VOLUME FOUR HUNDRED AND SIXTY-SIX

METHODS IN ENZYMOLOGY

Biothermodynamics, Part B

METHODS IN ENZYMOLOGY

Editors-in-Chief

JOHN N. ABELSON AND MELVIN I. SIMON

Division of Biology
California Institute of Technology
Pasadena, California, USA

Founding Editors

SIDNEY P. COLOWICK AND NATHAN O. KAPLAN

VOLUME FOUR HUNDRED AND SIXTY-SIX

METHODS IN ENZYMOLOGY

Biothermodynamics, Part B

EDITED BY

MICHAEL L. JOHNSON
University of Virginia Health System
Department of Pharmacology
Charlottesville, Virginia, USA

GARY K. ACKERS AND JO M. HOLT
Emeritus Department of Biochemistry and Molecular Biophysics
Washington University School of Medicine
St. Louis, Missouri, USA

ELSEVIER

AMSTERDAM • BOSTON • HEIDELBERG • LONDON
NEW YORK • OXFORD • PARIS • SAN DIEGO
SAN FRANCISCO • SINGAPORE • SYDNEY • TOKYO
Academic Press is an imprint of Elsevier

Academic Press is an imprint of Elsevier
525 B Street, Suite 1900, San Diego, CA 92101-4495, USA
30 Corporate Drive, Suite 400, Burlington, MA 01803, USA
32 Jamestown Road, London NW1 7BY, UK

First edition 2009

ISBN: 978-0-12-374776-1
ISBN: 0076-6879

Printed and bound in United States of America
09 10 11 12 10 9 8 7 6 5 4 3 2 1

CONTENTS

Contributors

Mitsuru Akita
Satellite Venture Business, Ehime University, 3 Bunkyo-cho, Matsuyama, Japan, and Faculty of Agriculture, and The United Graduate School of Agricultural Science, Ehime University, 3-5-7 Tarumi, Matsuyama, Japan

Andrei T. Alexandrescu
Department of Molecular and Cell Biology, University of Connecticut, Storrs, Connecticut, USA

Aileen Y. Alontaga
Department of Biochemistry and Molecular Biology, The University of Kansas Medical Center, Kansas City, Kansas, USA

Kathryn M. Armstrong
Department of Chemistry and Biochemistry, University of Notre Dame, Notre Dame, Indiana, USA

Brian M. Baker
Walther Cancer Research Center, and Department of Chemistry and Biochemistry, University of Notre Dame, Notre Dame, Indiana, USA

Sergio Barranco-Medina
Biochemistry and Physiology of Plants, W5-134, Bielefeld University, Bielefeld, Germany

Wlodzimierz Bujalowski
The Sealy Center for Structural Biology, Sealy Center for Cancer Cell Biology, and Department of Obstetrics and Gynecology; Department of Biochemistry and Molecular Biology, The University of Texas Medical Branch at Galveston, Galveston, Texas, USA

Adam P. Christensen
Department of Biochemistry, Roy J. and Lucille A. Carver College of Medicine, University of Iowa, Iowa City, Iowa, USA

Patricia L. Clark
Department of Chemistry and Biochemistry, University of Notre Dame, Notre Dame, Indiana, USA

Robyn L. Croke
Department of Molecular and Cell Biology, University of Connecticut, Storrs, Connecticut, USA

Tanneke den Blaauwen
Molecular Cytology, Swammerdam Institute for Life Sciences, University of Amsterdam, Amsterdam, The Netherlands

Enrico Di Cera
Department of Biochemistry and Molecular Biophysics, Washington University School of Medicine, St. Louis, Missouri, USA

Karl-Josef Dietz
Biochemistry and Physiology of Plants, W5-134, Bielefeld University, Bielefeld, Germany

Carmen Domene
Physical and Theoretical Chemistry Laboratory, Department of Chemistry, University of Oxford, Oxford, United Kingdom

Arnold J. M. Driessen
Department of Molecular Microbiology, Groningen Biomolecular Sciences and Biotechnology Institute and the Zernike Institute for Advanced Materials, University of Groningen, The Netherlands

Aron W. Fenton
Department of Biochemistry and Molecular Biology, The University of Kansas Medical Center, Kansas City, Kansas, USA

Michael G. Fried
Department of Molecular and Cellular Biochemistry, Center for Structural Biology, University of Kentucky, Lexington, Kentucky, USA

John R. Froehlig
Department of Biochemistry, Roy J. and Lucille A. Carver College of Medicine, University of Iowa, Iowa City, Iowa, USA

Simone Furini
Department of Medical Surgery and Bioengineering, University of Siena, Siena, Italy

Brian E. Gloor
Department of Chemistry and Biochemistry, University of Notre Dame, Notre Dame, Indiana, USA

Lisa M. Gloss
School of Molecular Biosciences, Washington State University, Pullman, Washington, USA

Gerald R. Grimsley
Department of Molecular and Cellular Medicine, Texas A&M Health Science Center, College Station, Texas, USA

Andrew Hagarman
Department of Chemistry, Drexel University, Philadelphia, Pennsylvania, USA

Lance M. Hellman
Department of Molecular and Cellular Biochemistry, Center for Structural Biology, University of Kentucky, Lexington, Kentucky, USA

Anne Hinderliter
Department of Chemistry and Biochemistry, University of Minnesota Duluth, Duluth, Minnesota, USA

Rainbo Hultman
Department of Biochemistry, Roy J. and Lucille A. Carver College of Medicine, University of Iowa, Iowa City, Iowa, USA

Hitoshi Inoue
The United Graduate School of Agricultural Science, Ehime University, 3-5-7 Tarumi, Matsuyama, Japan

Maria J. Jezewska
The Sealy Center for Structural Biology, Sealy Center for Cancer Cell Biology, and Department of Obstetrics and Gynecology; Department of Biochemistry and Molecular Biology, The University of Texas Medical Branch at Galveston, Galveston, Texas, USA

Sarah Johnson
Department of Pharmaceutical Sciences, College of Pharmacy, University of Nebraska Medical Center, Omaha, Nebraska, USA

Alexej Kedrov
Department of Molecular Microbiology, Groningen Biomolecular Sciences and Biotechnology Institute and the Zernike Institute for Advanced Materials, University of Groningen, The Netherlands

Irine Khutsishvili
Department of Pharmaceutical Sciences, College of Pharmacy, University of Nebraska Medical Center, Omaha, Nebraska, USA

Kristofer Knutson
Department of Chemistry and Biochemistry, University of Minnesota Duluth, Duluth, Minnesota, USA

Alexey S. Ladokhin
Department of Biochemistry and Molecular Biology, University of Kansas Medical Center, Kansas City, Kansas, USA

Jonathan R. LaRochelle
Department of Molecular and Cell Biology, University of Connecticut, Storrs, Connecticut, USA

Hui-Ting Lee
Department of Pharmaceutical Sciences, College of Pharmacy, University of Nebraska Medical Center, Omaha, Nebraska, USA

Conggang Li
Department of Chemistry, University of North Carolina at Chapel Hill, Chapel Hill, North Carolina, USA

George I. Makhatadze
Department of Biochemistry and Molecular Biology, Penn State University College of Medicine, Hershey, Pennsylvania, USA, and Center for Biotechnology and Interdisciplinary Studies, Rensselaer Polytechnic Institute, Troy, New York, USA

Luis A. Marky
Department of Pharmaceutical Sciences, College of Pharmacy, University of Nebraska Medical Center, Omaha, Nebraska, USA

Manana Melikishvili
Department of Molecular and Cellular Biochemistry, Center for Structural Biology, University of Kentucky, Lexington, Kentucky, USA

Andrew C. Miklos
Department of Chemistry, University of North Carolina at Chapel Hill, Chapel Hill, North Carolina, USA

Jesse Murphy
Department of Chemistry and Biochemistry, University of Minnesota Duluth, Duluth, Minnesota, USA

Rhonda A. Newman
Department of Biochemistry, Roy J. and Lucille A. Carver College of Medicine, University of Iowa, Iowa City, Iowa, USA

Susan E. O'Donnell
Department of Biochemistry, Roy J. and Lucille A. Carver College of Medicine, University of Iowa, Iowa City, Iowa, USA

C. Nick Pace
Department Biochemistry and Biophysics, Texas A&M University, and Department of Molecular and Cellular Medicine, Texas A&M Health Science Center, College Station, Texas, USA

Gary J. Pielak
Lineberger Comprehensive Cancer Center; Department of Chemistry, and Department of Biochemistry and Biophysics, University of North Carolina at Chapel Hill, Chapel Hill, North Carolina, USA

Kurt H. Piepenbrink
Department of Chemistry and Biochemistry, University of Notre Dame, Notre Dame, Indiana, USA

J. Martin Scholtz
Department Biochemistry and Biophysics, Texas A&M University, and Department of Molecular and Cellular Medicine, Texas A&M Health Science Center, College Station, Texas, USA

Katrina L. Schweiker
Department of Biochemistry and Molecular Biology, Penn State University College of Medicine, Hershey, Pennsylvania, USA, and Center for Biotechnology and Interdisciplinary Studies, Rensselaer Polytechnic Institute, Troy, New York, USA

Reinhard Schweitzer-Stenner
Department of Chemistry, Drexel University, Philadelphia, Pennsylvania, USA

Madeline A. Shea
Department of Biochemistry, Roy J. and Lucille A. Carver College of Medicine, University of Iowa, Iowa City, Iowa, USA

Sarah R. Sheftic
Department of Molecular and Cell Biology, University of Connecticut, Storrs, Connecticut, USA

Tzilhav Shem-Ad
Department of Life Sciences and the Zlotowski Center for Neurosciences, Ben-Gurion University of the Negev, Beer Sheva, Israel

Jana K. Shen
Department of Chemistry and Biochemistry, University of Oklahoma, Norman, Oklahoma, USA

Jonathan B. Soffer
Department of Chemistry, Drexel University, Philadelphia, Pennsylvania, USA

Martin Thompson
Department of Chemistry, Michigan Technological University, Houghton, Michigan, USA

Krastyu G. Ugrinov
Department of Chemistry and Biochemistry, University of Notre Dame, Notre Dame, Indiana, USA

Daniel Verbaro
Department of Chemistry, Drexel University, Philadelphia, Pennsylvania, USA

Jason A. Wallace
Department of Chemistry and Biochemistry, University of Oklahoma, Norman, Oklahoma, USA

Travis J. Witt
Department of Biochemistry, Roy J. and Lucille A. Carver College of Medicine, University of Iowa, Iowa City, Iowa, USA

Yujia Xu
Department of Chemistry, Hunter College-CUNY, New York, USA

Ofer Yifrach
Department of Life Sciences and the Zlotowski Center for Neurosciences, Ben-Gurion University of the Negev, Beer Sheva, Israel

Nitzan Zandany
Department of Life Sciences and the Zlotowski Center for Neurosciences, Ben-Gurion University of the Negev, Beer Sheva, Israel

Preface

Unfortunately, thermodynamics is commonly either poorly taught or not at all in graduate or undergraduate departments of chemistry, biochemistry, etc. My first exposure to thermodynamics was a graduate level physics department course where all of thermodynamics was covered in a single lecture. Steam engines come to mind when I think of my first chemistry department thermodynamics course. Such courses have caused a large fraction of biochemical researchers have the impression that thermodynamic approaches are archaic and, at best, ancillary to the central issues of biochemistry. Sadly, thermodynamics has seldom been fused with developments in molecular biology, structural analysis, or computational chemistry. Another reason for this narrow and insular perception is that thermodynamics is frequently equated with a single experimental technique (i.e., calorimetry). However, all of these perceptions are far from accurate.

The importance of thermodynamics is its use as a "logic tool." One of many quintessential examples of such a use of thermodynamics is Wyman's theory of linked functions. This volume is the second in a continuing series which foster and develop this vision of how thermodynamics can be an important tool for the study of biological systems.

Michael L. Johnson

METHODS IN ENZYMOLOGY

VOLUME 123. Vitamins and Coenzymes (Part H)
Edited by FRANK CHYTIL AND DONALD B. MCCORMICK

VOLUME 124. Hormone Action (Part J: Neuroendocrine Peptides)
Edited by P. MICHAEL CONN

VOLUME 125. Biomembranes (Part M: Transport in Bacteria, Mitochondria, and Chloroplasts: General Approaches and Transport Systems)
Edited by SIDNEY FLEISCHER AND BECCA FLEISCHER

VOLUME 126. Biomembranes (Part N: Transport in Bacteria, Mitochondria, and Chloroplasts: Protonmotive Force)
Edited by SIDNEY FLEISCHER AND BECCA FLEISCHER

VOLUME 127. Biomembranes (Part O: Protons and Water: Structure and Translocation)
Edited by LESTER PACKER

VOLUME 128. Plasma Lipoproteins (Part A: Preparation, Structure, and Molecular Biology)
Edited by JERE P. SEGREST AND JOHN J. ALBERS

VOLUME 129. Plasma Lipoproteins (Part B: Characterization, Cell Biology, and Metabolism)
Edited by JOHN J. ALBERS AND JERE P. SEGREST

VOLUME 130. Enzyme Structure (Part K)
Edited by C. H. W. HIRS AND SERGE N. TIMASHEFF

VOLUME 131. Enzyme Structure (Part L)
Edited by C. H. W. HIRS AND SERGE N. TIMASHEFF

VOLUME 132. Immunochemical Techniques (Part J: Phagocytosis and Cell-Mediated Cytotoxicity)
Edited by GIOVANNI DI SABATO AND JOHANNES EVERSE

VOLUME 133. Bioluminescence and Chemiluminescence (Part B)
Edited by MARLENE DELUCA AND WILLIAM D. MCELROY

VOLUME 134. Structural and Contractile Proteins (Part C: The Contractile Apparatus and the Cytoskeleton)
Edited by RICHARD B. VALLEE

VOLUME 135. Immobilized Enzymes and Cells (Part B)
Edited by KLAUS MOSBACH

VOLUME 136. Immobilized Enzymes and Cells (Part C)
Edited by KLAUS MOSBACH

VOLUME 137. Immobilized Enzymes and Cells (Part D)
Edited by KLAUS MOSBACH

VOLUME 138. Complex Carbohydrates (Part E)
Edited by VICTOR GINSBURG

VOLUME 139. Cellular Regulators (Part A: Calcium- and Calmodulin-Binding Proteins)
Edited by ANTHONY R. MEANS AND P. MICHAEL CONN

VOLUME 140. Cumulative Subject Index Volumes 102–119, 121–134

VOLUME 141. Cellular Regulators (Part B: Calcium and Lipids)
Edited by P. MICHAEL CONN AND ANTHONY R. MEANS

VOLUME 142. Metabolism of Aromatic Amino Acids and Amines
Edited by SEYMOUR KAUFMAN

VOLUME 143. Sulfur and Sulfur Amino Acids
Edited by WILLIAM B. JAKOBY AND OWEN GRIFFITH

VOLUME 144. Structural and Contractile Proteins (Part D: Extracellular Matrix)
Edited by LEON W. CUNNINGHAM

VOLUME 145. Structural and Contractile Proteins (Part E: Extracellular Matrix)
Edited by LEON W. CUNNINGHAM

VOLUME 146. Peptide Growth Factors (Part A)
Edited by DAVID BARNES AND DAVID A. SIRBASKU

VOLUME 147. Peptide Growth Factors (Part B)
Edited by DAVID BARNES AND DAVID A. SIRBASKU

VOLUME 148. Plant Cell Membranes
Edited by LESTER PACKER AND ROLAND DOUCE

VOLUME 149. Drug and Enzyme Targeting (Part B)
Edited by RALPH GREEN AND KENNETH J. WIDDER

VOLUME 150. Immunochemical Techniques (Part K: *In Vitro* Models of B and T Cell Functions and Lymphoid Cell Receptors)
Edited by GIOVANNI DI SABATO

VOLUME 151. Molecular Genetics of Mammalian Cells
Edited by MICHAEL M. GOTTESMAN

VOLUME 152. Guide to Molecular Cloning Techniques
Edited by SHELBY L. BERGER AND ALAN R. KIMMEL

VOLUME 153. Recombinant DNA (Part D)
Edited by RAY WU AND LAWRENCE GROSSMAN

VOLUME 154. Recombinant DNA (Part E)
Edited by RAY WU AND LAWRENCE GROSSMAN

VOLUME 155. Recombinant DNA (Part F)
Edited by RAY WU

VOLUME 156. Biomembranes (Part P: ATP-Driven Pumps and Related Transport: The Na, K-Pump)
Edited by SIDNEY FLEISCHER AND BECCA FLEISCHER

VOLUME 173. Biomembranes [Part T: Cellular and Subcellular Transport: Eukaryotic (Nonepithelial) Cells]
Edited by SIDNEY FLEISCHER AND BECCA FLEISCHER

VOLUME 174. Biomembranes [Part U: Cellular and Subcellular Transport: Eukaryotic (Nonepithelial) Cells]
Edited by SIDNEY FLEISCHER AND BECCA FLEISCHER

VOLUME 175. Cumulative Subject Index Volumes 135–139, 141–167

VOLUME 176. Nuclear Magnetic Resonance (Part A: Spectral Techniques and Dynamics)
Edited by NORMAN J. OPPENHEIMER AND THOMAS L. JAMES

VOLUME 177. Nuclear Magnetic Resonance (Part B: Structure and Mechanism)
Edited by NORMAN J. OPPENHEIMER AND THOMAS L. JAMES

VOLUME 178. Antibodies, Antigens, and Molecular Mimicry
Edited by JOHN J. LANGONE

VOLUME 179. Complex Carbohydrates (Part F)
Edited by VICTOR GINSBURG

VOLUME 180. RNA Processing (Part A: General Methods)
Edited by JAMES E. DAHLBERG AND JOHN N. ABELSON

VOLUME 181. RNA Processing (Part B: Specific Methods)
Edited by JAMES E. DAHLBERG AND JOHN N. ABELSON

VOLUME 182. Guide to Protein Purification
Edited by MURRAY P. DEUTSCHER

VOLUME 183. Molecular Evolution: Computer Analysis of Protein and Nucleic Acid Sequences
Edited by RUSSELL F. DOOLITTLE

VOLUME 184. Avidin-Biotin Technology
Edited by MEIR WILCHEK AND EDWARD A. BAYER

VOLUME 185. Gene Expression Technology
Edited by DAVID V. GOEDDEL

VOLUME 186. Oxygen Radicals in Biological Systems (Part B: Oxygen Radicals and Antioxidants)
Edited by LESTER PACKER AND ALEXANDER N. GLAZER

VOLUME 187. Arachidonate Related Lipid Mediators
Edited by ROBERT C. MURPHY AND FRANK A. FITZPATRICK

VOLUME 188. Hydrocarbons and Methylotrophy
Edited by MARY E. LIDSTROM

VOLUME 189. Retinoids (Part A: Molecular and Metabolic Aspects)
Edited by LESTER PACKER

Volume 206. Cytochrome P450
Edited by Michael R. Waterman and Eric F. Johnson

Volume 207. Ion Channels
Edited by Bernardo Rudy and Linda E. Iverson

Volume 208. Protein–DNA Interactions
Edited by Robert T. Sauer

Volume 209. Phospholipid Biosynthesis
Edited by Edward A. Dennis and Dennis E. Vance

Volume 210. Numerical Computer Methods
Edited by Ludwig Brand and Michael L. Johnson

Volume 211. DNA Structures (Part A: Synthesis and Physical Analysis of DNA)
Edited by David M. J. Lilley and James E. Dahlberg

Volume 212. DNA Structures (Part B: Chemical and Electrophoretic Analysis of DNA)
Edited by David M. J. Lilley and James E. Dahlberg

Volume 213. Carotenoids (Part A: Chemistry, Separation, Quantitation, and Antioxidation)
Edited by Lester Packer

Volume 214. Carotenoids (Part B: Metabolism, Genetics, and Biosynthesis)
Edited by Lester Packer

Volume 215. Platelets: Receptors, Adhesion, Secretion (Part B)
Edited by Jacek J. Hawiger

Volume 216. Recombinant DNA (Part G)
Edited by Ray Wu

Volume 217. Recombinant DNA (Part H)
Edited by Ray Wu

Volume 218. Recombinant DNA (Part I)
Edited by Ray Wu

Volume 219. Reconstitution of Intracellular Transport
Edited by James E. Rothman

Volume 220. Membrane Fusion Techniques (Part A)
Edited by Nejat Düzgüneş

Volume 221. Membrane Fusion Techniques (Part B)
Edited by Nejat Düzgüneş

Volume 222. Proteolytic Enzymes in Coagulation, Fibrinolysis, and Complement Activation (Part A: Mammalian Blood Coagulation Factors and Inhibitors)
Edited by Laszlo Lorand and Kenneth G. Mann

Volume 223. Proteolytic Enzymes in Coagulation, Fibrinolysis, and Complement Activation (Part B: Complement Activation, Fibrinolysis, and Nonmammalian Blood Coagulation Factors)
Edited by Laszlo Lorand and Kenneth G. Mann

Volume 224. Molecular Evolution: Producing the Biochemical Data
Edited by Elizabeth Anne Zimmer, Thomas J. White, Rebecca L. Cann, and Allan C. Wilson

Volume 225. Guide to Techniques in Mouse Development
Edited by Paul M. Wassarman and Melvin L. DePamphilis

Volume 226. Metallobiochemistry (Part C: Spectroscopic and Physical Methods for Probing Metal Ion Environments in Metalloenzymes and Metalloproteins)
Edited by James F. Riordan and Bert L. Vallee

Volume 227. Metallobiochemistry (Part D: Physical and Spectroscopic Methods for Probing Metal Ion Environments in Metalloproteins)
Edited by James F. Riordan and Bert L. Vallee

Volume 228. Aqueous Two-Phase Systems
Edited by Harry Walter and Göte Johansson

Volume 229. Cumulative Subject Index Volumes 195–198, 200–227

Volume 230. Guide to Techniques in Glycobiology
Edited by William J. Lennarz and Gerald W. Hart

Volume 231. Hemoglobins (Part B: Biochemical and Analytical Methods)
Edited by Johannes Everse, Kim D. Vandegriff, and Robert M. Winslow

Volume 232. Hemoglobins (Part C: Biophysical Methods)
Edited by Johannes Everse, Kim D. Vandegriff, and Robert M. Winslow

Volume 233. Oxygen Radicals in Biological Systems (Part C)
Edited by Lester Packer

Volume 234. Oxygen Radicals in Biological Systems (Part D)
Edited by Lester Packer

Volume 235. Bacterial Pathogenesis (Part A: Identification and Regulation of Virulence Factors)
Edited by Virginia L. Clark and Patrik M. Bavoil

Volume 236. Bacterial Pathogenesis (Part B: Integration of Pathogenic Bacteria with Host Cells)
Edited by Virginia L. Clark and Patrik M. Bavoil

Volume 237. Heterotrimeric G Proteins
Edited by Ravi Iyengar

Volume 238. Heterotrimeric G-Protein Effectors
Edited by Ravi Iyengar

VOLUME 239. Nuclear Magnetic Resonance (Part C)
Edited by THOMAS L. JAMES AND NORMAN J. OPPENHEIMER

VOLUME 240. Numerical Computer Methods (Part B)
Edited by MICHAEL L. JOHNSON AND LUDWIG BRAND

VOLUME 241. Retroviral Proteases
Edited by LAWRENCE C. KUO AND JULES A. SHAFER

VOLUME 242. Neoglycoconjugates (Part A)
Edited by Y. C. LEE AND REIKO T. LEE

VOLUME 243. Inorganic Microbial Sulfur Metabolism
Edited by HARRY D. PECK, JR., AND JEAN LEGALL

VOLUME 244. Proteolytic Enzymes: Serine and Cysteine Peptidases
Edited by ALAN J. BARRETT

VOLUME 245. Extracellular Matrix Components
Edited by E. RUOSLAHTI AND E. ENGVALL

VOLUME 246. Biochemical Spectroscopy
Edited by KENNETH SAUER

VOLUME 247. Neoglycoconjugates (Part B: Biomedical Applications)
Edited by Y. C. LEE AND REIKO T. LEE

VOLUME 248. Proteolytic Enzymes: Aspartic and Metallo Peptidases
Edited by ALAN J. BARRETT

VOLUME 249. Enzyme Kinetics and Mechanism (Part D: Developments in Enzyme Dynamics)
Edited by DANIEL L. PURICH

VOLUME 250. Lipid Modifications of Proteins
Edited by PATRICK J. CASEY AND JANICE E. BUSS

VOLUME 251. Biothiols (Part A: Monothiols and Dithiols, Protein Thiols, and Thiyl Radicals)
Edited by LESTER PACKER

VOLUME 252. Biothiols (Part B: Glutathione and Thioredoxin; Thiols in Signal Transduction and Gene Regulation)
Edited by LESTER PACKER

VOLUME 253. Adhesion of Microbial Pathogens
Edited by RON J. DOYLE AND ITZHAK OFEK

VOLUME 254. Oncogene Techniques
Edited by PETER K. VOGT AND INDER M. VERMA

VOLUME 255. Small GTPases and Their Regulators (Part A: Ras Family)
Edited by W. E. BALCH, CHANNING J. DER, AND ALAN HALL

VOLUME 256. Small GTPases and Their Regulators (Part B: Rho Family)
Edited by W. E. BALCH, CHANNING J. DER, AND ALAN HALL

VOLUME 257. Small GTPases and Their Regulators (Part C: Proteins Involved in Transport)
Edited by W. E. BALCH, CHANNING J. DER, AND ALAN HALL

VOLUME 258. Redox-Active Amino Acids in Biology
Edited by JUDITH P. KLINMAN

VOLUME 259. Energetics of Biological Macromolecules
Edited by MICHAEL L. JOHNSON AND GARY K. ACKERS

VOLUME 260. Mitochondrial Biogenesis and Genetics (Part A)
Edited by GIUSEPPE M. ATTARDI AND ANNE CHOMYN

VOLUME 261. Nuclear Magnetic Resonance and Nucleic Acids
Edited by THOMAS L. JAMES

VOLUME 262. DNA Replication
Edited by JUDITH L. CAMPBELL

VOLUME 263. Plasma Lipoproteins (Part C: Quantitation)
Edited by WILLIAM A. BRADLEY, SANDRA H. GIANTURCO, AND JERE P. SEGREST

VOLUME 264. Mitochondrial Biogenesis and Genetics (Part B)
Edited by GIUSEPPE M. ATTARDI AND ANNE CHOMYN

VOLUME 265. Cumulative Subject Index Volumes 228, 230–262

VOLUME 266. Computer Methods for Macromolecular Sequence Analysis
Edited by RUSSELL F. DOOLITTLE

VOLUME 267. Combinatorial Chemistry
Edited by JOHN N. ABELSON

VOLUME 268. Nitric Oxide (Part A: Sources and Detection of NO; NO Synthase)
Edited by LESTER PACKER

VOLUME 269. Nitric Oxide (Part B: Physiological and Pathological Processes)
Edited by LESTER PACKER

VOLUME 270. High Resolution Separation and Analysis of Biological Macromolecules (Part A: Fundamentals)
Edited by BARRY L. KARGER AND WILLIAM S. HANCOCK

VOLUME 271. High Resolution Separation and Analysis of Biological Macromolecules (Part B: Applications)
Edited by BARRY L. KARGER AND WILLIAM S. HANCOCK

VOLUME 272. Cytochrome P450 (Part B)
Edited by ERIC F. JOHNSON AND MICHAEL R. WATERMAN

VOLUME 273. RNA Polymerase and Associated Factors (Part A)
Edited by SANKAR ADHYA

Volume 328. Applications of Chimeric Genes and Hybrid Proteins (Part C: Protein–Protein Interactions and Genomics)
Edited by Jeremy Thorner, Scott D. Emr, and John N. Abelson

Volume 329. Regulators and Effectors of Small GTPases (Part E: GTPases Involved in Vesicular Traffic)
Edited by W. E. Balch, Channing J. Der, and Alan Hall

Volume 330. Hyperthermophilic Enzymes (Part A)
Edited by Michael W. W. Adams and Robert M. Kelly

Volume 331. Hyperthermophilic Enzymes (Part B)
Edited by Michael W. W. Adams and Robert M. Kelly

Volume 332. Regulators and Effectors of Small GTPases (Part F: Ras Family I)
Edited by W. E. Balch, Channing J. Der, and Alan Hall

Volume 333. Regulators and Effectors of Small GTPases (Part G: Ras Family II)
Edited by W. E. Balch, Channing J. Der, and Alan Hall

Volume 334. Hyperthermophilic Enzymes (Part C)
Edited by Michael W. W. Adams and Robert M. Kelly

Volume 335. Flavonoids and Other Polyphenols
Edited by Lester Packer

Volume 336. Microbial Growth in Biofilms (Part A: Developmental and Molecular Biological Aspects)
Edited by Ron J. Doyle

Volume 337. Microbial Growth in Biofilms (Part B: Special Environments and Physicochemical Aspects)
Edited by Ron J. Doyle

Volume 338. Nuclear Magnetic Resonance of Biological Macromolecules (Part A)
Edited by Thomas L. James, Volker Dötsch, and Uli Schmitz

Volume 339. Nuclear Magnetic Resonance of Biological Macromolecules (Part B)
Edited by Thomas L. James, Volker Dötsch, and Uli Schmitz

Volume 340. Drug–Nucleic Acid Interactions
Edited by Jonathan B. Chaires and Michael J. Waring

Volume 341. Ribonucleases (Part A)
Edited by Allen W. Nicholson

Volume 342. Ribonucleases (Part B)
Edited by Allen W. Nicholson

Volume 343. G Protein Pathways (Part A: Receptors)
Edited by Ravi Iyengar and John D. Hildebrandt

Volume 344. G Protein Pathways (Part B: G Proteins and Their Regulators)
Edited by Ravi Iyengar and John D. Hildebrandt

Volume 345. G Protein Pathways (Part C: Effector Mechanisms)
Edited by Ravi Iyengar and John D. Hildebrandt

Volume 346. Gene Therapy Methods
Edited by M. Ian Phillips

Volume 347. Protein Sensors and Reactive Oxygen Species (Part A: Selenoproteins and Thioredoxin)
Edited by Helmut Sies and Lester Packer

Volume 348. Protein Sensors and Reactive Oxygen Species (Part B: Thiol Enzymes and Proteins)
Edited by Helmut Sies and Lester Packer

Volume 349. Superoxide Dismutase
Edited by Lester Packer

Volume 350. Guide to Yeast Genetics and Molecular and Cell Biology (Part B)
Edited by Christine Guthrie and Gerald R. Fink

Volume 351. Guide to Yeast Genetics and Molecular and Cell Biology (Part C)
Edited by Christine Guthrie and Gerald R. Fink

Volume 352. Redox Cell Biology and Genetics (Part A)
Edited by Chandan K. Sen and Lester Packer

Volume 353. Redox Cell Biology and Genetics (Part B)
Edited by Chandan K. Sen and Lester Packer

Volume 354. Enzyme Kinetics and Mechanisms (Part F: Detection and Characterization of Enzyme Reaction Intermediates)
Edited by Daniel L. Purich

Volume 355. Cumulative Subject Index Volumes 321–354

Volume 356. Laser Capture Microscopy and Microdissection
Edited by P. Michael Conn

Volume 357. Cytochrome P450, Part C
Edited by Eric F. Johnson and Michael R. Waterman

Volume 358. Bacterial Pathogenesis (Part C: Identification, Regulation, and Function of Virulence Factors)
Edited by Virginia L. Clark and Patrik M. Bavoil

Volume 359. Nitric Oxide (Part D)
Edited by Enrique Cadenas and Lester Packer

Volume 360. Biophotonics (Part A)
Edited by Gerard Marriott and Ian Parker

Volume 361. Biophotonics (Part B)
Edited by Gerard Marriott and Ian Parker

Volume 381. Oxygen Sensing
Edited by Chandan K. Sen and Gregg L. Semenza

Volume 382. Quinones and Quinone Enzymes (Part B)
Edited by Helmut Sies and Lester Packer

Volume 383. Numerical Computer Methods (Part D)
Edited by Ludwig Brand and Michael L. Johnson

Volume 384. Numerical Computer Methods (Part E)
Edited by Ludwig Brand and Michael L. Johnson

Volume 385. Imaging in Biological Research (Part A)
Edited by P. Michael Conn

Volume 386. Imaging in Biological Research (Part B)
Edited by P. Michael Conn

Volume 387. Liposomes (Part D)
Edited by Nejat Düzgüneş

Volume 388. Protein Engineering
Edited by Dan E. Robertson and Joseph P. Noel

Volume 389. Regulators of G-Protein Signaling (Part A)
Edited by David P. Siderovski

Volume 390. Regulators of G-Protein Signaling (Part B)
Edited by David P. Siderovski

Volume 391. Liposomes (Part E)
Edited by Nejat Düzgüneş

Volume 392. RNA Interference
Edited by Engelke Rossi

Volume 393. Circadian Rhythms
Edited by Michael W. Young

Volume 394. Nuclear Magnetic Resonance of Biological Macromolecules (Part C)
Edited by Thomas L. James

Volume 395. Producing the Biochemical Data (Part B)
Edited by Elizabeth A. Zimmer and Eric H. Roalson

Volume 396. Nitric Oxide (Part E)
Edited by Lester Packer and Enrique Cadenas

Volume 397. Environmental Microbiology
Edited by Jared R. Leadbetter

Volume 398. Ubiquitin and Protein Degradation (Part A)
Edited by Raymond J. Deshaies

Volume 399. Ubiquitin and Protein Degradation (Part B)
Edited by Raymond J. Deshaies

VOLUME 400. Phase II Conjugation Enzymes and Transport Systems
Edited by HELMUT SIES AND LESTER PACKER

VOLUME 401. Glutathione Transferases and Gamma Glutamyl Transpeptidases
Edited by HELMUT SIES AND LESTER PACKER

VOLUME 402. Biological Mass Spectrometry
Edited by A. L. BURLINGAME

VOLUME 403. GTPases Regulating Membrane Targeting and Fusion
Edited by WILLIAM E. BALCH, CHANNING J. DER, AND ALAN HALL

VOLUME 404. GTPases Regulating Membrane Dynamics
Edited by WILLIAM E. BALCH, CHANNING J. DER, AND ALAN HALL

VOLUME 405. Mass Spectrometry: Modified Proteins and Glycoconjugates
Edited by A. L. BURLINGAME

VOLUME 406. Regulators and Effectors of Small GTPases: Rho Family
Edited by WILLIAM E. BALCH, CHANNING J. DER, AND ALAN HALL

VOLUME 407. Regulators and Effectors of Small GTPases: Ras Family
Edited by WILLIAM E. BALCH, CHANNING J. DER, AND ALAN HALL

VOLUME 408. DNA Repair (Part A)
Edited by JUDITH L. CAMPBELL AND PAUL MODRICH

VOLUME 409. DNA Repair (Part B)
Edited by JUDITH L. CAMPBELL AND PAUL MODRICH

VOLUME 410. DNA Microarrays (Part A: Array Platforms and Web-Bench Protocols)
Edited by ALAN KIMMEL AND BRIAN OLIVER

VOLUME 411. DNA Microarrays (Part B: Databases and Statistics)
Edited by ALAN KIMMEL AND BRIAN OLIVER

VOLUME 412. Amyloid, Prions, and Other Protein Aggregates (Part B)
Edited by INDU KHETERPAL AND RONALD WETZEL

VOLUME 413. Amyloid, Prions, and Other Protein Aggregates (Part C)
Edited by INDU KHETERPAL AND RONALD WETZEL

VOLUME 414. Measuring Biological Responses with Automated Microscopy
Edited by JAMES INGLESE

VOLUME 415. Glycobiology
Edited by MINORU FUKUDA

VOLUME 416. Glycomics
Edited by MINORU FUKUDA

VOLUME 417. Functional Glycomics
Edited by MINORU FUKUDA

Volume 418. Embryonic Stem Cells
Edited by Irina Klimanskaya and Robert Lanza

Volume 419. Adult Stem Cells
Edited by Irina Klimanskaya and Robert Lanza

Volume 420. Stem Cell Tools and Other Experimental Protocols
Edited by Irina Klimanskaya and Robert Lanza

Volume 421. Advanced Bacterial Genetics: Use of Transposons and Phage for Genomic Engineering
Edited by Kelly T. Hughes

Volume 422. Two-Component Signaling Systems, Part A
Edited by Melvin I. Simon, Brian R. Crane, and Alexandrine Crane

Volume 423. Two-Component Signaling Systems, Part B
Edited by Melvin I. Simon, Brian R. Crane, and Alexandrine Crane

Volume 424. RNA Editing
Edited by Jonatha M. Gott

Volume 425. RNA Modification
Edited by Jonatha M. Gott

Volume 426. Integrins
Edited by David Cheresh

Volume 427. MicroRNA Methods
Edited by John J. Rossi

Volume 428. Osmosensing and Osmosignaling
Edited by Helmut Sies and Dieter Haussinger

Volume 429. Translation Initiation: Extract Systems and Molecular Genetics
Edited by Jon Lorsch

Volume 430. Translation Initiation: Reconstituted Systems and Biophysical Methods
Edited by Jon Lorsch

Volume 431. Translation Initiation: Cell Biology, High-Throughput and Chemical-Based Approaches
Edited by Jon Lorsch

Volume 432. Lipidomics and Bioactive Lipids: Mass-Spectrometry–Based Lipid Analysis
Edited by H. Alex Brown

Volume 433. Lipidomics and Bioactive Lipids: Specialized Analytical Methods and Lipids in Disease
Edited by H. Alex Brown

Volume 434. Lipidomics and Bioactive Lipids: Lipids and Cell Signaling
Edited by H. Alex Brown

Volume 435. Oxygen Biology and Hypoxia
Edited by Helmut Sies and Bernhard Brüne

Volume 436. Globins and Other Nitric Oxide-Reactive Protiens (Part A)
Edited by Robert K. Poole

Volume 437. Globins and Other Nitric Oxide-Reactive Protiens (Part B)
Edited by Robert K. Poole

Volume 438. Small GTPases in Disease (Part A)
Edited by William E. Balch, Channing J. Der, and Alan Hall

Volume 439. Small GTPases in Disease (Part B)
Edited by William E. Balch, Channing J. Der, and Alan Hall

Volume 440. Nitric Oxide, Part F Oxidative and Nitrosative Stress in Redox Regulation of Cell Signaling
Edited by Enrique Cadenas and Lester Packer

Volume 441. Nitric Oxide, Part G Oxidative and Nitrosative Stress in Redox Regulation of Cell Signaling
Edited by Enrique Cadenas and Lester Packer

Volume 442. Programmed Cell Death, General Principles for Studying Cell Death (Part A)
Edited by Roya Khosravi-Far, Zahra Zakeri, Richard A. Lockshin, and Mauro Piacentini

Volume 443. Angiogenesis: *In Vitro* Systems
Edited by David A. Cheresh

Volume 444. Angiogenesis: *In Vivo* Systems (Part A)
Edited by David A. Cheresh

Volume 445. Angiogenesis: *In Vivo* Systems (Part B)
Edited by David A. Cheresh

Volume 446. Programmed Cell Death, The Biology and Therapeutic Implications of Cell Death (Part B)
Edited by Roya Khosravi-Far, Zahra Zakeri, Richard A. Lockshin, and Mauro Piacentini

Volume 447. RNA Turnover in Bacteria, Archaea and Organelles
Edited by Lynne E. Maquat and Cecilia M. Arraiano

Volume 448. RNA Turnover in Eukaryotes: Nucleases, Pathways and Analysis of mRNA Decay
Edited by Lynne E. Maquat and Megerditch Kiledjian

Volume 465. Liposomes, Part G
Edited by Nejat Düzgüneş

Volume 466. Biothermodynamics, Part B
Edited by Michael L. Johnson, Gary K. Ackers, and Jo M. Holt

CHAPTER ONE

Using NMR-Detected Backbone Amide ^{1}H Exchange to Assess Macromolecular Crowding Effects on Globular-Protein Stability

Andrew C. Miklos,* Conggang Li,* *and* Gary J. Pielak*,†,‡

Contents

Abstract

The biophysical properties of proteins in the crowded intracellular environment are expected to differ from those for proteins in dilute solution. Crowding can be studied *in vitro* through addition of polymers at high concentrations. NMR-detected amide ^{1}H exchange is the only technique that provides equilibrium stability data for proteins on a per-residue basis under crowded conditions. We describe the theory behind amide ^{1}H exchange and provide a detailed

* Department of Chemistry, University of North Carolina at Chapel Hill, Chapel Hill, North Carolina, USA
† Department of Biochemistry and Biophysics, University of North Carolina at Chapel Hill, Chapel Hill, North Carolina, USA
‡ Lineberger Comprehensive Cancer Center, University of North Carolina at Chapel Hill, Chapel Hill, North Carolina, USA

Methods in Enzymology, Volume 466
ISSN 0076-6879, DOI: 10.1016/S0076-6879(09)66001-8

description of the experiments used to quantify globular protein stability at the residue level under crowded conditions. We also discuss the detection of weak interactions between the test protein and the crowding molecules.

1. Introduction

The macromolecular components of a cell's interior include proteins, DNAs, and RNAs with a combined concentration of up to 400 g/L, and comprise up to 44% of a cell's volume (Zimmerman and Trach, 1991). These macromolecules cannot occupy the same space at one time, leading to excluded volume effects (Zhou *et al.*, 2008). As a result, macromolecules should favor compact states in cells. Other nonspecific interactions between these macromolecules can also affect the system. We call this combination of excluded volume effects and nonspecific interactions macromolecular crowding. Macromolecular crowding can affect protein stability (Charlton *et al.*, 2008), folding (Ai *et al.*, 2006), and aggregation (Munishkina *et al.*, 2008) compared to these processes in dilute solution. The direction and magnitude of any change depends on the relative strengths of the non-specific interactions and excluded volume effects. Quantifying these effects is daunting because both are weak and only become important at high macromolecule concentration, as in cells. In addition, it can be difficult to detect the target protein because the crowding agent's weight concentration can be up to two orders-of-magnitude higher than that of the target protein.

Solution-state, high-resolution nuclear magnetic resonance (NMR) spectroscopy is a tool that can overcome these difficulties. Most importantly, detection is limited to NMR-active nuclei. Because of this property, we can use isotopic enrichment of the target protein not only to remove background signals due to crowding agents but also to simplify protein spectra (Serber and Dötsch, 2001). For the work described here, the isotope is ^{15}N. NMR-detected backbone amide ^{1}H exchange is a technique that allows residue-level stability determination (Englander and Kallenbach, 1983). For these experiments, ^{15}N–^{1}H heteronuclear single quantum correlation (HSQC) spectra are obtained (Bodenhausen and Ruben, 1980; Kay *et al.*, 1992). These spectra correlate the chemical shift of an ^{15}N nucleus with its attached proton. The result is a crosspeak for each nonproline backbone amide bond and for most nitrogen-containing side chains. We focus on the backbone amide resonances. The volume of a crosspeak corresponds to the concentration of the N–H group. Knowledge of the ^{15}N and ^{1}H chemical shift assignments for these residues is a prerequisite for amide ^{1}H exchange. As described in Sections 2 and 3, the exchange of the amide proton for a deuteron can be used to assess protein stability.

2. Globular Protein Stability

Global protein stability is defined as the free modified standard state energy change, $\Delta G^{0'}$, for the following reaction

$$N \rightleftarrows D \tag{I}$$

where N represents the native, biologically active structural state and D represents the denatured (or nonnative) thermodynamic state (Lumry *et al.*, 1966). This simple two-state reaction applies to many single-domain globular proteins. The strongest evidence for this behavior is the correspondence between the indirectly measured van't Hoff denaturation enthalpy and the directly measured calorimetric denaturation enthalpy (Privalov and Khechinashvili, 1974). The two-state nature is further confirmed when all probes used to monitor reversible denaturation yield similar thermodynamic values (e.g., Huang and Oas, 1995).

A variety of probes, including circular dichroism spectropolarimetry (e.g., Doyle *et al.*, 1996), absorbance (e.g., Cohen and Pielak, 1994) and fluorescence (e.g., Betz and Pielak, 1992) spectroscopies, and differential scanning calorimetry (e.g., Cohen and Pielak, 1994), are used to assess the equilibrium constant for global denaturation, $K^{0'}_{\mathrm{eq(den)}} = [\mathrm{D}]/[\mathrm{N}]$, in dilute solution. Values of $\Delta G^{0'}_{\mathrm{den}}$ can then be obtained from $K^{0'}_{\mathrm{eq(den)}}$ and the Gibbs equation

$$\Delta G^{0'}_{\mathrm{den}} = -RT\ln\left(K^{0'}_{\mathrm{eq(den)}}\right) \tag{1.1}$$

where R is the gas constant and T is the absolute temperature. Backbone amide ^{1}H exchange can be used to assess $K^{0'}_{\mathrm{eq(den)}}$ at the level of individual amino acid residues of a protein.

3. Mechanism and Limits of Amide ¹H Exchange

Consider a short unstructured peptide in aqueous solution. Its amide protons react reversibly with hydroxide ions and then H_2O, with the net result that the amide protons exchange with protons from H_2O. The kinetics of exchange are pseudo first order, and can be catalyzed by both acid and base (Englander and Kallenbach, 1983) (Fig. 1.1). When placed into D_2O, the amide protons exchange for deuterons. NMR-detected amide ^{1}H exchange exploits this reaction in two ways. First, ^{2}H is not detected in ^{1}H NMR. Second, the concentration of ^{2}H from the D_2O is hundreds- to thousands-fold higher than any exchangeable ^{1}H, making the exchange essentially irreversible. These properties make exchange experiments amenable for study by using NMR.

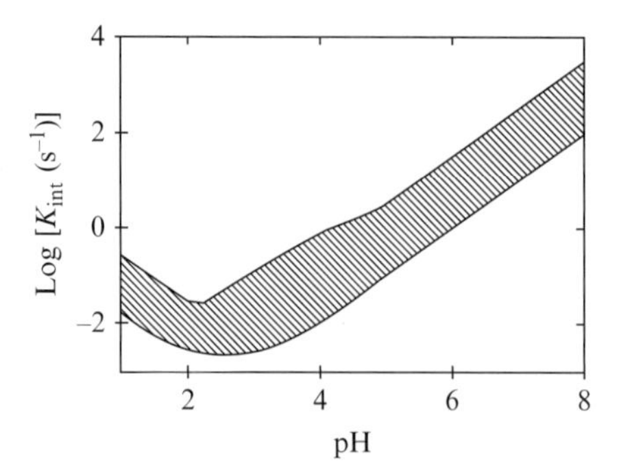

Figure 1.1 pH dependence of k_{int}. The shaded area corresponds to the range of k_{int} values for chymotrypsin inhibitor 2 (PDB 2CI2) at 37 °C. Rate constants are calculated by using the SPHERE program (Zhang, 1995).

To start exchange, the peptide or protein of interest is usually lyophilized from H_2O and dissolved in a solution containing D_2O. Using NMR, the exponential decay in the volume of each amide ^{1}H crosspeak can be used to obtain a rate constant for exchange, k_{obs}, provided the reaction is slow enough to be monitored. Approaches to measuring k_{obs} under crowded conditions are discussed in Section 6. Once measured, k_{obs} values can be correlated to protein stability.

Hvidt and Nielsen (1966) connected protein stability to amide ^{1}H exchange rates, using arguments that parallel those for global stability. Each amide region in the protein can be in one of two states: open or closed. Amide ^{1}H exchange only occurs from the open state, such that the irreversible exchange reaction can be described as

$$\mathrm{cl}-{}^1\mathrm{H} \underset{k_{cl}}{\overset{k_{op}}{\rightleftarrows}} \mathrm{op}-{}^1\mathrm{H} \xrightarrow{k_{int}} \mathrm{op}-{}^2\mathrm{H} \underset{k_{op}}{\overset{k_{cl}}{\rightleftarrows}} \mathrm{cl}-{}^2\mathrm{H} \qquad \text{(II)}$$

where cl and op are the open and closed states, k_{op} and k_{cl} are the associated rate constants, and k_{int} is the exchange rate from the open state. The open states are ensembles whose subpopulations range from small, low amplitude fluctuations of the native state to rare, globally unfolded forms. As shown by Frost and Pearson (1953), the overall rate constant, k_{obs} for such a reaction is

$$k_{obs} = \frac{(k_{op} + k_{cl} + k_{int}) - \sqrt{(k_{op} + k_{cl} + k_{int})^2 - 4k_{op}k_{int}}}{2} \qquad (1.2)$$

With the addition of some assumptions, the equation can be simplified to link k_{obs} and stability. First, the protein is assumed to be stable, that is,

$k_{cl} \gg k_{op}$. Given this assumption, there are two limiting extremes. First, if $k_{int} \gg k_{cl}$, then the exchange is said to occur under an EX1 limit, and the equation simplifies to

$$k_{obs} = k_{op} \tag{1.3}$$

The EX1 limit is typically associated with less stable or slow folding proteins (Ferraro *et al.*, 2004) and does not provide information about stability. The EX2 limit occurs when $k_{cl} \gg k_{int}$, that is, when exchange from the open state is the rate determining step. In this instance the equation simplifies to:

$$k_{obs} = \frac{k_{op}}{k_{cl}} k_{int} \quad \text{and} \quad \frac{k_{op}}{k_{cl}} = K_{op}^{0'} \tag{1.4}$$

where $K_{op}^{0'}$ is the equilibrium constant for opening the amide backbone at that residue. The local stability is defined as

$$\Delta G_{op}^{0'} = -RT\ln\left(K_{op}^{0'}\right) \tag{1.5}$$

These $\Delta G_{op}^{0'}$ values need not be uniform for a two-state unfolding model. Local unfolding events that are not indicative of full denaturation can occur, and have $\Delta G_{op}^{0'}$ values that are smaller than $\Delta G_{den}^{0'}$ (Englander and Kallenbach, 1983). Figure 1.2 shows representative $\Delta G_{op}^{0'}$ data, demonstrating the effect of 300 g/L poly(vinylpyrrolidone) as a crowding agent on the stability of chymotrypsin inhibitor 2 (Charlton *et al.*, 2008). Some residues only exchange when the entire protein is unfolded, and for these global unfolding events, $\Delta G_{op}^{0'} = \Delta G_{den}^{0'}$. To determine $\Delta G_{op}^{0'}$, $K_{op}^{0'}$ must be extracted from a list of k_{ex} values using Eq. (1.4), which requires determination of k_{int} values for each residue.

The base-catalyzed amide ^{1}H exchange rate in an unstructured peptide, k_{int}, is directly proportional to the hydroxide-ion concentration. The minimum rate constant for exchange, $\sim 1\ s^{-1}$, occurs between pH 2 and 4, and increases to $\sim 10^{10}\ s^{-1}$ at pH 12 (Fig. 1.1). The rate is also proportional to the amide proton's acidity, K_a ($\sim 10^{-18}$) (Molday and Kallen, 1972). The nonadditivity of peptide backbone solvation (Avbelj and Baldwin, 2009) means that K_a, and hence the exchange rate of a particular amide ^{1}H, depends on primary structure. Molday *et al.* (1972) conducted the classical studies on the exchange of amide ^{1}H for ^{2}H in D_2O. Their results have been refined in Englander's laboratory, and an easy to use computer program for calculating these values is available on-line from the Roder Lab (http://www.fccc.edu/research/labs/roder/sphere/sphere.html) (Zhang, 1995). Our approach to estimate k_{int} values under crowded conditions is discussed in Section 5.5.

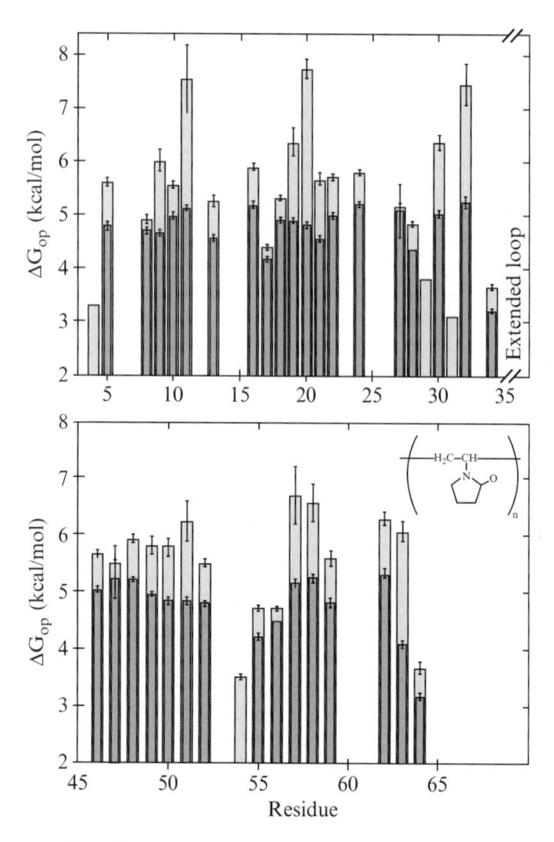

Figure 1.2 Macromolecular crowding with PVP40 stabilizes the I29A/I37H variant of CI2 relative to dilute solution. Values of $\Delta G^{0'}_{op}$ in 300 g/L PVP40 (cyan) and dilute solution (green) are plotted versus residue number. The height of each bar represents the average from three trials. The error bars represent the standard deviation. *Conditions*: 700–800 μM variant protein, 50 mM acetate buffer in D_2O, pH 5.4 at 37 °C. The inset shows the backbone structure of PVP. Reprinted with permission from Charlton *et al.* (2008). Copyright 2008 American Chemical Society. (See Color Insert.)

4. Requirements for Candidate Systems

The ability to acquire a high-resolution solution NMR spectrum of a globular protein depends on its rotational diffusion, which is determined by the protein's molecular weight and by the viscosity of the sample (Pielak *et al.*, 2008). As molecular weight and viscosity increase, the quality of NMR spectra decrease. Techniques exist for examining larger proteins, but conventional heteronuclear studies of amide ^1H exchange are usually restricted to monomeric globular proteins up to ~40 kDa in size. Crowded solutions decrease this value because of their increased viscosities. Luckily, the relationship between viscosity and rotational diffusion in macromolecular crowding conditions can lead to a higher diffusion rate than the Stokes–Einstein–Debye equation predicts (Kuttner *et al.*, 2005; Pielak *et al.*, 2008), provided there are only weak interactions between the crowder and the test protein. With the proper conditions, exchange experiments are possible under crowded conditions.

NMR tubes come in a variety of sizes, but typical 5-mm diameter tubes require a volume of ~0.4–1.0 mL, and measuring the diminution of an amide ^{1}H crosspeak accurately requires a protein concentration of ~1 m*M*. Given these parameters, each exchange experiment will require ~10 mg of protein. Of course, the protein must remain soluble at that concentration for the duration of the experiment.

The instrument time required for an exchange experiment is always a compromise. Longer acquisition times allow quantification of larger $\Delta G^{0'}_{op}$ values, but also risk precipitation of protein. We find that 24 h strikes a balance between these factors. To ensure accurate determination of k_{obs}, only crosspeak volumes that decay by a factor of two are analyzed. This requirement gives a lower limit for k_{obs} value of $\sim 8 \times 10^{-6}$ s^{-1}. The largest observable k_{obs} values must also be considered.

It takes 60 min to acquire a high-quality HSQC spectrum, and we require 3–4 points above the baseline to perform a regression. These requirements set the upper limit for k_{obs} values at $\sim 1 \times 10^{-4}$ s^{-1}. Using another experiment called SOFAST (Schanda and Brutscher, 2005), we can reduce the acquisition time to 7 min per spectrum, which moves the upper limit to $\sim 3 \times 10^{-3}$ s^{-1}. For our experiments (pH 5.4, 37 °C), this detection window corresponds to $\Delta G^{0'}_{op}$ values between 1 and 8 kcal/mol. Because the observation window is related to both k_{int} and protein stability, the window can be altered by mutagenesis, temperature changes, and pH alterations. With these time and stability constraints in mind, the crowding agent must also be selected with care.

There are two kinds or macromolecular crowding agents, polydisperse polymers and natural proteins. Whichever crowding agent is chosen, it must have the same key property as the test protein; solubility over the length of time required to collect the data. In addition, the crowding agent must be reasonably pure and should be inexpensive. Consider a crowder that is 99% pure by weight. A 300-g/L exchange sample will contain impurities at weight concentrations rivaling that of the test protein. Expense is a factor for the crowding agent because tens to hundreds of grams are necessary for characterization and to perform detailed exchange studies. Many polydisperse polymers satisfy these requirements.

Polydisperse crowders include synthetic polymers, and naturally occurring polymers such as polysaccharides. One advantage of choosing synthetic polymers is that they are often available in a range of molecular weights. For instance, poly(vinylpyrrolidone) (PVP) is available in molecular weights from 10 kDa to 1.3 MDa. This property allows exploration of volume exclusion with regards to molecular weight without changing the underlying monomer. A second advantageous property is that a model for the monomer is often available and can be used to probe the importance of the crowder's polymeric nature. We used *N*-ethylpyrrolidone as the model for PVP to show that the polymeric nature of PVP was the factor that increased chymotrypsin inhibitor 2 stability (Charlton *et al.*, 2008). In fact, we were able to show that the monomer model destabilized the protein. Polymeric crowding, however, does come with disadvantages.

One potential disadvantage is that synthetic polymers are polydisperse. Measuring polydispersity is discussed in Section 5.1. One might overcome this problem by using disordered proteins, such as FlgM (Dedmon *et al.*, 2002), which are random coil-like, but are monodisperse. Another disadvantage involves physiological significance. If one is only interested in understanding stability, synthetic polymers are fine, but cells do not contain synthetic polymers. Instead, they are crowded with monodisperse proteins, most of which are globular.

Globular proteins are the most physiologically relevant crowding agents. Furthermore, they all approximate the same shape, which scales rather smoothly with the number of residues between about 50 and 200 amino acids in length (Richards, 1977). They also have drawbacks. For instance, the surface features of proteins can vary wildly because they are composed of 20 types of subunits. The result is that it is difficult to vary the size of the crowder while maintaining its surface properties. Second, at concentrations up to 300 g/L, most globular proteins will aggregate, which means the crowding molecules will tend to interact with each other and with the test protein. Knowing all the requirements for a successful amide ^{1}H exchange experiment, controls and characterizations can be performed to find an acceptable candidate system.

5. Preliminary Experiments

Before any system can be used to assess protein stability under crowded conditions, a number of exploratory experiments must be performed. These experiments will ensure that the results obtained from an amide ^{1}H experiment are valid and indicative of a real change in protein stability upon addition of crowder.

5.1. Polymer characterization

Unlike most natural proteins, synthetic polymers are not monodisperse. As a result of the polymerization process, synthetic polymers comprise a mixture of different molecular weights. Properties such as the number average molecular weight, weight average molecular weight, and polydispersity must be determined before using such polymeric crowding agents.

The number average molecular weight (M_N) is an unweighted average of the molecular weight of all polymers in a mixture. The weight average molecular weight (M_W) is weighted by the molar mass of each species.

$$M_N = \frac{\sum_i N_i \cdot W_i}{\sum_i N_i}; \quad M_W = \frac{\sum_i N_i \cdot W_i^2}{\sum_i N_i \cdot W} \tag{1.6}$$

where N_i is the number of particles of species i and W_i is the molecular weight of species i. Polydispersity is the ratio of M_W to M_N, and indicates the spread in the size distribution. There are many ways to quantify these averages. We will focus on viscometry and quasi-elastic light scattering. Viscometry can be used because there is a correlation between intrinsic viscosity, $[\eta]$, and molecular weight (Rubinstein and Colby, 2003). The intrinsic viscosity is defined as:

$$[\eta] = \left(\frac{\eta - \eta_0}{c \cdot \eta_0} \right)_{c \to 0} \tag{1.7}$$

where η is measured solution viscosity, η_0 is dilute solution viscosity, and c is solute concentration. PVP has been characterized in this way (Molyneux, 1983). Light scattering, on the other hand, is an absolute method for determining molecular weight of macromolecules; it requires no standards. Furthermore, when used with size exclusion chromatography, it is possible to obtain M_W, M_N, and hence, the polydispersity in a single experiment (Wyatt, 1993). Once suitable crowding agents have been identified and characterized, the interactions between the test protein and crowding agent must be considered.

5.2. Aggregation studies

To avoid wasting time and ^{15}N-enriched protein, some preliminary experiments can be performed with unenriched samples. First, it is important to know whether the protein aggregates in the presence of the crowder, because aggregation confounds analysis of hydrogen exchange experiments. Specifically, aggregation and amide ^{1}H exchange both reduce the amide ^{1}H signal, and cannot be separated. Preliminary aggregation studies are performed by making 1-mL solutions containing 1-mM test protein with varying concentrations of crowding agent up to the highest concentration desired in the experiment. The samples are stored in sealed tubes at the desired temperature. After 24 h, the tubes are centrifuged at 16,000×g for 5 min and inspected for pellets. Although precipitation is easy to detect, there are more subtle problematic interactions, the detection of which requires more sophisticated methods.

5.3. Protein–crowder interactions

Soluble aggregates will not be detected in the centrifugation experiment. Furthermore, high concentrations of crowding agent can promote nonspecific interactions between the crowding agent and the test protein. These interactions can induce structural changes, especially in loop regions (Charlton *et al.*, 2008). NMR can be used to detect soluble aggregates, nonspecific interactions, and structural changes. To determine the presence of soluble aggregates, HSQC spectra are analyzed for alterations in crosspeak volume and linewidth (width at half-peak height) under nonexchange conditions as functions of time.

An increase in crosspeak linewidth in the presence of crowding agents results from an increase in the rotational correlation time of the protein (Li and Pielak, 2009). Such an effect is induced by the increase in the apparent molecular weight of the test protein from an increase in viscosity, aggregation, or from binding to a larger species. In our study of the stability of chymotrypsin inhibitor 2 in 300 g/L of 40 kDa PVP, we assessed the aggregation state of chymotrypsin inhibitor 2 by monitoring changes in linewidths and crosspeak volumes with time (Charlton *et al.*, 2008). Neither broadening nor volume changes were observed. These data are consistent with the monomeric nature of the protein in both dilute and crowded solutions. However, the width of the crosspeaks increased in 40 kDa PVP compared to dilute solution as a result of increased viscosity and, perhaps, protein–crowder interactions. Analysis of chemical shift changes in crowded condition will provide information about changes in protein structure and help identify weak protein-crowder interaction.

NMR chemical shifts are highly sensitive, empirical indicators of the chemical environment of the nucleus being studied. Therefore, changes in this environment induced by binding or alteration in protein structure can result in significant chemical shift changes and even crosspeak disappearance due to severe line-broadening. We found that 300 g/L 40 kDa PVP induced changes in ^{1}H and ^{15}N backbone chemical shifts in the loops and turns of chymotrypsin inhibitor 2. Such small changes are expected because crowding causes compaction, and these regions are not maximally compact (Charlton *et al.*, 2008). This observation is consistent with other studies showing that crowding can force unstructured regions into more compact states (Perham *et al.*, 2007; Stagg *et al.*, 2007). Of course, the chemical shift changes could also reflect weak interactions between 40 kDa PVP and these surface areas. The existence of weak interactions can be assessed quantitatively by using a more advanced experiment.

We have shown that nonspecific interactions between the test protein and crowding agents can be characterized by using the product of the transverse and longitudinal relaxation rates, R_1 and R_2, respectively (Li and Pielak, 2009). Their product is sensitive to nonspecific binding brought about by high crowder concentrations, but is insensitive to the crowder-induced increase in viscosity. This method was first tested in a model system comprising chymotrypsin inhibitor 2 in 200-g/L bovine serum albumin (BSA) (Fig. 1.3). Chymotrypsin inhibitor 2 not only interacts with BSA but also forms a small amount of homodimer in BSA. This method also shows that chymotrypsin inhibitor 2 also interacts weakly with PVP (Miklos, Li, Sharaf, and Pielak, unpublished). Having established the suitability of the protein-crowder combination for NMR-detected amide ^{1}H exchange experiments, exchange limits must be examined.

5.4. Exchange limit determination

As mentioned in Section 3, there are two exchange limits: EX1 and EX2. The EX2 limit is necessary to correlate exchange rates to stabilities. Two generally accepted methods for determining the exchange limit are analysis of the pH dependence of k_{obs} and NOESY crosspeak analysis.

The pH dependence of k_{obs} can be used if pH changes do not affect protein stability. pH meter readings from D_2O-containing solutions should be listed as pH_{read} because pD differs from pH (Schowen *et al.*, 1982), but we forego this convention here. For most proteins, physiologically relevant conditions exist in the base-catalyzed region for amide ^{1}H exchange

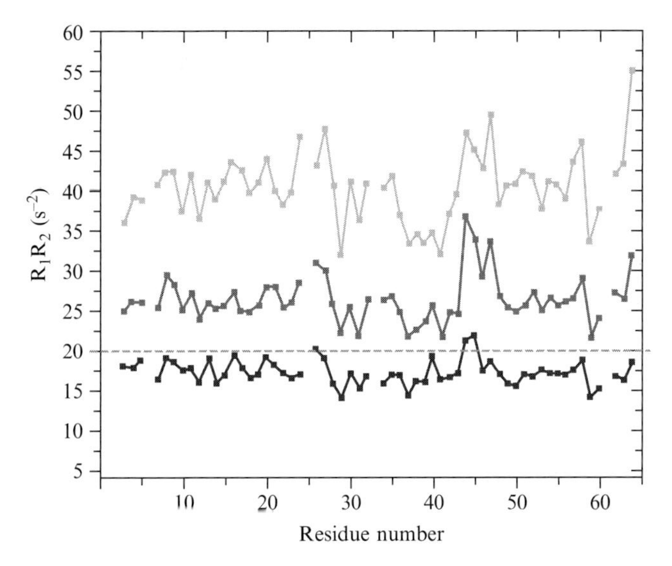

Figure 1.3 Histograms of R_1R_2 values as a function of residue number for chymotrypsin inhibitor 2 at 25 °C in 42% glycerol (blue) and 200 g/L BSA at pH 5.4 (red) and pH 6.8 (green). The rigid limit is also shown. Reprinted with permission from Li and Pielak (2009). Copyright 2009 American Chemical Society. (See Color Insert.)

(Fig. 1.1). For EX2 exchange, the exchange event (represented by k_{obs}) is rate determining (Eq. (1.4)). As a result, a plot of log k_{obs} of a given residue versus pH will yield a line of unitary slope with a nonzero intercept. The same result is expected for a plot of log k_{obs} for individual residues for one value of pH versus a second value. In this instance, the intercept will equal the difference in pH values.

When the test protein exchanges via the EX1 limit, and rate of opening (k_{op}) is not pH dependent, k_{int} is not the rate determining step for exchange. The pH should not affect log k_{obs}. Likewise, a plot of log k_{obs} for individual residues for one value of pH versus a second value will result in a line of unitary slope and an intercept of zero.

Using the pH dependence of k_{obs} to determine the exchange limit is well-established for dilute solution studies (Neira *et al.*, 1997; Qu and Bolen, 2003) and has yielded success in macromolecular crowding conditions (Charlton *et al.*, 2008) in which the stability of the protein remained constant over the pH range studied. The size of some crowding agents depends on pH, which can change the stability of the target protein. In this instance, another technique can be used.

Nuclear Overhauser enhancement spectroscopy (NOESY) creates a correlation between NMR-active nuclei that are spatially close to one another. Wagner (1980) showed that the exchange limit can be determined by analyzing the disappearance of an NOE signal from a partially exchanged sample for which exchange has been halted. Time-resolved NOESY-detected exchange experiments are now possible (Neira *et al.*, 1997). For the purposes of this experiment, there are three types of crosspeaks of interest: an amide–amide crosspeak corresponding to the combined decay of two amide protons with rate $k_{obs(A,\ B)}$, and two amide–aliphatic crosspeaks corresponding to each proton's individual decay, with rates $k_{obs(A)}$ and $k_{obs(B)}$. For the EX1 limit,

$$k_{obs(A,B)} = k_{obs(A)} = k_{obs(B)} \tag{1.8}$$

For an EX2 limit, the exchanges should be independent events, and

$$k_{obs(A,B)} = k_{obs(A)} + k_{obs(B)} \tag{1.9}$$

To determine the exchange limit, the relationship between these exchange rates must be analyzed.

An exchange sample is prepared with ^{15}N-enriched protein. Consecutive ^{15}N-filtered ^{1}H–^{1}H NOESY spectra (Kim and Szyperski, 2003; Kupce and Freeman, 2003, 2004) are then obtained over the course of 12 h. The ^{15}N filtering removes a significant portion of signals from C_{α} protons and other aliphatic protons. The resulting spectra are analyzed to identify NOE crosspeaks in the amide–amide and amide–aliphatic regions that correspond to amide backbone proton resonances. Crosspeak volumes are plotted as a function of time, and a first-order rate constant is obtained from fits to a single exponential function. This amide–amide "linked" decay is then compared to the individual proton decays from the amide–aliphatic crosspeaks to determine exchange limit. Our results for chymotrypsin inhibitor 2 are shown in Table 1.1. The data show that the k_{obs} value for a given amide–amide crosspeak corresponds to the sum of the k_{obs} values of the corresponding amide–aliphatic crosspeaks. Furthermore, the correspondence between the rate constants measured from the NOESY spectra and the HSQC spectra lend confidence to the conclusion that chymotrypsin inhibitor 2 exchanges under the EX2 limit in 50 m*M* sodium acetate buffer at 37 °C, pH 5.4. This conclusion also agrees with a previous study (Neira *et al.*, 1997). Having established an EX2 limit, one final control experiment is needed.

5.5. Intrinsic exchange rate

The value of the intrinsic exchange rate (k_{int}) depends not only on the acidity and solvation of the amide nitrogens (Avbelj and Baldwin, 2009) but may also be affected by the behavior of water (LeMaster *et al.*, 2006). High concentrations of crowding agents could alter these parameters either directly or indirectly by affecting the pK_a of H_2O. If this were to occur, k_{int} values from SPHERE would no longer be applicable to crowded conditions.

We attempted to measure the effect of 40 kDa PVP on k_{int} (Charlton *et al.*, 2008) using methods described by Bai *et al.* (1993). The experiment involved the exchange of L-alanyl-L-alanine under crowded conditions, but

Table 1.1 Determination of exchange limit by NOESY

Residue	NOESY k_{obs} ($s^{-1} \times 10^{-5}$)	HSQC k_{obs} ($s^{-1} \times 10^{-5}$)
17	65.8	73.6
18	35.4	28.6
Sum(17, 18)	101.2	102.2
17/18 crosspeak	77.0	N/A
19	4.60	5.30
20	3.39	3.30
Sum(19, 20)	7.99	8.60
19/20 crosspeak	8.41	N/A
58	5.51	7.06
59	4.93	5.33
Sum(58, 59)	10.44	12.39
58/59 crosspeak	10.4	N/A

Calculated exchange rates for chymotrypsin inhibitor 2 based on NOESY and HSQC exchange experiments (50 m*M* acetate buffer, pH 5.4, 37 °C). N/A, not applicable.

our results suggests a strong interaction between the peptide and PVP, obviating the measurement, so we turned to an alternative method.

Chymotrypsin Inhibitor 2 contains an extended loop between residues 33 and 44 that is not maximally compacted. We assumed this loop mimics an unstructured peptide, and resolved to measure k_{int} through determination of the rapid exchange rates of these residues (Charlton *et al.*, 2008). A water-saturation transfer experiment, CLEANEX-PM (Hwang *et al.*, 1998) was used. In contrast to the protocol described in Section 6, CLEANEX-PM does not require fully deuterated solutions, and can measure rates for exchanges that are too fast (3–55 s^{-1}) to be detected using traditional amide ^{1}H exchange (8×10^{-6}–3×10^{-3} s^{-1}).

Two experiments were performed. A control sample of lyophilized protein dissolved in buffer was analyzed to determine dilute solution k_{int} values. These values were compared to values from SPHERE (Zhang, 1995). A second sample of lyophilized protein dissolved in buffered crowding solution was then analyzed. The presence of the crowding agent did not affect k_{int} (Charlton *et al.*, 2008). With the knowledge that the intrinsic exchange rate is not changed by the experimental setup, amide ^{1}H exchange experiments can proceed.

6. A Protocol for Amide ^{1}H Exchange

There are two methods to prepare a sample for amide ^{1}H exchange. Buffer exchange can be achieved by using centrifugal filter devices, but the viscosity of crowded solutions makes buffer exchange difficult. Alternatively, lyophilized protein can be added to prepared solutions. The latter method is greatly preferred for the sake of time and convenience. For these reasons, we focus on samples made from lyophilized protein. Two samples are prepared. The first sample is used to adjust the shims. This sample contains a low ionic strength solution (<150 m*M*) with 10–20% D_2O (to lock the spectrometer) at the desired pH, and contains 1 m*M* protein, but has no crowding agent. The second sample is an exchange sample, and comprises 100% D_2O buffered solution at the same ionic strength and pH with crowding agent at the desired concentration. Enough lyophilized protein to bring the exchange sample to a final concentration of 1 m*M* is brought to the spectrometer. The protein is not mixed into the sample until later.

After shim optimization, the lyophilized protein is added to the exchange sample and mixed. This complete sample is then immediately spun in a microcentrifuge at 16,000×*g* for 1 min to remove any insoluble aggregates. The supernatant is transferred to an NMR tube, which is immediately placed in the magnet. The 90° pulse width is optimized by examining the first increment of a ^{15}N–^{1}H HSQC spectrum (Bodenhausen and Ruben, 1980; Kay *et al.*, 1992). Multiple (20–24) sequential $^{15}N-^1H$ HSQC spectra are obtained using the established parameters. Upon completion, the sample tube is again examined for aggregates and the pH is checked. The HSQC data are then processed, making sure to use the same processing parameters for each spectrum, to yield a list of residue numbers with crosspeak volumes as a function of time. To determine k_{obs}, each peak decay plot is fitted to a three parameter exponential function. Some crosspeaks overlap, which creates difficulties. To overcome this problem, both peaks are measured together and a five parameter, double exponential function can be used. Care must be taken to determine which of the two peaks is the faster-exchanging species. We find that the uncertainties in k_{obs} from curve fitting are of equal magnitude to those from reproducing the experiment. To convert the k_{obs} values into $K_{op}^{0'}$ values, the k_{int} value for each residue must be known.

Section 5.5 describes the methods for determining whether k_{int} values calculated from SPHERE can be used for the crowded solutions. If this is not possible, k_{int} values in crowded conditions can be quantified by using the technique described by Bai *et al.* (1993). An alternative would be to adjust k_{int} values as determined by the CLEANEX-PM experiment. This method has not been tested, and its use assumes that the alterations to k_{int} in the loop region of a protein uniformly apply to all residues.

7. Summary and Future Directions

Amide ^{1}H exchange provides a valuable tool to investigate the stability of proteins. NMR techniques add to its utility by providing residue-level interrogation. When coupled with *in vitro* experiments using macromolecular crowding agents, valuable information about the combination of excluded volume effects and nonspecific interactions can be obtained. Although we have focused on macromolecular crowding, the methods should be applicable to any solvent additive, whether stabilizing (Charlton *et al.*, 2008) or destabilizing (Bai *et al.*, 1995). *In vitro* studies should provide insight as to how crowded environments like cell interiors affect protein properties. To understand macromolecular crowding fully, however, it will be necessary to design experiments that isolate the excluded volume effect from nonspecific interactions in a more quantitative way.

The ultimate goal is to use NMR-detected amide ^{1}H exchange to quantify protein stability in living cells. Although a qualitative study has been reported using NMR spectroscopy (Inomata *et al.*, 2009), the only quantitative studies of in-cell protein stability have used mass spectrometry to detect exchange (Ghaemmaghami and Oas, 2001) or fluorescence detection of urea denaturation (Ignatova and Gierasch, 2004). The development of amide

[1]H exchange techniques to quantify $\Delta G_{op}^{0'}$ inside cells will open the way for residue-level interrogation of protein stability in cells, and bring us closer to understanding proteins as they function in their natural environment.

ACKNOWLEDGMENTS

We thank Elizabeth Pielak, Oscar Millet, and Mohona Sarkar for helpful comments.

REFERENCES

Ai, X., Zhou, Z. H., Bai, Y., and Choy, W.-Y. (2006). ^{15}N NMR spin relaxation dispersion study of the molecular crowding effects on protein folding under native conditions. *J. Am. Chem. Soc.* **128,** 3916–3917.

Avbelj, F., and Baldwin, R. L. (2009). Origin of the change in solvation enthalpy of the peptide group when neighboring peptide groups are added. *Proc. Natl. Acad. Sci. USA* **106,** 3137–3141.

Bai, Y., Milne, J. S., Mayne, L., and Englander, S. W. (1993). Primary structure effects on peptide group hydrogen exchange. *Proteins: Struct. Funct. Genet.* **17,** 75–86.

Bai, Y. W., Sosnick, T. R., Mayne, L., and Englander, S. W. (1995). Protein folding intermediates: Native-state hydrogen exchange. *Science* **269,** 192–197.

Betz, S. F., and Pielak, G. J. (1992). Introduction of a disulfide bond into cytochrome *c* stabilizes a compact denatured state. *Biochemistry* **31,** 12337–12344.

Bodenhausen, G., and Ruben, D. J. (1980). Natural abundance nitrogen-15 NMR by enhanced heteronuclear spectroscopy. *Chem. Phys. Lett.* **69,** 185–189.

Charlton, L. M., Barnes, C. O., Li, C., Orans, J., Young, G. B., and Pielak, G. J. (2008). Macromolecular crowding effects on protein stability at the residue level. *J. Am. Chem. Soc.* **130,** 6826–6830.

Cohen, D. S., and Pielak, G. J. (1994). Stability of yeast iso-1-ferricytochrome *c* as a function of pH and temperature. *Protein Sci.* **3,** 1253–1260.

Dedmon, M. M., Patel, C. N., Young, G. B., and Pielak, G. J. (2002). FlgM gains structure in living cells. *Proc. Natl. Acad. Sci. USA* **99,** 12681–12684.

Doyle, D. F., Waldner, J. C., Parikh, S., Alcazar-Roman, L., and Pielak, G. J. (1996). Changing the transition state for protein (un)folding. *Biochemistry* **35,** 7403–7411.

Englander, S. W., and Kallenbach, N. R. (1983). Hydrogen exchange and structural dynamics of proteins and nucleic acids. *Q. Rev. Biophys.* **16,** 521–655.

Ferraro, D. M., Lazo, N. D., and Robertson, A. D. (2004). EX1 hydrogen exchange and protein folding. *Biochemistry* **43,** 587–594.

Frost, A. A., and Pearson, R. G. (1953). *Kinetics and Mechanism.* Wiley, New York, NY.

Ghaemmaghami, S., and Oas, T. G. (2001). Quantitative protein stability measurement *in vivo. Nat. Struct. Biol.* **8,** 879–882.

Huang, G. S., and Oas, T. G. (1995). Structure and stability of monomeric λ repressor: NMR evidence for two-state folding. *Biochemistry* **34,** 3884–3892.

Hvidt, A. A., and Nielsen, S. O. (1966). Hydrogen exchange in proteins. *Adv. Protein Chem.* **21,** 287–386.

Hwang, T.-L., van Zijl, P. C. M., and Mori, S. (1998). Accurate quantitation of water-amide exchange rates using the phase-modulated CLEAN chemical EXchange (CLEANEX-PM) approach with a fast-HSQC (FHSQC) detection scheme. *J. Biomol. NMR* **11,** 221–226.

Ignatova, Z., and Gierasch, L. M. (2004). Monitoring protein stability and aggregation *in vivo* by real-time fluorescent labeling. *Proc. Natl. Acad. Sci. USA* **101,** 523–528.
Inomata, K., Ohno, A., Tochio, H., Isogai, S., Tenno, T., Nakase, I., Takeuchi, T., Futaki, S., Ito, Y., Hiroaki, H., and Shirakawa, M. (2009). High-resolution multi-dimensional NMR spectroscopy of proteins in human cells. *Nature* **458,** 106–109.
Kay, L., Keifer, P., and Saarinen, T. (1992). Pure absorption gradient enhanced heteronuclear single quantum correlation spectroscopy with improved sensitivity. *J. Am. Chem. Soc.* **114,** 10663–10665.
Kim, S., and Szyperski, T. (2003). GFT NMR, a new approach to rapidly obtain precise high-dimensional NMR spectral information. *J. Am. Chem. Soc.* **125,** 1385–1393.
Kupce, E., and Freeman, R. (2003). Projection-reconstruction of three-dimensional NMR spectra. *J. Am. Chem. Soc.* **125,** 13958–13959.
Kupce, E., and Freeman, R. (2004). Projection-reconstruction technique for speeding up multidimensional NMR spectroscopy. *J. Am. Chem. Soc.* **126,** 6429–6440.
Kuttner, Y. Y., Kozer, N., Segal, E., Schreiber, G., and Haran, G. (2005). Separating the contribution of translational and rotational diffusion to protein association. *J. Am. Chem. Soc.* **127,** 15138–15144.
LeMaster, D. M., Anderson, J. S., and Hernandez, G. (2006). Role of native-state structure in rubredoxin native-state hydrogen exchange. *Biochemistry* **45,** 9956–9963.
Li, C., and Pielak, G. J. (2009). Using NMR to distinguish viscosity effects from nonspecific protein binding under crowded conditions. *J. Am. Chem. Soc.* **131,** 1368–1369.
Lumry, R., Biltonen, R., and Brandts, J. F. (1966). Validity of the "two-state" hypothesis for conformational transitions of proteins. *Biopolymers* **4,** 917–944.
Molday, R. S., and Kallen, R. G. (1972). Substituent effects on amide hydrogen-exchange rates in aqueous-solution. *J. Am. Chem. Soc.* **94,** 6739–6745.
Molday, R. S., Englander, S. W., and Kallen, R. G. (1972). Primary structure effects on peptide group hydrogen exchange. *Biochemistry* **11,** 150–158.
Molyneux, P. (1983). *Water-soluble synthetic polymers: Properties and behavior.* CRC Press, Boca Raton, CA.
Munishkina, L. A., Ahmad, A., Fink, A. L., and Uversky, V. N. (2008). Guiding protein aggregation with macromolecular crowding. *Biochemistry* **47,** 8993–9006.
Neira, J. L., Itzhaki, L. S., Otzen, D. E., Davis, B., and Fersht, A. R. (1997). Hydrogen exchange in chymotrypsin inhibitor 2 probed by mutagenesis. *J. Mol. Biol.* **270,** 99–110.
Perham, M., Stagg, L., and Wittung-Stafshede, P. (2007). Macromolecular crowding increases structural content of folded proteins. *FEBS Lett.* **581,** 5065–5069.
Pielak, G. J., Li, C., Miklos, A. C., Schlesinger, A. P., Slade, K. M., Wang, G., and Zigoneanu, I. G. (2008). Protein NMR under physiological conditions. *Biochemistry* **48,** 226–234.
Privalov, P. L., and Khechinashvili, N. N. (1974). A thermodynamic approach to the problem of stabilization of globular protein structure: A calorimetric study. *J. Mol. Biol.* **86,** 665–684.
Qu, Y. X., and Bolen, D. W. (2003). Hydrogen exchange kinetics of RNase A and the urea: TMAO paradigm. *Biochemistry* **42,** 5837–5849.
Richards, F. M. (1977). Areas, volumes, packing, and protein structure. *Annu. Rev. Biophys. Bioeng.* **6,** 151–176.
Rubinstein, M., and Colby, R. (2003). *Polymer Physics.* Oxford University Press, New York, NY.
Schanda, P., and Brutscher, B. (2005). Very fast two-dimensional NMR spectroscopy for real-time investigation of dynamic events in proteins on the time scale of seconds. *J. Am. Chem. Soc.* **127,** 8014–8015.
Schowen, B. K., Schowen, R. L., and Daniel, L. P. (1982). Solvent isotope effects on enzyme systems. *Methods Enzymol.* **87,** 551–606.

Serber, Z., and Dötsch, V. (2001). In-cell NMR spectroscopy. *Biochemistry* **40,** 14317–14323.

Stagg, L., Zhang, S.-Q., Cheung, M. S., and Wittung-Stafshede, P. (2007). Molecular crowding enhances native structure and stability of α/β protein flavodoxin. *Proc. Natl. Acad. Sci. USA* **104,** 18976–18981.

Wagner, G. (1980). A novel application of nuclear Overhauser enhancement (NOE) in proteins: Analysis of correlated events in the exchange of internal labile protons. *Biochem. Biophys. Res. Commun.* **97,** 614–620.

Wyatt, P. J. (1993). Light-scattering and the absolute characterization of macromolecules. *Anal. Chim. Acta* **272,** 1–40.

Zhang, Y.-Z. (1995). *Protein and peptide structure and interactions studied by hydrogen exchange and NMR*. University of Pennsylvania.

Zhou, H. X., Rivas, G. N., and Minton, A. P. (2008). Macromolecular crowding and confinement: Biochemical, biophysical, and potential physiological consequences. *Annu. Rev. Biophys.* **37,** 375–397.

Zimmerman, S. B., Trach, S. O. (1991). Estimation of Macromolecule Concentrations and Excluded Volume Effects for the Cytoplasm of *Escherichia coli*. *J. Mol. Biol.* **222,** 599–620.

CHAPTER TWO

Fluorescence Spectroscopy in Thermodynamic and Kinetic Analysis of pH-Dependent Membrane Protein Insertion

Alexey S. Ladokhin*

Contents

* Department of Biochemistry and Molecular Biology, University of Kansas Medical Center, Kansas City, Kansas, USA

Methods in Enzymology, Volume 466

ISSN 0076-6879, DOI: 10.1016/S0076-6879(09)66002-X

Abstract

Experimental determination of the free energy stabilizing the structure of membrane proteins in their native lipid environment is undermined by a lack of appropriate methods and suitable model systems. Here, we demonstrate how fluorescence correlation spectroscopy can be used to characterize thermodynamics of pH-triggered bilayer insertion of nonconstitutive membrane proteins (e.g., bacterial toxins, colicins). The experimental design is guided by the appropriate thermodynamic scheme which considers two independent processes: pH-dependent formation of a membrane-competent form and its insertion into the lipid bilayer. Measurements of a model protein annexin B12 under conditions of lipid saturation demonstrate that protonation leading to the formation of the membrane-competent state occurs near membrane interface. Lipid titration experiments demonstrate that the free energy of transfer to the intermediate interfacial state is especially favorable, while the free energy of final insertion is modulated by interplay of hydrophobic and electrostatic interactions on the bilayer interface. The general principles of kinetic measurements along the insertion pathway containing interfacial intermediate are discussed and practical examples emphasizing appropriate fitting and normalization procedures are presented.

Abbreviations

ANX	Annexin B12
ANX-143C-NBD ANX-144C-NBD	NBD labeled single-cysteine mutants T143C and S144C of ANX
NBD	7-nitrobenz-2-oxa-1,3-diazol-4-yl
LUV	large unilamellar vesicles
POPC	palmitoyloleoylphosphatidylcholine
POPG	palmitoyloleoylphosphatidylglycerol
25PC:75PG and 75PC:25PG	mixtures of POPC and POPG that contain a molar percentage of corresponding lipid specified by the number
TM	transmembrane
IF	interfacial
F_{MC}	fraction of membrane-competent ANX
ΔG	free energy of transfer from water to membrane

φ	membrane surface potential
FCS	fluorescence correlation spectroscopy

1. Introduction: Co-Translational Versus Post-Translational Membrane Protein Insertion

Folding and stability of membrane proteins remains one of the most elusive problems in physical biochemistry. Generally, membrane protein folding and bilayer insertion is managed by complex multiprotein translocon complexes (Alder and Johnson, 2004; Johnson and van Waes, 1999; Rapoport *et al.*, 2004). For nonconstitutive proteins (e.g., bacterial toxins (Collier and Young, 2003; Miller *et al.*, 1999; Oh *et al.*, 1999; Senzel *et al.*, 2000; Shatursky *et al.*, 1999) and colicins (Parker *et al.*, 1990; Tory and Merrill, 1999; Zakharov *et al.*, 1999)), however, such insertion is achieved spontaneously, in response to changes in environment. For example, acidification of the endosome causes conformational change in endocytosed diphtheria toxin T-domain, resulting in its insertion into the membrane and translocation of its own N-terminus with the attached catalytic domain into the cytoplasm, possibly through a pore in the lipid bilayer (Oh *et al.*, 1999). Neither translocon-assisted nor spontaneous membrane insertion of proteins is well understood on a molecular level. Despite the obvious structural and mechanistic differences between the two, recent thermodynamic evidence indicates that the underlying physicochemical principles for these processes are likely to be the same (Bowie, 2005; Hessa *et al.*, 2005; White and von Heijne, 2005). Thus, deciphering these principles with the help of spontaneously inserting proteins is relevant to the larger problems of membrane protein folding and stability. The situation with membrane proteins is not too different from that of water-soluble proteins, which also can have assisted and spontaneous folding. Although it is likely that most proteins in the cell require molecular chaperones for efficient folding, the main thermodynamic principles that govern folding of soluble proteins were established from thorough studies of spontaneously folding proteins. Similarly, spontaneous insertion of nonconstitutive proteins can yield important insights into the much broader problem of membrane protein folding and stability, critical for predicting structure from sequence. Here, we present an overview of experimental approaches which utilize fluorescence spectroscopy to characterize kinetics and thermodynamics of pH-triggered insertion of membrane proteins into lipid bilayers.

2. Challenges of Thermodynamic Analysis of Membrane Protein Folding/Insertion

2.1. Lack of additivity of electrostatic and hydrophobic interactions on membrane interfaces

Since most biological membrane are negatively charged, it is important to understand how electrostatic and hydrophobic interactions add up in the interfacial region. While hydrophobic (Fernandez-Vidal *et al.*, 2007; Wimley and White, 1996) and electrostatic (Murray *et al.*, 1997) interactions between peptides lipid bilayers are reasonably well understood *separately*, their *relative interplay* in macromolecules with both charged and hydrophobic segments has not received the attention it deserves, and it is often assumed that the two components of the free energy are additive, although without any substantial evidence. Several studies on various systems, including antimicrobial peptides (Ladokhin and White, 2001) and ion-channel blockers (Posokhov *et al.*, 2007), demonstrate the general lack of additivity of free energy components originating from these interactions. Although an empirical rule has been derived from these results allowing the prediction of the free energy of membrane interactions of hydrophobic peptides and segments of proteins with mixed anionic/zwitterionic membranes (Ladokhin and White, 2001), the nature of this phenomenon or the general applicability of such a rule are not clear. The balance of electrostatic and hydrophobic interactions affects membrane insertion of many membrane-active peptides and proteins (Bradshaw, 2000; Heller *et al.*, 1998; Poklar *et al.*, 1999; Wieprecht *et al.*, 2000), most strikingly colicin E1(Zakharov *et al.*, 1996). The latter study presents convincing evidence for an optimal anionic lipid concentration (which is close to physiological values) for channel formation by colicin E1.

2.2. Experimental exploration of transmembrane insertion

Direct experimental exploration of the folding and stability of membrane proteins has been hindered by their insolubility. But because membrane proteins are equilibrium structures, their folding and stability can be examined by studying various aspects of the membrane interactions of peptides and constructing thermodynamic pathways, involving partitioning of unfolded chain, interfacial folding, transbilayer insertion, and association of TM helices. These pathways are not intended to depict the actual biological assembly process, but rather to provide a thermodynamic context within which assembly of membrane protein must proceed. A detailed and in-depth description of these approaches can be found in reviews of White and coworkers (White *et al.*, 1998, 2001).

While various model peptides were extremely useful experimental models for studying interfacial binding and folding (Fernandez-Vidal *et al.*, 2007;

Hristova and White, 2005; Ladokhin and White, 1999, 2001; Wimley and White, 1996; Wimley *et al.*, 1998), deciphering the energetics of transbilayer insertion turned out to be much more elusive. A systematic study of membrane interactions of a designed peptide TMX-3 (Ladokhin and White, 2004) demonstrated that the interfacial folded state is the most thermodynamically stable state for a self-inserting peptide. This trend cannot be reversed by an increase in sequence hydrophobicity, since the latter results in peptide precipitation in solution (rendering thermodynamic analysis impossible) prior to any noticeable increase in insertion. To circumvent this limitation one has to switch from peptides to more complex protein systems to study the energetics of membrane insertion such as annexin B12 and diphtheria toxin T-domain. Both of these proteins undergo a reversible pH-dependent membrane insertion (Ladokhin and Haigler, 2005; Ladokhin *et al.*, 2004) which makes them attractive models for thermodynamic and kinetic studies using fluorescence correlation spectroscopy (FCS) and other spectroscopic approaches (Ladokhin *et al.*, 2002; Posokhov *et al.*, 2008a,b).

3. FCS and Protein–Membrane Interactions

FCS measures intensity fluctuations of a small number of fluorescent molecules diffusing through a small focal volume. Due to recent technical developments, FCS is gaining popularity in biological studies (Haustein and Schwille, 2003; Hess *et al.*, 2002; Schwille *et al.*, 1999). Several groups have utilized it to study interfacial binding and transmembrane insertion of peptides and proteins (Posokhov *et al.*, 2008a,b; Rhoades *et al.*, 2006; Rusu *et al.*, 2004). Rusu *et al.* (2004) compared FCS results to those available from other measurements and found them to be in excellent agreement, opening the doors for quantitative application of FCS to membrane protein binding.

3.1. FCS experiment

The description of FCS experimental setup, calibration, data collection and multiple biological applications can be found in these reviews (Chen *et al.*, 2008; Garcia-Saez and Schwille, 2008; Haustein and Schwille, 2004, 2007). Here, we present several aspects of FCS measurements specific to thermodynamic studies of protein–membrane interactions. Our experiments were conducted using commercially available instrumentation utilizing confocal microscopy technology (e.g., MicroTime 200 from PicoQuant, Berlin, Germany or Alba FCS workstation from ISS, Champaign, IL) as described in detail before (Posokhov *et al.*, 2008a). For a model membrane system we highly recommend the use of large unilamellar vesicles (LUV) with a diameter of ~0.1 μm, which can be prepared by extrusion (Mayer *et al.*, 1986).

The use of the smaller sonicated vesicles (SUV) should be avoided, as they are metastable and known to produce anomalous binding (Greenhut *et al.*, 1986; Ladokhin *et al.*, 2000; Seelig and Ganz, 1991), due perhaps to distorted lipid packing associated with high surface curvature. Protein should be labeled with a relatively bright and photostable fluorophore (Alexa series of dyes from Invitrogen, Carlsbad, CA is a popular choice) at a unique position. Normally we used single cysteine mutants (ANX Q4C or DTT N235C) labeled either with Alexa488 or Alexa647 meleimide. The labeling site is chosen so as not to interfere with membrane interactions. The samples typically contained 0.5–6 n*M* of dye-labeled protein, which falls on the linear range for FCS concentration measurements, determined to be 6×10^{-8}–5×10^{-11} *M* for ANX-Alexa647 and 10^{-7}–4×10^{-10} *M* for ANX-Alexa488.

3.2. FCS data analysis

The autocorrelation function for single diffusing species can be described with the following equation (Hess *et al.*, 2002):

$$G(\tau) = \frac{1}{N} \times g(\tau) = \frac{1}{N} \cdot \left(1 + \frac{T}{1-T} \cdot e^{-\tau/\tau_P}\right) \cdot \left(\frac{1}{1+\tau/\tau_D}\right) \cdot \left(\frac{1}{1+\tau/S^2\tau_D}\right)^{1/2} \quad (2.1)$$

where $\boldsymbol{N}$ is the average number of the fluorescent molecules in the focus volume and τ_D is the correlation time of the particles. The correlation time represents the diffusion time through the focus volume and equals $\tau_D = \omega^2/4D$, where ω^2 is the square of the radius of the laser focus and D is the diffusion constant. S is the ratio of the distances from the center of the laser beam focus in the radial and axial directions, respectively. T is the fraction of fluorophores in the triplet state and τ_P is the phosphorescence lifetime. Because τ_P is normally several orders of magnitude shorter than the diffusion time of biological macromolecules, the phosphorescence term does not interfere with the determination of protein interactions with membranes and is ignored in our subsequent discussion.

The measured correlation function $G(\tau)$ of a multicomponent system is a weighted sum of the autocorrelation functions of each component $g_i(\tau)$ with amplitudes A_i (Clamme *et al.*, 2003):

$$G(\tau) = \sum_{i=1}^{M} A_i \cdot g_i(\tau) = \frac{\sum_{i-1}^{M} q_i^2 N_i \, g_i(\tau)}{\left[\sum_{i-1}^{M} q_i N_i\right]^2} \quad (2.2)$$

where N_i is the mean particle number and q_i is the ratio of the fluorescence yield of the ith component to that of the first component. For thermodynamic studies of membrane binding only two diffusing species need to be considered: the fluorescently labeled proteins (*index* P) and bilayer vesicles with bound fluorescently labeled proteins (*index* V):

$$G(\tau) = A_P \cdot g_P(\tau) + A_V \cdot g_V(\tau). \tag{2.3}$$

In our studies of membrane binding we have labeled membrane-inserting proteins ANX (Posokhov *et al.*, 2008a) and diphtheria toxin T-domain (Posokhov *et al.*, 2008b) with Alexa fluorophores which are rather convenient, as their quantum yield is normally insensitive to the changes in environment so that $q_i = 1$ (at least for the systems we have measured so far). Although not crucial, this further simplifies quantitative analysis as the amplitudes in Eq. (2.3) will now depend only on the numbers of fast and slow moving particles in the focal volume (N_P and N_V , respectively):

$$A_P = \frac{N_P}{(N_V + N_P)^2} \quad A_V = \frac{N_V}{(N_V + N_P)^2} \tag{2.4}$$

3.3. Determination of membrane partitioning from FCS data

The goal of these binding studies is to determine the fraction of the protein associated with the membrane: $f_B = [P]_{bound}/[P]_{total}$. There are two principal ways of extracting quantitative information on membrane binding from the FCS data (Eqs. (2.5) and (2.6)), both of which start with the fitting of the data collected in a sample containing free and vesicle-bound protein with Eq. (2.3). Information on the fraction of free and bound ANX can be extracted from the values of weighting factors, depending on certain assumptions.

3.3.1. Infinite dilution regime and FCS measurements at lipid saturation

The infinite dilution regime refers to the conditions of an overwhelming molar excess of lipid over protein ($\sim 10^6$) so that each vesicle contains no more than a single bound protein. It also requires that the protein be the monomer, or, at least, that the number of subunits in the oligomer is known. In this case, the number of particles associated with the slow mobility will be equal to the number of vesicle-bound protein molecules and the fraction of bound protein will be defined by the following expression (Clamme *et al.*, 2003; Posokhov *et al.*, 2008a):

$$f_B = \frac{A_V}{A_V + A_P} \tag{2.5}$$

Because of the extremely high sensitivity of the FCS experiment, reliable data can be obtained using subnanomolar concentrations of protein, suggesting that the infinite dilution regime can be achieved using a millimolar range of lipid concentrations. The main advantage of this approach is that a single autocorrelation curve, measured at a particular lipid concentration, contains all the information needed to calculate free and bound protein fractions. The example of such an experiment presented in Fig. 2.1 reveals a progressive shift of the autocorrelation curve upon lowering the pH, consistent with progressive binding of ANX to vesicles. The diffusion time measured at pH 7.2 coincided with that of ANX in the absence of vesicles, while one measured at pH 4.6 coincided with that of a labeled vesicle. Thus, the limiting curves in Fig. 2.1 can be assigned to a totally free and totally membrane-bound ANX. This assignment allowed us to determine the fractions of free and bound protein at intermediate pH using the global (or linked) analysis approach (Knutson *et al.*, 1983). Specifically, the diffusion times for free and bound ANX were linked for all four curves, while the fractions of free and bound ANX for the two middle curves were allowed to float. The resulting global fit (solid curves) is rather good and provides accurate determination of the bound fractions. Note that the samples are not required to have exactly the same protein concentration and the fractions of free and bound protein can be calculated from the normalized autocorrelation curves.

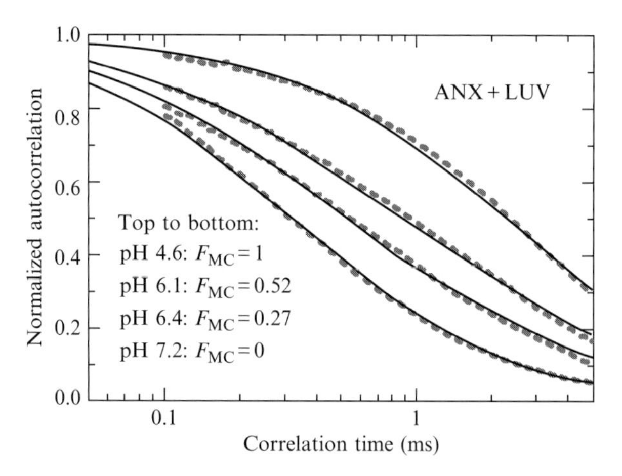

Figure 2.1 Example of binding measurement using FCS performed under conditions of the "infinite dilution regime" satisfied at extremely high lipid excess over fluorescently labeled protein ($\sim 10^6$). Acidification results in a progressive shift of mobility due to a sixfold increase of the correlation time resulting from binding of labeled ANX to the vesicles. Quantitative determination of partitioning is achieved by a global analysis (solid lines) of fluorescence autocorrelation curves (dashed lines) that links the two correlation times for all curves and allows free fitting of the preexponential amplitudes (see Eq. (2.3)). All data are normalized to the same number of fluorescent particles in the focal volume for better visual representation (Note that the absolute values of amplitudes are not important under these conditions, but only their relative contributions (Eq. (2.5)). Reproduced from Posokhov *et al.* (2008a) with permission.

3.3.2. Lipid titration FCS measurements

If the desired experimental conditions do not satisfy the requirements of the infinite dilution regime, the fraction of bound protein can still be calculated from the FCS experiment, in this case by comparing measurements in the presence and in the absence of vesicles. The autocorrelation collected in the presence of LUV is used to determine the amplitude, associated with the free protein (A_P) by fitting the data with Eq. (2.3). The autocorrelation for the sample containing the same amount of protein, but no LUV, will be fitted with Eq. (2.1) to determine the total number of proteins/peptides in the focal volume (N). The fraction of the bound protein will be equal (Rusu *et al.*, 2004):

$$f_B = 1 - A_P N \tag{2.6}$$

An example of such an analysis to membrane interactions of ANX (Posokhov *et al.*, 2008a) is presented in Fig. 2.2: symbols are experimental autocorrelation curves and the solid lines are fitting results with Eq. (2.3) (or Eq. (2.1) in case of no LUV sample). Addition of increasing concentrations of LUV results in progressive membrane binding and reduction in concentration of free ANX. This leads to an overall reduction in the number of fluorescent particles, since under these conditions a single vesicle caries more than a single labeled protein. The average number of particles in the

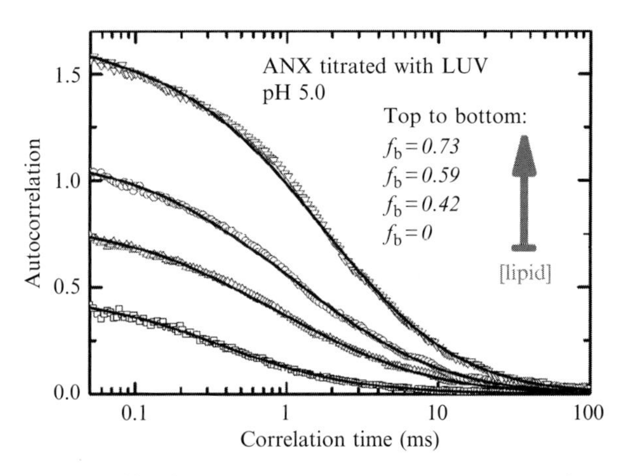

Figure 2.2 Example of lipid titration measurement using FCS performed outside of the "infinite dilution regime." Correlation curves (symbols) were measured for samples containing 1 nM ANX and no LUV or LUV with 1, 2, and 4 μM of total lipid LUV (bottom-to-top). Progressive vesicle binding of fluorescently labeled ANX results in a reduction of the number of fluorescent particles in the focal volume and a corresponding increase in total autocorrelation amplitude. Solid curves correspond to fit with Eq. (2.3) (or Eq. (2.1), for sample containing no LUV). The fractions of bound ANX f_b are calculated by comparing absolute amplitudes in the presence and absence of LUV (Eq. (2.6)). Reproduced from Posokhov *et al.* (2008a) with permission.

focal volume is inversely proportional to the autocorrelation amplitude. Therefore, the amplitude is expected to increase with increasing lipid concentration, which is indeed observed (Fig. 2.2). The quantitative analysis of binding is based on fitting individual curves to a two-component autocorrelation function Eq. (2.3) (solid lines), and extracting the fractions of bound ANX by comparing relative amplitudes in the presence and absence of LUV (Eq. (2.6)).

Although the lipid titration approach based on Eq. (2.6) is more general than the analysis allowed under the conditions of infinite dilution regime (Eq. (2.5)), it is not without drawbacks. First of all, since it relies on comparison of amplitudes in two protein samples, one in the absence and one in the presence of LUV, it is essential that the samples have exactly the same concentration of the protein. This is not a trivial task to accomplish, as membrane samples are notorious for variances in protein concentration due to precipitation and absorption to the surfaces. Another limitation of this approach is an increasing uncertainty at saturation of binding, as the fraction of the free protein is reduced drastically, and, as a result, the autocorrelation amplitudes do not change appreciably. The obvious limitation of the alternative approach based on Eq. (2.5) lies in the stringent requirements of the infinite dilution regime, which makes measurements extremely difficult when lipid concentration is below 50 μM. Nevertheless, we have demonstrated that both approaches are in excellent agreement when comparable conditions are used (Posokhov *et al.*, 2008a).

4. Thermodynamic Schemes for Analysis of Membrane Partitioning

The general definitions of thermodynamic parameters, such as the mole-fraction partitioning coefficient and the free energy of transfer, as well as an overview of the principles and methods used for determining the energetics of the partitioning of peptides into bilayer membranes are discussed in detail by White and coworkers (White *et al.*, 1998). In this section, we present the extension of the ideas developed in thermodynamic studies of peptides and demonstrate their application to more complex and challenging systems, such as nonconstitutive membrane proteins (e.g., bacterial toxins, colicins, and annexins). One of the main distinctions of proteins, as compared to peptides, is that they seldom follow a simple partitioning mode and the description of their membrane interactions requires a more complex thermodynamic scheme, which accounts for the existence of membrane-competent and membrane-incompetent protein conformations. In Section 4.1, we use the example of pH-dependent membrane insertion of ANX to illustrate this type of analysis.

4.1. Free energy of pH-triggered membrane protein insertion

The thermodynamic scheme describing acid-induced membrane interactions of such proteins as ANX or diphtheria toxin T-domain considers two independent processes (Fig. 2.3): pH-dependent formation of a membrane-competent form and its interaction with the lipid bilayer. (We would like to point out that FCS measurements as described here are not designed to distinguish the interfacially bound (Fig. 2.7, I-state) and transmembrane-inserted (T-state) protein conformations. This can be done, however, using other fluorescence techniques, such as steady-state and lifetime quenching topology methods (Ladokhin *et al.*, 2002; Posokhov and Ladokhin, 2006).) By combining these topology measurements with FCS-measured changes in autocorrelation during lipid titration, one can determine the free energy ΔG of both transmembrane insertion and interfacial penetration (Posokhov *et al.*, 2008a).

First, the fraction of bound protein (in this case ANX) as a function of lipid concentration is estimated depending on experimental conditions discussed above using either Eq. (2.5) or (2.6). The resulting titration isotherm (Fig. 2.4) is fitted to a mole-fraction partitioning equation (Ladokhin *et al.*, 2000; White *et al.*, 1998), modified to account for the presence of membrane-incompetent protein species (Posokhov *et al.*, 2008a):

$$f_B = F_{MC} \cdot \frac{K_x \cdot [L]}{[W] + K_x \cdot [L]} \tag{2.7}$$

where F_{MC} is a fraction of membrane-competent ANX (determined at lipid saturation), [L] is lipid concentration, [W] is water concentration (55.3 M), and K_x is mole-fraction partitioning coefficient which is used to determine the free energy of binding $\Delta G = -RT \ln K_x$.

4.2. pH-dependent formation of membrane-competent state

The next step in thermodynamic analysis involves characterization of the protonation required for the formation of membrane-competent protein conformation (Fig. 2.3) and can be accomplished by FCS measurements conducted under lipid saturation. Such measurements usually satisfy the conditions for infinite dilution regime described in Section 3.3.1, which simplifies their experimental implementation and analysis. A typical example of such an experiment is presented in Fig. 2.1, for which fractions of membrane-bound ANX are estimated using Eq. (2.5). Because this experiment is performed under conditions of lipid saturation, the fraction of bound protein corresponds to the fraction of membrane-competent form F_{MC} at this pH (Fig. 2.5). The plots of F_{MC} versus pH, referred to as titration profiles of membrane binding, contain information on the protein protonation required for membrane association (Fig. 2.1) and can be fitted to the following equation:

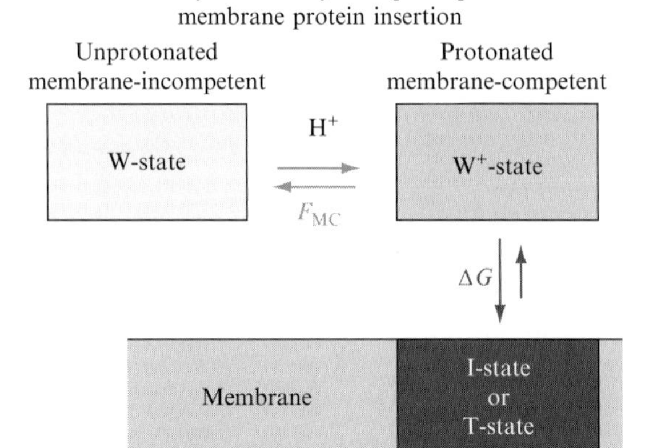

Free energy of membrane partitioning (ΔG):

$$\Delta G = -RT \cdot \ln(K_x) = -RT \cdot \ln\left(\frac{f_B/[L]}{f_F/[W]}\right)$$

K_x = mole-fraction partitioning coefficient

f_B = membrane-bound protein fraction

f_F = free protein fraction

[L] = lipid concentration

[W] = water concentration (55.3 M)

Simple partitioning: $f_F + f_B = 1$

Partitioning in the presence of membrane-incompetent form:

$f_F + f_B = F_{MC} = 1 - F_{MI}$

F_{MI} = fraction of membrane-incompetent protein

F_{MC} = fraction of membrane-competent protein

($f_B = F_{MC}$ and $f_F = 0$ when $[L] = \infty$)

$$f_B([L]) = F_{MC} \cdot \frac{K_x \cdot [L]}{[W] + K_x \cdot [L]}$$

Figure 2.3 Thermodynamic scheme employed in the analysis of the energetics of pH-triggered membrane interactions of proteins (Posokhov *et al.*, 2008a), such as Annexin B12 and diphtheria toxin T-domain. Environment acidification leads to the formation of a membrane-competent form (horizontal arrows), which interacts with the lipid bilayer (vertical arrows). Each of the two processes is characterized by an independent equilibrium parameter: pH-dependent fraction of the membrane-competent form (F_{MC}) and the free energy of membrane partitioning of this form (ΔG). Both parameters, F_{MC} and ΔG, can be experimentally determined via FCS methodology using the principal equations presented in the lower panel, as described in the text. Note that a transmembrane T-state and an interfacial I-state are distinguished in a separate membrane topology experiment as described in these publications (Ladokhin *et al.*, 2002; Posokhov and Ladokhin, 2006; Posokhov *et al.*, 2008a).

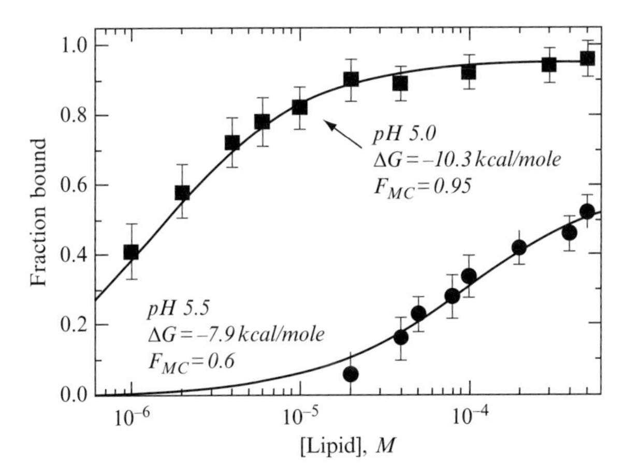

Figure 2.4 Typical membrane-binding isotherms used in determination of free energy of partitioning. Data for ANX binding to 75PC:25PG LUV was measured at pH 5.5 (circles) and pH 5 (squares). Solid curves are the fitting results using Eq. (2.7) with the corresponding parameters presented on the graph. Note that both parameters, the fraction of the membrane-competent form F_{MC} and the free energy of membrane partitioning ΔG, are pH-dependent. Reproduced from Posokhov *et al.* (2008a) with permission.

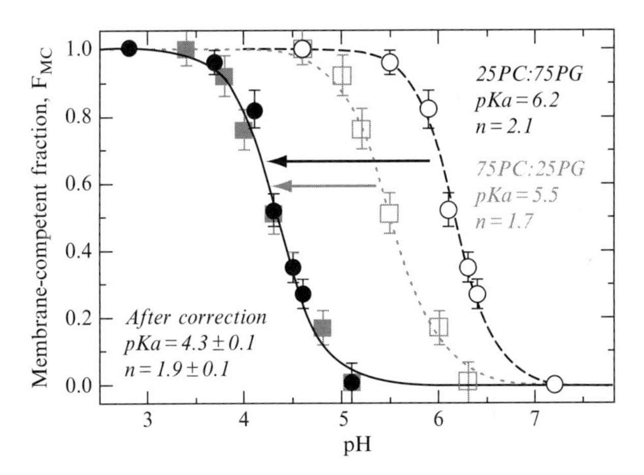

Figure 2.5 Analysis of pH-dependent formation of membrane-competent state of ANX. Squares and circles correspond to 75PC:25PG and 25PC:75PG lipid compositions, respectively. The same data sets are plotted using either uncorrected bulk pH values (open symbols) or corrected pH values (Eq. (2.9)) near the membrane interfaces with different surface potentials (closed symbols). Corrected pH-dependences for the formation of the membrane-competent form measured with the two lipid compositions coincide and can be fitted with a single curve using Eq. (2.8) (black line). The resulting parameters suggest that the transition to a membrane-competent form occurs near membrane interface with a p*Ka* of 4.3 and requires protonation of two residues ($n \sim 2$). Reproduced from Posokhov *et al.* (2008a) with permission.

$$F_{MC} = \frac{1}{1 + 10^{n(\mathrm{pH}-\mathrm{pK})}} \tag{2.8}$$

where pK is a negative logarithm of the dissociation constant, and n is the Hill coefficient.

In general, the following two limiting cases should be considered for interpreting pH-triggered membrane association: (a) protonation occurs predominantly in the bulk phase of solution and (b) protonation occurs predominantly on the interface. The first scenario will result in titration profiles being completely independent of the lipid bilayer, while in the second one a relatively strong dependence on the surface potential is expected. Protonation of ANX falls under the latter case, as the corresponding apparent pK's differ substantially: 5.5 ± 0.2 for 75PC:25PG and 6.2 ± 0.1 for 25PC:75PG (Posokhov *et al.*, 2008a). This difference for the apparent protonation behavior in solution (presumed to be lipid independent) is surprising only at first glance. In reality, formation of the membrane-competent form is likely to occur near the membrane interface, where local proton concentration is different from that in the bulk solution. Moreover, local proton concentration will be different for the two lipid compositions because of the difference in the surface potential. One can correct for such variation using the equation relating pH values at membrane interfaces and in the bulk of solution with the values of membrane surface potential (φ) (Fernandez and Fromherz, 1977; Posokhov *et al.*, 2008a):

$$\mathrm{pH}_{\mathrm{interface}} = \mathrm{pH}_{\mathrm{bulk}} - \frac{F\varphi}{RT} \tag{2.9}$$

where F and R are Faraday and gas constants, respectively, and T is the absolute temperature. Surprisingly, this rather simple correction gives excellent results and the combined dataset (Fig. 2.5 solid symbols) can be fitted with Eq. (2.6) (black line) and the following parameters are obtained: $\mathrm{pK} = 4.3 \pm 0.1$; $n = 1.9 \pm 0.1$.

5. Kinetic Analysis of Membrane Protein Insertion

5.1. Interface-directed membrane insertion

To enter the membrane, a nonconstitutive membrane protein must pass through the interfacial region of the bilayer, a region that must play an important role in protein refolding (White *et al.*, 2001). The general concept that emerges from the peptide studies is that the interfacial region of the bilayer can (a) provide special accommodation for large elements of protein structure undergoing refolding and (b) serve as a catalyst for the conformational change resulting in transbilayer insertion. Extension of the peptide results to more complex proteins suggests the importance of transient interfacial unfolded states. Indeed, kinetically populated or stable intermediates have been reported for a number of proteins including colicin E1 (Zakharov *et al.*, 1999), diphtheria toxin T-domain (Chenal *et al.*, 2002; Ladokhin *et al.*, 2004; Wang *et al.*, 1997), and ANX (Ladokhin *et al.*, 2002; Posokhov *et al.*, 2008a). All this evidence allows one to put forward the following hypothesis of interface-directed membrane insertion: an important component of a membrane insertion pathway for nonconstitutive membrane proteins is an obligatory interfacial intermediate state, characterized by high content of secondary structure and few native contacts. Formation of this intermediate and subsequent transbilayer insertion is mediated by a subtle balance of hydrophobic and electrostatic interactions between proteins and the membrane interface. Below we discuss the applications of fluorescence spectroscopy for kinetic measurements of ANX insertion along the pathway from aqueous W-state to interfacial intermediate I-state and finally to transbilayer inserted T-state (Fig. 2.7).

5.2. Spectroscopic approaches to study insertion pathways containing intermediates

5.2.1. On the interpretation of quantitative kinetic analysis

The exact functional form of kinetic equation in a pathway containing intermediate states can be rather complex and generally nonalgebraic (Moore and Pearson, 1981). Working with membrane systems further complicates kinetic analysis, as it limits the usable ranges of concentrations of interacting species. Therefore, it is customary to utilize empirical exponential decay (Eq. (2.10a)), or closely related to it, exponential association functions (Eq. (2.10b)) for quantitative analysis of a certain spectroscopic signal S that changes with the time passed since initiation of reaction, t (Compton *et al.*, 2006; Constantinescu and Lafleur, 2004; Ladokhin, 1999; Ladokhin *et al.*, 2002; Zakharov *et al.*, 1999):

$$S(t) = A_0 + A_1 \cdot e^{-k_1 t} + A_2 \cdot e^{-k_2 t} \tag{2.10a}$$

$$S(t) = A_0 + A_1 \cdot (1 - e^{-k_1 t}) + A_2 \cdot (1 - e^{-k_2 t}) \tag{2.10b}$$

where A_i are the amplitudes of the corresponding kinetic component with apparent rates of k_i. A_0 can represent a constant value corresponding to the final (Eq. (2.10a)) or initial signal (Eq. (2.10b)), or it can be interpreted as an amplitude associated with an infinitely slow, $k_0 = \infty$, to an infinitely fast component with $k_0 = 0$, in Eqs. (10a) and (10b), respectively.

Fitting with exponential functions, however, often leads to poorly defined fits (a common problem for fitting with any nonorthogonal functions), which further affects the reliability of apparent kinetic rates obtained from such analysis. Given all these limitations, one should be particularly aware of the danger of overinterpretation when describing complex kinetic data. Nevertheless, certain conclusions can be attempted, as long as the underlying assumptions are clearly stated. For a general reaction containing a single intermediate (e.g., W $\leftrightarrow$ I $\leftrightarrow$ T), the formation of the final state follows a biexponential law, in which the apparent kinetic parameters are complex functions of the four kinetic rates for each of the direct and reverse reaction steps. The assumption that under the conditions of the experiment the reverse reactions can be neglected simplifies the mathematical expressions and results in the apparent kinetic rates coinciding with the two rates for each step of the direct reaction (Moore and Pearson, 1981).

5.2.2. Spectroscopic measurements along W $\rightarrow$ I $\rightarrow$ T pathway

The time course of the change in spectroscopic linear-response parameter, $S(t)$, along a W (water-soluble) $\rightarrow$ I (interfacial intermediate) $\rightarrow$ T (transmembrane) pathway is as follows:

$$S(t) = S_W \cdot f_W(t) + S_I \cdot f_I(t) + S_T \cdot f_T(t) \tag{2.11}$$

where S_W, S_I, and S_T are the intensities and f_W, f_I, and f_T are the fractional populations of the W, I, and T states, respectively. It should be noted that not all spectroscopic signals satisfy Eq. (2.11), but only those that are linear response parameters (for more discussion see Ladokhin *et al.*, 2000; White *et al.*, 1998). For example, intensity measured at any constant wavelength and at any orientation of polarizers (or without them) is an appropriate linear-response parameter, while position of spectral maximum or ratio of intensities at any two wavelengths are not. The time-dependencies of the fractional populations are given by Ladokhin *et al.* (2002) and Moore and Pearson (1981):

$$f_W(t) = e^{-k_{wi}t} f_I(t) = \frac{k_{wi}}{k_{it} - k_{wi}} (e^{-k_{wi}t} - e^{-k_{it}t})$$

$$f_T(t) = 1 - f_W(t) - f_I(t) \tag{2.12}$$

where k_{wi} is the rate of the W $\rightarrow$ I transition and k_{it} the rate of the I $\rightarrow$ T transition. The underlying assumption is that the reverse rates can be neglected compared to direct rates, that is, rate for W $\leftarrow$ I and for I $\leftarrow$ T transition is much smaller than that for W $\rightarrow$ I and I $\rightarrow$ T transition, respectively. This assumption significantly simplifies mathematical expression of the kinetics and allows for easy comparison of the model with empirical fitting functions. Note for example, that under these conditions the apparent rate constants k_1 and k_2 obtained by fitting with Eqs. (2.10a) or (2.10b) coincide with the transition rates k_{wi} and k_{it} for the W $\rightarrow$ I $\rightarrow$ T pathway (Eqs. (2.11) and (2.12)).

5.3. Example of kinetic analysis of membrane insertion of ANX

We illustrate the application of kinetic analysis outlined above using the transmembrane insertion of the ANX observed upon acidification in the absence of Ca^{+2}. The crystallographic structure of the soluble W-state is known (Luecke *et al.*, 1995) and the final inserted T-state was shown to form a water-filled pore by site-directed EPR studies (Langen *et al.*, 1998). A helix-loop-helix region of ANX stretching between residues 134 and 162 (Fig. 2.6) undergoes a conformational change and becomes a single transmembrane helix in the inserted state with a residues 134 and 162 on the *cis* and *trans* side of the membrane, respectively (Ladokhin *et al.*, 2002; Posokhov *et al.*, 2008a). The kinetics of this process is complex and demonstrates the existence of the interfacial intermediate, which can be stabilized by high content of anionic lipids in the membrane (Ladokhin *et al.*, 2002; Posokhov *et al.*, 2008a). At moderate contents of anionic lipids ($\sim$25%) this intermediate can be observed kinetically by following the fluorescence of environment-sensitive probe NBD attached at single cysteine residues introduced in the middle of the TM segment in positions 143 (Fig. 2.6, dotted line) or 144 (dashed line). The trace corresponding to the latter is most indicative of intermediate, as the fluorescence intensity undergoes a rapid increase followed by slow decrease (the decrease of intensity at zero time corresponds to closing of the shutters during vesicle injection and is ignored during the analysis). For ANX-143C-NBD the intermediate can be identified from the nonexponential character of intensity increase: a rapid jump followed by slower increase.

Given together the data in Fig. 2.6A indicate a rapid formation of the membrane-associated intermediate state in which the probes attached at positions 143 and 144 are located in a similar environment, and hence similar intensity, followed by slow formation of the final T-state. Because this state is a water-filled pore comprised of several helical segments of the monomeric ANX (Ladokhin and Haigler, 2005; Langen *et al.*, 1998), the slow changes in intensity are position dependent. For the pore-exposed residue 144 the intensity is decreased, while for residue 143, which is facing the lipid, an additional increase is observed. We would like to point out that the intensities in the initial W-state are different, which is consistent with the degree of exposure expected from the crystallographic structure (Fig. 2.6, upper panel). This difference is confirmed by the variation in the lifetime of NBD dye being longer for the ANX-143C-NBD in solution

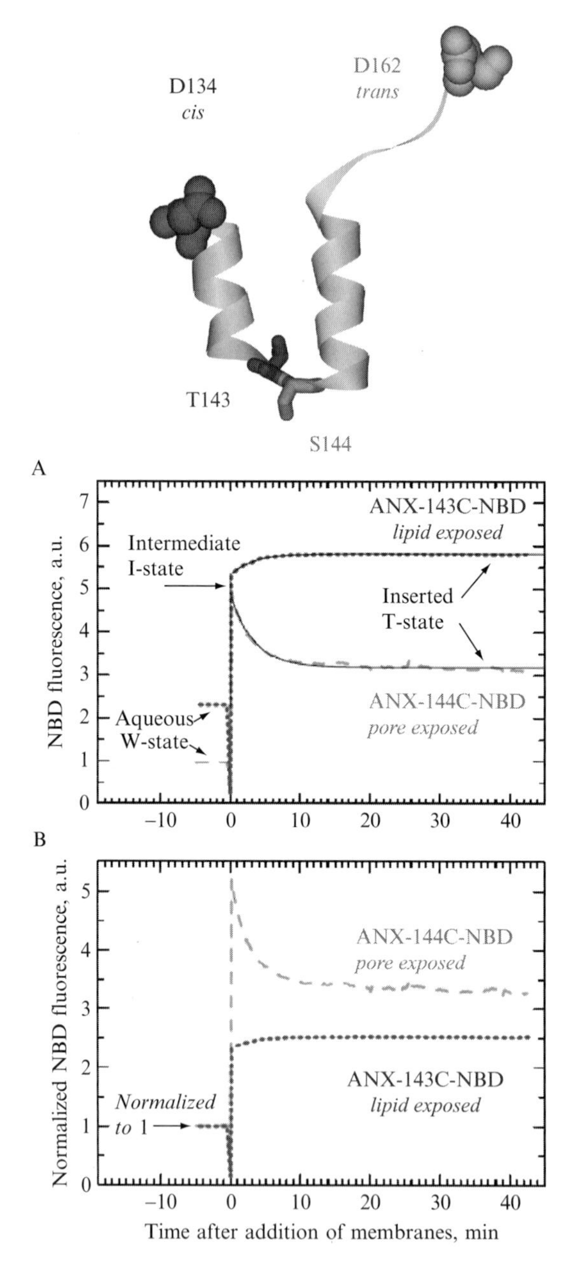

Figure 2.6 Example of kinetic study of pH-triggered membrane protein insertion pathway containing interfacial intermediate. A ribbon structure represents a segment of ANX flanked by residues 134 and 162 which is a helix-loop-helix in a soluble W-state with a known three-dimensional structure (the rest of the protein structure is omitted for clarity). At low pH in the absence of $Ca2^{+}$, ANX inserts into the lipid

than for ANX-144C-NBD (Ladokhin, unpublished results). Note that the interpretation of relative changes in intensity is more obscured, when the traces are normalized to the same value of signal in W-state (Fig. 2.6B). Unfortunately, this rather common practice of comparing different mutants or proteins contradicts physical principles (see Eq. (2.11)) and can lead to misinterpretation of results (e.g., one would not be able to identify similar environments for both 143 and 144 positions for I-state, based on Fig. 2.6B).

Because the initial binding event occurred faster than the dead-time of the hand-mixing experiment (<10 s), we could not recover the value of the rate of formation of the I-state and fitted the data with a simplified model containing a step and a single exponential component (Eq. (2.10b) with $A_2 = 0$). To more reliably determine the rate of insertion we have fitted both datasets in a linked analysis (Knutson *et al.*, 1983), where a single remaining kinetic constant was shared, and amplitudes were allowed to float independently for the two mutants. The resulting curves (solid lines) adequately describe the data and the rate of T-state formation was found to be $k_{it} = 0.005\ s^{-1}$. Because the k_{wi} rate is more than an order of magnitude higher, the kinetic analysis is rather robust. When the two rates are becoming similar, the reliability of fit decreases and other restraints should be applied (e.g., comparison of kinetics collected in different lipid systems Ladokhin *et al.*, 2002). One possibility would be to use additional experiments which are sensitive only to binding, but not insertion. Another would be the use of a double-kinetic approach (Beechem, 1997), which relies on a combination of stopped-flow kinetics and fluorescence lifetime measurements.

bilayer and forms a water-filled pore in which this segment becomes a single transmembrane helix with residues 134 and 162 on *cis* and *trans* sides, respectively (Ladokhin *et al.*, 2002; Langen *et al.*, 1998). The time-dependencies of the change in fluorescence of environment-sensitive probe NBD, attached to single-cysteine mutants T143C and S144C, are shown as dotted and dashed lines, respectively. Membrane interactions were initiated at zero time. Both panels contain the same data normalized to the same amount of the labeled protein (A) or to the same initial intensity (B). The latter normalization obscures correct interpretation of fluorescence changes and precludes proper identification of the intermediate state (see Eq. (2.11) and text for details). The analysis of properly normalized fluorescence kinetics (A) reveals fast formation of the intermediate I-state in which the environment of the probes attached at residues 143 and 144 is similar. The variation in probes environment in the final T-state is consistent with their different exposure to the pore and is confirmed by fluorescence lifetime measurements. Solid lines represent linked fits of the two kinetic datasets using Eq. (2.10b) as described in the text.

6. Perspectives: Annexin B12 as a Model for Thermodynamic and Kinetic Analysis of Membrane Protein Insertion, Folding and Misfolding

Different stages of the pH-triggered insertion pathway of ANX, along with their corresponding timescales, are summarized in Fig. 2.7. A fast transition from the water-soluble W-state to an interfacial intermediate

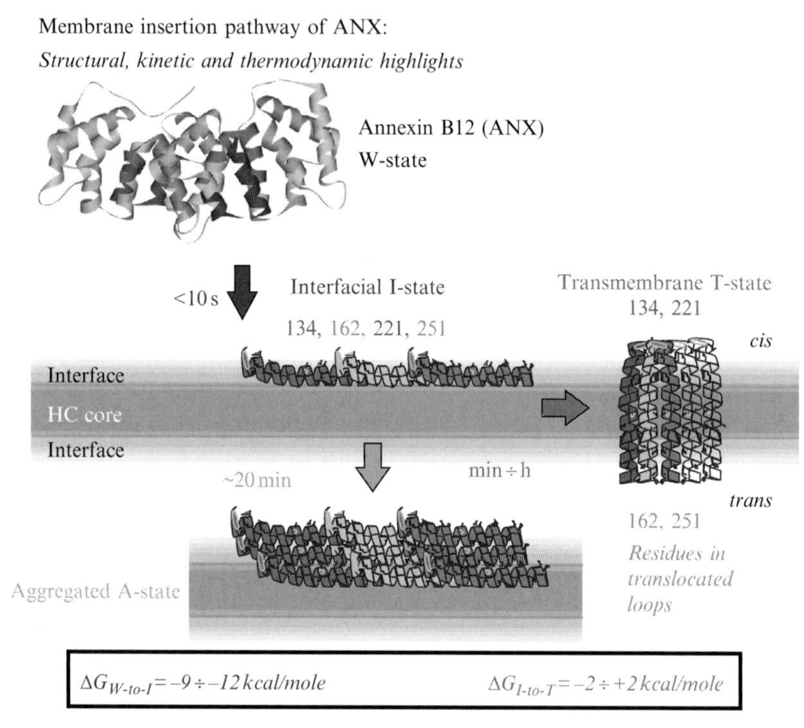

Figure 2.7 pH-triggered membrane insertion pathway of ANX, consisting of a water-soluble W-state, final transmembrane T-state, interfacial intermediate I-state, and aggregated A-state. The initial binding and unfolding (W-to-I transition) occurs very fast (Ladokhin *et al.*, 2002), while the final insertion (I-to-T transition), resulting in translocation of several helical hairpins (Posokhov *et al.*, 2008a), may take somewhere from minutes to hours, depending on the lipid composition (see also Fig. 2.6 for an example of kinetic experiment). Excess of anionic lipids will stabilize the insertion intermediate (Ladokhin *et al.*, 2002; Posokhov *et al.*, 2008a), which at high concentrations is prone to aggregation on the timescale of tens of minutes (Ladokhin, 2008). The main gain in free energy occurs upon interfacial penetration ($\sim -9 \div -12$ kcal/mol), while the free energy of final insertion is modulated by an interplay of hydrophobic and electrostatic interactions in a rather narrow free energy range ($-2 \div +2$ kcal/mol) (Posokhov *et al.*, 2008a).

I-state (<10 s; Ladokhin *et al.*, 2002) is followed by a slower insertion to a final TM T-state. The rate of the latter process is strongly slowed by a high content of anionic lipids, presumably *via* stabilization of an I-state with Coulombic attraction between anionic lipids and cationic protein groups. If enough protein lingers in the I-state, it starts aggregating (A-state), while the properly inserted T-state shows no signs of aggregation in the same concentration range. At the present time we do not understand what forces contribute to this selective aggregation of the noninserted ANX, but the nature of the interfacial zone of the lipid bilayer is bound to play a role. Importantly, all other transitions involving W-, I-, and T-states are reversible and do not involve oligomerization (Ladokhin and Haigler, 2005; Ladokhin *et al.*, 2004), which simplifies thermodynamic and kinetic analysis. Application of the FCS methodology had permitted accurate determination of the free energy of transfer of ANX from solution into the lipid bilayer. Remarkably, the main gain in free energy occurs already on the first step of the insertion pathway, that is, during formation of interfacial intermediate state ($-9 \div -12$ kcal/mol). The free energy for the final insertion changes from mildly favorable to mildly unfavorable ($+2 \div -2$ kcal/mol) and depends on the lipid composition (Ladokhin, 2008; Posokhov *et al.*, 2008a). All this makes ANX an attractive template for further thermodynamic and kinetic studies of pH-triggered membrane insertion using a combination of spectroscopic approaches described here with the site-directed mutagenesis of key residues involved in insertion and refolding.

One of the reasons that studies of folding and stability of membrane proteins lag far behind those of their soluble counterparts is a shortage of appropriate experimental models of folding, and what is perhaps even more challenging, of *misfolding*. As illustrated in Fig. 2.7, ANX can also serve as a model for a misfolding process in which hindered insertion leads to aggregation on membrane interfaces (Ladokhin, 2008; Posokhov *et al.*, 2008a). In the cell, a misfolded(-inserted) state of *constitutive* membrane proteins is believed to be quickly disposed of via a tightly set protein degradation machinery (Sanders and Myers, 2004). Moreover, even a large fraction of the inserted protein is degraded in order to minimize the risk of accumulating any misinserted protein. The evidence for the interfacial state of ANX being prone to aggregation sheds some light on the possible reasons why this may be the case. If aggregation on membrane interfaces is also shared by misinserted constitutive proteins, the cell must act quickly to dispose of them or else deal with much tougher aggregates. We will follow the example of Sanders and Myers, who with following quote drew attention to the dangers attendant on folding membrane protein (Charles Spurgeon, sermon at Newington, 1889, quoted after (Sanders and Myers, 2004)): "*Where there is danger there should be prudent haste. Quick! Pilgrim, be quick and tarry not in the place of danger*" ... or you might end up aggregated on membrane interfaces.

ACKNOWLEDGMENTS

We are grateful to Mr. M. A. Myers for his editorial assistance. This research was supported by NIH GM-069783.

REFERENCES

Alder, N. N., and Johnson, A. E. (2004). Cotranslational membrane protein biogenesis at the endoplasmic reticulum. *J. Biol. Chem.* **279,** 22787–22790.

Beechem, J. M. (1997). Picosecond fluorescence decay curves collected on millisecond time scale: Direct measurement of hydrodynamic radii, local/global mobility, and intramolecular distances during protein-folding reactions. *Methods Enzymol.* **278,** 24.

Bowie, J. U. (2005). Border crossing. *Nature* **433,** 367–369.

Bradshaw, J. P. (2000). Phosphatidylglycerol promotes bilayer insertion of salmon calcitonin. *Biophys. J.* **72,** 2180–2186.

Chen, H., Farkas, E. R., and Webb, W. W. (2008). Chapter 1: *In vivo* applications of fluorescence correlation spectroscopy. *Methods Cell Biol.* **89,** 3–35.

Chenal, A., Savarin, P., Nizard, P., Guillain, F., Gillet, D., and Forge, V. (2002). Membrane protein insertion regulated by bringing electrostatic and hydrophobic interactions into play. A case study with the translocation domain of the diphtheria toxin. *J. Biol. Chem.* **277,** 43425–43432.

Clamme, J. P., Azoulay, J., and Mely, Y. (2003). Monitoring of the formation and dissociation of polyethylenimine/DNA complexes by two photon fluorescence correlation spectroscopy. *Biophys. J.* **84,** 1960–1968.

Collier, R. J., and Young, J. A. (2003). Anthrax toxin. *Annu. Rev. Cell Dev. Biol.* **19,** 45–70.

Compton, E. L., Farmer, N. A., Lorch, M., Mason, J. M., Moreton, K. M., and Booth, P. J. (2006). Kinetics of an individual transmembrane helix during bacteriorhodopsin folding. *J. Mol. Biol.* **357,** 325–338.

Constantinescu, I., and Lafleur, M. (2004). Influence of the lipid composition on the kinetics of concerted insertion and folding of melittin in bilayers. *Biochim. Biophys. Acta* **1667,** 26–37.

Fernandez, M. S., and Fromherz, P. (1977). Lipoid pH indicators as probes of electrical potential and polarity in micelles. *J. Phys. Chem.* **81,** 1755–1761.

Fernandez-Vidal, M., Jayasinghe, S., Ladokhin, A. S., and White, S. H. (2007). Folding amphipathic helices into membranes: Amphiphilicity trumps hydrophobicity. *J. Mol. Biol.* **370,** 459–470.

Garcia-Saez, A. J., and Schwille, P. (2008). Fluorescence correlation spectroscopy for the study of membrane dynamics and protein/lipid interactions. *Methods* **46,** 116–122.

Greenhut, S. F., Bourgeois, V. R., and Roseman, M. A. (1986). Distribution of cytochrome b_5 between small and large unilamellar phospholipid vesicles. *J. Biol. Chem.* **261,** 3670–3675.

Haustein, E., and Schwille, P. (2003). Ultrasensitive investigations of biological systems by fluorescence correlation spectroscopy. *Methods* **29,** 153–166.

Haustein, E., and Schwille, P. (2004). Single-molecule spectroscopic methods. *Curr. Opin. Struct. Biol.* **14,** 531–540.

Haustein, E., and Schwille, P. (2007). Fluorescence correlation spectroscopy: Novel variations of an established technique. *Annu. Rev. Biophys. Biomol. Struct.* **36,** 151–169.

Heller, W. T., Waring, A. J., Lehrer, R. I., and Huang, H. W. (1998). Multiple states of β-sheet peptide protegrin in lipid bilayers. *Biochemistry* **37,** 17331–17338.

Hess, S. T., Huang, S., Heikal, A. A., and Webb, W. W. (2002). Biological and chemical applications of fluorescence correlation spectroscopy: A review. *Biochemistry* **41,** 697–705.

Hessa, T., Kim, H., Bihlmaler, K., Lundin, C., Boekel, J., Andersson, H., Nilsson, I., White, S. H., and von Heijne, G. (2005). Recognition of transmembrane helices by the endoplasmic reticulum translocon. *Nature* **433,** 377–381.

Hristova, K., and White, S. H. (2005). An experiment-based algorithm for predicting the partitioning of unfolded peptides into phosphatidylcholine bilayer interfaces. *Biochemistry* **44,** 12614–12619.

Johnson, A. E., and van Waes, M. A. (1999). The translocon: A dynamic gateway at the ER membrane. *Annu. Rev. Cell Dev. Biol.* **15,** 799–842.

Knutson, J. R., Beechem, J. M., and Brand, L. (1983). Simultaneous analysis of multiple fluorescence decay curves: A global approach. *Chem. Phys. Lett.* **102,** 501–507.

Ladokhin, A. S. (1999). Evaluation of lipid exposure of tryptophan residues in membrane peptides and proteins. *Anal. Biochem.* **276,** 65–71.

Ladokhin, A. S. (2008). Insertion intermediate of annexin B12 is prone to aggregation on membrane interfaces. *Biopolym. Cell* **24,** 101–104.

Ladokhin, A. S., and Haigler, H. T. (2005). Reversible transition between the surface trimer and membrane-inserted monomer of annexin 12. *Biochemistry* **44,** 3402–3409.

Ladokhin, A. S., and White, S. H. (1999). Folding of amphipathic α-helices on membranes: Energetics of helix formation by melittin. *J. Mol. Biol.* **285,** 1363–1369.

Ladokhin, A. S., and White, S. H. (2001). Protein chemistry at membrane interfaces: Non-additivity of electrostatic and hydrophobic interactions. *J. Mol. Biol.* **309,** 543–552.

Ladokhin, A. S., and White, S. H. (2004). Interfacial folding and membrane insertion of a designed helical peptide. *Biochemistry* **43,** 5782–5791.

Ladokhin, A. S., Jayasinghe, S., and White, S. H. (2000). How to measure and analyze tryptophan fluorescence in membranes properly, and why bother? *Anal. Biochem.* **285,** 235–245.

Ladokhin, A. S., Isas, J. M., Haigler, H. T., and White, S. H. (2002). Determining the membrane topology of proteins: Insertion pathway of a transmembrane helix of annexin 12. *Biochemistry* **41,** 13617–13626.

Ladokhin, A. S., Legmann, R., Collier, R. J., and White, S. H. (2004). Reversible refolding of the diphtheria toxin T-domain on lipid membranes. *Biochemistry* **43,** 7451–7458.

Langen, R., Isas, J. M., Hubbell, W. L., and Haigler, H. T. (1998). A transmembrane form of annexin XII detected by site-directed spin labeling. *Proc. Natl. Acad. Sci. USA* **95,** 14060–14065.

Luecke, H., Chang, B. T., Mailliard, W. S., Schlaepfer, D. D., and Haigler, H. T. (1995). Crystal structure of the annexin XII hexamer and implications for bilayer insertion. *Nature* **378,** 512–515.

Mayer, L. D., Hope, M. J., and Cullis, P. R. (1986). Vesicles of variable sizes produced by a rapid extrusion procedure. *Biochim. Biophys. Acta* **858,** 161–168.

Miller, C. J., Elliott, J. L., and Collier, R. J. (1999). Anthrax protective antigen: Prepore-to-pore conversion. *Biochemistry* **38,** 10432–10441.

Moore, J. W., and Pearson, R. G. (1981). Kinetics and mechanism. Wiley, New York.

Murray, D., Ben-Tal, N., Honig, B., and McLaughlin, S. (1997). Electrostatic interaction of myristoylated proteins with membranes: Simple physics, complicated biology. *Structure* **5,** 985–989.

Oh, K. J., Senzel, L., Collier, R. J., and Finkelstein, A. (1999). Translocation of the catalytic domain of diphtheria toxin across planar phospholipid bilayers by its own T domain. *Proc. Natl. Acad. Sci. USA* **96,** 8467–8470.

Parker, M. W., Tucker, A. D., Tsernoglou, D., and Pattus, F. (1990). Insights into membrane insertion based on studies of colicins. *Trends Biochem. Sci.* **15,** 126–129.

Poklar, N., Fritz, J., Macek, P., Vesnaver, G., and Chalikian, T. V. (1999). Interaction of the pore-forming protein equinatoxin II with model lipid membranes: A calorimetric and spectroscopic study. *Biochemistry* **38,** 14999–15008.

Posokhov, Y. O., and Ladokhin, A. S. (2006). Lifetime fluorescence method for determining membrane topology of proteins. *Anal. Biochem.* **348,** 87–93.

Posokhov, Y. O., Gottlieb, P. A., Morales, M. J., Sachs, F., and Ladokhin, A. S. (2007). Is lipid bilayer binding a common property of inhibitor cysteine knot ion-channel blockers? *Biophys. J.* **93,** L20–L22.

Posokhov, Y. O., Rodnin, M. V., Lu, L., and Ladokhin, A. S. (2008a). Membrane insertion pathway of annexin B12: Thermodynamic and kinetic characterization by fluorescence correlation spectroscopy and fluorescence quenching. *Biochemistry* **47,** 5078–5087.
Posokhov, Y. O., Rodnin, M. V., Das, S. K., Pucci, B., and Ladokhin, A. S. (2008b). FCS study of the thermodynamics of membrane protein insertion into the lipid bilayer chaperoned by fluorinated surfactants. *Biophys. J.* **95,** L54–L56.
Rapoport, T. A., Goder, V., Heinrich, S. U., and Matlack, K. E. S. (2004). Membrane-protein integration and the role of the translocation channel. *Trends Cell Biol.* **14,** 568–575.
Rhoades, E., Ramlall, T. F., Webb, W. W., and Eliezer, D. (2006). Quantification of alpha-synuclein binding to lipid vesicles using fluorescence correlation spectroscopy. *Biophys. J.* **90,** 4692–4700.
Rusu, L., Gambhir, A., McLaughlin, S., and Radler, J. (2004). Fluorescence correlation spectroscopy studies of Peptide and protein binding to phospholipid vesicles. *Biophys. J.* **87,** 1044–1053.
Sanders, C. R., and Myers, J. K. (2004). Disease-related misassembly of membrane proteins. *Annu. Rev. Biophys. Biomol. Struct.* **33,** 25–51.
Schwille, P., Korlach, J., and Webb, W. W. (1999). Fluorescence correlation spectroscopy with single-molecule sensitivity on cell and model membranes. *Cytometry* **36,** 176–182.
Seelig, J., and Ganz, P. (1991). Nonclassical hydrophobic effect in membrane binding equilibria. *Biochemistry* **30,** 9354–9359.
Senzel, L., Gordon, M., Blaustein, R. O., Oh, K. J., Collier, R. J., and Finkelstein, A. (2000). Topography of diphtheria toxin's T domain in the open channel state. *J. Gen. Physiol.* **115,** 421–434.
Shatursky, O., Heuck, A. P., Shepard, L. A., Rossjohn, J., Parker, M. W., Johnson, A. E., and Tweten, R. K. (1999). The mechanism of membrane insertion for a cholesterol-dependent cytolysin: A novel paradigm for pore-forming toxins. *Cell* **99,** 293–299.
Tory, M. C., and Merrill, A. R. (1999). Adventures in membrane protein topology: A study of the membrane-bound state of colicin E1. *J. Biol. Chem.* **274,** 24539–24549.
Wang, Y., Malenbaum, S. E., Kachel, K., Zhan, H. J., Collier, R. J., and London, E. (1997). Identification of shallow and deep membrane-penetrating forms of diphtheria toxin T domain that are regulated by protein concentration and bilayer width. *J. Biol. Chem.* **272,** 25091–25098.
White, S. H., and von Heijne, G. (2005). Do protein-lipid interactions determine the recognition of transmembrane helices at the ER translocon? *Biochem. Soc. Trans.* **33,** 1012–1015.
White, S. H., Wimley, W. C., Ladokhin, A. S., and Hristova, K. (1998). Protein folding in membranes: Determining the energetics of peptide-bilayer interactions. *Methods Enzymol.* **295,** 62–87.
White, S. H., Ladokhin, A. S., Jayasinghe, S., and Hristova, K. (2001). How membranes shape protein structure. *J. Biol. Chem.* **276,** 32395–32398.
Wieprecht, T., Apostolov, O., Beyermann, M., and Seelig, J. (2000). Membrane binding and pore formation of the antibacterial peptide PGLa: Thermodynamic and mechanistic aspects. *Biochemistry* **39,** 442–452.
Wimley, W. C., and White, S. H. (1996). Experimentally determined hydrophobicity scale for proteins at membrane interfaces. *Nat. Struct. Biol.* **3,** 842–848.
Wimley, W. C., Hristova, K., Ladokhin, A. S., Silvestro, L., Axelsen, P. H., and White, S. H. (1998). Folding of β-sheet membrane proteins: A hydrophobic hexapeptide model. *J. Mol. Biol.* **277,** 1091–1110.
Zakharov, S. D., Heymann, J. B., Zhang, Y.-L., and Cramer, W. A. (1996). Membrane binding of the colicin E1 channel: Activity requires an electrostatic interaction of intermediate magnitude. *Biophys. J.* **70,** 2774–2783.
Zakharov, S. D., Lindeberg, M., and Cramer, W. A. (1999). Kinetic description of structural changes linked to membrane import of the colicin E1 channel protein. *Biochemistry* **38,** 11325–11332.

CHAPTER THREE

Evaluating the Energy-Dependent "Binding" in the Early Stage of Protein Import into Chloroplasts

Mitsuru Akita*,†,‡ *and* Hitoshi Inoue*

Contents

Abstract

During protein import into chloroplasts, precursor proteins are synthesized in the cytosol with an amino-terminal extension signal and irreversibly bind to chloroplasts under stringent energy conditions, such as low levels of GTP/ATP and low temperature, to form the early translocation intermediates. Whether the states of the early-intermediates that are formed under different energy conditions are similar has not been well studied. To evaluate the early intermediate states, we analyzed how precursor proteins within the early intermediates behave by employing two different approaches, limited proteolysis and

* The United Graduate School of Agricultural Science, Ehime University, 3-5-7 Tarumi, Matsuyama, Japan
† Faculty of Agriculture, Ehime University, 3-5-7 Tarumi, Matsuyama, Japan
‡ Satellite Venture Business, Ehime University, 3 Bunkyo-cho, Matsuyama, Japan

Methods in Enzymology, Volume 466

ISSN 0076-6879, DOI: 10.1016/S0076-6879(09)66003-1

site-specific cross-linking. Our results indicate that three different combinations of three different early intermediate stages are present and that the extent of precursor translocation differs between these stages based upon temperature as well as hydrolysis of GTP and ATP. Furthermore, the transition from the second to the third stage was only observed by increasing the temperature. This transition is also accompanied by the hydrolysis of ATP and the movement of the transit peptide. These results suggest the presence of temperature-sensitive and temperature-insensitive ATP-hydrolyzing steps during the early stages of protein import.

1. Introduction

Chloroplasts and mitochondria are semiautonomous organelles bound by double-membranes that are thought to have evolved from symbiotic events (Gray, 1992). Although both organelles contain their own set of genetic machinery that is used to express their own genomes, most proteins that are localized to these organelles are encoded by the nuclear genome and are imported into these organelles by translocons embedded in their double membranes. The majority of these proteins are synthesized as precursor proteins with amino-terminal extension sequences that specifically target the proteins to these organelles and are removed upon import. Both organelles share various features that support the symbiotic theory, but the protein import systems of each organelle differ in many aspects (for a review, see Wickner and Schekman, 2005). Although ATP is required for protein import into both organelles, a different set of translocon components is involved and different energy sources are required (e.g., membrane potential in the mitochondrial system and GTP in the chloroplastic system).

With regard to the chloroplastic protein import system, most of the proteins destined for chloroplasts are synthesized as precursors and imported through translocons embedded in the outer and inner envelope membranes, which are known as the Toc (translocon at the outer membrane of chloroplasts) and Tic (translocon at the inner membrane of chloroplasts) complexes, respectively (for reviews, see Hörmann *et al.*, 2007; Inaba and Schnell, 2008; Jarvis, 2008). Based on the energy requirement, import of these precursor proteins can be divided into at least three steps *in vitro* (Olsen and Keegstra, 1992): (1) an energy-independent interaction between the precursors and the surface of the chloroplast; (2) an energy-dependent irreversible binding of the precursors to the chloroplasts, which is known as docking, to form the early translocation intermediates; and (3) the complete translocation of the precursors through the envelope membranes.

Complete translocation takes place in the presence of high levels of ATP (more than 1 m*M*) at higher temperatures. On the other hand, docking is only observed under stringent energy conditions, such as low levels of ATP (less than 0.1 m*M*) (Olsen *et al.*, 1989) and low temperatures. Docking is also observed at low temperatures in the presence of high levels of ATP, whereas complete translocation takes place at a higher temperature (Leheny and Theg, 1994; Rensink *et al.*, 2000). A low level of GTP enhances the ATP-dependent docking step; however, GTP alone cannot be used as a substitute for ATP (Olsen and Keegstra, 1992; Olsen *et al.*, 1989; Young *et al.*, 1999). Although docking has exclusively been observed *in vitro*, it is recognized as a significant stage connecting the events between the recognition of precursors at the surface of chloroplasts and complete translocation. Because the translocation of precursor proteins is arrested at the docking stage, various translocon components in close proximity to precursor proteins have been captured along with the precursors and identified (Akita *et al.*, 1997; Kouranov and Schnell, 1997; Ma *et al.*, 1996; Nielsen *et al.*, 1997; Perry and Keegstra, 1994; Schnell *et al.*, 1994).

Because studies on docking have focused on the energy conditions needed to form the early translocation intermediates (Olsen and Keegstra, 1992; Olsen *et al.*, 1989; Theg *et al.*, 1989; Young *et al.*, 1999), direct comparisons of the state of the early translocation intermediates formed under different energy conditions have been overlooked, despite the importance of the docking stage in understanding the chloroplast protein import system. As a first step toward addressing this issue, we investigated the state of the precursor proteins within the early translocation intermediates formed under different energy conditions as follows: first, a docking/import assay system that could accommodate a recombinant precursor protein under strict energy conditions was established (Inoue *et al.*, 2008); second, the extent of precursor translocation within the early translocation intermediates was determined by limited proteolysis (Inoue and Akita, 2008b); and last, the surrounding environment of the precursor proteins within the early translocation intermediates was assessed by site-specific cross-linking (Inoue and Akita, 2008a). Throughout our work, we shed light on the mechanism of docking by revealing many important findings. In this chapter, we not only describe what we have discovered but also how we designed and developed these assay systems in order to turn this rather "outdated" topic into an exciting and challenging one.

2. The *In Vitro* Chloroplastic Import Assay Using Recombinant Precursor Proteins

Some studies on the early stages of protein import into chloroplasts have already been carried out, and their outcomes have been widely accepted (Olsen and Keegstra, 1992; Olsen *et al.*, 1989; Theg *et al.*, 1989;

Young *et al.*, 1999). To reevaluate the early translocation intermediates found under these circumstances, strict management of the energy conditions during the docking reaction is essential. To overcome this issue, we developed the *in vitro* import assay system, allowing docking to be regarded as the sole event taking place between the chloroplasts and the precursor protein. To this end, we used two strategies to strictly control the energy conditions in the assay system:

1. Removal of exogenous factors that would affect the docking reaction.
2. Manipulation of the energy conditions in the assay system.

To execute the first strategy, nonradiolabeled recombinant precursor proteins overexpressed in *Escherichia coli* were utilized as import substrates and an *in vitro* import assay system suitable for those precursors was established (Inoue *et al.*, 2008). Radiolabeled precursor proteins have been exclusively used in the previous studies on protein import into chloroplasts. The recombinant precursor proteins we prepared did not contain any energy sources or cytosolic factors that might be present in the *in vitro* transcription–translation system frequently used to prepare precursor proteins. To execute the second strategy, the chloroplast energy pools were manipulated before and/or during the docking and import reactions by controlling the ATP levels in the assay system.

2.1. Preparations for recombinant precursor proteins

For the precursor protein, we chose the precursor form of the small subunit of ribulose-1,5-bisphosphate carboxylase oxygenase (Rubisco) (prSS) from the pea plant (*pisum sativum*), fused with a double or quadruple HA tag (HAHA or 4HA, respectively) and a hexahistidine tag (H) at its carboxy terminus (Fig. 3.1). These tags are valuable for detecting and purifying precursor proteins. Furthermore, the precursor protein was manipulated to carry a single cysteine residue at various positions (Fig. 3.1), which allowed the precursor to be site-specifically modified for the applications described in this article.

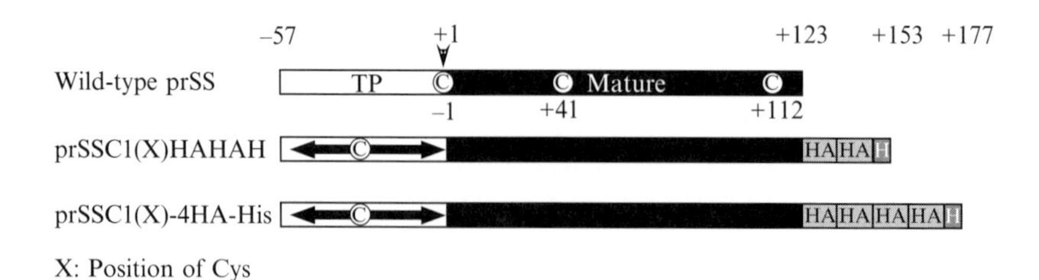

Figure 3.1 Schematic of the precursor proteins described in this article. TP indicates transit peptide.

The gene for pea wild-type prSS was cloned into the pET21a expression vector (Merck, Darmstadt, Germany). From this plasmid, a gene for a mutant form of prSS carrying the above-mentioned tags and a single cysteine residue was prepared by site-directed mutagenesis (Inoue and Akita, 2008a,b; Inoue *et al.*, 2008). To overexpress the precursor protein, 0.4 m*M* isopropyl 1-thio-β-D-galactopyranoside (IPTG) was added to the culture of *Escherichia coli* BL21(DE3) cells carrying the expression plasmid when it reached the mid-logarithmic growth phase, followed by a further incubation for 4 h at 37 °C. Harvested *E. coli* cells were washed once with W-buffer (50 m*M* Tris–HCl, pH 7.5, 10% glycerol) and resuspended in the same buffer. After the cells were ruptured using an ultra-sonicator (UD-201, TOMY SEIKO, Tokyo, Japan), the soluble and insoluble fractions were separated by centrifugation at 20,000*g* for 20 min at 4 °C. Recombinant precursor proteins were recovered from inclusion bodies, which occupied the majority of the precipitate (Inoue *et al.*, 2008). Therefore, the precipitate was washed with W-buffer and H_2O several times and then solubilized in buffer containing 8 *M* urea (S-buffer: 8 *M* urea, 25 m*M* HEPES–KOH, pH 7.5, 50 m*M* KCl, 2 m*M* $MgCl_2$). After the protein was quantified by the Lowry method (Lowry *et al.*, 1951) using bovine serum albumin (BSA) for the standards, the precursor proteins were diluted in S-buffer to a concentration of 0.1 m*M*. If necessary, the precursor protein was purified under denaturing conditions using Ni^{+2}-NTA-agarose (Qiagen, Hilden, Germany) to capture the precursor and then eluted in an imidazole-containing buffer, as previously described (Inoue and Akita, 2008b). The purified precursor protein was precipitated with 10% trichloroacetic acid (TCA) to remove imidazole. Precipitates were washed with acetone and ethanol, followed by solubilization in S-buffer as described above.

2.2. Modification of the precursor

The precursor protein was modified with the biotinylation reagent, biotin-maleimide (Sigma-Aldrich, St. Louis, MO) to allow for its detection, or with a photoreactive cross-linker, 4-(*N*-maleimido)benzophenone (MBP) (Sigma-Aldrich), to facilitate cross-linking (Inoue and Akita, 2008a,b). Both reagents contain a sulfhydryl (SH)-reactive maleimide group; therefore, a precursor protein containing a single cysteine residue could be site-specifically modified. Each reagent was dissolved in dimethylsulfoxide (DMSO) to create a 50-m*M* solution, which was diluted 20-fold with 0.1 m*M* precursor protein in S-buffer and incubated on ice for 1–4 h in the dark. The modification reaction was quenched by adding 5 m*M* dithiothreitol (DTT), followed by incubation on ice for an additional 15 min. The modified precursor protein was TCA precipitated and solubilized in S-buffer as described above.

2.3. The docking and the import assay

Pea plants (*Pisum sativum* cv. tsurunashiendou) were grown for 8–10 days with a cycle of 14 h-light at 24 °C/10 h-dark at 20 °C. Intact chloroplasts were isolated from these plants as described previously (Bruce *et al.*, 1994). Isolated chloroplasts suspended in import buffer (I-buffer: 50 m*M* HEPES–KOH, pH 8.0, 330 m*M* sorbitol) at a concentration of 2 mg of chlorophyll/ml were incubated on ice in the dark for 1 h and then diluted twice with 5 μM nigericin in I-buffer and incubated for 10 min at 25 °C. The chloroplasts were further diluted fivefold and energized with I-buffer containing 1 mg/ml BSA, 25 m*M* DTT, 5 m*M* $MgCl_2$, 0.1 m*M* GTP, and 0.1 m*M* ATP (docking) or 2.5 m*M* ATP (import) and incubated at 4 °C (docking) or 25 °C (docking or import) for 5 min in the dark (Inoue and Akita, 2008a,b; Inoue *et al.*, 2008). To initiate the docking and import reactions, 10 μM of the precursor protein in S-buffer was in the energized chloroplast suspension to make a 200-n*M* precursor solution and to maintain a urea concentration of 160 m*M*. Because the precursor aggregates were recovered with the chloroplasts when an excess of precursor was applied to the docking/import reaction (Inoue *et al.*, 2008), the combination of precursor and urea concentration was critical. The mixture was incubated at 4 °C (docking) or 25 °C (docking or import) in the dark for 5 min unless otherwise indicated. Intact chloroplasts were recovered by centrifugation (1500*g* for 5 min at 4 °C) through a cushion of 40% Percoll in I-buffer and washed once with I-buffer. If the docked precursor was used for the chase experiment, chloroplasts isolated from the docking reaction were suspended (0.2 mg of chlorophyll/ml) in I-buffer containing 5 m*M* $MgCl_2$ and either 0.1 m*M* ATP (for docking chase) or 2.5 m*M* ATP (for import chase) and incubated at 25 °C for 5 min (docking chase) or 20 min (import chase). Chloroplasts were then recovered as described above.

Recovered chloroplasts were dissolved in sample buffer and the proteins were separated by Tris–glycine buffer SDS–PAGE (Laemmli, 1970). After electrophoresis, immunoblotting was performed as described previously (Inoue *et al.*, 2008). Precursor proteins were immunoblotted with an anti-HA monoclonal antibody (Sigma-Aldrich) followed by the alkaline phosphatase (AP)-conjugated antimouse secondary antibodies. Biotinylated precursor proteins were identified by AP-conjugated streptavidin (Invitrogen, Carlsbad, CA). Color development of the AP conjugate was performed using nitro blue tetrazorium (NBT) and 5-bromo-4-chloro-3-indoyl phosphate (BCIP), as described previously (Inoue *et al.*, 2008).

Of the various prSS mutants, the docking and import assays were carried out using a precursor carrying a cysteine at residue −52, prSSC1(−52)HAHA, as a representative. The docking reaction was performed with this precursor in the presence of 0.1 m*M* ATP and 0.1 m*M* GTP and incubated for 5 min at 25 °C (Fig. 3.2, lane 3). The import assay was also carried out

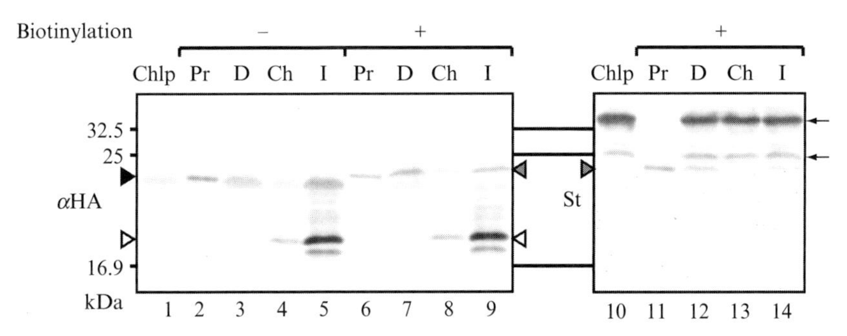

Figure 3.2 The docking and import assays using a recombinant protein. A mutant precursor protein, prSSC1(−52)HAHAH (Pr), either without (lane 2) or with (lanes 6 and 11) biotinylation, was used for the docking reaction (D; lanes 3, 7, and 12) at 25 °C and the import reaction (I; lanes 5, 9, and 14). Chloroplasts (Chlp; lanes 1 and 11) docked by the precursor (lanes 3, 7, and 12) were reisolated and used in the import chase assay (Ch; lanes 4, 8, and 13). After the assay, chloroplastic proteins were separated by Tris–glycine SDS–PAGE (Laemmli, 1970), followed by immunoblotting with either an anti-HA mouse monoclonal antibody (αHA) (lanes 1–9) or streptavidin (St) (lanes 10–14). Positions of prSS, biotinylated prSS, and mSS are indicated by the filled, shaded, and open arrowheads, respectively. Ten percent of the precursor that was used in the reactions was loaded onto the gel (lanes 2, 6, and 11). The arrows indicate two streptavidin-reactive endogenous proteins in the chloroplast.

with the same precursor in the presence of 2.5 m*M* ATP for 20 min at 25 °C (Fig. 3.2, lane 5). Approximately 10% of the precursor added to the mixture was recovered with chloroplasts after the docking reaction. To check whether the precursor recovered with chloroplasts maintained its import competency, an import chase reaction was performed by incubating the precursor-docked chloroplasts in the presence of 2.5 m*M* ATP for 20 min at 25 °C (Fig. 3.2, lane 4). Most of the predocked precursor disappeared and a smaller size band was observed at the same position as the processed mature Rubisco small subunit (mSS) (Fig. 3.2, lane 5) after the chase reaction. The efficiency of docking and import was not affected when the biotinylated precursor was used in the assays (Fig. 3.2, lanes 6–9). Because biotin was introduced at the amino terminus of the transit peptide, which is removed after import, the processed band was not detected by streptavidin (Fig. 3.2, lanes 13 and 14). Taken together, a recombinant precursor protein bound specifically to chloroplasts, regardless of whether it was modified by biotin, and the docked precursor maintained import competency during the procedure. Upon biotinylation, different parts of the precursors could be monitored by the position of the biotinylated cysteine residue and by the HA tag at the carboxy terminus (Inoue and Akita, 2008b). Furthermore, we were able to establish a docking/import system that avoided any exogenous factors that could be present in other precursor preparations by using a nonradiolabeled recombinant precursor protein.

2.4. Manipulation of chloroplast energy pools

ATP pools in chloroplasts are controlled in three ways: (A) generating ATP; (B) inhibiting ATP generation, or (C) degrading ATP with various factors, as illustrated in Fig. 3.3. Below is a brief description of various modes of action for each factor:

A. ATP generation
  a1. *Light*: ATP is synthesized by F_oF_1-ATPase embedded in the thylakoid membrane through the use of the proton gradient (ΔpH) formed by photosynthesis.
  a2. *The dihydroxyacetone phosphate (DHAP) system (DHAP + oxaloacetate (OAA) + Pi)*: OAA that has been taken in is oxidized to malate in a reaction catalyzed by malate:NADP$^+$ oxidoreductase to produce NADPH, which drives the Calvin–Benson cycle in reverse to generate ATP (Theg *et al.*, 1989).
  a3. *Adenine nucleotide translocator*: Exogenously added ATP is taken up by the ADP/ATP antiport (Neuhaus and Wagner, 2000).
  a4. *Nucleoside diphosphate kinase (NDPK)*: Because NDPK transfers the γ-Pi from an NTP to an NDP of another nucleoside, the addition of ATP or GTP would be assumed to be taken up by chloroplasts, which subsequently convert it to GTP or ATP, respectively (Hasunuma *et al.*, 2003). Because no inhibitor for chloroplastic NDPK has been currently reported, we could not control the activity of NDPK.

B. Inhibition of ATP generation
  b1. *Darkness*: Photosynthesis does not occur in the dark.
  b2. *Nigericin*: Nigericin uncouples electron transport through the thylakoid membrane, resulting in the inhibition of ATP synthesis by F_oF_1-ATPase (Theg *et al.*, 1989).

C. ATP degradation
  c1. *Apyrase*: External NTPs are degraded to NDPs, which are further degraded to NMPs by apyrase (Olsen and Keegstra, 1992).
  c2. *Glycerate*: Glycerate in the stromal space is the substrate for glycerate kinase, which dephosphorylates ATP (Olsen *et al.*, 1989).

Prior to the docking and import assays, chloroplasts were always incubated with nigericin and kept in the dark, as described above. If necessary, chloroplasts were further treated by other methods, such as apyrace, glycerate, the DHAP system, or a combination of apyrase and the DHAP system.

2.5. Docking under various energy conditions

Once the docking/import assay system was developed, we attempted to determine how precursor docking was affected by different energy conditions. The docking reaction was performed under different combinations of

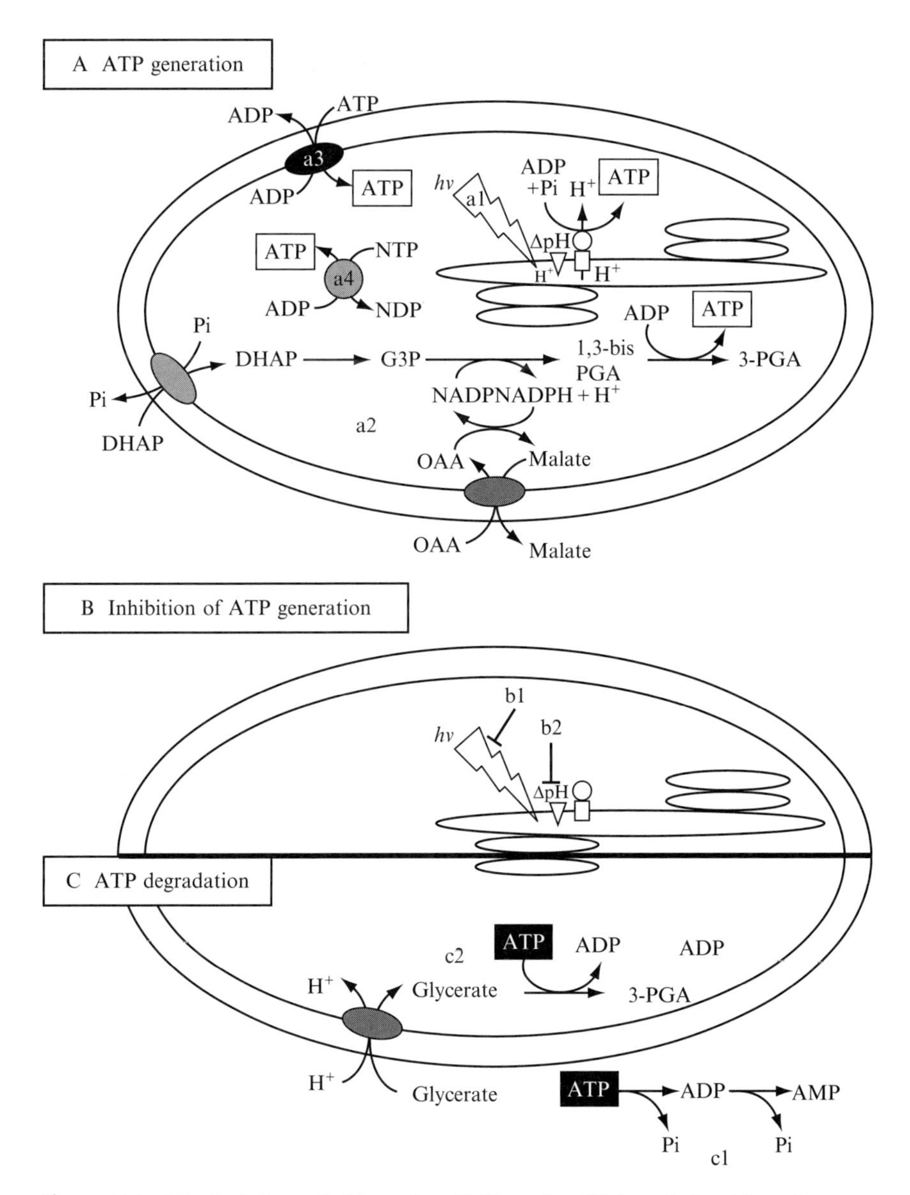

Figure 3.3 Manipulation of chloroplast ATP pools. ATP pools in chloroplasts are controlled by generating ATP (A), inhibiting ATP generation (B), or degrading ATP (C). Abbreviations: G3P, glyceraldehyde 3-phosphate; 1,3-bis PGA, 1,3-bis-phosphoglyceric acid; 3-PGA, 3-phosphoglyceric acid.

GTP and ATP concentrations and temperature (Fig. 3.4A). The docked precursor was seen on the blots (Fig. 3.4), and the docking rate (docked precursor (Fig. 3.4A, lanes 2–8 and 10–16) against precursor used in the

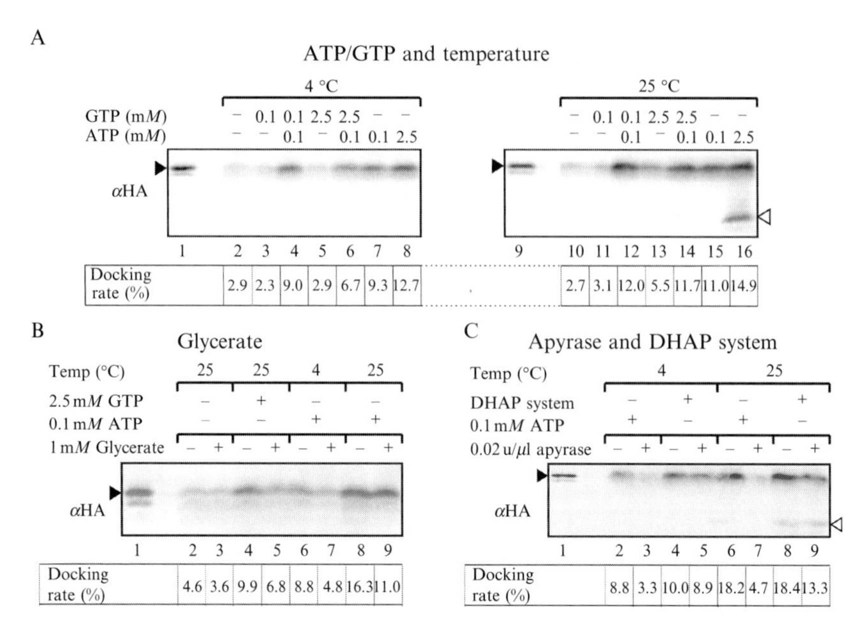

Figure 3.4 The docking assay under different energy conditions. A biotinylated precursor, prSSC1(−52)HAHAH was used in the docking reaction under the indicated conditions. Chloroplastic proteins were resolved by Tris–glycine buffer SDS–PAGE, followed by immunoblotting with αHA. The numbers shown below the blots are the docking rate calculated from the ratio of the intensity of the docked precursor against the intensity of the precursor used in the docking reaction. Ten percent of the amount of precursor used for the reactions was loaded onto the gel (Panel A, lanes 1 and 9; Panels B and C, lane 1). The positions of prSS and mSS are indicated by the filled and open arrowheads, respectively.

docking reaction (Fig. 3.4A, lanes 1 and 9)) was calculated from the intensity of the bands. The intensity of the bands was quantified from the digitized images of the blots using UN-SCAN-IT gel ver. 5.1 (Silk Scientific, Orem, UT). Regardless of the presence of GTP, the amount of the precursor that docked at 4 °C increased about threefold over the basal level in the presence of 0.1 m*M* ATP (Fig. 3.4A; compare lane 2 with lanes 4, 6, and 7). A further increase was observed when the ATP concentration was increased to 2.5 m*M* (Fig. 3.4A, compare lane 8 with 7). Furthermore, in the presence of the same NTP conditions, more precursor was docked at 25 °C than at 4 °C (Fig. 3.4A; compare lanes 4, 7, and 8 with lanes 12, 14, and 15). In the presence of GTP alone, the amount of docked precursor was comparable to the basal level (Fig. 3.4A, lanes 3, 5, and 11), except when the docking reaction was performed in the presence of 2.5 m*M* GTP at 25 °C (Fig. 3.4A, lane 13). In contrast to previous studies (Olsen and Keegstra, 1992; Olsen *et al.*, 1989; Young *et al.*, 1999), GTP enhancement of

ATP-dependent docking was not observed (Fig. 3.4A, compare lanes 4 with 7 and 12 with 15). This discrepancy was probably due to the difference in precursor preparations. Unlike our "pure" assay system, their precursor preparation might contain additional factor(s) from the *in vitro* translation system used to enhance ATP-dependent docking. The docking mechanism observed in the presence of 2.5 m*M* GTP at 25 °C is still unknown but might be explained by the combination of endogenous ATP and other populations of ATP converted from the excess amount of GTP by NDPK (Fig. 3.3A, a4), although there is no evidence to support the involvement of NDPK.

Therefore, the contribution of endogenous ATP to docking was investigated by depleting the internal ATP by treating chloroplasts with 1 m*M* glycerate prior to the docking reaction (Fig. 3.3C, c2). Although the contribution of endogenous ATP to docking has been shown in a previous study (Olsen and Keegstra, 1992), we used our pure system to reevaluate this question. Based on the glycerate treatment, the amount of docked precursor in the presence of NTP was reduced to 60–70% (Fig. 3.4B), indicating that the internal NTP also contributed to docking, although endogenous ATP alone (Fig. 3.4B, lane 3) was insufficient to support docking in the presence of 0.1 m*M* ATP (Fig. 3.4B, lanes 6 and 8). This result raised the question of whether docking used internal ATP. To address this question, chloroplasts were incubated with apyrase (0.02 units/μl) and the DHAP system (composed of 0.1 m*M* DHAP, 1 m*M* OAA, and 20 m*M* Na–Pi buffer, pH 8.0) (Fig. 3.3A, a2) prior to initiating the docking reaction with the addition of precursor proteins. Using this treatment, external NTP was depleted and internal ATP was generated at the same time. Regardless of the temperature, the amount of the docked precursor reached the basal level if the chloroplasts were treated with apyrase, even in the presence of 0.1 m*M* ATP (Fig. 3.4C, lanes 3 and 7). When internal ATP was generated by the DHAP system, the precursor docked to chloroplasts even when the external ATP was depleted by apyrase (Fig. 3.4C, lanes 5 and 9), indicating that docking was supported by internal ATP. Furthermore, ATP generation by the DHAP system was enough to drive the complete translocation, as shown in previous studies (Olsen and Keegstra, 1992; Theg *et al.*, 1989).

3. Limited Proteolysis of Docked Precursor Proteins

The results in the previous section only demonstrated the amount of precursor that docked to the chloroplasts under different energy conditions. However, these results did not give any indication of the state of the precursor proteins in the early translocation intermediates formed under these different energy conditions. We attempted to use limited proteolysis to solve this issue, as illustrated in Fig. 3.5. If the chloroplasts were treated

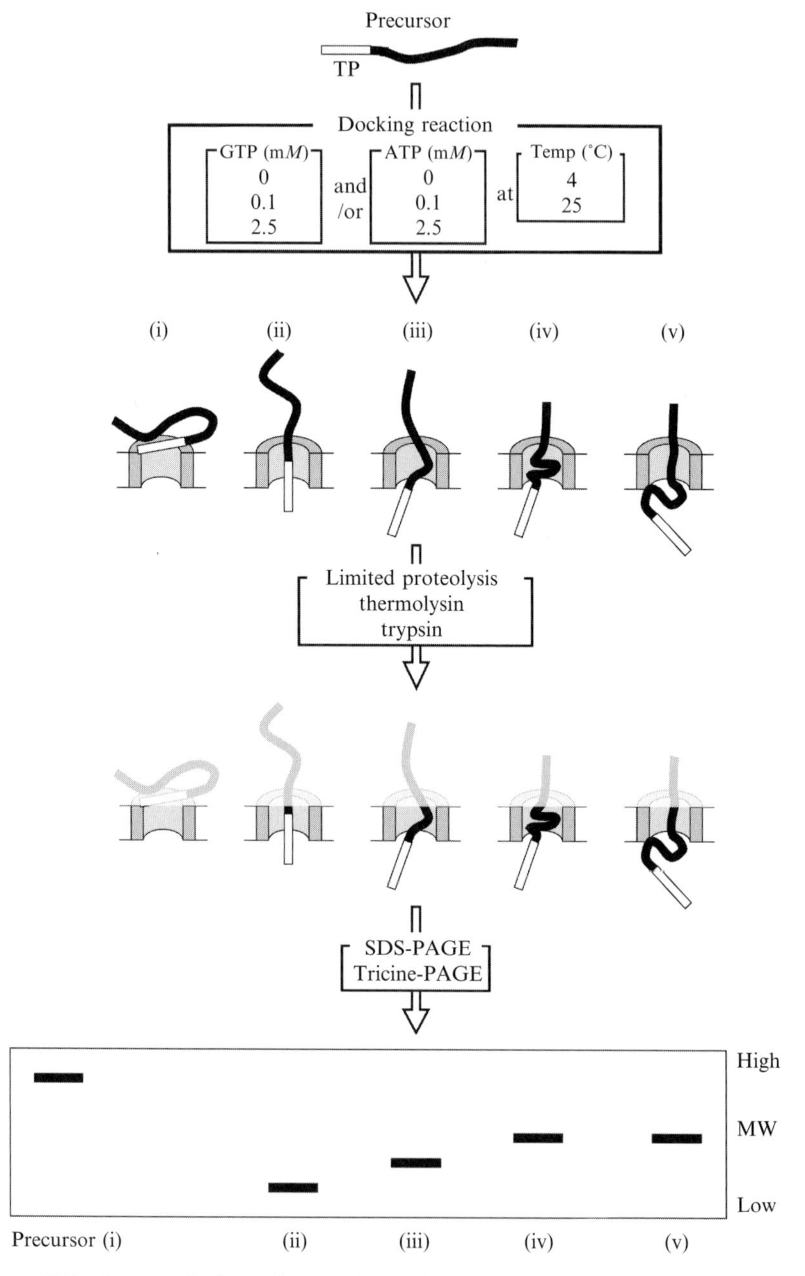

Figure 3.5 Strategy for limited proteolysis. The precursor docked to chloroplasts may be present in the following forms: (i) stationary at the surface; (ii) with the transit peptide (TP) partially translocated through the outer envelope; (iii) with the TP fully translocated; (iv) with the TP fully translocated but held in the same place as form iii; and (v) with the entire precursor penetrating further into the chloroplasts than form iii.

with a limited amount of protease, then the proteins on the surface of the chloroplasts and those exposed to the outside would be partially degraded; thus, the presence of protease-resistant fragments protected by the outer envelope would be expected. Applying this strategy, the state of the precursor proteins would be interpreted by the extent to which they are translocated within the early intermediates.

3.1. Proteolytic fragments

Precursor proteins biotinylated at residue −52 were mixed with chloroplasts in the presence of 0.1 m*M* ATP and 0.1 m*M* GTP and incubated at 4 °C for 5 min. After being reisolated, chloroplasts (1 mg of chlorophyll/ml in I-buffer) were treated with a limited amount thermolysin (0.8 mg/mg of chlorophyll in 4 m*M* $CaCl_2$) or trypsin (0.1 mg/mg of chlorophyll in 1 m*M* $CaCl_2$) for 30 min on ice in the dark. Proteolysis was quenched by diluting the reaction mixture fourfold with I-buffer containing protease inhibitors, which was either 12 m*M* EDTA for thermolysin-treated samples or a protease inhibitor cocktail containing 1 m*M* phenylmethylsulfonyl fluoride (PMSF), 0.05 mg/ml N^{α}-*p*-tosyl-L-lysine chloromethyl ketone (TLCK), 2 μg/ml aprotinin, and 0.1 mg/ml soybean trypsin inhibitor for trypsin-treated samples (Jackson *et al.*, 1998). The quenching reaction was performed by incubating the mixture on ice for 10 min in the dark. After chloroplasts were recovered through 40% Percoll in I-buffer containing the same protease inhibitors, chloroplasts were washed once with I-buffer containing inhibitors. The final precipitates were suspended in a sample buffer for Tris–glycine buffer SDS–PAGE (Laemmli, 1970) containing 1 m*M* EDTA (thermolysin-treated samples) or 1 m*M* PMSF (trypsin-treated samples), immediately followed by boiling for 10 min. Under these proteolytic conditions, the integral outer envelope membrane protein Toc75 remained intact, whereas Toc34, which has a large amino-terminal cytosolic domain, was degraded (Inoue and Akita, 2008b), indicating that thermolysin and trypsin degraded the outer surface of chloroplasts. Proteolytic fragments were separated by Tris–Tricine buffer SDS–PAGE (Schägger and von Jagow, 1987) and were detected on the blots probed with AP-conjugated streptavidin (Fig. 3.6A). Although this PAGE system is superior for separating smaller size polypeptides, the biotinylated precursor, when docked to chloroplasts (Fig. 3.4 and Fig. 3.6A, lane 2), was hindered by one of two streptavidin-reactive chloroplast proteins (Fig. 3.2, lanes 10, 12–14 and Fig. 3.6A, lanes 5 and 6). Therefore, docking should be assessed separately, as shown in Fig. 3.4. In a precursor- (Fig. 3.6A, compare lane 3 with 6) and protease- (Fig. 3.6A, compare lane 2 with 3) dependent manner, four fragments with molecular masses between 7.8 and 12 kDa were observed (Fig. 3.6A, lane 3). Those fragments were not detected using the anti-HA monoclonal antibody (data not shown). The similarly sized

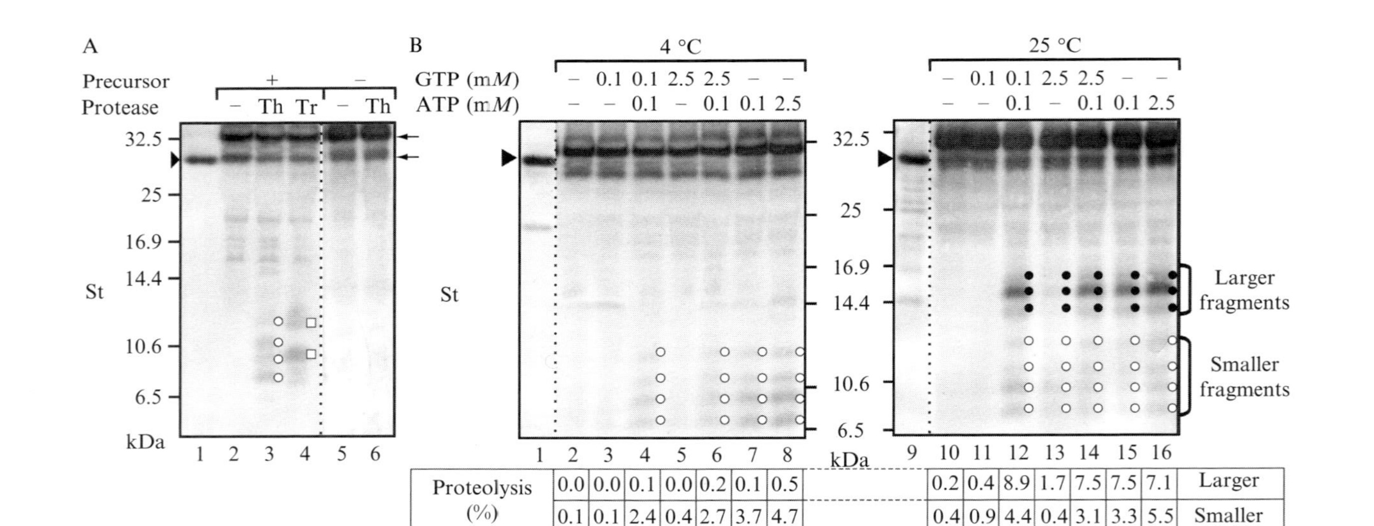

Proteolysis (%)	2	3	4	5	6	7	8	10	11	12	13	14	15	16	
	0.0	0.0	0.1	0.0	0.2	0.1	0.5	0.2	0.4	8.9	1.7	7.5	7.5	7.1	Larger
	0.1	0.1	2.4	0.4	2.7	3.7	4.7	0.4	0.9	4.4	0.4	3.1	3.3	5.5	Smaller

Figure 3.6 Proteolytic fragements produced from the docked precursor. (A) After the precursor was docked (lane 1), chloroplasts were treated with thermolysin (lane 3) or trypsin (lane 4). Chloroplastic proteins were resolved by Tris–Tricine buffer SDS–PAGE (Schägger and von Jagow, 1987), followed by immunoblotting with streptavidin (St). Open circles indicate the fragments generated by thermolysin, whereas open squares indicate those generated by trypsin. The arrows indicate two streptavidin-reactive endogenous proteins in the chloroplast. (B) The precursor was docked under different energy conditions, as shown in Fig. 3.4 (lanes 2–8 and 10–16), and treated with a limited amount of thermolysin. Smaller fragments are indicated by open circles, whereas larger fragments are indicated by filled circles. The numbers shown below the blots are the yields of the larger and smaller fragments calculated from the ratio of the intensities of the proteolytic fragments to the intensity of the precursor used in the docking reaction. Ten percent of the amount of precursor used for the reactions was loaded onto the gel (Panel A, lane 1; Panels B and C, lanes 1 and 9). The position of prSS is indicated by the filled arrowhead.

fragments were also observed when chloroplasts were treated with trypsin (Fig. 3.6A, lane 4) or when other precursor proteins biotinylated at residues −33 or +4 were used (data not shown). Taken together, we concluded that these were the proteolytic fragments generated from the amino-terminal part of the precursor protein.

When the docking reaction was performed in the presence of 0.1 m*M* ATP and 0.1 m*M* GTP at 25 °C, three larger sized proteolytic fragments of molecular masses between 13.5 and 15 kDa were generated from a docked precursor biotinylated at −52 (Fig. 3.6B, lanes 12–16), in addition to those fragments of 7.8–12 kDa (Fig. 3.6B, lanes 4, 6, 7, and 8). This indicated that the precursor was further translocated during the docking reaction at 25 °C. Based on the sizes of the proteolytic fragments, which were not affected by temperature during the docking reaction, the entire part of the transit peptide (about 5.8 kDa) was not exposed on the outer surface of the chloroplast. Therefore, the entire transit peptide was translocated through the outer envelope membrane (Fig. 3.5, iv). In addition, both the smaller and the larger proteolytic fragments were recovered in the membrane fractions after the chloroplasts were lysed (Inoue and Akita, 2008b).

To further examine the state of the docked precursor, the ratios of the smaller and larger proteolytic fragments to the amount of precursor used in the reaction were calculated (Fig. 3.6B) and compared with the docking rate, as shown in Fig. 3.4. Based upon these numbers, less than half of the docked precursor was recovered as proteolytic fragments at 4 °C, which indicated the complete degradation of the precursor. On the other hand, much more of the docked precursor was not recovered as proteolytic fragments at 25 °C. To generate the different sizes of proteolytic fragments that were observed, at least three separate precursor populations had to be present within the early translocation intermediates: (1) a completely degradable precursor; (2) a precursor that produced the smaller proteolytic fragments; and (3) a precursor that produced the larger proteolytic fragments. These precursor populations may represent the different stages of early translocation intermediates, Stages I, II, and III. In the Stage I intermediate, the precursor was docked to the chloroplasts and fully accessible to the proteases (Fig. 3.5, i). In Stage II, 70–110 amino acids of the precursor were inserted into the chloroplasts (Fig. 3.5, iii). In Stage III, an additional 20 amino acids of the precursor was further penetrated into the chloroplasts (Fig. 3.5, iv or v).

3.2. Transition between Stages II and III intermediates

Because the larger proteolytic fragments were exclusively observed when the docking reaction was performed at 25 °C in the presence of ATP, we wondered whether further penetration of the precursor from Stage II to III intermediates was achieved only by a temperature shift from 4 to 25 °C or

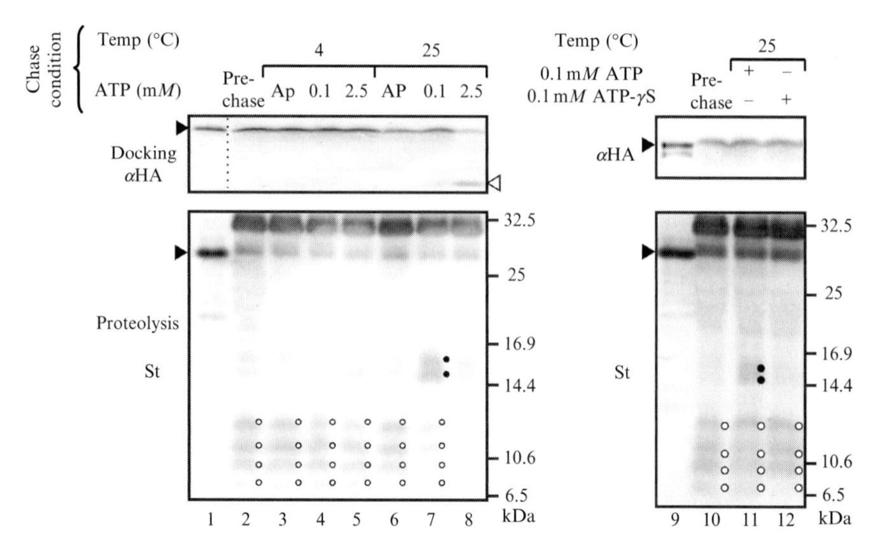

Figure 3.7 ATP hydrolysis is required for the transition of intermediates from Stage II to III. Precursor-docked chloroplasts in the presence of 0.1 m*M* ATP at 4 °C (lanes 2 and 10) were incubated in the presence of apyrase (Ap; lanes 3 and 6) or with different concentrations of ATP (lanes 4, 5, 7, and 8) or ATP-γS (lane 4) at either 4 °C (lanes 3–5) or 25 °C (lanes 6–8, 9, and 10) for 5 min to chase the docked precursor. The docked precursor was separated by Tris–glycine SDS–PAGE, followed by immunoblotting with αHA. The fragments generated by thermolysin treatment were separated by the Tris–Tricine buffer SDS–PAGE, followed by immunoblotting with strepatavidin (St). Ten percent of the amount of precursor used in the reactions was loaded onto the gel (Panel A, lanes 1 and 9; Panels B and C, lane 1). The positions of prSS and mSS are indicated by the filled and open arrowheads, respectively.

by a combination of this temperature shift and ATP. The precursor was docked at 4 °C in the presence of 0.1 m*M* ATP (Fig. 3.7, lane 2) and then chased by increasing the temperature to 25 °C under different ATP concentrations (Fig. 3.7, lanes 3–8). The amount of docked precursor observed during the chase reaction did not change except when incubated at 25 °C in the presence of 2.5 m*M* ATP. Complete translocation took place at that particular combination of temperature and ATP concentration (Fig. 3.7A, lane 8). Larger proteolytic fragments were only observed when 0.1 m*M* ATP was used during the docking chase reaction (Fig. 3.7A, lane 7), indicating that a low level of ATP is required for transition between the intermediate stages.

Previous studies indicated the hydrolysis of ATP and GTP is essential for precursor docking (Olsen and Keegstra, 1992; Olsen *et al.*, 1989; Young *et al.*, 1999). The GTP hydrolysis is used by the two Toc components, Toc34 and Toc159 (Kessler *et al.*, 1994). We wondered whether the hydrolysis of ATP was essential for this transition. During the chase reaction at 25 °C, the slowly hydrolyzed ATP analog, adenosine-5′-*O*-(3-thiotriphosphate) (ATP-γS) (Fig. 3.7, lane 4) was substituted for ATP (Fig. 3.7, lane 3). No precursor was released during the chase reaction (Fig. 3.7, compare lane 2 with lanes 3 and 4), and after limited proteolysis, there was no discernable change in the proteolytic fragments (Fig. 3.7, compare lane 2 with 3). These observations indicated that ATP hydrolysis was required for the transition of the early intermediates from Stage II to III. Taken together, there are two individual ATP-hydrolyzing steps, one of which is a temperature-insensitive step required for the Stage II intermediate and the other a temperature-sensitive step required for the Stage III intermediate. Currently, no ATP hydrolyzing factor has been identified in the intermembrane space other than Hsp70 (Hsp70-IAP) (Schnell *et al.*, 1994). Therefore, it is assumed that Hsp70-IAP may be involved in both of the ATP hydrolysis steps or that it may be involved in only one of the steps, with unidentified ATP-hydrolyzing components involved in the other step.

4. The Behavior of Transit Peptide During the Transition

The limited proteolytic study revealed the presence of different early translocation intermediates that differ in the extent of precursor translocation. However, results from limited proteolysis did not solve the issue of whether the precursors' surrounding environment differs at each stage. Although further penetration of the precursor is observed during the transition from Stage II to III, the question of whether the amino-terminal part of the precursor is held in place while the remainder of the precursor has penetrated (Fig. 3.5, iv) or the entire amino terminus of the precursor has penetrated (Fig. 3.5, v) remains unsolved. To solve this issue, we employed a site-specific, photoreactive cross-linking strategy. Multiple single-cysteine mutants (prSSC1(X)-4HA-His, X: the position of cysteine residue) were modified with MBP. MBP is a photocross-linker with a 10-Å arm, which introduces a benzophenone moiety to the SH-group of the cysteine residue and thus generates a reactive species that can be inserted into nearby bonds upon UV irradiation (Dormán and Prestwitch, 1994). Modified precursors were used in the docking reaction in the presence of 0.1 m*M* ATP and incubated at 4 °C. After reisolation of the chloroplasts, the chloroplast suspension was halved. Each aliquot was used in the docking chase experiment in the presence of 0.1 m*M* ATP and incubated at either 4 °C or 25 °C. After recovery of the chloroplasts, the chloroplast suspension (1 mg of chlorophyll/ml) was irradiated by a UV illuminator at 302 nm for 5 min on ice at a distance of 1–2 cm to initiate the cross-linking reaction (Inoue and Akita, 2008b). Cross-linked products were separated by Tris–glycine buffer SDS–PAGE (Laemmli, 1970) and detected by immunoblotting with an anti-HA monoclonal antibody. Various cross-linked products were observed when a membrane-localized precursor modified with MBP at position −15 was used (Fig. 3.8A) (Inoue and Akita, 2008a). Four of the

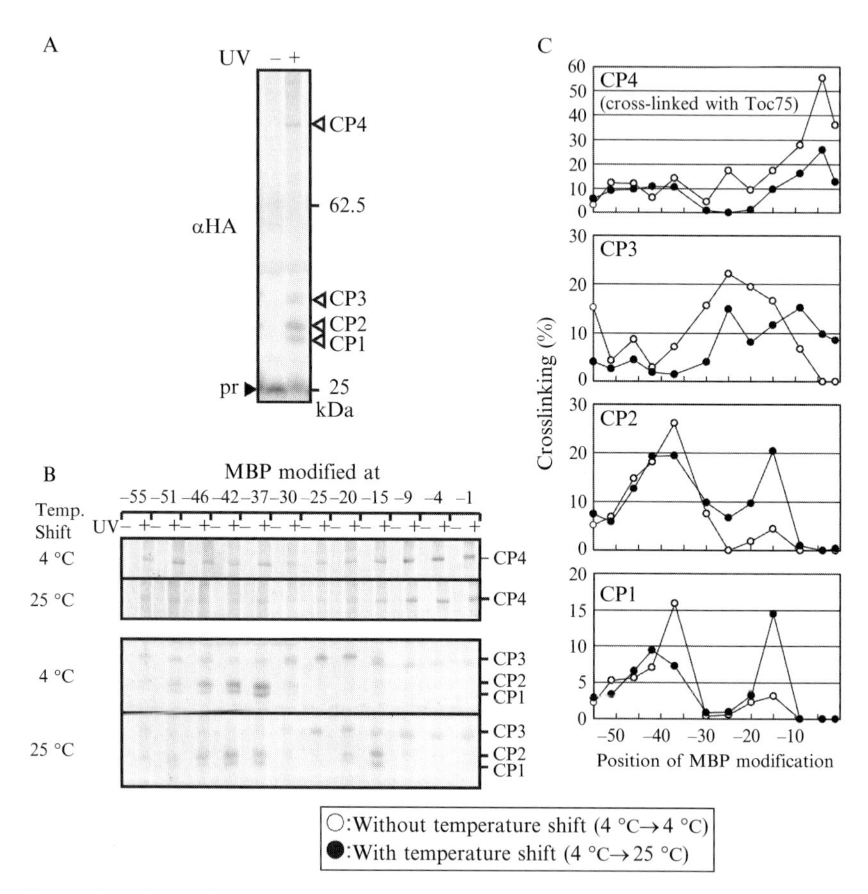

Figure 3.8 Cross-linked products generated with MBP-modified precursor. (A) Precursor protein modified with MBP at position −15 was docked to chloroplasts in the presence of 0.1 m*M* ATP at 25 °C for 5 min. Isolated chloroplasts were UV irradiated. Chloroplasts were separated by Tris–glycine buffer SDS–PAGE, followed by immunoblotting with αHA. Four major cross-linked products were designated as CP1–4, indicated by open triangles. (B) Precursors modified with MBP at various positions were docked to chloroplasts in the presence of 0.1 m*M* ATP at 4 °C for 5 min. After reisolation, the chloroplasts were incubated in the presence of 0.1 m*M* ATP at either 4 °C or 25 °C for 5 min, followed by the UV irradiation (+). (C) Yields of CP1–4 produced from the various mutant precursors are shown. Yields were calculated from the intensity of the precursor without UV irradiation and the intensity of each cross-linked products.

major cross-linked products were designated as CP1–4 and were examined to determine whether the temperature shift affected the yields (Fig. 3.8B and C). The position of the MBP modification in the precursor that yielded the highest amount of cross-linked products was shifted around 20 amino acids toward the carboxy terminus by the temperature shift, for example, −37 to −15 for CP1 and CP2 and −25 to −4 for CP3 (Fig. 3.8B and C). Stages I and II intermediates were found in the early intermediates that were prepared without the temperature shift (Fig. 3.7, lane 4), whereas Stages II and III intermediates were found with the temperature shift (Fig. 3.7, lane 7). Therefore, the shift of the MBP-modified position indicates that the transit peptide does not stay in place (Fig. 3.5, iii). Instead, the entire precursor protein penetrated further into the chloroplasts (Fig. 3.5, v), possibly by the action of a temperature-sensitive ATP hydrolyzing factor.

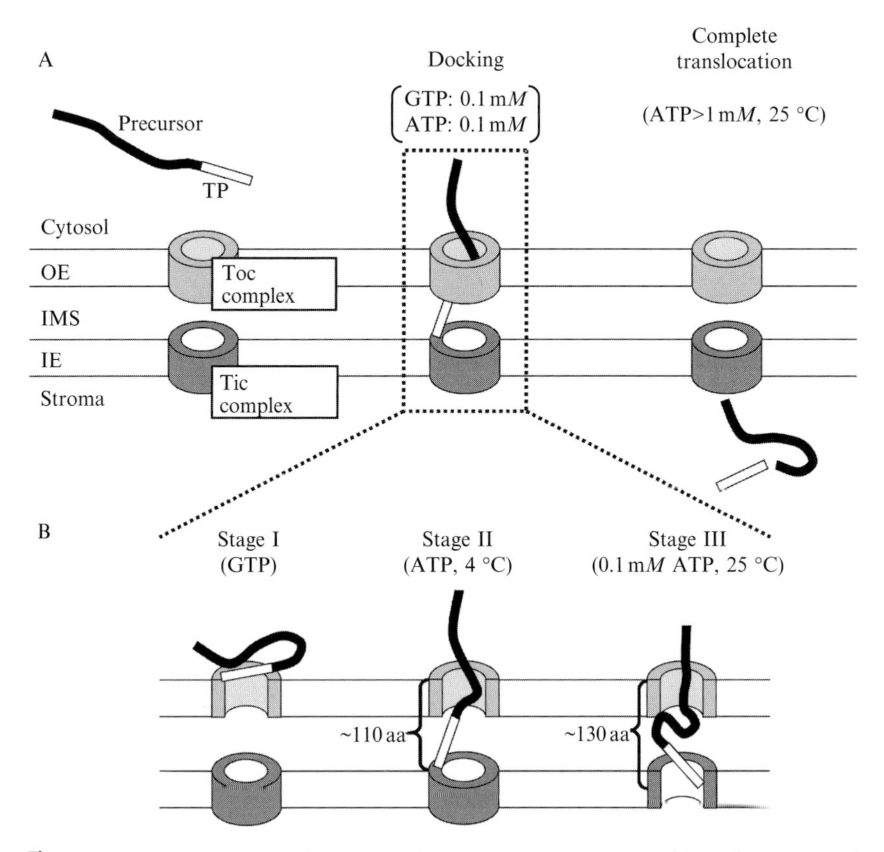

Figure 3.9 Outline of Toc/Tic-dependent protein import into chloroplasts (A) and the docking model deduced from our results (B). Abbreviations: IE, inner envelope membrane; IMS, intermembrane space; OE, outer envelope membrane; TP, transit peptide.

We have made some attempts to identify the cross-linking partner of the precursor. So far, CP4 has been proved to be the product of cross-linking between the precursor and Toc75 (Inoue and Akita, 2008a).

5. Conclusions

Protein translocation is regarded as a set of cooperative actions between a precursor and the components that comprise the translocon. In our studies, we focused on the behavior of the precursor protein within the early translocation intermediates and demonstrated that at least three translocation intermediate stages are present during the early process of protein import into chloroplasts. The specific energy conditions affect which intermediate stages are present, as illustrated in Fig. 3.9. Because we did not investigate the behavior of each component; however, we still do not have any information on the supercomplex of each intermediate, and whether any of the components are assembled into or dissociate from the intermediate during the transition from the intermediate stage remains unknown. Our observations that we could identify the different sets of intermediate states enable us to address this question in future studies by isolating intermediates formed under different energy conditions.

Acknowledgments

This work was supported in part by the Presidents's discretionary budget of Ehime University (to M. A.), Grants-in-Aid for Scientific Research (C) from JSPS (21580415) (to M. A), and Grants-in-Aid for JSPS Fellows (to H. I.). A portion of this work was undertaken at the Integrated Center for Sciences, Ehime University.

References

Akita, M., Nielsen, E., and Keegstra, K. (1997). Identification of protein transport complexes in the chloroplastic envelope membranes via chemical crosslinking. *J. Cell Biol.* **136,** 983–994.

Bruce, B. D., Perry, S., Froehlich, J., and Keegstra, K. (1994). *In vitro* import of proteins into chloroplasts. *In* "Plant Molecular Biology Manual," (S. B. Gelvin and R. A. Schilperoort, eds.) **J1,** pp. 1–15. Kluwer Academic Publishers, Boston.

Dormán, G., and Prestwitch, G. D. (1994). Benzophenone photophores in biochemistry. *Biochemistry* **33,** 5661–5673.

Gray, M. W. (1992). The endosymbiont hypothesis revisited. *Int. Rev. Cytol.* **141,** 233–357.

Hasunuma, K., Yabe, N., Yoshida, Y., Ogura, Y., and Hamada, T. (2003). Putative functions of nucleoside diphosphate kinase in plants and fungi. *J. Bioenerg. Biomembr.* **35,** 57–65.

Hörmann, E., Soll, J., and Bölter, B. (2007). The chloroplast protein import machinery: A review. *Methods Mol. Biol.* **390,** 179–193.

Inaba, T., and Schnell, D. J. (2008). Protein trafficking to chloroplasts: One theme, many variations. *Biochem. J.* **413,** 15–28.

Inoue, H., and Akita, M. (2008a). The transition of early translocation intermediates in chloroplasts is accompanied by the movement of the targeting signal on the precursor protein. *Arch. Biochem. Biophys.* **477,** 232–238.

Inoue, H., and Akita, M. (2008b). Three sets of translocation intermediates are formed during the early stage of protein import into chloroplasts. *J. Biol. Chem.* **283,** 7491–7502.

Inoue, H., Ratnayake, R. M., Nonami, H., and Akita, M. (2008). Development and optimization of an *in vitro* chloroplastic protein import assay using recombinant proteins. *Plant Physiol. Biochem.* **46,** 541–549.

Jackson, D. T., Froehlich, J. E., and Keegstra, K. (1998). The hydrophilic domain of Tic110, an inner envelope membrane component of the chloroplast protein translocation apparatus, faces the stromal compartment. *J. Biol. Chem.* **273,** 16583–16588.

Jarvis, P. (2008). Targeting of nucleus-encoded proteins to chloroplasts in plants. *New Phytol.* **179,** 257–285.

Kessler, F., Blobel, G., Patel, H. A., and Schnell, D. J. (1994). Identification of two GTP-binding proteins in the chloroplast protein import machinery. *Science* **266,** 1035–1039.

Kouranov, A., and Schnell, D. J. (1997). Analysis of the interactions of preproteins with the import machinery over the course of protein import into chloroplasts. *J. Cell Biol.* **139,** 1677–1685.

Laemmli, U. K. (1970). Cleavage of structural proteins during the assembly of the head of bacteriophage T4. *Nature* **227,** 680–684.

Leheny, E. A., and Theg, S. M. (1994). Apparent inhibition of chloroplast protein import by cold temperatures is due to energetic considerations not membrane fluidity. *Plant Cell* **6,** 427–437.

Lowry, O. H., Rosebrough, N. J., Farr, A. L., and Randall, R. J. (1951). Protein measurement with the Folin phenol reagent. *J. Biol. Chem.* **193,** 265–275.

Ma, Y., Kouranov, A., LaSala, S. E., and Schnell, D. J. (1996). Two components of the chloroplast protein import apparatus, IAP86 and IAP75, interact with the transit sequence during the recognition and translocation of precursor proteins at the outer envelope. *J. Cell Biol.* **134,** 315–327.

Neuhaus, H. E., and Wagner, R. (2000). Solute pores, ion channels, and metabolite transporters in the outer and inner envelope membranes of higher plant plastids. *Biochim. Biophys. Acta.* **1465,** 307–323.

Nielsen, E., Akita, M., Davila-Aponte, J., and Keegstra, K. (1997). Stable association of chloroplastic precursors with protein translocation complexes that contain proteins from both envelope membranes and a stromal Hsp100 molecular chaperone. *EMBO J.* **16,** 935–946.

Olsen, L., and Keegstra, K. (1992). The binding of precursor proteins to chloroplasts requires nucleoside triphosphates in the intermembrane space. *J. Biol. Chem.* **267,** 433–439.

Olsen, L. J., Theg, S. M., Selman, B. R., and Keegstra, K. (1989). ATP is required for the binding of precursor proteins to chloroplast. *J. Biol. Chem.* **264,** 6724–6729.

Perry, S. E., and Keegstra, K. (1994). Envelope membrane proteins that interact with chloroplastic precursor proteins. *Plant Cell* **6,** 93–105.

Rensink, W. A., Schnell, D. J., and Weisbeek, P. J. (2000). The transit sequence of ferredoxin contains different domains for translocation across the outer and inner membrane of the chloroplast envelope. *J. Biol. Chem.* **275,** 10265–10271.

Schägger, H., and von Jagow, G. (1987). Tricine-sodium dodecyl sulfate-polyacrylamide gel electrophoresis for the separation of proteins in the range from 1 to 100 kDa. *Anal. Biochem.* **166,** 368–379.

Schnell, D. J., Kessler, F., and Blobel, G. (1994). Isolation of the components of the chloroplast import machinery. *Science* **266,** 1007–1012.

Theg, S. M., Bauerle, C., Olsen, L. J., Selman, B. R., and Keegstra, K. (1989). Internal ATP is the only energy requirement for the translocation of precursor proteins across chloroplastic membranes. *J. Biol. Chem.* **264,** 6730–6736.

Wickner, W., and Schekman, R. (2005). Protein translocation across biological membranes. *Science* **310,** 1452–1456.

Young, M. E., Keegstra, K., and Froehlich, J. E. (1999). GTP promotes the formation of early-import intermediates but is not required during the translocation step of protein import into chloroplasts. *Plant Physiol.* **121,** 237–244.

CHAPTER FOUR

Use of DNA Length Variation to Detect Periodicities in Positively Cooperative, Nonspecific Binding

Manana Melikishvili, Lance M. Hellman, *and* Michael G. Fried

Contents

Abstract

The experiments described here demonstrate ways in which DNA length can be used as an experimental variable for the characterization of positively cooperative, sequence nonspecific DNA binding. Examples are drawn from recent studies of the interactions of O^6-alkylguanine DNA alkyltransferase (AGT) with duplex DNAs (Melikishvili *et al.* (2008). Interactions of human O^6-alkylguanine-DNA alkyltransferase (AGT) with short double-stranded DNAs. *Biochemistry* **47,** 13754–13763). Oscillations in binding density and apparent binding site size (S_{app}) are predicted by models in which a single cooperative assembly forms on each DNA molecule and in which enzyme molecules bind full-length binding sites, but not partial ones. These oscillations provide an accurate, DNA-length independent measure of the occluded binding site size (the length of DNA that one protein molecule occupies to the exclusion of others). In addition, length-dependent oscillations in association constant (K) and cooperativity (ω) reveal the degree to which substrate length can influence these parameters.

Department of Molecular and Cellular Biochemistry, Center for Structural Biology, University of Kentucky, Lexington, Kentucky, USA

Methods in Enzymology, Volume 466
ISSN 0076-6879, DOI: 10.1016/S0076-6879(09)66004-3

1. Introduction

Many proteins bind DNA with low sequence discrimination and at least moderate cooperativity. Important among these are single-strand binding proteins (Villemain and Giedroc, 1993), bacterial nucleoid-structuring proteins (Bouffartigues *et al.*, 2007; Peterson *et al.*, 2007), eukaryotic chromosome-structuring proteins (Watanabe, 1986), and DNA-repair proteins (Lee *et al.*, 2004; Rasimas *et al.*, 2007). In addition, several sequence-specific DNA binding proteins have been found to interact with nontarget sequences cooperatively and with low sequence discrimination. Examples include some transcription-regulatory proteins (human glucocorticoid receptor (Hard *et al.*, 1990) and NFkB (Phelps *et al.*, 2000), *Xenopus laevis* TFIIIA (Daly and Wu, 1989), *Escherichia coli* CAP (Saxe and Revzin, 1979; Takahashi *et al.*, 1979), and some restriction endonucleases (Taylor *et al.*, 1991)). The importance of cooperative, nonspecific interactions was recognized early (Latt and Sober, 1967; McGhee and von Hippel, 1974) and this recognition has stimulated the continuing development of theory to account for observable features of nonspecific protein–DNA assemblies (Bujalowski *et al.*, 1989; Epstein, 1978; Latt and Sober, 1967; McGhee and von Hippel, 1974; Nechipurenko and Gursky, 1986; Tsodikov *et al.*, 2001; Ucci and Cole, 2004; Wolfe and Meehan, 1992).

Epstein (1978) predicted that for highly cooperative binding to short DNA lattices, the fraction of residues occupied at binding saturation would oscillate with increasing lattice length. This reflects the expectation that over a wide concentration range, the protein will occupy all available full-length binding sites, but not partial sites for which it will have much-lower affinity (Fig. 4.1A). A second expectation is that for short DNAs, the probability that >1 cooperative assembly will form on a given DNA molecule will be small, so gaps between cooperative units will not contribute significantly to the binding density. When these conditions are met, the saturating stoichiometry, n_{max}, is the largest integer $\leq N/s$, where N is the number of residues (bp or nt) in the DNA and s is the number of bp (nt) that the protein occupies to the exclusion of others. If saturated complexes contain n_{max} proteins, we can define an apparent binding site size $S_{app} = N/n_{max}$ bp (or nt)/protein. For short DNAs, values of S_{app} are expected to fall between s and $2s - 1$ and to vary in a saw-tooth oscillation that diminishes with increasing N (Fig. 4.1B). Because minima in S_{app} occur whenever N/s is an integer, this oscillation should have a frequency equal to s. These features are shown schematically in Fig. 4.1B, for the example of a protein that occludes 4 bp. Oscillation in S_{app} with DNA length provides a measure of the occluded binding site size (s) and a basis for analysis of the dependence of cooperativity and association constants on binding density.

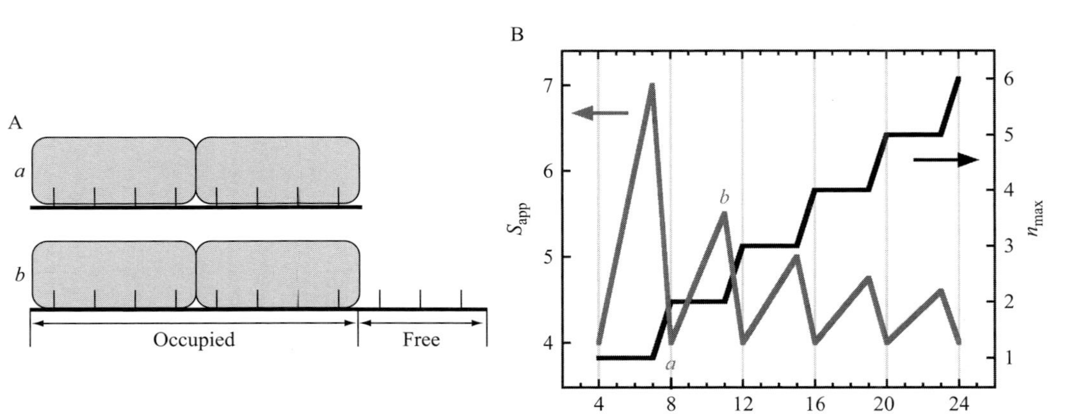

Figure 4.1 Contiguous protein binding to linear DNAs. (Panel A) Saturating stoichiometry $n_{max} = 2$. The protein (gray lozenge) occupies 4 lattice sites to the exclusion of other protein molecules ($s = 4$). *Above*: the number of lattice sites is an integral multiple of s and S_{app} is at a minimum ($S_{app} = 4$ residues). Values of S_{app} and n_{max} that correspond to this distribution are indicated by the letter *a* in panel B. *Below*: the number of lattice sites is not an integral multiple of s, and $S_{app} = 5.5$ residues, greater than its minimum value. Values of S_{app} and n_{max} that correspond to this distribution are indicated by the letter *b* in panel B. The lower distribution is one of four possible arrangements of two contiguously bound proteins on this lattice; the requirement for contiguous binding results in the same number of free sites for all arrangements. (Panel B) Dependence of apparent binding site size (S_{app}) and limiting stoichiometry (n_{max}) on lattice length N, for a protein with occluded site size (s) of 4 residues. Gray vertical lines indicate increments in N that are integral multiples of s, resulting in minima in S_{app}. The gray letters (*a*, *b*) indicate local minimum and maximum values of S_{app} that correspond to diagrams *a* and *b* in panel A.

In addition, because the proportion of binding sites near DNA ends decreases with increasing DNA length, analysis of length-dependent binding offers the potential to detect binding perturbations that are due to DNA ends.

2. Protein and DNA Preparations

The system that we have chosen to demonstrate DNA length-dependent binding is that of human O^6-alkylguanine-DNA alkyltransferase (AGT) interacting with short duplex DNA molecules. AGT is a well-characterized DNA-repair protein (Daniels *et al.*, 2004; Duguid *et al.*, 2005; Margison and Santibáñez-Koref, 2002; Pegg, 2000) that binds undamaged DNA with little sequence or base-composition specificity and moderate cooperativity (Fried *et al.*, 1996; Melikishvili *et al.*, 2008; Rasimas *et al.*, 2007). It binds the minor groove face of B-form DNA with little bending (Daniels *et al.*, 2004) or torsional deformation (Adams *et al.*, 2009). Recombinant human AGT protein (tagged with His_6 at its C-terminal end) was expressed in *E. coli* and purified to apparent homogeneity according to published protocols (Daniels *et al.*, 2004).

Oligodeoxyribonucleotides and their complements (sequences shown in Table 4.1) were synthesized by the Macromolecular Core Facility of the Penn State College of Medicine or were purchased from Midland Certified Reagent Company. Single-stranded DNA concentrations were measured spectrophotometrically, using extinction coefficients calculated by the nearest-neighbor method (Cantor and Tinoco, 1965; Cavaluzzi and Borer, 2004). Duplex DNAs were obtained by mixing a labeled oligonucleotide with a 1.05-fold molar excess of its unlabeled complement. Samples dissolved in 10 m*M* Tris (pH 8.0 at 20 ± 1 °C), 1 m*M* EDTA buffer, were heated to 90 °C for 3 min and cooled to 20 °C over 2 h. Duplex formation was monitored by nondenaturing PAGE (Maniatis and Efstratiadis, 1980).

3. Stoichiometry Analyses

The starting point of this analysis is the determination of stoichiometries (n) for saturated protein–DNA complexes. Accurate knowledge of n is crucial for measurements that depend on binding density. Although many methods will work, the best, in our opinion, are ones in which stoichiometry is inferred directly from a property of the complex measured at equilibrium, under the solution conditions used for the affinity measurements. Approaches that can meet these criteria include sedimentation equilibrium (SE) (Daugherty and Fried, 2005; Laue, 1995; Ucci and Cole, 2004), multiangle light scattering (Folta-Stogniew, 2006; Kendrick *et al.*, 2001)

Table 4.1 Duplex oligodeoxynucleotides

Length (bp)	Sequence
11	5′-TTT TTG TTT TT-3′ 3′-AAA AAC AAA AA-5′
12	5′-GAC TGA CTG ACT-3′ 3′-CTG ACT GAC TGA-5′
14	5′-GAC TGA CTG ACT GA-3′ 3′-CTG ACT GAC TGA CT-5′
16	5′-GAC TGA CTG ACT GAC T-3′ 3′-CTG ACT GAC TGA CTG A-5′
18	5′-GGA ACC TTG GAA CCT TGG-3′ 3′-CCT TGG AAC CTT GGA ACC-5′
21	5′-TGA AGT CCA AAG TTC AGT CCC-3′ 3′-CT TCA GGT TTC AAG TCA GGG A-5′
22	5′-CGC CAA CCC GCT GCC TAT CGT T-3′ 3′-GCG GTT GGG CGA CGG ATA GCA A-5′
24	5′-AAA AAA AAA AAA AAA AAA AAA AAA-3′ 3′-TTT TTT TTT TTT TTT TTT TTT TTT-5′
24	5′-GGG GGG GGG GGG GGG GGG GGG GGG-3′ 3′-CCC CCC CCC CCC CCC CCC CCC CCC-5′
26	5′-GAC TGA CTG ACT GAC TGA CTG ACT GA-3′ 3′-CTG ACT GAC TGA CTG ACT GAC TGA CT-5′
30	5′-GTG CCG CCA ACC CGC TGC CTA TCG TTA TAC-3′ 3′-CAC GGC GGT TGG GCG ACG GAT AGC AAT ATG-5′
41	5′-GCA ACG CAA TTA ATG TGA GTT AGC TCA CTC ATT AGG CAC CC-3′ 3′-CGT TGC GTT AAT TAC ACT CAA TCG AGT GAG TAA TCC GTG GG-5′

Modified from Melikishvili *et al.* (2008), with permission.

and tracer techniques that allow simultaneous measurement of the relative concentrations of protein and DNA (Fierer and Challberg, 1995; Fried and Crothers, 1983). Less direct methods such as continuous variation can be useful too (Huang, 1982; Huang *et al.*, 2003). If nonequilibrium methods such as nitrocellulose filter-binding (Stockley, 2009; Woodbury and von Hippel, 1983) or electrophoretic mobility shift assay (EMSA; Fried and Daugherty, 1998; Hellman and Fried, 2007) are used to detect binding, care should be taken to ensure that results are not affected by dissociation or disproportionation during periods of disequilibrium.

Shown in Fig. 4.2A is a subset of the SE data used for establishing limiting stoichiometries of AGT–DNA complexes. Because the SE method has been thoroughly discussed (Cole, 2004; Daugherty and Fried, 2005;

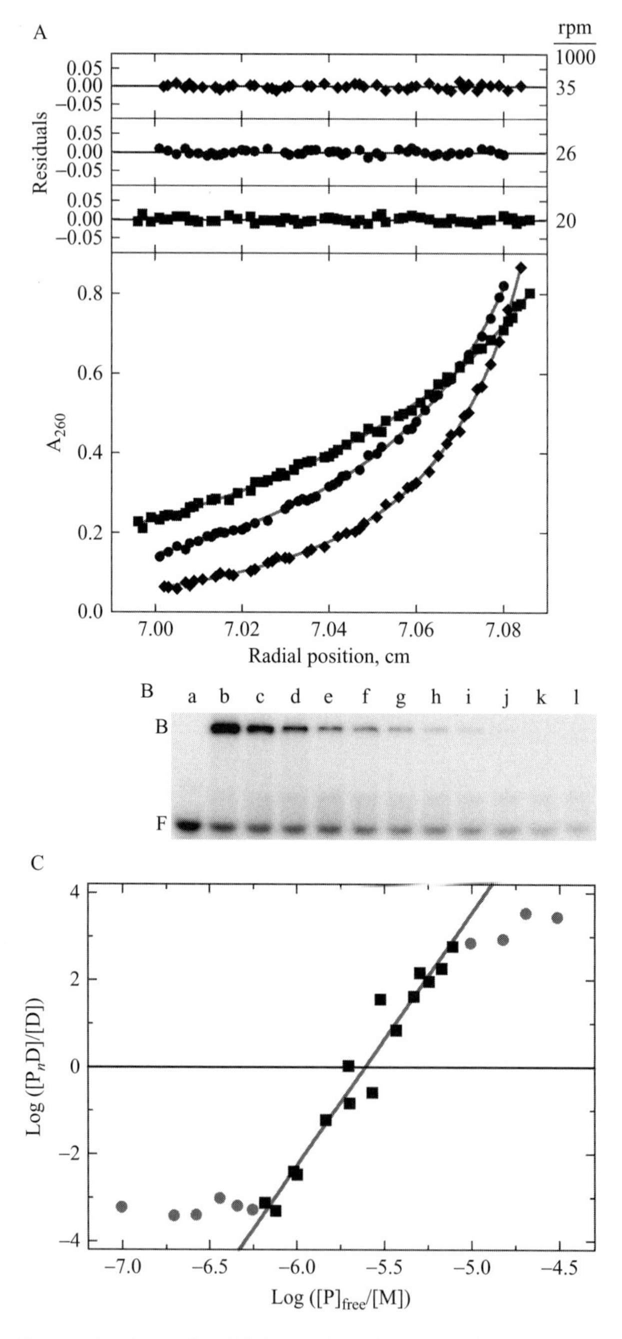

Figure 4.2 Determination of stoichiometries of saturated AGT–DNA complexes. (Panel A) Sedimentation equilibrium data for solutions containing AGT and 26 bp DNA, taken at 20 ± 1 °C. Samples contained DNA (5×10^{-7} *M*) and AGT

Laue, 1995) only relevant details are summarized here. For DNAs with small numbers of protein-binding sites, strong, positively cooperative binding can be described by an all-or-none mechanism, $n\mathrm{P} + \mathrm{D} \rightleftarrows \mathrm{P}_n\mathrm{D}$, in which free protein (P) and DNA (D) are in equilibrium with saturated complex ($\mathrm{P}_n\mathrm{D}$) and intermediates with protein stoichiometries $<n$ are not present in significant concentrations. The radial distribution of absorbance at SE for such a system is given by Eq. (4.1).

$$A(r) = \alpha_{\mathrm{P}}\exp\left[\sigma_{\mathrm{P}}(r^2 - r_{\mathrm{o}}^2)\right] + \alpha_{\mathrm{D}}\exp\left[\sigma_{\mathrm{D}}(r^2 - r_{\mathrm{o}}^2)\right] + \alpha_{\mathrm{P}_n\mathrm{D}}\exp\left[\sigma_{\mathrm{P}_n\mathrm{D}}(r^2 - r_{\mathrm{o}}^2)\right] + \varepsilon \quad (4.1)$$

here $A(r)$ is the absorbance at radial position r and α_{P}, α_{D}, and $\alpha_{\mathrm{P}_n\mathrm{D}}$ are absorbances of protein, DNA, and protein–DNA complex at the reference position, r_{o}. and ε is a baseline offset that accounts for radial position-independent differences in the absorbances of different cell assemblies. The reduced molecular weights of AGT protein, DNA, and protein–DNA complexes are given by $\sigma_{\mathrm{P}} = M_{\mathrm{P}}(1 - \bar{v}_{\mathrm{P}}\rho)z^2/(2RT)$, $\sigma_{\mathrm{D}} = M_{\mathrm{D}}(1 - \bar{v}_{\mathrm{D}}\rho)z^2/(2RT)$, and $\sigma_{\mathrm{P}_n\mathrm{D}} = (nM_{\mathrm{P}} + M_{\mathrm{D}})(1 - \bar{v}_{\mathrm{P}_n\mathrm{D}}\rho)z^2/(2RT)$. Here M_{P} and M_{D} are the molecular weights of protein and DNA, n is the protein: DNA ratio of the complex; ρ is the solvent density, z is the rotor angular velocity, R is the gas constant, and T is the temperature (K). The partial specific volume of AGT ($\bar{v}_{\mathrm{P}} = 0.744\,\mathrm{mL/g}$) was calculated by the method of Cohn and Edsall (1943), using partial specific volumes of amino acids tabulated by Laue *et al.* (1992). The partial specific volume of double-stranded NaDNA at 0.1 *M* NaCl

---

(1.45×10^{-5} *M*) in 10 m*M* Tris (pH 7.6), 1 m*M* DTT, 1 m*M* EDTA, 100 m*M* NaCl buffer. Radial scans acquired at 20,000 rpm (■), 26,000 rpm (●), and 35,000 rpm (♦) are shown. The data are offset vertically to improve clarity. The smooth curves correspond to a global fit of Eq. (4.1) to a dataset including these scans and others obtained at two additional AGT concentrations (1.82×10^{-5} *M* and 2.18×10^{-5} *M*). The small residuals, nearly symmetrically distributed about zero (upper panels) indicate that the cooperative $n\mathrm{P} + \mathrm{D} \leftrightarrows \mathrm{P}_n\mathrm{D}$ model is consistent with the observed mass distributions in these samples. (Panel B) Serial dilution analysis of an AGT complex formed with ^{32}P-labeled $\mathrm{dA}_{24}\bullet\mathrm{dT}_{24}$. Samples were resolved by electrophoresis and detected by autoradiography using an imaging plate. Sample a: 24-mer DNA (1.10×10^{-7} *M*) only. Sample b: 24-mer DNA (1.1×10^{-7} *M*) plus AGT (5.36×10^{-6} *M*). Samples c–l are sequential 1.33-fold dilutions of sample b. All samples were equilibrated in buffer (10 m*M* Tris (pH 7.6), 100 m*M* NaCl, 1 m*M* DTT, 0.05 mg/mL bovine serum albumin) for 30 min at 20 ± 1 °C prior to electrophoresis on a 10% native polyacrylamide gel (Hellman and Fried, 2007; Melikishvili *et al.*, 2008). (Panel C) Graph of the dependence of $\log[\mathrm{P}_n\mathrm{D}]/[\mathrm{D}]$ on $\log[\mathrm{P}]_{\mathrm{free}}$ for the AGT complex formed with $\mathrm{dA}_{24}\bullet\mathrm{dT}_{24}$. Data from the experiment shown in panel B and others that provide additional [AGT] values. The line represents a least squares fit to the data ensemble for the range about the mid-point of the reaction ($-6.18 \leq \log([\mathrm{AGT}]/\mathrm{M}) \leq -5.11$). *Symbols*: the points used in the fit are indicated by (■) other points in the dataset are indicated by closed circles (●). The slope equals 5.81 ± 0.34 for this subset of the data. From Melikishvili *et al.* (2008), with permission.

(0.540 mL/g) was estimated by interpolation of the data of Cohen and Eisenberg (1968). The partial specific volumes of protein–DNA complexes were estimated using Eq. (4.2).

$$\bar{\nu}_{\mathrm{P}_n\mathrm{D}} = \frac{nM_{\mathrm{P}}\bar{\nu}_{\mathrm{P}} + nM_{\mathrm{D}}\bar{\nu}_{\mathrm{D}}}{nM_{\mathrm{P}} + M_{\mathrm{D}}} \tag{4.2}$$

Equation (4.1) was used in global analysis of datasets obtained over a range of macromolecular concentrations and rotor speeds (Johnson *et al.*, 1981), implemented in the HETEROANALYSIS program (Cole, 2004). In this method, the values of α_{P}, α_{D}, $\alpha_{\mathrm{P}_n\mathrm{D}}$, and ε are unique to each sample but the value of n is optimized over all datasets.

Additional stoichiometry estimates were obtained by analysis of serial dilution series, with protein distributions resolved by native electrophoresis (Fig. 4.2B), detected using ^{32}P-labeled DNAs (Fried and Crothers, 1981; Garner and Revzin, 1981). While this is a rapid and sensitive method to detect protein–nucleic acid interactions, samples are not at equilibrium during the electrophoresis step. Factors affecting the stability of protein–DNA complexes during gel electrophoresis have been discussed elsewhere (Adams and Fried, 2007; Fried and Bromberg, 1997; Hellman and Fried, 2007). For this molecular system, control experiments in which the electrophoresis intervals were changed from 30 min to 2 h resulted in $<5\%$ decrease in the mole fractions of bound DNA, over a wide range of AGT: DNA (not shown), indicating that little dissociation was taking place during electrophoresis under these standard conditions.

The association constant for the binding of n AGT molecules to one of DNA is $K_n = [\mathrm{P}_n\mathrm{D}]/[\mathrm{D}][\mathrm{P}]^n_{\mathrm{free}}$. Separating variables and taking logarithms gives a relationship that predicts a linear dependence of $\log[\mathrm{P}_n\mathrm{D}]/[\mathrm{D}]$ on $\log[\mathrm{P}]_{\mathrm{free}}$, with slope equal to the stoichiometry, n.

$$\log\frac{[\mathrm{P}_n\mathrm{D}]}{[\mathrm{D}]} = n\log[\mathrm{P}]_{\mathrm{free}} + \log K_n \tag{4.3}$$

In a serial dilution experiment, mass action changes $[\mathrm{P}_n\mathrm{D}]/[\mathrm{D}]$ while the ratio of $[\mathrm{P}]_{\mathrm{total}}$ to $[\mathrm{D}]_{\mathrm{total}}$ remains unchanged. This feature is useful since quantitation of ^{32}P-DNA establishes the dilution factor at each step and thus $[\mathrm{P}]_{\mathrm{total}}$ for each sample. The free protein concentration at each dilution step can be estimated using $[\mathrm{P}]_{\mathrm{free}} = [\mathrm{P}]_{\mathrm{input}} - n[\mathrm{P}_n\mathrm{D}]$ starting with an initial value of n; Eq. (4.3) is then used to calculate a new value of n from the binding distribution. This value is used to calculate a new estimate of $[\mathrm{P}]_{\mathrm{free}}$ and the calculations repeated recursively. With typical data, values of n converge within 3 or 4 calculation cycles. An example of this analysis is shown in Fig. 4.2C.

The dependence of saturating stoichiometry (n) on DNA length (N) is shown in Fig. 4.3A. The slope of this graph, 0.22 ± 0.1, is consistent with a mean binding site size of 4.5 ± 0.2 bp. This is slightly larger than the minimum of 4 bp that can be estimated from the 4:1 stoichiometry observed with 16 bp DNA (Melikishvili *et al.*, 2008). Values of S_{app} oscillate with DNA length (Fig. 4.3B). Fitting the oscillation with a cosine function shows that the period and phase of the oscillation are maintained for 11 bp $\leq N \leq$ 30 bp. The oscillatory period, 4.05 ± 0.02 bp is compatible with contiguous binding models in which each protein molecule occludes 4 bp of DNA. The regularity of this oscillation, which occurs despite the

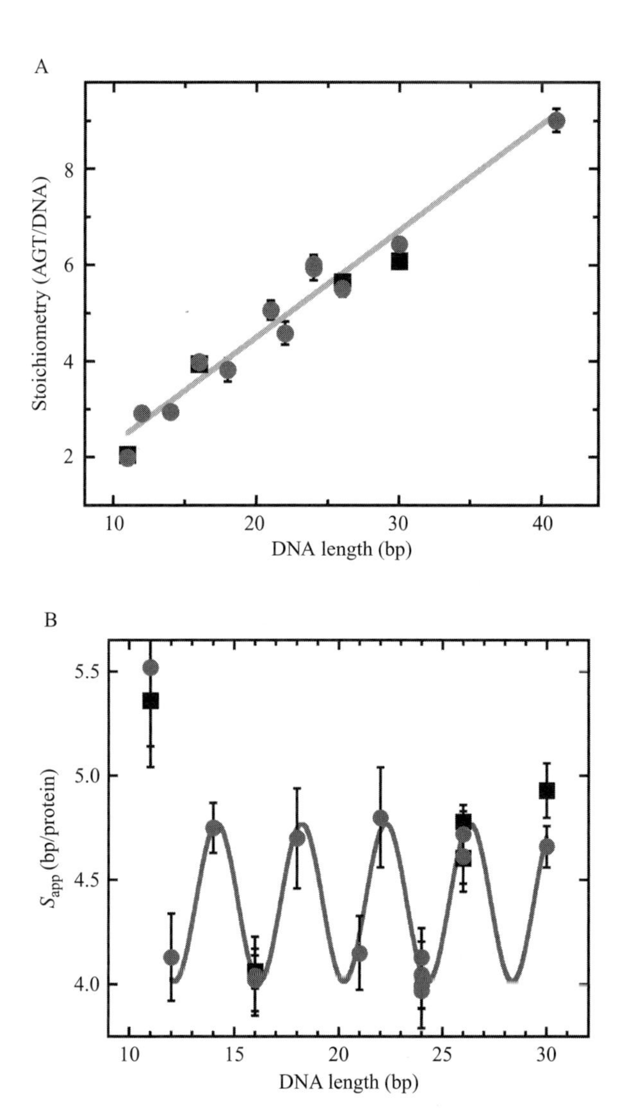

Figure 4.3 (Continued)

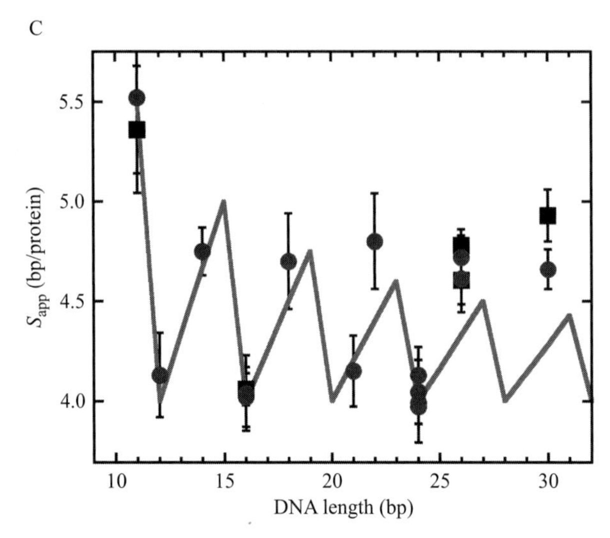

Figure 4.3 Stoichiometry and apparent binding site size (S_{app}) as functions of DNA length (N). (Panel A) Stoichiometries determined by sedimentation equilibrium (■) and serial dilution (●) analyses. Error bars (in some cases obscured by data points) represent 95% confidence intervals. (Panel B) AGT forms a binding motif with a 4-bp periodicity. S_{app} calculated using $S_{app} = N/n$, where N is DNA length in base pairs and n is the number of protein molecules bound to a DNA molecule. S_{app} values determined from sedimentation equilibrium (■) and serial dilution (●) analyses. The error bars correspond to 95% confidence limits. The smooth curve is the least-squares fit of the equation $S_{app} = A \cos(BN) + C$ in which A is the amplitude of the oscillation, B the displacement angle in degrees/bp, N is length in bp and C is an offset equal to the mean value of S_{app}. This fit returned $A = -0.37 \pm 0.04$, $B = 88.7° \pm 0.5°$ and $C = 4.39 \pm 0.02$. This value of B indicates that successive binding sites are separated by $360°/(88.7° \pm 0.5°) = 4.05 \pm 0.02$ bp along the DNA contour. (Panel C) Comparison of experimental and theoretical S_{app} values as a function of DNA length. Experimental S_{app} values determined from sedimentation equilibrium data are indicated by filled squares (■), values obtained from EMSA experiments are indicated by closed circles (●). The error bars correspond to 95% confidence limits. Theoretical S_{app} values (solid line) were calculated using $S_{app} = N/n_{max}$ where n_{max} is the largest integer $\leq N/4$. Panels B and C modified from Melikishvili *et al.* (2008), with permission.

fact that the DNAs used vary in sequence and base composition, is consistent with models in which the repeating structure of this cooperative assembly does not depend on these features.

For the shortest DNAs, experimental S_{app} values coincide with theoretical values calculated using $s = 4$ bp. This agreement is best when $N/4$ is an integer. The maxima in theoretical S_{app} decrease with increasing DNA length as the proportion of the total DNA length within an unoccupied partial binding site decreases. Intriguingly, this decrease is not reproduced by the experimental data. Our working hypothesis is that the maximal values of experimental S_{app} reflect increasingly degenerate binding as DNA length increases. Such degeneracy may result in population-average stoichiometries $<n_{max}$ and hence larger than expected values of S_{app}. At present we do not know the maximum DNA length over which oscillations in S_{app} can be detected. However, we expect that the range of length over which oscillations can be observed will depend on the magnitude of the binding cooperativity, as this determines the distribution of proteins between contiguous binding sites (adjacent to sites occupied by other bound proteins) and sites that are not contiguous. Tests of this prediction are underway.

4. Affinity and Cooperativity as Functions of DNA Length

Changes in intrinsic association constant (K) or in binding cooperativity (ω) may accompany changes in binding density. When suitable binding data are available, these parameters can be evaluated using the short-lattice variant of the McGhee–von Hippel relation (McGhee and von Hippel, 1974; Tsodikov *et al.*, 2001) (Eq. (4.4)).

$$\begin{aligned}\frac{v}{[\mathrm{P}]} &= K(1-sv)\left(\frac{(2\omega-1)(1-sv)+v-R}{2(\omega-1)(1-sv)}\right)^{s-1}\\ &\quad\left(\frac{1-(s+1)v+R}{2(1-sv)}\right)^{2}\left(\frac{N-s+1}{N}\right)\\ R &= \left((1-(s+1)v)^{2}+4\omega v(1-sv)\right)^{1/2}\end{aligned} \tag{4.4}$$

here, v is the binding density (protein molecules/bp), K is the equilibrium association constant for binding a single site, ω is the cooperativity parameter, N is the length of the DNA in base pairs, and s is the occluded site size. The cooperativity parameter ω is the equilibrium constant for moving a protein from an isolated site to a singly contiguous one or from a singly contiguous site to a doubly contiguous one (McGhee and von Hippel, 1974). The model embodied in this equation is one in which proteins are assumed not to bind to fractional sites of length $< s$ base pairs located within or at the ends of DNA molecules (Tsodikov *et al.*, 2001).

Shown in Fig. 4.4A are Scatchard plots of data obtained from serial dilution and direct titration experiments. The smooth curves represent least squares fits of Eq. (4.4) to binding data, for which values of s were calculated from stoichiometries determined independently. Tests using fixed values of K show that K and ω are weakly anticorrelated (an increase in one parameter produces a slight decrease in the other). However, the quality of the fits, measured by the χ^2 parameter, become much worse if either parameter varies from its most probable value by more than its 95% confidence limits (results not shown). Thus, while the values of these parameters cannot be considered unique, the data shown in this example are good enough to allow us to parse binding affinity into reasonably well-defined estimates of K and ω.

Values of K and ω oscillate with increasing DNA length (Fig. 4.4B). The fact that these oscillations are not in phase suggests that they are not the result of fitting correlations between K and ω (see above). The variation of ω with DNA length is regular in amplitude and period; it has the same frequency but opposite phase to that of S_{app}. Thus, in this example, the most compact complexes (with smallest S_{app}) have relatively stronger protein–protein interactions than do less compact assemblies. One possible interpretation of these results is that optimal protein juxtaposition is lost as binding density decreases (S_{app} increases). The baseline for the oscillation of ω increases uniformly for 11 bp $\leq$ DNA length $\leq$ 30 bp. This is an expected consequence of binding linear substrates. Proteins at the ends of a cooperative array must interact with a single neighbor while proteins in the center of the array interact with two. With increasing array length, the proportion

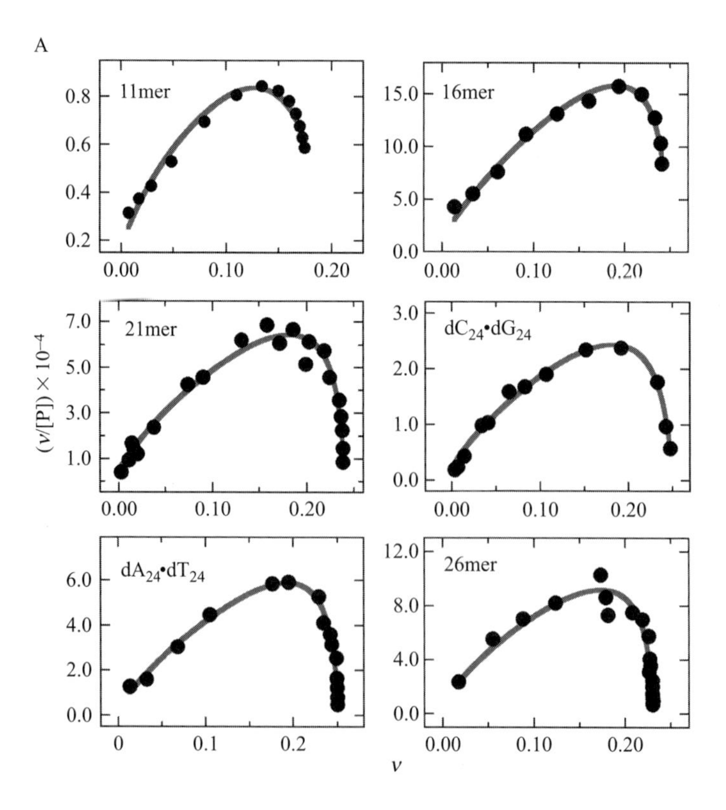

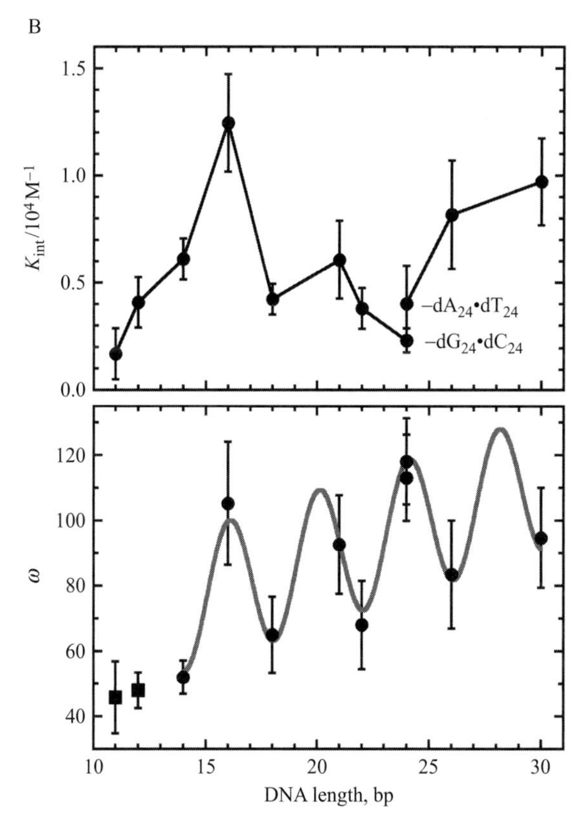

Figure 4.4 Affinity and cooperativity as functions of DNA length. (Panel A) Scatchard plots. DNAs (individual concentrations differed but all were close to 2×10^{-8} *M*) were titrated with AGT protein ($0 \leq$ [AGT] $\leq 5.2 \times 10^{-5}$ *M*) in buffer consisting of 10 m*M* Tris (pH 7.6), 1 m*M* DTT, 1 m*M* EDTA, 100 m*M* NaCl. Free and bound DNA species were resolved by native electrophoresis (EMSA) and mole fractions were quantitated as described for Fig. 4.2. Each dataset shown is derived from 2 or 3 independent titrations. The smooth curves are nonlinear least squares fits of Eq. (4.4) to the data. (Panel B) Dependence of *K* and ω on DNA length (*N*). The data are derived in part from the experiments shown in panel A. The error bars correspond to 95% confidence limits estimated for each parameter. The points corresponding to *K*-values for $dA_{24}•dT_{24}$ and $dG_{24}•dC_{24}$ templates are labeled. Panels A and B modified from Melikishvili *et al.* (2008), with permission.

of proteins that are in the center (i.e., not bound at array ends) increases, with corresponding increase in ω averaged over the whole assembly.

In contrast to the pattern seen with ω, oscillations in *K* are irregular in amplitude and phase. This suggests that factors other than DNA length and binding density contribute significantly to the strength of these protein–DNA contacts. The difference in *K* values for $dG_{24}•dC_{24}$ and $dA_{24}•dT_{24}$

templates (points designated in Fig. 4.4B) suggests that DNA sequence has an important influence on affinity. Consistent with this notion, the two DNAs that give local maxima in K (the 16-mer and the 26-mer) have repeating sequences that provide a guanine at every fourth residue on each strand (shown in Table 4.1). This frequency, which is the same as that of AGT in the cooperative complex, may allow the enzyme to make unusually favorable contacts with these substrates. We expect that a set of homopolymer DNAs, such as poly dA•poly dT, differing in length, will offer the best chance to observe any changes in K that depend on binding density, without confounding contributions from sequence-specific effects.

Short synthetic DNAs are available in quantity, high purity and at reasonable cost from commercial sources. These features make them substrates of choice for studies of protein–DNA interactions. The experiments described here demonstrate the ease with which DNA length can be used as an experimental variable and some of the results that can be derived from such studies. Although these experiments have focused on duplex DNA, the same logic can be applied to single-stranded substrates (Rasimas *et al.*, 2007). In addition, the experiments shown here were performed with a protein that has little DNA sequence specificity, but it seems likely that a similar approach could be used to study cooperative interactions between proteins that bind specific sequences or structures. Finally, the fact that equilibrium constants and cooperativity parameters vary with DNA length suggests that dynamic features of cooperative assemblies might depend on DNA length as well. This possibility calls for further study.

ACKNOWLEDGMENTS

We gratefully acknowledge valuable discussions with Drs. Jack Correia, Michael Johnson, Jacob Lebowitz, James Cole, and Nichola Garbett. This work was supported by NIH grant GM-070662 to M. G. F.

REFERENCES

Adams, C., and Fried, M. G. (2007). Analysis of protein-DNA equilibria by native gel electrophoresis. *In* "Protein Interactions: Biophysical Approaches For The Study of Multicomponent Systems," (P. Schuck, ed.), pp. 417–446. Academic Press, New York.

Adams, C. A., Melikishvili, M., Rodgers, D. W., Rasimas, J. J., Pegg, A. E., and Fried, M. G. (2009). Topologies of complexes containing O^6-alkylguanine-DNA alkyltransferase and DNA. *J. Mol. Biol.* **389,** 248–263.

Bouffartigues, E., Buckle, M., Badaut, C., Travers, A., and Rimsky, S. (2007). H-NS cooperative binding to high-affinity sites in a regulatory element results in transcriptional silencing. *Nat. Struct. Mol. Biol.* **14,** 441–448.

Bujalowski, W., Lohman, T. M., and Anderson, C. F. (1989). On the cooperative binding of large ligands to a one-dimensional homogeneous lattice: The generalized three-state lattice model. *Biopolymers* **28,** 1637–1643.
Cantor, C. R., and Tinoco, I., Jr. (1965). Absorption and optical rotatory dispersion of seven trinucleoside diphosphates. *J. Mol. Biol.* **13,** 65–77.
Cavaluzzi, M. J., and Borer, P. N. (2004). Revised UV extinction coefficients for nucleoside-5′-monophosphates and unpaired DNA and RNA. *Nucl. Acids Res.* **32,** e13.
Cohen, G., and Eisenberg, H. (1968). Deoxyribonucleate solutions: sedimentation in a density gradient, partial specific volumes, density and refractive density increments and preferential interactions. *Biopolymers* **6,** 1077–1100.
Cohn, E. J., and Edsall, J. T. (1943). Proteins, Amino Acids and Peptides as Ions and Dipolar Ions. *In* "Proteins, Amino Acids and Peptides as Ions and Dipolar Ions," (E. J. Cohn and J. T. Edsall, eds.), pp. 370–381, 428–431. Reinhold, New York.
Cole, J. L. (2004). Analysis of heterogeneous interactions. *Methods Enzymol.* **384,** 212–232.
Daly, T. J., and Wu, C. W. (1989). Cooperative DNA binding by Xenopus transcription factor IIIA. Use of a 66-base pair DNA fragment containing the intragenic control region of the 5 S RNA gene to study specific and nonspecific interactions. *J. Biol. Chem.* **264,** 20394–20402.
Daniels, D. S., Woo, T. T., Luu, K. X., Noll, D. M., Clarke, N. D., Pegg, A. E., and Tainer, J. A. (2004). DNA binding and nucleotide flipping by the human DNA repair protein AGT. *Nat. Struct. Mol. Biol.* **11,** 714–720.
Daugherty, M. A., and Fried, M. G. (2005). Protein-DNA interactions at sedimentation equilibrium. *In* "Modern Analytical Ultracentrifugation: Techniques and Methods," (D. Scott, ed.), pp. 195–209. Royal Society of Chemistry, Oxford.
Duguid, E. M., Rice, P. A., and He, C. (2005). The structure of the human AGT protein bound to DNA and its implications for damage detection. *J. Mol. Biol.* **350,** 657–666.
Epstein, I. R. (1978). Cooperative and non-cooperative binding of large ligands to a finite one-dimensional lattice. A model for ligand-oligonucleotide interactions. *Biophys. Chem.* **8,** 327–339.
Fierer, D. S., and Challberg, M. D. (1995). The stoichiometry of binding of the herpes simplex virus type 1 origin binding protein, UL9, to OriS. *J. Biol. Chem.* **270,** 7330–7334.
Folta-Stogniew, E. (2006). Oligomeric states of proteins determined by size-exclusion chromatography coupled with light scattering, absorbance, and refractive index detectors. *Methods Mol. Biol.* **328,** 97–112.
Fried, M. G., and Bromberg, J. L. (1997). Factors that affect the stability of protein-DNA complexes during gel electrophoresis. *Electrophoresis* **18,** 6–11.
Fried, M. G., and Crothers, D. M. (1981). Equilibria and kinetics of lac repressor-operator interactions by polyacrylamide gel electrophoresis. *Nucl. Acids Res.* **9,** 6505–6525.
Fried, M. G., and Crothers, D. M. (1983). CAP and RNA polymerase interactions with the *lac* promoter: Binding stoichiometry and long range effects. *Nucl. Acids Res.* **11,** 141–158.
Fried, M. G., and Daugherty, M. A. (1998). Electrophoretic analysis of multiple protein-DNA interactions. *Electrophoresis* **19,** 1247–1253.
Fried, M. G., Kanugula, S., Bromberg, J. L., and Pegg, A. E. (1996). DNA binding mechanisms of O^6-alkylguanine-DNA alkyltransferase: stoichiometry and effects of DNA base composition and secondary structures on complex stability. *Biochemistry* **35,** 15295–15301.
Garner, M. M., and Revzin, A. (1981). A gel electrophoresis method for quantifying the binding of proteins to specific DNA regions: Application to components of the Escherichia coli lactose operon system. *Nucl. Acids Res.* **9,** 3047–3060.
Hard, T., Dahlman, K., Carlstedt-Duke, J., Gustafsson, J.-A., and Rigler, R. (1990). Cooperativity and specificity in the interactions between DNA and the glucocorticoid receptor DNA-binding domain. *Biochemistry* **29,** 5358–5364.

Hellman, L. M., and Fried, M. G. (2007). Electrophoretic mobility shift assay (EMSA) for detecting protein-nucleic acid interactions. *Nat. Protoc.* **2,** 1849–1861.

Huang, C. Y. (1982). Determination of binding stoichiometry by the continuous variation method: The Job plot. *Methods Enzymol.* **87,** 509–525.

Huang, C. Y., Zhou, R., Yang, D. C., and Boon Chock, P. (2003). Application of the continuous variation method to cooperative interactions: mechanism of Fe(II)-ferrozine chelation and conditions leading to anomalous binding ratios. *Biophys. Chem.* **100,** 143–149.

Johnson, M. L., Correia, J. J., Yphantis, D. A., and Halvorson, H. R. (1981). Analysis of data from the analytical ultracentrifuge by non-linear least squares techniques. *Biophys. J.* **36,** 575–588.

Kendrick, B. S., Kerwin, B. A., Chang, B. S., and Philo, J. S. (2001). Online size-exclusion high-performance liquid chromatography light scattering and differential refractometry methods to determine degree of polymer conjugation to proteins and protein-protein or protein-ligand association states. *Anal. Biochem.* **299,** 136–146.

Latt, S. A., and Sober, H. A. (1967). Protein-nucleic acid interactions. II. Oligopeptide-polyribonucleotide binding studies. *Biochemistry* **6,** 3293–3306.

Laue, T. M. (1995). Sedimentation equilibrium as a thermodynamic tool. *Methods Enzymol.* **259,** 427–452.

Laue, T. M., Shah, B. D., Ridgeway, T. M., and Pelletier, S. L. (1992). Computer-aided interpretation of analytical sedimentation data for proteins. *In* "Analytical Ultracentrifugation in Biochemistry and Polymer Science," (S. E. Harding, A. J. Rowe, and J. C. Harding, eds.), pp. 90–125. The Royal Society of Chemistry, Cambridge, England.

Lee, C. Y., Bai, H., Houle, R., Wilson, G. M., and Lu, A. L. (2004). An *Escherichia coli* MutY mutant without the six-helix barrel domain is a dimer in solution and assembles cooperatively into multisubunit complexes with DNA. *J. Biol. Chem.* **279,** 52653–52663.

Maniatis, T., and Efstratiadis, A. (1980). Fractionation of low molecular weight DNA or RNA in polyacrylamide gels containing 98% formamide or 7 *M* urea. *Methods Enzymol.* **65,** 299–305.

Margison, G. P., and Santibáñez-Koref, M. F. (2002). O^6-Alkylguanine-DNA alkyltransferase: role in carcinogenesis and chemotherapy. *BioEssays* **24,** 255–266.

McGhee, J., and von Hippel, P. H. (1974). Theoretical aspects of DNA-protein interactions: Co-operative and non-co-operative binding of large ligands to a one-dimensional homogeneous lattice. *J. Mol. Biol.* **86,** 469–489.

Melikishvili, M., Rasimas, J. J., Pegg, A. E., and Fried, M. G. (2008). Interactions of human O6-alkylguanine-DNA alkyltransferase (AGT) with short double-stranded DNAs. *Biochemistry* **47,** 13754–13763.

Nechipurenko, Y. D., and Gursky, G. V. (1986). Cooperative effects on binding of proteins to DNA. *Biophys. Chem.* **24,** 195–209.

Pegg, A. E. (2000). Repair of O(6)-alkylguanine by alkyltransferases. *Mutat. Res.* **462,** 83–100.

Peterson, S. N., Dahlquist, F. W., and Reich, N. O. (2007). The role of high affinity non-specific DNA binding by Lrp in transcriptional regulation and DNA organization. *J. Mol. Biol.* **369,** 1307–1317.

Phelps, C. B., Sengchanthalangsy, L. L., Shiva Malek, S., and Ghosh, G. (2000). Mechanism of B DNA binding by Rel/NFk-B dimers. *J. Biol. Chem.* **275,** 24392–24399.

Rasimas, J. J., Kar, S. R., Pegg, A. E., and Fried, M. G. (2007). Interactions Of human O^6-alkylguanine-DNA alkyltransferase (AGT) with short single-stranded DNAs. *J. Biol. Chem.* **282,** 3357–3366.

Saxe, S. A., and Revzin, A. (1979). Cooperative binding to DNA of catabolite activator protein of *Escherichia coli*. *Biochemistry* **18,** 255–263.

Stockley, P. G. (2009). Filter-binding assays. *Methods Mol. Biol.* **543,** 1–14.

Takahashi, M., Blazy, B., and Baudras, A. (1979). Non-specific interactions of CRP from *E. coli* with native and denatured DNAs: control of binding by cAMP and cGMP and by cation concentration. *Nucl. Acids Res.* **7,** 1699–1712.

Taylor, J. D., Badcoe, I. G., Clarke, A. R., and Halford, S. E. (1991). EcoRV restriction endonuclease binds all DNA sequences with equal affinity. *Biochemistry* **30,** 8743–8753.

Tsodikov, O. V., Holbrook, J. A., Shkel, I. A., and Record, M. T., Jr. (2001). Analytic binding isotherms describing competitive interactions of a protein ligand with specific and nonspecific sites on the same DNA oligomer. *Biophys. J.* **81,** 1960–1969.

Ucci, J. W., and Cole, J. L. (2004). Global analysis of non-specific protein-nucleic interactions by sedimentation equilibrium. *Biophys. Chem.* **108,** 127–140.

Villemain, J. L., and Giedroc, D. P. (1993). Energetics of arginine-4 substitution mutants in the N-terminal cooperativity domain of T4 gene 32 protein. *Biochemistry* **32,** 11235–11246.

Watanabe, F. (1986). Cooperative interaction of histone H1 with DNA. *Nucl. Acids Res.* **14,** 3573–3585.

Wolfe, A. R., and Meehan, T. (1992). Use of binding site neighbor-effect parameters to evaluate the interactions between adjacent ligands on a linear lattice. Effects on ligand-lattice association. *J. Mol. Biol.* **223,** 1063–1087.

Woodbury, C. P., Jr., and von Hippel, P. H. (1983). On the determination of deoxyribonucleic acid-protein interactions parameters using the nitrocellulose filter-binding assay. *Biochemistry* **22,** 4730–4737.

CHAPTER FIVE

The Impact of Ions on Allosteric Functions in Human Liver Pyruvate Kinase

Aron W. Fenton *and* Aileen Y. Alontaga

Contents

Abstract

Experimental designs used to monitor the magnitude of an allosteric response can greatly influence observed values. We report here the impact of buffer, monovalent cation, divalent cation, and anion on the magnitude of the allosteric regulation of the affinity of human liver pyruvate kinase (hL-PYK) for substrate, phosphoenolpyruvate (PEP). The magnitudes of the allosteric activation by fructose-1,6-bisphosphate (Fru-1,6-BP) and the allosteric inhibition by alanine are independent of most, but not all buffers tested. However, these magnitudes are dependent on whether Mg^{2+} or Mn^{2+} is included as the divalent cation. In the presence of Mn^{2+}, any change in $K_{app\text{-}PEP}$ caused by Fru-1,6-BP is minimal. hL-PYK activity does not appear to require monovalent cation. Monovalent cation binding in the active site impacts PEP affinity with minimum influence on the magnitude of allosteric coupling. However, Na^+ and Li^+ reduce the magnitude of the allosteric response to Fru-1,6-BP, likely due to mechanisms outside of the active site. Which anion is used to maintain a constant monovalent cation concentration also influences the magnitude of the allosteric

Department of Biochemistry and Molecular Biology, The University of Kansas Medical Center, Kansas City, Kansas, USA

Methods in Enzymology, Volume 466
ISSN 0076-6879, DOI: 10.1016/S0076-6879(09)66005-5

response. The value of determining the impact of ions on allosteric function can be appreciated by considering that representative structures used in comparative studies have often been determined using protein crystals grown in diverse buffer and salt conditions.

1. Introduction

A review of recent literature demonstrates that there is a broad renewed interest in understanding the molecular basis of allosteric regulation. In our own studies, we prefer to use a strict definition of allostery as the energetic coupling between two binding events on the same protein (Fenton, 2008). However, many other definitions of allosteric regulation are in widespread use. The commonality of these definitions is the propagation of an effect due to a local perturbation (most often ligand binding) through the protein. In turn, this results in change in protein structure and/or function at a site that is distinct (and often distant) from the original site of perturbation. Here, we use protein structure to refer to the combined characteristics of the average protein conformation and protein dynamics. Protein dynamics is, in turn, defined as fluctuations within a stable protein without external perturbations (i.e., a change in average conformation upon ligand binding is not included in a reference to "dynamic"). Of the two components of structural descriptions, crystallographic techniques have produced a wealth of understanding about protein conformation. Although protein dynamics have been recognized for many years (Cui and Karplus, 2008; Frauenfelder *et al.*, 1988), the number of techniques that can monitor protein dynamics have been limited until recently. It follows that much of the discussion in the latest literature surge focuses on how protein dynamics (rather than more historical focuses on conformation) contribute to propagation of an allosteric signal through a protein (Formaneck *et al.*, 2006; Gunasekaran *et al.*, 2004; Kern and Zuiderweg, 2003; Popovych *et al.*, 2006; Swain and Gierasch, 2006; Volkman *et al.*, 2001).

Studies of both protein conformation and dynamics (and even evolution; Lockless and Ranganathan, 1999; Pendergrass *et al.*, 2006; Suel *et al.*, 2003) approach the structure/function relationship of allosteric regulation from the structural side. In contrast and, building on the groundwork established by Weber (1972), Reinhart (1983, 1988, 2004) has outlined the application of allosteric linkage to quantitatively characterize allosteric functions in enzymes undergoing steady-state turnover. Therefore, the theoretical basis of allosteric linkage between two ligand-binding events to one protein has now been thoroughly discussed for both systems that operate at equilibrium and those that function at steady state. The focus of the current chapter highlights technical aspects of data acquisition necessary to facilitate a linkage analysis. Specifically, the binding affinity of one ligand (which typically requires titration using a concentration range of this ligand) is determined over a concentration range of a second ligand (see below). If one or both ligands carry charge, a variation in the concentration of that ligand will be accompanied by a pH gradient, an ionic strength gradient, and/or a counterion concentration gradient, depending on experimental design. Any one of these gradients can modify protein function independent of or in concert with modifications caused by ligand binding.

We are interested in understanding the molecular basis of the allosteric regulation of human liver pyruvate kinase (L-PYK for liver isozymes in general; hL-PYK for the human protein). Pyruvate kinase (PYK) catalyzes the conversion of phosphoenolpyruvate (PEP) and MgADP to pyruvate and MgATP as the last step of glycolysis. There are four different mammalian PYK isozymes. L-PYK and the isozyme expressed in erythrocytes (R-PYK) are products of a single gene due to the use of different translation start sites and differ only by the additional 31 amino acids located at the N-terminus of R-PYK (Noguchi *et al.*, 1987); the muscle (M_1-PYK) and "fetal" (M_2-PYK) forms of PYK are gene products from a second gene. Historically accepted allosteric regulatory features of L-PYK include (1) allosteric activation of PEP affinity by fructose-1,6-bisphosphate (Fru-1,6-BP), (2) allosteric inhibition of PEP affinity by alanine (Ala), and (3) allosteric inhibition of PEP affinity by ATP. It is also historically accepted that binding of allosteric effectors to L-PYK alters the affinity of the enzyme for PEP without altering the catalytic rate (k_{cat}) or the affinity of the enzyme for MgADP (Blair, 1980; Hall and Cottam, 1978; Munoz and Ponce, 2003). We find the best definition for this "K-type" allostery is a comparison of how one ligand binds in the absence versus in the presence of a second ligand (Fenton, 2008). This definition facilitates a linked-equilibrium analysis of this regulatory phenomenon.

PEP binding in the active site of PYK is in large part due to coordination with both a divalent and a monovalent cation (Fig. 5.1) (Larsen *et al.*, 1997). It has been historically accepted that both ions are required for enzymatic activity, as is the divalent cation-bound form of ADP (Boyer, 1962; Nowak and Suelter, 1981). Given the inherent challenges due to changing counterion concentrations associated with the ligand titrations (introduced above), the ion requirements of PYK further complicate a study of the allosteric properties of hL-PYK. Therefore, this chapter presents experimental designs that have been useful in exploring how ion types (and to a less extent concentration) impact observed allosteric responses of hL-PYK to Fru-1,6-BP and Ala.

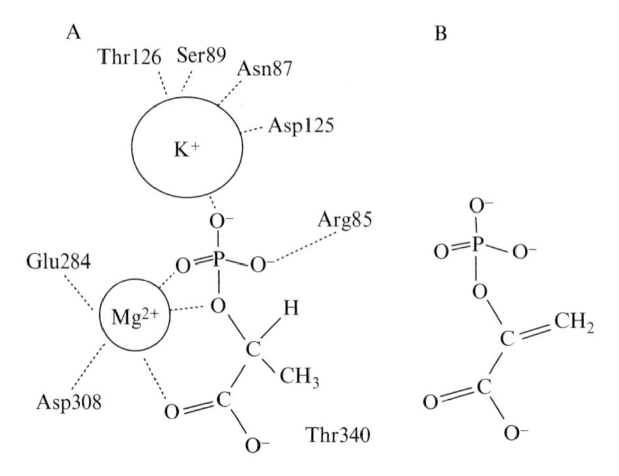

Figure 5.1 (A) Schematic of the coordination between K^+, protein-bound Mg^{2+}, and phospholactate in the PEP-binding site (the active site) as determined in M_1-PYK (Larsen *et al.*, 1997), but shown with L-PYK numbering (residue types are identical). Coordinating interactions are indicated by dashed lines. (B) Comparative structure of PEP.

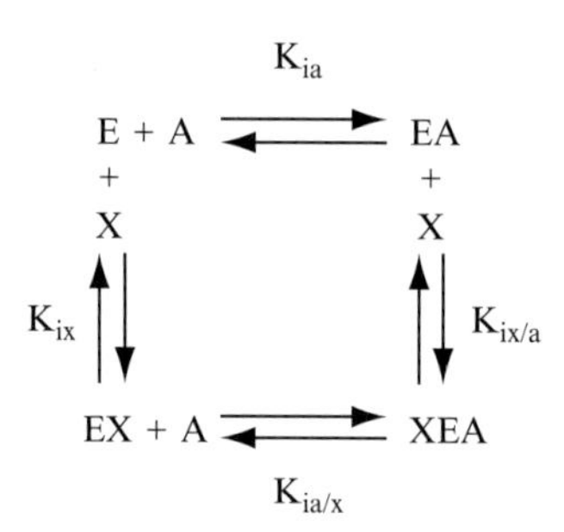

Figure 5.2 Thermodynamic energy cycle for a protein or enzyme (E) which binds one substrate (A) and one allosteric effector (X). Each enzyme complex may be a single protein conformation, an equilibrium of a limited number of conformational substates, or an ensemble of conformational substates (a dynamic structure).

2. General Strategy to Assess Allosteric Coupling

Figure 5.2 represents the energy cycle that describes allostery. The relationship between dissociation constants that defines the allosteric coupling constant (Q_{ax}) is (Weber, 1972)

$$Q_{ax} = K_{ia}/K_{ia/x} = K_{ix}/K_{ix/a}. \tag{5.1}$$

If $Q_{ax} > 1$, the allosteric effector (X) causes increased affinity of the protein for A. If $Q_{ax} < 1$, the allosteric effector causes decreased affinity of the protein for A. If $Q_{ax} = 1$, there is no allosteric coupling between A and X. One established method to determine Q_{ax} is to monitor the affinity (or apparent affinity, K_{app}, derived from initial velocity techniques, see Reinhart, 1983) of the protein for one ligand as a function of the concentration of the second ligand (Di Cera, 1995; Reinhart, 2004). On a log–log plot, the allosteric coupling is the difference between the upper and lower plateaus (Fig. 5.3). Because the allosteric coupling is a comparison of dissociation (or affinity) constants, it is independent of either the substrate affinity in the absence of effector or the effector affinity in the absence of substrate. Therefore, when an allosteric system is perturbed (e.g., introduction of mutations or modification of ligand), the varied experimental conditions may alter the ligand affinities, allosteric coupling or both (Fenton *et al.*, 2003; Williams *et al.*, 2006). This is graphically exemplified in Fig. 5.3, in which curves A, B, and C share a common Q_{ax} value. Curve D represents a condition with an altered value of Q_{ax}.

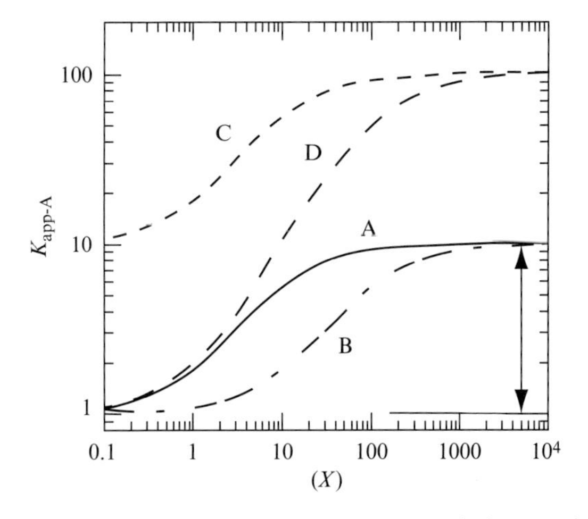

Figure 5.3 *Model data.* Model data demonstrate potential changes that could result from modifying the allosteric effector, mutating or covalently modifying the protein, and/or changing temperature, pH, or other solution conditions. Curve A is the reference line. The allosteric coupling (Q_{ax}) for curve A is represented by the double headed arrow. Although compared to A, B has a 10-fold decrease in effector affinity in the absence of substrate and C has a 10-fold decrease in substrate affinity in the absence of effector, A, B, and C have equivalent allosteric coupling. D has a 10-fold change in allosteric coupling as compared to A (reproduced with permission from Fenton, 2008).

3. PYK Assay

Throughout this chapter, several components of the assay are independently varied. Therefore, a "standard assay" will be described here and used as the control in comparative studies. Variations from the standard assay will be detailed in the relevant sections. We have previously demonstrated the pH dependency of the allosteric functions of hL-PYK (Fenton and Hutchinson, 2009); with the exception of studies that vary divalent cation, pH will be maintained at pH 7.5 in the current chapter. Creation of the E130K mutation, protein expression, and protein purification (ammonium sulfate fractionation and DEAE-cellulose column) were as previously described (Fenton and Hutchinson, 2009; Fenton and Tang, 2009). Purified protein was stored in 10 m*M* MES (pH 6.8), 5 m*M* $MgCl_2$, 10 m*M* KCl, and 2 m*M* DTT and characterized within 2 days after the last step of purification. When monovalent or divalent cations were varied, purified hL-PYK and lactate dehydrogenase were first desalted with Sephadex G-50 (Sigma) resin into the respective assay buffer before use.

Activity measurements in the "standard assay" were carried out at 30 °C using a lactate dehydrogenase-coupled assay. Reactions contained 50 m*M* bicine (pH 7.5), 10 m*M* $MgCl_2$, 2 m*M* (K)ADP, 0.1 m*M* (2Na)EDTA, 0.18 m*M* (2Na)NADH, and 19.6 U/ml lactate dehydrogenase. (K)PEP and effector concentrations were varied as indicated. Stock solutions of PEP and effectors were adjusted to pH 7.5 with KOH before addition and serial dilutions of ligands were in KCl to maintain constant K^+ concentration. K^+ concentrations from additions of KOH and from counterions of ligands were summed and KCl was supplemented to a total K^+ concentration of 150 m*M* in all assays. Due to the high affinity for Fru-1,6-BP, only low concentrations of (Na)Fru-1,6-BP are required; the Na^+ contributed by additions of this ligand are considered negligible. A solution of Ala in water has a pH near pH 7.5. Therefore, there is essentially no change in cation concentrations or ionic strength over a concentration range of Fru-1,6-BP or Ala. Independent of other sources of ions, the cumulative Na^+ concentration (in all assays in this chapter) derived from EDTA, NADH (and when applicable Fru-1,6-BP), ranges from 0.56 to 0.65 m*M* in the absence and saturating presence of Fru-1,6-BP, respectively. To minimize pipetting error, all assay components except effector and PEP were combined in a cocktail. Therefore, three additions were made to each well of a 96-well plate using a 12-channel pipettor: (1) cocktail, (2) effector, and (3) PEP. All volumes added were greater than 70 μl. The enzymatic reaction was initiated with PEP and monitored at 340 nm over time in a UV-Star flat-bottomed 96-well plate (Greiner Bio-one) containing a total reaction volume of 350 μl. All activity readings were collected using a Molecular

Devices Spectramax Plus384 spectrophotometer. 30 °C approaches the minimum temperature (ambient +5°) requirement of the plate reader used for semi-high-throughput data collection.

The use of monovalent cations other than K^+ and Na^+ (these two are the cations available in commercial sources of PEP and ADP) required ion exchange. This exchange was performed by mixing (K)PEP or (K)ADP with Dowex 50WX4 resin (Sigma) in a ratio of 1 mol (K)ligand to 4 kg (wet weight) of washed resin. Once ligand and resin were suspended in a minimum volume of water, the suspension was allowed to equilibrate at 4 °C for 30 min and then the acidic form of the ligand was recovered using either a syringe filter or Buchner funnel/filter paper. The acidic form of ligands were lyophilized to near dryness and stored at −20 °C. Immediately before use, lyophilized samples were resuspended in water. pH was adjusted to 7.5 with measured quantities of the desired base. PEP and ADP concentrations were determined using the PYK/lactate dehydrogenase assay and monitoring the absolute A_{340} difference (similar to Czof and Lamprecht, 1974). Dilutions of cation–PEP and cation effector were in the respective cation–Cl.

Data were fit to appropriate equations using the nonlinear least-squares fitting analysis of Kaleidagraph (Synergy) software. As previously noted, the dependency of enzymatic activity on PEP concentration is biphasic (Fenton and Hutchinson, 2009). $K_{\text{app-PEP}}$ values were obtained by fitting initial rates obtained from kinetic assay to

$$v = \frac{V_{\max}[\text{PEP}]^{n_{\text{H}}}}{(K_{\text{app-PEP}})^{n_{\text{H}}} + [\text{PEP}]^{n_{\text{H}}}} + c[\text{PEP}], \qquad (5.2)$$

where $V_{\max}$ is the maximum velocity associated with the low PEP phase, $K_{\text{app-PEP}}$ is the concentration of substrate that yields a rate equal to one-half the $V_{\max}$, and n_{H} is the Hill coefficient associated with the low PEP (i.e., high PEP affinity) phase. The response of activity to increasing concentrations of PEP is biphasic, although a maximum activity for the second phase has not been obtained (Fenton and Hutchinson, 2009). We have speculated multiple potential mechanisms that might give rise to the biphasic response. The protein may actually bind PEP to active sites with different affinities, such that the tetramer functions as a dimer of dimer. In this mechanism, c would be the $V_{\max}/K_{\text{app-PEP}}$ for the low-affinity PEP phase (Fenton and Hutchinson, 2009). Due to the experimental design aimed at holding cation concentrations constant, Cl^- concentration decreases as PEP concentration increases. Dilution of Cl^-, change in total anion concentration over the PEP concentration range, and/or PEP binding with low affinity to some site other than the active site may give rise to the second phase. In this second speculation, c is the linear response expected due to nonspecific effects caused by changing concentrations of anions.

$K_{\text{app-PEP}}$ values obtained from Eq. (5.2) were plotted as a function of effector concentration and fit to (Reinhart, 1983, 1988, 2004):

$$K_{\text{app-PEP}} = K_{\text{a}}\left(\frac{K_{\text{ix}} + [\text{Effector}]}{K_{\text{ix}} + Q_{\text{ax}}[\text{Effector}]}\right), \tag{5.3}$$

where $K_{\text{a}} = K_{\text{app-PEP}}$ when $[\text{Effector}] = 0$, K_{x} is the dissociation constant for effector when $[\text{PEP}] = 0$, and Q_{ax} is the coupling constant between PEP and the effector. Q_{ax} is related to the coupling free energy, ΔG_{ax} between PEP and effector by (Johnson and Reinhart, 1994)

$$\Delta G_{\text{ax}} = -RT\ln(Q_{\text{ax}}). \tag{5.4}$$

To gain some appreciation for the reproducibility of data collection, Fig. 5.4 presents nine data sets for Ala inhibition and seven data sets for Fru-1,6-BP activation collected using seven different protein preparations, and collected over 24 months. Fit parameters obtained from simultaneously fitting all data to Eq. (5.3) are presented in Table 5.1. These parameters are in good agreement with those previously reported from the same assay conditions, with the exception that HEPES was previously used as the buffer (Fenton and Tang, 2009). In Table 5.1, each parameter is shown ± the error estimate determined by this simultaneous fit. In addition, each data set was fit individually. The standard deviation of the equivalent parameters determined by these individual fits is also presented in Table 5.1. Of the parameters, evaluations of $K_{\text{ix-FBP}}$ are the least reproducible amongst data sets.

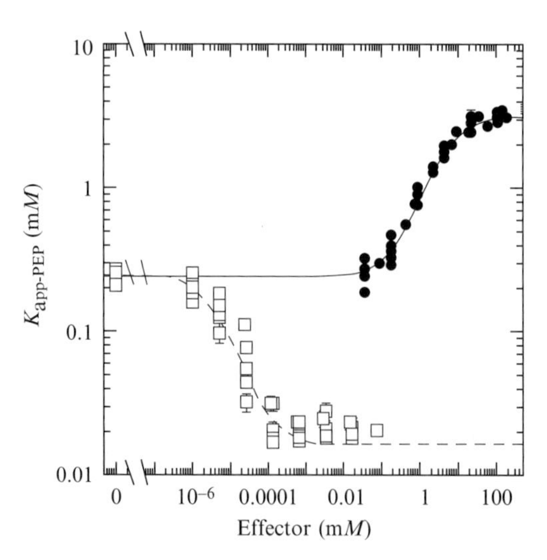

Figure 5.4 $K_{\text{app-PEP}}$ as a function of Ala (●) or Fru-1,6-BP (□) as determined in the standard assay. Lines represent fits to Eq. (5.3). When error bars are not apparent they are smaller than the symbols. Fit parameters are listed in Table 5.1.

Table 5.1 Fit parameters for data obtained in the standard assay as presented in Fig. 5.4

Parameter[a]	Fit value from simultaneous fit all sets	Standard deviation of fit values from individual data sets[b]
$K_{a\text{-PEP}}$ (mM)	0.244 ± 0.003	0.03
$K_{ix\text{-FBP}}$ (μM)	0.068 ± 0.004	0.03
$K_{ix\text{-Ala}}$ (mM)	0.29 ± 0.01	0.05
$Q_{ax\text{-FBP}}$	15.2 ± 0.4	2.7
$Q_{ax\text{-Ala}}$	0.074 ± 0.001	0.01

[a] From Eq. (5.3).
[b] $n = 9$ for Ala data and $n = 7$ for Fru-1,6-BP data.

4. Buffers

Useful buffers should not interact with the study system. However, many common buffers are sulfate derivatives. Sulfate can mimic phosphate in some settings. Since hL-PYK binds phosphorylated ligands at both the active site and the Fru-1,6-BP-binding site, sulfate-based buffers may have the potential to interact with this enzyme. Consistent with this cautionary consideration, buffer effects on L-PYK properties have previously been noted by several authors (Blair, 1980).

Initial tests of the impact of buffers were performed by replacing an effector in the standard assay with the buffer being tested (noted as "effector buffer" to distinguish from the assay buffer). The impact of effector buffers on $K_{app\text{-PEP}}$ was determined in different assay buffers (Fig. 5.5A and B). The use of Tris/HCl as the effector buffer caused an increase in $K_{app\text{-PEP}}$ independent of whether titrations were in HEPES/KOH or bicine/KOH assay buffer. Likewise, increasing concentrations of MOPS/KOH causes $K_{app\text{-PEP}}$ to decrease in both HEPES/KOH and bicine/KOH assay buffers. In bicine/KOH assay buffer, HEPES/KOH also causes $K_{app\text{-PEP}}$ to decrease. Considered in isolation, this data might indicate that a number of buffers interact with hL-PYK and, therefore, modify function of the protein. Since the use of bicine as the effector buffer did not impact $K_{app\text{-PEP}}$ when assayed in HEPES assay buffer, bicine was chosen as the assay buffer in the control assay. However, a finding that many different buffers interact with the same protein seems questionable. Therefore, experiments were designed to challenge this interpretation.

Within the experimental design of the standard assay, the uncontrolled variables are the concentration of Cl^- and the total anionic concentration across the concentration of effector (or effector buffer in this instance). One experimental design that would allow changes in buffer type with minimal

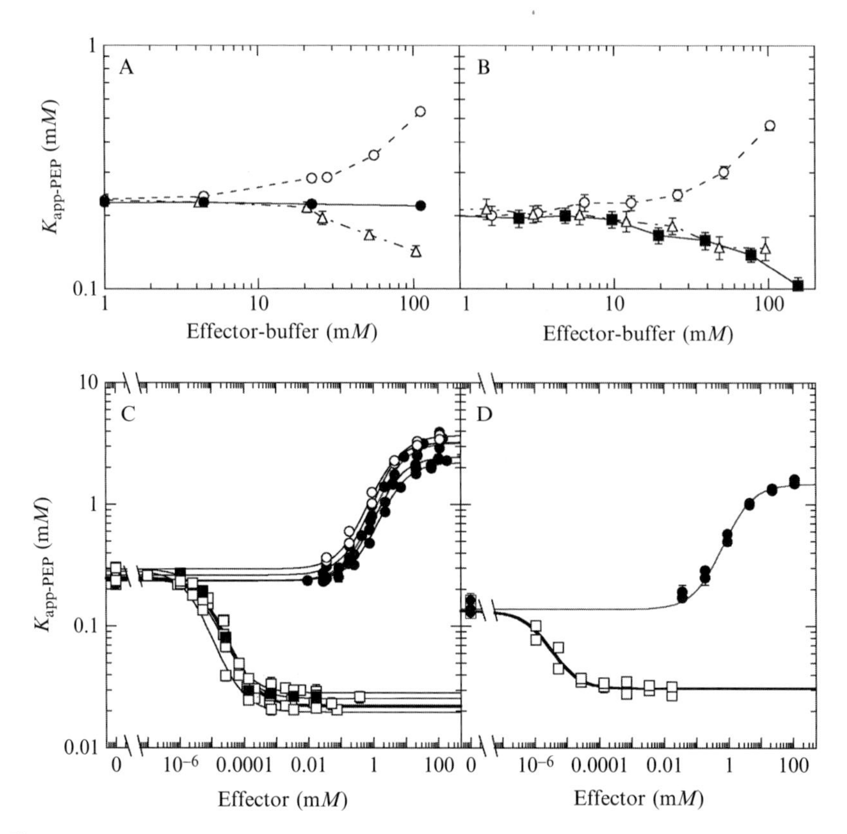

Figure 5.5 The impact of buffer type on $K_{app\text{-}PEP}$ and allosteric function of wild-type hL-PYK. (A) $K_{app\text{-}PEP}$ determined over a concentration range of Tris/HCl (○), MOPS/KOH (△), or bicine/KOH (●) in HEPES/KOH assay buffer. (B) $K_{app\text{-}PEP}$ determined over a concentration range of Tris/HCl (○), MOPS/KOH (△), or HEPES/KOH (■), in bicine/KOH assay buffer. In A and B, effector buffer replaces effector in the standard assay; lines reflect data trends. (C) $K_{app\text{-}PEP}$ as a function of Ala (circles) or Fru-1,6-BP (squares) determined in bicine/KOH, EPPS/KOH, HEPES/KOH, Tris/bicine, and Tris/HCl. With the exception of Tris/HCl (open squares and filled circles), no effort is made to distinguish data for other buffers (closed squares and open circles). (D) $K_{app\text{-}PEP}$ as a function of Ala (●) or Fru-1,6-BP (□) determined in MOPS/KOH. Lines in C and D represent fits to Eq. (5.3). Note different y-axis and x-axis ranges in A and B compared with C and D. When error bars are not apparent they are smaller than the symbols.

changes in Cl^- concentration is a comparison of the impact of allosteric effectors on $K_{app\text{-}PEP}$ in different assay buffers. Figure 5.5C shows that data collected in bicine/KOH, EPPS/KOH, HEPES/KOH, and Tris/bicine are equivalent within error. Tris/bicine was added since assays with Tris/HCl have higher Cl^- than other buffers systems. (However, 50 mM of Tris requires high concentrations of bicine—213 mM—to obtain pH 7.5;

therefore, the total ionic strength of this assay buffer exceeded that of other buffer systems.) At 50 m*M* Tris/HCl (Fig. 5.5A and B), the effect on $K_{app\text{-}PEP}$ is minimal. This minor shift is also apparent in the $K_{app\text{-}PEP}$ obtained when Tris/HCl is the assay buffer (Fig. 5.5C). The only buffer that resulted in considerable differences from the data obtained in other buffers was MOPS (Fig. 5.5D). Therefore, estimations of the magnitudes of allosteric functions obtained in all buffers tested, with the exception of MOPS, are comparable (e.g., results reported in Table 5.1 for the control assay versus those reported in HEPES; Fenton and Tang, 2009).

The contrast between the apparent impact of titrating effector buffers on $K_{app\text{-}PEP}$ (Fig. 5.5A and B) and the finding that changing the type of assay buffer does not alter estimates of $K_{a\text{-}PEP}$ or allosteric functions (Fig. 5.5C and D) can be explained if Cl^- inhibits $K_{a\text{-}PEP}$. This can be appreciated by considering that titrations of $K_{app\text{-}PEP}$ with Tris/HCl (Fig. 5.5A and B) increases Cl^- over the concentration range of the effector buffer, while titrations with anionic buffers decrease Cl^- over this range. Given this new information, it seems likely that previously reported buffer effects on L-PYK properties (Blair, 1980) were at least in part due to uncontrolled Cl^- concentration. Due to the potential that Cl^- ions might impact PEP affinity, further studies of the effects of anions are discussed below.

5. Divalent Cation

During the initial characterization of PEP dephosphorylation in 1934, it was recognized that Mg^{2+} was required for enzymatic activity of PYK (Boyer, 1962). However, the role of the divalent cation in allosteric regulation remains poorly understood. This lack of understanding is likely in part due to the complex nature of this system. The complexity of a protein that binds both a free divalent cation and the divalent metal complex of ADP is well recognized. However, it is less appreciated that both Fru-1,6-BP and PEP can bind divalent metal in solution (McGilvery, 1965; Nowak and Lee, 1977). Likewise, the influence that monovalent cation concentrations have on the affinity of ADP for divalent cation have been reported (Martell and Smith, 1974–1989), yet little is understood about the impact of monovalent cation concentrations on the divalent cation interaction with hL-PYK, Fru-1,6-BP, or PEP. The potential synergistic effect in the binding of divalent and monovalent cations to the protein (Reed and Cohn, 1973; Suelter *et al.*, 1966) and the potential of competitive binding of these ligands to the enzyme add even more layers of complexity.

Due to this complexity, the experimental design used here was to replace Mg^{2+} with other divalent metals rather than to titrate the divalent metal concentration (i.e., $MgCl_2$, $MnCl_2$, or $CaCl_2$). Even this approach required additional modifications to the standard assay since saturating concentration of Mn^{2+} is limiting at pH 7.5; comparisons of the allosteric regulation elicited by both Fru-1,6-BP and Ala in either Mg^{2+} or Mn^{2+} were made at pH 6.5, 6.8, and 7.0 in 50 m*M* HEPES buffer.

Over the narrow pH range from 6.5 to 7.0, $K_{a\text{-PEP}}$ increases in the presence of Mg^{2+}, but not Mn^{2+} (Fig. 5.6). At any one pH, the magnitude of the allosteric response to Ala is dependent on which divalent cation is present; this magnitude is greater when Mg^{2+} serves as the required metal compared to when Mn^{2+} serves this role. The magnitude of the response to Ala increases as pH increases when Mn^{2+} is present, but only minimally increases when Mg^{2+} is present. The allosteric response to Fru-1,6-BP is present when Mg^{2+} is the divalent cation. The magnitude of this response

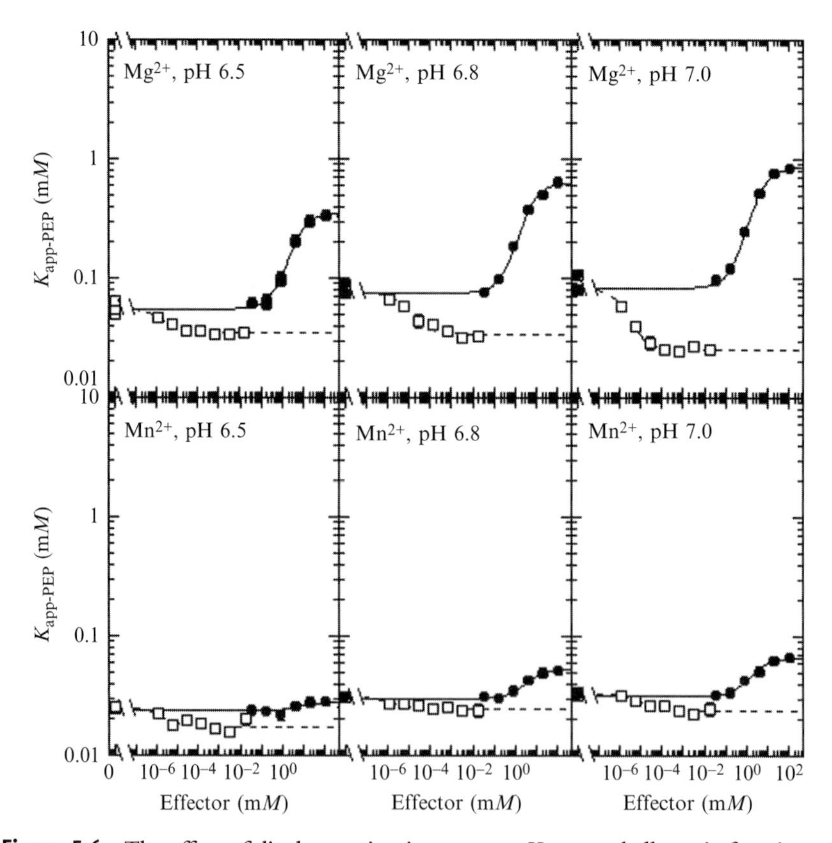

Figure 5.6 The effect of divalent cation ion type on $K_{a\text{-PEP}}$ and allosteric function of wild-type hL-PYK at pH 6.5, 6.8, and 7.0. $K_{app\text{-PEP}}$ was determined over a concentration range of Ala (●) or Fru-1,6-BP (□) and in 50 m*M* HEPES buffer. Lines represent fits to Eq. (5.3). When error bars are not apparent they are smaller than the symbols. No activity was observed in the presence of Ca^{2+}. Even at pH 6.5, 10 m*M* Ni^{2+} or Co^{2+} are oxidized and become insoluble and are therefore not studied.

increases with increasing pH. However, the ability of Fru-1,6-BP to elicit an allosteric response is questionable when Mn^{2+} is the divalent cation, at any of the three buffer pH values.

The role of divalent cation in the allosteric regulation of a PYK isozyme has most thoroughly been characterized to date in the Fru-1,6-BP activation of yeast PYK (Bollenbach and Nowak, 2001a,b; Mesecar and Nowak, 1997a,b), allowing a comparison with the current results. Despite the complexities caused by multiple binding events involving divalent cation, this ligand was varied in studies of yeast PYK. In the yeast enzyme, the interactions between the protein/divalent cation and PEP are different depending on whether Mg^{2+} or Mn^{2+} fills the divalent requirement. Therefore, the interaction between PEP and divalent cation may not simply be electrostatic in nature. In addition, Mg^{2+} or Mn^{2+} mediates the allosteric coupling between PEP and Fru-1,6-BP in the presence or absence of the second substrate, MgADP. In other words, Fru-1,6-BP binding to protein is coupled directly to divalent metal affinity; PEP binding to protein is coupled directly to divalent metal affinity; but Fru-1,6-BP binding and PEP binding are not directly coupled in the absence of the divalent metal cation. Collectively, results reported for the yeast enzyme are not entirely consistent with the results shown in Fig. 5.6; primarily, Mn^{2+} supports allosteric activation by Fru-1,6-BP in the yeast PYK isozyme, but supports no, or only a marginal allosteric response by Fru-1,6-BP in hL-PYK.

The marginal response of the $K_{app\text{-}PEP}$ for L-PYK to Fru-1,6-BP in the presence of Mn^{2+} has also been reported for the rat L-PYK isozyme (Blair and Walker, 1984). Additionally reported in the rat L-PYK system, PEP has the ability to impact Fru-1,6-BP binding even in the absence of divalent cation. This second result for the rat protein is in contrast to the expectation based on studies of yeast PYK. Therefore, the role of divalent cation in the regulation of L-PYK by Fru-1,6-BP may be dissimilar to the role of this metal in the same regulation of yeast PYK.

6. Monovalent Cation

M_1-PYK holds the historic honor as being the first enzyme demonstrated to require a monovalent cation for activity (Kachmar and Boyer, 1953; Nowak and Suelter, 1981; Page and Di Cera, 2006). This conclusion was based on the observation that the activity present when sodium salts of ADP and PEP were used was only 1.5% of that obtained in the presence of K^+ (Boyer, 1962; Kachmar and Boyer, 1953); 8% by later estimations (Kayne, 1971). More recent studies suggest that K^+ induces a conformational change in the active site that switches substrate binding from a ordered mechanism in the absence of K^+ (ADP cannot bind until after

PEP binds) to a random mechanism in the presence of K^+ (Oria-Hernandez *et al.*, 2005). Monovalent requirements for activity have also been noted in an *Escherichia coli* PYK isozyme (Valentini *et al.*, 1979). Much less is known about the role of the monovalent cation in the allosteric regulation of PYK isozymes in general, and specifically of the hL-PYK isozyme.

PEP, ADP, and Fru-1,6-BP carry negative charges at physiological pH. All three ligands are commercially sold as salts of sodium and/or potassium. Since potassium is likely the physiological ion and due to the historical acceptance that Na^+ does not support enzymatic activity, the potassium salts of anionic ligands have been used in our control assay. Also, due to this dependency on potassium for activity, titrations of ligands accompanied by a potassium counterion (and after adjustment of pH to the desired 7.5 using measured additions of KOH) have been with a serial dilution of the titrant made in a solution of KCl, such that the final K^+ concentration in all assays across the titration is constant at 150 m*M*.

To determine the role of the monovalent cation, potassium salts of PEP, ADP, and KOH were replaced with the sodium salts. Dilutions of Na-PEP and Na-effector were in NaCl. Na^+ supports activity in hL-PYK; activity with Na^+ as the monovalent cation is ~40–50% that of the activity observed when K^+ is present. Since activity is observed when Na^+ is present, it is unclear if Na^+ supports activity or if activity is present even in the absence of monovalent cation. In addition, PEP affinity is slightly decreased in the presence of Na^+ versus K^+ (Fig. 5.7). It is apparent that Na^+ supports both Fru-1,6-BP activation and Ala inhibition of $K_{app\text{-}PEP}$. However, the magnitude of the allosteric activation elicited by Fru-1,6-BP is greatly reduced when K^+ is substituted with Na^+.

In the presence of Na^+ versus K^+, different allosteric responses are observed. However, based on this data alone, it is unclear if only one or both cations elicit a response. Furthermore, if only one cation elicits a response, it is unclear whether K^+ elicits a response and the Na^+ data represent no response, or *vice versa* (i.e., which data might be a control with no effect from monovalent cation). To gain some insight into this question, the allosteric response to both Fru-1,6-BP and Ala were determined over different concentrations of Na^+ or K^+ (Fig. 5.8) Using this approach, the impact of varying Cl^- and varying ionic strength should be consistent whether Na^+ or K^+ is varied. When the data in Fig. 5.8 are examined in isolation, the lack of change over ion concentration (i.e., $Q_{ax\text{-}FBP}$ determined in Na^+ and the $Q_{ax\text{-}Ala}$ determined in K^+) might support that these conditions reflect the absence of an effect from monovalent cation (i.e., reference data). The reservation to this interpretation is that changes in the total ionic strength that accompany the ion concentration might counter effects caused by the monovalent cation to result in an apparent lack of response over the ion concentration. As monovalent cation concentration is decreased, $Q_{ax\text{-}FBP}$ determined in K^+ approaches that

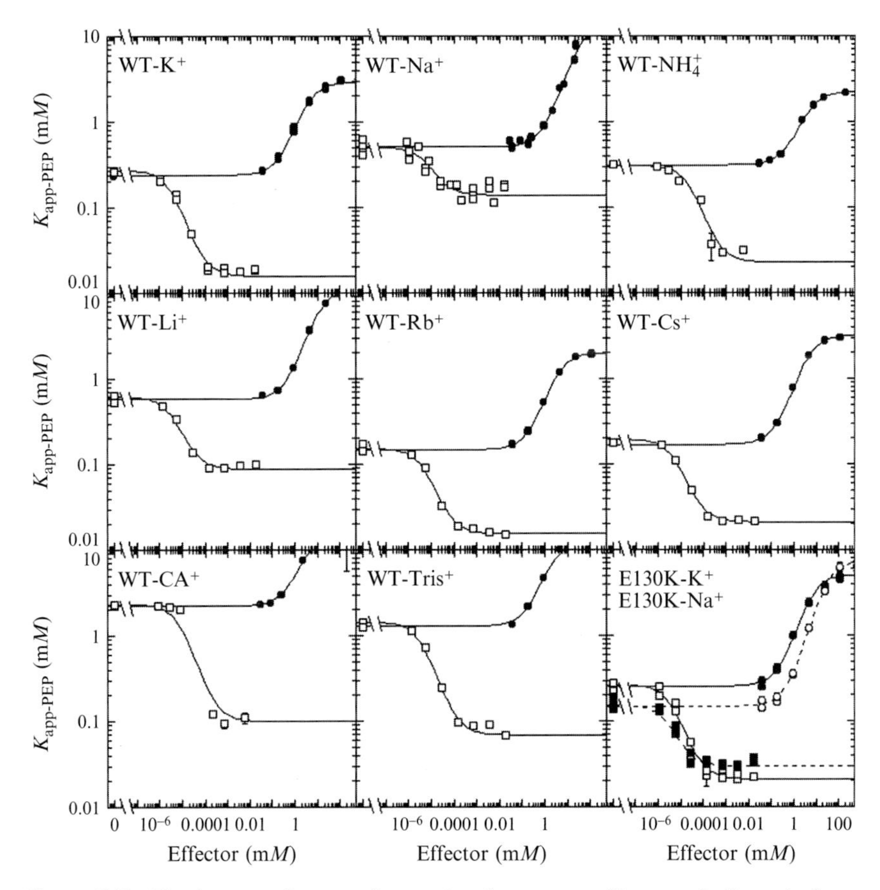

Figure 5.7 The impact of monovalent cation ion type on $K_{a\text{-PEP}}$ and allosteric function. $K_{app\text{-PEP}}$ was determined over a concentration range of Ala (●) or Fru-1,6-BP (□) for wild-type hL-PYK (WT) in the presence of a number of cations, including cyclohexylammonium (CA^+). Wild type/Na^+ includes data collected both with commercial sodium salts of substrates and with sodium substrates obtained using ion exchange. $K_{app\text{-PEP}}$ was also determined for E130K over a concentration range of Ala (●) or Fru-1,6-BP (□) in the presence of K^+ (solid line) or of Ala (○) or Fru-1,6-BP (■) in the presence of Na^+ (dashed line) Cl^- was used as the anion. Lines represent fits to Eq. (5.3). When error bars are not apparent they are smaller than the symbols.

determined in Na^+. However, the opposite is true when considering $Q_{ax\text{-Ala}}$, that is, $Q_{ax\text{-Ala}}$ determined in Na^+ approaches that determined in K^+. Therefore, based on this data alone, it is interesting to speculate that activation by Fru-1,6-BP might be impacted by K^+, whereas the inhibition by Ala might be impacted by Na^+. However, upon considering results with additional monovalent cations, these results must be reconsidered (below).

Based on sequence alignments and the observation that some PYK isozymes do not require K^+ for activity, it was found that replacing a

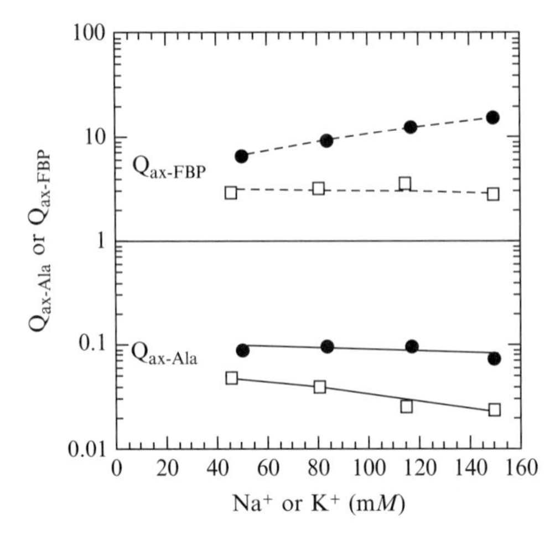

Figure 5.8 The influence of monovalent cation concentration on Q_{ax} values. $Q_{ax\text{-}FBP}$ (dashed lines) and $Q_{ax\text{-}Ala}$ (solid lines) for wild-type hL-PYK were determined over concentration ranges of either Na^+ (□) or K^+ (●). PEP and effectors were diluted in salt to maintain constant monovalent cation concentration. The minimum concentration of monovalent cations is determined by the contribution from substrates. Increasing concentration of monovalent cation is accompanied by increasing Cl^- concentration. However, changes in Cl^- concentration should be consistent whether Na^+ or K^+ is varied. Lines reflect data trends. When error bars are not apparent they are smaller than the symbols.

glutamic acid residue near the active site of M_1-PYK with lysine removes the requirement of this isozyme for K^+ (Laughlin and Reed, 1997). Creating the same mutation in hL-PYK (E130K) is predicted to both fulfill the role of the active site monovalent cation and to prevent binding of monovalent cations in the active site. Note that this consideration assumes some level of similarity between the catalytic mechanism of hL-PYK and that of M_1-PYK (this assumption is not easily tested given the level of hL-PYK activity in the presence of Na^+; therefore, there is no hL-PYK comparative control without activity). Consistent with K^+ and E130K fulfilling the same functional role in determining the $K_{a\text{-PEP}}$, this value for the wild-type protein and the E130K mutant protein are very similar when measured in the presence of K^+ (Fig. 5.7). This result can be used as evidence that the impact of the K^+ on $K_{a\text{-PEP}}$ is primarily a specific effect due to binding at the active site. In contrast, $K_{a\text{-PEP}}$ is modified by introduction of E130K when measurements are performed in the presence of Na^+. Therefore, if Na^+ binds in the active site, it does not perform the same functional role in determining $K_{a\text{-PEP}}$ as the E130K mutant or K^+. In addition, the magnitude of the Fru-1,6-BP allosteric response for E130K measured in Na^+ continues to be reduced compared to that measured in K^+. Therefore, it also seems likely that the impact of Na^+ on the magnitude of Fru-1,6-BP allostery results from a mechanism involving Na^+ binding outside of the active site.

Even after efforts to determine data trends as a function of varying K^+ and Na^+ and an effort to remove specific cation binding in the active site via introduction of the E130K mutant, a convincing argument cannot be made for which allosteric response to Fru-1,6-BP and Ala represents no influence by monovalent cation (i.e., a standard for comparison). Therefore, studies were expanded to consider effects of other monovalent cations. The use of monovalent cations besides K^+ and Na^+ required ion exchange to replace K^+ from both ADP and PEP with the respective cation. Note that data collected with these alternate cations were collected once and were not repeated. Activity was observed in the presence of all cations tested. Compared to V_{max} activity in the presence of K^+, observed activities were Tris (25%), NH_4^+ (140%), cyclohexylammonium (25%), Li^+ (25%), Cs^+ (12%), and Rb^+ (75%). Since Tris and cyclohexylammonium are much larger than the single atom cations bound in crystal structures, these larger ions are not likely to bind in the active site. Therefore, it appears that hL-PYK does not require a monovalent cation for activity. However, it should also be noted that the potential sources of contaminating monovalent cation, including NH_4^+ carryover with protein, are sufficiently numerous to prevent an estimate of the level of contamination by nonspecified monovalent cations.

Considering allosteric responses of wild-type hL-PYK in the presence of the various monovalent cations tested (Fig. 5.7), three groups can be distinguished. Data collected in K^+, NH_4^+, Rb^+, and Cs^+ show similar PEP affinities and allosteric responses. Na^+ and Li^+ constitute a second group with (as compared to the first group) reduced affinity for PEP and a reduced magnitude of the allosteric response to Fru-1,6-BP. The final group results from the use of either Tris or cyclohexylammonium as the monovalent cation; this group displays the lowest affinity for PEP, but a magnitude of allosteric activation by Fru-1,6-BP similar to that of the first group. If we return to the reasoning that Tris or cyclohexylammonium are not likely to bind in the active site, then it seems reasonable to conclude that the allosteric similarities in the first and third group represent no impact by the monovalent cation on this function. Therefore, cations in the first group increase PEP affinity, but do not play a role in the allosteric mechanism. Combining this explanation with conclusions from the study of the E130K mutant, we may suggest that the impact of Na^+ and Li^+ on the allosteric response to Fru-1,6-BP are not a result of these ions binding in the active site.

This final explanation also requires reconsideration of the effects observed in Fig. 5.8. Based on the above argument, K^+ modifies PEP affinity, but does not alter allosteric responses. It follows that the impact of increasing K^+ on $Q_{ax\text{-}FBP}$ (Fig. 5.8) is a result of changes in total ionic strength and/or anion concentration (a conclusion also supported from the anion data presented below). The same total ionic strength/anion effects must be included with increasing Na^+. Since Na^+ reduce the allosteric response for Fru-1,6-BP (discussed above), it seems likely that the lack of $Q_{ax\text{-}FBP}$ response to increasing NaCl (Fig. 5.8) is due to compensating effects result from ionic strength/anion enhancement and Na^+ reduction.

The collective findings in this section support that (1) hL-PYK does not likely require a monovalent cation for activity (based on the consideration that Tris and cyclohexylammonium are not expected to bind in the active site), (2) monovalent cation binding in the active site impacts PEP affinity without altering the magnitude of allosteric coupling, and (3) Li^+ and Na^+ impact the magnitude of allosteric coupling by Fru-1,6-BP due to mechanisms that occur outside of the active site cation-binding site.

7. Anion

By maintaining constant K^+, Cl^- concentration is left as the uncontrolled variable in the standard assay described above. Total anion concentration also decreases over the PEP concentration range; at 150 m*M* constant K^+, 143 m*M* Cl^- decreases to 109 m*M* while PEP (also an anion) concentration increases from 0 to 10 m*M*. The requirement for monovalent cation for activity (as assumed at the onset of this study) prevented a simplistic study that varied salt concentration. Instead, the experimental design most approachable with this system was to replace the KCl used to dilute ligands with other potassium salts. In this design, even though the anion concentration varies inversely to the concentration of PEP, the variation in anion concentration is the same for all monovalent anions (but not divalent anions).

In assays free of Cl^-, the magnitude of the allosteric inhibition elicited by Ala appears relatively independent of anion type, when this anion is NO_3^-, CH_3COO^-, or SO_4^{2-} (Fig. 5.9A). In contrast, NO_3^- causes a large increase in $K_{a\text{-}PEP}$. Both the CH_3COO^- and SO_4^{2-} decreased PEP affinity in the saturating presence of Fru-1,6-BP (Note that use of SO_4^{2-} changes anionic concentration relative to assays with Cl^-). The use of SO_4^{2-} also causes a noticeable decrease in the observed Fru-1,6-BP affinity.

Replacing Cl^- with Br^- or I^- increases $K_{a\text{-}PEP}$ consistent with the ionic radius of the anion (Fig. 5.9B); F^- caused precipitation at the required concentrations. The affinity for PEP at high concentrations of Ala also increases consistent with the ionic radius of the anion. However, the affinity for PEP at high concentrations of Fru-1,6-BP seems relatively independent of whether Cl^-, Br^-, or I^- is used. Results with this series of anions might be used to support that the anion effects are specific. However, before

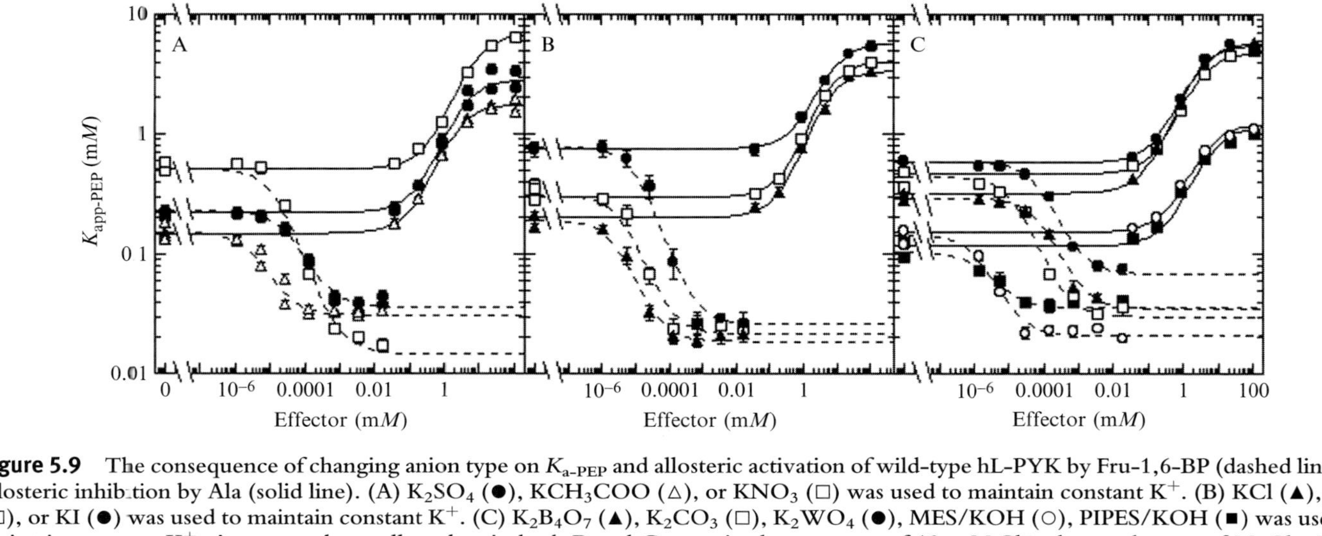

Figure 5.9 The consequence of changing anion type on $K_{a\text{-}PEP}$ and allosteric activation of wild-type hL-PYK by Fru-1,6-BP (dashed line) or allosteric inhibition by Ala (solid line). (A) K_2SO_4 (●), KCH_3COO (△), or KNO_3 (□) was used to maintain constant K^+. (B) KCl (▲), KBr (□), or KI (●) was used to maintain constant K^+. (C) $K_2B_4O_7$ (▲), K_2CO_3 (□), K_2WO_4 (●), MES/KOH (○), PIPES/KOH (■) was used to maintain constant K^+. Assays used to collect data in both B and C were in the presence of 10 m*M* Cl^-, due to the use of $MgCl_2$. Lines represent fits to Eq. (5.3). When error bars are not apparent they are smaller than the symbols.

making this conclusion, consider that PYK is a fluorokinase (Tietz and Ochoa, 1958). Therefore, monoatomic anions likely bind in the active site. If these ions bind in the active site competitively with PEP and this is the only specific anion-binding event, then the anions are not expected to alter allosteric coupling (i.e., when PEP binds, the competing anion will no longer be bound). Therefore, since anion type also alters allosteric function, a single, specific interaction in the active site is not supported. Effects caused by other anions (besides Cl^-, Br^-, and I^-) are also consistent with this conclusion. However, these data have not eliminated the possibility that anions specifically bind to a limited number of sites on the protein outside of the active site. Anion binding has not been elucidated in currently reported structures of PYK isozymes.

In as much as we can consider that the allosteric responses of the E130K mutant protein are not impacted by monovalent cation, effects on this mutant protein elicited by increasing concentration of potassium salts of various anions can be assigned to the combined effect of changes in ionic strength and anion concentration. Therefore, allosteric effects elicited by Ala and Fru-1,6-BP were determined over a concentration range of potassium-anion salts (Fig. 5.10). In these assays, PEP and effectors were diluted

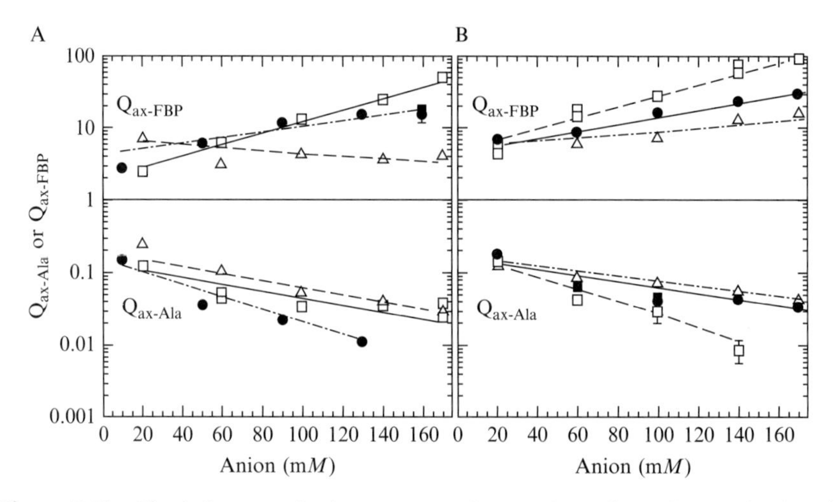

Figure 5.10 The influence of anion concentration on Q_{ax} values. $Q_{ax\text{-}FBP}$ (top) and $Q_{ax\text{-}Ala}$ (bottom) for E130K hL-PYK were determined over concentration ranges of (A) NO_3^- (□), SO_4^{2-} (●), or CH_3COO^- (△) and (B) Cl^- (△), Br^- (●), or I^- (□). All anions were added as the potassium salt. PEP and effectors were diluted in water, such that ionic strength was not constant. Unlike data collected in Fig. 5.9, Mg^{2+} was added as the respective Mg-anion salt; minimum concentration of anion was determined by this addition. Increasing concentration of anion is accompanied by increasing K^+ concentration. However, the use of the E130K mutation is expected to minimize the impact of the monovalent cation. Lines reflect data trends. When error bars are not apparent they are smaller than the symbols.

in water with no attempt to maintain constant ionic strength. Data trends over the ion series are consistent with the wild-type data presented in Fig. 5.9. Acetate is unique in that as the concentration of this anion increases, the allosteric activation by Fru-1,6-BP decreases, rather than increases as with all other anions; although not detailed here, potassium acetate also caused unique changes of $K_{app\text{-}PEP}$ and V_{max} relative to other anions, a finding that might indicate a unique interaction of acetate with the protein. Over the concentration range of the remaining anions, Cl^- appears to have one of the smallest effects both on the allosteric coupling elicited by Ala and on that elicited by Fru-1,6-BP; other anions elicited greater changes on one of the two allosteric effects. Therefore, Cl^- was used as a counterion in the standard assay.

To further characterize the impact that anions have on the observed allosteric effects in the absence of varying ionic strength, at least one noninteracting anion must be identified as a standard for comparison. However, what criteria should be used to identify such an anion is unclear. Data groups considered in conjunction with knowledge of the characterized binding site provided bases for discussion of monovalent cations; no data groups of the anion data are obvious and no anion-binding site has been detailed in PYK structures. Confidence that buffers are noninteracting is based on reproducible results independent of buffer type; no such anion-independent reproducibility was identified in the anion series used, including a screen of additional anions (Fig. 5.9C). Therefore, the results presented support that the type of anion present impacts the observed magnitude of the allosteric response. Although more descriptive conclusions cannot currently be made, this is the first indication that anions impact the allosteric response in hL-PYK.

8. Concluding Remarks

The role of PYK in glycolysis was discovered early in the history of metabolism (Boyer, 1962). This early identification can be viewed both as a benefit and as a hindrance to future studies of individual PYK isozymes. The benefits include decades of studies that have detailed many structural and functional aspects of this enzyme. However, current understanding relies on many studies (including much of the current understanding about catalysis; Boyer, 1962) that were completed before the relationship of isozymes was fully recognized. As a result, many properties have often been assumed to be common to all PYK isozymes with little or no verification. Our original assay conditions took into consideration these well accepted "knowns" regarding PYK function (Fenton and Hutchinson, 2009), including the assumption that K^+ was one of a limited number of monovalent cations that could activate PYK isozymes. However, such monovalent requirements were a result of studies primarily using the M_1-PYK isozyme, not an L-PYK isozyme. With knowledge that monovalent cations are not required for activity, but monovalent cation (and anion) type impacts V_{max}, $K_{app\text{-}PEP}$ and the magnitude of allosteric responses, much of the divergence in kinetic parameters previously noted (Blair, 1980) may be due to uncontrolled levels of Na^+ and Cl^- in the assay systems.

It should be emphasized that in a broader study of allostery, characterizations of the impact of various ions on allosteric function should not be viewed purely as an academic endeavor. As one demonstration of the utility of understanding the impact of ions on allosteric function, consider that much of the structural comparisons relevant to allosteric systems use structures determined by X-ray crystallography. To obtain crystal growth, different enzyme complexes are often crystallized using different buffer, salt, and/or pH conditions. Using hL-PYK as an example, we have shown here that changing buffer, divalent cation, monovalent cation or anion type and/or concentration can modify the allosteric response. In addition, we have previously demonstrated that the allosteric functions in this enzyme are pH dependent (Fenton and Hutchinson, 2009). The dependence of allosteric function on buffer/pH/ions requires a consideration of which structural comparisons are viable in an effort to understand changes associated with allosteric functions. In the specific example of PYK, structural comparisons used to explain conformational changes associated with Fru-1, 6-BP regulation have been with structures crystallized with Mn^{2+} bound in the active site. The difficulty in interpreting such comparisons now becomes apparent since hL-PYK does not respond to Fru-1,6-BP in the presence of Mn^{2+}. Furthermore, many past structural comparisons have been made between isozymes from different sources (including different species). In the case of PYK, the variable monovalent cation-independent activity catalyzed by rabbit M_1-PYK (Boyer, 1962; Kachmar and Boyer, 1953; Kayne, 1971) versus hL-PYK likely implies functional differences between the two enzymes. With variable functions between two isozymes, a structural comparison between such isozymes is at best a questionable approach to study allosteric function.

As a second example of the utility of understanding the impact of ions on allosteric function, NMR measurements have been used to determine the distances between cations in PYK (Nowak and Suelter, 1981; Villafranca and Raushel, 1982). These differences can be considered for each of the different complexes of an allosteric cycle (Fig. 5.2). Since Mg^{2+} does not have a paramagnetic signal, Mn^{2+} has been used as the divalent cation. Multiple monovalent cation (including Tl^+) have been used as a monovalent cation with paramagnetic signal (Loria and Nowak, 1998; Villafranca and Raushel, 1982). Again, we can now appreciate that changing which monovalent or divalent cations are added to the assay system modifies the magnitudes of the allosteric response. Furthermore, the difference between divalent function in L-PYK and yeast PYK (discussed above) demonstrates that these magnitudes are likely unique to individual isozymes. Therefore, to facilitate data interpretation of physical measurements such as changes in NMR-determined distances between monovalent and divalent cations, a complete characterization of allosteric responses with the alternative ions is required. Note with this consideration there is no current knowledge that directly ties physical changes with functional changes (e.g., does an effector induced change of a discrete distance between the divalent cation and K^+ result in a different magnitude of allosteric coupling depending on which divalent cation is present?).

Through the results presented here, it can be appreciated that even very small changes in the system (Mg^{2+} to Mn^{2+}; K^+ to Na^+; or Cl^- to Br^-) can alter the observed magnitude of the allosteric response. This conclusion highlights the necessity for quantitative studies of allosteric regulation at a functional level (Fenton, 2008; Reinhart, 2004). Information gained through these functional studies can, in turn, be used to guide structure/function correlations.

ACKNOWLEDGMENTS

We would like to thank Josh Smith and Michael Chopade for their aid in exchanging substrate ions. This chapter was supported by NIH grant DK78076.

REFERENCES

Blair, J. B. (1980). Regulatory properties of hepatic pyruvate kinase. *In* "The Regulation of Carbohydrate Formation and Utilization in Mammals" (C. M. Veneziale, ed.), pp. 121–151. University Park Press, Baltimore, MD.

Blair, J. B., and Walker, R. G. (1984). Rat liver pyruvate kinase: Influence of ligands on activity and fructose 1,6-bisphosphate binding. *Arch. Biochem. Biophys.* **232,** 202–213.

Bollenbach, T. J., and Nowak, T. (2001a). Kinetic linked-function analysis of the multiligand interactions on Mg(2+)-activated yeast pyruvate kinase. *Biochemistry* **40,** 13097–13106.

Bollenbach, T. J., and Nowak, T. (2001b). Thermodynamic linked-function analysis of Mg (2+)-activated yeast pyruvate kinase. *Biochemistry* **40,** 13088–13096.

Boyer, P. D. (1962). Pyruvate kinase. *In* "The Enzymes" (P. D. Boyer, *et al.*, eds.), Vol. 4, pp. 95–113. Academic Press, New York, NY.

Cui, Q., and Karplus, M. (2008). Allostery and cooperativity revisited. *Protein Sci.* **17,** 1295–1307.

Czof, R., and Lamprecht, W. (1974). Pyruvate, phosphoenolpyruvate and D-glycerate-2-phosphate. *In* "Methods of Enzymatic Analysis" (H. U. Bergmeyer, ed.), Vol. 3, pp. 1446–1451. Verlag Chemie/Academic Press, Weinheim/New York, NY.

Di Cera, E. (1995). Thermodynamic Theory of Site-Specific Binding Processes in Biological Macromolecules. Cambridge University Press, New York, NY.

Fenton, A. W. (2008). Allostery: An illustrated definition for the 'second secret of life'. *Trends Biochem. Sci.* **33,** 420–425.

Fenton, A. W., and Hutchinson, M. (2009). The pH dependence of the allosteric response of human liver pyruvate kinase to fructose-1,6-bisphosphate, ATP, and alanine. *Arch. Biochem. Biophys.* **484,** 16–23.

Fenton, A. W., and Tang, Q. (2009). An activating interaction between the unphosphorylated N-terminus of human liver pyruvate kinase and the main body of the protein is interrupted by phosphorylation. *Biochemistry* **48,** 3816–3818.

Fenton, A. W., *et al.* (2003). Identification of substrate contact residues important for the allosteric regulation of phosphofructokinase from *Escherichia coli*. *Biochemistry* **42,** 6453–6459.

Formaneck, M. S., *et al.* (2006). Reconciling the "old" and "new" views of protein allostery: A molecular simulation study of chemotaxis Y protein (CheY). *Proteins* **63,** 846–867.

Frauenfelder, H., *et al.* (1988). Conformational substrates in proteins. *Annu. Rev. Biophys. Biophys. Chem.* **17,** 451–479.

Gunasekaran, K., *et al.* (2004). Is allostery an intrinsic property of all dynamic proteins? *Proteins* **57,** 433–443.

Hall, E. R., and Cottam, G. L. (1978). Isozymes of pyruvate kinase in vertebrates: Their physical, chemical, kinetic and immunological properties. *Int. J. Biochem.* **9,** 785–793.

Johnson, J. L., and Reinhart, G. D. (1994). Influence of MgADP on phosphofructokinase from *Escherichia coli*. Elucidation of coupling interactions with both substrates. *Biochemistry* **33,** 2635–2643.

Kachmar, J. F., and Boyer, P. D. (1953). Kinetic analysis of enzyme reactions. II. The potassium activation and calcium inhibition of pyruvic phosphoferase. *J. Biol. Chem.* **200,** 669–682.

Kayne, F. J. (1971). Thallium (I) activation of pyruvate kinase. *Arch. Biochem. Biophys.* **143,** 232–239.

Kern, D., and Zuiderweg, E. R. (2003). The role of dynamics in allosteric regulation. *Curr. Opin. Struct. Biol.* **13,** 748–757.

Larsen, T. M., *et al.* (1997). Ligand-induced domain movement in pyruvate kinase: Structure of the enzyme from rabbit muscle with Mg^{2+}, K^{+}, and L-phospholactate at 2.7 Å resolution. *Arch. Biochem. Biophys.* **345,** 199–206.

Laughlin, L. T., and Reed, G. H. (1997). The monovalent cation requirement of rabbit muscle pyruvate kinase is eliminated by substitution of lysine for glutamate 117. *Arch. Biochem. Biophys.* **348,** 262–267.

Lockless, S. W., and Ranganathan, R. (1999). Evolutionarily conserved pathways of energetic connectivity in protein families. *Science* **286,** 295–299.

Loria, J. P., and Nowak, T. (1998). Conformational changes in yeast pyruvate kinase studied by 205Tl+ NMR. *Biochemistry* **37,** 6967–6974.

Martell, A. E., and Smith, R. M. (1974). Critical Stability Constants. Plenum Press, New York, NY–1989.

McGilvery, R. W. (1965). Fructose 1,6-diphosphate. Acidic dissociation constants, chelation with magnesium, and optical rotatory dispersion. *Biochemistry* **4,** 1924–1930.

Mesecar, A. D., and Nowak, T. (1997a). Metal-ion-mediated allosteric triggering of yeast pyruvate kinase. 1. A multidimensional kinetic linked-function analysis. *Biochemistry* **36,** 6792–6802.

Mesecar, A. D., and Nowak, T. (1997b). Metal-ion-mediated allosteric triggering of yeast pyruvate kinase. 2. A multidimensional thermodynamic linked-function analysis. *Biochemistry* **36,** 6803–6813.

Munoz, M. E., and Ponce, E. (2003). Pyruvate kinase: Current status of regulatory and functional properties. *Comp. Biochem. Physiol. B Biochem. Mol. Biol.* **135,** 197–218.

Noguchi, T., *et al.* (1987). The L- and R-type isozymes of rat pyruvate kinase are produced from a single gene by use of different promoters. *J. Biol. Chem.* **262,** 14366–14371.

Nowak, T., and Lee, M. J. (1977). Reciprocal cooperative effects of multiple ligand binding to pyruvate kinase. *Biochemistry* **16,** 1343–1350.

Nowak, T., and Suelter, C. (1981). Pyruvate kinase: Activation by and catalytic role of the monovalent and divalent cations. *Mol. Cell. Biochem.* **35,** 65–75.

Oria-Hernandez, J., *et al.* (2005). Pyruvate kinase revisited: The activating effect of K^+. *J. Biol. Chem.* **280,** 37924–37929.

Page, M. J., and Di Cera, E. (2006). Role of Na^+ and K^+ in enzyme function. *Physiol. Rev.* **86,** 1049–1092.

Pendergrass, D. C., *et al.* (2006). Mining for allosteric information: Natural mutations and positional sequence conservation in pyruvate kinase. *IUBMB Life* **58,** 31–38.

Popovych, N., *et al.* (2006). Dynamically driven protein allostery. *Nat. Struct. Mol. Biol.* **13,** 831–838.

Reed, G. H., and Cohn, M. (1973). Electron paramagnetic resonance studies of manganese (II)–pyruvate kinase–substrate complexes. *J. Biol. Chem.* **248,** 6436–6442.

Reinhart, G. D. (1983). The determination of thermodynamic allosteric parameters of an enzyme undergoing steady-state turnover. *Arch. Biochem. Biophys.* **224,** 389–401.

Reinhart, G. D. (1988). Linked-function origins of cooperativity in a symmetrical dimer. *Biophys. Chem.* **30,** 159–172.

Reinhart, G. D. (2004). Quantitative analysis and interpretation of allosteric behavior. *Methods Enzymol.* **380,** 187–203.

Suel, G. M., *et al.* (2003). Evolutionarily conserved networks of residues mediate allosteric communication in proteins. *Nat. Struct. Biol.* **10,** 59–69.

Suelter, C. H., *et al.* (1966). Studies on the interaction of substrate and monovalent and divalent cations with pyruvate kinase. *Biochemistry* **5,** 131–139.

Swain, J. F., and Gierasch, L. M. (2006). The changing landscape of protein allostery. *Curr. Opin. Struct. Biol.* **16,** 102–108.

Tietz, A., and Ochoa, S. (1958). Fluorokinase and pyruvic kinase. *Arch. Biochem. Biophys.* **78,** 477–493.

Valentini, G., *et al.* (1979). Monovalent cations requirement of the fructose 1,6-bisphosphate-activated pyruvate kinase from *E. coli*. *Ital. J. Biochem.* **28,** 345–361.

Villafranca, J. J., and Raushel, F. M. (1982). The monovalent cation site of pyruvate kinase and other enzymes: NMR investigations. *Fed. Proc.* **41,** 2961–2965.

Volkman, B. F., *et al.* (2001). Two-state allosteric behavior in a single-domain signaling protein. *Science* **291,** 2429–2433.

Weber, G. (1972). Ligand binding and internal equilibria in proteins. *Biochemistry* **11,** 864–878.

Williams, R., *et al.* (2006). Differentiating a ligand's chemical requirements for allosteric interactions from those for protein binding. Phenylalanine inhibition of pyruvate kinase. *Biochemistry* **45,** 5421–5429.

CHAPTER SIX

Conformational Stability of Cytochrome *c* Probed by Optical Spectroscopy

Reinhard Schweitzer-Stenner, Andrew Hagarman, Daniel Verbaro, *and* Jonathan B. Soffer

Contents

Abstract

Over the last 50 years cytochrome *c* has been used as a model system for studying electron transfer and protein folding processes. Recently, convincing evidence has been provided that this protein is also involved in other biological processes such as the apoptosis and α-synuclein aggregation. Numerous lines of evidence suggest that the diversity of the functional properties of cytochrome *c* is linked to its conformational plasticity. This chapter introduces circular dichroism and absorption spectroscopy, as an ideal tool to explore this protein's conformational in solution. Besides assisting in distinguishing different conformations and in obtaining the equilibrium thermodynamics of the transitions between them, the two spectroscopies can also be used to explore details of heme-protein interaction, for example, the influence of the external electric field on the prosthetic heme group.

Department of Chemistry, Drexel University, Philadelphia, Pennsylvania, USA

Methods in Enzymology, Volume 466
ISSN 0076-6879, DOI: 10.1016/S0076-6879(09)66006-7

1. Introduction

Cytochrome *c* is a comparatively simple and small protein (MW ~ 12.4 kDa) which is known for its role as electron transfer mediator in the mitochondria (Edman, 1979). This protein contains a single heme group, the central iron atom of which is coordinated to a histidine (H18) and the sulfur atom of a methionine (M80) side chain (Fig. 6.1). The native state of cytochrome *c*, adopted at physiological pH values and room temperature, is well characterized in structural and functional terms (Berghuis and Brayer, 1992; Bushnell *et al.*, 1990; Edman, 1979; Louie and Brayer, 1990; Moore and Pettigrew, 1990), which made it an ideal object for protein folding studies (Edman, 1979; Englander *et al.*, 1998; Pinheiro *et al.*, 1997). However, it has become increasingly unclear whether the native state of the protein is the most physiologically relevant one. Spectroscopic studies have provided evidence for cytochrome *c* to undergo a conformational change upon its binding to cytochrome *c* oxidase

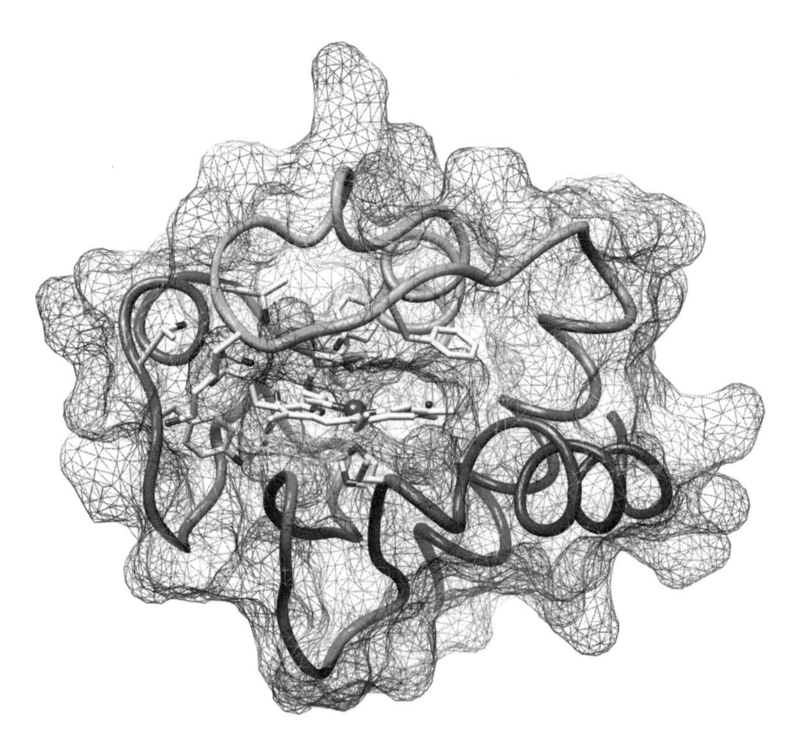

Figure 6.1 Environment of the functional heme *c* group in horse heart cytochrome *c*. The structure was reported by Bushnell *et al.* (1990), the figure was produced with the VMD software (Humphrey *et al.*, 1996). (See Color Insert.)

(Döpner *et al.*, 1999; Garber and Margoliash, 1994; Weber *et al.*, 1987). More importantly, it has become increasingly clear that cytochrome *c* changes its structure upon binding to negatively charged lipid membranes, with cardiolipin (CL) being the most effective target (Belikova *et al.*, 2006; Gorbenko *et al.*, 2006; Heimburg *et al.*, 1991, 1999; Kagan *et al.*, 2005; Kapralov *et al.*, 2007; Rytōman and Kinnunen, 1994; Rytōman *et al.*, 1992). The protein's interaction with CL containing liposomes has attracted considerable interest over the last 15 years, because CL bound cytochrome *c* can be considered as the precursor state of a recently discovered involvement of the protein in apoptosis (Green and Kroemer, 2004; Jiang and Wang, 2004), which has been found to be initiated by the binding of cytochrome *c* to Apaf-1. Subsequently, this leads to the aggregation of the latter, which is a prerequisite for the activation of caspase (Jiang and Wang, 2004; Purring-Koch and McLendon, 2000). Convincing experimental evidence has been provided for the notion that the binding to Apaf-1 requires a conformational change of cytochrome *c* (Jemmerson *et al.*, 1999). While the involvement of structural changes in liposome–cytochrome *c* and Apaf-1–cytochrome *c* interactions can be considered as established, the details of the respective changes of the secondary and tertiary structure remain unclear.

Conformational changes of cytochrome *c* can also occur in solution. In contrast to the reduced species, oxidized ferricytochrome *c* exhibits a pronounced conformational flexibility. Five different states of ferricytochrome *c*, numbered I–V, were identified by analysis of the protein's pH-dependence (different nomenclature is used in the literature, we used the classical nomenclature of Theorell and Åkessen, 1941). The corresponding reaction scheme can be written as follows:

$$I \underset{}{\overset{pK_{12}^{eff}}{\rightleftarrows}} II + n_1H^+ \overset{pK_{23}^{eff}}{\rightleftarrows} III + n_2H^+ \overset{pK_{34}^{eff}}{\rightleftarrows} IV_{a,b} + n_3H^+ \overset{pK_{45}^{eff}}{\rightleftarrows} V_{a,b} + n_4H^+ \qquad (I)$$

All p*K*-values of this scheme are most likely apparent values, representing multiple, cooperative deprotonation steps. n_j ($j = 1, 2, \ldots$) denote the number of released protons. State I is the acidic unfolded state, which is a classical statistical coil at low and a molten globule at high ion concentration (Dyson and Beattie, 1982). State II is a partially folded state, adopted between pH 2 and 3, where at least three different ligation and spin substates coexist (Cohen and Pielak, 1994; Indiani *et al.*, 2000). State III is the native state, observed at neutral pH. State III undergoes a transition into state IV at pH > 8.0. The transition from III to IV involves the replacement of the methionine ligand by a lysine residue (Barker and Mauk, 1992; Battistuzzi *et al.*, 1999a; Döpner *et al.*, 1998; Rossel *et al.*, 1998). Its apparent p*K*-value lies between 8.5 and 9.5 and depends on ionic strength and temperature. This transition is generally thought to cause a more open heme crevice and reduce the reduction potential. For yeast cytochrome *c*, Mauk, Hildebrandt, and their coworkers discovered that state III, in fact, undergoes two parallel transitions with very similar p*K*-values, which involve replacements of M80 by K79 and K73 (Döpner *et al.*, 1998). The corresponding substates were termed IV_a and IV_b. Blouin *et al.* (2001) showed that the III → IV transition is even more complex for horse heart cytochrome *c*, in that it involves three IV states. Two of them correspond to IV_a and IV_b of cytochrome *c*, which are separated from the third one by a rather high-energy barrier. An NMR based structure analysis of state IV_b has been reported by the Bertini group (they investigated the mutant K72AK79AC102T of yeast *iso* 1-cytochrome, the K79 mutation eliminates the trimethylated lysine residue and makes the protein more "horse heart" like) (Assfalg *et al.*, 2003). The authors' data show that most of the native structure (secondary and tertiary) is actually maintained in the mutant. This notion particularly applies to the proximal side of the protein. The distal side, however, was shown to be highly flexible which allows the replacement of M80 by K73 and causes an increased exposure of the heme to the solvent. Upon increasing the pH to 12, two new states (V_a and V_b) are formed in which lysine is replaced by another ligand, possibly a hydroxide ion (Döpner *et al.*, 1998). This second alkaline transition causes a substantial decrease of the redox potential (Barker and Mauk, 1992).

This chapter provides an overview over how absorption and CD spectroscopy can be used to identify and characterize different conformational states of cytochrome *c* in solution. Characterization, here, means the elucidation of how the protein interacts with its environment and the determination of thermodynamic parameters reflecting the energetics of conformational transitions. Most of the data and some of the analyses presented here have recently been reported in a variety of publications.

The chapter is organized as follows. Section 2 provides a very brief theoretical description of light absorption and circular dichroism (CD), which should facilitate the understanding of how both spectroscopic tools are used in the subsequent paragraphs. Section 3 introduces ultra-violet circular dichroism (UVCD), particularly vacuum ultra-violet (VUV) synchrotron radiation circular dichroism (SRCD) spectroscopy as the most reliable way of using CD spectroscopy for the secondary structure analysis of proteins. Section 4 focuses on how visible CD and absorption spectroscopy can be used to probe electronic and vibronic symmetry perturbations of the heme group by the electrostatic potential in the heme pocket. Subsequently, Section 5 demonstrates how to utilize visible CD and absorption spectroscopy for discriminating between different nonnative states and how the temperature and pH dependence of specific spectroscopic observables can be utilized for a thermodynamic analysis. Some new insights about the energy landscape of cytochrome *c* have emerged from this analysis.

2. Basic Theory of Absorption and Circular Dichroism Spectroscopy

The absorption of light occurs upon the reaction of electromagnetic radiation with matter. Macroscopically, this interaction causes the attenuation of the intensity of light, which for nonaggregating molecules in the gas phase of a liquid can be described by the Beer–Lambert law:

$$\frac{I(\tilde{\nu})}{I_0(\tilde{\nu})} = \mathrm{e}^{-A(\tilde{\nu})} \tag{6.1a}$$

where $I_0(\tilde{\nu})$ and $I(\tilde{\nu})$ are the intensities of light at the wavenumber[1] $\tilde{\nu}$ before and after the passage through the medium. $A(\tilde{\nu})$ is called the absorbance and is written as

$$A(\tilde{\nu}) = \varepsilon(\tilde{\nu})cd \tag{6.1b}$$

where c is the concentration of the substance in solution, d is the length of the path which the light travels in the medium and ε is the extinction coefficient. The latter is generally expressed in units of M^{-1} cm^{-1} and reflects the capability of the medium to interact with the electromagnetic radiation field. For electronically allowed transitions, ε can generally be described by a superposition of Voigtian profiles:

$$\varepsilon(\tilde{\nu}) = \sum_l \frac{1}{\sqrt{2\pi}\sigma_l} \int_{-\infty}^{\infty} \frac{f_l \Gamma_l / 2\pi}{(\tilde{\nu}'_l - \tilde{\nu})^2 + \Gamma_l^2} \mathrm{e}^{-\left((\tilde{\nu} - \tilde{\nu}'_l)/2\sigma_l^2\right)} \mathrm{d}\tilde{\nu} \tag{6.2a}$$

where

$$f_l = A_\varepsilon \tilde{\nu}_l |\langle l | \vec{\mu} | g \rangle|^2 = A_\varepsilon \tilde{\nu}_l \left| \int \psi_l^\star \vec{\mu} \psi_g d^3 r \right|^2 \tag{6.2b}$$

is the oscillator strength of the transition associate with the dipole $\vec{\mu}$ induced by the electromagnetic radiation field. $\psi_{l,g}$ are the electronic wavefunctions of the states $|g\rangle$ and $|l\rangle$. A_ε is a constant, the value of which depends on the chosen unit system. The transition dipole moment is defined as

$$\vec{\mu} = \sum_i e\vec{r}_i \tag{6.3}$$

where e is the elementary charge and $\vec{r}_i$ is the coordinate of the ith electron with respect to an arbitrary coordinate system. The transition dipole moment couples the ground state $|g\rangle$ to the excited state $|l\rangle$ by absorbing a photon from the radiation field. $\tilde{\nu}_l$ is the wavenumber corresponding to the energy difference between $|l\rangle$ and $|g\rangle$. Γ_l and σ_l are the half-halfwidths

[1] Wavenumbers rather than wavelengths or frequencies will be used throughout this chapter.

of the Lorentzian and Gaussian profiles associated with the excitation of state $|l\rangle$. While the former reflects the natural lifetime and dephasing of the excited state (Friedman *et al.*, 1977), the latter contains a static term describing the inhomogeneous broadening and a dynamic temperature dependent term which reflects the coupling of low frequency modes to the $|g\rangle \rightarrow |l\rangle$ transition (Cupane *et al.*, 1995). For molecules, the summation runs over all excited vibronic states accessible by electronic dipole transitions.

Circular dichroism is a measure of the difference between the extinction coefficients for left (L)- and right (R)-handed circular polarized light:

$$\Delta\varepsilon(\tilde{\nu}) = \varepsilon_{\mathrm{L}}(\tilde{\nu}) - \varepsilon_{\mathrm{R}}(\tilde{\nu}) \tag{6.4}$$

The difference cannot be accounted for by Eqs. (6.2a) and (6.2b) because the electronic transition dipole moment would just be the same for both rotations of the field vector. However, the electromagnetic field also induces a magnetic dipole moment which might couple $|g\rangle$ and $|l\rangle$ as well. It can be written as follows:

$$\vec{m} = \frac{e}{2m_{\mathrm{e}}}\sum_i \vec{L}_i \tag{6.5}$$

where m_{e} is the mass of a resting electron and $\vec{L}_i$ is the operator associated with the angular momentum of the *i*th electron. If one considers contributions from both the electronic and the magnetic transition dipole moment, the effective oscillator strength is written as

$$f_l^{\mathrm{L,R}} = A_\varepsilon \tilde{\nu}_l(\langle l|\vec{\mu}|g\rangle + \mathrm{Im}\langle l|\vec{m}|g\rangle)(\langle l|\vec{\mu}|g\rangle - \mathrm{Im}\langle l|\vec{m}|g\rangle) = f_l + G_l \pm 2R_l \tag{6.6}$$

where we consider the fact that the magnetic moment is imaginary. The term

$$R_l = \mathrm{Im}(\langle l|\vec{\mu}|g\rangle\langle l|\vec{m}|g\rangle) \tag{6.7}$$

is called the rotational strength of the transition. G_l is the oscillator strength of the magnetic transition. The magnetic dipole moment exhibits rotational symmetry so that the respective transition requires circular polarized light. Substituting f_l in Eqs. (6.2a) and (6.2b) by $f_l^{\mathrm{L,R}}$ yields the extinction profiles for left and right handed circular light. The CD spectrum is the difference of these two profiles:

$$\Delta\varepsilon(\tilde{\nu}) = \sum_l \frac{1}{\sqrt{2\pi}\sigma_l}\int_{-\infty}^{\infty} \frac{R_l\Gamma_l/2\pi}{(\tilde{\nu}'_l - \tilde{\nu})^2 + \Gamma_l^2}\,\mathrm{e}^{-\left((\tilde{\nu}_l - \tilde{\nu}'_l)/2\sigma_l^2\right)}\,\mathrm{d}\tilde{\nu} \tag{6.8}$$

If the considered transitions are well separated (i.e., the respective band profiles do not overlap), the shapes of the corresponding absorption and CD band profiles are identical. R_l can be positive or negative, which gives rise to positive and negative Cotton bands in the CD spectrum. Generally, the above condition is not met for molecules because an absorption band is composed of a series of subbands resulting from transitions into different vibronic states assignable to vibrational modes coupled to the corresponding electronic transition. However, in the absence of strong interstate non-Condon coupling (Franck–Condon limit) the electronic and magnetic transition dipole moments and their relative orientation can be considered as independent on the involved vibrational states (Cupane *et al.*, 1995; Stallard *et al.*, 1983; Wellman *et al.*, 1965). As a consequence, the shapes of corresponding CD and absorption profiles associated with the same electronic transition should be identical irrespective of the underlying vibronic structure.

The situation changes if the absorption profiles of different electronic transitions overlap substantially and if, for example, the rotational strengths of the two transitions have opposite signs. As described in more detail in the next chapter, this is exactly the case for split optical bands of native cytochrome *c*. Figure 6.2A shows how the actual strength and shape of a couplet produced by two overlapping Cotton bands of opposite sign depends on the difference, Δ, between their peak positions and the halfwidths, σ, of the corresponding Gaussian profiles. Apparently, a decrease of Δ/σ yields a decreasing couplet signal, owing to the destructive interference between the positive and negative rotational strengths of the two transitions. The signal disappears if $\Delta \rightarrow 0$. It thus follows that the rotational strength of a transition cannot be estimated just from the integration of a CD band if two or more bands overlap. Another possible scenario is displayed in Fig. 6.2B, namely, the overlap of two transitions with the same oscillator strengths and rotational strengths of the same sign (positive), but of substantially different magnitude. In the case depicted in Fig. 6.2B ($\Delta = \sigma$), the two transitions are not resolved in the absorption band. However, the CD profile of the two overlapping transitions is clearly asymmetric and its peak position does not coincide with the peak of the absorption band. Together, this demonstrates that a comparison of corresponding CD and absorption band profiles can facilitate the identification of overlapping bands. This insight will be used in the subsequent paragraphs for analyzing various absorption and CD bands of cytochrome *c*.

Generally, the rotational strength of a transition between different electronic states in a molecule is interpreted as reflecting its chirality. Besides indicating that a molecule is achiral, the absence of any circular dichroism can also be due to the 1:1 racemic mixture of two enantiomers, each of which being chiral. A molecule is achiral if it contains an inversion center or a plane of symmetry.

For macromolecules like proteins, CD spectra often solely probe the chirality of local "color centers" or chromophores in a very complex and symmetry lowering environment. The functional group of heme proteins is a classical example. In most cases, the functional group is a protoporphyrin IX derivative (heme group) with an iron as the central metal atom (Fig. 6.3). This metal atom is coordinated to one or two ligands provided by the protein environment. If one neglects the influence of peripheral substituents a planar porphyrin macrocycle exhibits D_{4h} symmetry, which is of course

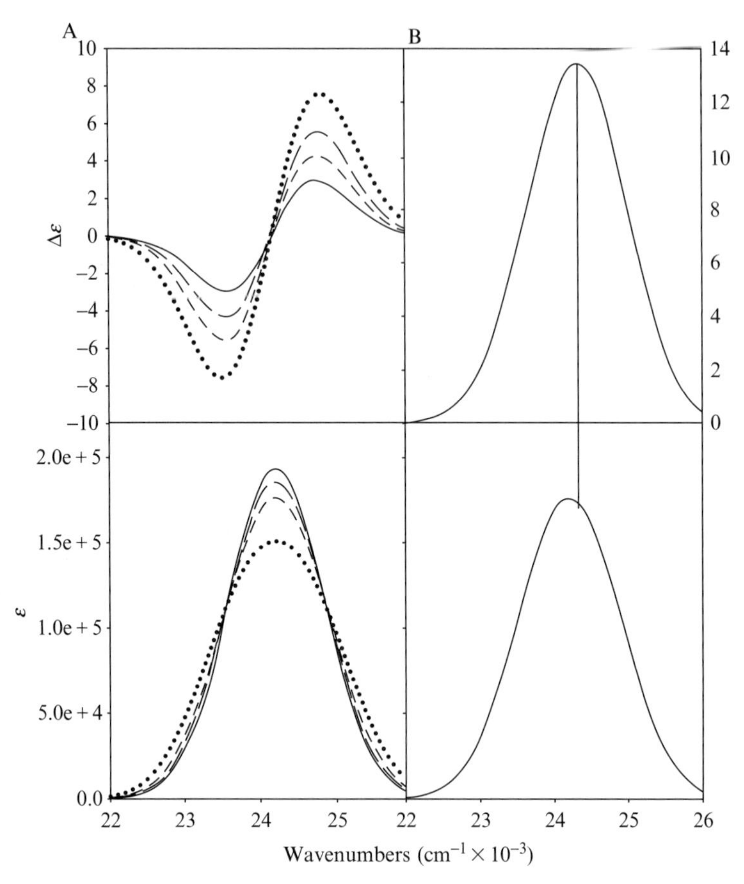

Figure 6.2 (A) CD couplet of two nearly degenerate electronic dipole-allowed transitions with opposite rotational strength calculated for different Δ/σ ratios: 0.5 (solid), 0.75 (medium dashed), 1.0 (short dashed), 1.3 (dotted). (B) CD and absorption spectrum of a split transition with $R_x/R_y = 0.5$ and $\Delta/\sigma = 1$.

achiral. Hence, one would expect that electronic transitions associated with the heme group show no or very weak CD activity (induced by the symmetry lowering power of the substituents). However, the CD spectrum of the B-band region, which results from a transition into the second lowest excited electronic state of the heme's π-electron system, is quite significant (Blauer *et al.*, 1993b; Dragomir *et al.*, 2007; Hsu and Woody, 1971; Kiefl *et al.*, 2002; Urry, 1965; Urry and Doty, 1965). This occurrence demonstrates that the protein environment induces chirality by lowering the symmetry of the heme group. The physical basis of this observation is that transition dipole moments associated with electronic transitions of side chains (i.e., the aromatic rings of F, Y, and W residues) and of the protein backbone can induce a magnetic dipole moment at the heme group

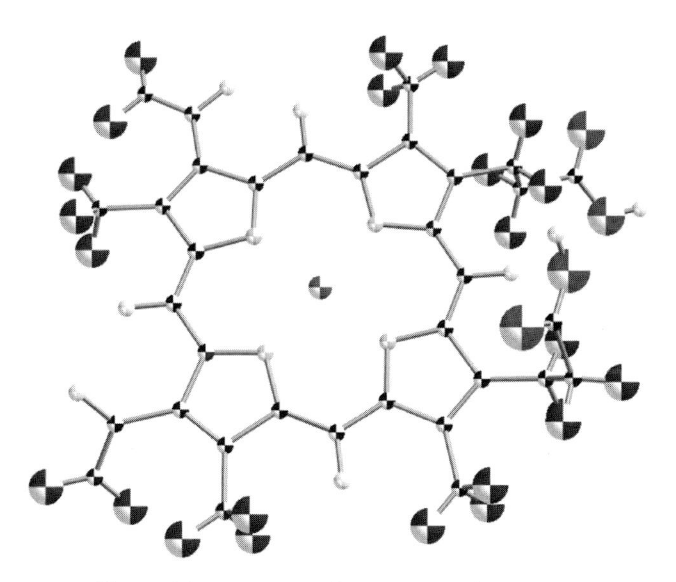

Figure 6.3 Structure of Fe-protoporphyring IX.

(Blauer *et al.*, 1993b; Hsu and Woody, 1971), which must be added to its intrinsic magnetic dipole moment (van Holde *et al.*, 2006):

$$\vec{m}_{\text{total}} = \sum_j \left(\vec{m}_0 + i\pi \vec{R}_j \times \frac{\vec{\mu}_j}{\lambda_j} \right) \tag{6.9}$$

where λ_j is the wavelength of the *j*th electronic transition, $\vec{\mu}_j$ is the dipole moment of the *j*th group, $\vec{R}_j$ is the distance vector from the heme to this group, and $\vec{m}_0$ is the intrinsic magnetic moment of the heme. If this perturbation produces a magnetic dipole vector, which is no longer perpendicular to the heme plane, rotational strength is induced into the electronic transition of the heme group. The very same argument can be made for electronic transitions in protein side chains and backbones (Blauer *et al.*, 1993b). For the latter this leads to very structure sensitive CD spectra in the far-UV region, which can be used for secondary structure analysis.

3. Secondary Structure Analysis of Cytochrome *c* Using Ultra-Violet Circular Dichroism Spectroscopy

For the past 50 years, far UV circular dichroism spectroscopy has been established as a diagnostic tool to determine the structure and dynamics of proteins in solution (Woody, 2004). Since CD is diagnostic of chirality, it is a particularly suitable tool to explore the secondary structure of proteins (Sreerama and Woody, 1993). Secondary structures of proteins such as α-helices, β-sheets, polyproline II, turns, etc., exhibit characteristic backbone dihedral angles (φ, ψ), which are indicated in the Ramachandran plot shown in Fig. 6.4. The amide backbone $\pi \rightarrow \pi^{\star}$ and n $\rightarrow \pi^{\star}$ electronic transitions occur in the ultra-violet (UV) regime. These transitions give rise to circular dichroism, owing to excitonic coupling between $\pi \rightarrow \pi^{\star}$ as well as $\pi \rightarrow \pi^{\star}$ and n $\rightarrow \pi^{\star}$ transitions in different peptides of a polypeptide chain (Tinoco, 1962). Polarization effects have recently been shown to contribute as well (Woody, 2009). This coupling depends on the dihedral angles so that different secondary structures yield significantly different CD spectra (Sreerama and Woody, 1993). Proteins containing large fractions of α-helices, such as Myoglobin, yield a UV-CD spectrum with minima at 222 and 208 nm and an intense maximum at 192 nm. Proteins such as Concavlin A, which contains a substantial amount of β-sheet, yield a UV-CD spectrum with a minimum at 210 nm and a maximum at 190 nm, the relative intensities of which are less than those from α-helices. Lysozyme, a protein with substantial fractions of both, α-helices and β-sheets, yields a UV-CD spectrum characteristic of a superposition of the respective spectra. The amount of α-helix present in a given protein can be roughly estimated by comparing the ellipticity at 222 nm. (Greenfiled, 1996) Secondary structure prediction software, such as the neural network analysis (K2D) (Andrade *et al.*, 1993; Böhm *et al.*, 1992; Merolo *et al.*, 1994), rigid regression (Contin program) (Provencher and Glöckner, 1981), variable selection (VARSLC program) (Manavalan and Johnson, 1987), and the self-consistent method

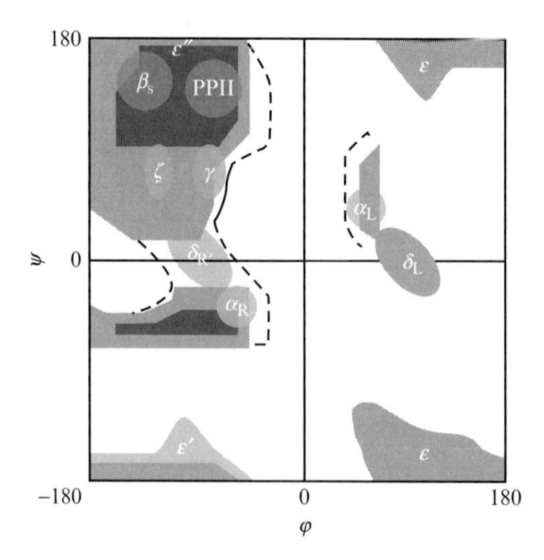

Figure 6.4 Ramachandran plot of backbone dihedral angles in proteins.

(SELCON program) (Sreerama and Woody, 1993, 1994a,b) have been developed for a more quantitative analysis of secondary structure composition, though the results varied (Black and Wallace, 2007). Spectra obtained from conventional instruments are limited in this regard, due to the confines of the Xenon arc lamp light source, which loses a significant amount of intensity below 190 nm.

Though being developed almost 30 years ago (Snyder and Rowe, 1980; Sutherland *et al.*, 1980), only over the last 10 years (Wallace, 2000) synchrotron radiation has been used to measure VUV CD spectra, due to the fact that this light source is still very intense at wavelengths below 190 nm and thus allows for higher energy transitions such as $n \rightarrow \sigma^{\star}$ and $n' \rightarrow \pi^{\star}$ to be probed. Valuable spectral information is gained by extending the recording of CD spectra into the VUV (<190 nm) region (Fig. 6.5). For instance, spectra from α-helices show an apparent shoulder on the positive maximum, located at 180 nm, which is difficult to measure properly using conventional instruments. The characteristics of CD spectra observed in the VUV enhance the capability of a quantitative secondary structure analysis using prediction software. Most recently, Wallace and coworkers developed a comprehensive secondary structure analysis program known as Dichroweb (Whitmore and Wallace, 2004, 2008), which is available online. Dichroweb allows the use of many of the most popular structural determination algorithms and several reference sets encompassing all ranges of wavelengths. This online server allows the user to input a

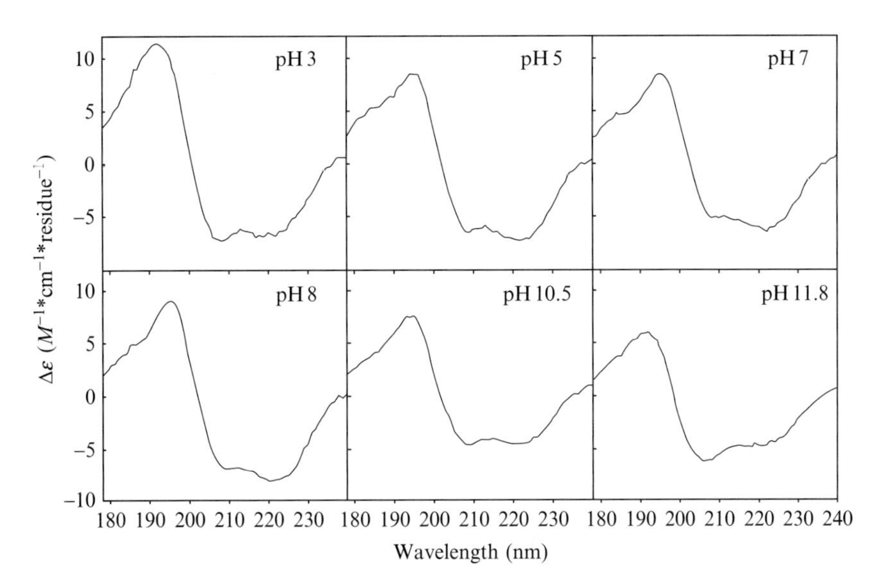

Figure 6.5 SRCD spectra of oxidized horse heart cytochrome *c* taken at indicated pH values plotted from 178 to 240 nm in Δε units per residue.

variety of different (a) file formats from various instrument manufacturers, (b) units, and (c) wavelength scales, with matching outputs, making it very user-friendly. Along with secondary structure fractions, the universal goodness-of-fit parameters is provided with the output file, which allows a judgment of their significance. To appropriately quantify secondary structure using UV-CD, exact protein concentrations and path lengths of cells must be known, and the spectra must be background corrected for solvent effects. The reader is herewith referred to the work of Greenfield, and Wallace and coworkers (Greenfiled, 1996; Wallace, 2000; Whitmore and Wallace, 2004, 2008), which include more details about Dichroweb and analyses using UV-CD for secondary structure determination.

Taking advantage of the opportunity to measure SRCD spectra at beamline U11 at Brookhaven National Laboratories in the National Synchrotron Light Source building, the VUV-CD spectra of oxidized cytochrome *c* was measured in the region between 175 and 240 nm from pH from 3 to 12 at room temperature in a 1-m*M* MOPS (3-(*N*-morpholino)propanesulfonic acid) buffer. A rather low ionic strength was chosen to avoid any influence of anion binding (Banci *et al.*, 1998; Battistuzzi *et al.*, 1999b; Feng and Englander, 1990; Moench *et al.*, 1991; Shah and Schweitzer-Stenner, 2008). As indicated in Section 1, this pH range covers the states II–V of the protein. The SRCD spectra can be seen in Fig. 6.5. The online Dichroweb software was employed to analyze these spectra. The Contin analysis algorithm (Provencher and Glöckner, 1981) was utilized with reference set 6. Protein concentrations were determined from the Soret band absorption of the protein (Bushnell *et al.*, 1990) solution in a 1-mm quartz cell (Hellma) by using an extinction coefficient of 10.6×10^4 (M^{-1} cm^{-1}). From the crystal structure of yeast ferricytochrome *c*, Louie and Brayer (1990) determined 40% α-helix, 22% β-turn, and negligible β-sheet structure. A practically identical result was obtained for horse heart cytochrome *c* (Bushnell *et al.*, 1990). The analysis of the SCRD spectra recorded at pH 7 and 8, at which the native state III is predominantly populated, yielded 44% α-helix and 23% β-turn, which is in very good agreement with crystallographic values. At pH 5, at which the protein is still believed to be in its native state (Dyson and Beattie, 1982), the values were almost identical, but indicate a small loss ($\sim$3%) of α-helical content. In state II (pH 3) both the α-helix and β-turn content decreased by $\sim$3% compared to that obtained for the native state measurement. This modest decrease indicates that this state still exhibits a rather intact secondary structure, which would be consistent with the notion that it is a molten globule (Indiani *et al.*, 2000). For pH 10.5, at which the protein is almost exclusively in state IV in a low ionic strength solution (Verbaro *et al.*, 2009), the α-helix content was determined to be 35.5%, while the β-turn fraction is comparable with that of the native state. In state V, probed at pH 11.8, the α-helix and β-turn contents were 31% and 20%, respectively. All these changes of the secondary structure are modest, which shows that the structural changes induced by the conformational transitions of ferricytochrome *c* involve mostly changes of the tertiary structure, some of which will be discussed in more detail below.

4. Visible CD and Absorption Spectroscopy of Native Cytochrome *c*

Visible CD has frequently been used, particularly in protein folding experiments, to probe the intactness of the heme environment in heme proteins, particularly predominantly for cytochrome *c*. Very early investigations of cytochrome *c* in 1960s used optical rotary dispersion studies (Urry, 1965; Urry and Doty, 1965), before CD-spectroscopy became the standard method. With respect to cytochrome *c*, visible CD measurements focused on the B-band region between 22,000 and 26,000 cm^{-1} because of its superior intensity (Blauer *et al.*, 1993b; Hsu and Woody, 1971; Kiefl *et al.*, 2002; Myer, 1968; Myer and Pande, 1978; Nantes *et al.*, 2001; Pinheiro *et al.*, 1997). Here, the CD spectra of the B- and Q-band regions are presented and discussed. Figures 6.6–6.9 show the CD spectra of the B- and Q-band region of native oxidized and reduced cytochrome *c* (horse heart and yeast) measured at nearly neutral pH and high ionic strength taken in our own laboratory (Dragomir *et al.*, 2007). For the oxidized state, very pronounced couplets are obtained at the positions of the B_0 and Q_0 band. The B-band couplet of the reduced protein is substantially different from that of the oxidized species. The corresponding Q-band region shows a fine-structure of comparatively sharp lines which does not reflect the rather broad and unresolved Q_v band of the absorption spectrum.

It is surprising that limited efforts have been made to understand these spectra. The first real theoretical approach to rationalize the B-band circular dichroism of myoglobin and hemoglobin was undertaken by Hsu and Woody (1971). Their CD spectra show positive Cotton bands rather than couplets. These authors calculated the rotational strength of the B-band transition in terms of coupled oscillators involving the $\pi \rightarrow \pi^\star$ transitions of the heme macrocycle and of nearby aromatic amino acid side chains. Blauer *et al.* (1993a) extended this approach by additionally considering the coupling of heme transitions to far-UV backbone and side chain electronic transitions, as well as intrinsic rotational strength of the heme group induced by nonplanar deformations of the heme group and applied it to explain the B-band CD spectrum of the heme undecapeptide of horse heart cytochrome *c*. Kiefl *et al.* (2002) showed that heme ruffling can contribute significantly to the rotational strength of the B-band of carbonmonoxy myoglobin.

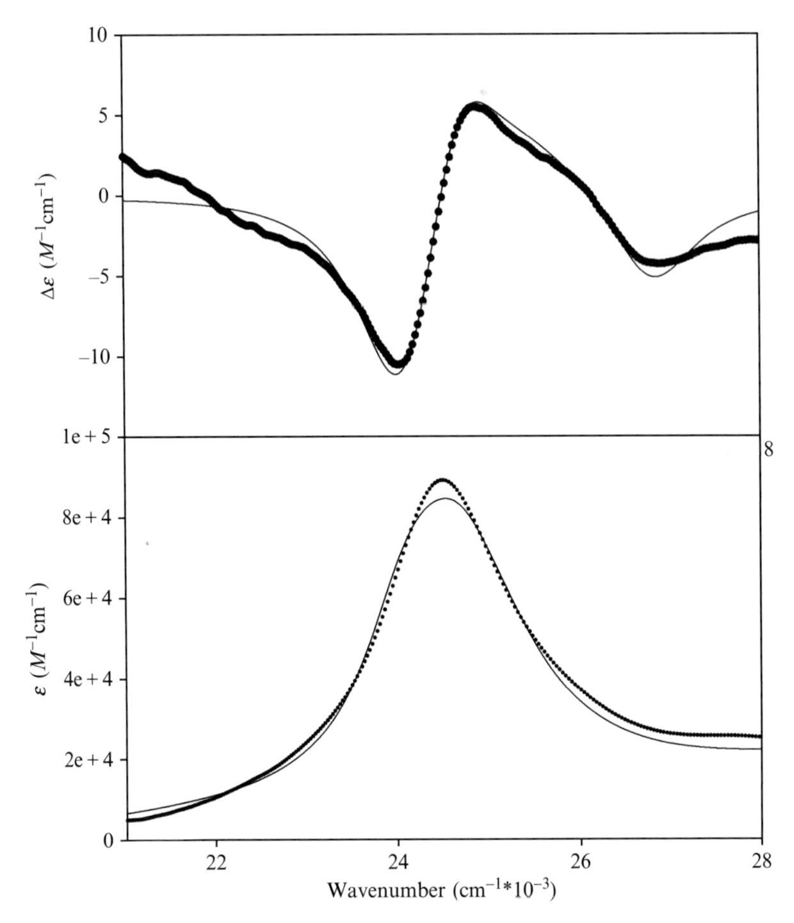

Figure 6.6 CD and absorption spectra of the B-band region of oxidized horse heart cytochrome *c* measured at neutral pH. The solid lines result from a simulation described in the text (data from Schweitzer-Stenner, 2008).

Though providing a deeper understanding of the rotational strength of the B-band displayed by the visible CD spectra of heme proteins, the theoretical approaches outlined above do not directly explain the couplet in the corresponding spectrum of ferricytochrome *c* and the somewhat skewed CD spectrum of the reduced protein, for which the CD maximum clearly does not coincide with the absorption maximum. An explanation was only recently suggested (Dragomir *et al.*, 2007), which was then substantiated in a subsequent paper (Schweitzer-Stenner, 2008). In what follows, the basic concept of the theory presented in this chapter will be briefly outlined. For the sake of the readability of this chapter for a broader

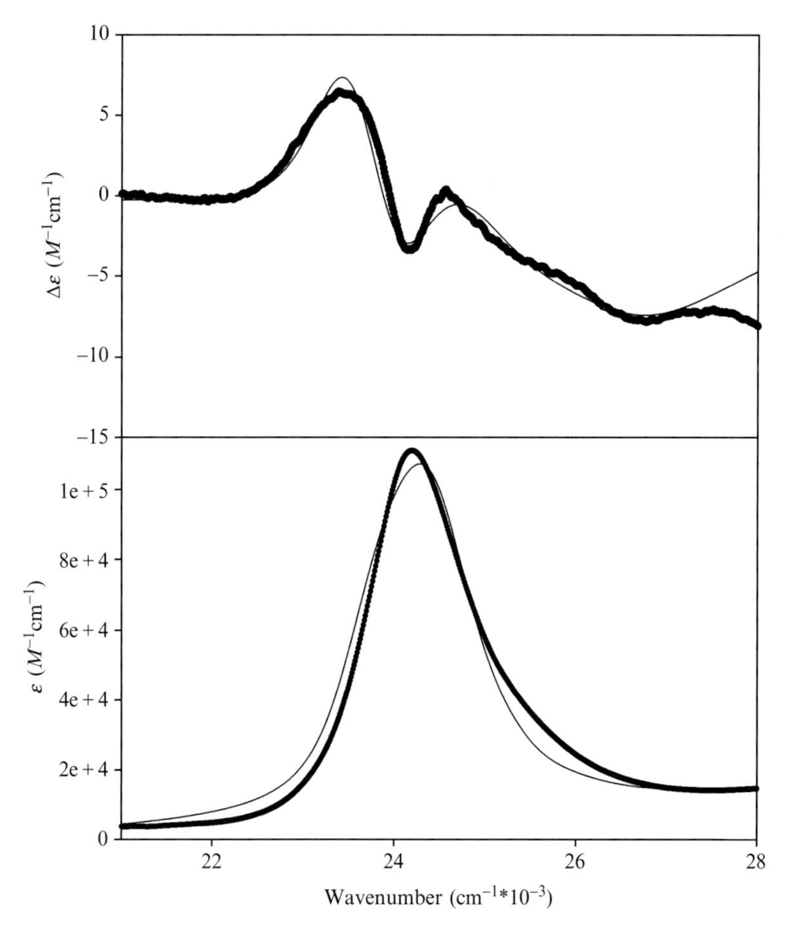

Figure 6.7 CD and absorption spectra of the B-band region of reduced horse heart cytochrome *c* measured at neutral pH. The solid lines result from a simulation described in the text (data from Schweitzer-Stenner, 2008).

readership, the mathematics is reduced to a minimum and the interested reader is referred to the above cited paper of Schweitzer-Stenner for details.

Assuming an unperturbed D_{4h} symmetry for the π-electron system of the heme macrocycle, the absorption spectrum can be understood in terms of only two electronic transitions from a totally symmetric ground state (symmetry A_{1g}) into 2 twofold degenerate states $|B_{x,y}\rangle$ and $|Q_{x,y}\rangle$, the respective symmetry of which is E_u (Fig. 6.10) (Gouterman, 1959). The energies of the $0 \rightarrow 0$ transitions between the vibrational ground states of the electronic ground state and the excited state coincide with the maxima of the B_0 and Q_o bands. For reasons described elsewhere, the oscillator

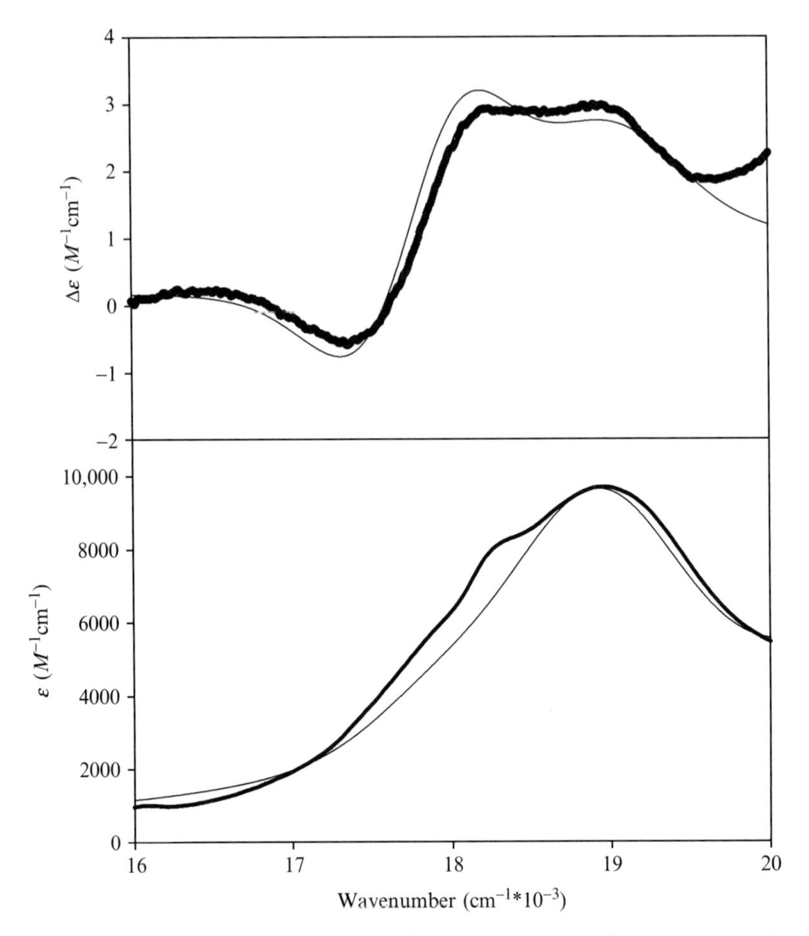

Figure 6.8 CD and absorption spectra of the Q-band region of oxidized horse heart cytochrome *c* measured at neutral pH. The solid lines result from a simulation described in the text (data from Schweitzer-Stenner, 2008).

strength of the Q-state transition is, by nearly an order of magnitude, smaller than that of the B-state transition (Gouterman, 1959; Schweitzer-Stenner, 1989). The weak oscillator strength and additional broadening of the band profile makes the Q_0-band barely detectable in the spectrum of ferricytochrome *c*. Both, the Q- and the B-band region exhibit vibronic side bands termed Q_v and B_v, which reflect $0 \rightarrow 1$ transitions into the first excited state of various vibrational modes of the heme macrocycle (Cupane *et al.*, 1995; Levantino *et al.*, 2005; Schweitzer-Stenner, 1989; Stallard *et al.*, 1983). The spectrum of the oxidized state additionally contains some weak bands in the "red" and "near IR" region below 15,000 cm^{-1} which are assignable to

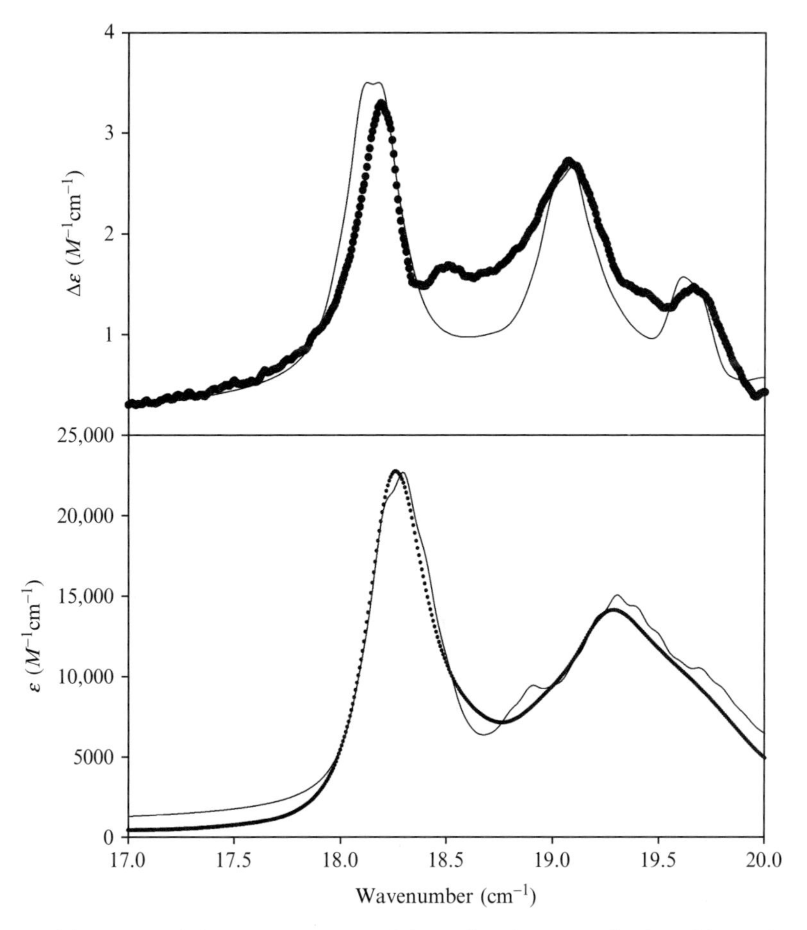

Figure 6.9 CD and absorption spectra of the Q-band region of reduced horse heart cytochrome *c* measured at neutral pH. The solid lines result from a simulation described in the text (data from Schweitzer-Stenner, 2008).

charge transfer processes (Eaton and Hochstrasser, 1968). One of these bands (the so-called 695 nm band) will be discussed in more detail in Section 4.

If the symmetry of the heme were indeed D_{4h}, the CD spectrum would not exhibit couplets, because all the transitions into different vibronic states of the same electronic state have the same rotational strength. Hence, the profiles of the CD and the absorption spectrum should be identical. This consideration led to the conclusion that the CD spectra of both oxidation states reflect a splitting of the excited states, owing to the symmetry lowering perturbations caused by the protein environment

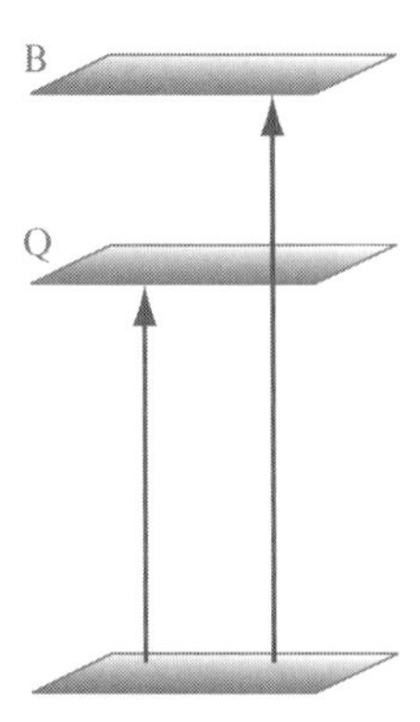

Figure 6.10 Schematic representations of electronic transitions of the heme macrocycle contributing to the visible absorption spectrum.

(Dragomir *et al.*, 2007). In this case, the absorption and CD profiles for a distinct electronic transition l can best be described by the following equations:

$$\varepsilon_l(\tilde{\nu}) = \sum_{\rho=x,y}\sum_{q}\frac{1}{\sqrt{2\pi\sigma_{l,p}}}\int_{-\infty}^{\infty}\frac{f^{\text{eff}}_{l,p,q}\Gamma_{l,p}/2\pi}{(\tilde{\nu}'_{l,p,q}-\tilde{\nu})+\Gamma^2_{l,p}}\,\mathrm{e}^{(\tilde{\nu}_{l,p,q}-\tilde{\nu}'_{l,p,q})/2\sigma^2_{l,p}}\,\mathrm{d}\tilde{\nu}$$

$$\Delta\varepsilon_l(\tilde{\nu}) = \sum_{\rho=x,y}\sum_{q}\frac{4}{\sqrt{2\pi\sigma_{l,p}}}\int_{-\infty}^{\infty}\frac{R^{\text{eff}}_{l,p,q}\Gamma_{l,p}/2\pi}{(\tilde{\nu}'_{l,p,q}-\tilde{\nu})^2+\Gamma^2_{l,p}}\,\mathrm{e}^{(\tilde{\nu}_{l,p,q}-\tilde{\nu}'_{l,p,q})/2\sigma^2_{l,p}}\,\mathrm{d}\tilde{\nu} \tag{6.10}$$

The first sum runs over the two split components x and y of the lth transition. For Q and B, the respective coordinate axes are generally identified with the Fe–N bonds of the heme center (Fig. 6.3). The second sum runs over all states of vibrations, which are vibronically coupled to the lth transition. $f^{\text{eff}}_{l,p,q}$ and $R^{\text{eff}}_{l,p,q}$ are effective oscillator and rotational strengths which can be calculated by multiplying f_l and R_l with factors that reflect the vibronic coupling between the qth vibration and lth electronic transition. Details of the formalism can be found elsewhere (Schweitzer-Stenner, 2008; Schweitzer-Stenner *et al.*, 2007a). Vibronic coupling parameters can be obtained by a simultaneous analysis of absorption, CD and resonance Raman data.

The couplets depicted in Figs. 6.6–6.8 result from band splitting combined with rotational strengths of opposite sign for the x and y components of the corresponding B- and Q-state transitions. The fine structure depicted by the Q-band region of ferrocytochrome c (Fig. 6.9) reflects the fact that only one of the two components of the g → Q_0 and the g → Q_v transitions carry substantial rotational strength, which reduces the effective halfwidths

Table 6.1 B- and Q-band splitting values of horse heart ferri- and ferrocytochrome *c* at neutral pH and room temperature

	1 = B, ferri[c]	1 = Q, ferri[c]	1 = B, ferro[c]	L = Q, ferro[d]
Δ^{l}_{elect} [a]	−505	31	126	60
Δ^{l}_{vib} [b]	−46	−168	390	50

[a] Splitting due to electronic coupling.
[b] Splitting due to vibronic coupling.
[c] Taken from Schweitzer-Stenner (2008).
[d] Taken from Levantino *et al.* (2005).
The splitting values are given in units of cm^{-1}.

of corresponding transitions by a factor 2 and thus greatly enhances the spectral resolution (Dragomir *et al.*, 2007). As a consequence, the CD spectrum looks very much like the low temperature absorption spectrum of ferrocytochrome *c* measured at 10 K (Levantino *et al.*, 2005; Schweitzer-Stenner and Bigman, 2001; Schweitzer-Stenner *et al.*, 2006). A vibronic coupling theory based approach yielded simulations, which are depicted as solid lines in Figs. 6.6–6.9 (Schweitzer-Stenner, 2008). The total band splitting values used for these simulations are listed in Table 6.1. All these values arise from a combination of electronic and vibronic perturbations, which are also listed in Table 6.1. Here, we focus exclusively on the electronic contribution to the B-band splitting.

As shown by Manas *et al.* (1999), the electronic contribution to the B-band splitting is caused by the interaction between the heme macrocycle and the internal electric field created by the distribution of charges in the heme cavity. This is reflected by the following equation:[2]

$$\begin{aligned}\Delta\tilde{\nu}_{electr} = \nu^{elec}_{lx} - \nu^{elec}_{ly} &= \langle l|\tilde{\mu}|g\rangle^2 E^2(\sin^2\theta - \cos^2\theta) \\ &+ \frac{e}{2}\left\{\frac{\partial E_x}{\partial x} - \frac{\partial E_y}{\partial y}\right\}\langle l|x^2 - y^2|l\rangle\end{aligned} \tag{6.11}$$

where E denotes the internal electric field, θ is the angle between the electric field and electronic dipole vector $\tilde{\mu}_{lx}$ ($\tan\theta = \mu_x/\mu_y$) and a line connecting two methine carbons of the heme group (Fig. 6.3), $\tilde{\nu}_l$ is the wavenumber value corresponding to the $0 \rightarrow 0$ position of the B-band in the absence of any splitting. The first term on the right hand side of Eq. (6.11) describes a quadratic Stark effect, which causes different shifts of $\tilde{\nu}_{lx}$ and ν^{elec}_{ly} if $\theta \neq \pi/4$. The second term accounts for the interaction between the quadrupole moment of the heme and the external field and

[2] The corresponding equation of eq. (6.11) in the paper of Schweitzer-Stenner *et al.* (2007a,b) exhibits a denominator for the quadrupole term, which is incorrect and has therefore been omitted.

reflects the difference between the field gradients along the x- and y-axis of the chosen coordinate system. The second term would disappear if the electric field were uniform. The combination of dipole and quadrupole term is additive for the Q- and subtractive for the B-band transition.

As shown by Manas *et al.* (Manas *et al.*, 1999, 2000; Schweitzer-Stenner, 2008) and confirmed by Schweitzer-Stenner (2008), the quadratic Stark effect dominates the B-band splitting. The Q-band splitting, however, is nearly exclusively governed by the quadrupole term (Levantino *et al.*, 2005), owing to its much weaker transition dipole moment. By comparing the electronic contributions to B- and Q-band splitting, the pure Stark-splitting of the B-band was isolated. The respective values for ferri- and ferrocytochrome *c* are listed in Table 6.1. Apparently, electronic splitting is substantially smaller in the reduced than in the oxidized state. For the oxidized state, the splitting is larger for horse heart than for yeast cytochrome *c*, while it is the other way round for the reduced state.

The Stark splitting can be used to estimate the electric field at the heme by using the equation (Manas *et al.*, 1999):

$$E = \frac{1}{|\langle B|\vec{\mu}\,|g\rangle|}\sqrt{\frac{\Delta E^{B}_{\text{Srart}}}{\gamma(\sin^2\theta - \cos^2\theta)}} \tag{6.12}$$

where $\gamma = 0.0057\ \text{cm}^{1/2}/(\text{MV D})$. The corresponding units of the transition dipole moment and the electric field are Debye (D) and MV/cm, respectively. Since values for θ are available from computational studies on ferrocytochrome *c* (55° for horse heart and 58° for yeast) (Manas *et al.*, 1999), Eq. (6.12) can be used to calculate the electric field for the corresponding redox state by invoking the experimentally determined values for the transition dipole moment, namely 11.4 and 10.4 D for horse heart and yeast[3] (Schweitzer-Stenner, 2008), which yields 27 and 29 MV/cm for horse heart and yeast, respectively. These values are somewhat larger than what Manas *et al.* (1999) obtained from theoretical calculations (14 and 15 MV/cm), but the value for horse heart cytochrome *c* is practically identical with the field strength, which Geissinger *et al.* (1997) observed from spectral hole burning experiments on the same cytochrome *c* species with a free base porphyrin.

An estimation of the electric field is more difficult for the ferristate, since no independent information about θ is available. However, a maximal value for E(ferri) can be estimate by assuming that the 300 cm^{-1} upshift of the B-band upon oxidation is entirely due to a quadratic Stark effect. This would yield a field strength of 40 MV/cm. If one combines this value with the experimentally obtained band splitting values listed in Table 6.1 in

[3] Data for yeast cytochrome *c* are not shown in this chapter.

Eq. (6.12), a value of 35° is obtained. This value is consistent with the opposite signs of the couplets exhibited by the CD B-band profiles of ferri- and ferrocytochrome. As argued by Schweitzer-Stenner (2008) the opposite signs reflect an exchange of the x- and y-component of the B-band transition occurring upon oxidation, which can only occur if θ becomes smaller than 45° in the oxidized state. By invoking a uniform electric field, this author estimated that the difference between the above determined electric field strengths of the two redox states would account approximately for a -55 kJ/mol (oxidized–reduced) contribution to the Gibbs energy of the redox reaction. This value is actually very close to the sum of the Gibbs energies associated with electrostatic heme–protein interactions (charge–charge and charge-induced dipole), for which Churg and Warshel (1986) calculated a value of -67.9 kJ/mol. This value is ca. 62% of the solvation energy contribution to the oxidation potential, which stabilizes the oxidized over the reduced state of the heme.

Churg and Warshel (1986) compared the different solvation energies of the two redox states of cytochrome *c* and microperoxidase 8. They found the above electrostatic heme–protein interactions to be approximately a factor of 70 weaker in microperoxidase 8, which should give rise to a very small band splitting, which would not be detectable with CD spectroscopy. In agreement, our recently measured CD spectrum of the very similar system AcM-microperoxidase 11 revealed a positive Cotton signal for the B-band region, which perfectly coincides with the absorption band (Verbaro *et al.*, 2009).

As described in more detail in the next chapter, the CD spectra of native and nonnative states differ considerably, in that the latter all show positive Cotton bands, which seem to coincide with the position of the respective absorption bands. As an example, Fig. 6.11 exhibits absorption and CD spectra of the B-band region of ferricytochrome *c* measured at pH 10 and low ionic strength. The alkaline state IV is predominantly populated under these conditions (Verbaro *et al.*, 2009). The coincidence between absorption and CD seem to suggest the absence of band splitting. However, an overlap of the absorption with the properly scaled CD spectrum shows that the latter is much narrower than the former. This effect has not been realized in earlier studies reporting similar spectra (Hagarman *et al.*, 2008). These data indicate that the CD band is assignable to only one component of the B-state transition (B_x and B_y). A similar phenomenon has been earlier discovered for deoxymyoglobin and myoglobin cyanide (Schweitzer-Stenner *et al.*, 2007a). In contrast to the CD spectrum, both components carry some electronic dipole strength, though to a different extent. We are in process of analyzing these band profiles further. It is clear, however, that at variance with earlier statements (Hagarman *et al.*, 2008), the CD and absorption spectra of the B-band of state IV do indicate band splitting.

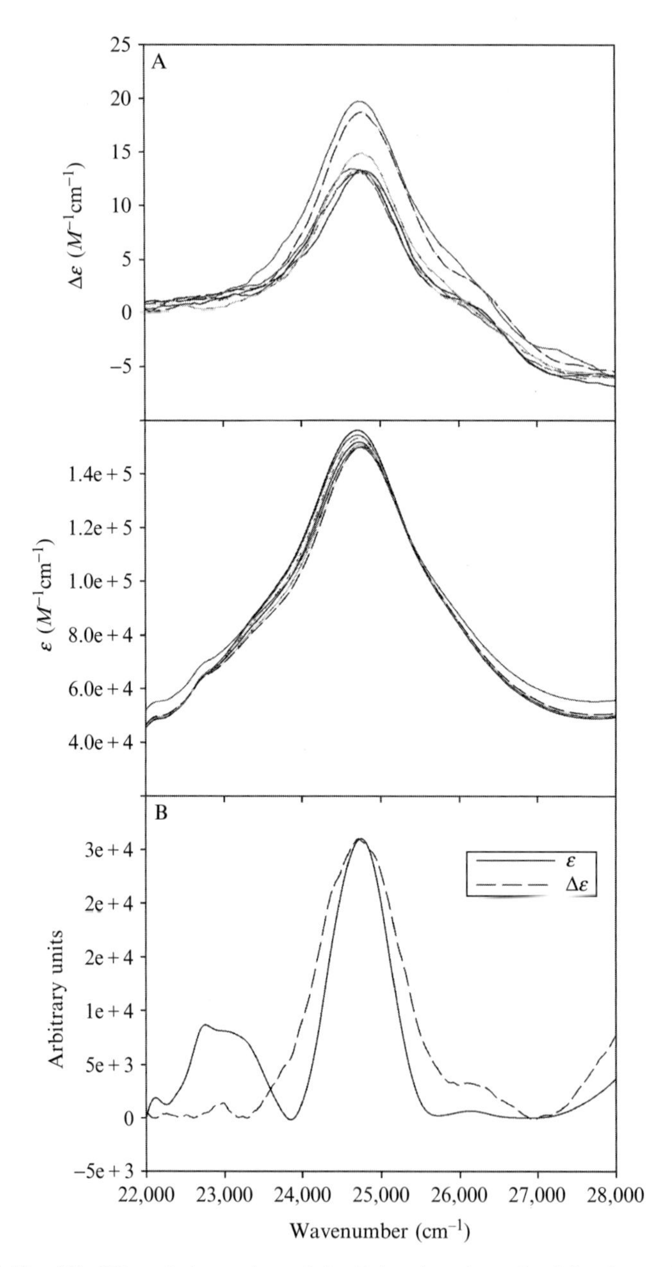

Figure 6.11 (A) CD and absorption of the B-band region of oxidized cytochrome *c* measured at pH 10 (state IV) at low ionic strength (<0.01 *M*) and room temperature. (B) Overlap of the CD and absorption spectrum for the comparison of the respective band profiles.

5. Nonnative States of Ferricytochrome *c* Probed by Visible CD and Absorption Spectroscopy

In contrast to the reduced state, the oxidized state of cytochrome *c* undergoes a series of transitions if the pH is changed between 1 and 12 as visualized by the reaction scheme I in Section 1. The native state III is generally thought to thermally unfold via the two-step mechanism

$$\mathrm{III} \overset{K_{\mathrm{III},1}}{\rightleftarrows} \mathrm{III_h} \overset{K_{\mathrm{III},2}}{\rightleftarrows} \mathrm{III_u} \qquad \text{(II)}$$

where $\mathrm{III_h}$ is a thermodynamic intermediate and $\mathrm{III_u}$ is the thermally unfolded state. Several lines of evidence seem to suggest that $\mathrm{III_h}$ should be identified with the alkaline state(s) IV, which becomes populated at high temperatures because the effective p*K*-value of the alkaline transition shifts down with increasing temperature (Battistuzzi *et al.*, 1999a; Taler *et al.*, 1995). In what follows, results are presented which lead to a different interpretation.

Figures 6.12 and 6.13 depict two different sets of visible CD and absorption spectra of the B-band region of bovine heart ferricytochrome *c* measured as a function of temperature. For the spectra in Fig. 6.12, the protein was dissolved in a 0.1-*M* Tris–HCl buffer, the pH of which was 7.0 at the lowest temperature (5 °C, 278 K). With rising temperature, the pH decreased to reach 6.3 at 333 K and 5.6 at 363 K. Hence, the system was moved out of the range of the above mentioned p*K*-shift of the alkaline transition with increasing temperature. Figure 6.13 shows the CD and absorption spectra of the protein dissolved in 1 m*M* MOPS buffer (3-(*N*-morpholino)propanesulfonic acid) buffer. The pH-value of this buffer (7.0 at 5 °C (278 K)) exhibits only minor changes as a function of temperature. The pH of the buffer was 7.0 at 5 °C (278 K). Apparently, the pH dependences of the CD spectrum are very similar for both buffers. As reported earlier (Hagarman *et al.*, 2008), the couplet disappears at high temperatures and is replaced by a positive Cotton band, which coincides with the absorption spectrum. Figures 6.14 and 6.15 depict the temperature dependences of the $\Delta\varepsilon$ values measured at the minimum and maximum positions (denoted $\Delta\varepsilon_{\mathrm{min}}$ and $\Delta\varepsilon_{\mathrm{max}}$) of the respective 278 K spectra. In both cases $\Delta\varepsilon_{\mathrm{min}}$ increases slightly with temperature between 278 and 343 K and steeply increases toward higher temperatures. $\Delta\varepsilon_{\mathrm{max}}$, however, first decreases with rising temperature and starts to increase (steeply) only at 343 K. Thus, these graphs suggest the population of an intermediate state around 338 K (65 °C) for both buffers. For the protein dissolved in Tris–HCl buffer, this intermediate cannot be identified with the alkaline state IV for reasons outlined above.

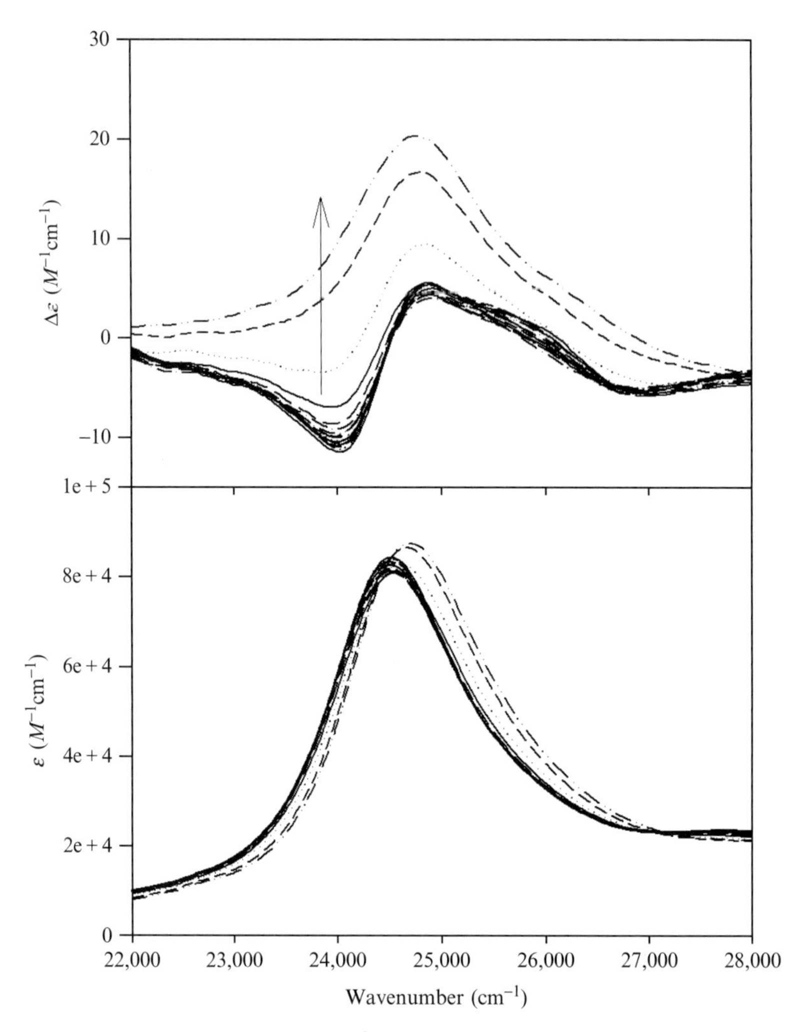

Figure 6.12 CD and absorption spectra of the B-band region of oxidized bovine heart cytochrome *c* dissolved in a 0.1-*M* Tris–HCl buffer (pH 7 at room temperature) measured as a function of temperature between 5 and 90 °C. Arrows indicate changes in temperature (data from Hagarman *et al.*, 2008).

The population of an intermediate state is further corroborated by the temperature dependence of the $\Delta\varepsilon$ (44964 cm^{-1}) value obtained from the far UV-CD spectrum of bovine heart cytochrome *c*, also shown in Fig. 6.14. For the former, the transition from the native to the intermediate state causes a decrease of its value followed by a steep increase above 340 K. The UV-CD values increases towards a saddle point between 330 and 340 K, from where it increases further.

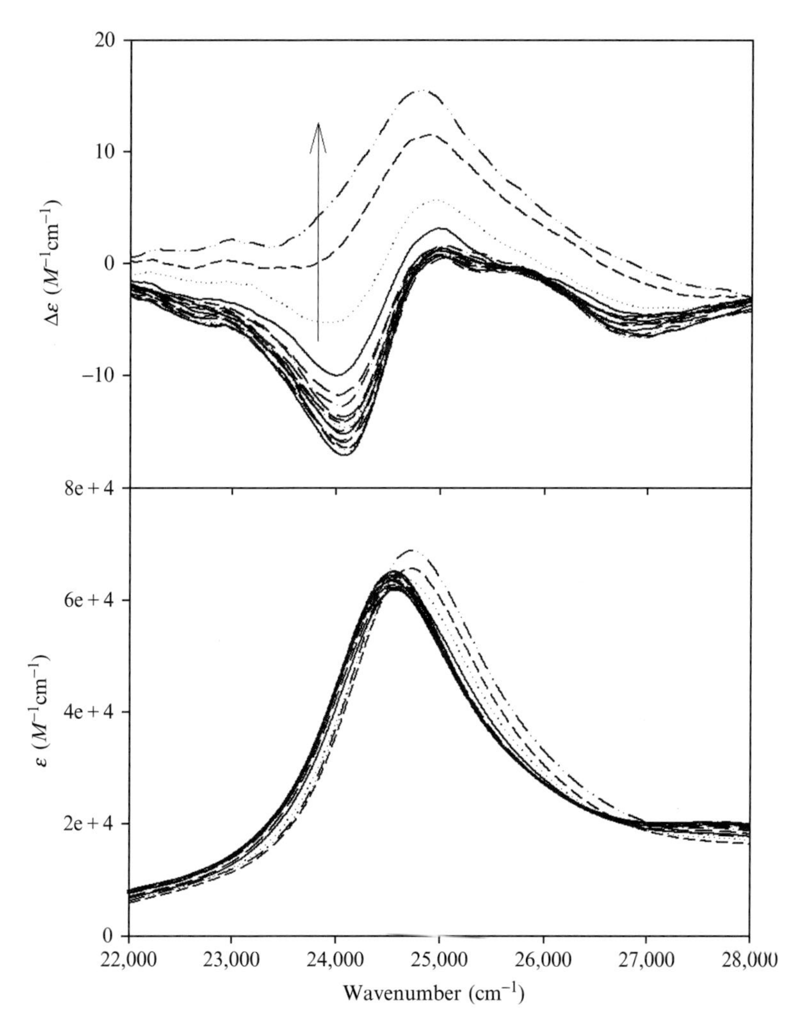

Figure 6.13 CD and absorption spectra of the B-band region of oxidized bovine heart cytochrome *c* dissolved in a 1-m*M* MOPS buffer (pH 7 at room temperature) measured as a function of temperature between 5 and 90 °C. Arrows indicate changes in temperature (data from Hagarman *et al.*, 2008).

The existence of state III_h had been postulated based on the finding that the 695 nm charge transfer band of ferricytochrome *c* is eliminated at higher temperatures before the onset of protein unfolding (Schejter and George, 1964). This band is generally considered as a marker of the integrity of the Fe-M80 linkage. The alkaline transition eliminates this band as well which reflects the replacement of the M80 by either K73 or K79 (Döpner *et al.*, 1998; Rossel *et al.*, 1998). This transition maintains a low spin state of the

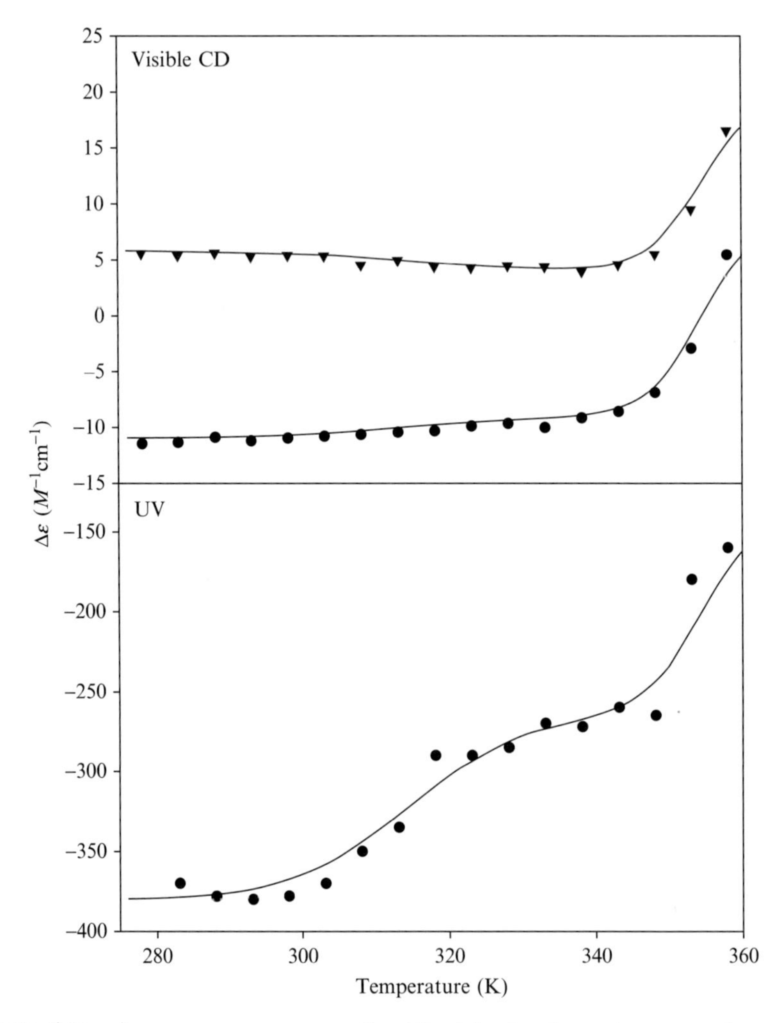

Figure 6.14 Δε versus temperature of oxidized bovine horse heart cytochrome *c* between 278 and 363 K. Upper panel: Δε obtained from the CD spectra in Fig. 6.12 at 24876 cm^{-1} (triangles) and 24,010 cm^{-1} (filled circles). Lower panel: Δε obtained from the corresponding ECD spectra at 44964 cm^{-1}. The experimental data were taken from Hagarman *et al.* (2008). The solid lines result from fits described in the text.

heme iron, as does the III → III_h transition. These similarities and a NMR study by Navon and coworkers led to the conclusion that III_h and IV are identical (Taler *et al.*, 1995). Now the question arises whether the intermediate obtained from our CD data is identical with the III_h state probed by the disappearance of the 695 nm band. To answer this question, the 695 nm band (absorption and CD) of bovine heart ferricytochrome *c* dissolved in

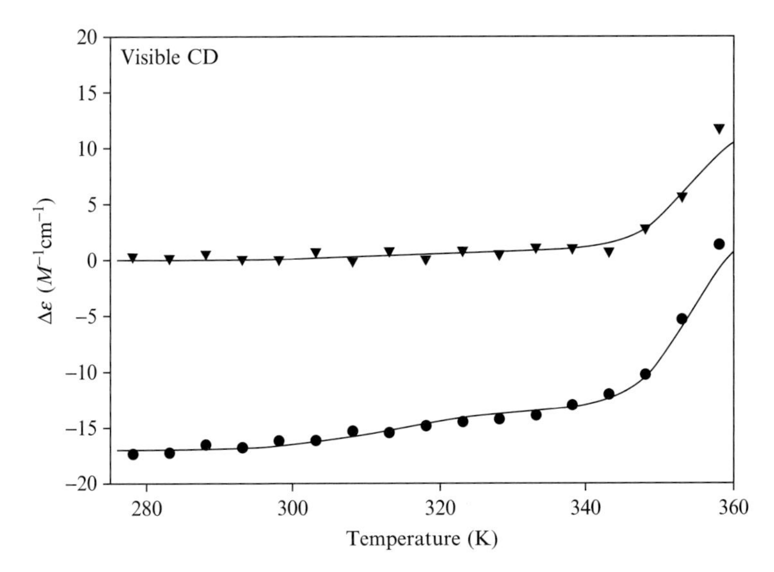

Figure 6.15 Δε versus temperature of oxidized bovine horse heart cytochrome *c* between 278 and 363 K. The Δε values obtained from the CD spectra in Fig. 6.13 at 24,876 cm^{-1} (triangles) and 24,010 cm^{-1} (filled circles). The experimental data were taken from Hagarman *et al.* (2008). The solid lines result from fits described in the text.

0.1 *M* Tris–HCl buffer was also measured. As shown in Fig. 6.16, the band's intensity decreases with increasing temperature, as expected. This seems to indicate that state III_h does not exhibit a 695-nm band, as stated by Hagarman *et al.* (2008). However, as we will argue below, a more thorough thermodynamic analysis leads to a different conclusion.

Thus far, this chapter demonstrated how the CD spectrum of the B-state transition of heme proteins could be used to compare different conformations of the protein in solution. Most of the results described in the section have been recently published (Hagarman *et al.*, 2008). In what follows, a simple thermodynamic model is presented by which they can be consistently analyzed.

Based on reaction scheme II the following formalism can be used to describe the temperature dependence of Δε at neutral pH:

$$\Delta\varepsilon = \frac{\Delta\varepsilon_0 + \Delta\varepsilon_h e^{-G^h/RT} + \Delta\varepsilon_u e^{-G^u/RT}}{Z} \tag{6.13}$$

where G^h and G^u are the Gibbs energies of the states III_h and III_u relative to the folded state III, R is the gas constant and T the absolute temperature. Z is the partition sum, which is written as

$$Z = 1 + e^{-G^h/RT} + e^{-G^u/RT} \tag{6.14}$$

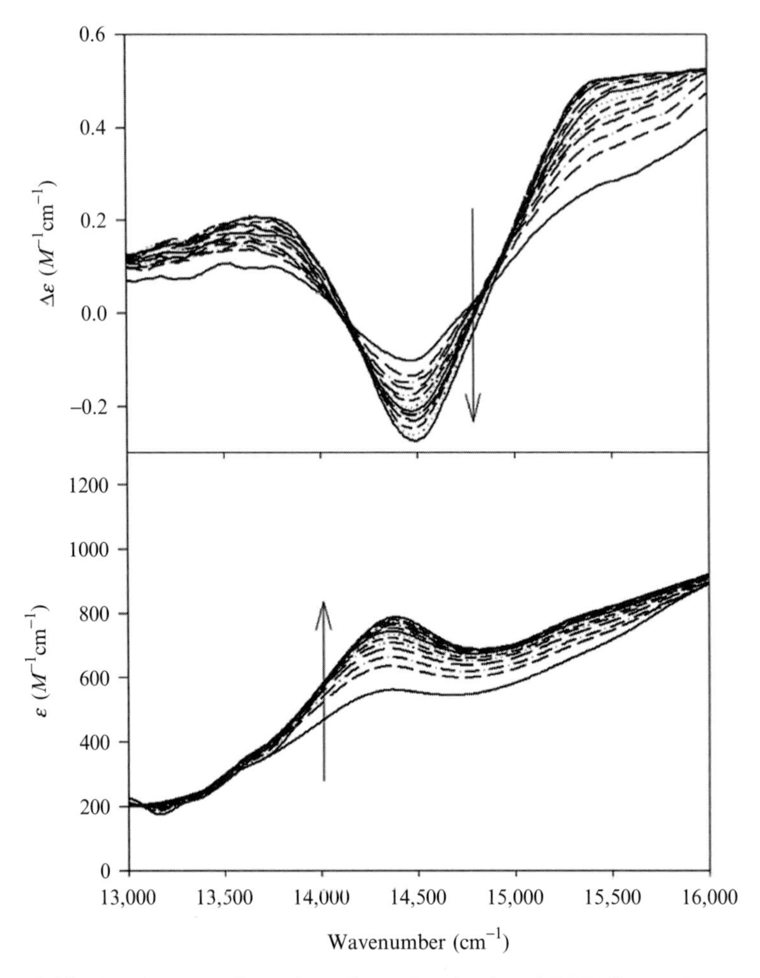

Figure 6.16 Temperature dependent absorption (top) and ECD (bottom) spectra of bovine heart ferricytochrome *c* measured in the charge transfer band region from 13,000 to 16,000 cm^{-1}. The protein was dissolved in a pH 7 (room temperature) Tris–HCl buffer with a 5-m*M* concentration. Arrows indicate changes in temperature.

We consider the III ⇔ IV transition as noncooperative, so that

$$G^{h} = H^{h} - TS^{h} \tag{6.15}$$

with temperature independent enthalpy and entropy. In our fit to the data, the transition temperature:

$$T_{h} = \frac{H^{h}}{S^{h}} \tag{6.16}$$

is used as free parameter. However, this approach does not work for the $III_h \Leftrightarrow III_u$ transition which indicates some cooperativity of the underlying unfolding/folding process in that a fit with a noncooperative approach as utilized for $III \Leftrightarrow III_h$ does not reproduce the slope of the $\Delta\varepsilon(T)$ graphs above 340 K. Cooperativity could be empirically accounted for by employing a Hill function (i.e., substituting $e^{-G^u/RT}$ by $e^{-nG^u/RT}$ in Eqs. (6.13) and (6.14), n: empirical Hill coefficient). To avoid the use of physically not very well-defined parameters, however, it is more appropriate to use the approach of Uchiyama *et al.* (2002), who considered a temperature dependent enthalpy, which in first order can be accounted for by

$$H^u = H_0^u + \delta c_p (T - T_u) \tag{6.17}$$

where δc_p is the heat capacity difference between the two states at constant pressure and T_u is the transition temperature for the $III_h \Leftrightarrow III_u$ transition. Then, we can write the Gibbs energy as

$$G^u = H_0^u + \delta c_p (T - T_u) + T\left(S^u(T_u) + \delta c_p \ln\left(\frac{T}{T_u}\right)\right) \tag{6.18}$$

The solid lines in Figs. 6.14 and 6.15 result from a consistent fit of the above formalism to the experimental data. Consistence here means that the very same thermodynamic parameters were used to fit $\Delta\varepsilon(T)$ and $\varepsilon(T)$ data obtained with the same buffer. To minimize the ambiguity, we took ΔH and δc_p values from the calorimetric studies of Uchiyama *et al.* for the $III_h \Leftrightarrow III_u$ transition and modified them only slightly for the purpose of optimization. The thermodynamic parameters used for the fits are listed in Table 6.2.

To subject the 695 nm band in Fig. 6.16 to the same thermodynamic analysis, it has to be decomposed into its (three) subbands termed S2, S3, and S4.[4] This is not an easy task because the band lies on the low energy wing of the much more intense Q-band. The band shapes of the 695 nm band were fitted by superimposing three Gaussian band profiles and a quadratic polynominal baseline with our spectral decomposition program, MULTIFIT (Jentzen *et al.*, 1995). A consistent fitting of the band profiles measured at different temperatures is complicated by the fact that the band positions and halfwidths can be temperature dependent (Cupane *et al.*, 1995; Gilch *et al.*, 1996). To minimize ambiguities, the profiles measured at the lowest and highest temperatures were first analyzed. If band positions and halfwidths of subbands were found to be different, the respective values at the remaining temperatures were interpolated and used as fixed values in the fitting. Figure 6.17 shows the spectral decomposition of the profile

[4] Bands named as S1 and S5 are flanking the 695 nm band. While they were considered in the band analysis they are not assignable to the same transition as the 695 nm band.

Table 6.2 Thermodynamic parameters obtained from fitting the $\Delta\varepsilon_{max}(T)$ and $\Delta\varepsilon_{min}(T)$ graphs in Fig. 6.14 (bovine ferricytochrome *c*, MOPs buffer, pH 7), Fig. 6.15 (Tris–HCl buffer, pH 7 at room temperature), and Fig. 6.21 (Bis-Tris–HCl buffer, pH 10.5 and 11.5)

	pH 7 MOPs	pH 7 Tris–HCl	pH 10.5	pH 11.5
H^{h} (kJ/mol)	100	100	100	200
H^{u} (kJ/mol)	300	250	200	200
T_{h} (K)	315	315	320	330
T_{u} (K)	355	355	350	355
c_{p} (kJ/mol K)	−2.8	−2.9	−2.8	−2.8

measured at 293 K, as an example. Table 6.3 provides a list of spectral parameters, which reveal that only the wavenumber positions were found to be (linearly) dependent on the temperature. Figure 6.18 depicts the integrated intensities of the three subbands as a function of temperature. The intensity dependence of the subbands is clearly different, which indicates that they are assignable to different conformational substates with slightly different optical and/or thermodynamical properties (Schweitzer-Stenner *et al.*, 2007b).

In the study of Schweitzer-Stenner *et al.*, the temperature dependence of the 695 nm band was analyzed by fitting the temperature dependence of the extinction coefficients measured at three different wavenumber positions of the band profile to a model that additionally considered the population of another state termed III_{l} at temperatures below 290 K (Schweitzer-Stenner *et al.*, 2007b). However, this effect much more pronounced at more acidic pH values (6.0) and can be neglected at the more neutral pH values (at these temperatures) used for our experiments. Hence, one would expect that it is sufficient to fit a simple two-state model reflecting the III $\Leftrightarrow$ III_{h} transition to the 695 nm data in Fig. 6.21. However, at least two of the $\varepsilon(T)$ curves in Fig. 6.16 seem to indicate a biphasic behavior. Thus, a consistent fitting of all three curves has to be based on the following reaction scheme:

$$\begin{array}{ccccc} III_{2} & \leftrightarrow & III_{2h} & \leftrightarrow & III_{2u} \\ b & & b & & b \\ III_{3} & \leftrightarrow & III_{3h} & \leftrightarrow & III_{3u} \\ b & & b & & b \\ III_{4} & \leftrightarrow & III_{4h} & \leftrightarrow & III_{4u} \end{array} \qquad \text{(III)}$$

where III_{j}, III_{jh}, and III_{ju} denote the native, intermediate, and unfolded state of the *j*th substate. The partition sum for this scheme reads as

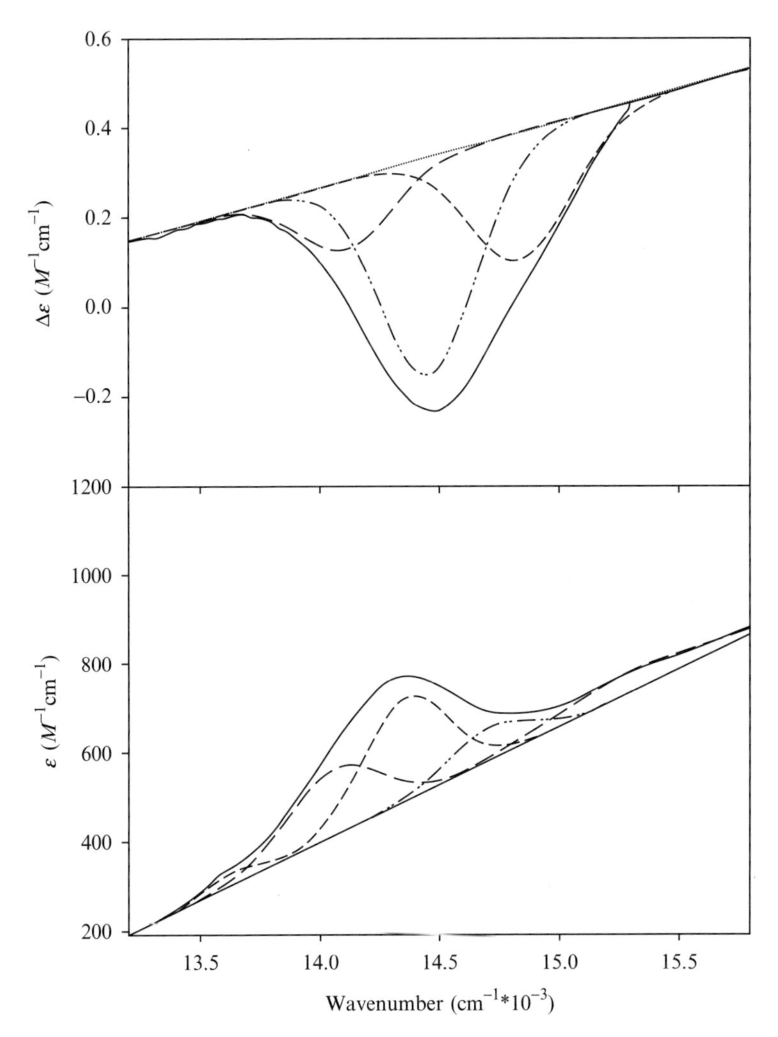

Figure 6.17 Decomposition of the absorption (bottom) and CD (top) band profile of the 695 nm transition of bovine ferricytochrome *c* into three subbands. The spectra were measured at pH 7 with a 0.1-*M* Tris–HCl buffer (data from Verbaro *et al.*, 2009).

$$Z = \sum_{j=2}^{4} e^{-G_j^0/RT}\left[1 + e^{-G_j^h/RT}\left(1 + e^{-G_j^u/RT}\right)\right] \tag{6.19}$$

where G_j^0, G_j^h, and G_j^u are the Gibbs energies of *j*th substates for the indicated thermodynamic states. The Gibbs energy of the III_2 can be chosen as reference point, so that $G_2^0 = H_2^0 = S_2^0 = 0$. The $\varepsilon_j(T)$ graph for each substate is thus described by:

Table 6.3 Wavenumber positions of the subbands obtained from the spectral decomposition of the 695 nm band of bovine ferricytochrome *c* at pH 7

	$\tilde{\nu}$ (cm^{-1} × 10^{-3})	
	278 K	363 K
S2	14.1	13.9
S3	14.4	14.2
S4	14.77	14.49

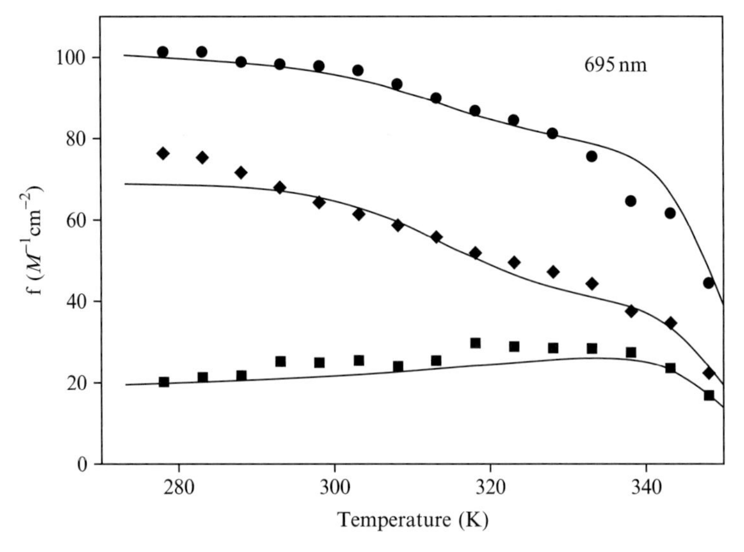

Figure 6.18 Oscillator strengths of the three subbands of the 695 nm absorption band of bovine heart ferricytochrome *c* as function temperature. Filled circles, S3; triangles, S4; open circles, S2. The solid lines represent a fit described in the text.

$$\varepsilon_j(T) = \frac{\sum_{j=2}^{4}\left[\varepsilon_j^0 e^{-G_j^0/RT} + \varepsilon_j^h e^{-(G_j^0 - G_j^h)/RT}\right]}{Z} \qquad (6.20)$$

Since scheme (III) describes a thermodynamic cycle, changes of the thermodynamic parameters for a distinct subband j affect the calculated graphs for all subbands in a qualitative and quantitative way. The thermodynamic parameters obtained from the fits to the $\Delta\varepsilon(T)$ graphs in Fig. 6.14 were used. Only the enthalpies H_3^0 and H_4^0 and the extinction coefficients had to be used as free parameters. The entropic parts of G_3^0 and G_4^0 were not explicitly considered, since they cannot be sufficiently discriminated from the extinction ratios $\varepsilon_4^0/\varepsilon_2^0$ and $\varepsilon_3^0/\varepsilon_2^0$. Hence, the values obtained for the latter must be considered as reflecting both, differences between the oscillator strengths and the entropies of the respective substates. Fitting Eq. (6.20) to the datasets in Fig. 6.18 yielded quite acceptable results, as visualized by the solid lines. A minor deviation from the S3 data at low temperature most likely reflects the population of the aforementioned low temperature state, which has not been considered. Altogether, it can be concluded that the temperature dependences of the B-band (CD and absorption) and of the 695 nm absorption can all be fitted by the same model with a consistent set of parameters.

The model, which emerges from the current analysis, is somewhat different from what Schweitzer-Stenner *et al.* recently suggested based on an analysis of the temperature dependence of the 695 nm band of horse heart ferricytochrome *c* in phosphate buffer (Schweitzer-Stenner *et al.*, 2007b). These authors based their analysis on the temperature dependence of three ε values of the band profile rather than on integrated intensities. They came to the conclusion that state III_h does not exhibit a 695 nm band, which is in agreement with earlier studies particularly by Schejter and coworkers. Here, the combined analysis of B-band absorption and CD as well as of the 695 nm band subbands of bovine ferricytochrome *c* reveal that state III_h is still very similar to state III in that its Fe-M80 bond is still present (though weakened) and its B-band CD still shows a pronounced couplet. For our data, the decay of the 695 nm band can be clearly correlated with the population of III_u. Performing the experiments with bovine rather than with horse heart cytochrome *c* had the advantage that the former is less prone to aggregation than the latter so that the 695 nm band could be measured at higher temperatures.

Further evidence for the notion that III_h and IV are thermodynamically and structurally distinct is provided by a comparison of the respective CD spectra. Figure 6.19 exhibits the Soret band CD spectrum of bovine heart ferricytochrome *c* dissolved in a pH 10.5 solution (0.1 *M* Bis/Tris buffer) as a function of temperature. The room temperature spectrum exhibits a positive Cotton band, which is clearly different from the rather symmetric couplet obtained for the Tris–HCl buffer solution at $T = 338$ K, which mostly represents state III_h. The latter indicates that the heme environment of the III_h state is very similar to that of the native state, whereas major changes occur upon the III → IV alkaline transition. If one subjects the state III_h to the same type of vibronic analysis as the spectrum of the native state as described in the preceding chapter, one obtains that III and III_h differ in terms of the respective rotational strength of B_x and B_y, but exhibit practically the same band splitting (Hagarman *et al.*, 2008).

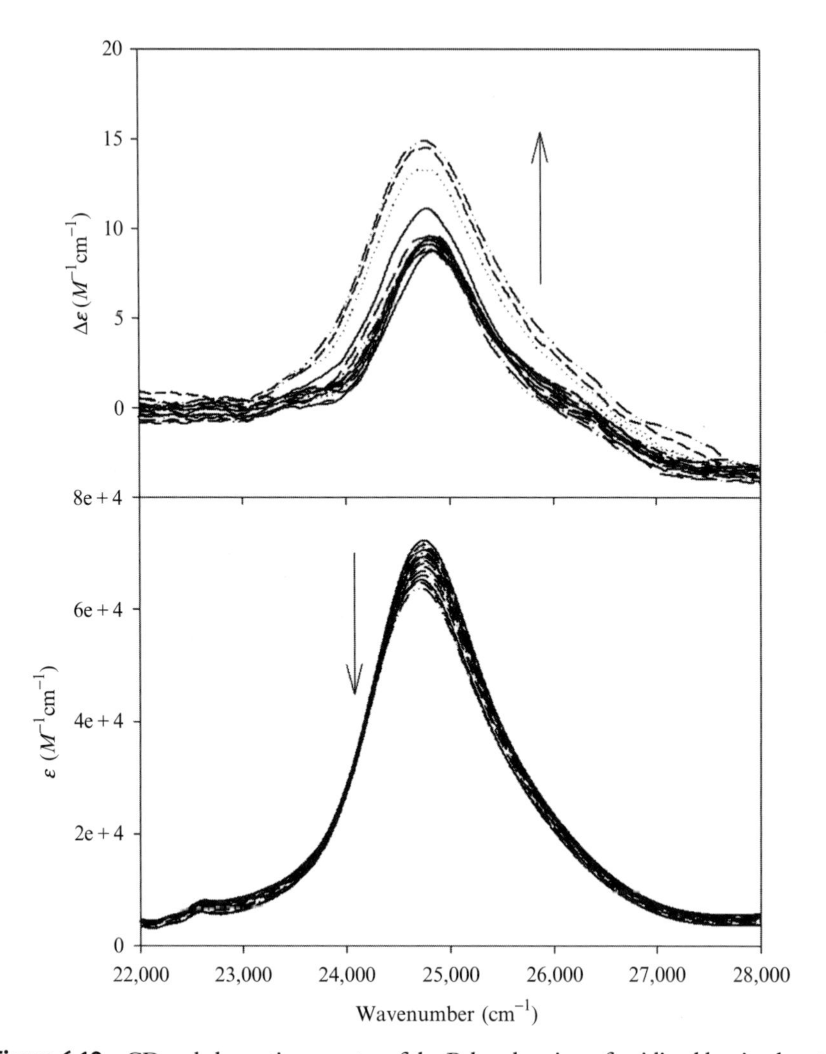

Figure 6.19 CD and absorption spectra of the B-band region of oxidized bovine heart cytochrome *c* dissolved in a 0.1-*M* Bis–Tris–HCl buffer (pH 10.5 at room temperature) measured as a function of temperature between 5 and 90 °C. Arrows indicate changes in temperature (data from Hagarman *et al.*, 2008).

For the sake of completeness, Fig. 6.20 also exhibits the visible CD and absorption spectrum of the B-band region of ferricytochrome *c* at pH 11.5. The work of Döpner *et al.* (1998) suggests a substantial fraction of state V to be populated at this pH. At room temperature, the CD spectrum looks very much like the spectrum measured at pH 10.5 (state IV), but this changes dramatically above 320 K, when the $\Delta\varepsilon_{max}$—value measured at pH 11.5

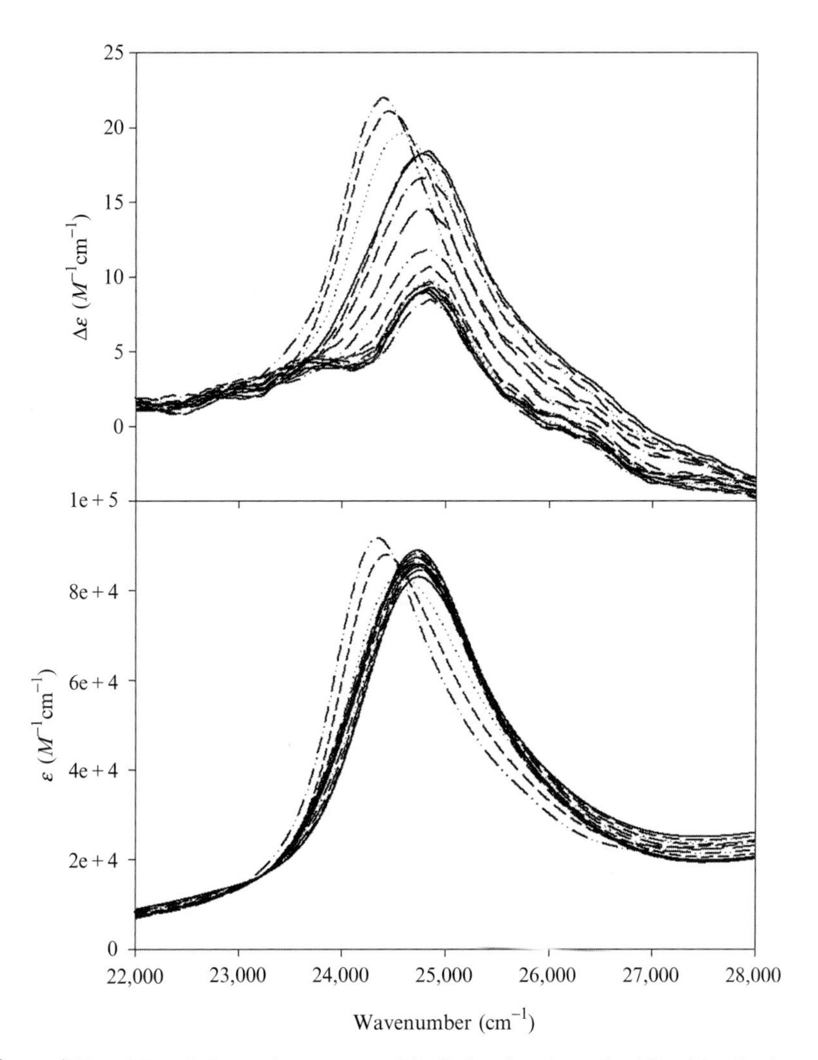

Figure 6.20 CD and absorption spectra of the B-band region of oxidized bovine heart cytochrome *c* dissolved in a 0.1-*M* Bis–Tris–HCl buffer (pH 11.5 at room temperature) measured as a function of temperature between 5 and 90 °C (data from Hagarman *et al.*, 2008).

exhibits a much more pronounced increase with rising temperature. The subsequent decrease above 350 K reflects the significant downshift of the CD band. The absorption behaves concomitantly. Figure 6.21 exhibits the temperature dependence of $\Delta\varepsilon_{max}$ for both the spectra recorded at pH 10.5 and 11.5.

The $\Delta\varepsilon(T)$ graphs measured at pH 10.5 and 11.5 were also subjected to fits based on reaction scheme II, which, for the pH 10.5 data, led to the

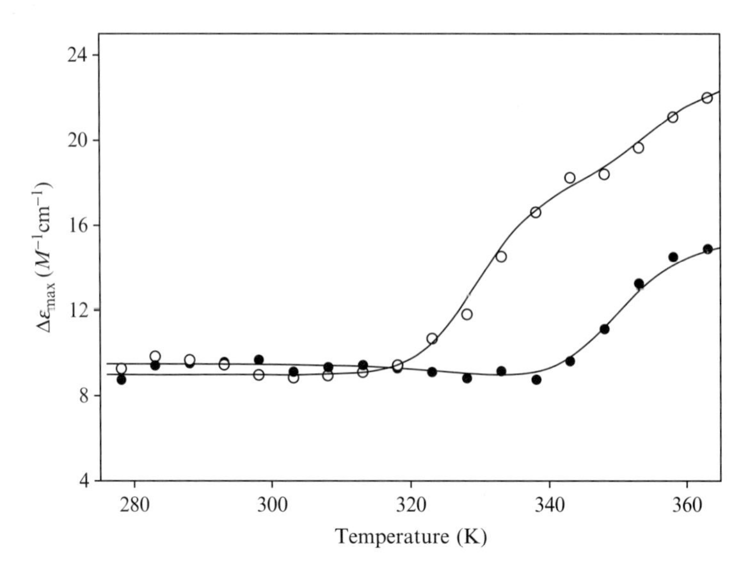

Figure 6.21 $\Delta\varepsilon$ versus temperature of oxidized bovine horse heart cytochrome *c* between 278 and 363 K. The $\Delta\varepsilon$ values obtained from the CD spectra in Fig. 6.18 (state IV, filled circles) and 19 (state V, open circles) at the position of the Cotton band maximum. The experimental data were taken from Hagarman *et al.* (2008). The solid lines result from fits described in the text.

respective solid line in Fig. 6.21. The population of an intermediate is difficult to discern from the data, though its existence becomes more obvious if one inspects the temperature dependence of the Kuhn anisotropy. The Cotton band in the unfolded state IV_u is by a factor 3/4 lower less intense than the CD band of the III_u state populated at neutral pH and high temperatures. This difference seems to make sense in that we expect IV_u to be even more unstructured than III_u, which would lead to a less ordered heme environment.

The attempt to fit reaction scheme II to the data acquired at pH 11.5 had only limited success in that the rather steep increase of $\Delta\varepsilon(T)$ at 320 K could not be reproduced. This steepness indicates that the observed changes at this temperature are already reflecting the transition into an unfolded state and that the slight variation of $\Delta\varepsilon(T)$ could be interpreted as indicating the occupation of an intermediate. The reaction scheme was therefore extended to

$$V \rightleftarrows V_h \rightleftarrows V_{u1} \rightleftarrows V_{u2} \qquad \text{(IV)}$$

For the transition into V_{u1}, the same cooperative model was utilized to model the transitions $III_h \Leftrightarrow III_u$ and $IV_{u1} \Leftrightarrow IV_{u2}$. This model yielded an excellent fit to the experimental data (Fig. 6.21). Interestingly, the $\Delta\varepsilon_{max}$ value for V_{u1} is similar to that obtained for IV_u and $V_{u.}$ This similarity suggests that, for example, III_u might actually not be the final stat of the unfolding process and that the transition into III_u might be followed by another transition into the final unfolded state at higher temperatures. Generally, this state is not accessible to the optical measurements used for the current study, owing to the onset of aggregation.

Most of the above data used for the thermodynamic analysis have been obtained for proteins dissolved in a solution with intermediate to high ionic strength (buffer concentration $> 0.05\ M$). Normally, not much attention is paid to the ionic composition in spectroscopic studies, even though the latter is known to affect the equilibrium constants of the alkaline transitions and the redox potential of the protein (Battistuzzi *et al.*, 1999a,b, 2001). Recently, Shah and Schweitzer-Stenner showed that an increase of HPO_4^{2-} ion concentration causes a substantial increase of the 695 nm band absorptivity, which is indicative of a strengthening of the Fe-M80 bond (Shah and Schweitzer-Stenner, 2008). This result demonstrates that anion binding to the positive patch of the protein affects the heme-protein linkage. In what follows, the influence of phosphate ions on the energy landscape of the alkaline transitions is analyzed in more detail.

The 695 nm band has been used at a very early stage to probe the III $\Leftrightarrow$ IV transition (Davis *et al.*, 1973). The respective data, which were measured at comparatively high ionic strength, clearly suggest a monophasic titration, which can be rationalized with a Henderson–Hasselbach equation and a single, effective p*K*-value. The latter itself reflects two consecutive processes, that is, the protonation of a basic residue (most likely a lysine side chain) and a subsequent conformation transition. However, Wallace and coworkers showed by a very meticulous analysis of the alkaline titration of horse heart cytochrome *c*, that, in fact, at least two different protonation sites are involved (Blouin *et al.*, 2001). Recently, Verbaro *et al.* investigated the 695 nm band titration of horse heart ferricytochrome *c* at high (0.05 *M*) and low phosphate ion concentration (1 m*M*) (Verbaro *et al.*, 2009). Their astonishing results, which are briefly described in the following, reveal again the usability of CD and absorption spectroscopy for the thermodynamic analysis of heme proteins.

Figure 6.22 exhibits the integrated intensities of two 695 nm subband of horse heart ferricytochrome *c* as a function of pH between 7.0 and 10.0 for the above mentioned phosphate ion concentrations. The titration curves of the "high ionic strength" sample are clearly monophasic, as expected, and can be fitted by assuming a single protonation step, the p*K*-values of which are listed in Table 6.4. The titration curves measured at low anion concentrations, however, are all biphasic, indicating the involvement of two protonatable sites. One is tempted to assume that the two p*K*-values should be assigned to the two structural isomers of state IV. However, as shown by Verbaro *et al.*, this assignment does not explain a biphasic titration if one

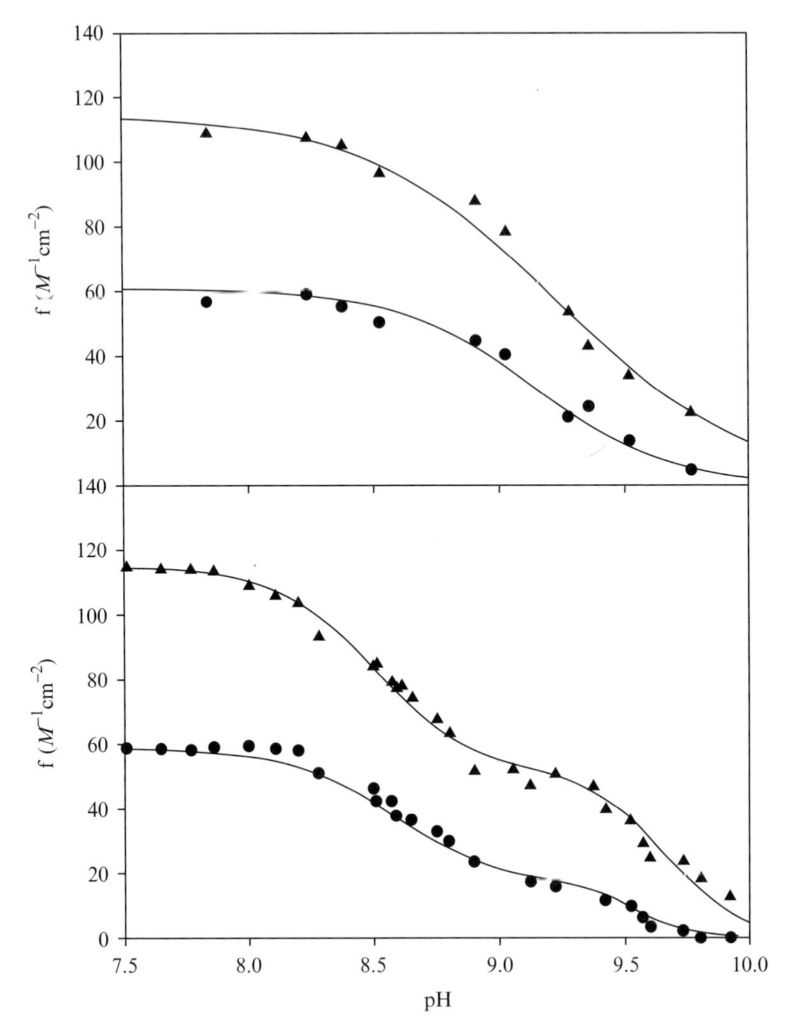

Figure 6.22 Oscillator strengths of the subbands (filled circles: S3 and filled triangles: S4) of the 695 nm absorption band as function of pH between 7 and 10. The respective spectra were obtained for the protein dissolved in 50 mM (top panel) and 0.5 m*M* (lower panel) phosphate buffer. The solid lines represent fits described in the text. Taken from Verbaro *et al.* (2009) and modified.

assumes that the two isomers are in thermodynamic equilibrium (Verbaro *et al.*, 2009). The authors instead invoked a model, which can be visualized by the following reaction scheme:

Table 6.4 p*K*-values and Hill coefficients obtained from the fits to the titration curves of 695 nm subbands plotted in Fig. 6.22 (taken from Verbaro *et al.*, 2009)

	High ionic strength		Low ionic strength			
	pK	n	pK_1	n_1	pK_2	n_2
S3	9.2	1.14	9.6	2.2	8.5	2.86
S4	9.1	1.6	9.5	2.0	8.6	4.0

$$\begin{array}{ccccc} & & K_2' & & \\ & III & \leftrightarrow & III_1^* & \\ K_1 & \updownarrow & & \updownarrow & K_1' \\ & III_2^* & \leftrightarrow & IV & \\ & & K_2 & & \end{array} \tag{V}$$

where $\mathrm{III}_1^\star$ and $\mathrm{III}_2^\star$ denote two thermodynamic intermediate states which are populated by the protonation of two different sites. If the difference between the pK-values pK_1 and pK_2 is large, only one of these states will be substantially populated and V becomes essentially a two-step mechanism. Based on V, the titration curves for the oscillator strength f_j(pH) of the jth subband can be described by

$$f_j(\mathrm{pH}) = \left\{ \frac{f_j^{\mathrm{III}}}{1 + \left(\frac{K_1}{[\mathrm{H_3O^+}]}\right)^{n_1} + \left(\frac{K_2}{[\mathrm{H_3O^+}]}\right)^{n_2} + \frac{K_1^{n_1} K_2^{n_2}}{[\mathrm{H_3O^+}]^{n_1+n_2}}} \right.$$
$$+ f_j^{\mathrm{III}^\star} \frac{1}{1 + \left(\frac{K_1}{[\mathrm{H_3O^+}]}\right)^{n_1} + \left(\frac{[\mathrm{H_3O^+}]^2}{K_2}\right)^{n_2} + \frac{K_1^{n_1}}{K_2^{n_2}}} \tag{6.21}$$
$$\left. + \frac{1}{1 + \left(\frac{K_2}{[\mathrm{H_3O^+}]}\right)^{n_2} + \left(\frac{[\mathrm{H_3O^+}]^2}{K_1}\right)^{n_1} \frac{K_2^{n_2}}{K_1^{n_1}}} \right\}$$

Equation (6.21) is based on the assumption that the two protonation steps are independent of each other, which implies that $K_1 = K_1'$ and $K_1 = K_2'$, n_1 and n_2 are Hill coefficients which describe the cooperativity of the involved transitions, f_j^{III} and $f_j^{\mathrm{III}\star}$ are the intrinsic oscillator strengths of the states III and III★ ($\mathrm{III}_1^\star$ and $\mathrm{III}_2^\star$ cannot be discriminated by the titration and are assumed to exhibit the same oscillator strength). The titration curves for the rotational strengths can be described by the same formalism. The solid lines in Fig. 6.22 reflect the result of the fitting, which were achieved by identical fitting parameters for corresponding f(pH) and R(pH) graphs. All parameters are listed in Table 6.4.

The existence of a thermodynamic intermediate of the III $\rightarrow$ IV transition of horse heart cytochrome *c* has recently been reported by Weinkam *et al.* (2008) These authors used the selected deuteration of side chains in combination with FTIR spectroscopy to probe structural changes involved in the alkaline transition. The experiments were carried out at high phosphate (50 m*M*) and sodium chloride (200 m*M*) concentration, so that their results are not directly comparable with those of Verbaro *et al.* (2009). The authors' results led them to subdivide the alkaline transition into two steps with midpoints at 8.8 and 10.2. They attributed the former to a transition from state III into an intermediate 3.5, whereas the second step was assigned to the formation of state(s) IV. The III $\rightarrow$ 3.5 transition involves only conformational changes of the M80 ligand, but not any of the lysine ligands. There are several differences between their results and those of Verbaro *et al.* Weinkam *et al.* claimed that the III $\rightarrow$ 3.5 transition leads to a complete disappearance of the 695 nm band, which can therefore not be interpreted as reflecting a loss of the Fe-M80 coordination, which they considered as still existing in 3.5. The intermediate state III* inferred from Fig. 6.16 still exhibits considerable 695 nm band intensity and rotational strength. Moreover, one has to consider the finding of Blouin *et al.*, who showed that the monophasic titration curve of the 695 nm band observed at high phosphate ion concentration can still be decomposed into two protonation steps, which we could now clearly resolve at low anion concentration (Blouin *et al.*, 2001). This result indicates that the state III_1^* also becomes populated with high anion concentration, though in a very small pH range and to a much lesser extent than observed at low anion concentration (Verbaro *et al.*, 2009). One might therefore speculate that the state 3.5 reported by Weinkam *et al.* could be identified with state III_2^*, which can become populated at high anion concentrations, because of the reduced gap between pK_1 and pK_2.

6. Summary and Outlook

In this chapter, we presented three applications of circular dichroism spectroscopy for the investigation of heme proteins. While we focused on cytochrome *c*, similar investigations can be carried with other heme proteins (Schweitzer-Stenner *et al.*, 2007a). The first application is synchotron radiation circular dichroism which provides a broader range of spectroscopic information than conventional far UV-CD spectroscopy for the determination of the secondary structure of proteins. Here, SRCD spectroscopy was used to compare the secondary structure of native and nonnative states of cytochrome *c*. The second application exploited the fact that visible CD spectroscopy can be used to identify the splitting of the Soret band transition of the heme chromophore, which is predominantly caused by the internal electric field in the heme pocket. The third application utilized visible (and in part also conventional UV CD) spectroscopy to explore conformational changes of ferricytochrome *c* in solution. In this context spectroscopic information was combined with thermodynamic analyses. For the future, we envisaging an even more quantitative use of visible CD spectroscopy by combining CD measurements with other tools of structure analysis (NMR spectroscopy, MD simulation) which could provide us with structural models for, for example, nonnative states of cytochrome *c* which could then be used to calculate CD spectra.

REFERENCES

Andrade, M. A., Chacón, P., Merolo, J. J., and Morán, F. (1993). Evaluation of secondary structure of proteins from UV circular dichroism spectra using an unsupervised learning neural network. *Protein Eng.* **6,** 383–390.

Assfalg, M., Bertini, I., Dolfi, A., Turano, P., Mauk, A. G., Rossel, F. I., and Gray, H. B. (2003). Structural model for an alkaline Form of ferricytochrome *c*. *J. Am. Chem. Soc.* **125,** 2913–2922.

Banci, L., Bertini, I., Redding, T., and Turano, P. (1998). Monitoring the conformational flexibility of cytochrome c at low ionic strength by H-NMR spectroscopy. *Eur. J. Biochem.* **1998,** 271–278.

Barker, P. D., and Mauk, A. G. (1992). pH-Linked conformational regulation of a metalloprotein oxidation-reduction equilibrium: Electrochemical analysis of the alkaline form of cytochrome c. *J. Am. Chem. Soc.* **114,** 3619–3624.

Battistuzzi, G., Borsari, M., Loschi, L., Martinelli, A., and Sola, M. (1999a). Thermodynamics of the alkaline transition of cytochrome c. *Biochemistry* **38,** 7900–7907.

Battistuzzi, G., Loschi, L., Borsari, M., and Sola, M. (1999b). Effects of nonspecific ion-protein interactions on the redox chemistry of cytochrome c. *JBIC.* **4,** 601–607.

Battistuzzi, G., Borsari, M., and Sola, M. (2001). Medium and temperature effects on the redox chemistry of cytochrome c. *Eur. J. Inorg. Chem.* **2001,** 2989–3004.

Belikova, N. A., Vladimirov, Y. A., Osipov, A. N., Kapralov, A. A., Tyurin, V. A., Potapovich, M. V., Basova, L. V., Peterson, J., Kurnikov, I. V., and Kagan, V. E. (2006). Peroxidase activity and structural transitions of cytochrome c bound to cardiolipin-containing membranes. *Biochemistry* **45,** 4998–5009.

Berghuis, A. M., and Brayer, G. D. (1992). Oxidation state-dependent conformational changes in cytochrome *c*. *J. Mol. Biol.* **223,** 959–976.

Black, K. M., and Wallace, C. J. A. (2007) *Biochem. Cell. Biol.* **85,** 366–374.

Blauer, G., Sreerama, N., and Woody, R. W. (1993a). Optical activity of hemoproteins in the Soret region—Circular dichroism of the heme undecapeptide of cytochrome c in aqueous solution. *Biochemistry* **32,** 6674–6679.

Blauer, G., Sreerama, N., and Woody, R. W. (1993b). Optical activity of hemoproteins in the soret region. Circular dichroism of the heme undecapeptide of cytochrome c in aqueous solution. *Biochemistry* **32,** 6674–6679.

Blouin, C., Guillemette, J. G., and Wallace, C. J. A. (2001). Resolving the individual components of a pH-induced conformational change. *Biophys. J.* **81,** 2331–2338.

Böhm, G., Muhr, R., and Jaenicke, R. (1992). *Protein Eng.* **5,** 191–195.

Bushnell, G. W., Louie, G. V., and Brayer, G. D. (1990). High-resolution three-dimensional structure of horse heart cytochrome c. *J. Mol. Biol.* **1990,** 214.

Churg, A. K., and Warshel, A. (1986). Control of the redox potential of cytochrome c and microscopic dielectric effects in proteins. *Biochemistry* **25,** 1675–1681.

Cohen, D. S., and Pielak, G. J. (1994). Stability of yeast iso-1 ferricytochrome c as a function of pH and temperature. *Protein Sci.* **3,** 1253–1280.

Cupane, A., Leone, M., Vitrano, E., and Cordone, L. (1995). Low temperature optical absorption spectroscopy: An approach to the study of stereodynamic properties of hemeproteins. *Eur. Biophys. J.* **23,** 385–398.

Davis, L. A., Schejter, A., and Hess, G. P. (1973). Alkaline isomerization of oxidized cytochrome c. *J. Biol. Chem.* **249,** 2624–2632.

Döpner, S., Hildebrandt, P., Rosell, F. I., and Mauk, A. G. (1998). The alkaline conformational transitions of ferricytochrome c studied by resonance Raman spectroscopy. *J. Am. Chem. Soc.* **120,** 11246–11255.

Döpner, S., Hildebrandt, P., Rosell, F. I., Mauk, A. G., von Walter, M., Soulimane, T., and Buse, G. (1999). The structural and functional role of lysine residues in the binding domain of cytochrome c for the redox process with cytochrome c oxidase. *Eur. J. Biochem.* **261,** 379–391.

Dragomir, I., Hagarman, A., Wallace, C., and Schweitzer-Stenner, R. (2007). Optical band splitting and electronic perturbations of the heme chromophore in cytochrome c at room temperature probed by visible electronic circular dichroism spectroscopy. *Biophys. J.* **92,** 989–998.

Dyson, H. J., and Beattie, J. K. (1982). Spin state and unfolding equilibria of ferricytochrome c in acidic solution. *J. Biol. Chem.* **257,** 2267–2273.

Eaton, W. A., and Hochstrasser, R. M. (1968). Singe crystal spectra of ferrimyoglobin in polarized light. *J. Chem. Phys.* **49,** 985–995.

Edman, E. T. (1979). A comparison of the structures of electron transfer proteins. *Biochim. Biophys. Acta* **549,** 107–144.

Englander, S. W., Sosnick, T. R., Mayne, L. C., Shtilerman, M., Qi, P. X., and Bai, Y. W. (1998). Fast and slow folding in cytochrome c. *Acc. Chem. Res.* **31,** 737–744.

Feng, Y., and Englander, S. W. (1990). Salt dependent structural change and ion binding in cytochrome c studied by two-dimensional proton NMR. *Biochemistry* **29,** 3505–3509.

Friedman, J. M., Rousseau, D. L., and Adar, F. (1977). Excited state lifetimes in cytochromes measured from Raman scattering data: Evidence for iron-porphyrin interactions. *Proc. Natl. Acad. Sci. USA* **74,** 2607–2611.

Garber, E. A. E., and Margoliash, E. (1994). Circular dichroism studies of the binding of mammalian and non-mammalian cytochromes c to cytochrome c oxidase, cytochrome c peroxidase, and polyanions. *Biochim. Biophys. Acta* **1187,** 289–295.

Geissinger, P., Kohler, B. E., and Woehl, J. C. (1997). Experimental determination of internal electric fields in ordered systems: Myoglobin and cytochrome c. *Synth. Met.* **84,** 937–938.

Gilch, H., Schweitzer-Stenner, R., Dreybrodt, W., Leone, M., Cupane, A., and Cordone, L. (1996). Conformational substates of the Fe^{2+}-His F8 linkage in deoxymyoglobin and hemoglobin probed in parallel by the Raman band of the Fe-His stretching vibration and the near-infrared absorption band III. *Int. J. Comp. Chem.* **59,** 301–313.

Gorbenko, G. P., Molotkovsky, J. G., and Kinnunen, P. K. J. (2006). Cytochrome c interaction with cardiolipin/phosphatidylcholine model membranes: Effect of cardiolipin protonation. *Biophys. J.* **90,** 4093.

Gouterman, M. (1959). Study of the effect of substitution on the absorption spectra of porphyrins. *J. Chem. Phys.* **30,** 1139.

Green, D. R., and Kroemer, G. (2004). The pathology of mitochondrial cell death. *Science* **305,** 626–629.

Greenfiled, N. J. (1996). Methods to estimate the conformtation of proteins and polypeptides from circular dichroism data. *Anal. Biochem.* **235,** 1–10.

Hagarman, A., Duitch, L., and Schweitzer-Stenner, R. (2008). The conformational manifold of ferricytochromec explored by visible and far-UV electronic circular dichroism spectroscopy. *Biochemistry* **47,** 9667–9677.

Heimburg, T., Hildebrandt, P., and Marsh, D. (1991). Cytochrome c-lipid interactions studied by resonance Raman and ^{31}P NMR spectroscopy. Correlation between the conformational change of the protein and the lipid bilayer. *Biochemistry* **30,** 9084–9089.

Heimburg, T., Angerstein, A., and Marsh, D. (1999). Binding of peripheral proteins to mixed lipid membranes: Effect of lipid demixing upon binding. *Biophys. J.* **76,** 2575–2586.

Hsu, M. C., and Woody, R. W. (1971). The origin of the heme cotton effects in myoglobin and hemoglobin. *J. Am. Chem. Soc.* **93,** 3515–3525.

Humphrey, W., Dalke, A., and Schulten, K. (1996). VMD—Visual molecular dynamics software. *J. Mol. Graph.* **14,** 33–38.

Indiani, C., de Sanctis, G., Neri, F., Santos, H., Smulevich, G., and Coletta, M. (2000). Effect of pH on axial ligand coordination of cytochrome c from *Methylophilus methylotrophus* and horse heart cytochrome c. *Biochemistry* **39,** 8234–8242.

Jemmerson, R., Liu, J., Hausauer, D., Lam, K.-P., Mondino, A., and Nelson, R. D. (1999). A conformational change in cytochrome c of apoptotic and necrotic cells is detected by monoclonal antibody binding and mimicked by association of the native antigen with synthetic phospholipid vesicles. *Biochemistry* **38,** 3599–3609.

Jentzen, W., Unger, E., Karvounis, G., Shelnutt, J. A., Dreybrodt, W., and Schweitzer-Stenner, R. (1995). Conformational properties of nickel(II) octaethylporphyrin in solution. 1. Resonance excitation profiles and temperature dependence of structure-sensitive Raman lines. *J. Phys. Chem.* **100,** 14184–14191.

Jiang, X., and Wang, X. (2004). Cytochrome c-mediated apoptosis. *Annu. Rev. Biochem.* **73,** 87–106.

Kagan, V. E., Tyurin, V. A., Jiang, J., Tyurin, V. A., Ritov, V. B., Amoscato, A., Osipov, A. N., Belikova, N. A., Kapralov, A. A., Kini, V., Vlasova, I. I., Zhao, Q., *et al.* (2005). Cytochrome c acts as a cardiolipin oxygenase required for release of proapoptic factors. *Nat. Chem. Biol.* **4,** 223–232.

Kapralov, A. A., Kurnikov, I. V., Vlasova, I. I., Belikova, N. A., Tyurin, V. A., Basova, L. V., Zhao, Q., Tyurina, Y. Y., Jiang, J., Bayir, H., Vladimirov, Y. A., and Kagan, V. E. (2007). The hierachy of structural transitions induced in cytochrome c by anionic phospholipids determines its peroxidase activation and selective peroxidation during apoptosis in cells. *Biochemistry* **46,** 14232–14244.

Kiefl, C., Sreerama, N., Haddad, R., Sun, L., Jentzen, W., Lu, Y., Qiu, Y., Shelnutt, J. A., and Woody, R. W. (2002). Heme distortions in sperm-whale carbonmonoxy myoglobin: Correlations between rotational strengths and heme distortions in MD-generated structures. *J. Am. Chem. Soc.* **124,** 3385–3394.

Levantino, M., Huang, Q., Cupane, A., Laberge, M., Hagarman, A., and Schweitzer-Stenner, R. (2005). The importance of vibronic perturbations in ferrocytochrome c spectra: A reevaluation of spectral properties based on low-temperature optical absorption, resonance Raman, and molecular-dynamics simulations. *J. Chem. Phys.* **123,** 054508.

Louie, G. V., and Brayer, G. D. (1990). High-resolution refinement of yeast iso-1-cytochrome c and comparisons with other eukaryotic cytochromes c. *J. Mol. Biol.* **214,** 527–555.

Manas, E. S., Vaderkooi, J. M., and Sharp, K. A. (1999). The effects of protein environment on the low temperature electronic spectroscopy of cytochrome c and microperoxidase-11. *J. Phys. Chem. B* **103,** 6334–6348.

Manas, E. S., Wright, W. W., Sharp, K. A., Friedrich, J., and Vanderkooi, J. M. (2000). The influence of protein environment on the low temperature electronic spectroscopy of Zn-substituted cytochrome c. *J. Phys. Chem. B.* **104,** 6932–6941.

Manavalan, P., and Johnson, W. C. Jr. (1987). *Anal. Biochem.* **167,** 76–85.

Merolo, J. J., Andrade, M. A., Prieto, A., and Morán, F. (1994). *Neurocomputing* **6,** 443–454.

Moench, S. J., Shi, T.-M., and Satterlee, J. S. (1991). Proton-NMR studies of the effects of ionic strength and pH on the hyperfine-shifted resonances and phenylalanine-82 environment of three species of mitochondrial ferricytochrome c. *Eur. J. Biochem.* **197,** 631–641.

Moore, G. W., and Pettigrew, G. W. (1990). Cytochrome c—Evolutionary, Structural and Physicochemical Aspects. Springer, Berlin.

Myer, Y. P. (1968). Conformation of cytochromes. III. Effect of urea, temperature, extrinsic ligands and pH variation on the conformation of horse heart ferricytochrome-c. *Biochemistry* **7,** 765–776.

Myer, Y. P., and Pande, A. (1978). Circular dichroism studies of hemoproteins and heme models. *In* "The Porphyrins," (D. Dolphin, ed.), **III,** pp. 271–322. Academic Press, New York.

Nantes, I. I., Zucchi, M. R., Nascimento, O. R., and Faljoni-Alario, A. (2001). Effect of heme iron valence state on the conformation of cytochrome c and its association with membrane interfaces. *J. Biol. Chem.* **276,** 153–158.

Pinheiro, T. J. T., Elöve, G., Watts, A., and Roder, H. (1997). Structural and kinetic description of cytochrome c unfolding induced by the interaction with lipid vesicles. *Biochemistry* **36,** 13122–13132.

Provencher, S. W., and Glöckner, J. (1981). *Biochemistry* **20,** 33–37.

Purring-Koch, C., and McLendon, G. (2000). Cytochrome c binding to Apaf-1: The effects of dAtp and ionic strength. *Proc. Natl. Acad. Sci. USA* **97,** 11928–11931.

Rossel, F. I., Ferrer, J. C., and Mauk, A. G. (1998). Proton-linked protein conformational switching: Definition of the alkaline conformational transition of yeast iso-l-ferricytochrome c. *J. Am. Chem. Soc.* **120,** 11234–11245.

Rytōman, M., and Kinnunen, P. K. J. (1994). Evidence for two distinct acidic phospholipid-binding sites in cytochrome c. *J. Biol. Chem.* **269,** 1770–1774.

Rytōman, M., Mustonen, P., and Kinnunen, P. K. J. (1992). Reversible, nonionic and pH-dependent association of cytochrome c with cardiolipin-phosphatidylcholine liposomes. *J. Biol. Chem.* **267,** 22243–22248.

Schejter, A., and George, P. (1964). The 695 μm band of ferricytochrome c and its relationship to protein conformation. *Biochemistry* **3,** 1045–1049.

Schweitzer-Stenner, R. (1989). Allosteric linkage-induced distortions of the prosthetic group in heme proteins as derived by the theoretical interpretation of the depolarization ratio in resonance Raman scattering. *Q. Rev. Biophys.* **22,** 381.

Schweitzer-Stenner, R. (2008). The internal electric field in cytochrome c explored by visible electronic circular dichroism spectroscopy. *J. Phys. Chem. B* **112,** 10358–10366.

Schweitzer-Stenner, R., and Bigman, D. (2001). Electronic and vibronic contributions to the band splitting in optical spectra of heme proteins. *J. Phys. Chem. B* **105,** 7064–7073.

Schweitzer-Stenner, R., Levantino, M., Cupane, A., Laberge, M., and Huang, Q. (2006). Functional relevant electric-field induced perturbations of the prosthetic group of yeast ferrocytochrome c mutants obtained from a vibronic analysis of low temperature absorption spectra. *J. Phys. Chem. B* **110,** 12155–12161.

Schweitzer-Stenner, R., Gorden, J. P., and Hagarman, A. (2007a). The asymmetric band profile of the soret band of deoxymyoglobin is caused by electronic and vibronic perturbations of the heme group rather than by a doming deformation. *J. Chem. Phys.* **127,** 135103.

Schweitzer-Stenner, R., Shah, R., Hagarman, A., and Dragomir, I. (2007b). Conformational substates of horse heart cytochrome c exhibit different thermal unfolding of the heme cavity. *J. Phys. Chem. B* **111,** 9603–9607.

Shah, R., and Schweitzer-Stenner, R. (2008). Structural changes of horse heart ferricytochrome c induced by changes of ionic strength and anion binding. *Biochemistry* **47,** 5250–5257.

Snyder, P. A., and Rowe, E. M. (1980). *Nucl. Instrum. Methods* **172,** 345–349.

Sreerama, N., and Woody, R. W. (1993). A self-consistent method for the analysis of protein secondary structure from circular dichroism. *Anal. Biochem.* **209,** 32–44.

Sreerama, N., and Woody, R. W. (1994a). *Biochemistry* **33,** 10022–10025.

Sreerama, N., and Woody, R. W. (1994b). *J. Mol. Biol.* 242.

Stallard, B. R., Champion, P. M., Callis, P. R., and Albrecht, A. C. (1983). Advances in calculating Raman excitation profiles by means of the transform theory. *J. Chem. Phys.* **78,** 712–722.

Sutherland, J. C., Desmond, E. J., and Takacs, P. Z. (1980). *Nucl. Instr. Meth.* **172,** 195–199.

Taler, G., Schejter, A., Navon, G., Vig, I., and Margoliash, E. (1995). The nature of the thermal equilibrium affecting the iron coordination of ferric cytochrome c. *Biochemistry.* **34,** 14209–14212.

Theorell, H., and Åkessen, Å. (1941). *J. Am. Chem. Soc.* **63,** 1804–1820.

Tinoco, I. (1962). Theoretical aspects of optical activity. *Adv. Chem. Phys.* **4,** 113–160.

Uchiyama, S., Hasegawa, J., Tanimoto, Y., Moriguchi, H., Mizutani, M., Igarashi, Y., Sambongi, Y., and Kobayashi, Y. (2002). Thermodynamic characterization of variants of mesophilic cytochrome c and its thermophilic counterpart. *Protein Eng.* **15,** 445.

Urry, D. W. (1965). Protein-heme interactions in heme-proteins: Cytochrome c. *Proc. Natl. Acad. Sci. USA* **54,** 640–648.

Urry, D. W., and Doty, P. (1965). On the conformation of horse heart ferri- and ferrocytochrome-c. *J. Am. Chem. Soc.* **87,** 2756–2758.

van Holde, K. E., Johson, W. C., and Ho, P. S. (2006). Principles of Physical Biochemistry. Pearson Prentice Hall, Upper Saddle River.

Verbaro, D., Hagarman, A., Soffer, J., and Schweitzer-Stenner, R. (2009). The pH dependence of the 695 nm charge transfer band reveals the population of an intermediate state of the alkaline transition of ferricytochrome *c* at low ion concentrations†. *Biochemistry* **48,** 2990–2996.

Verbaro, D., Hagarman, A., Kohli, A., and Schweitzer-Stenner, R. J. (2009). *Biol. Inorg. Chem.* DOI 10.1007/s00775-009-0574-9.

Wallace, B. A. (2000). Synchrotron radiation circular-dichroism spectroscopy as a tool for investigating protein structures. *J. Synchrotron Rad.* **7,** 289–295.

Weber, C., Michel, B., and Bosshard, H. R. (1987). Spectroscopic analysis of the cytochrome c oxidase-cytochrome *c* complexes: Circular dichroism and magnetic circular dichroism measurements reveal changes of cytochrome *c* heme geometry imposed by complex formation. *Proc. Natl. Acad. Sci. USA* **84,** 6687–6691.

Weinkam, P., Zimmermann, J., Sagle, L. B., Matsuda, S., Dawson, P. E., Wolynes, P. G., and Romesberg, F. E. (2008). Cahracterization of alkaline transitions in ferricytochrome *c* using carbon-deuterium infrared probes. *Biochemistry* **47,** 13470–13480.

Wellman, K. M., Laur, P. H. A., Briggs, W. S., Moscowitz, A., and Djerassi, C. (1965). Optical rotary dispersion studies. XCIX. Siperposed multiple cotton effects of saturated ketones and their significance in the circular dichroism measurement of (-)-menthone. *J. Am. Chem. Soc.* **87,** 66–72.

Whitmore, L., and Wallace, B. A. (2004). DICHROWEB, an online server for protein secondary structure analyses from circular dichroism spectroscopic data. *Nucleic Acids Res.* **32,** W668–W673.

Whitmore, L., and Wallace, B. A. (2008). *Biopolymers* **89,** 392–400.

Woody, R. W. (2004). The circular dichroism of protein folding intermediates. *Methods Enzymol.* **380,** 242–351.

Woody, R. W. (2009). Circular dichroism spectrum of peptides in the poly(pro)II conformation. *J. Am. Chem. Soc..* **131,** 8234–8245.

CHAPTER SEVEN

Examining Ion Channel Properties Using Free-Energy Methods

Carmen Domene* *and* Simone Furini†

Contents

Abstract

Recent advances in structural biology have revealed the architecture of a number of transmembrane channels, allowing for these complex biological systems to be understood in atomistic detail. Computational simulations are a powerful tool by which the dynamic and energetic properties, and thereby the function of these protein architectures, can be investigated. The experimentally observable properties of a system are often determined more by energetic than dynamics, and therefore understanding the underlying free energy (FE) of biophysical processes is of crucial importance.

* Physical and Theoretical Chemistry Laboratory, Department of Chemistry, University of Oxford, Oxford, United Kingdom
† Department of Medical Surgery and Bioengineering, University of Siena, Siena, Italy

Methods in Enzymology, Volume 466
ISSN 0076-6879, DOI: 10.1016/S0076-6879(09)66007-9

Critical to the accurate evaluation of FE values are the problems of obtaining accurate sampling of complex biological energy landscapes, and of obtaining accurate representations of the potential energy of a system, this latter problem having been addressed through the development of molecular force fields. While these challenges are common to all FE methods, depending on the system under study, and the questions being asked of it, one technique for FE calculation may be preferable to another, the choice of method and simulation protocol being crucial to achieve efficiency. Applied in a correct manner, FE calculations represent a predictive and affordable computational tool with which to make relevant contact with experiments. This chapter, therefore, aims to give an overview of the most widely implemented computational methods used to calculate the FE associated with particular biochemical or biophysical events, and to highlight their recent applications to ion channels.

1. Introduction

Membrane proteins are among the most challenging targets in structural biology, though technical advances in recent years are leading to a rapid increase in the rate at which membrane protein structures are solved (Carpenter *et al.*, 2008). Among the membrane proteins, the number of ion channel structures experimentally determined has been increasing steadily since the publication of the three-dimensional structure of the KcsA K^+ channel in 1998 (Doyle *et al.*, 1998). Out of the 15 unique ion channel structures available at the time this chapter is written, only examples of K^+ channels are available, as well as some channels that present nonselective conductance. The most important experimental technique for obtaining ion channel structures is X-ray crystallography, although a small percentage of the membrane protein structures available nowadays have been also obtained by nuclear magnetic resonance (NMR). A crystal structure is an average of many different conformations present in the crystal and produced during the experiment, which represents the best fit to the available experimental data. With the exception of the information contained in the B-factors, these three-dimensional models give only a very limited picture of the dynamics of the system. Thus, computational simulations are important for the insight they give into the dynamics and energetics associated with ion channel function. Figure 7.1 shows a representative example of a simulation system of a K^+ channel, KirBac1.1, inserted in a lipid bilayer.

Numerical simulation of ion channels faces various computational challenges. Firstly, even with large-scale computational resources and efficiently parallelized codes, it is still difficult to simulate long timescales. For instance, the average time required for the gating of an ion is too long relative to the timescales typically accessible by brute-force classical molecular dynamics

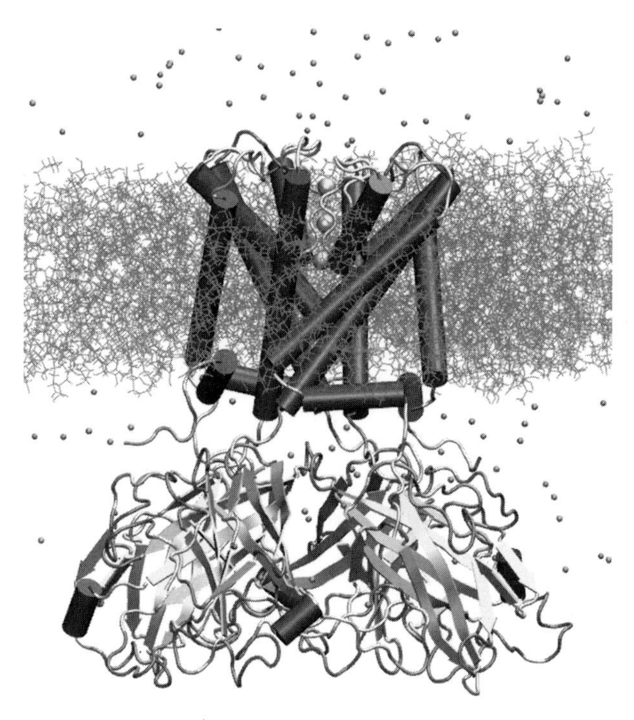

Figure 7.1 The KirBac1.1 K^+ channel structure (PDB 1P7B) embedded in a lipid bilayer with individual lipids rendered as gray chains. The transmembrane α-helices are rendered in purple and the extracellular domain is colored according to its structure (β-sheet: yellow; random coil: white; turns: cyan). K^+ ions inside the selectivity filter of the channel are represented as yellow spheres. Other ions in solution are yellow and orange spheres. (See Color Insert.)

(MD). Secondly, despite simulations having been described in the literature of over million atoms, the physical size of the system that can be computationally modeled still imposes a limit on what can be achieved through calculation.

2. Free-Energy Calculations

Atomistic simulations provide insight into the behavior of biological systems at atomic level. In MD simulations, the interactions between atoms are described by empirical potential functions, which are used to integrate the classical equations of motion. This process yields a trajectory of the system in time, from which structural and dynamical quantities as well as kinetic and thermodynamic properties can be calculated using the principles of statistical mechanics. Among the thermodynamic quantities, it is the free energy (FE) that provides a direct link between statistical mechanics and thermodynamics, and through which other thermodynamic quantities can be obtained. The FE is a measure of the probability of finding a system in a given state. Thus, it is a crucial descriptor when the atomic characteristics of a system are to be related to some experimental, macroscopic, property (Chandler, 1987; Hill, 1987). While the absolute free energy of a system can be difficult to calculate, the difference in free energy between two states is easier to obtain, and can provide sufficient information.

For example, in the context of ion channels, FE calculations can be applied to identify the molecular determinants of channel blockage, a problem of profound interest in pharmaceutical research. In this case, evaluation of how structural changes in the channel and the blocker affect binding affinities is performed by calculation of the FE, the binding affinities being a function of the FE difference between the compounds attached to the channel or free in solution.

Using statistical thermodynamics, one can relate the FE with MD or Monte Carlo (MC) trajectories. The FE difference, ΔA, between two configurations, 0 and 1, can be expressed as

$$\Delta A = A_1 - A_0 = -k_B T \ln \frac{P_1}{P_0}, \tag{7.1}$$

where k_B is the Boltzmann's constant, T is the temperature, and P_0 and P_1 are the probabilities to find the system in configurations 0 and 1, respectively. However, relating the FE with MD or MC trajectories through Eq. (7.1) is not straightforward; sampling the most probable configurations corresponding to the states of interest is required, and this may be hampered by long relaxation times of the system. Let us imagine that the starting configuration of the system is close to configuration 0, which is separated from configuration 1 by a FE barrier. Due to the exponential dependence in Eq. (7.1), even a small energy barrier is unlikely to be crossed in a regular MD simulation. The system will visit only local minima close to configuration 0, and configuration 1 will never be reached, preventing the calculation of the FE difference.

A plethora of techniques have been developed to improve the sampling of the configuration space, allowing the calculation of FE differences by atomistic simulations (for a review see, e.g., Chipot and Pohorille, 2007). Reviewing this vast field is out of the scope of this chapter. In the next sections, the intention is to describe and illustrate with examples those methods readily implemented in conventional software packages, and more frequently used to investigate the biophysics of ion channels.

3. Thermodynamic Integration

In thermodynamic integration (TI), two states of interest are connected by a pathway along which the FE profile is determined. Such a FE profile is often called a potential of mean force (PMF) because to obtain it, the mean force along the path is integrated. Constrained MD (den Otter and Briels, 1998) and adaptive biasing force MD (Darve and Pohorille, 2001) are variants of this technique as well as the popular steered MD method (Gullingsrud *et al.*, 1999). The latter belongs to nonequilibrium FE methods which has been recently reviewed in Cossins *et al.* (2009) where we refer the interested reader.

In TI the Hamiltonian, that is, the sum of the kinetic and potential energy of the physical system, can be defined as a function of a continuously varying parameter, λ. Supposing the states of interest to be labeled 0 and 1, λ might be defined such that when $\lambda = 0$ the system is in state 0, and when $\lambda = 1$ the system is in state 1. In practice λ is modeled at a series of discrete points λ_i. The system is allowed to fully relax at each of the points λ_i and the energy gradient is estimated at each point. The FE difference between two states $\Delta A^{\mathrm{TI}}(0 \rightarrow 1)$ is then expressed as

$$\Delta A^{\mathrm{TI}}(0 \rightarrow 1) = \int_{\lambda=0}^{\lambda=1} \left\langle \frac{\partial H}{\partial \lambda} \right\rangle_{\lambda} \mathrm{d}\lambda \approx \sum_{i=1}^{n-1} \left\langle \frac{\partial H}{\partial \lambda} \right\rangle_{\lambda_i}, \qquad (7.2)$$

where H is the Hamiltonian and $\langle\rangle_\lambda$ is an ensemble average at a particular value of λ.

In a TI calculation, the ensemble averages are determined using either molecular dynamics or Monte Carlo sampling. The energy derivative is approximated by a sum over a discrete number of points, which can introduce errors, although provided that enough points λ_i are chosen, accurate FE can be computed. Depending on the path taken to interconvert the system from $\lambda = 0$ to $\lambda = 1$ rather long dynamics may be necessary to achieve statistical accuracy.

As the FE is a state function, it is independent of the paths connecting the initial and final states. This allows the use of nonphysical paths, as it is only required that the path brings the system from the physical initial state to the physical final state. However, in practice, the choice of the path is restricted by numerical requirements. Equation (7.2) can be understood by recalling the relationship between work and FE from the second law of thermodynamics. An amount of work W is performed on the system, whereby the system is transformed from state 0 to state 1. The integral in Eq. (7.2) gives this work, and by the second law of thermodynamics, this amount of work obeys $W \geq \Delta A(0 \rightarrow 1)$, where the equality holds only if the transformation is carried out along a reversible path.

TI can be implemented very easily. Many simulations of the system at values of λ ranging from 0 to 1 should be carried out in order to perform the numerical integration. Thus, the phase space is explored in a deterministic and systematic manner. The accuracy of this method is strongly dependent upon an adequate sampling of the phase space, and thus the intermediate averages must be accurately calculated in order for the integration to yield the correct result.

Evaluation of the accuracy of a TI calculation can be obtained by comparing the values of two FE calculations. In the forward sampling, estimations of the gradient are calculated as the coefficient i increases, moving from λ_i to λ_{i+1}, while in the backward sampling estimations are calculated as the coefficient i decreases, moving from λ_{i+1} to λ_i. Given a small enough gap between the λ_i, and a long enough simulation, the size of the difference between the resultant integrals would be zero. In practice this is not the case, and the difference, known as the hysteresis error, is a measure of the accuracy of the calculation (Jorgensen and Ravimohan, 1985). Accuracy can also be measured by convergence between repeated runs of the calculation (Beveridge and Dicapua, 1989; Mitchell and McCammon, 1991). The simulation is said to be converged when the calculated properties are no longer significantly changing. Particular care must be taken in the vicinity of $\lambda = 0$ and $\lambda = 1$, where it is possible to encounter loss of numerical accuracy and singularities, referred as the end-points catastrophes (Beutler *et al.*, 1994). However, the presence of convergence or the absence of hysteresis are necessary but not sufficient indicators of precision in calculated FE (Mitchell and McCammon, 1991).

4. Free-Energy Perturbation

This approach to the calculation of FE differences is generally attributed to Zwanzig (1954). According to free-energy perturbation theory (FEP), the FE difference for going from state 0 to state 1 is obtained from the following equation, known as the *Zwanzig* equation:

$$\Delta A^{\mathrm{FEP}}(0 \to 1) = A_1 - A_0 = -k_{\mathrm{B}} T \ln \left\langle \exp\left(-\frac{H_1 - H_0}{k_{\mathrm{B}} T} \right) \right\rangle_0, \quad (7.3)$$

where T is the temperature, k_{B} is the Boltzmann's constant, H is the Hamiltonian, that is, the sum of the kinetic and potential energy of the system at state 0 and 1, and $\langle \rangle_0$ refers to the ensemble average over a simulation run for state 0.

Whereas in TI the free energy is calculated as the integral of the gradient of the free energy, calculated along a path, in FEP the free energy is calculated as a simple sum of differences between a series of intermediate states. Writing the Hamiltonians of the intermediate states as

$$H(\lambda_i) = \lambda_i H_1 + (1 - \lambda_i)H_0, \quad 0 \leq \lambda_i \leq \lambda_{i+1} \leq 1 \tag{7.4}$$

transforms Eq. (7.3) as follows:

$$\begin{aligned}\Delta A^{\mathrm{FEP}}(0 \rightarrow 1) &= \sum_{i=0}^{n-1} \Delta A^{\mathrm{FEP}}(\lambda_i \rightarrow \lambda_{i+1}) \\ &= \sum_{i=0}^{n-1} -k_{\mathrm{B}} T \ln\left\langle \exp\left(-\frac{H(\lambda_{i+1}) - H(\lambda_i)}{k_{\mathrm{B}} T}\right)\right\rangle_{\lambda_i}.\end{aligned} \tag{7.5}$$

Thus, the calculation is divided into the sum of a series of free-energy differences between intermediate states, for which the difference between the Hamiltonians is small. Efficient choice of the λ_i, to obtain sufficient overlap between states without unnecessary calculation, is key. Again, accuracy can be measured by hysteresis or convergence. Improvements in the accuracy of a single calculation can be obtained through double-wide sampling, in which differences are calculated simultaneously in a forward and a backward direction (Jorgensen and Ravimohan, 1985). A more reliable estimation of the error can possibly be obtained by comparing results of multiple independent simulations.

Since obtaining the ensemble average for each value of λ_i is an independent process, the method can be trivially parallelized by running each window using a different processor. One inconvenience of the technique is the fact that it is hard to know in advance what size of interval between the λ_i to choose. Pearlman's dynamically modified windows in some way mitigate this problem (Pearlman and Kollman, 1989).

Gradual transformation of one chemical species into another leads to the question of how to represent the geometries of the intermediate states. Two approaches have been devised to solve this problem, namely the single and dual topology methods. In the single topology method, the changes between initial and final states of a system are represented by modifying the atom types and internal coordinates, for example by incrementally scaling a bond length, atomic charge, or force-field parameter to generate a single hybrid molecule combining the properties of the initial and final states. If the number of atoms between the initial and final configurations differs, dummy atoms are utilized. By contrast, in the dual topology, the changes are specified in terms of a system where two complete versions of the initial and final states coexist. Atoms of the initial state do not interact with those of the final state throughout the simulation, while the remainder of the system interacts with a linear sum of both of the states, scaled so that as λ_i increases, the interaction with the final state gradually takes over from the interaction with the initial state.

In such a calculation, particular care must be taken in the vicinity of $\lambda = 0$ and $\lambda = 1$. Van der Waals forces between atoms are modeled using a potential with a singularity at distance zero, which is not removed by the linear scaling. This leads to a sharp change where the molecule appears or disappears at the extreme values of λ, referred to as the end-points catastrophe (Beutler *et al.*, 1994). This problem can be solved by use of a soft-core function which smoothes the appearance or disappearance of atoms at the ends of the transition (Beutler *et al.*, 1994). Of the two approaches, dual topology is likely to require greater conformational rearrangement of the system at particular values of λ than in the single topology method, since a greater number of atoms interact, and hence longer sampling is required to obtain the same accuracy. Dual topology calculations, however, have the advantage that more complex changes in topology can be accommodated more easily than in a single topology approach.

5. Umbrella Sampling

FEP and TI allow FE differences between two configurations of a molecular system to be computed, therefore enabling the calculation of properties such as binding affinities or channel selectivity. However, in other situations, a detailed knowledge of the FE profile is desired, for instance to describe the permeation of ions through a membrane protein, where the complete FE profile of the ion along the channel pore is required. Knowledge of the FE differences between a subset of configurations is not enough, since the energetic barriers between these configurations are crucial to estimate the channel permeation characteristics. Therefore, changes in the FE are calculated in terms of a function of the atomic coordinates of the system, $\xi(\underline{r}^{3N})$, which describes the state of the system, by means of the equation:

$$A(\xi) = -k_{\mathrm{B}} T \ln P(\xi). \tag{7.6}$$

Here, $P(\xi)$ is the probability to find the system in the state characterized by ξ, and the FE is calculated relative to some reference state. In the situation where one is just interested in FE differences, the FE of the reference state is irrelevant, as it vanishes once the difference is taken. The function ξ is referred to as the collective variable (CV), and is here assumed to be one-dimensional, although extension to higher dimensions is straightforward.

Defining the CVs is a critical step in the calculation of the FE. The methodologies used to compute the FE are optimized to accelerate the sampling along the CV chosen, assuming that once the CV is fixed, the other degrees of freedom of the system will reach equilibrium easily during the dynamics. In some cases, the CV necessary to describe a transition is obvious. In the example of ion permeation, the position of the ion along the channel pore is the natural choice. In other cases, the definition of an effective CV can be more complex and one has to proceed by trial and error.

In the umbrella sampling (US) technique, biasing potentials are used to improve the sampling of the CVs in MD simulations (Torrie and Valleau, 1977). In practice, the region of interest that needs to be sampled is divided into R sets, called windows, representing points along the CV. For each of these windows, a trajectory is obtained using MD. The Hamiltonian of the atomic system is then expressed by

$$H_i(\Gamma) = H_0(\Gamma) + W_i(\xi), \tag{7.7}$$

where $H_0(\Gamma)$ is the Hamiltonian of the unbiased system and $W_i(\xi)$ represents the biasing potential. The array Γ defines the microscopic state of the system, thus also the value of the CV, ξ. In an MD simulation of a N-particle system, Γ depends on the atomic three-dimensional coordinates, $X = (x_1, \ldots, x_{3N})$, and momenta, $P = (p_1, \ldots, p_{3N})$. The biased probability $P_i^{\mathrm{b}}(\xi)$ of finding the system *inthe* state ξ when the biasing potential $W_i(\xi)$ is added to the Hamiltonian is

$$\begin{aligned} P_i^{\mathrm{b}}(\xi) &= \frac{\int_\Gamma \delta(\xi(X) - \xi)\mathrm{e}^{-\{H_i(\Gamma)/k_{\mathrm{B}}T\}}\mathrm{d}\Gamma}{\int_\Gamma \mathrm{e}^{-\{H_i(\Gamma)/k_{\mathrm{B}}T\}}\mathrm{d}\Gamma} \\ &= \frac{\int_\Gamma \delta(\xi(X) - \xi)\mathrm{e}^{-\{H_0(\Gamma)/k_{\mathrm{B}}T\}}\mathrm{d}\Gamma}{\int_\Gamma \mathrm{e}^{-\{H_0(\Gamma)/k_{\mathrm{B}}T\}}\mathrm{d}\Gamma}\mathrm{e}^{-\beta W_i(\xi)}\frac{\int_\Gamma \mathrm{e}^{-\{H_0(\Gamma)/k_{\mathrm{B}}T\}}\mathrm{d}\Gamma}{\int_\Gamma \mathrm{e}^{-\{H_i(\Gamma)/k_{\mathrm{B}}T\}}\mathrm{d}\Gamma} \\ &= P^{\mathrm{u}}(\xi)\exp\left\{\frac{W_i(\xi) - F_i}{k_{\mathrm{B}}T}\right\}, \end{aligned} \tag{7.8}$$

where $P^{\mathrm{u}}(\xi)$ is the probability of finding the unbiased system in the state ξ, δ is the Dirac delta function, and F_i is a constant. Equation (7.8) shows that it is possible to estimate the unbiased probability, $P^{\mathrm{u}}(\xi)$, by simulating a system biased by the forcing potential $W_i(\xi)$. Once this unbiased probability is known, the FE can be computed using Eq. (7.6).

The value $W_i(\xi) - F_i$ in Eq. (7.8) is the excess FE in the biased system i with respect to the unbiased system. Therefore, from Eq. (7.8) the biased free energy can be expressed as

$$A_i^{\mathrm{b}}(\xi) = A(\xi) + W_i(\xi) - F_i. \tag{7.9}$$

According to Eq. (7.9), optimum sampling in the simulation of the biased system will be reached if ξ is equal to $-A(\xi)$. With such a choice of the biasing potential, the system will follow a diffusive trajectory along the path characterized by ξ, rendering an efficient sampling of the CV. Obviously, since $A(\xi)$ is not known *a priori*, it is not possible to adopt this strategy. Therefore, the biasing potentials are usually defined as

$$W_i(\xi) = \frac{k_i}{2}(\xi - \xi_1^0)^2, \quad i = 1, \ldots, R, \tag{7.10}$$

where ξ_i^0 is the center of the region analyzed by the *i*th simulation. Assuming this choice of biasing potential, the system will be forced to sample a region close to the center of the window, ξ_i^0. The harmonic force constant, k_i, the number of windows, R, and their spacing need to be defined carefully in order to efficiently sample the entire range of the CV ξ. Equation (7.8) is strictly valid only when complete statistics of ξ are achieved, that is, in the limit of an infinite number of trajectories being calculated. In reality, however, only a portion of the ξ space is sampled. For instance, if harmonic potentials like those in Eq. (7.10) are used, only a region around ξ_i^0 will be described by the MD trajectory. An estimate of the unbiased probability will be accurate in this region, but it will not be adequate far from the center of the window, ξ_i^0, where samples are not collected. Therefore, to obtain a full description of the behavior of the unbiased system, R simulations are run, with different biasing potentials, and the R different estimates of the unbiased probability, $P_i^{\mathrm{u}}(\xi)$ are then combined. This is usually accomplished using the weighted histogram analysis method (WHAM) (Kumar *et al.*, 1992). Figure 7.2 shows how the free energy estimated by US with the WHAM algorithm changes with different degree of overlap between windows.

6. Adaptive Biasing Force

In order to analyze how a system behaves with respect to a CV, a strategy to sample the system at different ξ values is required. In US, biasing potentials are used. As already discussed, sampling of the CV is optimized if the biasing potentials compensate for the underlying FE profile. Since the FE profile is not known *a priori*, one needs to make some assumptions in advance. A possible strategy is to estimate the FE by short US simulations and use this initial profile in a second set of US simulations, which will be more accurate(Mezei, 1987), a procedure which can be repeated iteratively. The adaptive biasing force (ABF) method uses a similar approach. The estimated FE derivative, computed for small intervals of the CV, is canceled

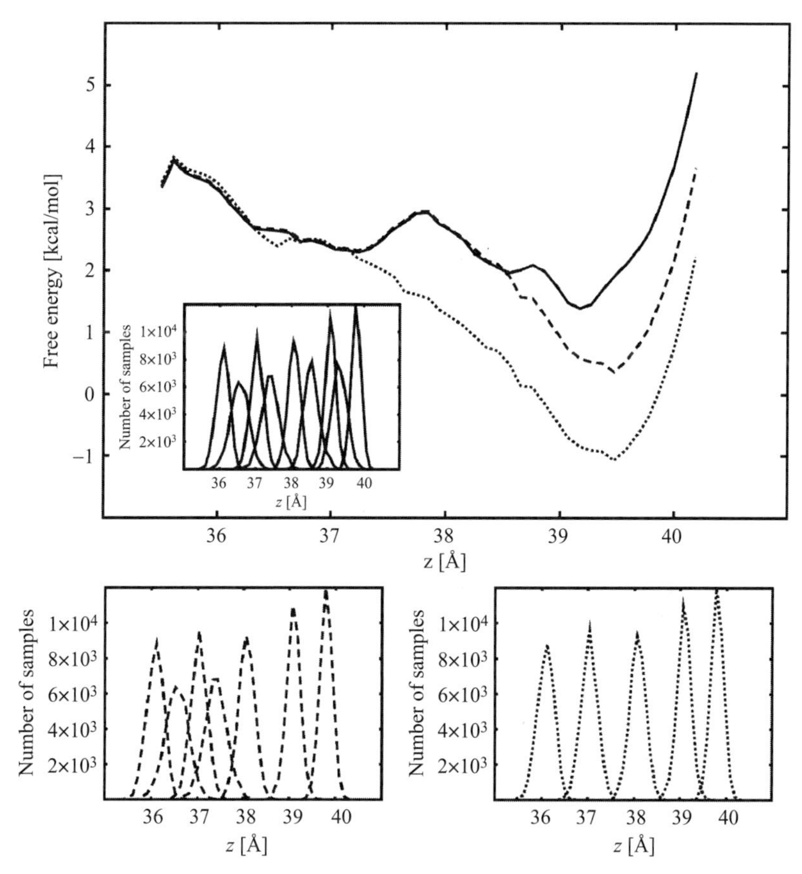

Figure 7.2 Free energy estimated with the US method using different number of windows and thus, of overlap between them. The main plot shows the free energy for moving a K^+ ion along the pore axis of the K^+ channel KcsA, from the cavity center ($z \sim 36$ Å) to the inside of the selectivity filter ($z \sim 39.5$ Å). The three curves (continuous, dashed, and dotted) refer to the window histograms shown in the respective plots. Note that the dashed curve is close to the continuous one as long as there is overlap between windows. They diverge at $z \sim 38.5$ Å, where there is no overlap between the windows represented by the dashed curve.

by the introduction of an ABF (Darve and Pohorille, 2001; Henin and Chipot, 2004). The idea behind the ABF method can be easily illustrated in the special case where the CV is a Cartesian coordinate of one of the N-atoms in the system. Recalling that the coordinates of the atomic system are denoted as $X = (x_1, \ldots, x_{3N})$, let the CV ξ be equal to x_1. Then

$$A(x_1) = -k_B T \ln P(x_1) \tag{7.11}$$

and the probability of finding the system in the state x_1:

$$P(x_1) \propto \int \exp\left\{-\frac{U(x_1,\ldots,x_{3N})}{k_B T}\right\} dx_2 \ldots dx_{3N}, \tag{7.12}$$

where U is the potential energy of the atomic system (note that the integral in Eq. 7.12 is not performed over x_1). Since we are interested in FE differences, the FE can be written as

$$A(x_1) = -k_B T \ln \int e^{-(U/k_B T)} dx_2 \ldots dx_{3N}. \tag{7.13}$$

Thus, the derivative of the FE with respect to x_1 is

$$\frac{dA(x_1)}{dx_1} = \left\langle \frac{\int e^{-(U/k_B T)} \frac{\partial U}{\partial x_1} dx_2 \ldots dx_{3N}}{\int e^{-(U/k_B T)} dx_2 \ldots dx_{3N}} \right\rangle_{x_1} = \left\langle \frac{\partial U}{\partial x_1} \right\rangle_{x_1} = -\langle F_{x_1} \rangle_{x_1}, \tag{7.14}$$

where F_{x_1} is the force acting in the direction of the CV and the average is calculated at a specific x_1 value. From an MD simulation, the force acting in the direction of the CV is obtained and the average value of this force is computed. As the MD simulation proceeds, the estimate of the average force is improved. If this force is then subtracted from the forces acting on the atomic system, the force acting in the direction of the CV will be almost negligible. Under such a condition, the system undergoes diffusive motion along the ξ-axis, thus providing efficient sampling. The derivative of the FE estimated by the FEP simulation can then be integrated, for example with the trapezoidal rule, to obtain the FE along the ξ-axis. This methodology can be extended to a generic n-dimensional CV (Henin and Chipot, 2004).

To calculate the FE by the ABF strategy, the first step is to define the CV, ξ, and to calculate analytically the derivative of the FE with respect to this CV. Two parameters then need to be defined: the limits of the CV, and the number of bins used to compute the FE profile, which is related to the resolution of the FE. The computational cost of the ABF simulation increases with the number of bins, since more values of the average force need to be computed. Using a low number of bins will render low-resolution FEs, and configurations characterized by different values of the FE will be clustered in the same average. During the MD simulation, the value of the force along the CV is stored in each bin, and an average value is computed at runtime. At the beginning of the simulation the average force may oscillate widely, so to avoid instability, the average force is not subtracted from the force of the atomic system until an adequate number of force samples have been collected. Once enough samples are collected, the average force on the CV is subtracted and the system is forced to escape the minima, ready to sample new regions of the configuration space. The simulation should proceed until uniform sampling of the region of interest of the CV is achieved.

7. Metadynamics

Metadynamics, developed by Laio and Parrinello (2002), is a method that encompasses several features of other techniques and can provide a unified framework for computing FEs, and for accelerating rare events, in this case, without the requirement of *a priori* knowledge of the energy surface. In this respect, the method provides an efficient scheme to sample the improbable but important configurations, such as transition states, during an MD simulation. In this approach, the reaction coordinate is discouraged to remain fluctuating around a local minimum, for example at the minimal geometry of the reactant state of a system, by adding a small Gaussian-shaped potential to the reaction coordinate every few dynamics steps. In this way, over time the potential well associated with the local minimum will gradually be filled, allowing the system to escape via the lowest transition state to the next (product or intermediate) well. In the same way, this next potential well will eventually also be filled by Gaussians and so forth. The time required to escape from a local minimum in the FE surface is determined by the depth of the well itself, and the number and volume of the Gaussians that are added to fill the well.

If too many Gaussians, or of too great size, are deposited in a short period of time, the reconstructed FE can be incorrect. However, if small enough Gaussians are added at suitable intervals, the FE surface of the system under study, including the transition states, can be obtained to arbitrary accuracy, by subtracting the sum of the added Gaussians, when the total FE surface has been flattened. At this point, the motion of the CV should be diffusive across its defined range.

Metadynamics exits in different versions: discrete or continuous, extended Lagrangian and direct. The original variant is known as discrete metadynamics. This algorithm requires the evaluation of the derivative of the FE with respect to the CV at every step. These are then evolved in a discrete fashion in the direction of the maximum gradient. In principle, an exact separation between the dynamics of the CVs and the dynamics of the normal microscopic variables could be achieved. Later, the direct version was introduced (Iannuzzi *et al.*, 2003) to facilitate the implementation of the algorithm in conjunction with molecular dynamics.

The external, history-dependent potential, $V_G(\xi(X), t)$, acting on the system at a time t is of the form

$$V_G(\xi(X), t) = \frac{w}{\tau_G}\int_0^t \exp\left(-\frac{|\xi(X) - \xi(X(t'))|^2}{2|\delta s|^2}\right) dt', \tag{7.15}$$

where $\xi(X)$ is a set of CVs which are a function of the atomic coordinates of the system, X, and are added to the potential according to the classical

Newtonian forces acting on the particles. δs and w are the width and height of the Gaussians, τ_G is the constant rate at which successive Gaussians are deposited, and $X(t')$ is the trajectory of the system. After a sufficiently long time $V_G(\xi(X), t)$ is assumed to provide an estimate of the underlying FE, $A(\xi)$:

$$\lim_{t \to \infty} V_G(\xi(X), t) \approx -A(\xi). \tag{7.16}$$

In other words, the FE, an equilibrium quantity, is estimated by non-equilibrium dynamics, in which the underlying potential is changed every time a new Gaussian is added.

If information on the nature of the FE landscape is unavailable, the scaling parameters are chosen by performing short coarse-grained dynamics without any biasing potential. The efficiency of the method in filling a well can be estimated by the number of Gaussians that are required to fill the well, which is proportional to $(1/\delta s)^n$, where n is the dimensionality of the problem. The characteristic parameters of the Gaussian, w and δs, have to be carefully chosen in order to obtain a balance between accuracy and sampling efficiency.

An analytical expression for the error as a function of the parameters of the method—namely the width and height of the Gaussians and the time interval of deposition, and of system-specific parameters such as the system size (S), temperature (T), and diffusion coefficient (D)—has also been reported (Laio and Parrinello, 2005):

$$\varepsilon = C(d)\sqrt{\frac{\delta s T S w}{D \tau_G}}, \tag{7.17}$$

where $C(d)$ is a constant that depends on the dimensionality d, that is, the number of CVs.

Correct choice of CVs is essential to this technique as the method suffers from serious hysteresis if the appropriate CVs are not selected. In some cases, there is a natural choice of the CVs dictated by the problem at hand; however on many occasions, a knowledge of the system beyond that which is initially available is required. Subsequently, selection of CVs is often an arbitrary process where experience plays a significant role, and in the absence of an *a priori* recipe for choosing the correct CVs, one has to proceed by trial and error. CVs are system-specific and therefore selecting them requires intuition and a good understanding of the physical or chemical process of interest. Primarily, they should describe all the slow events that are relevant to the process of interest and be capable of differentiating between the initial, intermediate, and final states. In some cases, from the dynamical behavior of the system observed in one simulation, the intuitively chosen reaction coordinate can be improved.

The possibility of tracing more than one or two collective degrees of freedom independently, such as bond lengths, bond and torsion angles, or coordination numbers, allows for the study of concerted reactions, although if too many CVs are used, filling the FE surface can take a very long time. The diversity of possible CVs that can be used in metadynamics leads to a potential range of applications of this method beyond that of either constrained molecular dynamics or umbrella sampling.

8. Applications of Free-Energy Methods to Study Ion Channel Properties

The literature is particularly rich with examples of applications of FE methods to the study of membrane proteins and ion channels in particular. In this section, just a few representative examples are reviewed with the aim of illustrating the applicability of the methods described previously.

8.1. Free-energy perturbation and ion permeation and selectivity

The selectivity filter of K^+ channels is the crucial structural element controlling permeation and selectivity. Four identical subunits are symmetrically disposed to form the central pore, through which ions travel. Each contains a conserved signature peptide sequence: the TVGYG motif. Key carbonyl oxygen atoms together with the side-chain hydroxyl oxygen of the threonine define four equally spaced ion-binding sites, S1 to S4, labeled from the extracellular to the intracellular region. Another binding site, S0, is described just above S1.

Results from extensive MD simulations and FEP calculations (Aqvist and Luzhkov, 2000) using the KcsA channel established the nature of the multiple ion conduction mechanism, showing that out of the 16 theoretically possible loading states of the filter, the most favorable pathway that connects subsets of these states in a cyclic fashion involves transitions between two main configurations, namely S1S3 and S2S4, with a FE difference of $\sim$5 kcal/mol. It was also possible to report that the channel has a preference for accommodating a total of three ions, two in the selectivity filter, and one in the central water-filled cavity. The FEP treatment provides a way to take into account only local structural perturbations without moving the entire group of ions and water through the pore.

To examine the selectivity properties, FEP calculations were performed in which two K^+ ions in the configurations S2S4 and S1S3 were simultaneously mutated to Na^+ and Rb^+, respectively, revealing that both states are selective for K^+. The values reported were in good agreement with selectivity measurements in KcsA (LeMasurier *et al.*, 2001). Subsequently, these authors addressed the protonation states of acidic groups located close to the channel selectivity filter and which are thought to play an important role in conduction (Cordero-Morales *et al.*, 2006). FEP calculations suggested that Glu71 in KcsA is most stable in its neutral form, while Asp80 is negatively charged. Other authors have also employed FEP calculations to study selectivity in K^+ channels, in particular KcsA, KirBac, and NaK channels, and other membrane proteins (Berneche and Roux, 2001; Domene and Furini, 2009; Noskov and Roux, 2005, 2007).

8.2. Umbrella sampling and conduction mechanisms

The average time required for the study and analysis of permeation of a single ion through a K^+ channel is much longer than can be simulated with brute-force MD. To avoid this limitation, the energetics of K^+ conduction through the KcsA K^+ channel was studied by means of umbrella sampling (Berneche and Roux, 2001). The conduction process was described to involve transitions between two main states with two or three K^+ ions. Ions travel along the pore axis of the channel in single file accompanied by a single water molecule between the K^+ ions. The largest FE barrier for this process was proposed to be of the order of 2–3 kcal/mol.

Recently, we have computed the energetics of ion conduction along two different pathways using the umbrella sampling technique. The first one (KWK) features the ion–water arrangement originally proposed by crystallographers (Morais-Cabral *et al.*, 2001; Zhou *et al.*, 2001) and supported by modeling studies(Berneche and Roux, 2001) with two K^+ ions separated by a water molecule (Fig. 7.3). The second pathway studied (KK) has ions at adjacent binding sites (Fig. 7.3). This alternative pathway entails the possible presence of vacancies, that is, neither K^+ ions nor water molecules in certain sites. We proposed that coexistence of several ion permeation mechanisms is energetically possible. Thus, conduction can be described as a more anarchic phenomenon than previously characterized by the concerted translocations of K^+–water–K (Furini and Domene, 2009).

In another study, using MD simulations, umbrella sampling, and FEP calculations, the stability of the SF, the selectivity of its binding sites, and permeation events of the low-K^+ KcsA structure in an explicit lipid bilayer were investigated (Domene and Furini, 2009) and compared with other structures described in the literature (Berneche and Roux, 2005; Domene *et al.*, 2008a; Furini *et al.*, 2009), thought to be in inactivated states. The X-ray structure of the KcsA channel obtained in the presence of low-K^+ concentration is thought to be representative of a K^+ channel in the C-type inactivated state. While the structural analysis of the low-K^+ KcsA static conformation suggests that pore-lining amide hydrogen atoms would

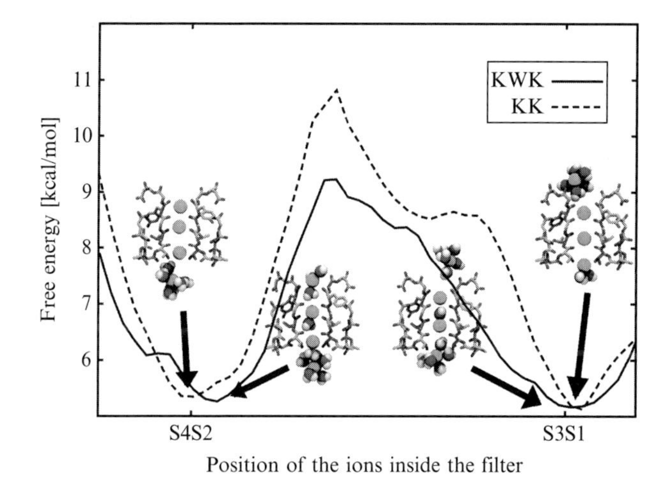

Figure 7.3 Free energy for the movement of the ions inside the selectivity filter of the K^+ channel KcsA in the KWK and KK conduction mechanisms.

prevent the permeation of ions, uncertainties remain about its stability under physiological conditions and its ion occupancy state. It was found that the low-K^+ KcsA structure is stable on the timescale of the MD simulations performed, and that ions preferably remain in S1 and S4. In the absence of ions, the selectivity filter evolves toward an asymmetric architecture, as already observed in other computations with metadynamics of the high-K^+ structure of KcsA and KirBac (and described later in this chapter). The low-K^+ KcsA structure is not permeable by Na^+, K^+, or Rb^+, and the selectivity of its binding sites is different from that of the high-K^+ structure.

The atomic structure of the K^+-depleted KcsA and the mechanisms of water permeation was also recently characterized by means of all-atom MD simulations, in conjunction with umbrella sampling and a nonequilibrium approach to simulate pressure gradients (Furini *et al.*, 2009). Previous studies had reported that KcsA has an osmotic permeability coefficient of 4.8×10^{-12} cm^3/s, giving it a significantly higher osmotic permeability coefficient than that of some membrane channels specialized in water transport. This high osmotic permeability is proposed to occur when the channel is depleted of K^+ ions, the presence of which slows down the water permeation process. Computations revealed that the crystallographic structure of the channel in high-K^+ concentration is not permeable by water molecules moving along the channel axis and an alternative permeation pathway was identified with a computed osmotic permeability in agreement with the experiments.

8.3. Adaptive biasing force and barriers of translocation

ABF simulations were carried out to investigate structural mechanisms of ion translocation and selectivity in the α_1-GlyR and α_7-nAChr ligand-gated ion channels (Ivanov *et al.*, 2007). These channels belong to the Cys-loop family of receptors whose major function is to bind the neurotransmitter and change the membrane potential through selective permeation of ions. nAChr selects cations and conducts Na^+ and K^+ ions and in some subunits also Ca^{2+}, where by contrast, GlyR conducts Cl^- anions. Together, they mediate the majority of fast synaptic transmission throughout the nervous system.

In this study, the translocation coordinate was chosen as the collective variable. The simulations were carried out in 10 windows of length 5 Å along the direction of the translocation, sufficient to cover the entire length of the transmembrane region of the two receptors. Within each window, the average force acting on the selected ion was accumulated in 0.1 Å sized bins. Smoothing within 0.2 Å was applied by averaging the content of the adjacent bins. Application of the adaptive bias was initiated only after accumulation of 800 samples in individual bins, subsequently to which the biasing force was introduced progressively in the form of a linear ramp. Production runs in each window were of ~3 ns, rendering a total simulation time of 150 ns.

In this manner, free energies were calculated for transport of Na^+ and Cl^- through each of the two channels, testing the hypothesis that hydrophobic effects in the transmembrane region are responsible for channel gating, this being confirmed for nAChr, but no significant hydrophobic barrier being identified for Cl^- in GlyR.

Another study is reported in the literature (Henin *et al.*, 2008) where the FE landscape characterizing the assisted transport of glycerol by the glycerol uptake facilitator, GlpF, is explored by means of ABF simulations. Intramolecular relaxation is proposed to play a crucial role on the diffusion properties of the permeant species. Reorientation and conformational equilibrium of glycerol in GlpF is proposed to represent the obstacle in the permeation process. To understand how the measured FE is affected by the choice of the reaction coordinate, both the center of mass of the entire channel and that of the selectivity filter were examined. A barrier of ~8.7 kcal/mol is found for the translocation of glycerol through the selectivity filter of this channel, in agreement with a value of 9.5 ± 1.5 kcal/mol quoted from experiments.

8.4. Metadynamics and gating at the selectivity filter of K^+ channels

In combination with umbrella sampling, the metadynamics method has been applied to study the conduction of K^+ ions through the KirBac channel, and the conformational changes and gating mechanisms associated

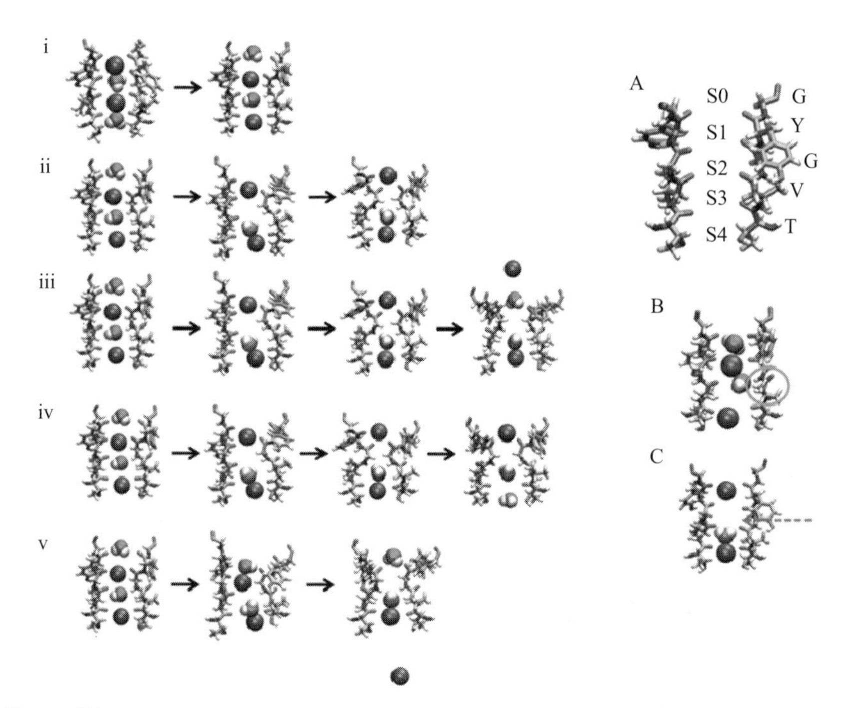

Figure 7.4 Metadynamics. (i–v) Sequence of movements of ions coupled to backbone conformational changes, during a metadynamics run. (A) Atomic detail of the SF in a representative conductive state. Labels S0 to S4 indicate the ion-binding sites along the TVGYG motif of the SF. (B) Representative snapshot of the SF where a Val carbonyl flips. This event is commonly observed in classical MD simulations of nanosecond time length. (C) Snapshot of the conformation of other two chains not shown in parts (i)–(v) for simplicity; S2 looks empty, although in fact that volume is occluded by the residues of the other two chains (those presented in diagrams (i)–(v)).

with the translocation of ions through the selectivity filter(Domene *et al.*, 2008b). Initially, the ions occupied canonical positions S1 and S3. Figure 7.4i–v shows some of the mechanisms for SF gating observed using the position of two K^+ ions along the axis of the pore as CVs. Unequivocally, the sequence of ion movements is coupled to changes in the backbone conformations. A new two-fold symmetric conformation is identified, which supports the existence of a physical gate or constriction in the selectivity filter of K^+ channels. This structure is likely associated with C-type inactivation. There are several interesting features which differentiate this structure from those obtained crystallographically at high- and low-K^+ concentration. For instance, only two out of the four chains that constitute the pore alter their conformation; the other two remain as in the high-K^+ channel. Figure 7.4A shows two opposite chains that resemble those of the channel in high-K^+ concentration and Fig. 7.4C shows two of the chains of the "locked" conformation observed in metadynamics. In this snapshot, it can be observed that S2 appears empty, although in reality the two other chains occlude this site as the valine carbonyl adopts a different orientation in two out of the four chains. However, this valine carbonyl flip is more pronounced than that observed in classical MD, and showed in Fig. 7.4B. In addition, the volume of S4 is greater, increasing the accessibility of water in this region. K^+ ions can be only found in S1 and S4. S3 is in general occupied by a water molecule and S2 is physically inaccessible.

A path-based CV method (Branduardi *et al.*, 2007) was also employed to confirm the mechanism that lead to this "locked" structure and a barrier for the conformational transition of ~5–6 kcal/mol was found.

This example, therefore, provides a very enlightening illustration of how metadynamics can be used to enhance the sampling as a function of a reduced number of predefined CVs, and reconstruct the FE surface of a *rare* process, in this particular case, the gating mechanism. In this context, it also illustrates the ability of the method to explore a transition mechanism, through the connection of the *a priori* known initial and final states, for example, the closed and open SF. More importantly, it also highlights the potential of metadynamics as a method by which new structures can be predicted.

9. Conclusions and Future Outlook

Improvements in computer hardware combined with development of new algorithms have led to the existence of a vast range of computational techniques available for use in the molecular modeling of biological processes. Energetics, rather than dynamics, often provides the key link between biological function and experimentally observable properties of a system, and hence understanding the underlying FE of biophysical processes is of crucial importance. At present, the techniques described in this chapter can be applied in a predictive and reliable fashion. However, several limitations still exist. Besides the importance and delicate task of adequately sampling the slow degrees of freedom, in charged systems, like the ones considered here, explicit inclusion of polarization could potentially be critical to render accurate FE differences. Yet polarization effects and any many-body interactions are in general neglected in common pair potential, or just considered in an average fashion in effective pair potentials because of their computational expense (Illingworth and Domene, 2009). The issue of system size accessible to simulation follows. A way to overcome this particular difficulty is to abandon the all-atom description, and use instead coarse-grained models. These are reduced representations, retaining only those characteristics of a molecule which are essential. However, this approach probably requires a detailed knowledge of the systems that is often unavailable. On the other hand, multiscale modeling which combines different levels of calculations shows a great deal of promise. It is important to stress that depending on the specifics of the problem, one technique may be preferable over the others, no one technique providing a catch-all solution. The challenge is to choose among these diverse and sometimes sophisticated techniques competently. This, in combination with critical analyses of the results, will certainly shed some light into the crucial issues of interest to biologists and biophysicists. In particular, in the field of ion channels, many questions concerning gating, permeation, and selectivity are still to be answered.

ACKNOWLEDGMENTS

C.D. thanks The Royal Society for a University Research Fellowship, the EPSRC, and The Leverhulme Trust.

REFERENCES

Aqvist, J., and Luzhkov, V. (2000). Ion permeation mechanism of the potassium channel. *Nature* **404,** 881–884.

Berneche, S., and Roux, B. (2001). Energetics of ion conduction through the K^+ channel. *Nature* **414,** 73–77.

Berneche, S., and Roux, B. (2005). A gate in the selectivity filter of potassium channels. *Structure* **13,** 591–600.

Beutler, T. C., Mark, A. E., Vanschaik, R. C., Gerber, P. R., and Vangunsteren, W. F. (1994). Avoiding singularities and numerical instabilities in free-energy calculations based on molecular simulations. *Chem. Phys. Lett.* **222,** 529–539.

Beveridge, D. L., and Dicapua, F. M. (1989). Free-energy via molecular simulation—Applications to chemical and biomolecular systems. *Annu. Rev. Biophys. Biophys. Chem.* **18,** 431–492.

Branduardi, D., Gervasio, F. L., and Parrinello, M. (2007). From A to B in free energy space. *J. Chem. Phys.* **126,** 054103.

Carpenter, E. P., Beis, K., Cameron, A. D., and Iwata, S. (2008). Overcoming the challenges of membrane protein crystallography. *Curr. Opin. Struct. Biol.* **18,** 581–586.

Chandler, D. (1987). Introduction to Modern Statistical Mechanics. Oxford University Press, New York, NY.

Chipot, C., and Pohorille, A. (2007). Free Energy Calculations. Springer-Verlag, Heidelberg.

Cordero-Morales, J. F., Cuello, L. G., Zhao, Y. X., Jogini, V., Cortes, D. M., Roux, B., and Perozo, E. (2006). Molecular determinants of gating at the potassium-channel selectivity filter. *Nat. Struct. Mol. Biol.* **13,** 311–318.

Cossins, B. P., Foucher, S., Edge, C. M., and Essex, J. W. (2009). Assessment of nonequilibrium free energy methods. *J. Phys. Chem. B* **113,** 5508–5519.

Darve, E., and Pohorille, A. (2001). Calculating free energies using average force. *J. Chem. Phys.* **115,** 9169–9183.

den Otter, W. K., and Briels, W. J. (1998). The calculation of free-energy differences by constrained molecular-dynamics simulations. *J. Chem. Phys.* **109,** 4139–4146.
Domene, C., and Furini, S. (2009). Dynamics, energetic and selectivity of the low-K^+ KcsA channel structure. *J. Mol. Biol.* **389,** 637–645.
Domene, C., Klein, M., Branduardi, D., Gervasio, F., and Parrinello, M. (2008a). Conformational changes and gating at the selectivity filter of potassium channels. *J. Am. Chem. Soc.* **130,** 9474–9480.
Domene, C., Klein, M. L., Branduardi, D., Gervasio, F. L., and Parrinello, M. (2008b). Conformational changes and gating at the selectivity filter of potassium channels. *J. Am. Chem. Soc.* **130,** 9474–9480.
Doyle, D. A., Cabral, J. M., Pfuetzner, R. A., Kuo, A. L., Gulbis, J. M., Cohen, S. L., Chait, B. T., and MacKinnon, R. (1998). The structure of the potassium channel: Molecular basis of K^+ conduction and selectivity. *Science* **280,** 69–77.
Furini, S., and Domene, C. (2009). Atypical mechanism of conduction in potassium channels? *Proc. Natl. Acad. Sci. USA* **106,** 16074–16077.
Furini, S., Beckstein, O., and Domene, C. (2009). Permeation of water through the KcsA K^+ channel. *Proteins* **74,** 437–448.
Gullingsrud, J. R., Braun, R., and Schulten, K. (1999). Reconstructing potentials of mean force through time series analysis of steered molecular dynamics simulations. *J. Comp. Phys.* **151,** 190–211.
Henin, J., and Chipot, C. (2004). Overcoming free energy barriers using unconstrained molecular dynamics simulations. *J. Chem. Phys.* **121,** 2904–2914.
Henin, J., Tajkhorshid, E., Schulten, K., and Chipot, C. (2008). Diffusion of glycerol through *Escherichia coli* aquaglyceroporin GlpF. *Biophys. J.* **94,** 832–839.
Hill, T. L. (1987). An Introduction to Statistical Thermodynamics. Dover Publications, New York, NY.
Iannuzzi, M., Laio, A., and Parrinello, M. (2003). Efficient exploration of reactive potential energy surfaces using Car-Parrinello molecular dynamics. *Phys. Rev. Lett.* **90,** 238302.
Illingworth, C. J., and Domene, C. (2009). Many-body effects and simulations of potassium channels. *Proc. R. Soc. A* **465,** 1701–1716.
Ivanov, I., Cheng, X. L., Sine, S. M., and McCammon, J. A. (2007). Barriers to ion translocation in cationic and anionic receptors from the Cys-loop family. *Proc. R. Soc. A* **129,** 8217–8224.
Jorgensen, W. L., and Ravimohan, C. (1985). Monte Carlo simulation of differences in free-energies of hydration. *J. Chem. Phys.* **83,** 3050–3054.
Kumar, S., Bouzida, D., Swendsen, R. H., Kollman, P. A., and Rosenberg, J. M. (1992). The weighted histogram analysis method for free-energy calculations on biomolecules. 1. The method. *J. Comput. Chem.* **13,** 1011–1021.
Laio, A., and Parrinello, M. (2002). Escaping free-energy minima. *Proc. Natl. Acad. Sci. USA* **99,** 12562–12566.
Laio, A., and Parrinello, M. (2005). Computing free energies and accelerating rare events with metadynamics. *In* "Conference on Computer Simulations in Condensed Matter Systems," (M. Ferrario, G. Ciccotti, and K. Binder, eds.), Erice, Italy, pp. 315–347.
LeMasurier, M., Heginbotham, L., and Miller, C. (2001). KcsA: It's a potassium channel. *J. Gen. Physiol.* **118,** 303–313.
Mezei, M. (1987). Adaptive umbrella sampling—Self-consistent determination of the non-Boltzmann bias. *J. Comput. Phys.* **68,** 237–248.
Mitchell, M. J., and McCammon, J. A. (1991). Free-energy difference calculations by thermodynamic integration—Difficulties in obtaining a precise value. *J. Comput. Chem.* **12,** 271–275.
Morais-Cabral, J. H., Zhou, Y., and MacKinnon, R. (2001). Energetic optimization of ion conduction by the K^+ selectivity filter. *Nature* **414,** 37–42.

Noskov, S. Y., and Roux, B. (2005). Ion selectivity in potassium channels 230th National Meeting of the American Chemical Society, Washington, DC, pp. 279–291.

Noskov, S. Y., and Roux, B. (2007). Importance of hydration and dynamics on the selectivity of the KcsA and NaK channels. *J. Gen. Physiol.* **129,** 135–143.

Pearlman, D. A., and Kollman, P. A. (1989). A new method for carrying out free-energy perturbation calculations—Dynamically modified windows. *J. Chem. Phys.* **90,** 2460–2470.

Torrie, G. M., and Valleau, J. P. (1977). Non-physical sampling distributions in Monte Carlo free-energy estimation—Umbrella sampling. *J. Comput. Phys.* **23,** 187–199.

Zhou, Y., Morais-Cabral, J. H., Kaufman, A., and MacKinnon, R. (2001). Chemistry of ion coordination and hydration revealed by a K^+ channel-Fab complex at 2.0 Å resolution. *Nature* **414,** 43–48.

Zwanzig, R. W. (1954). High-temperature equation of state by a perturbation method. 1. Nonpolar gases. *J. Chem. Phys.* **22,** 1420–1426.

CHAPTER EIGHT

Examining Cooperative Gating Phenomena in Voltage-Dependent Potassium Channels: Taking the Energetic Approach

Ofer Yifrach, Nitzan Zandany, *and* Tzilhav Shem-Ad

Contents

Department of Life Sciences and the Zlotowski Center for Neurosciences, Ben-Gurion University of the Negev, Beer Sheva, Israel

Methods in Enzymology, Volume 466
ISSN 0076-6879, DOI: 10.1016/S0076-6879(09)66008-0

Abstract

Allosteric regulation of protein function is often achieved by changes in protein conformation induced by changes in chemical or electrical potential. In multi-subunit proteins, such conformational changes may give rise to cooperativity in ligand binding. Conformational changes between open and closed states are central to the function of voltage-activated potassium (Kv) channel proteins, homotetrameric pore-forming membrane proteins involved in generating and shaping action potentials in excitable cells. Accessible to extremely high signal-to-noise ratio in functional measurements, combined with the availability of high-resolution structural data for different conformations of the protein, the Kv channel represents an excellent allosteric model system to further understand the aspects of synergism and cooperative effects in protein function. In this chapter, we demonstrate how the use of the simple law of mass action combined with thermodynamic mutant cycle energetic coupling analysis of Kv channel gating can be used to provide valuable information regarding (1) how cooperativity in Kv channel pore opening can be assessed; (2) how one can directly discriminate whether conformational transitions during Kv channel pore opening occur in a concerted or sequential manner; and (3) how mechanistically, the coupling between distant activation gate and selectivity filter functional elements of the prototypical *Shaker* Kv channel protein might be achieved. In addition to providing valuable insight into the function of this important protein, the conclusions reached at using high-order thermodynamic energetic coupling analysis applied to the Kv channel allosteric model system reveal much about the function of allosteric proteins, in general.

1. Introduction

Synergistic phenomena in protein function are of immense importance in many fundamental biological processes, such as signal transduction, protein allostery, and electrical signaling, to name but a few. Such phenomena usually rely on long-range communication between distal functional elements of the protein(s), manifested as nonadditive cooperative effects on protein function (LiCata and Ackers, 1995). The development of the conceptual framework of thermodynamic double-mutant cycle coupling analysis by Fersht and colleagues (Carter *et al.*, 1984) provides solid ground for a deeper understanding of synergism in protein function. This approach has allowed for evaluation of the nature and magnitude of the coupling free energy between two protein residues, two modules of a protein, two interacting proteins, or protein–DNA interactions, during various thermodynamic processes, including folding, binding, enzyme catalysis, or conformational transitions (Horovitz, 1996). Furthermore, this analytic approach was further extended to examine the complex hierarchical shells of cooperative effects involving additional protein residues, protein modules, or more interacting partners (Horovitz and Fersht, 1990).

Recently, it became clear that the membrane-spanning, voltage-activated potassium (Kv) channel, a homotetrameric protein, involved in generating electrical signals in the nervous system (Hille, 2001) can serve as an excellent allosteric model system to address issues related to nonadditive effects in protein function (Hidalgo and Mackinnon, 1995; Ranganathan *et al.*, 1996; Sadovsky and Yifrach, 2007; Yifrach and MacKinnon, 2002). Due to its extremely high signal-to-noise ratio in electrophysiological recording measurements, its fast and reliable readout, and the fact that this protein can be studied in its natural membrane environment, thus eliminating the need to purify the protein, mutant cycle coupling analysis can be applied to the study of the Kv channel protein to yield novel and general insight into synergistic phenomena in proteins (Ben-Abu *et al.*, 2009; Sadovsky and Yifrach, 2007; Zandany *et al.*, 2008). In addition, the wealth of structural data available on various K^+ channels (Doyle *et al.*, 1998; Jiang *et al.*, 2002), in particular on Kv channels (Long *et al.*, 2005, 2007), makes it possible to propose mechanistic explanations regarding how long-range communication in proteins might be achieved (Sadovsky and Yifrach, 2007).

In what follows, a rigorous description of high-order mutant cycle energetic coupling analysis and an introduction to the Kv channel model allosteric system are given, followed by a demonstration of how the use of this strategy, in the case of Kv channel pore opening, yielded valuable insight not only into the function of this important protein, but also into allosteric proteins, in general.

2. High-Order Thermodynamic Mutant Cycle Coupling Analysis

A leap forward in our understanding of synergistic effects on protein function came with the development, by Fersht and coworkers, of the conceptual framework of thermodynamic double-mutant cycle coupling analysis (Carter *et al.*, 1984). This analysis was later generalized and extended to examine the complex hierarchical shells of cooperative effects involving additional protein residues, protein modules, or other interacting partners (Horovitz and Fersht, 1990). Although description of this approach can be found in the seminal papers by Horovitz and Fersht (1990, 1992) and elsewhere (Horovitz, 1996), the following rigorous step-by-step portrayal of the method highlights, in an explicit manner, several important points to facilitate the practice of this conceptual framework.

2.1. Double-mutant cycle coupling analysis

Double mutant cycle analysis represents a formal method for quantifying the independence of two mutations in terms of their effects on the function of a protein (Horovitz and Fersht, 1990). Consider, for example, two protein residues, *i* and *j* (residues 1 and 2, respectively). Each residue may separately alter the function of the protein, although their effects may or may not be additive when the mutations are combined. To better explain this concept, consider the two-dimensional square in Fig. 8.1A describing the double-mutant cycle relating the wild-type (WT), single mutants (M1 or M2), and double-mutant (M1 and M2) proteins. In such a cycle, the free energy changes upon mutation of residue *i* in the presence (ΔG_1) or absence ($\Delta G_{1/2}$) of the native *j* position are compared (parallel vertical arrows along the cycle in Fig. 8.1A). If the free energy change of mutating residue *i* is independent of the presence or absence of residue *j* (i.e., $\Delta G_1 = \Delta G_{1/2}$),

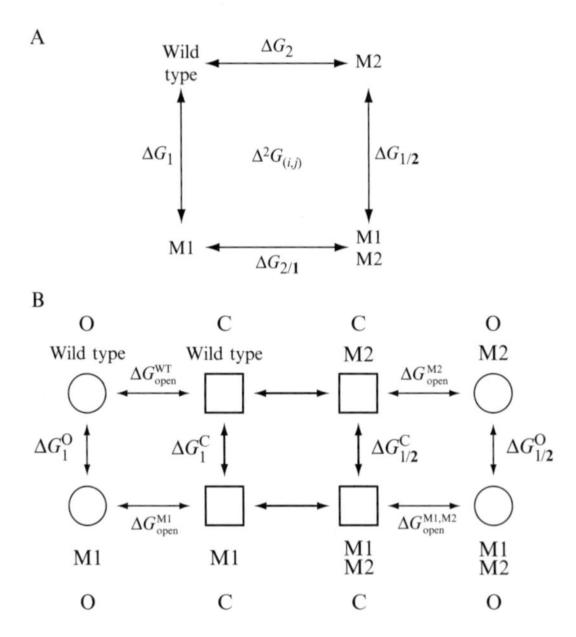

Figure 8.1 Double-mutant cycle analysis. (A) A two-dimensional thermodynamic construct used to measure the pairwise coupling free energy between residues *i* and *j* (residues 1 and 2, respectively). M1, M2, and M1M2 denote the two single and double mutants, respectively. (B) Double-mutant cycle analysis enables the measurement of state-dependent coupling. Shown for the particular case of voltage-dependent channel gating, the middle cycle in bold and the cycle comprising the outer four corners represent the double-mutant cycles used to measure the coupling free energies between residues *i* and *j* in the closed (C) and open (O) states, respectively. Refer to the main text body for notations' description.

then the two residues are not coupled and $\Delta^2 G_{(i,j)}$, the second-order (pairwise) coupling free energy between residues *i* and *j*, defined by Eq. (8.1), is equal to zero:

$$\Delta^2 G_{(i,j)} = \Delta G_1 - \Delta G_{1/2}. \tag{8.1}$$

This is the additive case, where the two residues tested do not interact. Nonadditivity in mutational effects is indicated by $\Delta^2 G_{(i,j)} \neq 0$, suggesting the two residues to be coupled. The extent of deviation of $\Delta^2 G_{(i,j)}$ from zero upon which a residue pair is defined as coupled (i.e., the threshold coupling value) is protein system specific and requires measurements of several such pairs to realize its value. From symmetry considerations along a closed thermodynamic cycle, it follows that

$$\Delta^2 G_{(i,j)} = \Delta G_1 - \Delta G_{1/2} = \Delta G_2 - \Delta G_{2/1}. \tag{8.2}$$

Obviously, the free energy change upon residue mutation (described by the different energy terms in Eq. 8.2) cannot be directly measured. However, this quantity can be evaluated using any thermodynamic property associated with the function of the protein under study. This can be realized by considering the thermodynamic construct depicted in Fig. 8.1B, describing a double-mutant cycle applied to the case of ion channel gating between the closed (C) and open (O) states. Although this example relates to the specific case of channel gating, its outcome, in essence, is general (see below). In this figure, the central bolded cycle is a double-mutant cycle to measure the coupling between residues *i* and *j* in the closed channel state ($\Delta^2 G_{(i,j)}(C)$), as given in Eq. (8.3):

$$\Delta^2 G_{(i,j)}(C) = \Delta G_1^{C} - \Delta G_{1/2}^{C}, \tag{8.3}$$

whereas, as before (Eq. 8.1), ΔG_1^{C} and $\Delta G_{1/2}^{C}$ represent the free energy change upon mutating residue *i* in the presence or absence of the native *j* position, respectively, in the closed channel state. These two quantities can be indirectly calculated using the respective left and right outward closed thermodynamic cycles, as follows:

$$\Delta G_1^{C} = \Delta G_{\text{open}}^{\text{WT}} + \Delta G_1^{O} - \Delta G_{\text{open}}^{\text{M1}}, \tag{8.4}$$

$$\Delta G_{1/2}^{C} = \Delta G_{\text{open}}^{\text{M2}} + \Delta G_{1/2}^{O} - \Delta G_{\text{open}}^{\text{M1,M2}}. \tag{8.5}$$

In Eqs. (8.4) and (8.5), ΔG_1^{O} and $\Delta G_{1/2}^{O}$ represent the free energy change upon mutating residue *i* in the open channel state in the presence or absence of the native *j* position, respectively. All other terms are the experimentally measurable free energy of channel opening of the four indicated wild-type or mutant channel proteins. These latter expressions represent the closed-to-open equilibrium gating reaction. Substituting the expressions for ΔG_1^C and $\Delta G_{1/2}^C$ into Eq. (8.3), followed by term rearrangement, yields the following expression for $\Delta^2 G_{(i,j)}(C)$:

$$\Delta^2 G_{(i,j)}(C) = (\Delta G_{\text{open}}^{\text{WT}} - \Delta G_{\text{open}}^{\text{M1}}) - (\Delta G_{\text{open}}^{\text{M2}} - \Delta G_{\text{open}}^{\text{M1,M2}}) + \Delta^2 G_{(i,j)}(O), \tag{8.6}$$

where $\Delta^2 G_{(i,j)}(O) (= \Delta G_1^O - \Delta G_{1/2}^O)$ is the pairwise coupling free energy between residues *i* and *j* in the open channel state, as delineated by the outer corners cycle in Fig. 8.1B. Equation (8.6) can be rearranged into a final form to yield an expression for $\Delta^2 G_{(i,j)}$ in terms of thermodynamically measurable quantities of the process being studied:

$$\begin{aligned} \Delta^2 G_{(i,j)} &= \Delta^2 G_{(i,j)}(C) - \Delta^2 G_{(i,j)}(O) \\ &= (\Delta G_{\text{open}}^{\text{WT}} - \Delta G_{\text{open}}^{\text{M1}}) - (\Delta G_{\text{open}}^{\text{M2}} - \Delta G_{\text{open}}^{\text{M1,M2}}). \end{aligned} \tag{8.7}$$

Inspection of Eq. (8.7) reveals the important point that pairwise coupling free energies calculated using double-mutant cycles are not absolute but rather state dependent. Therefore, both the magnitude and sign of the coupling free energy are important for double-mutant cycle analysis (Aharoni and Horovitz, 1997; Horovitz and Fersht, 1990; Sadovsky and Yifrach, 2007). Whereas the extent of deviation of $\Delta^2 G_{(i,j)}$ from zero determines the strength (magnitude) of the interaction between the tested residue pair, the sign designator, whether positive or negative, indicates in what state the coupling is stronger. In the example of channel gating discussed here, a positive sign indicates an interaction that is stronger in the closed channel state than in the open, while a minus sign indicates the opposite.

At this point, several general remarks should be clarified. First, although delineated here for the channel-gating reaction [closed (C) to open (O) equilibrium], this approach is appropriate for any thermodynamic process associated with the protein of interest, such as a protein stability denaturation reaction [folded (F) to unfolded (U) equilibrium], ligation reaction [bound (B) to unbound (U) equilibrium], or conformational transitions associated with allosteric proteins (tense (T) to relaxed (R) equilibrium). Thus, the ΔG_{open} experimentally determined variable of the different wild-type or mutant proteins comprising the cycle in the example of the Kv channel can be replaced by protein denaturation free energies, ligand-binding free energy, or free energies of allosteric transitions, respectively. Second, the double-mutant cycle described here for measuring coupling between two residues within a protein can, in principle, be performed between two different proteins (Hidalgo and MacKinnon, 1995). Third, although in the analysis described here we relate to a "mutation" in that traditional sense (i.e., an amino acid substitution), "mutations" can be interpreted in a broader sense to describe any change associated with the protein (e.g., changes associated with complete functional elements as a whole) where we want to evaluate whether a functional effect is dependent on a second change or not. Fourth, although high pairwise coupling free energies usually correlate with the spatial proximity of the tested residues (Schreiber and Fersht, 1995), indirect long-range effects cannot be excluded. Finally, although double-mutant cycles are more easier to understand and interpret in terms of free energies of the process under study, calculation of coupling based on equilibrium constants is also frequently employed (recall that $\Delta G = -RT\ln K$). In such a case, the following equation, analogous to Eq. (8.7), describes the magnitude of second-order coupling between residues i and j ($\Omega^{(2)}_{(i,j)}$):

$$\Omega^{(2)}_{(i,j)} = \frac{\Omega^{(2)}_{(i,j)}(C)}{\Omega^{(2)}_{(i,j)}(O)} = \frac{K^{\mathrm{WT}}_{\mathrm{open}} K^{\mathrm{M1M2}}_{\mathrm{open}}}{K^{\mathrm{M1}}_{\mathrm{open}} K^{\mathrm{M2}}_{\mathrm{open}}}. \tag{8.8}$$

It can be seen that in this case, additivity in mutational effects is manifested by a $\Omega^{(2)}_{(i,j)}$ value of 1, whereas non-additive (cooperative) effects are manifested by $\Omega^{(2)}_{(i,j)}$ values greater and smaller than 1 (indicating closed- and open-state stabilizing interactions, respectively). The relation between $\Omega^{(2)}_{(i,j)}$ and $\Delta^2 G_{(i,j)}$ is given by Eq. (8.9):

$$\Delta^2 G_{(i,j)} = -RT\ln\Omega^{(2)}_{(i,j)}. \tag{8.9}$$

Before proceeding to describe high-order thermodynamic coupling analysis, an additional remark concerning the appropriate reference state for a mutation should be made. While in many structure–function studies amino acids are usually substituted by residues with similar physicochemical properties, we would like to advocate the notion that in mutant cycle coupling analysis, the reference of choice is alanine. Mutation to the smallest residue after glycine eliminates the interaction between the residue pair under study without the possible formation of a new interaction, thus facilitating interpretation of the coupling data, as realized by Faiman and Horovitz (1996).

2.2. High-order mutant cycle coupling analysis

The double-mutant cycle analytic formalism described here in great detail was further extended by Horovitz and Fersht (1990) to examine the complex hierarchical shells of cooperative effects involving additional protein residues, protein modules, or other interacting partners. For example, one may wish to know what is the energetic effect of a third residue, k, on the coupling free energy that exists between the (i, j) residue pair (with $\Delta^3 G_{(i,j)k}$ describing the third-order coupling) or how a second interaction pair involving residues k and l affects the magnitude of coupling between the (i, j) residue pair (with $\Delta^4 G_{(i,j)(k,l)}$ defining the fourth-order coupling between the two interacting pairs). Likewise, fifth-order coupling between five protein residues defines the effect of a fifth residue on the magnitude of coupling between the two (i, j) and (k, l) residue interaction pairs, while sixth-order coupling describes the effect of a third interaction pair on the magnitude of coupling between two other pairs. In general, the nth-order coupling term, $\Delta^n G$, can be defined by the nth-order coupling of n-protein residues (Horovitz and Fersht, 1990). It can be realized from the description given in this paragraph that high-order coupling terms address the context dependency of a given interaction, thus delineating the complex hierarchy of cooperative, nonadditive interactions involving more residues. Below, we provide a detailed and practical description of how third- and fourth-order coupling, for example, can be measured and their intuitive meaning understood. Again, the seminal paper by Horovitz and Fersht (1990) should be addressed if one wishes a more generalized description of high-order thermodynamic coupling analysis.

If, in performing of double-mutant cycle analysis, a thermodynamic square is required to measure the coupling between two adjacent residues (Fig. 8.1A), then a thermodynamic cube (Fig. 8.2A, upper panel) is needed to study how a third residue affects the magnitude of coupling between two adjacent amino acids. Similarly, a four-dimensional thermodynamic construct (intuitively represented by double-mutant cycle of double-mutant cycles) is needed to measure the magnitude of coupling between two residue interaction pairs (Fig. 8.2B). To calculate third-order coupling free energy involving residues i, j, and k ($\Delta^3 G_{(i,j)k}$), the free energies of the functional property associated with the eight different proteins found at the corners of the thermodynamic cubic construct described in Fig. 8.2A (upper panel) should be measured. These proteins include the wild-type protein, the three single mutants, the three possible double-mutant combinations, and the triple mutant. In this construct, the front and back faces of the cube are double-mutant cycles used to measure the interaction between residues i and j in the absence ($\Delta^2 G_{(i,j)}$) and presence ($\Delta^2 G_{(i,j)/\mathbf{k}}$) of a third mutation (in residue k), respectively (Fig. 8.2A, lower panel). Subtracting the coupling free energies calculated for each of these cycles gives a quantitative measure of the effect of a third residue (k) on the magnitude of interaction between the (i, j) reside pair as follows:

$$\Delta^3 G_{(i,j)k} = \Delta^2 G_{(i,j)} - \Delta^2 G_{(i,j)/\mathbf{k}}. \tag{8.10}$$

From symmetry considerations involving closed thermodynamic constructs, it should be noted that $\Delta^3 G_{(i,j)k} = \Delta^3 G_{(i,k)j} = \Delta^3 G_{(k,j)i}$. In other words, for a given residue triad, the effect of any third residue on the magnitude of coupling between the other two residues is identical (reflected in the three different pairs of parallel faces of the cubic construct in Fig. 8.2A).

Fourth-order coupling free energies ($\Delta^4 G_{(i,j)(k,l)}$) can be calculated using a four-dimensional thermodynamic construct more intuitively represented as a double-mutant cycle of double-mutant cycles in Fig. 8.2B. This construct is comprised of 16 proteins including the wild-type protein,

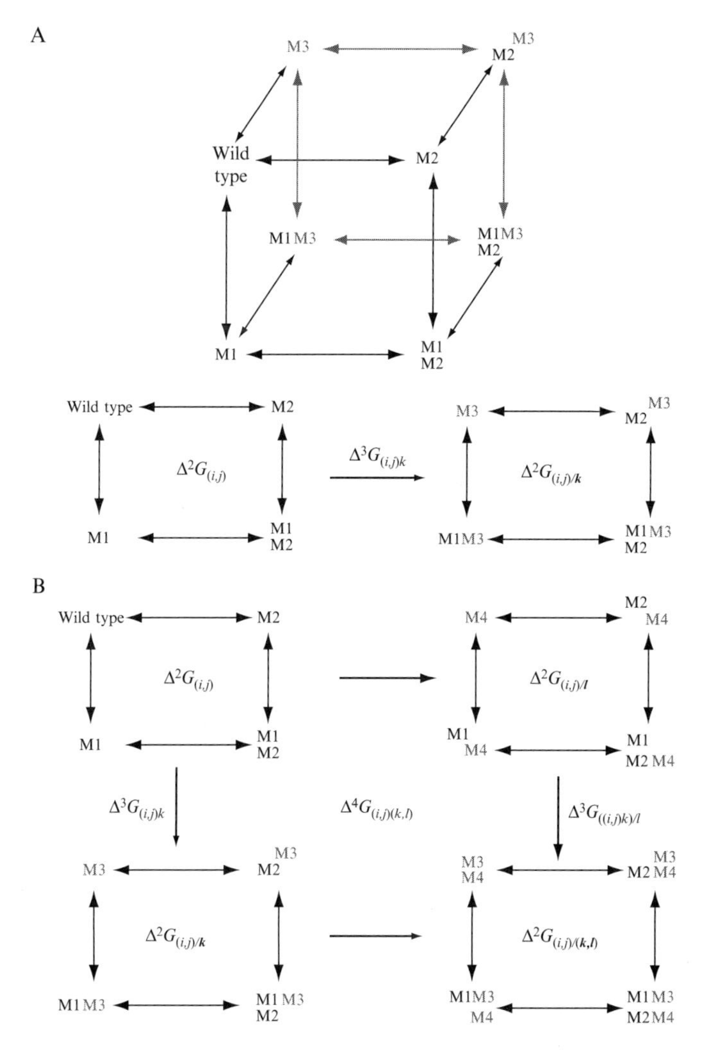

Figure 8.2 High-order thermodynamic mutant cycle coupling analysis. (A) A thermodynamic cube is used to calculate the effect of a third residue, k, on the interaction between residues i and j. The lower panel presents the two front and back cube faces used to calculate third-order coupling between the (i, j, k) residue triad ($\Delta^3 G_{(i,j)k}$). (B) A four-dimensional construct (i.e., double-mutant cycle of double-mutant cycles) is used to measure the effect of the interaction between the k and l residue pair on the magnitude of coupling between the i and j residue pair. Refer to the main text body for notations' description.

the 4 single mutants, the 6 double-mutant combinations, the 4 triple-mutant combinations, and the corresponding quadruple mutant. At the corners of this construct lay the double-mutant cycles used to measure the coupling free energy between residues i and j in the absence of any further mutation ($\Delta^2 G_{(i,j)}$, upper left corner) or in the presence of another single mutation ($\Delta^2 G_{(i,j)/\mathbf{k}}$ and $\Delta^2 G_{(i,j)/\mathbf{l}}$, corresponding to the respective lower left and upper right corners) or a double mutation ($\Delta^2 G_{(i,j)/(\mathbf{k},\mathbf{l})}$, corresponding to the lower right corner). $\Delta^4 G_{(i,j)(k,l)}$ is calculated according to the following equation:

$$\Delta^4 G_{(i,j)(k,l)} = \Delta^3 G_{(i,j)k} - \Delta^3 G_{((i,j)k)/\mathbf{l}} \qquad (8.11)$$

where $\Delta^3 G_{(i,j)k}$ is the third-order coupling between the residue pair (i, j) and residue k in the presence of the fourth native residue, l, while $\Delta^3 G_{((i,j)k)/\mathbf{l}}$ is the similar third-order coupling, only where the fourth residue l has been mutated (indicated by the bold face). It is worth noting that, as in the case of third-order coupling where a value for $\Delta^3 G_{(i,j)k}$ is calculated by subtracting two terms of the immediately lower one dimension ($\Delta^2 G_{(i,j)}$ terms corresponding to the parallel faces of the cube), fourth-order coupling is calculated by subtracting two third-order coupling terms representing differences between two three-dimensional cubes. A generalization of this principle to measure nth-order coupling involving n-protein residues can be found in Horovitz and Fersht (1990). The means to understand the meaning of high-order coupling using thermodynamic constructs, such as those described in Fig. 8.2, becomes less intuitive as the order of coupling one wishes to measure grows. Here again, from symmetry considerations, for any given four residues, fourth-order coupling is identical for any two pairs of interactions being tested for coupling.

The high-order thermodynamic coupling analysis outlined above highlights the requirement for measurements involving many wild-type and mutant proteins (8 in the case of a single third-order coupling and 16 in the case of a single fourth-order coupling). Such calculations, based on labor-intensive functional measurement of many mutant proteins, emphasize the need for a protein system exhibiting a high signal-to-noise ratio in functional measurements. One such unique protein system is the Kv channel protein, introduced in the section that follows.

3. The Voltage-Activated Potassium Channel Allosteric Model System

Kv channels are allosteric pore-forming proteins that open and close in response to changes in membrane potential (Bezanilla, 2000; Sigworth, 1994; Yellen, 1998). This form of gating regulates the flow of potassium ions across the membrane, a process underlying many fundamental biological processes, in particular the generation of nerve and muscle action potentials (Hille, 2001). Kv channels are homotetrameric modular proteins

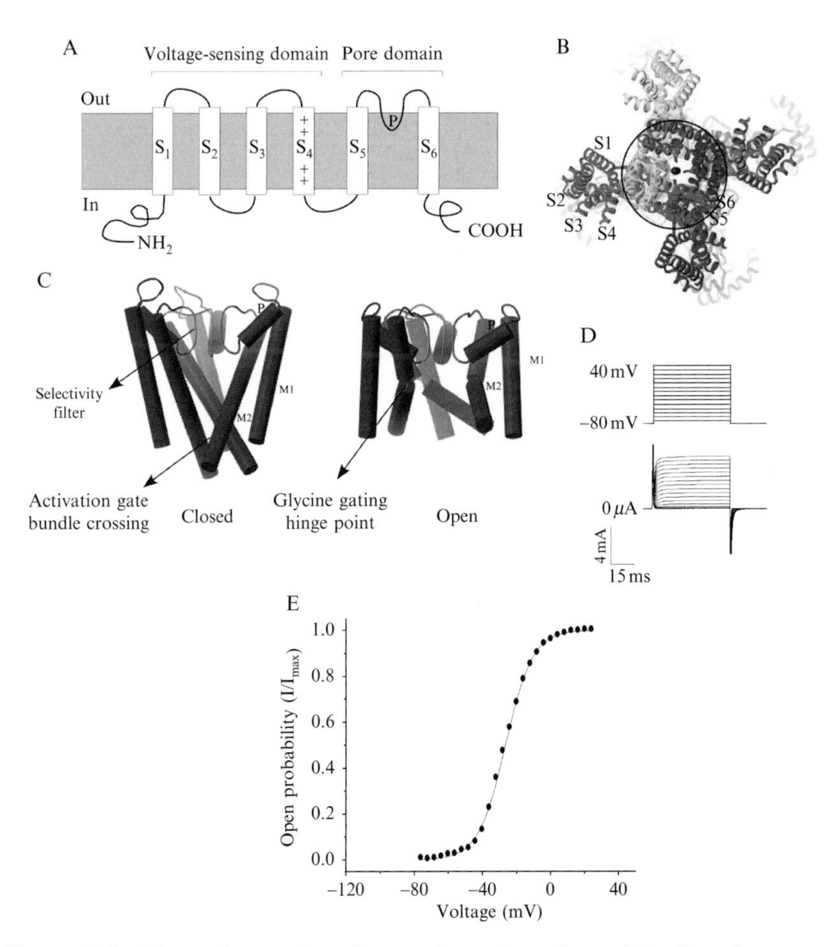

Figure 8.3 The voltage-activated potassium channel model allosteric system. (A) Membrane topology of Kv channels. The alternating basic residues of the S4 transmembrane helix are indicated by a "+" sign. (B) Crystallographic structure of the human Kv1.2 channel viewed from the intracellular side along the channel's fourfold symmetry axis. Each subunit is depicted in a different color. The black circle indicates the ion-conduction pore domain. Adapted with permission from Long *et al.*, 2005. (C) Helix-rod representations of the KcsA (closed) and MthK (open) pore structures, with M1, P, and M2 corresponding to the outer, pore, and inner helices, respectively. Adapted with permission from Yifrach, 2004. (D) Typical voltage protocol for wild-type *Shaker* Kv channel activation and traces of the elicited K^+ currents flowing through the membrane-expressed Kv channels. (E) Voltage-activation curve of the wild-type *Shaker* Kv channel. Smooth curve corresponds to a two-state Boltzmann function. (See Color Insert.)

comprised primarily of voltage-sensing and pore membrane-spanning domains (Fig. 8.3A). Whereas the voltage-sensing domain is composed of the first four transmembrane helices (S1–S4) of the protein, the pore domain composes the two last transmembrane helices (i.e., the S5 and S6 helices, corresponding to the respective M1 and M2 outer and inner helices of voltage-insensitive K^+ channels), along with the "selectivity filter"-containing P-loop element in between. The recent crystal structure of a voltage-dependent K^+ channel (Fig. 8.3B) reveals that the four channel subunits give rise to a central ion-conduction pore domain, composed of the pore segments, one from each subunit, whereas the four voltage-sensing domains lay at the periphery of the pore domain, in a seemingly loosely connected manner (Long *et al.*, 2005). In principle, the process of voltage-dependent gating is simple: Following membrane depolarization, structural rearrangements of the voltage-sensing domains are transmitted to the pore domain to allow channel opening (Sigworth, 1994).

The voltage sensitivity of Kv channels arises from voltage-induced displacement of alternating basic amino acids that are uniquely arranged along the S4 transmembrane helix of the voltage-sensing domain, a motif common to all voltage-activated cation channels (Fig. 8.3A) (Liman *et al.*, 1991; Noda *et al.*, 1984; Papazian *et al.*, 1991). In the prototypical *Shaker* voltage-dependent K^+ channel, for example, this so-called gating charge corresponds to nearly 12–14 electron charge units traversing the membrane electric field (Aggarwal and MacKinnon, 1996; Schoppa *et al.*, 1992). Additional contributions to the voltage sensitivity of Kv channels may arise, however, from cooperative interactions along the activation pathways of such channels. Indeed, cooperative interactions play a fundamental role in determining the voltage sensitivity of the Kv channel (Schoppa *et al.*, 1992; Smith-Maxwell *et al.*, 1998; Tytgat and Hess, 1992; Yifrach, 2004; Zagotta *et al.*, 1994). Based on steady-state and kinetic analyses (Schoppa and Sigworth, 1998; Zagotta *et al.*, 1994), a detailed mechanism of action has been suggested for the *Shaker* Kv channel in which transition of the four voltage-sensor domains occurs in a sequential but independent manner. Once such transitions between closed channel states are completed, a late, concerted pore-opening transition ensues.

Insight into the conformational transition associated with pore opening may be gained by comparing the crystal structures of voltage-independent K^+ channel pore domains in the closed (i.e., KcsA) (Doyle *et al.*, 1998) and open (i.e., MthK) (Jiang *et al.*, 2002) states (Fig. 8.3C). Following voltage-induced structural rearrangement of the voltage-sensing domains, the inner helix of the pore domain bends at a glycine hinge point (marked in red), so as to open the pore, and straighten, so as to close it (Fig. 8.3C) (Jiang *et al.*, 2002). This conformational change leads to the disassembly of the bundle crossing activation gate formed from the four inner helices of the different channel subunits and results in a dramatic rearrangement of intersubunit contacts (Fig. 8.3C).

Before proceeding, a concise practical description of how Kv channel opening is measured should be given. Principally, gating transitions of voltage-activated cation channels are usually studied by measuring voltage-activation relations of the channel using conventional electrophysiological recording techniques, such as the two-electrode voltage clamp technique. In such measurements, the transition of the channel from the closed to the opened state(s) is induced by stepwise changes in the membrane voltage, with elicited ionic currents resulting from ion flow through the open channels being recorded (Hille, 2001). Figure 8.3D describes a typical voltage protocol applied to the membrane of a *Xenopus laevis* oocyte expressing the "wild-type" *Shaker* Kv channel variant lacking the N-terminal inactivation domain, along with the resulting K^+ current traces. The probability of the channel to be open (*Po*) and its dependence on voltage is then inferred by plotting the normalized isochronal tail current amplitude as a function of voltage (Fig. 8.3E). This typical voltage-induced activation curve of a Kv channel is analogous to those ligand-binding curves describing the initial reaction velocity of an enzyme, as a function of substrate concentration. Channel-gating effects as a result of residue mutation, for example, are manifested by leftward or rightward shift of the activation curve along the voltage axis (as compared to the wild-type curve), indicating open or closed state stabilization effects upon mutation, respectively.

Activation curves of voltage-dependent ion channels are frequently parameterized using two phenomenological values, namely the slope factor, Z [proportional to the total nominal channel-gating charge of the channel that moves across the membrane electrical field upon depolarization (Z_T) and to the magnitude of cooperative interactions along the channel-gating reaction], and $V_{1/2}$, the half activation voltage, that is, that voltage at which the probability of the channel to be open is 0.5. Estimates for the values of these parameters can be obtained by fitting the actual activation data to a simple Boltzmann function, derived assuming a simple two-state voltage-dependent model for channel activation. According to this model, the equilibrium between the closed (C) and open (O) states of the channel is voltage dependent [$K_{open}(V)$] and is given by Eq. (8.12):

$$K_{open}(V) = \frac{[O]}{[C]} = K_{open} e^{Z_T FV/RT}, \tag{8.12}$$

where K_{open} is the chemical equilibrium constant for channel gating in the absence of voltage (at 0 mV), Z_T is as defined above and all the other constants have their usual thermodynamic meanings. The probability of channel opening $\{Po = ([O]/([O] + [C]))\}$ is then given by the Boltzmann function, as follows:

$$\frac{I}{I_{max}} \propto Po = (1 + e^{-ZF(V-V_{1/2})/RT})^{-1}, \tag{8.13}$$

where I/I_{max} is the normalized tail current amplitude of the channel and all other symbols are as defined above. One can easily show that the free energy of channel opening at 0 mV (ΔG_{open}) is given by

$$\Delta G_{\text{open}} = -RT \ln K_{\text{open}} = ZFV_{1/2}. \tag{8.14}$$

As we shall see in the following sections, the free energy of channel opening (ΔG_{open}) is the principal thermodynamic quantity upon which all calculations of coupling free energies, using the high-order thermodynamic mutant cycle coupling analysis, are based upon.

In the closing paragraph of this section, after describing mutant cycle coupling analysis and the Kv channel model system, we would like to stress how intuitively coupling between two channel residues can be realized in terms of the specific channel-gating measurements described above (Fig. 8.3E): In a simple manner, two residues would be suspected as non-coupled if the activation curve of the corresponding double mutant is shifted along the voltage axis, compared to the wild-type curve, by the sum of the shifts of the two single mutants (the additive case). The two residues are coupled if nonadditive gating effects are observed. Concrete examples of these two scenarios can be found in Fig. 8.4 in Yifrach and MacKinnon (2002).

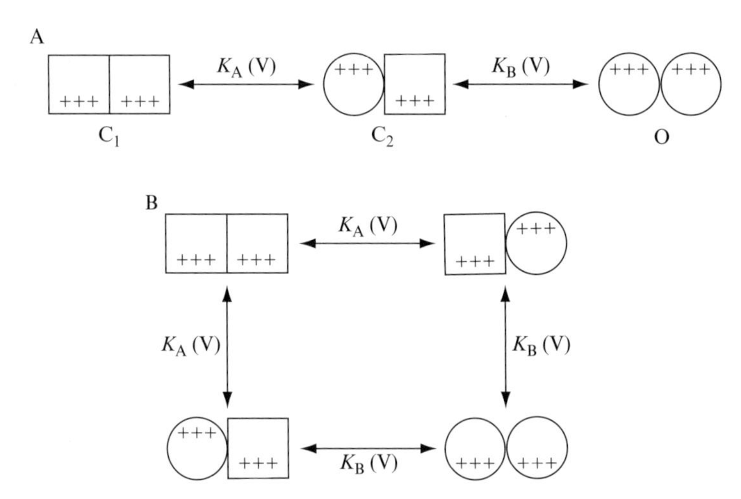

Figure 8.4 The classical KNF allosteric model applied to voltage-dependent gating. (A) Scheme of the different states considered by the KNF model. A homodimeric channel undergoes two sequential voltage-dependent subunit transitions from the closed (square) to the open (circle) state. A "+" sign indicates the channel-gating charge. $K_A(V)$ and $K_B(V)$ are the equilibrium constants for the first and second transitions, respectively. (B) Double-mutant cycle representation of the above KNF model. The "mutation" in this cycle is a transition of a subunit from the closed to the open conformation. Cooperativity arises when the free energy associated with this "mutation" depends on the conformational state of an adjacent subunit. Such cooperativity is a function of intersubunit interactions between the monomers in the different conformations and is manifested by K_B/K_A. Adapted with permission from Yifrach, 2004.

4. Deriving a Hill Coefficient for Assessing Cooperativity in Voltage-Dependent Ion Channels

Having introduced the Kv channel model system, we can now ask how cooperativity in voltage-dependent ion channel gating is assessed. In this section, we demonstrate that the two-state voltage-dependent Boltzmann equation (Eq. 8.13), frequently used to assess channel activation data, is in fact, analogous, to the original Hill equation (Hill, 1910). We outline a framework to understand the meaning of the Hill coefficient (n_H) for channel-gating transitions, and, using a particularly simple gating scheme for channel activation, demonstrate the relation of n_H to the magnitude and nature of cooperativity in channel-gating transitions (Yifrach, 2004). Using simple math, this section highlights the power of the simple thermodynamic law of mass action to obtain important insight into the Kv channel-gating reaction addressed here.

4.1. The analogy between voltage-dependent ion channels and allosteric enzymes

Gating transitions of voltage-dependent ion channels are usually described using a reductionist, two-state gating model that assumes a voltage-dependent equilibrium between the closed (C) and open (O) states of the channel [$K(V)$], according to Eq. (8.12). This equation can be transformed into

$$\log\frac{P}{1-P} = \log K + n_H \frac{Z_U F}{RT} V, \tag{8.15}$$

where n_H is the number of channel subunits and Z_U is the unitary channel-gating charge associated with one channel subunit ($Z_T = n_H Z_U$). Written in this form, it is worth noting the similarity of the equation to the original Hill equation (Hill, 1910), extensively used to estimate the magnitude of cooperativity in multisubunit allosteric enzymes:

$$\log\frac{\bar{Y}}{1-\bar{Y}} = \log K + n_H \log S. \tag{8.16}$$

Here, $\bar{Y}$, the fractional binding saturation function, is the fraction of sites occupied with the substrate (S), n_H is the Hill coefficient, and K is the apparent binding constant of the substrate to the enzyme. This equation was derived assuming an infinitely cooperative case, that is, where n substrate molecules bind simultaneously to the enzyme. In practice, however, by fitting experimental binding data with the Hill equation, a value between 1 and the total number of binding sites (n) is frequently observed for n_H. n_H, therefore, is an index of the cooperativity of ligand binding. The similarity between the two equations is not only in structure but also in essence. Conformational transitions within a protein may be driven by changes in chemical potential (log S, in the case of ligand-binding systems) or by changes in electrical potential $[(Z_U F/RT)V]$, in the case of voltage-gated ion channels. Given that the unitary gating charge of the channel (Z_U) is known, fitting channel activation data to the Hill form of the Boltzmann equation can yield a value for the Hill coefficient for channel-gating transitions (Yifrach, 2004). In a simple manner, this allows for comparison of the magnitude of steady-state cooperativity in channel-gating transitions between different subtypes of the same channel or between different cation channels, irrespective of differences in the nominal gating charge of the channels (Z_T). It should be stressed, however, that although it is common practice to compare Hill values of different proteins, such comparisons, as will be made evident below, is more informative when comparing channels that gate according to similar activation mechanisms.

4.2. Understanding the meaning of the Hill coefficient

The form of the Hill equation used to describe channel-gating transitions (Eq. 8.15) is based on the assumption that no intermediate states exist along the channel-gating reaction pathway, or in other words, that all four channel subunits switch simultaneously from the closed to the fully open state. This, of course, is far from reality for almost all channels studied thus far. How reliable is it then to estimate the magnitude of cooperativity in gating transitions using a two-state Hill equation? In analogy to ligand-binding allosteric systems (Levitzki, 1978), to allow for a precise understanding of the meaning of the Hill coefficient, we present below a general framework that enables one to derive expressions for n_H that depend on the model parameters of the particular channel-gating scheme addressed.

A general definition for a Hill coefficient for gating transitions may be derived from Eq. (8.15), as follows:

$$n_H(V) = \frac{RT}{Z_U F}\frac{\partial \log[P/(1-P)]}{\partial V} = \frac{RT}{Z_U F}\frac{\partial P/\partial V}{P/(1-P)}. \tag{8.17}$$

The above definition of the Hill coefficient makes intuitive sense: $n_H(V)$ is the scaled slope of an open probability function ($\partial P/\partial V$, corresponding to a particular gating scheme), relative to the slope of a reference open probability function $[P(1 - P)]$, derived for a gating scheme that assumes independent subunit transitions (Yifrach, 2004). Thus, a Hill coefficient value of 1 will be obtained for n_H if the slope of the open probability function ($\partial P/\partial V$) is equal to that of the independent, noncooperative case. Deviations of the Hill value from 1 would indicate the existence of cooperativity in channel-gating transitions, a reflection of coupling between channel subunits. Using this empirical definition of cooperativity in channel-gating transitions, one can derive explicit expressions for n_H for different channel-gating schemes.

Consider, for example, the following simple gating scheme for a homodimeric channel undergoing two sequential KNF-type subunit transitions (Koshland *et al.*, 1966), as reflected in Fig. 8.4A. In this gating scheme, it is assumed that the channel is open only when both subunits are in the open state (O state). Using the principle of microscopic reversibility and Boltzmann-type transitions for channel subunits (Eq. 8.12), one can derive expressions for P, $1 - P$, and $\partial P/\partial V$ for this gating scheme and substitute them into Eq. (8.17) to obtain the following expression for n_H, as a function of voltage:

$$n_H(V) = \frac{2 + 2K_A e^{(Z_U F/RT)V}}{1 + 2K_A e^{(Z_U F/RT)V}}. \quad (8.18)$$

Usually, however, gating isotherms of voltage-activated channels are parameterized based on the voltage midpoint of the gating transition ($V_{1/2}$), where $P = 1/2$, and on the slope of the gating isotherm around this midpoint (Z). The Hill coefficient at an open probability of 0.5 [n_H ($V = V_{1/2}$)] may be derived for such a gating model by combining Eq. (8.18) with a limiting condition derived from $P = 1/2$ to yield the following expression:

$$n_H(V = V_{1/2}) = \frac{1}{1 + [2/(K_B/K_A)](1 + \sqrt{1 + (K_B/K_A)})} + 1, \quad (8.19)$$

where K_A and K_B are the equilibrium constants for the first and second subunit transitions in the absence of voltage (at 0 mV), respectively. It can be seen that for a channel that gates according to the KNF model, n_H ($V = V_{1/2}$) is voltage independent and depends only on the K_A and K_B model parameters. The quantity K_B/K_A reflects how much the second transition of the channel is facilitated (or inhibited) by the first and is an intuitive measure of the magnitude of cooperativity in channel-gating transitions. In the context of the sequential gating model discussed here, this quantity is directly related to the magnitude of intersubunit interactions, as indicated by the double-mutant cycle representation of the KNF model (Fig. 8.4B). The "mutation" in this cycle is a transition of a subunit from the closed to the open conformation. Cooperativity arises when the free energy associated with this "mutation" depends on the conformational state of the adjacent subunit. Such cooperativity is a function of the magnitude of intersubunit interactions between the monomers in the different conformations and is manifested by K_B/K_A.

Examining Eq. (8.19) in the extreme case yields a logical outcome. At the limit where K_B/K_A is very high, a Hill coefficient of 2 (the total number of subunits in the model discussed here) is obtained, reflecting complete and concerted gating charge movement. In the case of negative coupling between subunits, that is, where K_B/K_A approaches zero ($K_A \gg K_B$), a value of 1 is obtained for n_H. This outcome argues against the strict association of negative cooperativity between protein subunits with Hill values smaller than 1. It is further interesting to note that an n_H value of 1.17 is obtained by substituting $K_A = K_B$ in Eq. (8.19). This result addresses the common misconception that a Hill coefficient value greater than 1 is always indicative of positive cooperativity along the reaction pathway, be it ligand binding or gating. In the particular case presented here, an apparent positive cooperativity is detected for n_H ($n_H > 1$), despite the fact that channel subunits switch from closed-to-open states in an independent manner (i.e. $K_A = K_B$). This specific outcome underscores the need for detailed analysis of the specific channel's gating pathway, by means of Eq. (8.17), to truly capture the meaning of the Hill coefficient.

To summarize, the general empiric definition for n_H derived for channel-gating transitions (Eq. 8.17) provides a useful framework for interpreting the meaning of the Hill coefficient. It follows that comparison between Hill values of different proteins, channels, or enzymes, is probably more informative, when the activation mechanisms or ligation pathways of the compared proteins are similar (Yifrach, 2004).

5. Direct Analysis of Cooperativity in Multisubunit Allosteric Proteins

The Hill analysis, originally deduced in 1910 to account for the sigmoidal kinetics of oxygen binding to hemoglobin, is model independent. It makes no assumption about the molecular mechanisms giving rise to cooperativity. Mechanistic models that account for cooperative phenomena and regulation of enzymes were later suggested by Monod and Koshland and their associates in the MWC and KNF models, respectively, presented in the early 1960s (Koshland *et al.*, 1966; Monod *et al.*, 1965). In Section 4, we demonstrated how the use of the general definition of the Hill coefficient (Eq. 8.17) can serve to link between the phenomenological measure of the Hill coefficient and specific channel-gating model parameters (Yifrach, 2004). The specific example considered allowed understanding of the meaning of the Hill coefficient and further highlighted the involvement of intersubunit interactions in determining cooperativity in channel gating. In the current section, we address the involvement of such intersubunit interactions, as assessed using high-order coupling analysis, in determining the sequential or concerted nature of conformational transitions underlying allosteric protein function.

5.1. The high-order coupling pattern of intersubunit interactions as a criterion for discerning concerted or sequential conformational transitions

Despite the general acceptance of intersubunit interactions as playing a fundamental role in determining the magnitude of cooperativity in ligand binding by an allosteric enzyme, structure–function analysis of such proteins through mutagenesis has proven unsatisfactory in providing information on the contribution of individual subunits and their interactions to cooperativity in protein function. This is because any point mutation introduced into a homooligomeric protein will be present in all subunits. Thus, we describe here a strategy to directly explore the nature and magnitude of cooperativity in multisubunit allosteric proteins that further allows for evaluation of the contribution of individual protein subunits and mutual interactions thereof to cooperativity in protein function (Zandany *et al.*, 2008). The approach adopted involves measuring the effects on protein function of all possible combinations of subunit mutations introduced into tandem-linked subunits of a homooligomeric protein (Fig. 8.5) and calculation of the hierarchy of intersubunit interaction energies, using high-order mutant cycle coupling analysis (Zandany *et al.*, 2008).

To gain insight into possible intersubunit interactions contributing to cooperativity in the function of a protein, a concatenated gene encoding the *n* identical subunits of the protein connected through flexible linkers, each harboring unique restriction sites, is employed (Fig. 8.5A). This tandem-subunit gene construct enables one to "cut and paste" mutated subunits in any combination desired (Yang *et al.*, 1997). Functional measurements using tandem-linked multisubunit proteins can, for instance, reveal, in a direct manner, the nature of conformational transition experienced by the allosteric protein of interest. For instance, a mutation that dramatically disrupts intersubunit contacts in tetrameric model protein can be separately introduced into one, two, three, or four protein subunits using a tandem tetrameric gene construct (Fig. 8.5A). For an MWC-type (tetrameric) allosteric protein involving concerted conformational transition of all four protein subunits (Monod *et al.*, 1965), it is predicted that the effect of a single mutated subunit on protein function would be very similar to the case where all four subunits are mutated. In other words, a mutation introduced into only one subunit should induce similar structural changes in all neighboring wild-type subunits. On the other hand, for a KNF-type allosteric enzyme undergoing sequential subunit transitions (Koshland *et al.*, 1966), a gradual effect on protein function is expected upon successive addition of mutated protein subunits.

An informative and discriminative approach to discerning the type of conformational transitions underlying the function of an allosteric protein may further come from considering the hierarchy of intersubunit

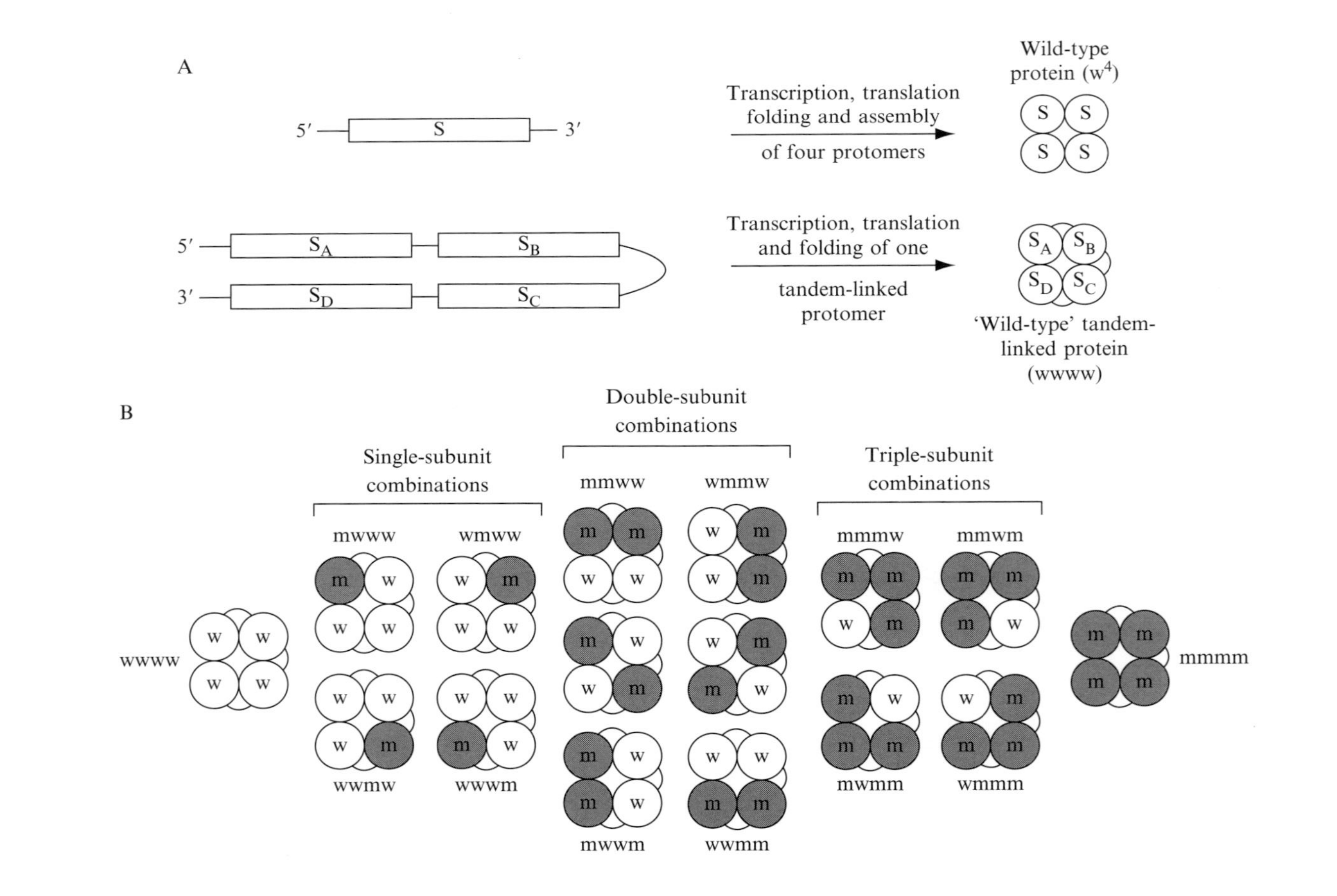
A
5′
S
3′
Transcription, translation
folding and assembly
of four protomers
Wild-type
protein (w4)
S
5′
3′
SA
SB
SD
SC
Transcription, translation
and folding of one
tandem-linked
protomer
'Wild-type' tandem-
linked protein
(wwww)
B
wwww
Single-subunit
combinations
mwww
wmww
wwmw
wwwm
Double-subunit
combinations
mmww
wmmw
mwwm
wwmm
Triple-subunit
combinations
mmmw
mmwm
mwmm
wmmm
mmmm

interactions contributing to cooperativity in protein function, as revealed by high-order thermodynamic coupling analysis (Horovitz and Fersht, 1990; Zandany *et al.*, 2008). When one considers the MWC and KNF allosteric mechanisms described above, different high-order intersubunit coupling profiles would be expected in each case. Whereas for a tetrameric KNF-type tetrameric enzyme, $\Delta^2 G_{(i,j)} < \Delta^3 G_{(i,j)k} < \Delta^4 G_{(i,j)(k,l)}$, for a MWC-type tetrameric enzyme, $\Delta^2 G_{(i,j)} = \Delta^3 G_{(i,j)k} = \Delta^4 G_{(i,j)(k,l)}$ (Note that the i, j, k and l notations refer, in this case, to the four protein subunits considered). These distinct intersubunit coupling patterns result from the concerted versus sequential nature of conformational changes predicted by the MWC and KNF models, respectively. Since, in the case of the MWC model, a functionally sensitive mutation in even one subunit triggers a similar structural change in all other subunits so as to achieve the full potential of the functional effect, mutations in the third and fourth subunits would not be expected to affect the magnitude of coupling between a subunit pair much further. On the other hand, the increase in binding affinity attained upon the successive ligand binding typical of KNF-type allosteric enzymes dictates increased contributions for high-order intersubunit coupling terms. Thus, differences in high-order intersubunit coupling profiles may further serve to discriminate between possible allosteric models. What follows is a detailed example of such an analysis applied to the case of Kv channel protein.

5.2. Direct demonstration for concerted pore opening in Kv channels

As outlined in Section 3, the late pore-opening transition of the Kv channel that follows voltage-sensing transitions occurs in a concerted manner. This assertion was indirectly deduced based on detailed steady state and transient kinetics measurements involving data fitting (Schoppa and Sigworth, 1998; Zagotta *et al.*, 1994). To directly probe for the nature of conformational transitions underlying activation gate opening in the archetypical *Shaker* Kv channel, we introduced the glycine to proline gating hinge mutation in a

Figure 8.5 Strategy for direct analysis of cooperativity in multisubunit allosteric proteins. (A) Schematic representations of monomeric and tandem tetrameric gene constructs giving rise to assembled tetrameric protein particles, one in which the subunits are not linked to each other (upper scheme) and a second, where the identical subunits are tandem-linked (lower scheme). The identical subunits encoded by the tandem tetrameric gene (arbitrarily designated A–D) are separated by unique restriction sites. Such a gene design allows the construction of all possible combinations of mutated subunits (B). Lower case letters "w" and "m" designate wild-type and mutated subunits, respectively. Adapted with permission from Zandany *et al.*, 2008.

tandem tetrameric version of the Kv channel. Such mutation (G466P) dramatically perturbs intersubunit contacts (Fig. 8.3C) (Ding *et al.*, 2005; Magidovich and Yifrach, 2004). Initially, we verified that the functional behavior of tandem tetrameric wild-type or mutant channels is identical to wild-type or mutant channels assembled from four separate monomers (Fig. 8.6A) (Zandany *et al.*, 2008). Next, we examined the effects on voltage-dependent gating of all four possible combinations of single-subunit (G466P) point mutations (mwww, wmww, wwmw, and wwwm), of all six possible double-subunit mutant combinations (mmww, mwmw, mwwm, wmmw, wmwm, and wwmm), and of all four triple-subunit mutant combinations (mmmw, mmwm, mwmm, and wmmm), where "w" is the wild-type and "m" is the mutant channel (Fig. 8.6B). The resulting voltage-activation curves of all three groups of single-, double-, and triple-subunit mutants were compared to those of the wild-type (wwww) and quadruple-mutant (mmmm) tandem tetrameric channels, as presented in Fig. 8.6B. As can be clearly seen, the activation curves of all four possible triple-subunit mutant combinations are comparable and similar to curves of the single-, double-, and quadruple-subunit mutant combinations. These steady-state channel-gating measurements strongly imply that movements underlying activation gate opening and closing occur in a concerted all-or-none fashion, with respect to the four Kv channel subunits.

As outlined above, the pattern of high-order intersubunit couplings of an oligomeric protein, obtained through the use of high-order mutant cycle coupling analysis, can further serve to discriminate between sequential and concerted conformational transitions of an allosteric protein. Following calculation of the free energy change of channel gating (Eq. 8.14) of the wild-type and the 15 possible combinations of subunit mutations (Fig. 8.5B), high-order intersubunit coupling free energy terms were calculated. As can be seen in Fig. 8.6C, the high-order intersubunit coupling profile obtained using the G466P gating hinge mutant of the *Shaker* Kv channel reveals that $\Delta^2 G_{(i,j)} \cong \Delta^3 G_{(i,j)k} \cong \Delta^4 G_{(i,j)(k,l)}$. This hierarchy of high-order intersubunit coupling is in line with a concerted opening of the Kv channel pore domain activation gate, as qualitatively realized above.

To summarize, in this section we have presented a general strategy for the direct analysis of cooperativity in multisubunit proteins that combines measurement of the effects on protein function of all possible combinations of mutated subunits with analysis of the hierarchy of intersubunit interactions, assessed using high-order mutant cycle coupling analysis (Zandany *et al.*, 2008). We showed that the pattern of high-order intersubunit coupling can serve as a discriminative criterion for defining concerted versus sequential conformational transitions underlying protein function. This strategy was applied to the particular case of the Kv channel protein to provide compelling evidence for a concerted all-or-none activation gate opening of the Kv channel pore domain.

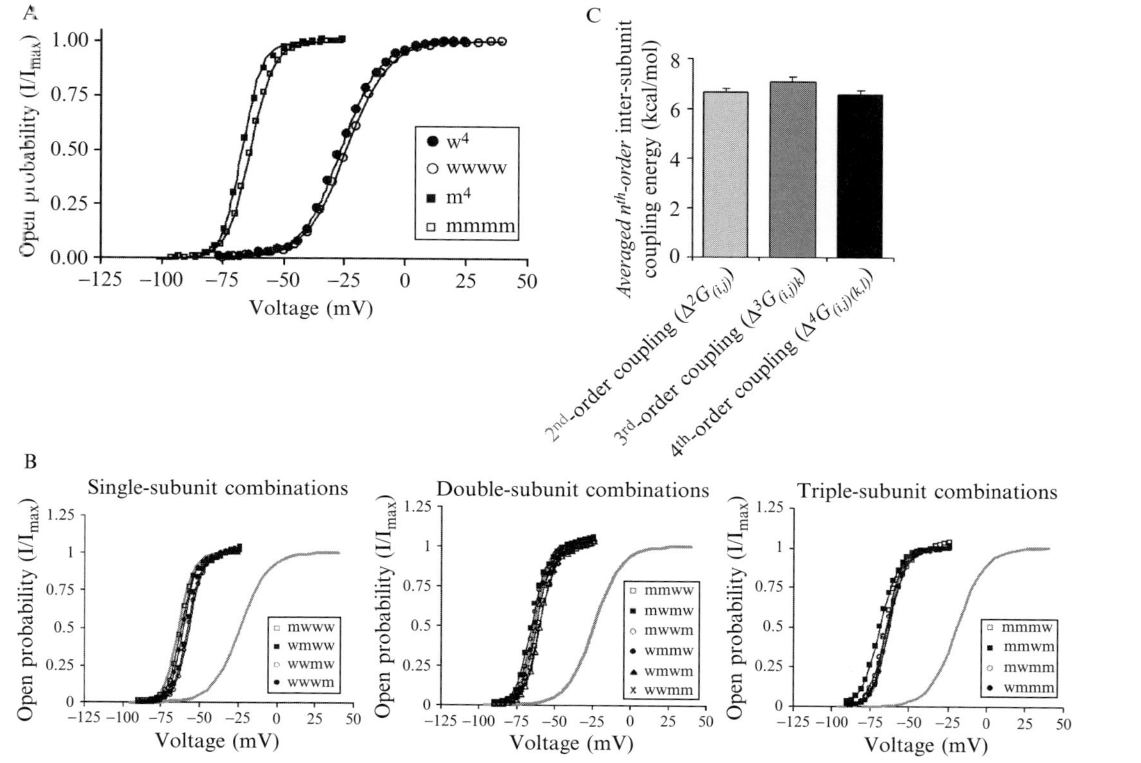

Figure 8.6 Kv channel pore domain opening occurs in a concerted fashion. (A) Voltage-activation curves of the monomeric and tandem tetrameric wild-type (w4 and wwww) or mutant (m4 and mmmm) channels. Lower case "w" and "m" represent the wild-type or G466P mutant *Shaker* Kv channel subunit, respectively. (B) The effect on voltage-dependent gating of all possible combinations of single-, double-, and triple-subunit mutations of the tandem tetrameric *Shaker* Kv channel. (C) A comparison of the average high-order second-, third-, and fourth-order intersubunit coupling free energies associated with six residue pairs along the Kv channel pore domain allosteric trajectory is shown. The different panels of this figure were adapted with permission from Zandany *et al.*, 2008.

6. Long-Range Energetic Coupling Mediated Through Allosteric Communication Trajectories

The classical MWC and KNF models of allostery addressed in the previous section provide deep insight into the role of conformational transitions in explaining cooperative phenomenon in proteins. Yet, knowing the type of conformational change a protein undergoes, be it sequential or concerted, still does not offer an atomic-resolution mechanistic explanation of how distant functional elements of that protein might be coupled. Information transfer between such elements may be achieved by propagation of conformational changes through the protein structure, induced by changes in either chemical or electrical potential. However, while the structures of stable conformational states of several allosteric proteins are known, the mechanism(s) by which ligand-induced structural changes propagate through the molecule remains elusive. It has been suggested that conformational changes may propagate by simple mechanical deformation of the protein structure along pathways of energetic connectivity, comprising adjacent amino acid positions in the tertiary structure (Lockless and Ranganathan, 1999; Suel *et al.*, 2003).

Indeed, such allosteric communication trajectory was revealed in the case of the Kv channel protein. Recent systematic energy perturbation analysis of the pore domain of the archetypical *Shaker* Kv channel revealed that gating-sensitive positions, that is, those positions where mutation dramatically affects the closed-to-open equilibrium of the channel, cluster around the channel activation gate and near the selectivity filter functional elements, when mapped onto the closed channel pore structure (Fig. 8.7A) (Yifrach and MacKinnon, 2002). Other gating-sensitive positions were found to lie along a pathway connecting these two elements (Fig. 8.7A). Through double-mutant cycle analysis (Horovitz and Fersht, 1990), it was shown that residues along this pathway are energetically coupled (Yifrach and MacKinnon, 2002), implying that the allosteric pathway connecting the activation gate and selectivity filter can be described as a trajectory of energetic connectivity, in accord with the mechanical view of signal propagation presented by Ranganathan and colleagues (Lockless and Ranganathan, 1999). In this section, we take advantage of the Kv channel model allosteric system and describe how the application of high-order thermodynamic energetic coupling analysis can be used to unravel the mechanism underlying experimentally determined features of energy transduction along allosteric communication trajectories (Sadovsky and Yifrach, 2007).

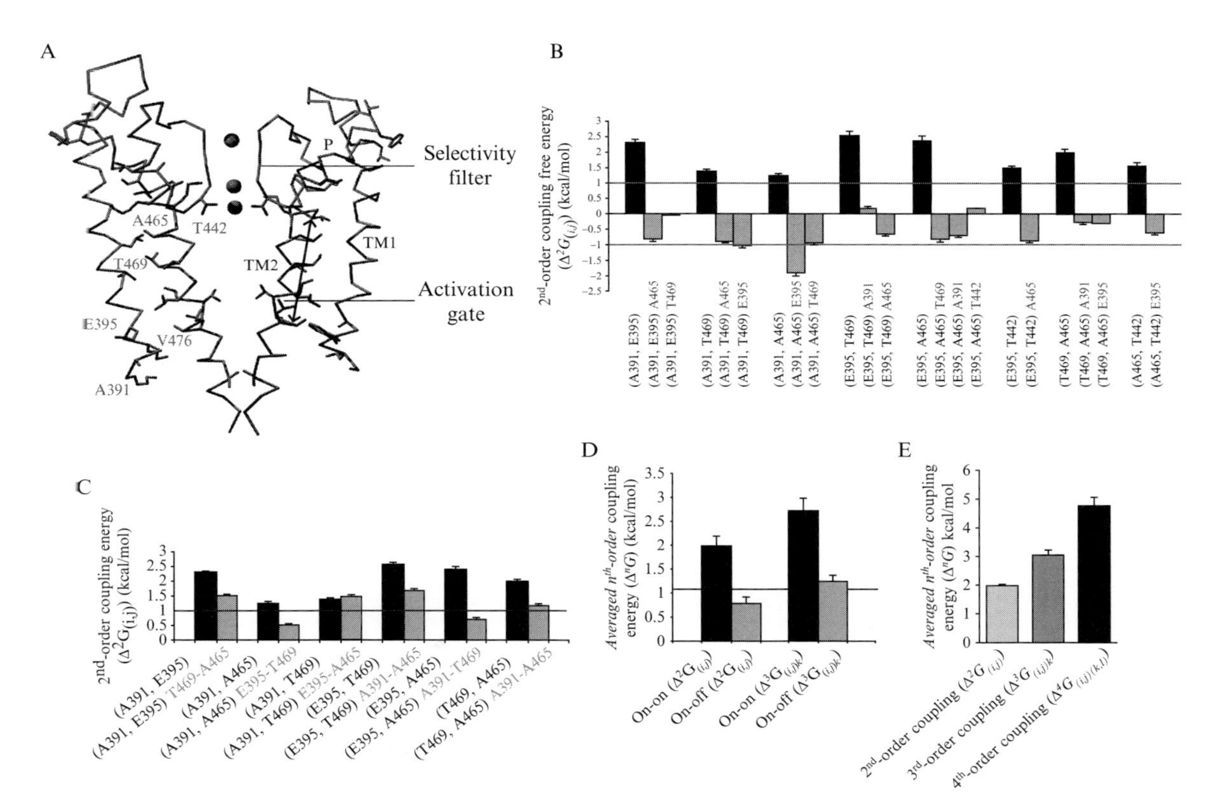

Figure 8.7 Emerging principles for long-range coupling along the Kv channel pore domain allosteric trajectory. (A) The allosteric communication network found in the pore domain of the *Shaker* Kv channel. Gating-sensitive positions of the *Shaker* Kv channel pore (marked in red) form a connected pattern when mapped onto the closed pore conformation of the KcsA K^+ channel (for clarity, only two

6.1. High-order coupling analysis of Kv channel allosteric communication trajectories: Emerging principles

We were interested in understanding what is unique about the relation between residues lining the allosteric trajectory of the Kv channel pore domain. Specifically, we asked what properties of these residues allow them to achieve the tight coupling observed between the selectivity filter and activation gate functional elements of the channel (Fig. 8.3C) (Panyi and Deutsch, 2006). To address this question, according to Horovitz and Fersht (1992), it is first necessary to describe ΔG_i^{open}, namely the contribution of a single residue, *i*, to the closed-to-open equilibrium of the channel. Equation (8.20) describes all possible free energy contributions to channel opening associated with residue *i* and considers the context of all possible structural interactions of this residue with other channel residues:

$$\Delta G_i^{\text{open}} = \Delta G_i^{\text{open,intrinsic}} + \sum_j \Delta^2 G_{(i,j)} + \sum_k \sum_j \Delta^3 G_{(i,j)k} + \sum_l \sum_k \sum_j \Delta^4 G_{(i,j)(k,l)} + \ldots\ldots \tag{8.20}$$

In the simple case, where residue *i* is not coupled to any other residue, $\Delta G_i^{\text{open}} = \Delta G_i^{\text{open,intrinsic}}$. If, however, residue *i* is coupled to other residues, then coupling terms up to the order *n*, where *n* corresponds to the total number of protein residues, must be considered. Although Eq. (8.20) should ideally extend to$\Delta^n G$, that is, the *n*th-order coupling of *n* channel residues, it is reasonable to assume that for most protein residues, the enormous complexity of energy parsing for any thermodynamic process can be usually reduced to only the first few coupling terms, involving only those residues

diagonal channel subunits are shown) (Sadovsky and Yifrach, 2007). The black arrow schematically represents a possible route for the allosteric trajectory of the Kv channel. (B) Third-order coupling along the Kv channel pore domain allosteric trajectory. Comparison of the magnitudes of coupling free energies ($\Delta^2 G_{(i,j)}$) between different trajectory-spanning residue pairs in the presence of a native (black bars) or mutated (gray bars) third trajectory-lining residue. The gray lines delineate the 1 kcal/mol cutoff line above which two residues are considered to be coupled. (C) Fourth-order coupling along the Kv channel pore domain allosteric trajectory. Comparison of the magnitude of coupling free energies ($\Delta^2 G_{(i,j)}$) between different trajectory-lining residue pairs in the presence (black bars) or absence (gray bars) of the indicated adjacent interactions. For the third- and fourth-order coupling energies, the respective single residue or residue pair affecting the targeted (*i*, *j*) interaction is indicated in gray. (D) The boundaries of the allosteric trajectory are sharply defined. Comparisons of the averaged second- and third-order coupling free energies between trajectory-lining residue pairs and residue pairs comprising a trajectory-lining residue and an adjacent off-trajectory residue are presented. (E) Allosteric communication trajectories exhibit increasingly stronger layers of cooperative interactions. Comparison of the average second-, third-, and fourth-order coupling free energies associated with six residue pairs along the allosteric trajectory. Adapted with permission from Sadovsky and Yifrach, 2007.

immediately adjacent to the residue in question (Schreiber and Fersht, 1995). By contrast, residues involved in long-range coupling would be expected to make significant contribution to higher-order coupling terms.

Using this coupling paradigm (Eq. 8.20) to address interactions between residues lining the allosteric trajectory of the *Shaker* K^+ channel (i.e., those gating-sensitive residues), including A391, E395, T469, A465, and T442 (Fig. 8.7A), we measured second-, third-, and fourth-order coupling terms and compared their magnitudes to those of off-trajectory residues. We will first focus on the results of pairwise coupling along the allosteric trajectory. Many such residue pairs, comprising different combinations of the residues indicated above, were tested for coupling. Our results revealed that all trajectory-lining residue pairs are highly coupled (Fig. 8.7B, black bars) in a manner that is context dependent. As can be seen in Fig. 8.7B, pairwise coupling all along the allosteric trajectory [different (i, j) residue pairs] is dramatically abolished upon mutation of a third, adjacent trajectory-lining residue, k (compare gray and black bars). This context sensitivity is manifested by the high third-order coupling values (averaged $\Delta^3 G_{(i,j)k}$ coupling value of ~3 kcal/mol). Similarly, the magnitude of pairwise coupling for several (i, j) trajectory-lining pairs is significantly reduced upon elimination of an adjacent neighboring (k, l) trajectory interaction pair (compare black and gray bars in Fig. 8.7C), again reflected in the high fourth-order coupling value observed for these interactions ($\Delta^4 G_{(i,j)(k,l)}$ of ~5 kcal/mol). The strong context sensitivity in high-order couplings indicates that allosteric trajectory residues must be functionally essential. Hence, all coupled residues lying along the allosteric communication trajectory are important for efficient long-range coupling between distant functional elements.

In the next stage, to address how well defined are the boundaries of allosteric communication networks, we compared the average magnitudes of second- and third-order coupling free energies along and "normal" to the allosteric trajectory, as defined between pairs of on-trajectory residues and pairs comprising an on- and an adjacent off-trajectory residue partner, respectively. As shown in Fig. 8.7D, the average magnitude of pairwise coupling free energy along the allosteric trajectory ($n = 12$) is much stronger than for pairwise interactions "normal" to the trajectory ($n = 9$), despite the average distance between residue pairs in the former group being almost twofold higher than that in the latter (Sadovsky and Yifrach, 2007). A similar trend is observed when third-order coupling free energies ($\Delta^3 G_{(i,j)k}$) along and "normal" to the allosteric trajectory are compared ($n = 5$ and 3, respectively). A third residue, k, affects the magnitude of coupling between any measured trajectory-lining pair (i, j) to a much greater extent if k itself resides along the trajectory. The anisotropic, direction-dependent coupling profile of trajectory-lining residues for both $\Delta^2 G_{(i,j)}$ and $\Delta^3 G_{(i,j)k}$ implies that the boundaries of the allosteric trajectory are well defined.

The above paragraph addressed how allosteric trajectories, evolved to transmit information, are insulated from the surrounding environment. In this paragraph, we elaborate on the information conductivity properties of such an allosteric trajectory. Efficient communication between distal functional elements would be achieved were all interactions along the allosteric trajectory highly coupled. To thus assess the degree of synergy between residues lining the allosteric trajectory, second-, third-, and fourth-order coupling free energies for six on-pathway interactions were calculated by quantifying the extent to which each pairwise interaction ($\Delta^2 G_{(i,j)}$) is affected by the presence of an adjacent trajectory-lining residue k ($\Delta^3 G_{(i,j)k}$) or interacting residue pair (k, l) ($\Delta^4 G_{(i,j)(k,l)}$). Strikingly, for all six interaction pairs considered, $\Delta^2 G_{(i,j)} < \Delta^3 G_{(i,j)k} < \Delta^4 G_{(i,j)(k,l)}$. Trajectory residue k affects the interaction between the (i, j) pair by 1 kcal/mol, on average, whereas the adjacent trajectory (k, l) interacting pair affects the interaction of the same (i, j) pair by 2.7 kcal/mol, on average (Fig. 8.7E). Residue coupling along the trajectory is, therefore, highly cooperative, with the presence of an adjacent trajectory-lining residue or interacting pair progressively increasing the strength of coupling of a given trajectory residue pair. Indeed, this highlights the uniqueness of allosteric trajectory-lining residues and may explain how such residues are strongly coupled to each other over distances reaching 18 Å (Fig. 8.7A). In other words, allosteric trajectories assume a hierarchical organization whereby increasingly stronger layers of cooperative residue interactions act to ensure efficient and cooperative long-range coupling between distal channel regions. Indeed, a linear relation between the magnitude of high-order coupling free energy along the trajectory and the Hill coefficient (n_H) for channel gating (see Section 4), was observed (Sadovsky and Yifrach, 2007).

To summarize, the high-order coupling analysis of voltage-dependent gating described here reveals additional features underlying allosteric communication trajectories in Kv channels. We find that all trajectory-lining residues are energetically coupled over long distances and are important for efficient energy transduction (Fig. 8.7B and C). The trajectory boundaries are, moreover, well defined, ensuring anisotropic information transfer along the trajectory pathway, rather than in other directions (Fig. 8.7D). Trajectory-lining residues also exhibit layers of cooperative interactions of increasing magnitude (Fig. 8.7E), implying that all trajectory-lining residues are highly interconnected, again, to ensure efficient information conductance. The magnitude of high-order coupling along the allosteric trajectory is related to cooperativity in channel gating (n_H), meaning that the stronger the magnitude of high-order cooperative interactions along the allosteric trajectory, the more cooperative is the channel's gating transition. Together, these features may explain how crosstalk between the Kv channel lower activation gate and the upper C-type inactivation gate, elegantly revealed by Panyi and Deutsch (2006), may occur.

7. Concluding Remarks

Synergistic phenomena in protein function usually rely on long-range coupling between distal functional elements of the protein(s), manifested as nonadditive (cooperative) effects on protein function. In this chapter, we have demonstrated how use of the simple thermodynamic law of microscopic reversibility and the conceptual framework of high-order mutant cycle energetic coupling analysis can be applied to study various synergistic phenomena associated with pore domain gating of the Kv channel model allosteric system. Following a description of the Kv channel model allosteric system, we delineated in a rigorous step-by-step manner, high-order thermodynamic coupling analysis. The reason for this detailed explanation is that, despite the deep insight this approach can provide into many protein-based processes, for various reasons, it is still not as widely used as would be expected. Thus, we hope that this detailed contribution dedicated to the description of this conceptual framework will influence the protein analysis practices of both young and senior scientists interested in biochemical and biophysical sciences. In applying this analysis to the particular case of Kv channel gating, we show that the pattern of high-order intersubunit coupling can serve as a discriminative criterion for defining concerted versus sequential conformational transitions underlying protein function. Furthermore, the high-order coupling analysis of the Kv channel allosteric communication trajectory assisted in revealing the principles underlying energetic coupling between the distant activation gate and selectivity filter functional elements of the channel pore domain. In both cases, the application of the high-order energetic coupling analysis to the Kv channel model system yielded valuable insight not only for this particular system but also for the function of allosteric proteins, in general.

Acknowledgments

This research was supported by a research grant from the ISRAEL SCIENCE FOUNDATION (127/08). O. Y. is an incumbent of the Belle and Murray Nathan Career Development Chair in Neurobiology.

References

Aggarwal, S. K., and MacKinnon, R. (1996). Contribution of the S4 segment to gating charge in the *Shaker* K^+ channel. *Neuron* **16,** 1169–1177.

Aharoni, A., and Horovitz, A. (1997). Detection of changes in pairwise interactions during allosteric transitions: Coupling between local and global conformational changes in GroEL. *Proc. Natl. Acad. Sci. USA* **94,** 1698–1702.

Ben-Abu, Y., Zhou, Y., Zilberberg, N., and Yifrach, O. (2009). Inverse coupling in leak and voltage-activated K^+ channel gates underlies distinct roles in electrical signaling. *Nat. Struct. Mol. Biol.* **16,** 71–79.

Bezanilla, F. (2000). The voltage sensor in voltage-dependent ion channels. *Physiol. Rev.* **80,** 555–592.

Carter, P. J., Winter, G., Wilkinson, A. J., and Fersht, A. R. (1984). The use of double mutants to detect structural changes in the active site of the tyrosyl-tRNA synthetase (*Bacillus stearothermophilus*). *Cell* **38,** 835–840.

Ding, S., Ingleby, L., Ahern, C. A., and Horn, R. (2005). Investigating the putative glycine hinge in *Shaker* potassium channel. *J. Gen. Physiol.* **126,** 213–226.

Doyle, D. A., Morais Cabral, J., Pfuetzner, R. A., Kuo, A., Gulbis, J. M., Cohen, S. L., Chait, B. T., and MacKinnon, R. (1998). The structure of the potassium channel: Molecular basis of K^+ conduction and selectivity. *Science* **280,** 69–77.

Faiman, G. A., and Horovitz, A. (1996). On the choice of reference mutant states in the application of the double-mutant cycle method. *Protein Eng.* **9,** 315–316.

Hidalgo, P., and MacKinnon, R. (1995). Revealing the architecture of a K^+ channel pore through mutant cycles with a peptide inhibitor. *Science* **268,** 307–310.

Hill, A. V. (1910). The possible effects of the aggregation of the molecules of haemoglobin on its oxygen dissociation curve. *J. Physiol. (Lond.)* **40,** IV–VII.

Hille, B. (2001). Ion Channels of Excitable Membranes. Sinauer Associates, Sunderland, MA.

Horovitz, A. (1996). Double mutant cycles: A powerful tool for analyzing protein structure and function. *Fold. Des.* **1,** R121–R126.

Horovitz, A., and Fersht, A. R. (1990). Strategy for analyzing the co-operativity of intramolecular interactions in peptides and proteins. *J. Mol. Biol.* **214,** 613–617.

Horovitz, A., and Fersht, A. R. (1992). Co-operative interactions during protein folding. *J. Mol. Biol.* **224,** 733–740.

Jiang, Y., Lee, A., Chen, J., Cadene, M., Chait, B. T., and MacKinnon, R. (2002). The open pore conformation of potassium channels. *Nature* **417,** 523–526.

Koshland, D. E. Jr., Nemethy, G., and Filmer, D. (1966). Comparison of experimental binding data and theoretical models in proteins containing subunits. *Biochemistry* **5,** 365–385.

Levitzki, A. (1978). Quantitative Aspects of Allosteric Mechanisms. Springer-Verlag, Berlin.

LiCata, V. J., and Ackers, G. K. (1995). Long-range, small magnitude nonadditivity of mutational effects in proteins. *Biochemistry* **34,** 3133–3139.

Liman, E. R., Hess, P., Weaver, F., and Koren, G. (1991). Voltage-sensing residues in the S4 region of a mammalian K^+ channel. *Nature* **353,** 752–756.

Lockless, S. W., and Ranganathan, R. (1999). Evolutionarily conserved pathways of energetic connectivity in protein families. *Science* **286,** 295–299.

Long, S. B., Campbell, E. B., and MacKinnon, R. (2005). Crystal structure of a mammalian voltage-dependent *Shaker* family K^+ channel. *Science* **309,** 897–903.

Long, S. B., Tao, X., Campbell, E. B., and MacKinnon, R. (2007). Atomic structure of a voltage-dependent K^+ channel in a lipid membrane-like environment. *Nature* **450,** 376–382.

Magidovich, E., and Yifrach, O. (2004). Conserved gating hinge in ligand- and voltage-dependent K^+ channels. *Biochemistry* **43,** 13242–13247.

Monod, J., Wymann, J., and Changeux, J.-P. (1965). On the nature of allosteric transitions: A plausible model. *J. Mol. Biol.* **12,** 88–118.

Noda, M., Shimizu, S., Tanabe, T., Takai, T., Kayano, T., Ikeda, T., Takahashi, H., Nakayama, H., Kanaoka, Y., and Minamino, N. (1984). Primary structure of *Electrophorus electricus* sodium channel deduced from cDNA sequence. *Nature* **312,** 121–127.

Panyi, G., and Deutsch, C. (2006). Cross talk between activation and slow inactivation gates of *Shaker* potassium channels. *J. Gen. Physiol.* **128,** 547–559.

Papazian, D. M., Timpe, L. C., Jan, Y. N., and Jan, L. Y. (1991). Alteration of voltage-dependence of *Shaker* potassium channel by mutations in the S4 sequence. *Nature* **349,** 305–310.

Ranganathan, R., Lewis, J. H., and MacKinnon, R. (1996). Spatial localization of the K^+ channel selectivity filter by mutant cycle-based structure analysis. *Neuron* **16,** 131–139.

Sadovsky, E., and Yifrach, O. (2007). Principles underlying energetic coupling along an allosteric communication trajectory of a voltage-activated K^+ channel. *Proc. Natl. Acad. Sci. USA* **104,** 19813–19818.

Schoppa, N. E., and Sigworth, F. J. (1998). Activation of *Shaker* potassium channels. III. An activation gating model for wild-type and V2 mutant channels. *J. Gen. Physiol.* **111,** 313–342.

Schoppa, N. E., McCormack, K., Tanouye, M. A., and Sigworth, F. J. (1992). The size of gating charge in wild-type and mutant *Shaker* potassium channels. *Science* **255,** 1712–1715.

Schreiber, G., and Fersht, A. R. (1995). Energetics of protein–protein interactions: Analysis of the barnase–barstar interface by single mutations and double mutant cycles. *J. Mol. Biol.* **248,** 478–486.

Sigworth, F. J. (1994). Voltage gating of ion channels. *Q. Rev. Biophys.* **27,** 1–40.

Smith-Maxwell, C. J., Ledwell, J. L., and Aldrich, R. W. (1998). Role of the S4 in cooperativity of voltage-dependent potassium channel activation. *J. Gen. Physiol.* **111,** 399–420.

Suel, G. M., Lockless, S. W., Wall, M. A., and Ranganathan, R. (2003). Evolutionarily conserved networks of residues mediate allosteric communication in proteins. *Nat. Struct. Biol.* **10,** 59–69.

Tytgat, J., and Hess, P. (1992). Evidence for cooperative interactions in potassium channel gating. *Nature* **359,** 420–423.

Yang, Y., Yan, Y., and Sigworth, F. J. (1997). How does the W434F mutation block current in *Shaker* potassium channels? *J. Gen. Physiol.* **109,** 779–789.

Yellen, G. (1998). The moving parts of voltage-gated ion channels. *Q. Rev. Biophys.* **31,** 239–295.

Yifrach, O. (2004). Hill coefficient for estimating the magnitude of cooperativity in gating transitions of voltage-dependent ion channels. *Biophys. J.* **87,** 822–830.

Yifrach, O., and MacKinnon, R. (2002). Energetics of pore opening in a voltage-gated K^+ channel. *Cell* **111,** 231–239.

Zagotta, W. N., Hoshi, T., and Aldrich, R. W. (1994). *Shaker* potassium channel gating. III. Evaluation of kinetic models for activation. *J. Gen. Physiol.* **103,** 321–362.

Zandany, N., Ovadia, M., Orr, I., and Yifrach, O. (2008). Direct analysis of cooperativity in multi-subunit allosteric proteins. *Proc. Natl. Acad. Sci. USA* **105,** 11697–11702.

CHAPTER NINE

Thermal Stability of Collagen Triple Helix

Yujia Xu

Contents

Abstract

Chief among the challenges of characterizing the thermal stability of the collagen triple helix are the lack of the reversibility of the thermal transition and the presence of multiple folding–unfolding steps during the thermal transition which rarely follows the simple two-state, all-or-none mechanism. Despite of the difficulties inherited in the quantitative depiction of the thermal transition of collagen, biophysical studies combined with proteolysis and mutagenesis approaches using full-chain collagens, short synthetic peptides, and recombinant collagen fragments have revealed molecular features of the thermal unfolding of the *subdomains* of collagen and led to a better understanding of the diverse biological functions of this versatile protein. The *subdomain* of collagen generally refers to a segment of the long, rope-like triple helical molecule that can unfold cooperatively as an independent unit whose properties (their size, location, and thermal stability) are considered essential for the molecular recognition during the self-assembly of collagen and during the interactions of collagen with other macromolecules. While the unfolding of segments of the triple helix at temperatures below the apparent melting temperature of the molecule has been used to interpret much of the features of the thermal unfolding of full-chain collagens, the thermal studies of short, synthetic

Department of Chemistry, Hunter College-CUNY, New York, USA

Methods in Enzymology, Volume 466
ISSN 0076-6879, DOI: 10.1016/S0076-6879(09)66009-2

peptides have firmly established the molecular basis of the subdomains by clearly demonstrating the close dependence of the thermal stability of a triple helix on the constituent amino acid residues at the X and the Y positions of the characteristic Gly-X-Y repeating sequence patterns of the triple helix. Studies using recombinant collagen fragments further revealed that in the context of the long, linear molecule, the stability of a segment of the triple helix is also modulated by long-range impact of the local interactions such as the interchain salt bridges. Together, the combined approaches represent a unique example on delineating molecular properties of a protein under suboptimal conditions. The related knowledge is likely not to be limited to the applications of collagen studies, but contributes to the understanding of the molecular properties and functions of protein in general.

1. Introduction

Professor Peter Privalov once stated: "The problem of stability of proteins is not as simple as it seems. Moreover, this is one of the most complicated and obscure problems of present protein physics." 20 years later, instruments are more sensitive; calculation procedures are simplified by using computer software packages; data on proteins are mounting, the question about the stability *per se* still remains obscure and complicated, especially for large proteins like collagens. Only in the ideal case when the transition between the fully folded, native state and the fully unfolded, denatured state is *reversible* and following the *all-or-none* mechanism, the stability of the protein can be clearly defined as the work required to cause the cooperative disruption of the *entire* structure of the molecule. Based on the thermodynamic principles, this work can be determined from the difference of the Gibbs free energy between the native state and the denatured sate. Unfortunately, only the denaturation transition of a small percentage of small, single-domained, globular proteins fall into this category. For proteins with more complex structure and more complicated folding reactions, the stability can often be characterized only using empirical parameters, such as the melting temperature (T_m). The T_m is loosely defined as the temperature at which the fraction of folded protein $f = 0.5$ during a thermal denaturation reaction. The T_m relates to the stability of proteins their ability to resist the heat-induced unfolding. But such empirical characterization on its own is often not very informative about the molecular properties or the extent of the conformational changes of the protein at temperatures other than the T_m. The functions of proteins often involve the conformational changes of only a part of the molecule and rarely require the complete unfolding of the full structure. It has been especially challenging to relate the stability of the protein to the properties of a specific region of the molecule during its functions.

Understanding the microunfolding of a region—a *subdomain*—of collagen is especially relevant to the understanding of the diverse biological functions of this versatile protein. Collagen triple helix consists of three polypeptide chains, each in polyproline II conformation, wrapped around each other to form a long, rope-like helical conformation (Rich and Crick, 1961). The close packing of the helix requires the characteristic Gly-X-Y repeating amino acid sequence, where Gly is the only residue buried at the center of the helix while the residues at the X and Y positions are largely exposed to solvent. The X and Y can be any amino acids (although Trp has never been found in native collagens) but are frequently occupied by Pro (at X) and Hyp (hydroxyproline, at Y). A *subdomain* of collagen generally refers to a segment of the long helix that can unfold cooperatively as an independent unit. The properties of the subdomains: their size, location, and thermal stability are considered essential for the molecular recognition during the self-assembly of collagen and during the interactions of collagen with other macromolecules (Kadler *et al.*, 1988; Makareeva *et al.*, 2006, 2008; Persikov *et al.*, 2005). So far, not much about the subdomains of collagen are clearly known. Studies using short, triple helical peptides have demonstrated that the stability of the triple helix is highly sensitive to the residues at the X and/or the Y positions. What remains unclear is how the stabilizing or destabilizing effects of the residues affect each other in the context of the long helix. The central questions remain: what factor(s) determines a segment of the helix to fold/unfold as a cooperative unit; and given the known stabilizing (or destabilizing) effects of each constituent amino acid residue can we identify the subdomains from their amino acid sequences. The related knowledge is likely not to be limited to the applications of collagen studies, but contributes to the understanding of the molecular properties and functions of protein by and large.

Thermodynamic characterization of collagen triple helix faces special challenges. First of all, the triple helix is a trimer, consisting of three long polypeptide chains. Second, it is large, with a linear, rather homogeneous backbone conformation and repeating amino acid sequences. The triple helix domain of the fibrillar collagen, the best characterized collagens including collagens type I, II, and III, often consist of more than 1000 residues (per chain) in none interrupted Gly-X-Y repeating amino acid sequences (Kielty *et al.*, 1993). In addition to these unique structural features of the triple helix, the thermodynamic studies of the collagen are further complicated by the fact that the denaturation reaction of collagen is often irreversible due to the exceedingly slow folding reaction and other complications.

Many studies have been devoted to the determination of the thermodynamic parameters of collagen despite the complications (Leikina *et al.*, 2002; Privalov, 1982; Privalov *et al.*, 1979). The discussion of the current chapter focuses on the applications of the stability studies on the characterization of the *subdomains* of collagen. Instead of rigorous theoretical analysis with complex equations and compounded assumptions, the studies described here are often empirical and rely on the combinations of several experimental observations. Albeit lacking quantitative details at times, these efforts have provided new insight on the molecular interactions deterministic of the cooperative behavior of the subdomains. The stability study using denaturants will also not be included here. Such studies of collagen have been problematic due to preferential bindings of the denaturant by certain residues. The intention here is not to discuss the experimental details of any particular method; we refer the readers to the original reference and reviews, but to use the study of the subdomains of collagen as an example to demonstrate the applications of biophysical investigations that can derive molecular insights on the properties of proteins under suboptimal conditions.

2. Methods

2.1. The thermal stability of full-chain collagen—An empirical treatment of an irreversible process

2.1.1. Characterization of the thermostability of protein

The thermostability of a protein is derived from the temperature-induced denaturation reactions, also known as the denaturation transition. Under the condition that the transition is *reversible* and following the *all-or-none* mechanism (also known as the *two-state* model), the denaturation reaction can be described by an equilibrium between the native (N) and the denatured (D) states, and the Gibbs free energy of denaturation (ΔG_D)can be determined from the equilibrium constant K_{eq}:

$$K_{eq} = \frac{[U]}{[N]} = \frac{\theta(x) - \theta_N}{\theta_D - \theta_N} = \frac{f}{1-f}; \qquad (9.1)$$

$$\Delta G_D^o = -RT \ln K_{eq},$$

where $\theta(x)$ is an experimentally observed parameter of the protein at conditions specified by x, θ_N, and θ_D are the same experimental observables characterizing, respectively, the pure native and pure denatured states; f is the fraction of folded. It follows from Eq. (9.1) that at the temperature (the T_m) when $f = 0.5$, $K_{eq} = 1$. Thus, in this case the T_m is rigorously defined as the temperature where the $\Delta G_D^o = 0$, and the denaturation reaction reaches an equilibrium.

The common choices of methods for the study of the thermal denaturation are the differential scanning calorimetry (DSC) or the optical methods. The DSC monitors the heat involved during the denaturation reaction; the optical observations monitor the temperature-induced changes of the optical signal at specific wavelength characteristic of the structure of the protein. For collagen, the optical observation is often carried out using optical rotatory dispersion (ORD) at 365 nm or more commonly, the circular dichroism (CD) spectroscopy at 225 nm. The characteristic CD spectrum of the triple helix consists of one positive peak at 225 nm and another deep, negative peak at ~219 nm (an example is given in Fig. 9.6). Since salts, solvents, and even the peptide backbone all have significant absorbance in the far-UV region, the precise measurements of the signal at 219 nm is often not practical and are rarely used for stability studies.

For a reversible, cooperative *all-or-none* transition, the heat-induced change from the native state to the denatured state often takes place within a narrow range of temperature, and presented by the characteristic sigmoidal curve when observed by the optical methods, or by a single heat-absorption peak by DSC. The T_m can be estimated conveniently from the half-way point of the sigmoidal curve or from the peak of the DSC thermograms. Within the transition zone, the melting curve can be converted to a plot of f versus T suing Eq. (9.1) and the enthalpy of the denaturation reaction, ΔH_D, can be determined using the van't Hoff equation (Privalov, 1979). For DSC experiments, the van't Hoff enthalpy can be determined using the following equation:

$$\Delta H^{eff}(T_m) = \frac{RT^2}{f(1-f)}\frac{\Delta C_p(T)}{Q_d} = \frac{4RT_m^2}{\Delta T_{1/2}}, \tag{9.2}$$

where ΔC_p is the excess heat capacity at temperature T, Q_d is the total heat of denaturation, and $\Delta T_{1/2}$ is the half-width of the peak (Privalov, 1979). In addition to the ΔH^{eff}, the total amount of heat absorbed during the transition can be estimated by integration of the heat-absorption peak, which gives a direct measure of the enthalpy of transition, the ΔH^{cal}. Any factors affecting the shape of the heat-absorption peak, or the deviation from the *two-state* assumption, will likely affect the estimated value of ΔH^{eff}, while that of ΔH^{cal} is independent of any assumptions about the transition and independent of the shape of the curve. In fact, the deviation from unity of the ratio $\Delta H^{cal}/\Delta H^{eff}$ is often used as an indication that the *two-state* model is not adequate for the description of the thermal denaturation.

When the association of more than one polypeptide chain is an integral part of the folding, like the folding of collagen triple helix which involves three peptide chains, the T_m is a function of the concentration of the protein. For simple cases of self-association, with the total concentration of the protein defined by the total molar concentration of the monomer unit: $C_T = [M] + n[M_n]$, where [M] and $[M_n]$ are the molar concentration of monomer and the fully folded protein (the n-mer), respectively, and n stands for degree of association. At the point where the fraction of folded $f = n[M_n]/C_T = 0.5$, the equilibrium constant of denaturation has the following form (Breslauer, 1987)

$$K_{uf}(T_m) = \frac{[M]^n}{[M_n]} = n\left(\frac{1}{2}C_T\right)^{n-1}. \quad (9.3)$$

This equation is not limited to self-association but applicable to other types of association reactions following the 1:1:1:. . .:1 stoichiometry.

It follows from Eq. (9.3) that under the assumptions that the ΔH_D and ΔS_D are independent of temperature, the T_m depends on C_T:

$$T_m = -\frac{\Delta H}{R}\left[\ln\left(\frac{3}{4}C_T^2 - \frac{1}{T'}\right)\right]^{-1}, \quad (9.4)$$

where T' is the temperature when $\Delta G_D = 0$. Despite the slightly more complex looking equations for an association system, one could generally derive the thermodynamic parameters from the temperature melt experiments so long as the reaction is *two state* and *reversible.*

The T_m can often be readily determined from thermal denaturation reaction as the temperature marking the half-way point of the transition even for cases when Eq. (9.1) does not apply—when the transition is not all-or-none. Under these conditions, the T_m should be treated as an empirical parameter representing the temperature at which the energy of heating is enough to cause about half of the protein to unfold. For cases where the thermal denaturation reaction is not reversible, the apparent values of T_m, the T_m^{app}, will vary with the equilibration time or with the heating rate. A linear extrapolation of the T_m^{app} to zero heating rate was often used to derive the *true equilibrium* T_m of the protein. However, such an extrapolation must be examined carefully for each case since the linear relationship may not exist (more discussion in Section 2).

2.1.2. The thermal denaturation of collagen

While the collagen is generally considered to be too large to follow an all-or-none transition, the melting curve of collagens often shows a characteristic single, sharp transition, whether observed by spectroscopic methods or by DSC (Fig. 9.1). It is not hard to reveal that this denaturation reaction is not a reversible one. The melting temperature strongly depends on the heating rate and the reverse melt by cooling revealed a very different transition from that of heating, even after extended period of equilibration of days or weeks. The slow refolding of the triple helix is a major factor for the lack of reversibility. The major kinetic barriers for the folding of the triple helix are the slow *cis–trans* isomerization of the imino acids (Bachinger, 1987; Bachinger *et al.*, 1978; Breslauer, 1987), which the collagen has a high quantity of, and the difficulties in nucleation—association of the three chains with the correct one residue stagger at the ends (Engel and Bachinger, 2000). Because of the repeating sequences and the linear conformation, partially formed helices with three

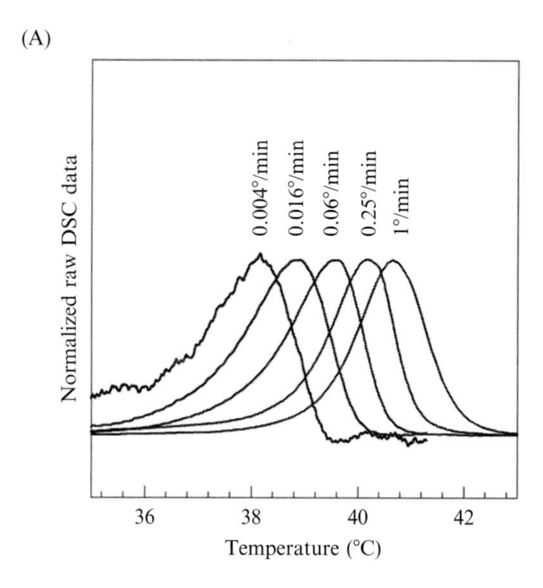

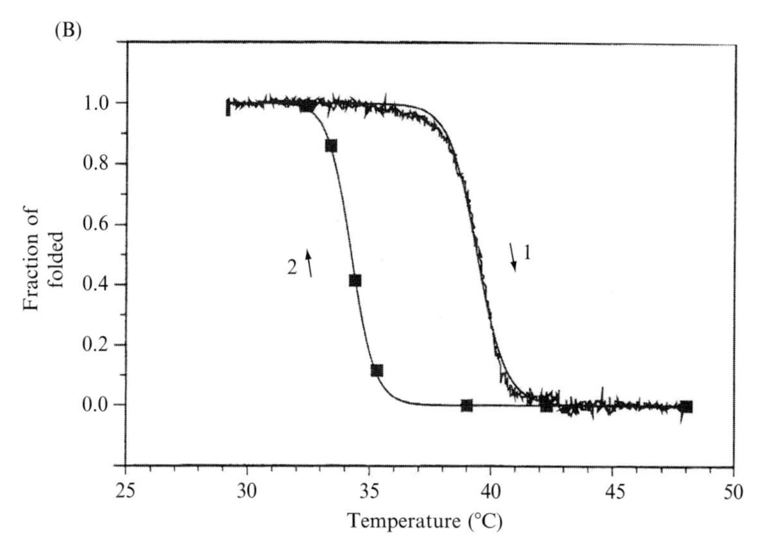

Figure 9.1 (A) The thermal unfolding of collagen measured by DSC at different heating rate in 0.2 *M* phosphate/0.5 *M* glycerol (pH 7.4). To simplify the visual comparison, the baselines were subtracted and the data were normalized to ensure the same height of the denaturation peak (Leikina *et al.*, 2002 with permission). (B) Thermal unfolding and refolding profiles of collagen III. The experimental unfolding curve was measured at a slow temperature increase of 1 °C/h (curve 1). Plateau values were measured after overnight recording of the refolding at the indicated temperatures (filled squares) (from Engel and Bachinger, 2000 with permission).

misaligned chains are common. These partially folded, or misfolded, triple helices are stabilized by lateral H-bonds which prevent the shifting of the chains with respect to one another. As a result, the misfolded helices have to completely unfold before they can eventually form correctly aligned helices (Engel and Bachinger, 2000). As expected, addition of proline isomerase was shown to facilitate the folding but only up to a degree (Bachinger, 1987; Bachinger *et al.*, 1978). Similarly, crosslinked triple helices fold much faster by avoiding the limitations in chain alignment, yet the complete folding still has to wait for the *cis–trans* isomerization of the numerous Pro and Hyp residues (Davis and Bachinger, 1993).

Despite of the complexity, experiments have clearly demonstrated that the renaturation of the unfolded chain is a reversible, thermodynamic driven process (Engel and Bachinger, 2000; Makareeva *et al.*, 2008). A recent study of human type I collagen indicated that, at least at temperatures <28 °C, the fully unfolded chains can refold into the triple helix which has the same melting behavior as the original ones (Makareeva *et al.*, 2008). The refolded fraction is small, and the refolding often takes weeks or longer. The folding study using crosslinked collagens has shown that even under the conditions that the optical signal of the triple helix can resume to the level before the denaturation, the reverse melt by cooling followed a very different transition than that of heating (Davis and Bachinger, 1993; Engel and Bachinger, 2000). These data support the conclusion that the denaturation/renaturation reaction of collagen is an intrinsically reversible equilibrium process; the apparent irreversibility of the thermal transition in the time scale of hours or days is simply the manifestation of the hysteresis of a cooperative equilibrium transition.

Even if the renaturation is reversible the equilibrium is not achievable in practice because of the slow refolding. The early approaches to such an irreversible process is to, well, treat it as an equilibrium process. This equilibrium approximation was justified by the observation that the heat-absorption peak of collagen monitored by DSC did not change significantly if the heating rate is slowed down to 0.1–0.5 K/min; and a very narrow range of the transition temperature was observed by studies using different techniques (Privalov, 1982). Thus, the irreversibility of collagen was considered to be a "secondary phenomenon." From these studies, it was found that the denaturation of collagen was accompanied by an unusually high enthalpy change as estimated by ΔH^{cal}. The molar heat of enthalpy of collagen is in the range of 960–1600 kcal/mol, comparing to $\sim$60–200 kcal/mol for typical globular proteins. This unusually high ΔH^{cal} is often considered as an indication of the high cooperativity of the denaturation transition of collagen. This highly cooperative denaturation of collagen is also reflected in the dramatic change within a narrow temperature range of all properties of collagen solutions sensitive to shape, conformation, or molecular weight of dissolved macromolecules (Privalov, 1982). Yet, this highly cooperative transition is not two state. When the effective enthalpy was calculated using Eq. (9.2), the ratio of $\Delta H^{eff}/\Delta H^{cal}$ was clearly not unity but close to be 1/10. This ratio was used to indicate the existence of microunfolding units of collagen (more discussion in Section 2.1.3).

While some considered the ΔH^{cal} is an adequate measure of the ΔH_D for collagen, since its estimation is "model independent" and is also independent of the heating rate, others argued that the total heat involved in an irreversible denaturation reaction is a poor presentation of the ΔH of an equilibrium process. Instead, a kinetic approach was suggested (Miles, 1993; Sanchez-Ruiz, 1992) to treat the DSC data. Since the denaturation reaction of collagen was dominated by the unfolding of the triple helix, the melting curve at one constant heating rate was fitted to determine the activation energy and the rate constant of the unfolding reaction. When carried out using collagens isolated from lens capsules, the unfolding rate was found to be 10^{-2}–10^{-4} s^{-1} depending on temperature, and the activation free energy to be in the range of ~860 kJ/mol. While it may be true that the folding of the triple helix from three *completely separated monomer chains* is slow and contribute little to the denaturation curve, the unfolding *per se* may not be completed during the course of a DSC experiment. Experimental observations of light scattering, viscosimetric, and optical rotation indicated that the separation of the three chains during the thermal denaturation took place only at the final stage of the melting process (Privalov, 1982). The refolding of partially unfolded helices is rather fast and can contribute directly to the melting curve. It seems that the thermal transition of collagen is neither equilibrium nor a kinetic-limited process, but fall uncomfortably somewhere in between of these two limiting conditions.

As for many other proteins with an irreversible thermal transition, the "true" equilibrium T_m^{eff} of collagen (the *effective* T_m) was determined from a linear extrapolation of T_m^{app} to zero heating rate (Privalov, 1982). This linear extrapolation has proven to be quite problematic for collagen as demonstrated in a more recent study (Leikina *et al.*, 2002). When the temperature melt of type I collagen was investigated at very slow heating rates by DSC, Leikin *et al.* have shown that the dependence of T_m^{app} on the heating rate is exponential rather than linear. Thus, the extrapolation to zero heating rate is mathematically incorrect and impossible. Combined with the data from gel electrophoresis and CD spectroscopy, the same study also reported that the true T_m of human type I collagen is a few degrees *lower* than the body temperature, contradicting to the studies using the linear extrapolation which had often concluded that the T_m of a collagen is usually a few degrees higher than the body temperature (Leikina *et al.*, 2002).

Despite the doubt about the equilibrium approximation of the thermal transition of collagen, some of the conclusions of the early studies are generally accepted, albeit not at the quantitative level. For lacking of a better method to handle the exponential dependence of T_m^{app} on the heating rate, the linear extrapolation of $T_{\mathrm{m}}^{\mathrm{app}}$ was often used as the only practical approach for an empirical estimation of the thermal stability of the collagen. It was noted in the early studies that the enthalpy of denaturation estimated from ΔH^{cal} for collagens from different species correlated well with the $T_{\mathrm{m}}^{\mathrm{eff}}$ from the *linear* extrapolation. The temperature dependence of enthalpy is usually a measure of the change of the heat capacity based on the Kirchhoff's law ($\mathrm{d}\Delta H/\mathrm{d}T = \Delta C_{\mathrm{p}}$), yet the ΔC_{p} for collagens is negligibly small as shown in Fig. 9.1 (Leikina *et al.*, 2002; Privalov *et al.*, 1979). In addition, the $T_{\mathrm{m}}^{\mathrm{eff}}$ of collagen was also found to correlate well with the physiological temperature of each species. Thus, it was concluded that this strong dependence of ΔH^{cal} on $T_{\mathrm{m}}^{\mathrm{eff}}$ is not caused by the ΔH_{D} as a function of temperature, but rather it reflects the dependence of the thermal stability of collagen on its chemical composition. Further studies along this line have demonstrated the clear relationship of the thermal stability of collagen on the content of the imino acids. For example, the imino content of Cod is about 15% and its $T_{\mathrm{m}}^{\mathrm{eff}}$ was found to be ~19 °C, while the imino acid content of rat is higher, ~21%, and its $T_{\mathrm{m}}^{\mathrm{eff}}$ was found to be 39 °C. This correlation of thermal stability of collagen with the content of imino acids has puzzled many researchers in the collagen field and has been at the center of many investigations.

2.1.3. Characterization of the subdomains—The conclusions of the DSC study of collagen

It was demonstrated in studies of proteins undergoing reversible transitions that the ratio of $\Delta H^{\mathrm{cal}}/\Delta H^{\mathrm{eff}}$ appeared to be a measure of the number of cooperative units of the protein during temperature-induced denaturation. Such an approach was successfully applied to identify the functional subdomains of another linear, helical molecule—the myosins—which consists of α-helix coiled coil (Privalov, 1982). The presence of multiple cooperative units is considered as one of the most probable explanations of the deviation of $\Delta H^{\mathrm{cal}}/\Delta H^{\mathrm{eff}}$ from unity, but any conclusions about the subdomains of a molecule must be supported by other independent evidences in addition to the calorimetric data. Several studies of type I, II, and III collagens have found the ratio of $\Delta H^{\mathrm{cal}}/\Delta H^{\mathrm{eff}}$ to be ~10 (per single chain) for collagens (Privalov, 1982; Davis and Bachinger, 1993). These collagens contain, on average, about 1000 amino acids per single chain in repeating Gly-X-Y sequence. Thus, it appears that each cooperative unit of collagen is about 125–100 amino acid long along the helix, and consists about 300 residues in trimer form. In some of the analyses, the collagen denaturation was treated as a monomeric protein (Privalov, 1982), largely based on the observations that the three chains do not separate completely as mentioned above. Among the three collagens only the type III is a "true" monomer since the three chains are crosslinked by disulfide bonds at the end of the C-termini. As shown by Breslauer, the equilibrium association reaction affects the ΔH^{eff} (under the *two-state* assumption) by a factor of $(2 + 2n)/4$, where n is the degree of association ($n = 1$ for monomeric protein). Thus, if the monomer–trimer association reaction is to be considered during the denaturation of collagen, the ΔH^{eff} would be twice as large as reported, so will be the size of the cooperative units of collagen. The real quantitative interpretation of the ratio of $\Delta H^{\mathrm{cal}}/\Delta H^{\mathrm{eff}}$, however, is obscured by the lack of reversibility of the denaturation reaction of collagen. It should be pointed out that the size of the cooperative units so estimated is a statistical estimate. It does not directly link to the opening of a segment of the helix with of 100 consecutive amino acid residues.

Hydrogen exchange study of collagen was often the method of choice when characterizing the local unfoldings. By estimating the exchange rate at different temperatures under the EX II limit, it was found that there were two types of slow exchanging hydrogens in collagen: the very slow ones that exchange with the solvent only when the temperature is close to $T_{\mathrm{m}}^{\mathrm{eff}}$, and the slow ones which exchange at lower temperature (Privalov *et al.*, 1979). By assuming the exchange behavior can be treated as an equilibrium process and using the van't Hoff equations for a *two-state* model, it was calculated that the slow exchange collagen corresponded to those undergoing local unfolding of about one Gly-X-Y triplet; while the very slow ones corresponding to the exchange through the unfolding of about 30 Gly-X-Y triplets ($\sim$100 amino acids). While this finding appears to support the calorimetry studies, both methods used the same questionable equilibrium approximation and the *two-state* assumption. So far, there have been little other experimental data substantiate this finding about the cooperative units of the collagen.

Alternatively using the kinetic limit, the ΔH^{eff} was interpreted as the activation energy of the rate limiting step of unfolding (Makareeva *et al.*, 2008). The fact that the ΔH^{eff} is only a fraction of ΔH^{cal} indicates the unfolding starts with a fraction of the helix. After the unfolding of these crucial regions, the rest of the helix unravels quickly in an energy downhill manner. Again, this estimated initial unfolding sites of the helix is a statistical measure and does not allow the identification of any particular segment of the helix. The 1/3 of helix could include several segments along the chain of the helix or a single segment about 1/3 long of the helix. These initial unfolding sites were also considered the sites that provide the essential stabilizing effects of the collagen.

2.2. The triple helical peptides—A model for reversible denaturation reaction of the triple helix

2.2.1. The reversible, two-state thermal transition of triple helical peptides

Synthetic peptides with 24–45 amino acid residues and the repeating Gly-X-Y sequences can form stable triple helix (Baum and Brodsky, 1999). The smaller sizes of these peptides appear to be more suitable for thermodynamic characterizations and for addressing some of the questions raised from the studies of full-chain collagen. The thermal transition of the peptides seems to follow the simple *all-or-none* mechanism, yet the equilibrium is still slow and limited by the slow *cis–trans* isomerization. Since the repeating sequence of GPO (O stands for 4R-hydroxy proline or Hyp) is often needed for stabilization, the peptides often have an even higher percentage of imino acids than that of natural collagens. As a result, the heating rate dependence and the lack of superposition of the heating and cooling transition curves are common in denaturation studies of peptides (Fig. 9.2A). This slow equilibration also contributed to other anomalies observed during the temperature melt experiments. At the average heating rate of ~0.1 °C/min, the estimated T_m appeared to be independent of the concentration of the peptides. This result contradicts the theoretical prediction for a monomer–trimer equilibrium shown in Eq. (9.4).

The helical conformation of the peptides is demonstrated by the similar CD spectra with that of the collagen (Long *et al.*, 1993). As for any other multimeric proteins, the folding of the molecule should also be characterized by the change of the molecular weight. When analytical ultracentrifugation study was carried out for the most widely used and also the most stable triple helical peptide $(GPO)_{10}$, the molecular weight average of the peptide agrees well with that of a trimer at 20 °C (Long *et al.*, 1993), consistent with the formation of the triple helix. Ideally, the equilibrium sedimentation could be used to further characterize the monomer–trimer equilibrium, or even the thermal transition at different temperatures. Unfortunately, for this case, the peptides appear to be too small for accurate determination of the K_{eq}. The slow equilibrium between the folded (trimer) and unfolded (monomer) state has also made such a study by analytical ultracentrifugation difficult.

The temperature melt experiments of the triple helical peptides usually reveal a single transition when studied by DSC or by optical techniques. The *two-state* transition was examined by several techniques. Analytical ultracentrifugation was used to monitor the change of molecular weight average of peptides $(GPP)_{10}$ and $(GPO)_{10}$ at different temperature (Y. Xu and B. Brodsky, unpublished data). For both peptides, the molecular weight average varies from nearly pure monomer to nearly completely trimer within a narrow range of temperature at the concentration of ~1 mg/ml (~0.03 m*M*). However, the transition zone shifted to lower temperature range comparing to that observed by CD, mainly due to the different time duration for equilibration. The ultracentrifugation could not discriminate the misfolded intermediates with misaligned chains from that of fully folded helices, since both have the same molecular weight. The accumulation of the misaligned chains was more or less ruled out by NMR studies of $(GPO)_{10}$ at a higher concentration. No unstructured ends of the helix were observed beyond the usual helix fray of one or two residues were observed (Li *et al.*, 1993). The close to *two-state* transition of the triple helical peptides is attributed to their smaller sizes. The stability of the triple helix closely depends on

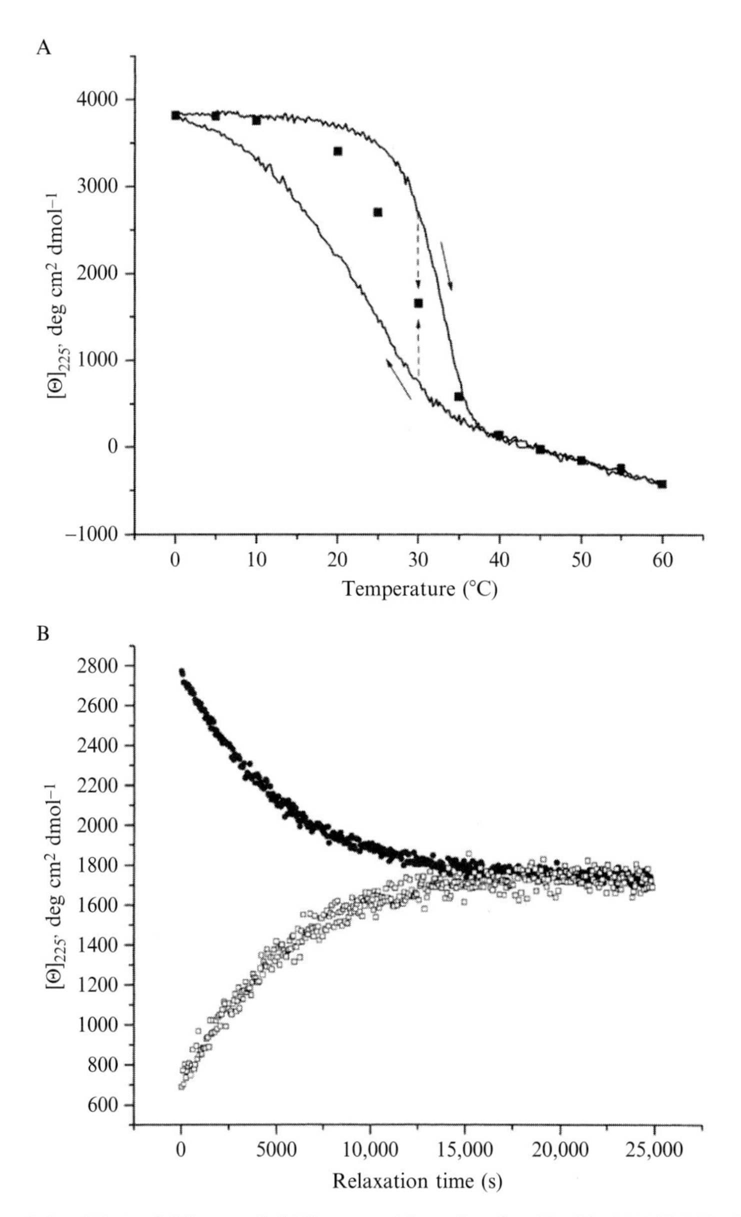

Figure 9.2 (A) Unfolding and folding transitions for the $(GPO)_3GPD(GPO)_4$ host–guest peptide as monitored by the CD at 225 nm. The arrows indicate the direction of heating at an average rate of 0.2 °C/min, while the squares indicate equilibrium values obtained through isothermal experiments (c = 1 mg/ml in PBS, pH 7.0). (B) Relaxation of the peptide when sample is transferred from 25 to 30 °C (filled squares) and from 35 to 30 °C (empty squares) (from Persikov *et al.*, 2004 with permission).

the size of the molecule. Peptides with less than seven GXY often do not form stable helix (Brodsky and Ramshaw, 1997). Thus, any intermediates with partially folded or misaligned chains are expected to be unstable and not to populate to a significant level during the denaturation reaction.

A fundamental difference of the denaturation transition of the triple helical peptides from that of the full-chain collagen is that the fully reversible condition can be achieved without hysteresis (Persikov *et al.*, 2004). Given sufficient time, the melting curve from heating and from cooling will merge together and remain constant afterwards without further change with time. The process of establishing the equilibrium can also be monitored in kinetic experiments (Fig. 9.2B). In the refolding experiment, the peptide sample was heated to 35 °C then transferred to 30 °C and the change of the CD signal was observed over a period of more than 10 h. The equilibrium value of the fraction of folded of this particular peptide was about 10% and ~40% at 35 and 30 °C, respectively. Upon transferring to 30 °C, a significant population of the peptide will start to fold until the fraction of folded reaches ~40%, which took more than 8 h. After that point, the CD signal, which is a measure of the fraction of folded, remained constant. The data for the unfolding reaction are nearly symmetric but opposite in direction. When the peptide was transferred from a state with ~70% folded at 25 °C back to 30 °C the fraction of folded of this sample reached the equilibrium value of 40% after about 8 h and remained constant afterwards. Thus, the equilibrium CD signal was reached after 8 h at 30 °C whether it was by heating or by cooling. As for any other equilibrium processes, the time required to reach the equilibrium depends on the change in temperature ($\Delta T = 5$ °C in this case) and the location of the final temperature on the melt curve. The equilibrium state dominated by folded species (e.g., $f > 85\%$) usually takes longer to establish than those dominated by unfolded species ($f < 30\%$) since the folding reaction is significantly slower than that of the unfolding reaction.

The *true equilibrium* T_{m} determined under the equilibrium condition demonstrated the expected concentration dependence of Eq. (9.4) (Fig. 9.3). The T_{m} was estimated by equilibrate the sample at each temperature for 4–20 h, till no more change of the CD signal, whether by heating or by cooling. The plot of $1/T_{\mathrm{m}}$ versus ln C_{T} fit well by Eq. (9.4), consistent with a trimerization reaction. The enthalpy change of the equilibrium denaturation determined using Eq. (9.4) appears to agree well with that of van't Hoff enthalpy estimated from the equilibrium melting curve. The enthalpy ΔH^{eq} is in the range of 300–400 kJ/mol, depending on the sequence and the size of the peptide. Since the two estimations of enthalpy are independent, the good agreement of the values confirms the *two-state* model and the true equilibrium nature of the transition.

The values of $T_{\mathrm{m}}^{\mathrm{app}}$ estimated from a temperature melt experiment with a heating rate too fast for a reversible denaturation, such as the ones estimated by CD at an average heating rate of ~0.3 °C/min, are consistently 4–8 °C

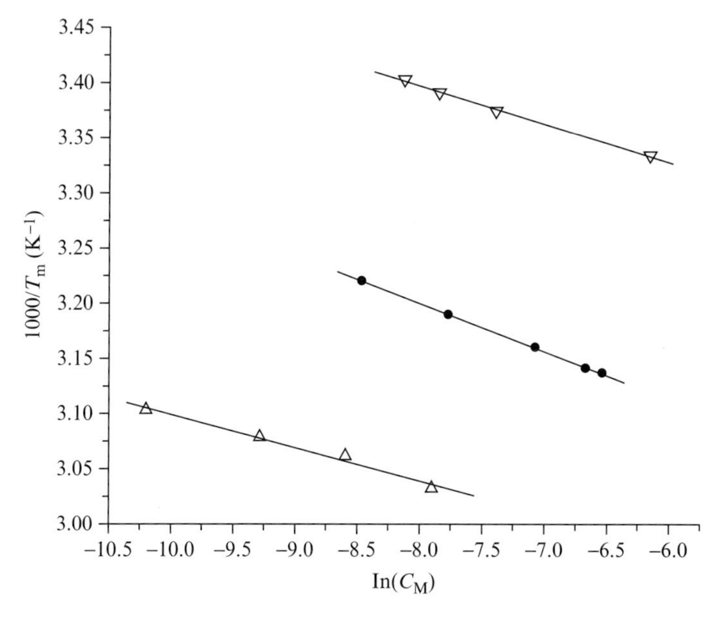

Figure 9.3 Concentration dependence of melting temperature for collagen-like peptides: $(GPO)_3GPD(GPO)_4$ (circles) and T1-892 (down triangles) in PBS (pH 7.0), and $(GPO)_{10}$(up triangles) in acetic acid, pH 2.9 (from Persikov *et al.*, 2004 with permission).

lower than that of the T_m^{eq}. However, in a rather systematic study, it was found that the order of stability of a group of host–guest triple helical peptides remains unchanged whether represented by T_m^{eq} or by T_m^{app}, as long as the T_m^{app} were determined using the same heating rate (A.V. Persikov and B. Brodsky unpublished data; Persikov *et al.*, 2004). The T_m^{app} was often used as a practical, empirical measure of the *relative* stability of peptides. The apparent values of ΔH estimated with fast heating rates, on the other hand, show significant deviations from that of ΔH^{eq} (Persikov *et al.*, 2004). Thus, cautions must be taken when the ΔH^{app} is used to infer any quantitative, structural information about the triple helix.

2.2.2. The propensity scale for triple helix

While the size of the synthetic peptides are too small to be used to investigate the subdomains—they often exhibit the all-or-none behavior during thermal transition as a single cooperative unit—studies using the peptides have provided the molecular basis for the subdomains of collagen. The significant effects of the X and the Y residues on the thermal stability of the triple helix have been documented by a systematic study using host–guest peptides $(GPO)_3GXY(GPO)_4$, where X and Y are the guest sites for X and Y residues, respectively (Persikov *et al.*, 2005). The study was summarized into the complete propensity scale for all 20 amino acids at the X or the Y positions. For practical concerns and also for historical reasons, the propensity was presented using the empirical parameter $T_{\mathrm{m}}^{\mathrm{app}}$ determined using CD spectroscopy at a heating rate of 0.3 °C/min. The most stabilizing tripeptide is GPO, with the Pro at X and Hyp at the Y positions; and the $T_{\mathrm{m}}^{\mathrm{app}}$ of the host peptide $(\mathrm{GPO})_8$ is about 47 °C. Substitution of the Pro at X position by another amino acid residue reduces the $T_{\mathrm{m}}^{\mathrm{app}}$ of the host peptide by 4–15 °C; and the substitution of Hyp at Y reduces the $T_{\mathrm{m}}^{\mathrm{app}}$ between 1.8 and ~20 °C.

Using the $T_{\mathrm{m}}^{\mathrm{app}}$, the stabilizing effects (more precisely, the destabilizing effects) of the X and Y residues appear to follow a simple *additive rule* (Persikov *et al.*, 2000, 2005). Thus, if replacing Pro by Glu at X and replacing Hyp by Gln reduce the $T_{\mathrm{m}}^{\mathrm{app}}$ of the host peptide by 4.4 and 6 °C, respectively, replacing the Pro-Hyp by Glu-Gln simultaneously will cause the decrease of the $T_{\mathrm{m}}^{\mathrm{app}}$ of the host peptide by ~10.4 °C. While the appropriateness of using the $T_{\mathrm{m}}^{\mathrm{app}}$ as a measure of propensity is open for debate, the additive rule implies that the effects of the X and Y residues on the triple helix is independent of each other and their impacts remain *local*. Studies using another set of host–guest peptides including a guest site of GXYGX′Y′ have revealed certain interchain interactions between the X and Y residues (Persikov *et al.*, 2002). The most significant of them is the interchain salt bridges formed between the charged residues Lys and Glu or Asp in the sequence GX**KGE**Y or GX**KGD**Y. An increase of 15–16 °C of the $T_{\mathrm{m}}^{\mathrm{app}}$ was reported when such a sequence was involved. A simple algorithm was derived based on the additive rule to calculate the $T_{\mathrm{m}}^{\mathrm{app}}$ for a peptide with n amino acid residues (Persikov *et al.*, 2005):

$$T_{\mathrm{m}} = T_{\mathrm{m}}^{0} - \sum_{i=2}^{n-1} \Delta T_{\mathrm{m}}^{\mathrm{GXY}} + \sum \Delta T_{\mathrm{m}}^{\mathrm{int}}, \tag{9.5}$$

where T_{m}^{0} is the "base" T_{m} for $(\mathrm{GPO})_n$ repeating sequence, $\Delta T_{\mathrm{m}}^{\mathrm{GXY}}$ is the destabilizing effects of the X and Y residues, and $\Delta T_{\mathrm{m}}^{\mathrm{int}}$ is the additional stabilizing effects such as the ones from the salt bridges of KGE or KGD sequences. A good agreement of the calculated T_{m} and the experimentally determined $T_{\mathrm{m}}^{\mathrm{app}}$ were found when tested on over 40 synthetic peptides (Persikov *et al.*, 2005).

The peptide results were applied to identify the thermally stable and labile regions along the helix of collagen. With the known amino acid sequence of collagens, a stability curve similar as the one shown in Fig. 9.4 were derived by several groups using different database for stability, but more or less the same calculation scheme (Davis *et al.*, 1989; Makareeva *et al.*, 2008; Persikov *et al.*, 2005). Typically, the "relative stability" was calculated for a window of 5–7 GXY units for each of the three chains (some collagen are heterotrimer) of the collagen, and the average of the three chain was plotted against the tripeptide number in collagen sequences.

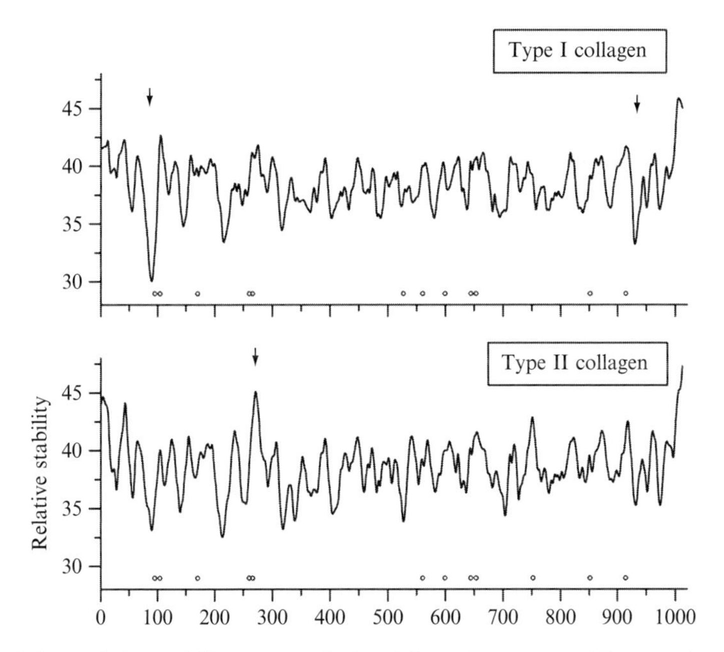

Figure 9.4 Relative stability curve calculated for collagen type I (heterotrimer) and collagen type II. The arrows indicate the low stability sites. The location of KGE/D sequences is shown by circles (from Persikov *et al.*, 2005 with permission).

The window was then shifted by 1 GXY triplet, and the calculation repeated. Several unstable regions can be identified from the "valleys" of the plot. Yet, it remains unclear if these *valley regions* will unfold as cooperative units under the physiological conditions. Despite the extensive peptide studies on the stabilizing/destabilizing effects of isolated X or Y residues and on the intrachain and interchain molecular interactions involving both X and Y residues in one Gly-X-Y tripeptide or in two adjacent tripeptides (Gly-X-Y–Gly-X′-Y′), how these interactions modulate each other in the context of the long helix of collagen remain largely unknown.

2.3. The microunfolding of a subdomain of the collagen is modulated by the long-range impacts of the local interactions

Recombinant collagen fragments with a size significantly larger than that of the synthetic peptides appear to be a logical choice of systems to investigate the long-range impact of the interactions between the X and the Y residues and to characterize the cooperative behavior of subdomains of collagen. From the example demonstrated below, it is clear that the combination of mutagenesis and the limited proteolytic digestion offers an effective way to characterize the subdomains of the triple helix at a level that is difficult to reach using either full-chain collagen or short, synthetic triple helical peptides. This approach takes advantage of the fact that when fully folded, the triple helix is resistant to chymotrypsin, pepsin, and a few other proteases (Bruckner and Prockop, 1981; Davis *et al.*, 1989; Makareeva *et al.*, 2008; Persikov *et al.*, 2005). Thus, the proteolytic digestion at one particular site can be used as a probe of the partial unfolding of the site. Given the linear conformation of the triple helix, the size and location of a microunfolding subdomain can be quite clearly characterized.

Several features of the recombinant fragment F877 make it a very interesting molecule to study the microunfolding of the triple helix (Xu *et al.*, 2008). The fragment F877 was used to model the 63-residue segment of the α_1-chain of type I collagen corresponding to residues 877–939, with the foldon domain of bacteriophage T4 fibritin at the C-terminus serving as the nucleation domain for folding (Fig. 9.5). This fragment was expressed in *Escherichia coli*. A short sequence containing two Cys residues, GlyProCys-CysGly, was added to crosslink the three chains through disulfide bonds and to further increasing the stability of the triple helix. When oxidized in folded triple helix conformation, the Cys residues form a set of interchain disulfide bonds—also known as the Cys-knot. The triple helix domains contain 99 amino acid residues with GXY repeating sequences and the repeating sequences of GPP at the ends to eliminate the effects of helix fray. Using the stability calculation described above, it is clear that the triple helix domain of F877 can be separated into two regions with quite different stability: a relatively stable 30-residue region at the N-terminal end (exclude

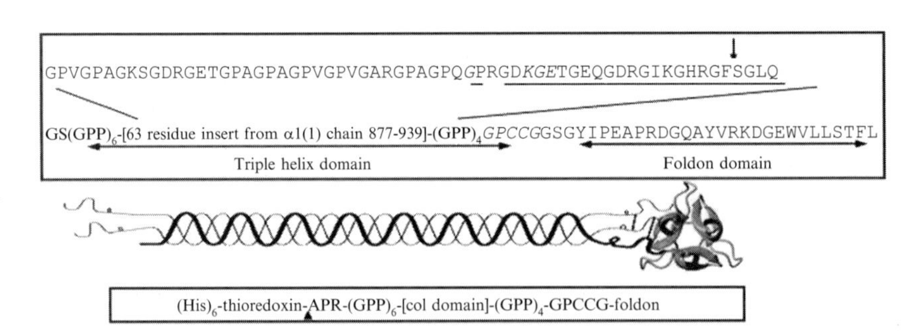

Figure 9.5 The F877 fragment. The amino acid sequence of the fragment F877 is shown in the upper panel. The 24-residue Pro-free region (see text) is underlined and KGE sequence (position 918–920, see text) is in italic. The Gly913 is underlined in italic. The chymotrypsin digestion site Phe-935 is marked by the arrow. The residues introduced to form the Cys-knot are shown in italics. The gene expression construct of the fragments is shown in the bottom panel. This fragment does not contain Hyp and are, thus, suitable for *E. coli* expression without needs for posttranslational modification. The His-tagged Thioredoxin was removed during the last purification procedure by thrombin digestion at the cleavage site APR marked by the triangle.

the GPP repeats), and a 24-residue *Pro-free region* between G916 and Q939 at the C-terminus. In addition to be free of any imino acids, Pro or Hyp, this Pro-free region contains several low propensity triplets like GIK and GFS. At the middle of the triple helix domain, and located right at the N-terminal edge of the Pro-free region, is a KGE sequence, which is expected to have a strong stabilizing effects through the interchain salt bridges.

The triple helix conformation of this fragment and its thermal unfolding were characterized using CD spectroscopy (Fig. 9.6). The melting curve revealed a single transition with $T_m \sim 40$ °C, which is higher than that of type I collagen due to the stabilizing effects of the foldon domain and the Cys-knot. However, when the salt bridges of the KGE sequence was removed by a Glu → Ala mutation at position 920, the melting curve of the E920A fragment revealed two transitions. The first transition occurs at a much lower temperature, between 15 and 20 °C, and involves the decrease of about 40% of the ellipticity at 225 nm. The second melting step takes place between 29 and 35 °C causing the complete unfolding of the triple helix domain. Similar two-step melt was also found for F877 at pH 3 when the salt bridges were eliminated by neutralization of the negatively charged carboxyl group of Glu920. Given the linear conformation of the triple helix, this 40% loss of CD signal during the first transition corresponds well to the unfolding of the Pro-free region. Chymotrypsin digestion targeted at the sole Phe935 located at the C-terminal end of the Pro-free region was used to further examine the unfolding of this region. As expected, when fully folded at 4 °C, both the fragments F877 and E920A are resistant to chymotrypsin (Fig. 9.6). At 15 °C, however, the F877 is still resistant, but the E920A can be digested after only 1 min of incubation. The increased susceptibility of E920A suggests an unfolded or loose conformation at the Phe935 at 15 °C. The similar two-stepped melting curve and the increased sensitivity to chymotrypsin at 15 °C were also found in another molecule, the G913S fragment, containing a Gly → Ser mutation at the nearby Gly913. The Gly substitution mutation is known to cause local conformational alterations and can, thus, induce breakage of the salt bridges at the KGE site. All these observations support the same conclusion: the disruption of the salt bridges, whether by titration, by conformational changes, or by the removal of the Glu, causes the opening of more than 20 amino acids C-terminal to the KGE sequence. In other words, the stabilizing effects of the interchain salt bridges of KGE are not limited to its immediate neighbors but can affect the conformation of the entire Pro-free region. Further studies are underway to investigate if the salt bridges have the same range of effects on N-terminal to its location.

The stabilizing effects of the KGE and KGD sequences are well documented in synthetic peptides (Chan *et al.*, 1997; Persikov *et al.*, 2005) and in bacteria collagen (Mohs *et al.*, 2007). The inclusion of a KGE or KGD sequence increases the melting temperature of peptides with 24–30 amino

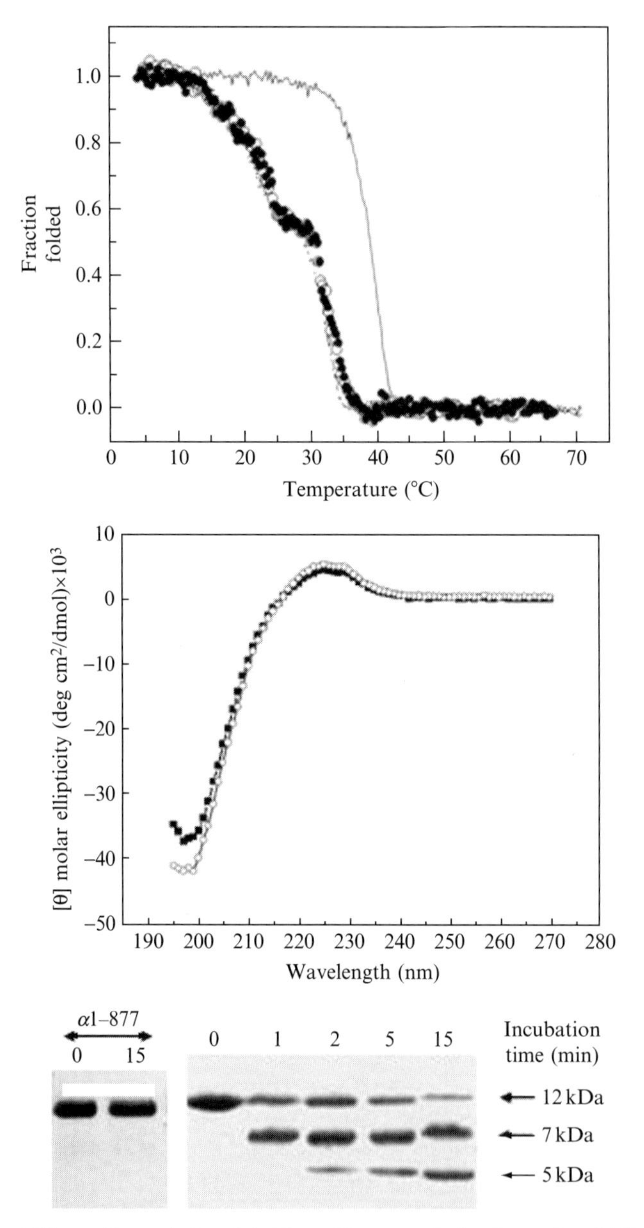

Figure 9.6 Effects of interchain salt bridges. *Upper panel*: the thermal unfolding of F877 (solid line), E920A (filled circle), and G913S (dotted line) all in PBS (pH 7), and F877 (open circle) in pH 3 buffer. The unfolding was monitored by the temperature-induced decrease of the CD signal at 225 nm, and normalized to the fraction folded of the triple helix. The heating rate is ~0.1 °C/min and the concentrations of the three fragments are 1 mg/ml. *Middle panel*: the CD scan of E920A at 4 °C (open circle) plotted together with that of F877 (filled square). *Lower panel*: chymotrypsin digestion experiment of F877 (left) and E920A (right) at 15 °C. The concentrations of the fragments were 1 mg/ml. The SDS-PAGE was performed with the presence of 20 m*M* DDT.

acid residues by ~15 °C; and their effects were considered the most essential stabilizing factors for bacteria collagen containing no Hyp. The study using the fragments has, for the first time, demonstrated experimentally the long-range impacts of the salt bridges. Knowledge of the long-range effects of molecular interactions of the triple helix is fundamental for our ability to infer the structures and molecular dynamics of a segment of the triple helix from its constituent sequence.

The above study of recombinant collagen fragments is a clear demonstration of the *microunfolding* of the triple helix. The unfolding of the 24-residue Pro-free region in E920A and G913S showed that a segment of the helix of about 20 residues with low intrinsic thermal stability can unfold as a cooperative unit. However, in fragment F877 this Pro-free region is kept in triple helix conformation by the salt bridges of the KGE sequence, and unfolds only with the rest of the helix. Thus, in addition to the propensity to helix conformation of the constituent residues, the stability and the cooperative unfolding of a subdomains of collagen is also modulated by certain long-range stabilizing factors like, but not limited to, the interchain salt bridges.

REFERENCES

Bachinger, H. P. (1987). The influence of peptidyl-prolyl *cis-trans* isomerase on the *in vitro* folding of type III collagen. *J. Biol. Chem.* **262,** 17144–17148.

Bachinger, H. P., Bruckner, P., Timple, R., and Engel, J. (1978). The role of *cis-trans* isomerization of peptide bonds in the coil to triple helix conversion of collagen. *Eur. J. Biochem.* **90,** 605–613.

Baum, J., and Brodsky, B. (1999). Folding of peptide models of collagen and misfolding diseases. *Curr. Opin. Struct. Biol.* **9,** 122–128.

Breslauer, K. J. (1987). Extracting thermodynamic data from equilibrium melting curves for oligonucleotide order–disorder transitions. *Methods Enzymol.* **259,** 221–242.

Brodsky, B., and Ramshaw, J. A. M. (1997). The collagen triple helix structure. *Matrix Biol.* **15,** 545.

Bruckner, P., and Prockop, D. J. (1981). Proteolytic enzymes as probes for the triple-helical conformation of procollagen. *Anal. Biochem.* **110**(2), 360–368.

Chan, V. C., Ramshaw, J. A. M., Kirkpatrick, A., Beck, K., and Brodsky, B. (1997). Positional preferences of ionizable residues in Gly-X-Y triplets of the collagen triple-helix. *J. Biol. Chem.* **272,** 31441–31446.

Davis, J. M., and Bachinger, H. P. (1993). Hysterisis in the triple helix-coil transtion of type III collagen. *J. Biol. Chem.* **268,** 25965–25972.

Davis, J. M., Boswell, B. A., and Bachinger, H. P. (1989). Thermal stability and folding of type IV procollagen and effect of peptidyl-prolyl cis-trans-isomerase on the folding os the triple helix *J. Biol. Chem.* **264**(15), 8956–8962.

Engel, J., and Bachinger, H. P. (2000). Cooperative equilibrium transitins couplde with a slow annealing step explain the sharpness and hysteresis of collagen folding. *Matrix Biol.* **19,** 235–244.

Kadler, K. E., Hojima, Y., and Prockop, D. J. (1988). Assembly of type I collagen fibrils de novo. Between 37 and 41 degrees C the process is limited by micro-unfolding of monomers. *J. Biol. Chem.* **263,** 10517–10523.

Kielty, C. M., Hopkinson, I., and Grant, M. E. (1993). The collagen family: Structure, assembly, and organization in the extracellular matrix. *In* "Connective Tissue and Its Heritable Disorders," (P. M. Royce and B. Steinmann, eds.), pp. 103–147. Wiley-Liss, New York, NY.

Leikina, E., Mertts, M. V., Kuznetsova, N., and Leikin, S. (2002). Type I collagen is thermally unstable at body temperature. *Proc. Natl. Acad. Sci. USA* **99**(3), 1314–1318.

Li, M.-H., Fan, P., Brodsky, B., and Baum, J. (1993). Two-dimensional NMR assignments and conformation of (Pro-Hyp-Gly)10 and a designed collagen triple-helical peptide. *Biochemistry* **32,** 7377–7387.

Long, C. G., Braswell, E., Zhou, D., Apigo, J., Baum, J., and Brodsky, B. (1993). Characterization of collagen-like peptides containing interruptions in the repeating Gly-X-Y sequence. *Biochemistry* **32,** 11688–11694.

Makareeva, E., Cabral, W. A., Marini, J. C., and Leikin, S. (2006). Molecular mechanism of alpha 1(I)-osteogenesis imperfecta/Ehlers-Danlos syndrome: Unfolding of an N-anchor domain at the N-terminal end of the type I collagen triple helix. *J. Biol. Chem.* **281**(10), 6463–6470.

Makareeva, E., Mertz, E. L., Kuznetsova, N. V., Sutter, M. B., DeRidder, A. M., Cabral, W. A., Barnes, A. M., McBride, D. J., Marini, J. C., and Leikin, S. (2008). Structural heterogeneity of type I collagen triple helix and its role in osteogenesis imperfecta. *J. Biol. Chem.* **283,** 4787–4798.

Miles, C. A. (1993). Kinetics of collagne denaturation in mammalian lens capsules studied by differential scanning calorimetry. *Int. J. Biol. Macromol.* **5,** 265–271.

Mohs, A., Silva, T., Yoshida, T., Amin, R., Lukomski, S., Inouye, M., and Brodsky, B. (2007). Mechanism of stabilization of a bacterial collagen triple helix in the absence of hydroxyproline. *J. Biol. Chem.* **282,** 29757.

Persikov, A. V., Ramshaw, J. A. M., Kirkpatrick, A., and Brodsky, B. (2000). Amino acid propensities for the collagen triple-helix. *Biochemistry* **39,** 14960–14967.

Persikov, A. V., Ramshaw, J. A. M., Kirkpatrick, A., and Brodsky, B. (2002). Peptide investigations of pairwise interactions in the collagen triple-helix. *J. Mol. Biol.* **316,** 385–394.

Persikov, A. V., Xu, Y., and Brodsky, B. (2004). Equilibrium thermal transitions of collagen model peptides. *Protein Sci.* **13**(4), 893–902.

Persikov, A. V., Ramshaw, J. A. M., and Brodsky, B. (2005). Prediction of collagen stability from amino acid sequence. *J. Biol. Chem.* **280,** 19343–19349.

Privalov, P. L. (1979). Stability of proteins. *Protein Chem.* **33,** 167–241.

Privalov, P. L. (1982). Stability of proteins. Proteins which do not present a single cooperative system. *Adv. Protein Chem.* **35,** 1–104.

Privalov, P. L., Tictopulo, E. I., and Tischenko, V. M. (1979). Stability and mobility of the collagen structure. *J. Mol. Biol.* **127,** 203–216.

Rich, A., and Crick, F. H. C. (1961). The molecular structure of collagen. *J. Mol. Biol.* **3,** 483–506.

Sanchez-Ruiz, J. M. (1992). Theoretical analysis of Lumry–Eyring models in differential scanning calorimetry. *Biophys. J.* **61,** 921–935.

Xu, K., Nowak, I., Kirchner, M., and Xu, Y. (2008). Recombinant collagen studies link the severe conformational changes induced by Osteogenesis imperfecta mutations to the disruption of a set of interchain salt-bridges. *J. Biol. Chem.* **283,** 34337–34344.

CHAPTER TEN

Electrostatic Contributions to the Stabilities of Native Proteins and Amyloid Complexes

Sarah R. Sheftic, Robyn L. Croke, Jonathan R. LaRochelle, *and* Andrei T. Alexandrescu

Contents

Abstract

The ability to predict electrostatic contributions to protein stability from structure has been a long-standing goal of experimentalists and theorists. With recent advances in NMR spectroscopy, it is possible to determine pK_a values of all ionizable residues for at least small proteins, and to use the pK_a shift between the folded and unfolded states to calculate the thermodynamic contribution from a change in charge to the change in free energy of unfolding. Results for globular proteins and for α-helical coiled coils show that electrostatic contributions to stability are typically small on an individual basis, particularly for surface-exposed residues. We discuss why NMR often suggests smaller electrostatic contributions to stability than X-ray crystallography or

Department of Molecular and Cell Biology, University of Connecticut, Storrs, Connecticut, USA

Methods in Enzymology, Volume 466
ISSN 0076-6879, DOI: 10.1016/S0076-6879(09)66010-9

site-directed mutagenesis, and discuss the type of information needed to improve structure-based modeling of electrostatic forces. Large pK_a shifts from random coil values are observed for proteins bound to negatively charged sodium dodecyl sulfate micelles. The results suggest that electrostatic interactions between proteins and charges on the surfaces of membrane lipid bilayers could be a major driving force in stabilizing the structures of peripheral membrane proteins. Finally, we discuss how changes in ionization states affect amyloid-β fibril formation and suggest that electrostatic repulsion may be a common destabilizing force in amyloid fibrils.

1. Introduction

Electrostatic forces play important roles in protein folding, stability, and association (Kumar and Nussinov, 2002), as well as enzymatic activity and conformational transitions (Graeff *et al.*, 2006; Kumar and Nussinov, 2002; Makhatadze *et al.*, 2003; Matousek *et al.*, 2007). A better understanding of electrostatic forces would shed new light on biological mechanisms ranging from structure formation to molecular recognition, and could lead to improved potential energy functions for protein structure prediction and design (Jones *et al.*, 2003; Kuhlman *et al.*, 2003; Ripoll *et al.*, 2005; Sheinerman and Honig, 2002; Sternberg *et al.*, 1987; Warshel and Papazyan, 1998).

Methods that can be used to study ionization equilibria include potentiometry (Thurlkill *et al.*, 2006), various spectroscopic techniques (Liu *et al.*, 2009), and X-ray crystallography (Berisio *et al.*, 1999). There are additional indirect methods that take advantage of properties of molecules that change with ionization state such as enzymatic activity, folding, or aggregation. Amongst these methods, NMR spectroscopy is unique in offering atomic-level resolution (Markley, 1975).

The protonated and unprotonated states of ionizable residues are almost always in fast exchange on the NMR timescale (Wuthrich, 1986), giving a population-averaged chemical shift of the charged and neutral species. At high pH the unprotonated form dominates the chemical shift. As the pH is lowered, the resonance position gradually shifts toward the chemical shift of the protonated form, eventually reaching an acidic chemical shift plateau. The pH at the midpoint of the sigmoidal titration curve is the pK_a value (Fig. 10.1A). The pK_a value is typically determined from a nonlinear least-squares fit of the NMR chemical shift versus pH data to the model specified by the Henderson–Hasselbalch equation (Markley, 1975; Matousek *et al.*, 2007). Although the majority of ionization equilibria are in fast exchange, we have seen some exceptions. Histidine resonances sometimes broaden at pH values near their pK_a, consistent with intermediate exchange between the protonated and unprotonated species

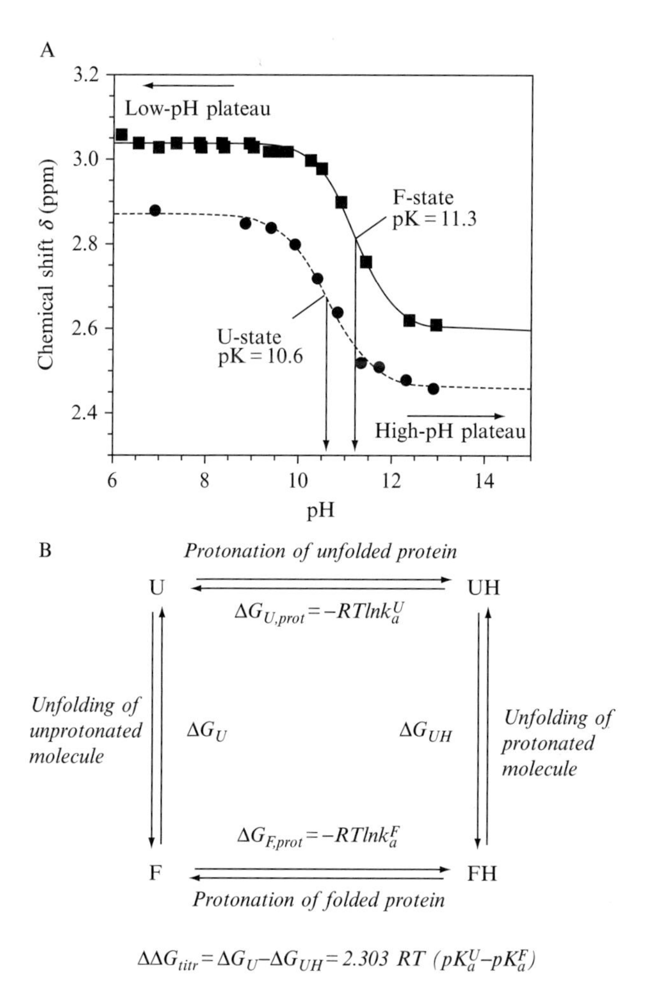

Figure 10.1 How the shift in pK_a values between folded and unfolded states relates to the change in the free energy of unfolding. (A) Titration data illustrating the pK_a shift for Lys8 between the folded and unfolded states of the GCN4p coiled-coil dimer (Matousek *et al.*, 2007). In this case, the positively charged state of Lys8 stabilizes the GCN4p structure; the side chain becomes protonated at a lower hydronium ion concentration in the folded (pK_a of 11.3) compared to the unfolded state (pK_a of 10.6). (B) The thermodynamic cycle that can be used to quantify the contribution of a change in charge to the change in stability (Anderson *et al.*, 1990; Bosshard *et al.*, 2004; Lumb and Kim, 1995). U is the unprotonated unfolded state, UH is the protonated unfolded state, F is the unprotonated folded state, FH is the protonated folded state, and R is the gas constant. $\Delta\Delta G_{\mathrm{titr}}$ needs to be multiplied by -1 for acidic groups, and by the number of magnetically equivalent monomers if used for an oligomeric protein. In the present example for Lys8 in the GCN4p coiled-coil dimer, $\Delta\Delta G_{\mathrm{titr}} = -1.9$ (kcal/mol dimer) at a temperature of 25 °C.

(Alexandrescu *et al.*, 1992). The titration of Arg25 of the GCN4p leucine zipper is in slow exchange on the NMR timescale, giving separate resonances for the protonated and unprotonated side chain (Matousek *et al.*, 2007). Arg25 is unusual in that it has a pK_a > 13 (e.g., a very low K_d for protons). For the vast majority of cases, however, ionization equilibria are in fast exchange on the NMR timescale.

This chapter describes NMR measurements of ionization constants and inferences obtained from these experiments into the roles that charges play in stabilizing folded protein structures. We compare NMR results with those from other techniques such as X-ray crystallography and site-directed mutagenesis. We briefly review NMR studies of the electrostatic properties of denatured states and evidence that charges in unfolded proteins can behave differently than predicted for a random coil conformation. Finally, we review recent work from our lab on the electrostatic properties of amyloidogenic proteins; in particular, the contributions of charges to the interactions of amyloidogenic proteins with micelles that present a membrane-like environment and the impact of ionization states on fibrillization kinetics.

2. Practical Aspects of pK_a Measurements by NMR

Typically about 10–15 spectra at different pH points are needed to precisely define a pK_a value. For more demanding experiments, it may be possible to obtain a pK_a from 6 to 8 pH points albeit with a greater experimental uncertainty. The most important parts of the titration curve are the midpoint, and the acidic and basic plateaus. When planning an experiment, it is therefore important to distribute most points around the expected pK_a value and to define the acidic and basic plateaus by collecting 2–3 pH points far from the pK_a.

Because of sample losses from pipetting and pH measurements, it is best to start the experiment with a larger sample volume than needed, for example, 0.7 ml for a standard 5 mm NMR tube. The pH adjustments are made using stock solutions of acid and base, typically at 1 *M* concentrations. The acid and base are delivered in 0.2–2.0 μl aliquots. Care must be taken to avoid inclusion of unwanted acid or base on the side of the pipette tip. To deliver very small amounts of acid or base, it is possible to touch the pipette tip to the surface of the stock solution and then blot the tip against a Kimwipe tissue, leaving only a trace of liquid. For titrations in D_2O, it is necessary to use DCl and NaOD when adjusting the sample pH to avoid introducing H_2O. For the same reason it helps to rinse the pH electrode with 99.8% D_2O before measurements. For experiments in D_2O, it is important to note that the deuterium isotope effect on pH is about

0.4 units. The isotope effect on pH, however, is approximately equal and opposite in sign to the isotope effect on a glass electrode. Therefore, the isotope effect is typically not corrected for when comparing pH measurements from D_2O and H_2O solutions (Markley, 1975).

To be thermodynamically defined, a pH titration should be microscopically reversible. This can be checked by collecting data points both as the pH is lowered and raised back to its initial value. In some cases, hysteresis occurs and the forward and backward curves are not superimposable. This may indicate irreversible changes in the protein but more often is a result of increasing salt concentration generated by the acid and base added to adjust the sample pH. Changes in salt concentration can change the value of the pK_a. A way to minimize this problem is to divide the sample into two halves and use them separately for the acidic and basic parts of the titrations, thus reducing the amount of salt introduced to the sample.

There are a number of experiments that can be used to follow NMR resonances during a pH titration. Proton 1D NMR experiments are typically useful only for histidine residues in small proteins. The aromatic $C\varepsilon_1$- and $C\delta_2$-protons of histidines resonate in a relatively isolated region of the aromatic NMR spectrum and undergo large downfield shifts of $\sim$1 ppm with decreasing pH (Alexandrescu *et al.*, 1988). Experiments to monitor histidine titrations are usually done in D_2O to avoid overlap from the amide protons present in H_2O. Isotopic labeling with ^{15}N and ^{13}C greatly increases the options for NMR experiments. For example, ^{13}C-filtered 1D NMR (Alexandrescu *et al.*, 1990; Matousek *et al.*, 2007) can be used to quickly and reliably follow the titration of histidine and tyrosine residues in H_2O, without interference from amide protons since the experiment only detects protons attached to ^{13}C. Moreover, ^{13}C-filtered 1D experiments can be done in H_2O avoiding any need for deuterium isotope effect corrections.

As with other applications, 2D and 3D NMR can be used to improve resolution (Wuthrich, 2003). The use of 2D and 3D NMR experiments for pH titrations often poses a tradeoff between obtaining chemical shift information as close as possible to the titrating group and obtaining sufficient resolution to separate the titration of amino acids of the same type. The total acquisition time for multidimensional NMR experiments can also become a factor if the sample is not stable at a given pH value. The side-chain H(CA)CO experiment, for example, can measure the side-chain carbonyls of Asp and Glu residues directly and provides chemical shift data for nuclei immediately adjacent to the titrating group (Yamazaki *et al.*, 1994). The experiment suffers from poor sensitivity, however, due to routing of magnetization through aliphatic carbons, which have unfavorable relaxation properties. While the H(CA)CO experiment provides chemical shift data adjacent to the titrating carboxylate groups, the limited side-chain dispersion of these groups may preclude the ability to distinguish between the different

Asp and Glu residues in the amino acid sequence. The problem becomes even more severe for unfolded proteins where the dispersion is smaller.

Backbone nitrogen chemical shift dispersion is typically excellent even for unfolded proteins (Shortle, 1996). Thus, a strategy for unfolded proteins is to obtain titration information from the "fingerprint" backbone ^{1}H–^{15}N crosspeaks in 2D ^{1}H–^{15}N HSQC spectra. It has been shown that the titration of ^{1}H–^{15}N HSQC crosspeaks in unfolded proteins is dominated by the residue itself, or residues that are nearest neighbors in the amino acid sequence. Residues more than 2 or 3 positions away in the sequence usually make only negligible contributions, so that ^{1}H–^{15}N HSQC spectra recorded as a function of pH provides a sensitive and straightforward way to obtain titration data for denatured states (Matousek *et al.*, 2007; Pujato *et al.*, 2006). For folded proteins, one usually needs to resort to more sophisticated experiments that give chemical shift information closer to the group becoming ionized. A number of 3D experiments are available that transfer magnetization from side chains to well-dispersed nitrogen resonances, even when the side-chain resonances are severely overlapped. These include 3D experiments such as HNCACB, CBCA(CO)NH, HBHA(CO)NH (Cavanagh *et al.*, 1996; Kay, 1995), and an experiment that transfers magnetization from Asp and Glu side chains to the backbone HN (Tollinger *et al.*, 2002). Most 3D NMR strategies for proteins are based on the detection of backbone HN protons and except for very stable structures will not work at high pH where exchange of the labile amide protons with solvent becomes severe (Croke *et al.*, 2008). If the amide protons are lost due to fast exchange, an alternative strategy is to use ^{13}C-based NMR experiments to resolve resonances according to multiple frequencies. We successfully used 2D constant-time ^{1}H–^{13}C HSQC experiments to follow the titration of the lysine residues in the folded form of the GCN4p leucine zipper fragment (Cavanagh *et al.*, 1996; Matousek *et al.*, 2007). For the unfolded form the dispersion was insufficient to resolve individual lysines, so we could only study these residues as a group. Selective labeling of the lysines would not have helped, because it would still not be possible to distinguish between different lysines in the amino acid sequence. 3D HCCH or ^{13}C NOESY-HSQC experiments may be useful in these cases to resolve signals by correlating them to different types of nuclei (Kay, 1995).

The data from multidimensional NMR can provide a number of different chemical shift handles on the titration of a given residue. In most cases, these different reporters give similar pK_a values but in some cases differences are observed. When choosing between multiple alternative pK_a values, we found that the most reliable approach is to consider the nuclei that are the closest in the structure to the titrating group and the nuclei that show the biggest chemical shift changes between the limiting acidic and basic plateau values.

NMR chemical shifts can be easily measured with very high accuracy relative to an internal reference compound such as DSS or TSP (Wishart *et al.*, 1995). The accuracy of standard pH meters is on the order of 0.05 or 0.1 pH units and this is usually the limiting factor determining the uncertainty of a pK_a value. Accurate calibrations of the pH electrode are essential. We typically measure the pH of the sample both before and after collecting an NMR spectrum, and take the average between these two values as the experimental pH. Typically, the difference between the pH measured before and after the NMR experiment is within 0.1–0.2 pH units, although larger differences are sometimes seen at basic pH.

There are two different types of pH electrodes commercially available. Glass electrodes have the advantage of having extensive literature on corrections for the effects of isotopes on the pH measurements and for the effects of denaturants such as urea (Acevedo *et al.*, 2002; Markley, 1975; Marti, 2005; Matousek *et al.*, 2007). A disadvantage of glass electrodes is that they are easily broken. Metal electrodes are "indestructible" but have the disadvantage that they only work for 1.5–2 years before they need to be replaced. Another disadvantage of metal electrodes is that the effects of isotopes or denaturants on pH measurements are not as well characterized as for glass electrodes. Glass electrodes have a particular disadvantage when working with SDS detergent micelles. The potassium salt of SDS is highly insoluble, so even a trace amount of KCl from using a glass electrode invariably leads to precipitation of some of the sample. To avoid precipitation, we exclusively use metal electrodes when studying proteins in the presence of detergents or lipids.

The data from a pH titration series are analyzed using a nonlinear least-squares fit to the modified Henderson–Hasselbalch equation (Dames *et al.*, 1998; Markley, 1975):

$$\delta = \delta_{\text{low}} - \frac{\delta_{\text{low}} - \delta_{\text{high}}}{1 + 10^{n(\text{p}K_a - \text{pH})}}, \tag{10.1}$$

where the pK_a is the ionization constant, δ_{low} is the low pH chemical shift plateau, δ_{high} is the high pH chemical shift plateau, and n is the apparent Hill coefficient. With these four parameters, it is possible to reconstruct the chemical shift of a titrating nucleus at any pH. The Hill coefficient describes the cooperativity of the titration (Markley, 1975). Values of $n < 1$ are associated with negative cooperativity. Values of $n > 1$ suggest positive cooperativity (Kaslik *et al.*, 1999; Markley, 1975). In practice, deviations of the n-value from unity can arise artifactually in cases where the data defining the titration are insufficient. It is prudent to use a statistical test such as the F-test on the chi-squared residuals, to see if a four-parameter fit with n included as a free variable is warranted by a significantly better fit to the titration data than when a three-parameter fit is used with n fixed to 1 (Shoemaker *et al.*, 1981).

3. Interpreting pK_a Values in Terms of Stability

Figure 10.1A shows an example of a pK_a shift between the folded and unfolded states of a protein. The foundation for interpreting pK_a shifts in terms of protein stability is the thermodynamic linkage cycle shown in Fig. 10.1B. The approach assumes that the ionization reaction can be separated from the other thermodynamic contributions to stability and relates the change in free energy from the protonation of an ionizable residue to the change in free energy of protein unfolding.

4. Importance of the Reference (Unfolded) State

To calculate the free energy contribution from the protonation of a given site to the stability of the folded state using thermodynamic linkage (Fig. 10.1B), it is necessary to have a pK_a value for the analogous reference state, for example, the unfolded protein in a folding transition (Bosshard *et al.*, 2004; Lumb and Kim, 1996; Matousek *et al.*, 2007). The pK_a for the unfolded state can be estimated from unstructured model peptides (Thurlkill *et al.*, 2006). It has been found, however, that short peptides can give a poor approximation for residues in an unfolded protein (Alexandrescu *et al.*, 1993; Lumb and Kim, 1996; Tan *et al.*, 1995; Tollinger *et al.*, 2003). Because of the long-range nature of electrostatic forces the ionization constants of residues in unfolded proteins can be affected by residues far away in the sequence, although these effects are more prominent at low ionic strength (Tan *et al.*, 1995). The larger hydrophobicity of a polypeptide compared to model compounds as well as residual structure that persists under denaturing conditions are additional factors that warrant using the unfolded state of a protein directly as the proper reference state (Croke *et al.*, 2008; Matousek *et al.*, 2007; Steinmetz *et al.*, 2007; Tan *et al.*, 1995; Tollinger *et al.*, 2003).

5. Results from Globular Proteins

There are a number of examples of residues in globular proteins with strongly perturbed pK_a values. In some cases, the impact on protein stability can be as large as 5 kcal/mol (Anderson *et al.*, 1990; Arbely *et al.*, 2009; Chiang *et al.*, 1996; Joshi *et al.*, 1997; Sali *et al.*, 1988; Schaller and Robertson, 1995; Wilson *et al.*, 1995; Yamazaki *et al.*, 1994). The contributions from charged residues can be either stabilizing (Anderson *et al.*, 1990; Chiang *et al.*, 1996; Schaller and Robertson, 1995) or destabilizing

(Wilson *et al.*, 1995). Residues with strongly perturbed ionization constants usually occur in unusual structural contexts such as salt bridges (Chiang *et al.*, 1996), buried in the hydrophobic interior (Joshi *et al.*, 1997), or at the ends of α-helix dipoles (Sali *et al.*, 1988). The majority of ionizable residues, however, and particularly side chains that are exposed to solvent appear to make little or no contribution to protein stability based on NMR-determined pK_a shifts (Bosshard *et al.*, 2004; Nakamura, 1996). This is thought to be because solvent-exposed surface charges in a folded protein experience a similar environment (dielectric constant) as in the unfolded state (Strickler *et al.*, 2006). The enthalpic advantage to forming an ion pair or hydrogen-bonded salt bridge on the surface of a protein would be offset by the loss in side-chain conformational entropy and by the desolvation of the interacting charges (Bosshard *et al.*, 2004). These reservations notwithstanding, a recent protein design study showed that the stabilities of five small proteins could be raised significantly by engineering surface electrostatic interactions (Strickler *et al.*, 2006). The conclusion of the study was that surface charge interactions are important determinants of protein stability (Strickler *et al.*, 2006).

Another consideration when interpreting the small pK_a shifts typically observed by NMR for proteins in solution is that to obtain titration data, the protein must remain folded over the range of pH values studied. This has the potential to bias results against proteins in which ion pairs make large contributions to stability. If we consider an attractive interaction between a negatively charged Glu residue and a positively charged Lys residue, the interactions would tend to shift the pK_a values for both residues toward the extremes of pH. The Lys residue would resist loss of a proton and its positive charge, and would have a pK_a raised above its 10.4 random coil value. The Glu residue would resist taking on a proton and losing its negative charge so that its pK_a would be lowered below the 4.2 value expected for a random coil. Many proteins would not survive such extremes of pH, so that these stabilizing interactions would be missed by NMR. Moreover, the larger the stabilization of a protein by an ion pair, the greater the chance that its disruption would lead to unfolding, a factor that may additionally mask stabilizing ion pairs.

6. Results from Coiled Coils

Coiled-coil oligomeric assemblies of α-helices are one of the best-studied systems to address the contributions of charges to protein stability. Inter- and intramolecular hydrogen-bonded salt bridges are common in the X-ray structures of coiled coils (Burkhard *et al.*, 2000; Glover and Harrison, 1995; Lee *et al.*, 2003; Lumb and Kim, 1996; O'Shea *et al.*, 1991).

By contrast, NMR shows that changes in the ionization states of coiled coils often have only a small impact on their stabilities (Dames *et al.*, 1999; Lumb and Kim, 1995; Marti and Bosshard, 2003; Marti *et al.*, 2000). The ionization of charged residues has a large effect on the stability of the nascent monomeric α-helix of the GCN4 trigger site but this intermediate structure is intrinsically unstable with a maximum α-helix population of only 50% at neutral pH (Steinmetz *et al.*, 2007). For stable native coiled-coil structures, the free energy of unfolding appears to be more invariant with changes in pH. We recently determined ionization constants of all 16 groups that titrate between pH 1 and 13 in the 33-residue leucine zipper fragment, GCN4p (Matousek *et al.*, 2007). We found that at pH 8 where the stability of the coiled coil is at a maximum, electrostatic contributions may account for up to $\sim$40% of the free energy change on unfolding (Matousek *et al.*, 2007). Although hydrophobic packing and hydrogen bonding have dominant roles, electrostatic interactions make significant albeit individually small contributions to the stability of the coiled coil. Of the 16 ionizable groups we looked at the largest individual contribution was -2 kcal/mol dimer from residue Lys8 (Fig. 10.1A). Lys8 forms a salt bridge with Glu11 in the X-ray structure of GCN4p. Many of the other charged residues that participated in salt bridges in the X-ray structures of GCN4p had much smaller contributions to protein stability, and in some cases made destabilizing contributions due to electrostatic repulsion (Lumb and Kim, 1995; Matousek *et al.*, 2007). To better understand how the ionization states of residues impact the stability of the coiled coil, continuum electrostatics calculations were done using the X-ray structure of the GCN4p fragment. The calculations reproduced the electrostatic properties of GCN4p very well at neutral pH. At the extremes of acidic and basic pH, however, the calculations greatly overestimated the contributions of charged residues to the stability of the folded structure. The work highlighted that in spite of considerable progress, some fundamental aspects of structure-based pK_a modeling remain unresolved (Matousek *et al.*, 2007). Below, we consider the origins of some of the discrepancies, and ways in which structure-based modeling of charges could be improved.

7. Comparison of NMR and Crystallographic Results

Ion pairs appear to be more prevalent in X-ray structures of proteins than in solution (Ibarra-Molero *et al.*, 2004; Lumb and Kim, 1995). A number of factors could contribute to this discrepancy. (1) Crystal structures are typically studied at cryogenic temperatures much lower than NMR. Although ionization enthalpies are typically small, these can result in pK_a shifts that are significant compared to the differences in pK_a values between F and U states (Bhattacharya and Lecomte, 1997). Low temperatures could also induce structuring of surface residues directly by reducing motion, or indirectly by affecting the properties of the water in the crystal (Halle, 2004). (2) The definition of pH and the dielectric effect are unclear in the solid state. Even though the protein crystals used for crystallography typically have a (v/v) water content of about 50% (Banaszak, 2000), hydrogen ion concentrations may not be uniform throughout the crystal making the definition of pH ambiguous. Similarly, the dielectric constant, which accounts for the environmental modulation of electrostatic forces, may be influenced by crystal lattice contacts that are absent in bulk solution. Differences in counterion concentrations between the solution and crystalline state could play additional roles in modulating electrostatic interactions (Linse *et al.*, 1995), as would osmotic effects associated with the precipitants and cryoprotectants used to grow crystals. (3) Calculations of electrostatic forces depend critically on the accuracies of the structures used for modeling. Small differences in structures between different crystal forms or even between monomers in a homo-oligomer can have a large impact on calculated electrostatic contributions. The conformations of surface residues in crystal structures can be subject to distortions from lattice packing such as freezing out of conformations that are mobile in solution, and this can lead to discrepancies between pK_a values calculated from crystal structures and observed by NMR in solution (Khare *et al.*, 1997). In the case of the GCN4p homodimer, we found that salt bridges seen in X-ray structures were more likely to give a significant pK_a shift by NMR if the interaction was present in multiple crystal structures (Matousek *et al.*, 2007). These observations indicate that ion-pair distances averaged over a family of X-ray structures are more representative of the structure in solution.

Ribonuclease A is unique in that pK_a values for the histidine residues are available from both solution NMR (Baker and Kintanar, 1996) and X-ray diffraction data for the crystalline structure collected between pH 5.2 and 8.8 (Berisio *et al.*, 1999). The pK_a of ribonuclease A residue His119, which is not involved in a salt bridge was identical in the crystal and solution (Baker and Kintanar, 1996; Berisio *et al.*, 1999). His12 and His48, which participate in ion pairs in the X-ray structure, gave pK_a values of 6.0 in solution but these were raised to 7.0 in the crystal (Baker and Kintanar, 1996; Berisio *et al.*, 1999). These observations suggest that the ion pairs are more difficult to break in the crystalline state than in solution.

8. Comparison of NMR and Mutagenesis: Nonadditivity of Ion Pairs

Mutagenesis of charged residues, verified to form salt bridges by X-ray crystallography, often destabilizes coiled coils more than predicted from NMR-derived pK_a shifts between the folded and unfolded states

(Lavigne *et al.*, 1996; Lumb and Kim, 1995, 1996). This is because a mutation does not involve just a change in charge but substitution of additional atoms that can affect interactions with the rest of the structure and with solvent (Lavigne *et al.*, 1996; Lumb and Kim, 1996). A classic example is provided by residue Glu22 of the GCN4p coiled coil (Lumb and Kim, 1995). The residue participates in the hydrogen-bonded salt bridges Glu22-Arg25 and Glu22-Lys27 in the X-ray structure of GCN4p (O'Shea *et al.*, 1991). Based on NMR data, Glu22 makes only small and unfavorable contributions to the stability of the folded structure in its charged form (Lumb and Kim, 1995; Matousek *et al.*, 2007). The NMR data for Lys27 and Arg25, however, show significant favorable contributions to the stability of the folded state. The results show that electrostatic contributions for partners in ion pairs need not be correlated. Discrepancies between pK_a values of interacting oppositely charged residues have been reported in a number of systems (Anderson *et al.*, 1990; Baker and Kintanar, 1996; Khare *et al.*, 1997), and in one example it was shown that the contribution to the stability of a protein from an ion pair depended on whether the interacting residues were reversed in the structure (Makhatadze *et al.*, 2003). The pK_a shift determines the free energy change for ionizing a residue in the folded compared to the unfolded state, but does not include contributions from the unsatisfied ion-pair partner (Ibarra-Molero *et al.*, 2004). Moreover, the ionization of a residue in an ion pair can affect interactions other than a point charge. This can include interactions of the charged group with solvent and residual interactions such as hydrogen bonds (Bosshard *et al.*, 2004). The context provided by the medium- and long-range interactions appears to make contributions to protein stability that can be as important as the ion pair itself.

9. Improving Structure-Based Modeling of pK_a Values

In spite of considerable progress, fundamental aspects of structure-based pK_a modeling remain unresolved. The accuracy of pK_a calculations depends critically on the quality of the structural information used for modeling. Modeling is typically done using an X-ray structure, usually characterized at a single pH value. More studies like the pioneering work on RNaseA (Berisio *et al.*, 2002) where X-ray structures were determined at multiple pH values would shed light on the types of structural changes that occur as the pH is changed. Much like studies of enzymes benefit from having structures of both the holo- and apo-states, crystallographic data for a number of pH values would facilitate structure-based modeling of electrostatics. NMR is also poised to provide information on how structure is affected by pH. For example with GCN4p, we found that the backbone and side-chain NMR resonances of nontitrating residues were conserved over the entire pH range 1–13 studied (Matousek *et al.*, 2007). This indicates that the overall structure is maintained with changing pH. The changes we detected using NOESY spectroscopy were restricted to the side chains of residues that participated in ion pairs. This suggests that pH-induced conformational changes are localized to the residues involved in electrostatic interactions, and that breaking an interaction causes an increase in disorder or lack of structure rather than a switch to an alternative conformation.

When available, a family of structures can improve modeling of electrostatic interactions by distinguishing invariant interactions from crystallographic noise. Also of note is that ensembles of NMR structures often model electrostatics more accurately than a single X-ray or NMR structure, probably because they provide a more realistic model of side-chain structures in solution as a group of conformations rather than a unique structure (Khare *et al.*, 1997; Whitten *et al.*, 2005). Another approach to incorporate conformational flexibility into modeling is to include structures from molecular dynamics simulations starting from an initial X-ray structure (van Vlijmen *et al.*, 1998; You and Bashford, 1995; Zhou and Vijayakumar, 1997).

10. Results with Micelle-Bound Proteins

Many of the proteins involved in amyloidogenic diseases such as amyloid-β (Aβ), and α-synuclein (αS) function as membrane-bound proteins, or may exert their cytotoxic effects through membrane-bound forms (Alexandrescu and Croke, 2008; Lashuel *et al.*, 2002). It is therefore important to study the properties of these proteins in membrane-like environments. The membrane mimetics most amenable for high-resolution structural studies are micelle monolayers since some of these form complexes sufficiently small for NMR (Coles *et al.*, 1998; Patil *et al.*, 2009; Ulmer *et al.*, 2005). To this end, we decided to characterize the pK_a values of Aβ in sodium dodecyl sulfate (SDS) micelles. The structure of the peptide bound to SDS micelles is known and other biophysical studies have been performed (Coles *et al.*, 1998).

The pK_a data for free Aβ(1–40) and for the peptide bound to SDS micelles are summarized in Table 10.1. Similar data have been obtained previously for a 1–28 residue fragment of Aβ (Ma *et al.*, 1999). We wanted to see if residues 29–40 in the naturally occurring 40-residue fragment Aβ(1–40) affected the pK_a values. Our data for Aβ(1–40) are in very good agreement with those previously reported for the 1–28 fragment. The two exceptions are His6 and Glu11 in the SDS-bound form of the peptide.

Table 10.1 Ionization constants for Aβ(1–40) and free energy contributions to the stability of the SDS micelle-bound structure[a]

Residue	pK_a free	pK_a SDS	$\Delta\Delta G_{titr}$ (kcal/mol)	Resonance	H_2O/D_2O[b]
N-term	8.11 ± 0.64	7.99 ± 0.15	0.16 ± 0.88	Asp1 (H$\beta_{1,2}$)	D_2O
Asp1	3.79 ± 0.40	5.43 ± 0.11	2.19 ± 0.56	Ala2 (H_N)	H_2O
Glu3	4.62 ± 0.34	6.17 ± 0.12	2.08 ± 0.48	Glu3 (H_N)	H_2O
His6	6.77 ± 0.03	7.87 ± 0.05	−1.47 ± 0.08	His6 (Hε_1)	D_2O
Asp7	4.20 ± 0.12	5.55 ± 0.09	1.81 ± 0.20	Asp7 (H_N/N)	H_2O
Tyr10	11.42 ± 0.22	12.30 ± 0.05	1.18 ± 0.30	Tyr10 (H$\varepsilon_{1,2}$)	D_2O
Glu11	4.40 ± 0.31	6.15 ± 0.13	2.34 ± 0.45	Glu11 (H_N)	H_2O
His13	6.67 ± 0.02	7.72 ± 0.07	−1.41 ± 0.10	His13 (Hε_1)	D_2O
His14	6.65 ± 0.02	7.82 ± 0.05	−1.57 ± 0.07	His14 (Hε_1)	D_2O
Lys16[c]	10.50 ± 0.24	>12.5	<−2.5	Lys(H$\varepsilon_{1,2}$–H$\delta_{1,2}$)	D_2O
Glu22	4.13 ± 0.07	6.12 ± 0.04	2.66 ± 0.11	Glu22 (H_N)	H_2O
Asp23	4.25 ± 0.10	5.52 ± 0.04	1.70 ± 0.14	Asp23 (N/H_N)	H_2O
Lys28[c]	10.50 ± 0.24	>12.5	<−2.5	Lys(H$\varepsilon_{1,2}$–H$\varepsilon_{1,2}$)	D_2O
C-term	3.72 ± 0.09	5.25 ± 0.02	2.41 ± 0.12	Ala40 (H_N/N)	H_2O

[a] All data were collected at a temperature of 20 °C in the absence of added salts. For experiments with micelles, the SDS concentration was 100 mM. Natural abundance Aβ(1–40) was from EZBiolab (Carmel, IN), and ^{15}N and ^{13}C/^{15}N-labeld Aβ(1–40) were from rPeptide (Bogart, GA).

[b] Describes whether the experiment was done in a 90% H_2O/10% D_2O or D_2O solvent. For the experiments in H_2O, data were obtained by following the indicated backbone amide resonances using ^{1}H–^{15}N HSQC experiments. At high pH these experiments no longer work because of rapid amide proton exchange. For experiments in D_2O we followed resonances either using 1D ^{1}H NMR for the aromatic residues, or 2D TOCSY for aliphatic protons. For experiments with micelles we used natural abundance SDS in H_2O and d_{25}-SDS in D_2O. No correction was made for the deuterium isotope effect on the pH or electrode.

[c] The values reported are for the unresolved side-chain resonances of Lys16 and Lys28. The protonated and unprotonated forms of the lysines are in slow exchange on the NMR timescale both in the free and the micelle-bound peptide, and their pK_a values were obtained by integrating the Hδ–Hε crosspeaks of the protonated and unprotonated species in 2D TOCSY spectra.

For His6, we think the chemical shift of the high pH form in the previous study (Ma *et al.*, 1999) was probably misassigned to formate, a common and almost ubiquitous impurity in NMR samples that resonates at 8.44 ppm. With our assignment for the high pH plateau of the Hε1 proton from His6 to 7.7 ppm, the pK_a value for His6 becomes close to the other two histidines. The other small change is that we obtain a pK_a of 6.14 for Glu11 from the pH dependence of its backbone amide proton chemical shift compared to a pK_a of 5.5 from the γ-methylene resonances of Glu11 (Ma *et al.*, 1999). Our correction for Glu11 brings its pK_a closer to those of other glutamates in Aβ (Table 10.1).

The pK_a values for free and micelle-bound Aβ together with the associated $\Delta\Delta G_{titr}$ values are summarized in Table 10.1. We see large and uniform pK_a shifts when the protein is bound to SDS corresponding to $\Delta\Delta G$ values of a magnitude between 1.2 and 2.7 kcal/mol at 20 °C. For groups that become positively charged when ionized, the contributions are favorable presumably reflecting attractive interactions with the sulfate head groups of SDS. For groups that become negatively charged on ionization the thermodynamic contributions are unfavorable, most likely due to electrostatic repulsion with negatively charged sulfate headgroups of the micelles. Another system we looked at is the unique histidine at position 50 of αS (Fig. 10.2). It is worth noting that αS prefers binding to membranes with a high proportion of negatively charged lipids, so the negatively charged SDS micelles in this case should be a better model than neutral DPC micelles (Eliezer *et al.*, 2001). Here as well, we see a large pK_a shift for His50 from 6.33 for the free protein to 8.06 for the protein bound to SDS micelles. The larger pK_a in the presence of SDS indicates His50 takes up a proton at smaller hydronium ion concentration, presumably because of electrostatic attraction between the resulting positively charged histidine and the negatively charged sulfate groups on the SDS micelle. The pK_a shift corresponds to an energetically favorable $\Delta\Delta G_{titr}$ of -2.4 kcal/mol at 37 °C. By contrast, in neutral DPC micelles the pK_a is raised only to 6.59, corresponding to a $\Delta\Delta G_{titr}$ of -0.4 kcal/mol. There appears to be a residual interaction, possibly involving a weak ion pair between Glu46 and His50 but its contribution to the stability of the folded structure is much smaller than that between His50 and the negatively charged micelle sulfate groups of SDS. Although we have not measured pK_a values for Aβ(1–40) bound to neutral DPC micelles, these have been reported previously for the 1–28 Aβ fragment (Ma *et al.*, 1999). For all 10 residues investigated, large pK_a shifts were observed when the 1–28 peptide was bound to negatively charged SDS micelles. By contrast, when bound to neutral DPC micelles the pK_a values for Aβ(1–28) were similar to those of the unstructured peptide in D_2O.

In summary, binding of proteins to SDS micelles appears to induce large shifts in pK_a values due to the interactions of polypeptide residues with the

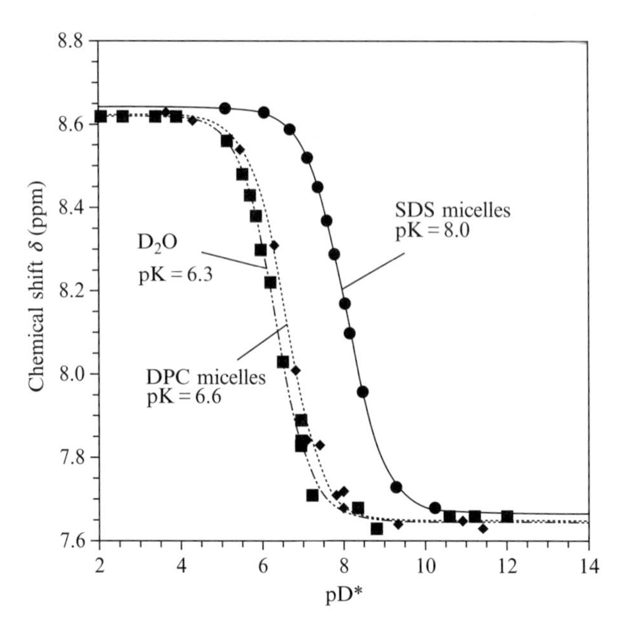

Figure 10.2 Titration curves of His50 from αS in various solutions. The data were obtained from 1D ^{1}H NMR experiments at a temperature of 37 °C. Samples of αS were obtained as previously described (Croke *et al.*, 2008). For the micelle samples the concentrations of deuterated SDS and DPC were 100 m*M* in a D_2O solution. The data in D_2O alone were collected in the presence of 100 m*M* NaCl while the protein in micelles had no added salt.

negatively charged surface of the micelle. The magnitudes of the shifts appear to be similar in the different systems we looked at (Aβ and αS), and while they show some dependence on residue type (Table 10.1), the pK_a shifts show little dependence on the location of the ionizable group in the sequence or structure (Table 10.1). This is important, because residues 1–14 are unstructured in micelle-bound Aβ, while residues 15–36 fold into an α-helical structure with a kink between residues 25 and 27 (Coles *et al.*, 1998). Moreover, paramagnetic relaxation enhancement spin-label experiments show that the segments 15–24 and 29–35 reside on the surface and in the interior of the micelle, respectively; while the unstructured segments 1–14 and 36–40 are solvent-exposed (Jarvet *et al.*, 2007). The pK_a shifts for Aβ appear to be insensitive to the differences in structural properties and to the positioning of the polypeptide with respect to the micelle (Table 10.1). At the same time, isolated amino acids do not experience the large pK_a shifts observed for Aβ in the presence and absence of SDS (Ma *et al.*, 1999), indicating that the observed titration behavior is specific to polypeptides.

Taken together these observations suggest that SDS micelles act as polyanions and provide a screen for the entire peptide that appears to be independent of structure and the degree of immersion of the polypeptides into the micelle. In other words, this screening appears to be nonspecific with respect to structural context. The interactions between negatively charged groups on the peptide and the negatively charged groups on the SDS micelles are energetically unfavorable due to electrostatic repulsion, while the interactions of positively charged residues with the micelle are favorable. By contrast, pK_a values are similar for the bound and free protein with neutral DPC micelles. This strongly suggests that electrostatic interactions formed between charged groups within the protein are small compared to those between charged groups from the protein and the micelle.

11. Results from Fibrillization Kinetics

Amyloid fibril formation follows nucleation kinetics (Alexandrescu and Rathgeb-Szabo, 1999; Harper and Lansbury, 1997). The reaction is initially characterized by a lag phase during which no aggregation is detected. The lag phase occurs because association of monomers is initially unfavorable due to the loss of rotational and translational freedom. Eventually, a sufficient number of monomers accrue to form a *critical nucleus*, and the enthalpy gain from incorporating additional monomers outweighs the increase in entropy from dissociation. Aggregation becomes energetically favorable and the reaction enters a growth phase. The growth phase continues until a plateau is reached that depends on the equilibrium between soluble species and the insoluble fibrils. A representative kinetic trace of Aβ(1–40) fibrillization detected by fluorimetry of the fibril-specific dye thioflavin T (ThT) (LeVine, 1993) is shown in Fig. 10.3A.

To gain insights into how electrostatics affect fibrillization, we looked at the effects of pH and single amino acid substitutions that replace ionizable residues. Representative data illustrating the effects of pH on the fibrillization of the wild-type Aβ(1–40) are shown in Fig. 10.3B. From these data, we can analyze the effects of pH on the lag phase of fibrillization (Fig. 10.3C) and on the plateau of the reaction (Fig. 10.3D). The kinetics of the growth phase showed no clear dependence with pH (not shown). The pH profile of the lag phase is similar to that previously observed for the solubility of Aβ. The lag phase is shortest between pH 3 and 5 where Aβ(1–40) is least soluble (LeVine, 1993). We could not obtain any data below pH 2 because the ThT dye becomes protonated at very acidic pH (not shown). In the region between pH 6.5 and 7.5, there is an apparent plateau in the pH dependence of the lag phase (Fig. 10.3C). We note, however, that above pH 8.5 we failed to detect any aggregation of Aβ (1–40) samples after 160 h so the lag phase must be longer than this (Fig. 10.3B, bottom trace). Indeed, when preparing Aβ we dissolve the

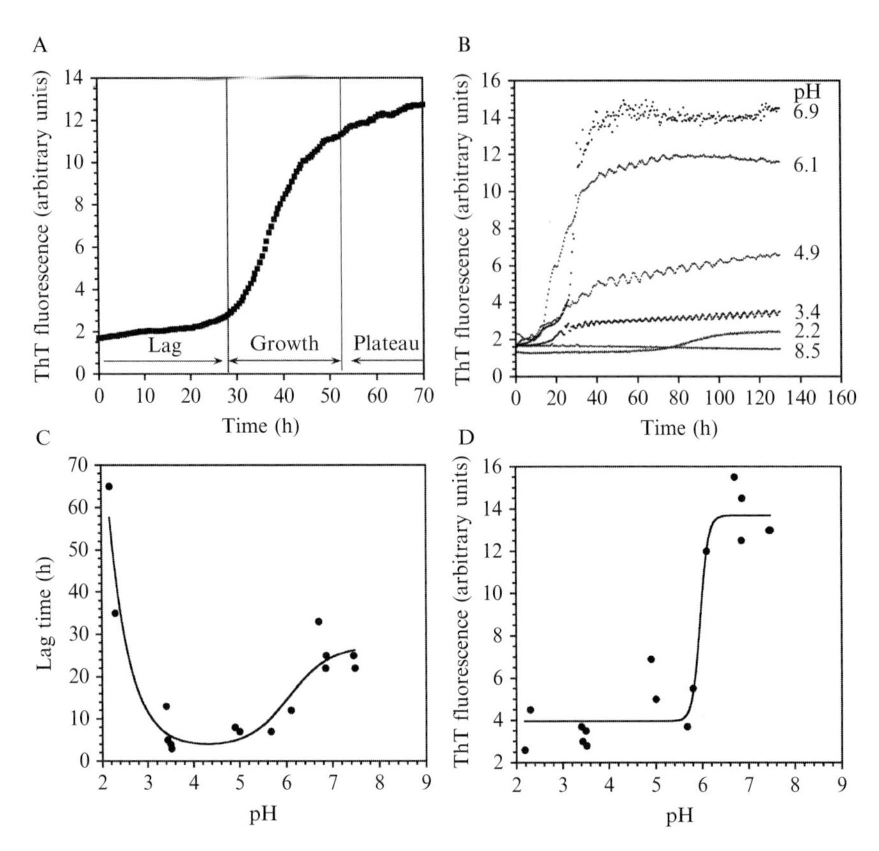

Figure 10.3 Dependence of Aβ(1–40) fibrillization kinetics on pH. (A) Representative Aβ(1–40) kinetic run at pH 7.45 showing the phases of a fibrillization reaction. (B) Kinetic runs for Aβ(1–40) at the pH values indicated. (C) Plot of fibrillization lag times at different pH values. The curve is a fit of the data to a function with two pK_a values of 1.6 $\pm$ 1.7 and 6.1 $\pm$ 0.4. (D) Plot of ThT fluorescence plateaus at different pH values. The data were fit to a Henderson–Hasselbalch equation yielding a pK_a of 6.0 $\pm$ 0.1 and a Hill coefficient of 5 $\pm$ 2. Samples were run in a 96-well plate on a Fluoroskan Ascent instrument. Each sample contained 46 μM Aβ(1–40), 15 μM ThT, 5 mM EDTA, and 0.02% sodium azide. All data were obtained at 37 °C with a shaking speed of 240 rpm. The following buffers were used: citric acid (pH 2–4), sodium acetate (pH 4–5), sodium citrate (pH 5–6), sodium HEPES (pH 6.5–7.5), and Tris–HCl (pH > 7.5).

samples in 10 mM NaOH (pH 10) to protect against aggregation. We looked at the effects of mutants on the pH dependence of the lag time but the results were difficult to interpret as many of the mutants gave significantly decreased lag times. Figure 10.3D shows the pH dependence of the ThT fluorescence plateau of wild-type Aβ at different pH values. The data define a single titration curve with an apparent pK_a value of 6.0. At its inflection point the curve is very steep and gives an apparent Hill coefficient of 5 ± 2, indicating positive cooperativity between multiple charged groups. This is not surprising considering that the repetitive fibril structure replicates charged groups along the length of its axis. Above pH 6 there is a marked increase in the ThT florescence plateau compared to samples at more acidic pH. This indicates that although the lag phase is slower above pH 6, more fibrils are formed compared to acidic pH. Since the ThT dye is only sensitive to fibrils, this does not imply more monomer at acidic pH. Indeed sedimentation experiments show that the solubility of the peptide is at a minimum between pH 5 and 3 (LeVine, 1993). A more consistent explanation for the decreased ThT fluorescence at low pH is that nonfibrillar, and therefore ThT-insensitive, aggregates become more prevalent under acidic conditions.

Figure 10.4A compares the pH dependence of the fluorescence plateau for wild-type Aβ(1–40) and mutants that replace charged residues. A triple mutant of Aβ(1–42) in which all three histidines are substituted by asparagines shows an inflection point similar to the wild type, indicating that the pK_a at 6.0 in the fluorescence plateau profile cannot be due to the histidines in Aβ. The closest remaining pK_a values are those of carboxylate groups (pK_a of 4.25 in a random coil) or the α-amino group at the N-terminus (pK_a of 8.0 in a random coil). Based on quenched hydrogen exchange and solid-state NMR data, residues ~10–36 form the fibril core structure in Aβ(1–40), whereas the first nine residues are disordered (Petkova *et al.*, 2004). This makes it unlikely that the pK_a at 6.0 could be from the N-terminus. Moreover, substitution of Glu22 with a glutamine abolishes the pK_a at 6.0 (red in Fig. 10.4A). As a control we looked at the equivalent mutation E11Q for a position at the edge of the fibril core structure and found that the substitution maintained the increase in the fluorescence profile observed near pH 6 for the wild type. Besides Glu11, His13, and His14, which we can exclude because their substitution retains the pK_a near 6, the only remaining ionizable residues in the region between residues 12 and 40 are Glu22, Asp23, and the α-carboxyl group at the C-terminus. Lys16 and Lys28 should have pK_a values above 10.4. We therefore assign the pK_a of 6.0 in the fluorescence plateau profile to Glu22. Based on the high-resolution structure of Aβ(1–42) protofibrils, the adjacent acidic residue Asp23 is involved in a salt bridge with Lys28 at this pH. If we properly consider the network of charges running along the axis of the Aβ fibrils (Fig. 10.4C), however, the carboxylate groups of Asp23 and the C-terminus could also contribute to the pK_a near 6.0.

Considering the pH dependence of the fluorescence plateau, a pK_a of 6.0 for a carboxylate, is about 1.75 pH units above that expected for an equivalent group in a random coil. This implies the charged state of Glu22 makes a destabilizing contribution to the fibril structure of about 2.5 kcal/mol per mole of Aβ monomer. Summed over the multiple

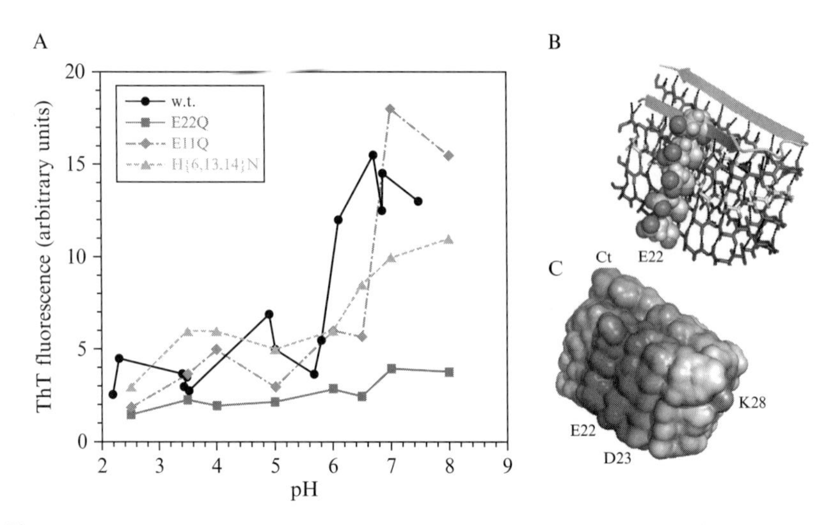

Figure 10.4 Roles of charged residues in the pH dependence of Aβ fibrillization. (A) Comparison of the ThT fluorescence plateau of Aβ variants as a function of pH. The lines are not fits but simply included to guide the eye. (B) Structure of the Aβ (1–42) protofilament (Luhrs *et al.*, 2005). The organization of five Aβ monomers in the protofilament is shown, and residue E22 is highlighted with CPK spheres. Note the side-chain carboxylates from E22 (red spheres) form an array of negative charges running along the length of the fibril. (C) Electrostatic surface of the structure in (B). Charged residues in the core of the fibril structure are labeled. The peptides [E11Q]-Aβ (1–40), [E22Q]-Aβ(1–40), and [H(6,13,14)N]-Aβ(1–42) were all from Anaspec (San Jose, CA). All peptides were dissolved according to a literature protocol (Hou *et al.*, 2004) and run through a 100-kDa molecular weight cutoff filter before use, to remove any aggregates at the start of the reactions. All other conditions were as described in the legend to Fig. 10.3. (See Color Insert.)

monomers that contribute to the structure, the electrostatic destabilization of the fibrils should be very large. The structural model of an Aβ protofibril (Fig. 10.4B) suggests why the charged state of Glu22 is destabilizing. The Glu22 carboxylate groups from individual monomers form an array of unmatched charges along the axis of the Aβ fibril that should lead to strong electrostatic repulsion between like charges. This is further emphasized in the electrostatic surface diagram of Fig. 10.4C. Electrostatic repulsion between the carboxylates of Glu22, Asp23 and the C-terminus may further destabilize the fibril structure.

We expect that electrostatic destabilization due to the propagation of like charges through the symmetry of the structure may be a common theme for amyloid fibrils. Preliminary results with the amyloidogenic proteins αS and amylin reinforce this view. We find that single amino acid substitutions can have large effects on the pH dependence of kinetic parameters for amylin and αS much like they do for Aβ. For the relatively simple amylin system where we can assign features in the pH profile to individual charged groups, we also see large shifts in pK_a from random coil values that suggest unfavorable electrostatic repulsion in the fibrils. The charge properties of fibrils may in part explain why salts and polyelectrolytes typically enhance fibrillization rates. Counterions may screen unfavorable charge interactions that inherently destabilize the fibrils structure (Calamai *et al.*, 2006). When the total of unfavorable charge interactions no longer supports the fibril structure the protein may associate into alternative soluble aggregates. The E22Q Aβ mutation is associated with the "Dutch" phenotype and increased toxicity of the Aβ(1–40) peptide (Wang *et al.*, 2000). It is tempting to speculate that the increased pathology associated with this mutation may result from its increased proclivity to form alternative soluble aggregates rather than fibrils at neutral pH (Fig. 10.4A).

Charges would appear to have important roles in modulating the conformational transitions of amyloidogenic proteins and their assembly into fibrils. A remaining challenge is to measure the pK_a values of ionizable groups directly from fibrils. We are exploring the possibility of obtaining such measurements using solid-state NMR or direct potentiometric titration measurements. The availability of these data would further improve our understanding of amyloidogenic assemblies, which are interesting both because of their biophysical complexity and because of their importance for human disease.

12. Conclusion

Recent NMR studies have shed new light on the contributions of ionization equilibria to protein stability. Charges can be stabilizing or destabilizing to a protein structure and need to be considered individually since the contributions for partners in an ion pair are often not additive. While some individual charges can have a large impact on protein stability, the typical contribution we found for surface-exposed residues is less than 1 kcal/mol. The magnitude of electrostatic contributions to folding stability is often overestimated in structure-based electrostatic calculations. This is in part due to a lack of information on the types of structural rearrangements that accompany changes in the charge states of residues. As the effects of ionization on structure become better characterized theoretical modeling of electrostatic interactions should improve. Large shifts in pK_a values are observed for the amyloidogenic proteins Aβ and αS when they become bound to negatively charged SDS micelles. The effects appear to be almost uniform, nonspecific with regard to structure, and are not seen with uncharged DPC micelles. The results suggest that in membrane bilayers composed of a heterogeneous mixture of lipid, the interactions between charged side chains and charged lipid headgroups are likely to be an important factor in stabilizing the structures of peripheral membrane proteins. We see large deviations from random coil values for the pK_a values of charged residues in amyloid fibrils. The shifts suggest that electrostatic repulsion between charges replicated along the length of the fibril is an important destabilizing force in the formation of these assemblies.

Macromolecular complexes such as membrane proteins, virus capsids, and amyloid fibrils represent a future frontier for investigations of electrostatic contributions to structural stability. Studies on these complexes could be important in establishing the mechanisms of subunit association and could suggest strategies to exploit electrostatic properties to control supramolecular assembly.

ACKNOWLEDGMENTS

A.T.A. was supported by a research grant from the American Parkinson Disease Association and by NSF grant MCB 02363116. NMR instrumentation was funded in part by grant 1S10RR016760 from the NIH-NCRR.

REFERENCES

Acevedo, O., *et al.* (2002). pH corrections in chemical denaturant solutions. *Anal. Biochem.* **306,** 158–161.

Alexandrescu, A. T., and Croke, R. L. (2008). NMR of amyloidogenic proteins. *In* "Protein Misfolding," (C. B. O'Doherty and A. C. Byrne, eds.), Nova Science Publishers, Hauppauge, NY.

Alexandrescu, A. T., and Rathgeb-Szabo, K. (1999). An NMR investigation of solution aggregation reactions preceding the misassembly of acid-denatured cold shock protein A into fibrils. *J. Mol. Biol.* **291,** 1191–1206.

Alexandrescu, A. T., *et al.* (1988). NMR assignments of the four histidines of staphylococcal nuclease in native and denatured states. *Biochemistry* **27,** 2158–2165.

Alexandrescu, A. T., *et al.* (1990). Chemical exchange spectroscopy based on carbon-13 NMR. Applications to enzymology and protein folding. *J. Magn. Reson.* **87,** 523–535.

Alexandrescu, A. T., *et al.* (1992). 1H-NMR assignments and local environments of aromatic residues in bovine, human and guinea pig variants of alpha-lactalbumin. *Eur. J. Biochem.* **210,** 699–709.

Alexandrescu, A. T., *et al.* (1993). Structure and dynamics of the acid-denatured molten globule state of alpha-lactalbumin: A two-dimensional NMR study. *Biochemistry* **32,** 1707–1718.

Anderson, D. E., *et al.* (1990). pH-induced denaturation of proteins: A single salt bridge contributes 3–5 kcal/mol to the free energy of folding of T4 lysozyme. *Biochemistry* **29,** 2403–2408.

Arbely, E., *et al.* (2009). Downhill versus barrier-limited folding of BBL 1: Energetic and structural perturbation effects upon protonation of a histidine of unusually low p*K*a. *J. Mol. Biol.* **387,** 986–992.

Baker, W. R., and Kintanar, A. (1996). Characterization of the pH titration shifts of ribonuclease A by one- and two-dimensional nuclear magnetic resonance spectroscopy. *Arch. Biochem. Biophys.* **327,** 189–199.

Banaszak, L. J. (2000). Foundations of Structural Biology. Academic Press, New York, NY.

Berisio, R., *et al.* (1999). Protein titration in the crystal state. *J. Mol. Biol.* **292,** 845–854.

Berisio, R., *et al.* (2002). Atomic resolution structures of ribonuclease A at six pH values. *Acta Crystallogr. D Biol. Crystallogr.* **58,** 441–450.

Bhattacharya, S., and Lecomte, J. T. (1997). Temperature dependence of histidine ionization constants in myoglobin. *Biophys. J.* **73,** 3241–3256.

Bosshard, H. R., *et al.* (2004). Protein stabilization by salt bridges: Concepts, experimental approaches and clarification of some misunderstandings. *J. Mol. Recognit.* **17,** 1–16.

Burkhard, P., *et al.* (2000). The coiled-coil trigger site of the rod domain of cortexillin I unveils a distinct network of interhelical and intrahelical salt bridges. *Struct. Fold. Des.* **8,** 223–230.

Calamai, M., *et al.* (2006). Nature and significance of the interactions between amyloid fibrils and biological polyelectrolytes. *Biochemistry* **45,** 12806–12815.

Cavanagh, J., *et al.* (1996). Protein NMR Spectroscopy: Principles and Practice. Academic Press, San Diego, CA.

Chiang, C. M., *et al.* (1996). The role of acidic amino acid residues in the structural stability of snake cardiotoxins. *Biochemistry* **35,** 9177–9186.

Coles, M., *et al.* (1998). Solution structure of amyloid beta-peptide(1–40) in a water-micelle environment. Is the membrane-spanning domain where we think it is? *Biochemistry* **37,** 11064–11077.

Croke, R. L., *et al.* (2008). Hydrogen exchange of monomeric {alpha}-synuclein shows unfolded structure persists at physiological temperature and is independent of molecular crowding in *Escherichia coli. Protein Sci.* **17,** 1434–1445.

Dames, S. A., *et al.* (1998). NMR structure of a parallel homotrimeric coiled coil. *Nat. Struct. Biol.* **5,** 687–691.

Dames, S. A., *et al.* (1999). Contributions of the ionization states of acidic residues to the stability of the coiled coil domain of matrilin-1. *FEBS Lett.* **446,** 75–80.

Eliezer, D., *et al.* (2001). Conformational properties of alpha-synuclein in its free and lipid-associated states. *J. Mol. Biol.* **307,** 1061–1073.

Glover, J. N., and Harrison, S. C. (1995). Crystal structure of the heterodimeric bZIP transcription factor c-Fos-c-Jun bound to DNA. *Nature* **373,** 257–261.

Graeff, R., *et al.* (2006). Acidic residues at the active sites of CD38 and ADP-ribosyl cyclase determine nicotinic acid adenine dinucleotide phosphate (NAADP) synthesis and hydrolysis activities. *J. Biol. Chem.* **281,** 28951–28957.

Halle, B. (2004). Biomolecular cryocrystallography: Structural changes during flash-cooling. *Proc. Natl. Acad. Sci. USA* **101,** 4793–4798.

Harper, J. D., and Lansbury, P. T. Jr. (1997). Models of amyloid seeding in Alzheimer's disease and scrapie: Mechanistic truths and physiological consequences of the time-dependent solubility of amyloid proteins. *Annu. Rev. Biochem.* **66,** 385–407.

Hou, L., *et al.* (2004). Solution NMR studies of the A beta(1–40) and A beta(1–42) peptides establish that the Met35 oxidation state affects the mechanism of amyloid formation. *J. Am. Chem. Soc.* **126,** 1992–2005.

Ibarra-Molero, B., *et al.* (2004). Salt-bridges can stabilize but do not accelerate the folding of the homodimeric coiled-coil peptide GCN4-p1. *J. Mol. Biol.* **336,** 989–996.

Jarvet, J., *et al.* (2007). Positioning of the Alzheimer Abeta(1–40) peptide in SDS micelles using NMR and paramagnetic probes. *J. Biomol. NMR* **39,** 63–72.

Jones, S., *et al.* (2003). Using electrostatic potentials to predict DNA-binding sites on DNA-binding proteins. *Nucleic Acids Res.* **31,** 7189–7198.

Joshi, M. D., *et al.* (1997). Complete measurement of the p*K*a values of the carboxyl and imidazole groups in *Bacillus circulans* xylanase. *Protein Sci.* **6,** 2667–2670.

Kaslik, G., *et al.* (1999). Properties of the His57-Asp102 dyad of rat trypsin D189S in the zymogen, activated enzyme, and alpha1-proteinase inhibitor complexed forms. *Arch. Biochem. Biophys.* **362,** 254–264.

Kay, L. E. (1995). Pulsed field gradient multi-dimensional NMR methods for the study of protein structure and dynamics in solution. *Prog. Biophys. Mol. Biol.* **63,** 277–299.

Khare, D., *et al.* (1997). p*K*a measurements from nuclear magnetic resonance for the B1 and B2 immunoglobulin G-binding domains of protein G: Comparison with calculated values for nuclear magnetic resonance and X-ray structures. *Biochemistry* **36,** 3580–3589.

Kuhlman, B., *et al.* (2003). Design of a novel globular protein fold with atomic-level accuracy. *Science* **302,** 1364–1368.

Kumar, S., and Nussinov, R. (2002). Close-range electrostatic interactions in proteins. *Chembiochem* **3,** 604–617.

Lashuel, H. A., *et al.* (2002). Neurodegenerative disease: Amyloid pores from pathogenic mutations. *Nature* **418,** 291.

Lavigne, P., *et al.* (1996). Interhelical salt bridges, coiled-coil stability, and specificity of dimerization. *Science* **271,** 1136–1138.

Lee, D. L., *et al.* (2003). Unique stabilizing interactions identified in the two-stranded alpha-helical coiled-coil: Crystal structure of a cortexillin I/GCN4 hybrid coiled-coil peptide. *Protein Sci.* **12,** 1395–1405.

LeVine, H. III (1993). Thioflavine T interaction with synthetic Alzheimer's disease beta-amyloid peptides: Detection of amyloid aggregation in solution. *Protein Sci.* **2,** 404–410.

Linse, S., *et al.* (1995). The effect of protein concentration on ion binding. *Proc. Natl. Acad. Sci. USA* **92,** 4748–4752.

Liu, L., *et al.* (2009). Direct measurement of the ionization state of an essential guanine in the hairpin ribozyme. *Nat. Chem. Biol.* **5,** 351–357.

Luhrs, T., *et al.* (2005). 3D structure of Alzheimer's amyloid-beta(1–42) fibrils. *Proc. Natl. Acad. Sci. USA* **102,** 17342–17347.

Lumb, K. J., and Kim, P. S. (1995). Measurement of interhelical electrostatic interactions in the GCN4 leucine zipper. *Science* **268,** 436–439.

Lumb, K. J., and Kim, P. S. (1996). Interhelical salt bridges, coiled-coil stability, and specificity of dimerization. *Science* **271,** 1136–1138.

Ma, K., *et al.* (1999). Residue-specific p*K*a measurements of the Abeta-peptide and mechanism of pH-Induced amyloid formation. *J. Am. Chem. Soc.* **120,** 8698–8706.

Makhatadze, G. I., *et al.* (2003). Contribution of surface salt bridges to protein stability: Guidelines for protein engineering. *J. Mol. Biol.* **327,** 1135–1148.

Markley, J. L. (1975). Observation of histidine residues in proteins by means of nuclear magnetic resonance spectroscopy. *Acc. Chem. Res.* **8,** 70–80.

Marti, D. N. (2005). Apparent p*K*a shifts of titratable residues at high denaturant concentration and the impact on protein stability. *Biophys. Chem.* **118,** 88–92.

Marti, D. N., and Bosshard, H. R. (2003). Electrostatic interactions in leucine zippers: Thermodynamic analysis of the contributions of Glu and His residues and the effect of mutating salt bridges. *J. Mol. Biol.* **330,** 621–637.

Marti, D. N., *et al.* (2000). Interhelical ion pairing in coiled coils: Solution structure of a heterodimeric leucine zipper and determination of p*K*a values of Glu side chains. *Biochemistry* **39,** 12804–12818.

Matousek, W. M., *et al.* (2007). Electrostatic contributions to the stability of the GCN4 leucine zipper structure. *J. Mol. Biol.* **374,** 206–219.

Nakamura, H. (1996). Roles of electrostatic interaction in proteins. *Q. Rev. Biophys.* **29,** 1–90.

O'Shea, E. K., *et al.* (1991). X-ray structure of the GCN4 leucine zipper, a two-stranded, parallel coiled coil. *Science* **254,** 539–544.

Patil, S. M., *et al.* (2009). Dynamic alpha-helix structure of micelle-bound human amylin. *J. Biol. Chem.* **284,** 11982–11991.

Petkova, A. T., *et al.* (2004). Solid state NMR reveals a pH-dependent antiparallel beta-sheet registry in fibrils formed by a beta-amyloid peptide. *J. Mol. Biol.* **335,** 247–260.

Pujato, M., *et al.* (2006). The pH-dependence of amide chemical shift of Asp/Glu reflects its p*K*a in intrinsically disordered proteins with only local interactions. *Biochim. Biophys. Acta* **1764,** 1227–1233.

Ripoll, D. R., *et al.* (2005). On the orientation of the backbone dipoles in native folds. *Proc. Natl. Acad. Sci. USA* **102,** 7559–7564.

Sali, D., *et al.* (1988). Stabilization of protein structure by interaction of alpha-helix dipole with a charged side chain. *Nature* **335,** 740–743.

Schaller, W., and Robertson, A. D. (1995). pH, ionic strength, and temperature dependences of ionization equilibria for the carboxyl groups in turkey ovomucoid third domain. *Biochemistry* **34,** 4714–4723.

Sheinerman, F. B., and Honig, B. (2002). On the role of electrostatic interactions in the design of protein–protein interfaces. *J. Mol. Biol.* **318,** 161–177.

Shoemaker, D. P., *et al.* (1981). Experiments in Physical Chemistry. McGraw-Hill, New York, NY.

Shortle, D. R. (1996). Structural analysis of non-native states of proteins by NMR methods. *Curr. Opin. Struct. Biol.* **6,** 24–30.

Steinmetz, M. O., *et al.* (2007). Molecular basis of coiled-coil formation. *Proc. Natl. Acad. Sci. USA* **104,** 7062–7067.

Sternberg, M. J., *et al.* (1987). Prediction of electrostatic effects of engineering of protein charges. *Nature* **330,** 86–88.

Strickler, S. S., *et al.* (2006). Protein stability and surface electrostatics: A charged relationship. *Biochemistry* **45,** 2761–2766.

Tan, Y. J., *et al.* (1995). Perturbed p*K*A-values in the denatured states of proteins. *J. Mol. Biol.* **254,** 980–992.

Thurlkill, R. L., *et al.* (2006). p*K* values of the ionizable groups of proteins. *Protein Sci.* **15,** 1214–1218.

Tollinger, M., *et al.* (2002). Measurement of side-chain carboxyl p*K*(a) values of glutamate and aspartate residues in an unfolded protein by multinuclear NMR spectroscopy. *J. Am. Chem. Soc.* **124,** 5714–5717.

Tollinger, M., *et al.* (2003). Site-specific contributions to the pH dependence of protein stability. *Proc. Natl. Acad. Sci. USA* **100,** 4545–4550.

Ulmer, T. S., *et al.* (2005). Structure and dynamics of micelle-bound human alpha-synuclein. *J. Biol. Chem.* **280,** 9595–9603.

van Vlijmen, H. W., *et al.* (1998). Improving the accuracy of protein p*K*a calculations: Conformational averaging versus the average structure. *Proteins* **33,** 145–158.

Wang, Z., *et al.* (2000). Toxicity of Dutch (E22Q) and Flemish (A21G) mutant amyloid beta proteins to human cerebral microvessel and aortic smooth muscle cells. *Stroke* **31,** 534–538.

Warshel, A., and Papazyan, A. (1998). Electrostatic effects in macromolecules: Fundamental concepts and practical modeling. *Curr. Opin. Struct. Biol.* **8,** 211–217.

Whitten, S. T., *et al.* (2005). Local conformational fluctuations can modulate the coupling between proton binding and global structural transitions in proteins. *Proc. Natl. Acad. Sci. USA* **102,** 4282–4287.

Wilson, N. A., *et al.* (1995). Aspartic acid 26 in reduced *Escherichia coli* thioredoxin has a p*K*a > 9. *Biochemistry* **34,** 8931–8939.

Wishart, D. S., *et al.* (1995). 1H, 13C and 15N chemical shift referencing in biomolecular NMR. *J. Biomol. NMR* **6,** 135–140.

Wuthrich, K. (1986). NMR of Proteins and Nucleic Acids. John Wiley & Sons, New York, NY.

Wuthrich, K. (2003). NMR studies of structure and function of biological macromolecules (Nobel lecture). *Angew. Chem. Int. Ed. Engl.* **42,** 3340–3363.

Yamazaki, T., *et al.* (1994). NMR and X-Ray evidence that the HIV protease catalytic aspartyl groups are protonated in the complex formed by the protease and a non-peptide cyclic urea-based inhibitor. *J. Am. Chem. Soc.* **116,** 10791–10792.

You, T. J., and Bashford, D. (1995). Conformation and hydrogen ion titration of proteins: A continuum electrostatic model with conformational flexibility. *Biophys. J.* **69,** 1721–1733.

Zhou, H. X., and Vijayakumar, M. (1997). Modeling of protein conformational fluctuations in p*K*a predictions. *J. Mol. Biol.* **267,** 1002–1011.

CHAPTER ELEVEN

Kinetics of Allosteric Activation

Enrico Di Cera

Contents

Abstract

Although enzyme inhibition results in most cases from the competing effect of a ligand with substrate, the ability of ligands to enhance enzyme function requires binding to a site distinct from the active site. This is the basis of allosteric activation of enzyme activity, documented most conspicuously in the vast family of enzymes activated by monovalent cations. In this chapter, we review the basic kinetic aspects of allosteric activation by taking into consideration the biologically relevant case of an enzyme E possessing a single site for substrate S and a single site for the allosteric effector L.

1. Linkage

The effect of a ligand L on the binding of a substrate S to the enzyme E is an example of linkage (Wyman, 1948, 1964; Wyman and Gill, 1990). Linkage is a thermodynamic concept derived from the first law (Di Cera, 1995) and reflects the reciprocity of effects that chemical components are subject to when they interact in a closed system. Steady state systems extend the properties of systems at equilibrium and add complexity to the rules governing communication. It is, therefore, instructive to start our discussion with the properties of a linkage scheme at equilibrium, as depicted in Scheme 11.1.

Department of Biochemistry and Molecular Biophysics, Washington University School of Medicine, St. Louis, Missouri, USA

Methods in Enzymology, Volume 466
ISSN 0076-6879, DOI: 10.1016/S0076-6879(09)66011-0

$$\begin{array}{ccc} & k_{-1,0} & \\ E & \leftrightarrows & ES \\ & k_{1,0}[S] & \\ k_A[L] \downarrow\uparrow k_{-A} & & k'_A[L] \downarrow\uparrow k'_{-A} \\ & k_{-1,1} & \\ EL & \leftrightarrows & ELS \\ & k_{1,1}[S] & \end{array}$$

Scheme 11.1

The protein is assumed to exist in two forms, one free (E) and the other bound to the allosteric effector (EL), featuring different values of kinetic rate constants for binding ($k_{1,0}$, $k_{1,1}$) and dissociation ($k_{-1,0}$, $k_{-1,1}$) of S. The parameters $K_A = k_A/k_{-A}$ and $K'_A = k'_A/k'_{-A}$ are the equilibrium association constants for L binding to E and ES, respectively. Detailed balance imposes a constraint among the rate constants in Scheme 11.1, that is, $k_{1,0}K'_A k_{-1,1} = k_{-1,0}K_A k_{1,1}$, which is a consequence of the first law of thermodynamics (Di Cera, 1995). The condition is equivalent to

$$\frac{K_0}{K_1} = \frac{K_A}{K'_A} \tag{11.1}$$

where $K_0 = k_{1,0}/k_{-1,0}$ and $K_1 = k_{1,1}/k_{-1,1}$ are the equilibrium association constants for S binding to E in the absence and presence of L, respectively. Mass law binding leads to the following expression for the binding probability of S to E:

$$\theta = \frac{[ES] + [ELS]}{[E] + [EL] + [ES] + [ELS]} = \frac{K[S]}{1 + K[S]} = \frac{K_0\left(\frac{1+K'_A[L]}{1+K_A[L]}\right)[S]}{1 + K_0\left(\frac{1+K'_A[L]}{1+K_A[L]}\right)[S]} \tag{11.2}$$

Equation (11.2) reflects the binding of S to E to a single site controlled by the allosteric effector L that likewise binds to a single site. All the terms in Eq. (11.2) are linear in [S] and [L] and the effect of L on S recognition is embodied by the linkage expression (Di Cera, 1995; Wyman, 1964)

$$K = K_0\left(\frac{1 + K'_A[L]}{1 + K_A[L]}\right) \tag{11.3}$$

for the apparent equilibrium association constant K as a function of [L]. In order for L to change the affinity of E for S, the allosteric effector must bind to the E and ES forms with different affinity. Enhanced binding ensues when $K'_A > K_A$, which implies $K_1 > K_0$ (see Eq. (11.1)). Allostery is, therefore, an example of linkage. The existence of a distinct site to which L binds with different affinities depending on the ligation state (E or ES) of the enzyme produces regulation of substrate binding. This regulation would not be possible if L were to bind to the active site and compete directly with S. In this case, $K'_A = 0$ and the affinity for S would decrease without limits with increasing [L]. Rigorously speaking, however, the case $K'_A = 0$ would not rule out binding of L to a site distinct from the active site. Therefore, the case of enhanced substrate binding to the enzyme can always be ascribed to an allosteric effect, but the case of decreased substrate binding may or may not be due to an allosteric effect and most often then not is simply due to direct competition of S and L at the active site.

2. Allosteric Activation at Steady State

The basic rules just described for an equilibrium system carry over to steady state. In this case, Scheme 11.1 is expanded to take into account the conversion of substrate S into product P, under the influence of the allosteric effector L, that is

$$
\begin{array}{ccccc}
\mathrm{E} & \underset{k_{1,0}[\mathrm{S}]}{\overset{k_{-1,0}}{\leftrightarrows}} & \mathrm{ES} & \xrightarrow{k_{2,0}} & \mathrm{E+P} \\
k_{A}x \downarrow\uparrow k_{-A} & & k'_{A}x \downarrow\uparrow k'_{-A} & & \\
\mathrm{EL} & \underset{k_{1,1}[\mathrm{S}]}{\overset{k_{-1,1}}{\leftrightarrows}} & \mathrm{ELS} & \xrightarrow{k_{2,1}} & \mathrm{EL+P}
\end{array}
$$

Scheme 11.2

Scheme 11.2 is analogous to the Botts–Morales scheme for the action of a modifier on substrate hydrolysis (Botts and Morales, 1953). The enzyme is assumed to exist in two forms, one free (E) and the other bound to the allosteric effector (EL), featuring different values of kinetic rate constants for binding ($k_{1,0}$, $k_{1,1}$), dissociation ($k_{-1,0}$, $k_{-1,1}$) and hydrolysis ($k_{2,0}$, $k_{2,1}$) of substrate S into product P. There are two components defining allosteric activation: the binding of L to its site and the transduction of this event into changes of the catalytic properties of the enzyme. The first component is a property of the free enzyme, independent of any substrate. The second component, on the other hand, depends specifically on the particular enzyme–substrate complex under consideration. There are two important consequences to be reminded here. First, structural determinants responsible for binding of the allosteric effector need not be the same as those responsible for transduction of this event into enhanced catalytic activity. Second, the extent of activation is energetically independent from the affinity of L. Activation is certainly not possible in the absence of L binding ($K_A = 0$ and $K'_A = 0$). However, absence of activation can also be caused by absence of transduction when L binding is present but does not change the values of the kinetic rate constants.

The function of interest in Scheme 11.2 is not the fractional saturation of the enzyme (see Eq. (11.2)), but rather the velocity of product formation at steady state, v, per unit active enzyme concentration, e_{tot}, along with the independent Michaelis–Menten parameters k_{cat} and $s = k_{cat}/K_m$. These parameters can be derived from a plot of the velocity of product generation expressed in units of active enzyme, v/e_{tot}, vs the substrate concentration, [S]. In this plot, the value of k_{cat} is the asymptotic value of v/e_{tot} as $[S] \to \infty$, and the value of s is the initial slope as $[S] \to 0$. It should be noted that K_m, the concentration of substrate giving half of the maximal velocity, is not an independent parameter because its definition requires knowledge of the value of k_{cat}. The values of s and k_{cat}, on the other hand, can be defined independently of each other.

The expression for v/e_{tot} depends on the total number of trajectories at steady state for Scheme 11.2, rather than the number of species. Each trajectory leading to a species must contain the product of three rate constants, as shown in Fig. 11.1, because of the four intermediates in Scheme 11.2 only three are independent. The sum of the trajectories toward each species defines the contribution of that species at steady state. The expression for the velocity of product formation at steady state is obtained with the elegant Hill diagram method (Hill, 1977) as (Di Cera, 2008; Page and Di Cera, 2006)

$$\frac{v}{e_{tot}} = \frac{k_{2,0}[ES] + k_{2,1}[ELS]}{[E] + [EL] + [ES] + [ELS]} = \frac{k_{2,0}\Sigma_{ES} + k_{2,1}\Sigma_{ELS}}{\Sigma_E + \Sigma_{EL} + \Sigma_{ES} + \Sigma_{ELS}} = \frac{\alpha[S] + \beta[S]^2}{\gamma + \delta[S] + \varepsilon[S]^2} \tag{11.4}$$

where Σ_{EX} is the sum of the trajectories toward species EX in Fig. 11.1 (EX = E, ES, EL, or ELS). An important property of Eq. (11.4), that distinguishes it from the equilibrium treatment of Scheme 11.1 leading to Eq. (11.2), is that the velocity of product formation is quadratic in [S], although the enzyme contains only a single site for S. This is a consequence of the difference in which terms are calculated for equilibrium and steady

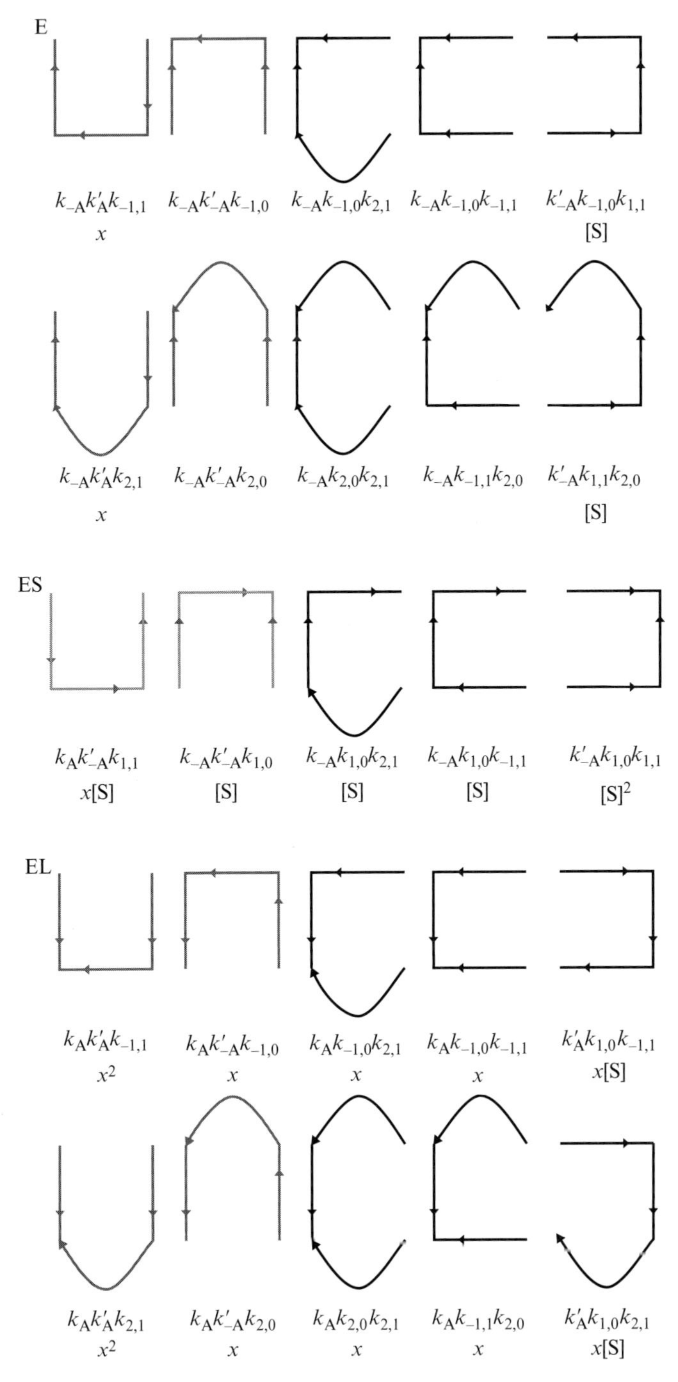

Figure 11.1 (Continued)

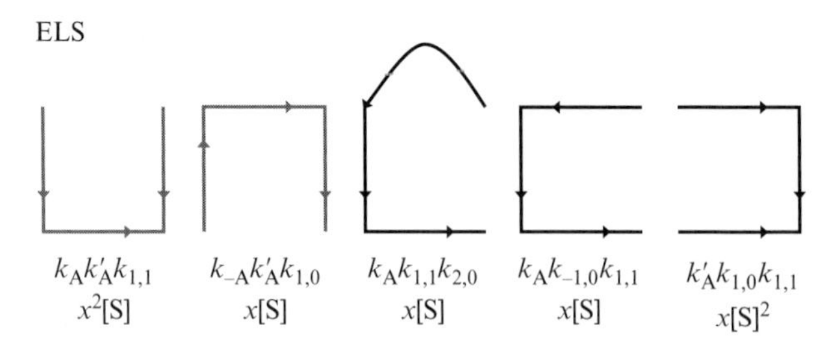

Figure 11.1 Hill diagrams depicting the trajectories toward each of the four species in the kinetic Scheme 11.1. Each trajectory contains the product of three rate constants in Scheme 11.1, because of the four species only three are independent due to mass conservation. Curved lines depict the irreversible reactions of product formation with rate constants $k_{2,0}$ and $k_{2,1}$. Trajectories in gray dominate under conditions where the rates of binding and dissociation of L are fast compared to all other rates. Combination of all trajectories gives the expressions for the coefficients in Eqs. (11.5a)–(11.5e).

state systems. Under the influence of L, an enzyme containing a single active site could in principle display an apparent cooperativity in substrate binding and hydrolysis. Glucokinase is a relevant example of such behavior (Kamata *et al.*, 2004) and many classical papers have documented the ability of monomeric enzymes to express complex kinetic behavior like hysteresis (Frieden, 1970), cooperativity (Botts and Morales, 1953) and allosteric regulation (Ainslie *et al.*, 1972). The coefficients in Eq. (11.4) can be found by application of the Hill diagram method (Hill, 1977) in Fig. 11.1 as follows (Di Cera, 2008; Page and Di Cera, 2006)

$$\frac{\alpha}{k_{-A}k'_{-A}} = (k_{2,0} + k_{2,1}K'_A x)(k_{1,0} + k_{1,1}K_A x) + \frac{k_{10}k_{11}K_{m,0}K_{m,1}}{k'_{-A}}(s_0 + s_1 K_A x) \tag{11.5a}$$

$$\frac{\beta}{k_{-A}k'_{-A}} = \frac{k_{1,0}k_{1,1}}{k_{-A}}(k_{2,0} + k_{2,1}k'_A x) \tag{11.5b}$$

$$\frac{\gamma}{k_{-A}k'_{-A}} = k_{1,0}K_{m,0}\left(1 + \frac{k_{1,1}K_{m,1}}{k'_{-A}} + \omega\frac{k_{1,1}}{k_{1,0}}K_A x\right)(1 + K_A x) \tag{11.5c}$$

$$\frac{\delta}{k_{-A}k'_{-A}} = (k_{1,0} + k_{1,1}k_A x)(1 + K'_A x) + k_{1,0}k_{1,1}K_{m,0} \times\left[\frac{1}{k_{-A}}(1 + \omega K_A x) + \frac{1}{k'_{-A}}\frac{K_A}{K'_A}(\omega + K'_A x)\right] \tag{11.5d}$$

$$\frac{\varepsilon}{k_{-A}k'_{-A}} = \frac{k_{1,0}k_{1,1}}{k_{-A}}(1 + K'_A x) \tag{11.5e}$$

where $K_{m,j} = (k_{-1,j} + k_{2,j})/k_{1,j}$ ($j = 0, 1$) is the Michaelis–Menten constant for the E and EL species. The coefficients contain terms quadratic in

$[L] = x$ although the enzyme has a single binding site for L. The situation is analogous to that just described for the substrate S. The independent parameters k_{cat} and s can be derived readily from Eq. (11.4) as

$$k_{cat} = \frac{\beta}{\varepsilon} = \frac{k_{2,0} + k_{2,1}K'_A x}{1 + K'_A x} \tag{11.6a}$$

$$\begin{aligned} s = \frac{\alpha}{\gamma} &= \frac{s_0 + s_1 K_A x}{1 + K_A x} + (\omega - 1)\frac{K_A x}{1 + K_A x}\frac{s_1 - \frac{k_{1,1}}{k_{1,0}}s_0}{1 + \frac{k_{1,1}K_{m,1}}{k'_{-A}} + \omega\frac{k_{1,1}}{k_{1,0}}K_A x} \\ &= \frac{s_0 + s_1 K_A x}{1 + K_A x} + (\omega - 1)\Lambda \end{aligned} \tag{11.6b}$$

Three independent parameters $k_{2,0}$, $k_{2,1}$, and K'_A are resolved from measurements of k_{cat} as a function of x, from which the binding affinity of L for the ES complex can be measured directly. On the other hand, measurements of s as a function of x only resolve two parameters because of the form of Eq. (11.6b). These parameters are $s_0 = (k_{1,0}k_{2,0})/(k_{-1,0} + k_{2,0})$ and $s_1 = (k_{1,1}k_{2,1})/(k_{-1,1} + k_{2,1})$, obtained respectively as the values of s in the absence or presence of saturating concentrations of allosteric effector. The ratio s_1/s_0 measures the allosteric transduction of the binding of L into the catalytic activity of the enzyme. As already mentioned, this ratio does not depend on the affinity of L for E. Therefore, although the effect of an allosteric modulator on the affinity of substrate can be predicted rigorously from its affinity for the E an ES forms of the enzyme (see Eq. (11.1)), the same information is neither necessary nor sufficient to predict the effect of the allosteric modulator on the enzyme activity. That is readily seen from Scheme 11.2 because the values of $k_{2,0}$ and $k_{2,1}$ are not subject to detailed balancing. Consequently, resolution of K_A, the important parameter measuring the binding affinity of L for the free enzyme E, cannot be secured from measurements of s_1, s_0, and knowledge of K'_A. The additional term $(\omega - 1)\Lambda$ that contains the independent parameters $\omega = (k_{-1,1} + k_{2,1}/k_{-1,0} + k_{2,0})(k_{-1,0}/k_{-1,1})$, $k_{1,1}K_{m,1}/k'_{-A}$, and $k_{1,1}/k_{1,0}$ complicates the task. When $(\omega - 1)\Lambda$ makes only a small contribution to the value of s, K_A can be estimated from the value of x at the midpoint of the transition of s from s_0 to s_1.

The velocity of product formation, Eq. (11.4), can be written in a form where its connection with the classical Michaelis–Menten equation is more readily appreciated

$$\frac{v}{e_{tot}} = \frac{\alpha[S] + \beta[S]^2}{\gamma + \delta[S] + \varepsilon[S]^2} = \frac{k_{cat}[S]}{K_m + [S]}\left(\frac{1}{1 + \frac{[S]}{K_m + [S]}(\omega - 1)\Theta}\right) \tag{11.7a}$$

or in the equivalent double-reciprocal form

$$\frac{e_{\rm tot}}{v} - \frac{K_{\rm m}}{k_{\rm cat}}\frac{1}{[{\rm S}]} + \frac{1}{k_{\rm cat}}[1 + (\omega - 1)\Theta] \tag{11.7b}$$

The additional term, $(\omega - 1)\Theta$, vanishes when $\omega = 1$ or $\Theta = 0$. The former condition was already identified by Botts and Morales and corresponds to the cases where substrate binds and dissociate rapidly from the enzyme, that is, $k_{-1,1} \gg k_{2,1}$ and $k_{-1,0} \gg k_{2,0}$, or in the interesting but rather peculiar circumstance where $k_{2,1}/k_{-1,1} = k_{2,0}/k_{-1,0}$. The expression for Θ is

$$\Theta = k_{1,1}\frac{\frac{1}{k'_{-{\rm A}}}\frac{K_A x}{1+K'_{\rm A}x}(k_{2,1} - k_{2,0}) + \frac{1}{k_{-{\rm A}}}K_m\left[(1 + \omega K_{\rm A}x)\Lambda - (s_1 - s_0)\frac{K_{\rm A}x}{1+K_{\rm A}x}\right]}{\frac{k_{11}K_{m,1}}{k'_{-{\rm A}}}(s_0 + s_1 K_{\rm A}x) + (s_0 + s_1\omega K_{\rm A}x)\left(1 + \frac{k_{1,1}}{k_{1,0}}K_{\rm A}x + \frac{k_{1,1}}{k_{-{\rm A}}}[S]\right)} \tag{11.8}$$

and the value of $\Theta = 0$ is obtained when binding and dissociation of L are fast compared to all other rates. In this case, the gray trajectories in Fig. 11.1 dominate the definitions of the coefficients (Eqs. (11.5a)–(11.5e)) and Eq. (11.4) reduces to the familiar Michaelis–Menten form

$$\frac{v}{e_{tot}} = \frac{\alpha[S]}{\gamma + \delta[S]} \tag{11.9}$$

with

$$k_{\rm cat} = \frac{\alpha}{\delta} = \frac{k_{2,0} + k_{2,1}K'_{\rm A}x}{1 + K'_{\rm A}x} \tag{11.10a}$$

$$s = \frac{\alpha}{\gamma} = \frac{s_0 + s_1 K_{\rm A}x}{1 + K_{\rm A}x} + (\omega - 1)\frac{K_{\rm A}x}{1 + K_{\rm A}x}\frac{s_1 - \frac{k_{1,1}}{k_{1,0}}s_0}{1 + \omega\frac{k_{1,1}}{k_{1,0}}K_{\rm A}x} \tag{11.10b}$$

The expression for $k_{\rm cat}$ is not affected by the drastic change in the form of v because it depends solely on the properties of the ES and ELS species. On the other hand, the form of s simplifies slightly but leaves resolution of $K_{\rm A}$ difficult even in the presence of Michaelis–Menten kinetics. Measurements of this important parameter must be carried out by means of other techniques or by resolving ω and $k_{1,1}/k_{1,0}$ from independent measurements of the individual kinetic rate constants.

3. Different Types of Activation (Type Ia, Type Ib, and Type II)

Scheme 11.2 in its general form applies to Type II activation (Di Cera, 2006; Page and Di Cera, 2006) where L promotes both substrate binding and catalysis. In this case, the steady state velocity of substrate hydrolysis should be measured accurately to confirm Michaelis–Menten kinetics. Departure from such behavior is indicative of binding and dissociation of L that take place on the same time scale as substrate binding and dissociation. If the enzyme obeys Michaelis–Menten kinetics over a wide range of solution conditions, then the likely explanation is that binding and dissociation of L are fast compared to all other rates. This is indeed the case for the activation of thrombin due to Na^+ binding (Di Cera, 2008). Measurements of k_{cat} and s as a function of $[Na^+]$ reveal a significant hyperbolic increase in both parameters from finite low values, $k_{2,0}$ and s_0, to higher values, $k_{2,1}$ and s_1. The midpoint of the transition in the k_{cat} versus $[Na^+]$ plot yields K_A', the equilibrium association constant for Na^+ binding to the enzyme–substrate complex (Fig. 11.2). The midpoint of the transition in the s versus $[Na^+]$ plot yields an approximate measure of K_A, the equilibrium association constant for Na^+ binding to the free enzyme.

Kinetic signatures of relevant types of activation are difficult to detect from inspection of the properties of s. On the other hand, the dependence of k_{cat} on [L] is of diagnostic value. A special case of Scheme 11.2 arises when binding of L is absolutely necessary for the binding of substrate S, which corresponds to Type Ia activation (Di Cera, 2006; Page and Di Cera, 2006). In this case, Scheme 11.1 is linear and contains only E, EL, and ELS.

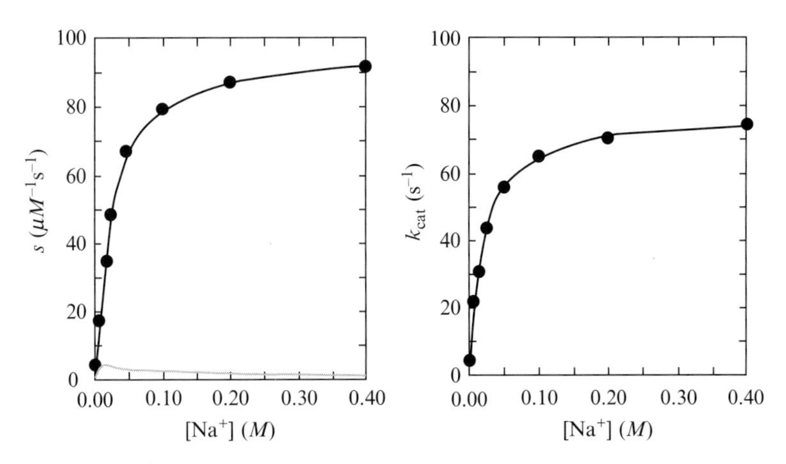

Figure 11.2 Na^+ dependence of the kinetic constants $s = k_{cat}/K_m$ (left) and k_{cat} (right) for the hydrolysis of a chromogenic substrate by thrombin. The data illustrate the signatures of Type II activation with both s and k_{cat} showing a marked Na^+ dependence and changing from low, finite values, to significantly higher values. Curves were drawn using Eqs. (11.10a) and (11.10b), with best-fit parameter values: (data at left) $s_0 = 2.3 \pm 0.1$ $\mu M^{-1}s^{-1}$, $s_1 = 99 \pm 3$ $\mu M^{-1}s^{-1}$, $K_A = 38 \pm 1$ M^{-1}; (data at right) $k_{2,0} = 4.7 \pm 0.2$ s^{-1}, $k_{2,1} = 78 \pm 2$ s^{-1}, $K_A' = 45 \pm 2$ M^{-1}. Also shown is the contribution of the additional term $(\omega - 1)\Lambda$ in Eq. (11.10b) (gray line, left). This term makes at most a 3% correction at low $[Na^+]$.

The enzyme always obeys Michaelis–Menten kinetics, regardless of whether binding and dissociation of L are fast or slow compared to binding and dissociation of substrate. The relevant expressions for k_{cat} and s are obtained from Eqs. (11.6a) and (11.6b) by taking the limits $k_{1,0} \rightarrow 0$, $k_{-1,0} \rightarrow 0$, $k_{2,0} \rightarrow 0$, and $K_A^{'} \rightarrow \infty$, so that

$$k_{cat} = k_{2,1} \tag{11.11a}$$

$$s = \frac{s_1 K_A x}{1 + K_A x} \tag{11.11b}$$

Equations (11.11a) and (11.11b) are significantly different from Eqs. (11.6a) and (11.6b). Type Ia activation always leads to Michaelis–Menten kinetics, a value of k_{cat} that is independent of [L] and a value of s that increases monotonically with [L] from 0 to s_1. Why is k_{cat} independent of [L]? The result seems counterintuitive given that in the absence of allosteric effector no binding of substrate and catalysis can occur. The reason for the constancy of k_{cat} is to be found in the fact that ELS is the only intermediate in the scheme bound to S and the value of k_{cat} is by definition the value of the velocity of substrate hydrolysis, per unit enzyme, under saturating conditions of substrate. Because ELS is the only substrate bound form in Scheme 11.2 when no ES is present, saturation with S will always drive the enzyme into the ELS conformation which will convert S into product with a rate constant k_{cat}. In this special case, the free enzyme E acts as a competitive inhibitor of the active form of the enzyme EL, that is the only species capable of interacting with substrate. Competitive inhibition only affects the value of K_m and leaves the value of k_{cat} unchanged. Among monovalent cation activated enzymes, diol dehydratase shows a k_{cat} that does not change among different monovalent cations that produce very different activation (Toraya *et al.*, 1971). This suggests that k_{cat} is constant although s changes widely, as predicted by Eqs. (11.11a) and (11.11b) for Type Ia activation. The activating effect of K^+ in this enzyme is therefore necessary to enable substrate binding.

Another special case of activation in Scheme 11.1 is when L is required for catalysis but not binding of substrate. In this case, denoted as Type Ib (Di Cera, 2006; Page and Di Cera, 2006), $k_{2,0} = 0$ in Scheme 11.2. There is no guarantee that the enzyme obeys Michaelis–Menten kinetics, unless binding and dissociation of L are fast compared to all other rates. The relevant expressions for k_{cat} and s are

$$k_{cat} = \frac{k_{2,1} K_A^{'} x}{1 + K_A^{'} x} \tag{11.12a}$$

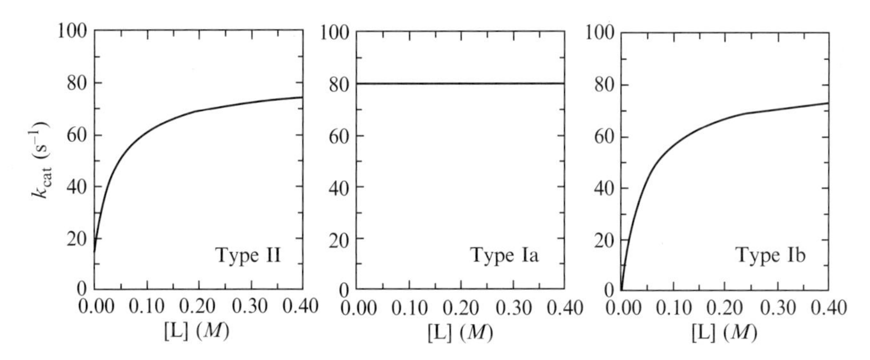

Figure 11.3 Dependence of k_{cat} on the concentratiaon of allosteric effector L is diagnostic for various cases of activation based on Scheme 11.2. In Type II activation, the value of k_{cat} changes from a finite low value to a higher value. The midpoint of the transition defines K_A', the binding constant for L binding to ES. In Type Ia activation, L is required for S binding and the value of k_{cat} is independent of [L]. In Type Ib activation, L is required for S hydrolysis only and the value of k_{cat} changes like in Type II activation, but goes to 0 when [L] = 0. Examples of these three types of activation are discussed in the text.

$$s = \frac{s_1 K_A x}{1 + K_A x} + (\omega - 1)\frac{K_A x}{1 + K_A x}\frac{s_1}{1 + \frac{k_{2,1} + k_{-1,1}}{k'_{-A}} + \omega \frac{k_{1,1}}{k_{1,0}} K_A x} \tag{11.12b}$$

Type Ib activation can easily be distinguished from Type Ia activation because k_{cat} increases with [L] from 0 to s_1. Furthermore, Type Ib activation can easily be distinguished from Type II activation because both k_{cat} and s are 0 in the absence of allosteric effector. An example of Type Ib activation was reported by Kachmar and Boyer for pyruvate kinase, a K^+-activated enzyme (Kachmar and Boyer, 1953). More recent measurements of the activating effect of K^+ on pyruvate kinase confirm Type Ib activation, with the value of k_{cat} dropping to 0 in the absence of K^+ (Laughlin and Reed, 1997; Oria-Hernandez *et al.*, 2005). This demonstrates that binding of K^+ in pyruvate kinase is strictly required only for catalysis and not for substrate binding to the active site. Figure 11.3 shows the different behavior of k_{cat} in the three possible activation cases.

4. Concluding Remarks

Allosteric activation of enzymes is a widespread phenomenon, prominently displayed by enzymes actived by monovalent cations like thrombin and pyruvate kinase (Di Cera, 2006; Page and Di Cera, 2006). A general linkage scheme can be invoked to explain the kinetics of activation with distinct cases (Type II, Types Ia and Ib) giving rise to kinetic signatures that are of diagnostic value. However, the mechanism of kinetic activation can be resolved unambiguously only by structural investigation (Di Cera, 2006). Although kinetic signatures of monovalent cation activation are different enough among Type I and Type II mechanisms, experimental errors often leave the exact mechanism of activation unresolved. For example, a Type II enzyme for which $k_{2,0}$ and s_0 are negligible relative to $k_{2,1}$ and s_1 is difficult to distinguish from a Type Ib enzyme. This case is approached by inosine monophosphate dehydrogenase from *Tritrichomonas foetus* where K^+ increases the k_{cat} > 100-fold from a very low basal level (Gan *et al.*, 2002). Likewise, a Type I enzyme capable of binding and hydrolyzing substrate at measurable rates even in the absence of monovalent cations would bear kinetic signatures indistinguishable from Type II activation. An example of such enzyme is the Na^+-activated Type I enzyme β-galactosidase from *E. coli*. Measurements of s and k_{cat} for the hydrolysis of *para*-nitrophenyl-β-D-galactopyranoside show Type II activation with values of $k_{2,0}$ and s_0 that are, respectively, twofold and 16-fold lower than $k_{2,1}$ and s_1 (Xu *et al.*, 2004). The value of the rigorous analysis presented in this chapter is greatly augmented when structural insights into the system confirm or clarify the predictions drawn from the interpretation of kinetic data.

ACKNOWLEDGMENT

This work was supported in part by NIH research grants HL49413, HL58141, HL73813 and HL95315.

REFERENCES

Ainslie, G. R. Jr., Shill, J. P., and Neet, K. E. (1972). Transients and cooperativity. A slow transition model for relating transients and cooperative kinetics of enzymes. *J. Biol. Chem.* **247,** 7088–7096.

Botts, J., and Morales, M. (1953). Analytical description of the effects of modifiers and of multivalency upon the steady state catalyzed reaction rate. *Trans. Faraday Soc.* **49,** 696–707.

Di Cera, E. (1995). Thermodynamic Theory of Site-Specific Binding Processes in Biological Macromolecules. Cambridge University Press, Cambridge, UK.

Di Cera, E. (2006). A structural perspective on enzymes activated by monovalent cations. *J. Biol. Chem.* **281,** 1305–1308.

Di Cera, E. (2008). Thrombin. *Mol. Aspects Med.* **29,** 203–254.

Frieden, C. (1970). Kinetic aspects of regulation of metabolic processes. The hysteretic enzyme concept. *J. Biol. Chem.* **245,** 5788–5799.

Gan, L., Petsko, G. A., and Hedstrom, L. (2002). Crystal structure of a ternary complex of Tritrichomonas foetus inosine 5′-monophosphate dehydrogenase: NAD+ orients the active site loop for catalysis. *Biochemistry* **41,** 13309–13317.

Hill, T. L. (1977). Free Energy Transduction in Biology. Academic Press, New York, NY.

Kachmar, J. F., and Boyer, P. D. (1953). Kinetic analysis of enzyme reactions. II. The potassium activation and calcium inhibition of pyruvic phosphoferase. *J. Biol. Chem.* **200,** 669–682.

Kamata, K., Mitsuya, M., Nishimura, T., Eiki, J., and Nagata, Y. (2004). Structural basis for allosteric regulation of the monomeric allosteric enzyme human glucokinase. *Structure* **12,** 429–438.

Laughlin, L. T., and Reed, G. H. (1997). The monovalent cation requirement of rabbit muscle pyruvate kinase is eliminated by substitution of lysine for glutamate 117. *Arch. Biochem. Biophys.* **348,** 262–267.

Oria-Hernandez, J., Cabrera, N., Perez-Montfort, R., and Ramirez-Silva, L. (2005). Pyruvate kinase revisited: The activating effect of K+. *J. Biol. Chem.* **280,** 37924–37929.

Page, M. J., and Di Cera, E. (2006). Role of Na^+ and K^+ in enzyme function. *Physiol. Rev.* **86,** 1049–1092.

Toraya, T., Sugimoto, Y., Tamao, Y., Shimizu, S., and Fukui, S. (1971). Propanediol dehydratase system. Role of monovalent cations in binding of vitamin B 12 coenzyme or its analogs to apoenzyme. *Biochemistry* **10,** 3475–3484.

Wyman, J. Jr. (1948). Heme proteins. *Adv. Protein Chem.* **4,** 407–531.

Wyman, J. Jr. (1964). Linked functions and reciprocal effects in hemoglobin: A second look. *Adv. Protein Chem.* **19,** 223–286.

Wyman, J., and Gill, S. J. (1990). Binding and Linkage. University Science Books, Mill Valley, CA.

Xu, J., McRae, M. A., Harron, S., Rob, B., and Huber, R. E. (2004). A study of the relationships of interactions between Asp-201, Na+ or K+ , and galactosyl C6 hydroxyl and their effects on binding and reactivity of beta-galactosidase. *Biochem. Cell Biol.* **82,** 275–284.

CHAPTER TWELVE

Thermodynamics of the Protein Translocation

Alexej Kedrov,* Tanneke den Blaauwen,† and Arnold J. M. Driessen*

Contents

Abstract

Many proteins synthesized in bacteria are secreted from the cytoplasm into the periplasm to function in the cell envelope or in the extracellular medium. The Sec translocase is a primary and evolutionary conserved secretion pathway in bacteria. It catalyzes the translocation of unfolded proteins across the cytoplasmic membrane via the pore-forming SecYEG complex. This process is driven by the proton motive force and ATP hydrolysis facilitated by the SecA motor protein. Current insights in the mechanism of protein translocation are largely based on elaborate multidisciplinary studies performed during the last three decades. To understand the process dynamics, the thermodynamic principles of

* Department of Molecular Microbiology, Groningen Biomolecular Sciences and Biotechnology Institute and the Zernike Institute for Advanced Materials, University of Groningen, The Netherlands
† Molecular Cytology, Swammerdam Institute for Life Sciences, University of Amsterdam, Amsterdam, The Netherlands

Methods in Enzymology, Volume 466
ISSN 0076-6879, DOI: 10.1016/S0076-6879(09)66012-2

translocation and the subunit interactions need to be addressed. Isothermal titration calorimetry has been widely applied to study thermodynamics of biological interactions, their stability, and driving forces. Here, we describe the examples that exploit this method to investigate key interactions among components of the Sec translocase and suggest further potential applications of calorimetry.

1. Introduction

1.1. Protein translocation

Protein translocation through biological membranes has remained in the focus of intensive multidisciplinary research for the last three decades, with the first major breakthrough done as early as in 1980, when a membrane-associated complex facilitating the translocation was isolated from the endoplasmic reticulum by Walter and Blobel (1980). Now, 30 years later we have a quite comprehensive knowledge of the mechanisms behind the translocation process, the key proteins that serve as molecular machines at different stages of the protein translocation reaction and the interactions between the protein subunits. While initial studies were done on protein translocation into the endoplasmic reticulum of eukaryotes, the secretion process of bacterial proteins shares a lot in common with those higher systems, and it is now clear that the pathways have remained highly conserved throughout evolution. Here, we focus on the bacterial Sec translocation pathway, the major route for the transport of proteins across and into the cytoplasmic membrane (Fig. 12.1A). The central component of this system that is also termed translocase is a membrane-embedded pore that consists of three protein subunits, called SecYEG. The heterotrimeric SecYEG complex is a structural and functional analogue of the Sec61 complex in eukaryotes. The SecYEG complex constitutes a central pore through which unfolded substrate polypeptide chains are translocated. The motor protein SecA provides the driving force for translocation utilizing multiple ATP hydrolysis cycles. Most secretory proteins in bacteria are translocated posttranslationally. These proteins are delivered to the translocase by a small chaperone protein SecB that in addition keeps the protein substrates in a translocation-competent state by preventing them from premature folding or aggregation. A wider set of cytoplasmic and membrane-associated proteins is involved in the translocation, but their functions appear nonessential or are understood only poorly. For a detailed description of the translocation pathways we would refer the reader to several comprehensive reviews that have been published recently (Driessen and Nouwen, 2008; Papanikou *et al.*, 2007).

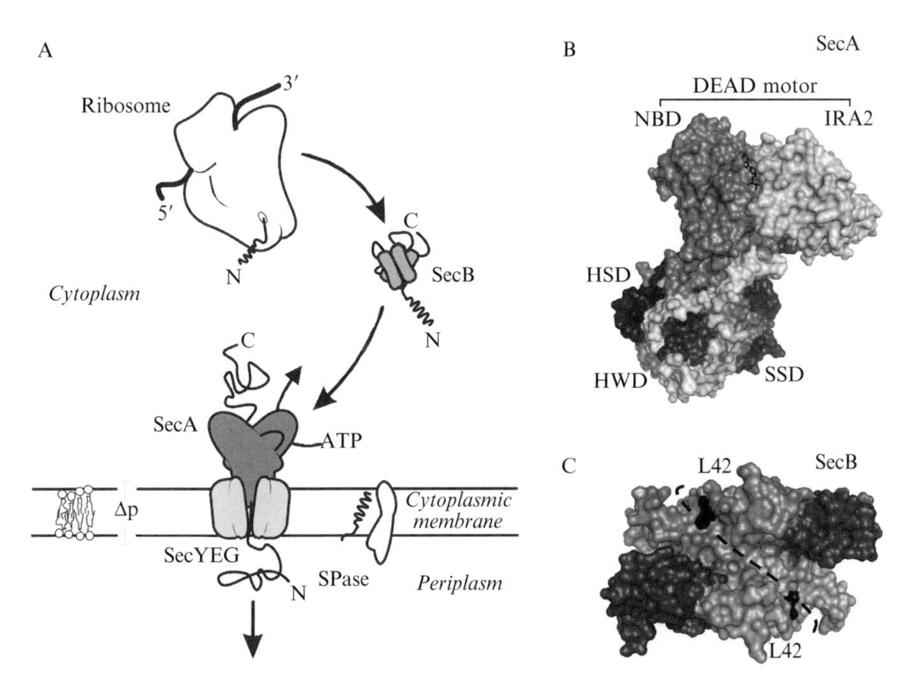

Figure 12.1 (A) Posttranslational protein export pathway via Sec translocase in bacteria (see text for the details). (B) Domain architecture of motor protein SecA of *Bacillus subtilis* (monomer is shown). ATPase motor domain (DEAD) consists of nucleotide-binding domain (NBD) and intramolecular regulator of ATPase (IRA2) capturing a nucleotide (ADP, shown in black lines). Interactions with the substrate polypeptide chain occur at substrate specificity domain (SSD), and helical Scaffold domain (HSD). The HSD and the helical wing domain (HWD) are also involved in interactions with SecYEG translocase. (C) Quaternary structure of the SecB chaperone protein. Four subunits form the functional protein with a hydrophobic groove exposed for the substrate binding (dashed line). Locations of L42 residues are shown in black.

The ubiquity of information on the translocation mechanism has been achieved through a combination of genetic, biochemical, and biophysical approaches, and the most recent structural data have further helped to consolidate many of those results (Zimmer *et al.*, 2008). While a few resolved structures of Sec proteins provide glimpses on possible protein conformations, dynamic aspects of the translocase function are commonly approached via conventional biochemical and biophysical techniques. Being able to dissect the complex translocation cycle into more elementary steps allows one to understand the basics of the process through a set of well-defined experiments. Here, we describe examples on the application of isothermal titration calorimetry (ITC) (Wiseman *et al.*, 1989) to investigate intermolecular interactions relevant to protein translocation. Such thermodynamic studies provide insights into the universal characteristics of the system, while the ability to perform the experiments over a wide range of conditions without modifications, such as labeling or immobilization, of studied macromolecules allows addressing the system in its physiologically relevant state.

1.2. Isothermal titration calorimetry

Applications of ITC technique have extensively increased upon the last 10 years, since advanced commercial instrumentation has become available on the market. Technical achievements made have been of particular importance for the study of biological systems characterized by a marginal stability of noncovalent interactions. A general ITC experiment represents titrating small volumes of a concentrated reactant X (often considered as a ligand) into a cell of a larger volume loaded with a reactant M (macromolecule). To ensure that the binding reaction proceeds in the whole cell volume stirring mixes the solution by the injector syringe during the experiment lapse. The system is thermally isolated from the environment and well equilibrated, so the heat effects associated with complex MX formation can be measured. Each injection is followed by a peak in the isotherm graph that is a manifestation of the heat released or consumed during the reaction. A gradual increase in the concentration of ligand added to the cell favors further complex formation until the reaction reaches its saturation. The associated heat effects decrease with subsequent titrations and eventually reach a steady level, which describes the simple dilution after each injection, normally below 0.1 μcal/sec. Those small heat effects can be accounted by control experiments when loading either the cell or syringe with the identical blank buffer solution and the effects should be subtracted from the total heats recorded.

When designing the ITC experiments one should pay particular attention to the concentration of both reactants. Firstly, a wrong guess on the required concentration range will immediately affect the calculations of the binding stoichiometry and thermodynamic parameters. Secondly, a precise analysis of ITC data requires that a sufficient number of data points are recorded along the binding reaction, and that the corresponding injections result in heat effects well above the level of dilution. The parameter C described in the seminal work of Wiseman *et al.* (1989) as

$$C = n[\mathrm{M_T}]K_\mathrm{a} \tag{12.1}$$

where n is the number of binding sites on the macromolecule in the cell, that is, the complex stoichiometry, K_a—association constant, and $\mathrm{M_T}$, the concentration of the macromolecule in the cell, should be chosen such that C lies above 10 to ensure a reliable analysis of ITC data. Low substrate concentrations would result in featureless isotherm traces. In later studies, that took advantage of technical improvements in ITC design, the ability of the technique to investigate rather weak interactions, where the C value decreases down to 1 or below was demonstrated (Tellinghuisen, 2005). However, it is generally recommended to set the cell concentration 10–50 times higher than the dissociation constant ($K_d = 1/K_a$) of the reaction. Alternatively, the following equation can be used to estimate the final ratio between the ligand X_T and the macromolecule in the cell (Tellinghuisen, 2005):

$$\left.\frac{[X_T]}{[M_T]}\right|_{\text{Final}} = \frac{6.4}{C^{0.2}} + \frac{13}{C}. \tag{12.2}$$

1.3. Data analysis

Integrating the area under each titration peak provides values of heat released or consumed upon the binding reaction, that is, the enthalpy change. From a set of heat values together with the concentrations of both reactants, the enthalpy of complex formation (ΔH) can be calculated, as well as the complex association/dissociation constant. The molar ratio of reactants at the midpoint of the reaction provides a precise estimate of the stoichiometry in the final complex (n). It is a remarkable advantage of ITC measurements that all three parameters can be derived from a single experiment data. The nonlinear least square algorithm used in the software (see below) fits the experimental data to the following ratio:

$$\frac{\Delta Q}{\Delta X_T} \approx \frac{dQ}{dX_T} = \frac{n\Delta HV}{2}\left(1 + \frac{1 - \frac{1}{M_T K_a} - \frac{X_T}{M_T}}{\left(\left(\frac{X_T}{M_T}\right)^2 - 2\frac{X_T}{M_T}\left(1 - \frac{1}{M_T K_a}\right) + \left(1 + \frac{1}{M_T K_a}\right)^2\right)^{1/2}}\right), \tag{12.3}$$

where V is the cell volume and $\Delta Q/\Delta X_T$ is the amount of heat released or consumed upon an injection as measured by ITC (Wiseman *et al.*, 1989). Data analysis packages provided with ITC instruments, such as the Origin-based software of MicroCal, not only allow one to perform the analysis immediately after the experiment, but also include options for different binding models and user-defined data weighting (Tellinghuisen, 2005).

While the values n, ΔH, and K_a are derived directly from the binding isotherm, the total free energy change (ΔG) and also entropy contribution (ΔS) are calculated from basic thermodynamics considerations, as

$$\Delta G = -RT \ln(K_a), \tag{12.4}$$

$$\Delta S = \frac{\Delta H - \Delta G}{T}, \tag{12.5}$$

where R is the gas constant, and T is the experimental temperature.

2. Example 1: SecA Nucleotide Binding

Molecular motors that interact with protein substrates and utilize the ATP hydrolysis for triggering unfolding, dissociation, or translocation can be found in many cellular pathways. SecA is a ubiquitous bacterial protein (cytoplasm concentration $\sim$0.1 mM) that drives the polypeptide translocation through the SecYEG channel (Vrontou and Economou, 2004) and may also facilitate substrate unfolding (Nouwen *et al.*, 2007). The atomic structures of SecA protein of different origins have been elucidated during the last decade (Hunt *et al.*, 2002; Osborne *et al.*, 2004; Papanikolau *et al.*, 2007), and recently also a structure of the complex of SecA with SecYEG channel was reported (Erlandson *et al.*, 2008; Zimmer *et al.*, 2008). The overall fold of the various SecA proteins is very similar with distinct structural domains that are responsible for the key activities of the protein, such as ATP hydrolysis (DEAD motor domain composed of nucleotide-binding domain, or NBD, and intramolecular regulator of ATPase 2, or IRA2), substrate binding (SSD, substrate specificity domain), or peptide movement into the SecYEG pore (HSD, helical Scaffold domain) (Fig. 12.1B). Certain regions of SecA are involved in dimerization, or in the interactions with the SecB chaperone and the SecYEG channel. Short flexible linkers between more structured domains ensure a high plasticity of the motor protein. Together with a large amount of data from biochemical and biophysical studies, the structural information now has led to a possible scenario of a large part of the ATPase-driven translocation reaction. The binding of ATP triggers a conformational change within the DEAD motor domain and this further propagates through the protein and allows a helical hairpin of the HSD to push the substrate polypeptide chain into SecYEG pore. In each ATP-dependent stroke about 25 amino acid residues of the unfolded polypeptide chain is translocated. Another 25 residues is translocated after the nucleotide hydrolysis and ADP dissociation from the motor protein which involves rebinding of the partially translocated polypeptide chain by SecA (van der Wolk *et al.*, 1997). The interaction with nucleotides that fuels the posttranslational protein translocation reaction represents primary steps of the cycle, and this has been addressed by studies of the thermodynamics of the binding reaction. In the experiments described below we exploited the ITC technique to study the interaction between motor protein SecA and the nucleotides ADP and ATPγS. The latter represents a nonhydrolysable homologue of ATP. Even in the absence of an atomic structure, such experiments shed light on thermodynamic and kinetic aspects of key biological interactions (den Blaauwen *et al.*, 1999).

2.1. Materials and methods

The isolation of SecA from overexpressing cells has been described in detail elsewhere. Shortly, His-tagged SecA from *Bacillus subtilis* was overexpressed in *Escherichia coli* and subsequently purified via Ni^{+}–NTA and anion exchange chromatography. A mutated SecA(K106N) protein with a defect in the nucleotide-binding site was prepared following the same protocol. Purified SecA was exchanged into TKM buffer system by ultrafiltration (Amicon, USA). TKM buffer contained 50 m*M* Tris–HCl, pH 8.0, 50 m*M* KCl, and 5 m*M* $MgCl_2$. Keeping both reactants in the identical buffer system is a key requirement for precise ITC measurements, and even minor discrepancies in the buffer composition may result in large dilution effects. Thus, ADP and ATPγS nucleotides were dissolved in the flow through of the SecA buffer exchange, and the pH of the solution was verified. The same flow through solution was used to fill the reference cell of the ITC setup. Prior to the ITC experiments, both the solutions of protein and nucleotides were thoroughly degassed by gentle stirring under vacuum. This prevents bubble formation during the experiment, thus ensuring a high signal-to-noise ratio.

Studies on the interaction between SecA and ADP using fluorescence methods previously showed that binding occurs with high affinity with an estimate for the dissociation constant of 0.15 μM (van der Wolk *et al.*, 1995). To keep the *C* parameter (see above) well above 10 in the ITC experiment we used a SecA concentrations of 3–10 μM (calculated for the monomer). About 1.4 mL SecA solution was loaded into the reaction cell of a MicroCal Omega titration calorimeter. The reference cell was loaded with blank TKM solution. A titration cycle included 20 injections, 5 μL each, of ADP or ATPγS solution at a concentration 100 and 300 μM, respectively, into the reaction cell. A smaller volume is commonly set for the initial injection to remove possible air bubbles from the tip of the syringe or the diluted reactant. This injection is excluded from the further analysis. Injections are spaced by 3–4 min intervals to ensure the system equilibration. Control experiments need to be performed under identical condition, except that the reaction cell is loaded with buffer only, and the corresponding dilution heats are to be subtracted from the nucleotide: SecA titration data.

Describing the system thermodynamics over a temperature range allows one to determine another term, ΔC_p, the change in heat capacity under constant pressure, as

$$\Delta C_p = \left(\frac{\delta \Delta H}{\delta T}\right)_p \approx \frac{\Delta \Delta H}{\Delta T}. \quad (12.6)$$

In the section below we show how this value can be used to provide further structural information on the system. The necessary ITC data were acquired from titrations of both nucleotides into a SecA solution at temperatures 20, 25, and 30 °C, while ATPγS was additionally titrated at 10 °C. Measured heat effects of binding were corrected for ionization enthalpy of Tris buffer system at studied temperatures.

2.2. Experimental part

The isotherms of SecA:nucleotide-binding reactions at 25 °C are shown in Fig. 12.2, where the power release in the reaction cell (μcal s^{-1}) is plotted versus time. Titrating either ADP or ATPγS into a SecA solution results in exothermic peaks that reflect the binding reaction, which approaches the

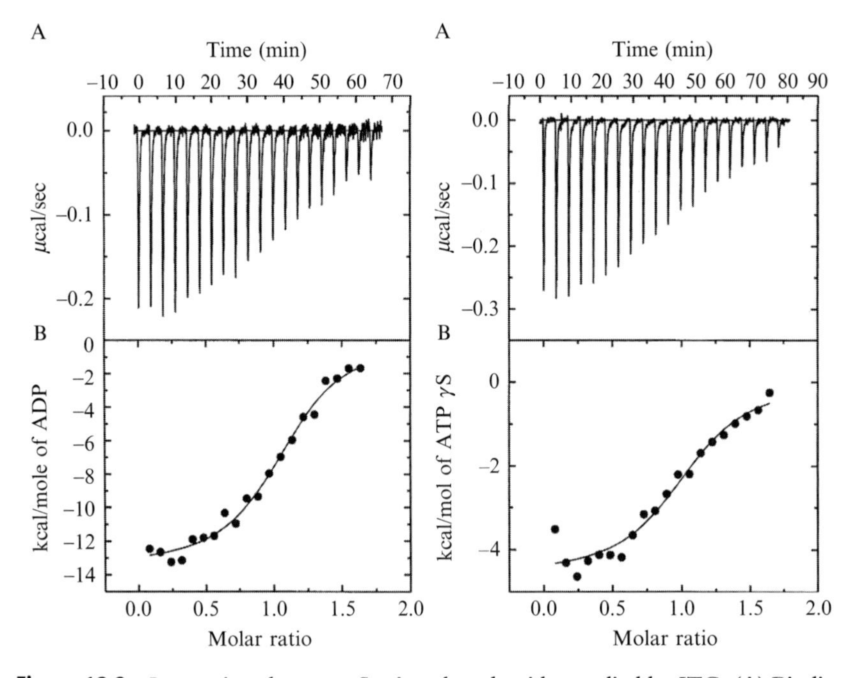

Figure 12.2 Interactions between SecA and nucleotides studied by ITC. (A) Binding isotherms of wild-type SecA to ADP (left panel) and ATPγS (right panel) at 25 °C. Energy released upon binding was calculated and plotted versus nucleotide:SecA molar ratio (B). The solid line represents a nonlinear least squares fit of the reaction heat per injection, with the assumption of a single high-affinity binding site per SecA monomer.

saturation upon 20 injections made. The enthalpy is calculated for each injection as the area under the peak, and its change along the experiment cycle shows a clear sigmoidal shape, as it is predicted for the binding reactions with C parameter between 10 and 1000.

The data are fitted to the model presuming a single, high-affinity binding site (Fig. 12.2). The stoichiometry of SecA:nucleotide complex is close to 1:1 for both ADP and ATPγS. The binding of ADP is driven by the negative enthalpy change of -14.4 kcal/mol, with K_d of 270 nM. Upon the reaction the entropy decreases by 18.25 cal/mol/grad that contributes 5.4 kcal/mol to the free energy of the reaction and largely compensates for the favorable enthalpic term. Binding of ATPγS is also driven by the negative, although smaller enthalpy change of -5.6 kcal/mol and is characterized by a K_d of 74 nM, that is, the affinity is almost fourfold higher than for SecA:ADP interactions. In contrast to the latter, binding of ATPγS is entropically favorable, adding as much as -2.9 kcal/mol to the total free energy change at 25 °C. Titrating nucleotides into the solution of the SecA (K106N) mutant does not show substantial heat effects suggesting that no binding reaction occurs. The mutation K106N is located in the NBD of the DEAD motor (Fig. 12.1B), so this SecA form is unable to bind nucleotides specifically and, correspondingly, does not support protein translocation (van der Wolk *et al.*, 1993). Thus, the characteristic isotherms recorded on functional SecA could be related to the specific interactions with nucleotides.

2.3. Temperature dependence

For the SecA:nucleotide interaction the contribution of the entropic term is specific for the particular nucleotide and also the values of the binding enthalpy change substantially, although the overall binding energies do not differ much between ADP and ATPγS. This is indicative for a change in the heat capacity of the system upon binding, which also can be determined from ITC experiments according to Eq. (12.6). The temperature dependencies of ΔG, ΔH, and $T\Delta S$ for both ADP and ATPγS binding to SecA were determined by titration experiments performed at temperatures ranging from 10 to 30 °C (Fig. 12.3). Although modern ITC instrumentation allows measurements at elevated temperatures up to 80 °C, one should also consider the stability of biological samples used when designing the experiment. Thus, already at 35 °C a partial unfolding of SecA is observed (den Blaauwen *et al.*, 1999) that contributes to the thermal effects in the system and overlaps with the heat of the nucleotide-binding reaction.

The SecA interaction with both ADP and ATPγS are characterized by an enthalpy change that decreases linearly with the temperature. However, this dependence is different between nucleotides, as can be seen from the slopes of ΔH versus T plots (Fig. 12.3). The calculated change in ΔC_p is

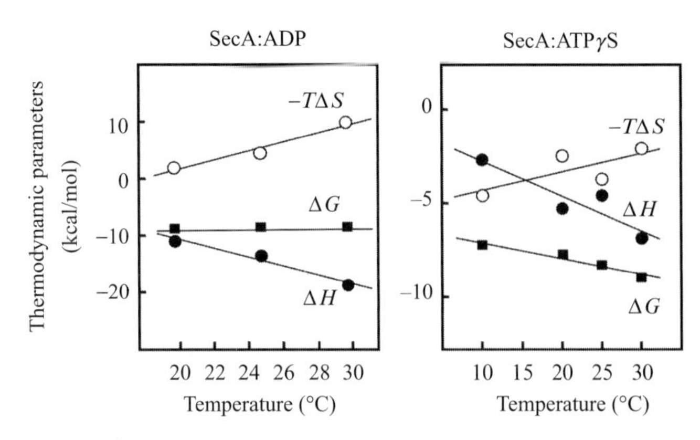

Figure 12.3 Temperature dependence of the thermodynamic parameters for binding of SecA to ADP (A) and ATPγS (B). Data points measured by titration calorimetry are included for ΔH. The heat capacity change ΔC_p associated with nucleotide binding to SecA was determined by linear regression analysis of the slope of the plot of ΔH versus temperature. Values of ΔG and $-T\Delta S$ are calculated for each temperature as described in the text.

negative suggesting a decrease in the solvent accessible area of SecA upon binding of the nucleotides. The binding of ADP induces a fourfold higher change in ΔC_p demonstrating that the protein undergoes a conformational change causing a change in water-accessible surface area. No substantial change in the SecA secondary structure upon ADP binding is observed by means of circular dichroism, so the decrease in the solvent accessible area is attributed to an altered organization of existing structural domains within the protein, rather than to folding effects. SecA substructures are suggested to come in closer contact thus partially shielding hydrophobic surfaces from the solvent. Likely, the NBD/IRA2 motor domain and the C-domain of SecA are involved in this movement as suggested by intrinsic tryptophan fluorescence experiments (den Blaauwen *et al.*, 1996). The intramolecular motion induced by ADP would result in a more compact state of the protein that is referred as the energy-loaded form, as its extension upon the functional cycle may provide a driving force for the substrate protein translocation. Although these conclusions are derived for SecA in solution, it may be extended to the case of protein translocation. The recently published structure of SecA in complex with ADP–BeF_x bound to SecYEG also shows a compact organization of SecA domains (Zimmer *et al.*, 2008), which only slightly differs from the SecA:ADP structure in solution (Osborne *et al.*, 2004).

In the shown example we describe a possible application of ITC to study interactions of proteins with small molecules, such as nucleotides. Because of the ubiquitous nature of molecular motors in the cell, the experiment may be easily extended to different biological systems (Inobe *et al.*, 2001;

Pretz *et al.*, 2006). Another conclusion here is that the calorimetry measurements may not only provide information on the binding thermodynamics, but also provide structural insights. When combined with other techniques, such as fluorescence spectroscopy, circular dichroism, or NMR, a detailed description of the system structural dynamics can be obtained.

3. Example 2: Probing SecB: Substrate Interactions

While a newly synthesized polypeptide chain protrudes from the ribosome exit tunnel, it will start to fold and eventually with the aid of molecular chaperones it will fold into its final compact structure. The presence of a signal sequence, such as in secretory preproteins reduces the folding rate. However, it may not suffice to ensure the efficient protein targeting to the translocase through the crowded cellular environment. The small chaperone protein SecB (Fig. 12.1C) recognizes such secretory proteins and targets them to the SecA/SecYEG complex. During this process, SecB also keeps these substrates in translocation-competent state, that is, a conformational state that lacks stably formed structure (Driessen, 2001). The signal sequence of secretory proteins is not essential for the interactions with SecB and so far no specific binding motif within substrate polypeptide chains has been elucidated. It is generally anticipated that SecB recognizes unfolded regions of proteins where hydrophobic side chains of amino acids are exposed to solvent. A long and deep hydrophobic groove at both sides of the SecB tetramer that was resolved by X-ray crystallography (Xu *et al.*, 2000) is most probably the site responsible for the substrate binding. Therefore, hydrophobic interactions likely play an important role in binding. SecB was shown to interact not only with its natural substrates, that is, secretory proteins, but also with a variety of polypeptides *in vitro*, which retain a "molten globule," but not fully unfolded state (Fekkes *et al.*, 1995; Hardy and Randall, 1991). Remarkably, the interaction with such model substrates may involve different number of subunits within SecB tetramer. Short polypeptides (up to 150 amino acids) interact with single SecB subunit, while the binding of longer proteins such as maltose- or galactose-binding proteins (MBP and GBP, respectively) requires all four subunits, so that the stoichiometry may range from 1:1 to 1:4.

To elucidate the role of nonpolar interactions between SecB and substrates the binding interface has been altered to produce a mutant form of the chaperone. Based on the three-dimensional structure of SecB (Dekker *et al.*, 2003; Xu *et al.*, 2000) a site for a point mutation was chosen that would interfere with substrate binding, but that would not affect the structural organization of SecB tetramer. Several residues have been suggested to participate in binding as their motion is dramatically reduced upon adding substrates (Crane *et al.*, 2006). SecB(L42R) mutant contains a positively charged residue at such a "hot-spot" of the hydrophobic groove and thus is expected to change its functional chaperone properties as the mutation likely interferes with protein binding. At the same time SecB (L42R) maintains the original quaternary structure, in contrast to previously described SecB variants that show severe defects in substrate binding (Muren *et al.*, 1999). The SecB(L42R) still supports protein translocation *in vitro*, but its holdase properties, that is, the ability to prevent substrate from folding/aggregation are substantially weakened (Bechtluft *et al.*, submitted). A study of the thermodynamics of the SecB:substrate interaction and a comparison between the wild-type and mutant form of SecB provides some details on the effect of the mutation. Below we describe examples of ITC experiments aimed to carry out such comparison.

3.1. Materials and methods

Overexpressed and purified SecB variants (wild-type and L42R mutant) were dialyzed versus potassium phosphate buffer. When necessary, the protein was additionally concentrated using MicroCon centrifuge concentrating units with a cut-off of 10 kDa. Bovine pancreas trypsin inhibitor (BPTI, 57 amino acids) and ribonuclease A (RNase A, 128 amino acids) were used as model substrates. To achieve their molten globule state, their cysteine residues were carboxymethylated to prevent formation of native disulphide bonds. Thus the proteins are captured in the reduced state and ceased from folding. BPTI and RNase A were denatured by Gdn-HCl and carboxymethylated in the presence of iodoacetamide as described elsewhere (Fekkes *et al.*, 1995; Houry *et al.*, 1994). In the following step, the excess of the iodoacetamide was removed and the buffer was replaced by dialysis using Slide-A-Lyzer cassettes (cut-off 2 kDa) or by size-exclusion chromatography on Superose 12 column. The carboxymethylation efficiency could be judged from a band shift on SDS–PAGE gels compared to the unmodified protein. To ensure an identical buffer composition and pH of the protein solutions, the same buffer was used to prepare both SecB and substrate samples for the ITC.

ITC measurements were performed with the ultrasensitive MicroCal ITC_{200} setup, which allowed to reduce the amount of material used in the experiment up to fivefold (less than 300 μL BPTI/RNase A, $\sim$40 μL SecB). The small volume of the ITC cell also requires less time for equilibration between two sequential injections, so the time necessary for a single experiment may be significantly reduced. Thus, this equipment allows a higher throughput too. Here, we use 2 min spacing for the SecB:substrate reaction, and 1 min for control titrations. Measurements were carried out in 100 m*M* potassium phosphate buffer, pH 7.4, which is characterized by a lower enthalpy of ionization compared to other wide-used buffer systems, such as Tris or HEPES (Doyle, 1999). Prior to each experiment the protein solutions were centrifuged for 10 min at 13,000 rpm, 4 °C to remove possible aggregates, and the protein concentration was determined in the supernatant. The calorimeter reaction cell was filled with the solution of a substrate, BPTI or RNase A, at concentrations varying between 10 and 30 μM, and the reference cell was filled with the buffer solution. SecB variants at concentrations between 200 and 500 μM were titrated into the reaction cell in 2 μL portions. To account for dilution heats accompanying each injection, control measurements were performed: (a) SecB was titrated into the buffer solution, and (b) the buffer solution was titrated into the cell loaded with the substrate. Although dilution heats did not exceed 0.1 μcal/sec, a monotonous decrease in heats from titration to titration was sometimes observed, so linear regression was performed on these data sets. In the following analysis these heats were subtracted from those recorded upon the binding reaction.

A rather low-binding constant K_a of SecB to nonnatural substrates (up to 10 μM^{-1}) is measured in ITC experiments (Panse *et al.*, 2000) which sets certain limitations on the experiment design. To achieve binding saturation the concentration of the substrate in the cell should be 20–50 lower than that of SecB loaded in the syringe. However, the substrate dilution is limited due to the drop of heat effects at low concentrations, while an increasing SecB concentration may result in the protein aggregation. Here, BPTI and RNase A were used at concentrations of 10–30 μM that normally ensures the binding saturation upon a single titration cycle, while the signal recorded at each titration is sufficiently high (Fig. 12.4A). However, the titrations may be continued after reloading the syringe, while leaving the same reactant in the cell if the saturation is not reached in one cycle. Corresponding isotherms are then "fused" using the software package that accompanies the instrument and the combined data set is analyzed as described earlier. This approach may be particularly useful when working at C values below 10, as the complete saturation is required for unambiguous data analysis. In experiments described here C values lay between 3 and 7 which is sufficient for calculating all thermodynamic parameters of the binding reaction and the stoichiometry n.

3.2. Experimental part

A typical trace recorded in an ITC experiment is shown in Fig. 12.4A. Titrating the concentrated SecB solution into the cell loaded with BPTI yields peaks that reflect the exothermal effect of the binding reaction. The peak amplitude reduces gradually with an increasing amount of SecB in the cell that manifests the progressive binding of available substrate molecules to the chaperone. At the end of the experiment saturation is reached as the

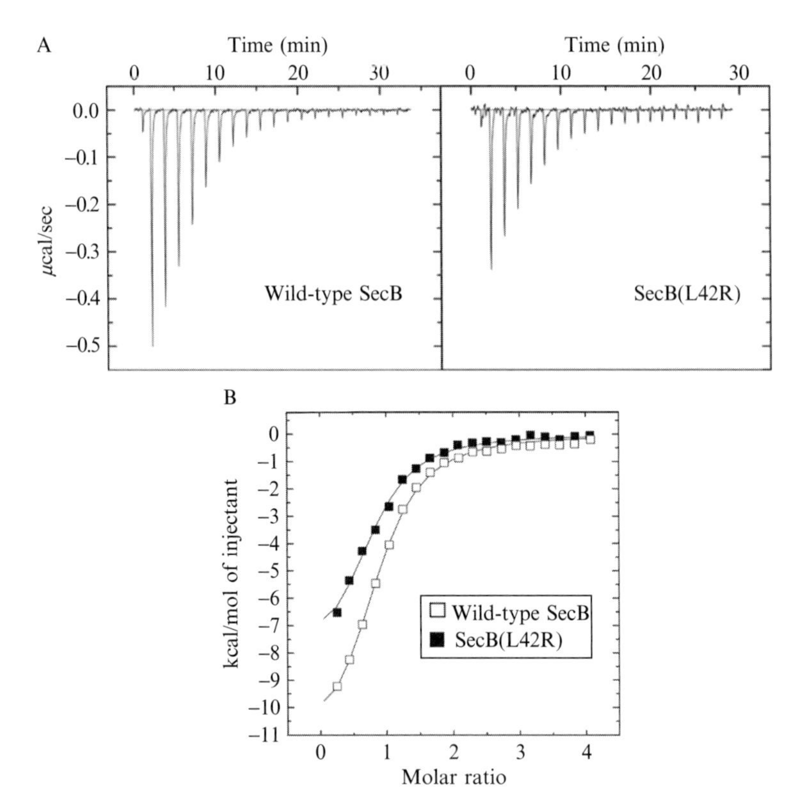

Figure 12.4 SecB:substrate interactions studied by isothermal titration calorimetry. (A) Binding isotherms of the titration of wild-type SecB and SecB(L42R) at concentrations 275 μM into a solution containing 13 μM BPTI at 15 °C. The area under each injection signal was integrated and plotted for both wild-type SecB (□) and SecB (L42R) mutant (■) versus SecB:substrate molar ratio (*C*). The solid lines represent a nonlinear least squares fit of the reaction heat for the injection.

peaks approach the levels described by the simple dilution. The experiment performed with the mutated SecB(L42R) results in an isotherm in which the peaks are substantially reduced, although the concentration of the protein and substrate are identical to those for wild-type SecB. Binding of RNase A to SecB variants demonstrates the same trend.

The difference in the substrate-binding energetics between SecB variants is seen in enthalpy plots (Fig. 12.4B). Nonlinear least square fits provide values of the binding stoichiometry n and thermodynamic parameters: ΔH, K_a, and, derived from those, the entropy ΔS (Table 12.1). As expected, both SecB variants bind one substrate molecule per monomer. A substantial negative enthalpy change accompanies the binding and favors the reaction. In agreement, the interaction between SecB with proteins was previously described as an enthalpy-driven process, while a decrease in

Table 12.1 Thermodynamic parameters and stoichiometry of SecB:substrate interactions studied by ITC

	SecB	BPTI 10 °C	BPTI 15 °C	BPTI 20 °C	RNase A 15 °C
N	Wt	1.0	1.1	1.2	0.9
	L42R	1.1	1.0	1.2	0.8
$K_a \times 10^5$ (M^{-1})	Wt	4.4	3.8	4.8	2.1
	L42R	4.3	2.6	2.2	1.2
ΔH (kcal/mol)	Wt	−11.0	−11.0	−10.4	−14.0
	L42R	−6.4	−9.0	−8.8	−10.0
$T\Delta S$ (kcal/mol)	Wt	−3.5	−3.7	−2.8	−7.5
	L42R	0.9	−1.8	−1.6	−3.7

entropy was observed upon the complex formation. The latter may be assigned to a drop in the substrate translational mobility and, most probably, a reduction in rotational degrees of freedom. Such an enthalpy–entropy compensation phenomenon is a common behavior for biological systems, characterized by weak intermolecular interactions (Dunitz, 1995).

Apparently, introducing a positively charged residue to the SecB:substrate interface reduces the enthalpic contribution to the binding energy. The loss in the enthalpy may be attributed to a change in the total charge carried by SecB at pH 7.4. Both studied substrates, BPTI and RNase A, are positively charged at the experimental conditions (+5.1 and +7.4, respectively), while wild-type SecB carries the net negative charge of −14.4. Adding a positive charge to the chaperone in SecB(L42R) mutant decreases the contribution of electrostatic interactions and affects the enthalpy of the reaction. Likely the association constant for native secretory proteins will be affected to a larger extent, since all four subunits are required for binding of such large proteins, and possible cooperativity between the binding sites will enhance the effect.

Remarkably, the mutation L42R not only decreases the enthalpic contribution to SecB:substrate binding, but also causes a lower entropy change. Assuming that neither the protein mobility, nor the solvent structure is substantially affected by the mutation, we suggest that the modified system retains a higher flexibility in the final bound state. The lack of hydrophobic contacts might results in loose docking of the substrate on the SecB surface. Thus some retained rotational mobility, when bound to the chaperone surface, may allow the substrates to acquire a partially folded state as suggested recently (Krishnan *et al.*, 2009). Altering the hydrophobic environment of the substrate-binding interface of SecB with a polar residue may not dramatically affect the binding affinity, but it interferes with the chaperone function as it allows folding of the substrate before it is delivered to the translocase complex. A set of biochemical assays support this phenomenon with the SecB(L42R) mutant protein and direct evidence for folding of the bound polypeptide chain was obtained by single-molecule optical tweezer experiments (Bechtluft *et al.*, submitted).

3.3. SecB:substrate interactions are temperature dependent

Previous studies on the interaction between SecB and protein substrates showed that the temperature has a substantial effect on the binding properties (Panse *et al.*, 2000). The enthalpy–entropy compensation occurs over a wide range of temperatures for the wild-type SecB, and the binding of proteins is accompanied by a reduction in the solvent accessible area. Performing ITC measurements over a range of temperatures provided further insight in the effect of the L42R mutation (Table 12.1). Remarkably, we do not observe a large change in the binding enthalpy when the temperature is increased from 10 to 20 °C, in contrast to the previous report on SecB:BPTI complex formation. Nevertheless, the effect of the mutation remains at all studied temperatures: SecB(L42R):substrate binding is characterized by lower values of ΔG and ΔH, so that substrate binding is reduced compared to the wild-type chaperone. While the enthalpy term does not show a clear trend with increasing temperature, the entropy change for the SecB mutant first approaches zero and then becomes negative. As a result, the association constant drops by a factor of two for the SecB(L42R) mutant, but remains almost constant for the wild-type chaperone (Table 12.1). When extended these results to 37 °C, that is, the growth temperature of *E. coli*, one may expect even further decline in the binding properties for SecB(L42R). Thus, the mutation affects both chaperone and binding properties of the protein.

4. Concluding Remarks

Even a general overview of protein translocation immediately shows the complexity of the process, as it is facilitated by several fine-tuned molecular machines, and regulated by dynamic interactions between those. With broad knowledge from previous biochemical and biophysical studies, together with recently acquired structural information we can now approach the dynamics of protein translocation, aiming to understand the details of the thermodynamics of the interactions among key components of the translocase. ITC is a valuable and powerful tool in this crusade. The examples presented here suggest that the method will be able to describe the energetics of each intermolecular interaction involved in the translocation process. While current studies have addressed molecular complexes formed in the cytoplasm, it is of particular interest to probe the interactions established on the membrane interface. The membrane-embedded pore SecYEG and its partners would be primary targets for those measurements, and surface plasmon resonance method has recently been used to study the translocase assembly using the membrane bound SecYEG complex immobilized on the sensor chip (de Keyzer *et al.*, 2003, 2005). Although ITC experiments can be performed on detergent-solubilized translocase, it will be a upcoming challenge to introduce the native lipid environment to the experimental setup. Latest studies using microcalorimetry methods, and in particular ITC, confirm that valuable information may be achieved from complex systems involving lipid membranes (Heerklotz, 2004). Nowadays ITC is routinely exploited to probe interactions of small molecules or even peptides with lipid membranes, and pioneer studies have recently addressed interactions of membrane-embedded transport proteins with their substrates (Sikora and Turner, 2005a,b). In addition, the recent miniaturization of the sample compartment of the ITC while maintaining the required sensitivity makes this technique more amendable to complex membrane-bound systems. Thus we foresee novel applications of microcalorimetry in general membrane protein biophysics and predict further advances in understanding molecular mechanisms of protein translocation from the thermodynamics point of view.

REFERENCES

Bechtluft, P., Kedrov, A., Slotboom, D. J., Nouwen, N., Tans, S. J., and Driessen, A. J. M. (submitted). Tight hydrophobic contacts with the SecB chaperone prevent folding of substrate proteins.

Crane, J. M., Suo, Y., Lilly, A. A., Mao, C., Hubbell, W. L., and Randall, L. L. (2006). Sites of interaction of a precursor polypeptide on the export chaperone SecB mapped by site-directed spin labeling. *J. Mol. Biol.* **363,** 63–74.

de Keyzer, J., van der Does, C., Kloosterman, T. G., and Driessen, A. J. M. (2003). Direct demonstration of ATP-dependent release of SecA from a translocating preprotein by surface plasmon resonance. *J. Biol. Chem.* **278,** 29581–29586.

de Keyzer, J., van der Sluis, E. O., Spelbrink, R. E., Nijstad, N., de Kruijff, B., Nouwen, N., van der Does, C., and Driessen, A. J. M. (2005). Covalently dimerized SecA is functional in protein translocation. *J. Biol. Chem.* **280,** 35255–35260.

Dekker, C., de Kruijff, B., and Gros, P. (2003). Crystal structure of SecB from *Escherichia coli*. *J Struct. Biol.* **144,** 313–319.

den Blaauwen, T., Fekkes, P., de Wit, J. G., Kuiper, W., and Driessen, A J. M. (1996). Domain interactions of the peripheral preprotein Translocase subunit SecA. *Biochemistry* **35,** 11994–12004.

den Blaauwen, T., van der Wolk, J. P., van der Does, C., van Wely, K. H., and Driessen, A. J. M. (1999). Thermodynamics of nucleotide binding to NBS-I of the *Bacillus subtilis* preprotein translocase subunit SecA. *FEBS Lett.* **458,** 145–150.

Doyle, M. L. (1999). Titrtaion Microcalorimetry. Wiley.

Driessen, A. J. (2001). SecB, a molecular chaperone with two faces. *Trends Microbiol.* **9,** 193–196.
Driessen, A. J. M., and Nouwen, N. (2008). Protein translocation across the bacterial cytoplasmic membrane. *Annu. Rev. Biochem.* **77,** 643–667.
Dunitz, J. D. (1995). Win some, lose some: Enthalpy–entropy compensation in weak intermolecular interactions. *Chem. Biol.* **2,** 709–712.
Erlandson, K. J., Miller, S. B., Nam, Y., Osborne, A. R., Zimmer, J., and Rapoport, T. A. (2008). A role for the two-helix finger of the SecA ATPase in protein translocation. *Nature* **455,** 984–987.
Fekkes, P., den Blaauwen, T., and Driessen, A. J. M. (1995). Diffusion-limited interaction between unfolded polypeptides and the *Escherichia coli* chaperone SecB. *Biochemistry* **34,** 10078–10085.
Hardy, S. J., and Randall, L. L. (1991). A kinetic partitioning model of selective binding of nonnative proteins by the bacterial chaperone SecB. *Science* **251,** 439–443.
Heerklotz, H. (2004). The microcalorimetry of lipid membranes. *J. Phys. Condens. Matter* **16,** R441–R467.
Houry, W. A., Rothwarf, D. M., and Scheraga, H. A. (1994). A very fast phase in the refolding of disulfide-intact ribonuclease A: Implications for the refolding and unfolding pathways. *Biochemistry* **33,** 2516–2530.
Hunt, J. F., Weinkauf, S., Henry, L., Fak, J. J., McNicholas, P., Oliver, D. B., and Deisenhofer, J. (2002). Nucleotide control of interdomain interactions in the conformational reaction cycle of SecA. *Science* **297,** 2018–2026.
Inobe, T., Makio, T., Takasu-Ishikawa, E., Terada, T. P., and Kuwajima, K. (2001). Nucleotide binding to the chaperonin GroEL: Non-cooperative binding of ATP analogs and ADP, and cooperative effect of ATP. *Biochim. Biophys. Acta* **1545,** 160–173.
Krishnan, B., Kulothungan, S. R., Patra, A. K., Udgaonkar, J. B., and Varadarajan, R. (2009). SecB-mediated protein export need not occur via kinetic partitioning. *J. Mol. Biol.* **385,** 1243–1256.
Muren, E. M., Suciu, D., Topping, T. B., Kumamoto, C. A., and Randall, L. L. (1999). Mutational alterations in the homotetrameric chaperone SecB that implicate the structure as dimer of dimers. *J. Biol. Chem.* **274,** 19397–19402.
Nouwen, N., Berrelkamp, G., and Driessen, A. J. (2007). Bacterial Sec-translocase unfolds and translocates a class of folded protein domains. *J. Mol. Biol.* **372,** 422–433.
Osborne, A. R., Clemons, W. M. Jr., and Rapoport, T. A. (2004). A large conformational change of the translocation ATPase SecA. *Proc. Natl. Acad. Sci. USA* **101,** 10937–10942.
Panse, V. G., Swaminathan, C. P., Surolia, A., and Varadarajan, R. (2000). Thermodynamics of substrate binding to the chaperone SecB. *Biochemistry* **39,** 2420–2427.
Papanikolau, Y., Papadovasilaki, M., Ravelli, R. B., McCarthy, A. A., Cusack, S., Economou, A., and Petratos, K. (2007). Structure of dimeric SecA, the *Escherichia coli* preprotein translocase motor. *J. Mol. Biol.* **366,** 1545–1557.
Papanikou, E., Karamanou, S., and Economou, A. (2007). Bacterial protein secretion through the translocase nanomachine. *Nat. Rev. Microbiol.* **5,** 839–851.
Pretz, M. G., Albers, S. V., Schuurman-Wolters, G., Tampe, R., Driessen, A. J. M., and van der Does, C. (2006). Thermodynamics of the ATPase cycle of GlcV, the nucleotide-binding domain of the glucose ABC transporter of sulfolobus solfataricus. *Biochemistry* **45,** 15056–15067.
Sikora, C. W., and Turner, R. J. (2005a). Investigation of ligand binding to the multidrug resistance protein EmrE by isothermal titration calorimetry. *Biophys. J.* **88,** 475–482.
Sikora, C. W., and Turner, R. J. (2005b). SMR proteins SugE and EmrE bind ligand with similar affinity and stoichiometry. *Biochem. Biophys. Res. Commun.* **335,** 105–111.
Tellinghuisen, J. (2005). Optimizing experimental parameters in isothermal titration calorimetry. *J. Phys. Chem. B* **109,** 20027–20035.

van der Wolk, J., Klose, M., Breukink, E., Demel, R. A., de Kruijff, B., Freudl, R., and Driessen, A. J. M. (1993). Characterization of a *Bacillus subtilis* SecA mutant protein deficient in translocation ATPase and release from the membrane. *Mol. Microbiol.* **8,** 31–42.

van der Wolk, J. P., Klose, M., de Wit, J. G., den Blaauwen, T., Freudl, R., and Driessen, A. J. M. (1995). Identification of the magnesium-binding domain of the high-affinity ATP-binding site of the *Bacillus subtilis* and *Escherichia coli* SecA protein. *J. Biol. Chem.* **270,** 18975–18982.

van der Wolk, J. P., de Wit, J. G., and Driessen, A. J. M. (1997). The catalytic cycle of the *Escherichia coli* SecA ATPase comprises two distinct preprotein translocation events. *EMBO J.* **16,** 7297–7304.

Vrontou, E., and Economou, A. (2004). Structure and function of SecA, the preprotein translocase nanomotor. *Biochim. Biophys. Acta* **1694,** 67–80.

Walter, P., and Blobel, G. (1980). Purification of a membrane-associated protein complex required for protein translocation across the endoplasmic reticulum. *Proc. Natl. Acad. Sci. USA* **77,** 7112–7116.

Wiseman, T., Williston, S., Brandts, J. F., and Lin, L. N. (1989). Rapid measurement of binding constants and heats of binding using a new titration calorimeter. *Anal. Biochem.* **179,** 131–137.

Xu, Z., Knafels, J. D., and Yoshino, K. (2000). Crystal structure of the bacterial protein export chaperone SecB. *Nat. Struct. Biol.* **7,** 1172–1177.

Zimmer, J., Nam, Y., and Rapoport, T. A. (2008). Structure of a complex of the ATPase SecA and the protein-translocation channel. *Nature* **455,** 936–943.

CHAPTER THIRTEEN

Thermodynamic Analysis of the Structure–Function Relationship in the Total DNA-Binding Site of Enzyme–DNA Complexes

Wlodzimierz Bujalowski*,†,‡ *and* Maria J. Jezewska*,†,‡

Contents

Abstract

Both helicases and polymerases perform their activities when bound to the nucleic acids, that is, the enzymes possess a nucleic acid-binding site. Functional complexity of the helicase or the polymerase action is reflected in the intricate structure of the total nucleic acid-binding site, which allows the

* Department of Biochemistry and Molecular Biology, The University of Texas Medical Branch at Galveston, Galveston, Texas, USA
† Department of Obstetrics and Gynecology, The University of Texas Medical Branch at Galveston, Galveston, Texas, USA
‡ The Sealy Center for Structural Biology, Sealy Center for Cancer Cell Biology, The University of Texas Medical Branch at Galveston, Galveston, Texas, USA

Methods in Enzymology, Volume 466
ISSN 0076-6879, DOI: 10.1016/S0076-6879(09)66013-4

enzymes to control and change their nucleic acid affinities during the catalysis. Understanding the fundamental aspects of the functional heterogeneity of the total nucleic acid-binding site of a polymerase or helicase can be achieved through quantitative thermodynamic analysis of the enzyme binding to the nucleic acids oligomers, which differ in their length. Such an analysis allows the experimenter to assess the presence of areas with strong and weak affinity for the nucleic acid, that is, the presence of the strong and the weak nucleic acid-binding subsites, determine the number of the nucleotide occlude by each subsite, and estimate their intrinsic free energies of interactions.

Abbreviations

DTT	dithiothreitol
εA	1,N^6-etheno adenosine

1. Introduction

Nonspecific interactions between enzymes involved in nucleic acid metabolism and the nucleic acids play a fundamental role in the transmission of genetic information from one cell generation to another. Both polymerases and helicases are essentially nonspecific nucleic acid-binding enzymes, that is, their interactions with the DNA or RNA have little dependence upon the specific nucleotide sequence (Baker *et al.*, 1987; Enemark and Joshua-Tor, 2008; Heller and Marians, 2007; Hubscher *et al.*, 2000; Joyce and Benkovic, 2004; Lohman and Bjornson, 1996; Morales and Kool, 2000; von Hippel and Delagoutte, 2002, 2003). Specific protein–nucleic acid interactions are predominantly engaged in carrying out a single physiological function, which is precisely defined by a short nucleotide sequence, for example, repressor–operator interactions. On the other hand, nonspecific interactions are primarily involved in metabolic pathways requiring multiple performances of the same reaction often on long stretches of the nucleic acid, regardless of the nucleotide sequence. In this context, recognition of specific nucleic acid conformational states, regardless of the nucleotide sequence of the DNA or RNA, is still a nonspecific protein–nucleic acid interaction. The intricate interactions between polymerases and the single-stranded conformation of the nucleic acids are well recognized, although not completely understood, particularly for different classes of polymerases, engaged in different metabolic pathways. On the other hand, there is still significant gap in our understanding of the corresponding interactions in the case of the helicases.

Helicases are a class of key enzymes that are involved in all major pathways of the DNA and RNA metabolism (Baker *et al.*, 1987; Enemark and Joshua-Tor, 2008; Heller and Marians, 2007; Lohman and Bjornson, 1996; von Hippel and Delagoutte, 2002, 2003). The enzymes are motor proteins, which catalyze the vectorial unwinding of the duplex DNA to provide an active, single-stranded intermediate, which is required, for example, in replication, recombination, translation, and repair processes. The unwinding reaction, as well as the mechanical translocation of the helicase along the nucleic acid lattice, is fueled by NTP hydrolysis (Ali and Lohman, 1997; Baker *et al.*, 1987; Enemark and Joshua-Tor, 2008; Galletto *et al.*, 2004a,b; Heller and Marians, 2007; Jankowsky *et al.*, 2002; Lohman and Bjornson, 1996; Lucius *et al.*, 2002; Nanduri *et al.*, 2002; von Hippel and Delagoutte, 2002, 2003). Helicases may also use the energy of NTP hydrolysis for activities in nucleic acid metabolism, different from duplex nucleic acid unwinding, for example, as molecular pumps (Bujalowski, 2003; Kaplan and O'Donnell, 2002; West, 1996). One of the fundamental elements of the helicase activity is the interaction with the ssDNA conformation of the nucleic acid. In fact, most of the helicases become active NTPases and are capable of free energy transduction only in the complex with the ssDNA or ssRNA (Ali and Lohman, 1997; Baker *et al.*, 1987; Enemark and Joshua-Tor, 2008; Galletto *et al.*, 2004a,b; Heller and Marians, 2007; Jankowsky *et al.*, 2002; Lohman and Bjornson, 1996; Lucius *et al.*, 2002; Nanduri *et al.*, 2002; von Hippel and Delagoutte, 2002, 2003). The allosteric effect of the ssDNA on the NTPase activity of the helicase and the mechanical translocation of the enzyme along the nucleic acid lattice provided first clues, although only intuitive ones, that interactions of the helicase with the ssDNA is far more complex than the simple recognition of a specific conformational state of the DNA or RNA (Amaratunga and Lohman, 1993; Bjorson *et al.*, 1996; Galletto *et al.*, 2004c; Jezewska *et al.*, 1996a, 1998a,b, 2000a,b, 2004; Lucius *et al.*, 2006; Marcinowicz *et al.*, 2007). Rather, they would more closely resemble the DNA or RNA polymerases with the intricate interplay among different areas of the total DNA-binding site of the enzyme with the nucleic acid.

In our discussion, we will first concentrate on the thermodynamic analyses of the *Escherichia coli* PriA helicase–ssDNA interactions, using spectroscopic approaches aimed at elucidation of the functional relationships in the PriA–ssDNA complex (Galletto *et al.*, 2004c; Jezewska *et al.*, 2000a,b; Lucius *et al.*, 2006). The method is based on the quantitative spectroscopic titration technique to obtain model-independent binding isotherms, which allows the experimenter to address structure–function relationships within the total DNA-binding site of the enzyme (Galletto *et al.*, 2004c; Jezewska *et al.*, 2000a,b; Lucius *et al.*, 2006). We will focus on the essential problem of obtaining thermodynamic parameters free of assumptions about the relationship between the observed signal and the degree of protein or nucleic acid saturation. The analogous approach will be next discussed in the case of the African swine fever virus (ASFV) polymerase X (pol X)–ssDNA complexes. ASFV pol X is engaged in the repair of the viral DNA. It is a small DNA polymerase whose complex interactions with the nucleic acid have been recently addressed. Although, in binding or kinetic studies, one generally monitors some spectroscopic signal (absorbance, fluorescence, fluorescence anisotropy, circular dichroism, NMR line width, or chemical shift), originating from either the protein or the nucleic acid, which changes upon the complex formation, we will discuss the quantitative analyses as applied to the use of fluorescence intensity. Fluorescence is the most widely used spectroscopic technique in energetics and dynamics studies of the protein–nucleic acid interactions in solution and its application will be illustrated in the case of the PriA–ssDNA complex (Bujalowski, 2006; Bujalowski and Jezewska, 2000; Bujalowski and Lohman, 1987; Galletto *et al.*, 2004c; Jezewska and Bujalowski, 1996; Jezewska *et al.*, 2000a,b; Lohman and Bujalowski, 1991; Lucius *et al.*, 2006). Nevertheless, the derived relationships are general and applicable to any physicochemical signal used to monitor the ligand–macromolecule interactions (Bujalowski, 2006; Bujalowski and Jezewska, 2000; Bujalowski and Lohman, 1987; Galletto *et al.*, 2004c; Jezewska and Bujalowski, 1996; Jezewska *et al.*, 2000a,b; Lohman and Bujalowski, 1991; Lucius *et al.*, 2006).

2. Thermodynamic Bases of Quantitative Equilibrium Spectroscopic Titrations

Determination of the binding isotherm is a fundamental first step in examining the energetics of the protein–nucleic acid association, or in general, any ligand–macromolecule interactions (Bujalowski, 2006; Bujalowski and Jezewska, 2000; Bujalowski and Lohman, 1987; Galletto *et al.*, 2004c; Jezewska and Bujalowski, 1996; Jezewska *et al.*, 2000a,b; Lohman and Bujalowski, 1991; Lucius *et al.*, 2006). Equilibrium-binding isotherm represents a direct relationship between the average total degree of binding (moles of ligand molecules bound per mole of a macromolecule) and the free ligand concentration (Bujalowski, 2006; Bujalowski and Jezewska, 2000; Lohman and Bujalowski, 1991). In the case of protein binding to a long, one-dimensional nucleic acid lattice, a more convenient parameter than the total average degree of binding is the total average binding density (moles of ligand bound per mole of bases or base pairs) (Bujalowski, 2006; Bujalowski and Lohman, 1987; Bujalowski *et al.*, 1989; Epstein, 1978; Lohman and Bujalowski, 1991; McGhee and von Hippel, 1974). The equilibrium-binding isotherm represents the direct relationship between the binding density and the free protein concentration (Bujalowski, 2006; Bujalowski and Jezewska, 2000; Bujalowski and Lohman, 1987; Bujalowski *et al.*, 1989; Epstein, 1978; Jezewska and Bujalowski, 1996; Lohman and Bujalowski, 1991; McGhee and von Hippel, 1974). The extraction of physically meaningful interaction

monitored through the changes of the nucleic acid fluorescence (Galletto *et al.*, 2004c; Jezewska *et al.*, 1998b, 2000a,b; Lucius *et al.*, 2006). Thus, in the examined case the nucleic acid is treated as the macromolecule and the PriA helicase is treated as the ligand.

The major task in examining the protein–nucleic acid interactions using a spectroscopic titration method is to convert the titration curve, in our case, the fluorescence titration curve, that is, a change in the monitored fluorescence as a function of the total concentration of the protein, into a model-independent, thermodynamic binding isotherm, which can then be analyzed, using an appropriate binding model to extract binding parameters. The thermodynamic basis of the method is that for the total nucleic acid concentration, N_T, the equilibrium distribution of the nucleic acid among its different states with a different number of bound protein molecules, N_i, is determined solely by the free protein concentration, L_F (Bujalowski, 2006; Bujalowski and Jezewska, 2000; Bujalowski and Lohman, 1987; Jezewska and Bujalowski, 1996; Lohman and Bujalowski, 1991). Therefore, at each L_F, the observed spectroscopic signal, F_{obs}, is the algebraic sum of the concentrations of the nucleic acid in each state, N_i, weighted by the value of the intensive spectroscopic property of that state, F_i. In general, a nucleic acid will have the ability to bind n protein molecules, hence the model-independent, "signal conservation" equation for the observed signal, F_{obs}, of a sample containing the ligand at a total concentration, L_T, and the nucleic acid at a total concentration, N_T, is given by (Bujalowski, 2006; Bujalowski and Jezewska, 2000; Bujalowski and Lohman, 1987; Jezewska and Bujalowski, 1996; Lohman and Bujalowski, 1991):

$$F_{obs} = F_F N_F + \sum F_i N_i \tag{13.1}$$

where F_F is the molar fluorescence of the free nucleic acid and F_i is the molar fluorescence of the complex, N_i, which represents the nucleic acid with i, bound protein molecules ($i = 1$ to n). In Eq. (13.1), all bound species of the nucleic acid are grouped according to the number of bound protein molecules. Another mass conservation equation relates N_F and N_i to N_T by

$$N_T = N_F + \sum N_i \tag{13.2}$$

Next, we define the partial degree of binding, Θ_i ("i" moles of protein bound per mole of nucleic acid), corresponding to all complexes with a given number "i" of bound protein molecules as

$$\Theta_i = \frac{iN_i}{N_T} \tag{13.3}$$

Therefore, the concentration of the nucleic acid with "i" protein molecules bound, N_i, is defined as

$$N_i = \left(\frac{\Theta_i}{i}\right) N_{\mathrm{T}} \tag{13.4}$$

Introducing Eqs. (13.3) and (13.4) into Eq. (13.1) provides a general relationship for the observed fluorescence, F_{obs}, as

$$F_{\mathrm{obs}} = F_{\mathrm{F}} N_{\mathrm{T}} + \left[\sum (F_i - F_{\mathrm{F}}) \left(\frac{\Theta_i}{i}\right)\right] N_{\mathrm{T}} \tag{13.5}$$

By rearranging Eq. (13.5), one can define the experimentally accessible quantity, ΔF_{obs}, that is, the fluorescence change normalized with respect to the initial fluorescence intensity of the free nucleic acid, as

$$\Delta F_{\mathrm{obs}} = \frac{F_{\mathrm{obs}} - F_{\mathrm{F}} N_{\mathrm{T}}}{F_{\mathrm{F}} N_{\mathrm{T}}} \tag{13.6}$$

and

$$\Delta F_{\mathrm{obs}} = \sum \left(\frac{\Delta F_{\mathrm{i}}}{i}\right) (\Theta_i). \tag{13.7}$$

The quantity, ΔF_{obs}, is the experimentally determined fractional fluorescence change observed at the selected total protein and nucleic acid concentrations, L_{T} and N_{T}. The quantity, $\Delta F_i/i$, is the average molar fluorescence change per bound protein in the complex containing "i" protein molecules. Because $\Delta F_i/i$ is an intensive molecular property of the protein–nucleic acid complex with "i" protein molecules bound, Eq. (13.7) indicates that ΔF_{obs} is only a function of the free protein concentration through the total average degree of binding, $\sum\Theta_i$. Therefore, for a given and specific value of ΔF_{obs}, the total average degree of binding, $\sum\Theta_i$, must be the same for any value of L_{T} and N_{T} (Bujalowski, 2006; Bujalowski and Jezewska, 2000; Bujalowski and Lohman, 1987; Jezewska and Bujalowski, 1996; Lohman and Bujalowski, 1991). Thus, if one performs a fluorescence titration of the nucleic acid with the protein, at different total nucleic acid concentrations, N_{T}, the same value of ΔF_{obs} at different N_{T}s indicates the same physical state of the nucleic acid, that is, the same degree of the nucleic acid saturation with the protein and the same $\sum\Theta_i$. Since $\sum\Theta_i$ is a unique function of the free protein concentration, L_{F}, then the value of L_{F}, at the same total average degree of binding, must also be the same. Expression (13.7) is rigorous and independent of any binding model (Bujalowski, 2006; Bujalowski and Jezewska, 2000; Bujalowski and Lohman, 1987; Jezewska and Bujalowski, 1996; Lohman and Bujalowski, 1991). An analogous expression can be derived for the case where the spectroscopic signal originates from the ligand and the binding analysis is performed using reverse titration method (Bujalowski and Jezewska, 2000; Bujalowski and Lohman, 1987; Lohman and Bujalowski, 1991).

Expression (13.7) indicates a very effective method of transforming the fluorescence titration curve into a thermodynamic binding isotherm

(Bujalowski, 2006; Bujalowski and Jezewska, 2000; Bujalowski and Lohman, 1987; Jezewska and Bujalowski, 1996; Lohman and Bujalowski, 1991). In optimal and minimal case, one can perform only two titrations at two different total concentrations of the macromolecule, that is, the nucleic acid, N_{T1} and N_{T2}. At the same value of ΔF_{obs}, the total average degree of binding, $\sum\Theta_i$, and the free protein concentrations, L_F, must be the same for both titration curves. Two hypothetical fluorescence titration curves are illustrated in Fig. 13.1A for the binding process where two ligand molecules bind to two different discrete binding sites on the macromolecule, characterized by intrinsic binding constants, K_1 and K_2, and the relative fluorescence changes, ΔF_1 and ΔF_2, respectively. The values of ΔF_{obs}, as defined by Eq. (13.7), are plotted as a function of the logarithm of the total protein concentration, L_T. At a higher nucleic acid concentration, a given relative fluorescence increase, ΔF_{obs}, is reached at higher protein concentrations, as more protein is required to saturate the nucleic acid at its higher concentration. A set of values of $(\sum\Theta_i)_j$ and $(L_F)_j$ for the selected "j" value of $(\Delta S_{obs})_j$ is obtained from these data in the following manner. One draws a horizontal line that intersects both titration curves at the same value of $(\Delta F_{obs})_j$ (Fig. 13.1A). The point of intersection of the horizontal line with each titration curve defines two values of the total protein concentration, $(L_{T1})_j$ and $(L_{T2})_j$, for which $(L_F)_j$ and $(\sum\Theta_i)_j$ are the same (Bujalowski, 2006; Bujalowski and Jezewska, 2000; Bujalowski and Lohman, 1987; Jezewska and Bujalowski, 1996; Lohman and Bujalowski, 1991). Then one has two mass conservation equations for the total concentrations of the protein, $(L_{T1})_j$ and $(L_{T2})_j$, as

$$(L_{T1})_j = (L_F)_j + \left(\sum\Theta_i\right)_j N_{T1} \tag{13.8a}$$

and

$$(L_{T2})_j = (L_F)_j + \left(\sum\Theta_i\right)_j N_{T2} \tag{13.8b}$$

from which one obtains that at a given $(\Delta F_{obs})_j$ the total average degree of binding and the free protein concentration are

$$\left(\sum\Theta_i\right)_j = \frac{(L_{T2})_j - (L_{T2})_j}{N_{T1} - N_{T2}} \tag{13.9}$$

and

$$(L_F)_j = (L_{Tx})_j - \left(\sum\Theta_i\right)_j (N_{Tx}) \tag{13.10}$$

where subscript x is 1 or 2 (Bujalowski, 2006; Bujalowski and Jezewska, 2000; Bujalowski and Lohman, 1987; Jezewska and Bujalowski, 1996; Lohman and Bujalowski, 1991). Performing a similar analysis along the titration curves at a selected interval of the observed signal change, one obtains model-independent values of $(L_F)_j$ and $(\sum\Theta_i)_j$ at any selected "j" value of $(\Delta S_{obs})_j$. Practically, the most accurate estimates of $(L_F)_j$ and $(\sum\Theta_i)_j$ are obtained in the region of the titration curves where the concentration of

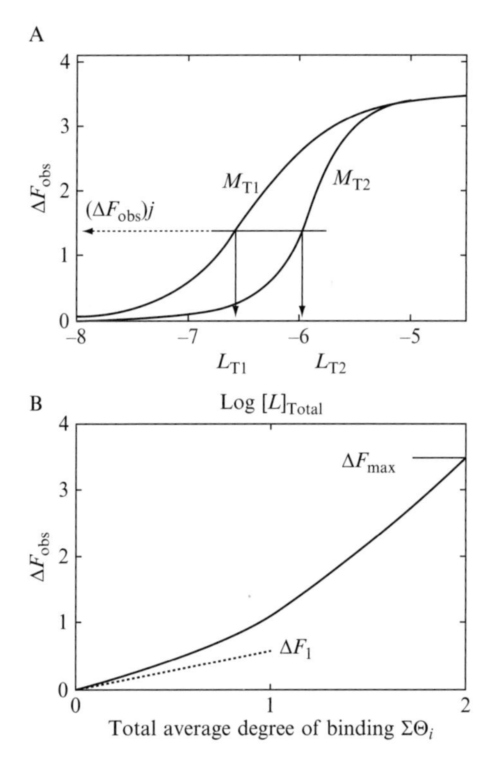

Figure 13.1 (A) Theoretical fluorescence titrations of a macromolecule with a ligand, obtained at two different macromolecule concentrations: (■) $M_1 = 1 \times 10^{-7}$ M; (□) $M_2 = 8 \times 10^{-7}$ M, respectively. The macromolecule has two discrete and different binding sites characterized by the intrinsic binding constants, $K_1 = 3 \times 10^7$ M^{-1} and $K_2 = 3 \times 0^6$ M^{-1}. The partition of the system, Z_D, is then $Z_D = 1 + K_1L_F + K_2L_F + K_1K_2L_F{}^2$. The binding to the first ligand induces the relative change of the macromolecule fluorescence, $\Delta F_1 = 0.5$, while the relative fluorescence change accompanying binding of the second ligand is $\Delta F_2 = 3$. The arrows indicate the total ligand concentrations, L_{T1} and L_{T2}, at the same selected value of the observed fluorescence change, marked by the dashed horizontal line, ΔF_{obs}, at which the total average degree of binding of the ligand, $\sum\Theta_i$, is the same at both titration curves. (B) Dependence of the relative fluorescence of the macromolecule, ΔF_{obs}, upon the total average degree of binding, $\sum\Theta_i$, of the ligand molecules. The short dashed line is an extrapolation of the initial slope of the plot, which indicates the value of $\Delta F_1 = 0.5$, characterizing the relative fluorescence change upon binding the first ligand molecules.

a bound protein is comparable to its total concentration, L_T. In our practice, this limits the accurate determination of the total average degree of binding and L_F to the region of the titration curves where the concentration of the bound ligand is at least ~10–15% of the L_T. Selection of suitable concentrations of the nucleic acid is of paramount importance for obtaining $(L_F)_j$ and $(\sum\Theta_i)_j$ over the largest possible region of the titration curves. The selection of the nucleic acid (macromolecule) concentrations is based on preliminary titrations that provide initial estimates of the expected affinity. Nevertheless, the accuracy of the determination of $(\sum\Theta_i)_j$ is mostly affected in the region of the high concentrations of the ligand, that is, where the binding process approaches the maximum saturation.

Usually, the maximum of the recorded fluorescence changes, that is, the saturation of the observed binding process, is attainable with adequate accuracy. What is mostly unknown is the maximum stoichiometry of the complex at saturation, as in the case of the PriA helicase (Jezewska *et al.*, 2000a,b; Lucius *et al.*, 2006). Another unknown is the relationship between the fluorescence change, ΔF_{obs}, and the total average degree of binding (Bujalowski, 2006; Bujalowski and Jezewska, 2000; Bujalowski and Lohman, 1987; Jezewska and Bujalowski, 1996; Jezewska *et al.*, 2000a; Lohman and Bujalowski, 1991; Lucius *et al.*, 2006). Both unknowns can be determined by plotting ΔF_{obs} as a function of $\sum\Theta_i$, as depicted in Fig. 13.1B, for the hypothetical binding model where two ligand molecules associate with two different, discrete sites on the macromolecule. The relative fluorescence change, accompanying the binding of the first high-affinity ligand, $\Delta F_1 = 0.5$, is lower than $\Delta F_2 = 3$, characterizing the binding of the second low-affinity ligand. As a result, the plot is clearly nonlinear. As mentioned above, in practice, the maximum value of $\sum\Theta_i$ cannot be directly determined, due to the inaccuracy at the high ligand concentration region. However, knowing the maximum increase of the nucleic acid fluorescence ΔF_{max}, one can perform a short extrapolation of the plot (ΔF_{obs} versus $\sum\Theta_i$) to this maximum value of the observed fluorescence change, which establishes the maximum stoichiometry of the formed complex at saturation (Bujalowski, 2006; Bujalowski and Jezewska, 2000; Bujalowski and Lohman, 1987; Jezewska and Bujalowski, 1996; Lohman and Bujalowski, 1991). Moreover, often the initial of the plot ΔF_{obs} versus $\sum\Theta_i$ can provide information on the value of the relative fluorescence change, ΔF_1, accompanying the association of the first ligand molecule. Such an estimate of ΔF_1 is illustrated in Fig. 13.1B.

3. Anatomy of the Total DNA-Binding Site in the PriA Helicase–ssDNA Complex

3.1. The site-size of the ssDNA-binding site proper of the PriA–ssDNA complex

The *E. coli* primosome is a multiprotein complex that catalyzes the DNA priming during the replication process (Jones and Nakai, 1999, 2001; Lee and Marians, 1987; Marians, 1999; Nurse *et al.*, 1990; Sangler and Marians, 2000). The PriA helicase is a key DNA replication enzyme in the *E. coli* cell that plays a fundamental role in the ordered assembly of the primosome (Jones and Nakai, 1999, 2001; Lee and Marians, 1987; Marians, 1999; Nurse *et al.*, 1990; Sangler and Marians, 2000). Originally, the PriA helicase was discovered as an essential factor during the synthesis of the complementary DNA strand of phage ΦX174 DNA (Lee and Marians, 1987). Current data indicate that the enzyme is involved not only in DNA replication but also in recombination and repair processes in the *E. coli* (Jones and Nakai, 1999, 2001). The native protein is a monomer with a molecular weight of 81.7 kDa and the monomer is the predominant form of the protein in solution (Galletto *et al.*, 2004c; Jezewska *et al.*, 1998b, 2000a, b; Jones and Nakai, 1999, 2001; Lee and Marians, 1987; Lucius *et al.*, 2006; Marians, 1999; Nurse *et al.*, 1990; Sangler and Marians, 2000).

The total site-size of a large protein ligand–DNA complex corresponds to the DNA fragment occluded by the protein, which includes nucleotides directly involved in interactions with the protein, its DNA-binding site proper and, in general, nucleotides that may not be engaged in direct interactions, or may interact differently with the protein matrix (Bujalowski, 2006; Bujalowski and Jezewska, 2000; Bujalowski and Lohman, 1987; Bujalowski *et al.*, 1989; Epstein, 1978; Jezewska and Bujalowski, 1996; Lohman and Bujalowski, 1991; McGhee and von Hippel, 1974). The latter are prevented from interacting with another protein molecule by the protruding protein matrix of the previously bound protein molecule over nucleotides adjacent to the DNA-binding site proper. Binding of the monomeric *E. coli* PriA helicase to the ssDNA provides an example of such a complex structure of the total ssDNA-binding site of the protein interacting with the nucleic acid (Galletto *et al.*, 2004c; Jezewska *et al.*, 1998b, 2000a,b; Lucius *et al.*, 2006). The most straightforward way of determining the total site-size of the ssDNA-binding site of a protein associating with the nucleic acid is to examine directly the protein binding to the homogeneous analog of the nucleic acid (Bujalowski, 2006; Bujalowski and Jezewska, 2000; Bujalowski and Lohman, 1987; Bujalowski *et al.*, 1989; Epstein, 1978; Jezewska and Bujalowski, 1996; Lohman and Bujalowski, 1991; McGhee and von Hippel, 1974). However, such an approach, although it does provide an accurate estimate of the site-size of the total DNA-binding site, does not provide any information about the structural and functional complexity of the total DNA-binding site itself. Another experimental strategy is to use a large series of ssDNA oligomers of well-defined length. This approach dramatically increases the resolution of the binding experiments and allows the experimenter to determine not only the total site-size of the protein–ssDNA complex, n, but also the number of nucleotides directly engaged in interactions with the protein, m, that is, the site-size of the DNA-binding site proper. Only when these two parameters are known, the correct statistical thermodynamic model may be formulated and the intrinsic binding parameters extracted (Bujalowski, 2006; Bujalowski and Jezewska, 2000; Bujalowski and Lohman, 1987; Jezewska and Bujalowski, 1996; Lohman and Bujalowski, 1991).

The fluorescent etheno-derivatives of homo-adenosine DNA polymer or oligomers seem to be one of the most suitable fluorescent analogs in quantitative examinations of the protein–DNA interactions (Baker *et al.*, 1978; Ledneva *et al.*, 1978; Tolman *et al.*, 1974). Binding of proteins to etheno-analogs is usually accompanied by a strong increase of the nucleic acid fluorescence (Chabbert *et al.*, 1987; Jezewska *et al.*, 1996b,c, 1998d, 2000a,b, 2001, 2003; Lucius *et al.*, 2006; Menetski and Kowalczykowski, 1985; Rajendran *et al.*, 1998, 2001). The applied excitation and emission wavelengths, $\lambda_{ex} = 325$ nm and $\lambda_{em} = 410$ nm, predominantly lead to the excitation of only the etheno-adenosine and observation of the nucleic acid fluorescence, without the excessive correction for the residual protein fluorescence. Thus the observed fluorescence change results exclusively from an increase of the quantum yield of the nucleic acid in the complex with the protein. The fluorescence change is usually very large (100–400%), greatly increasing the accuracy of the binding experiment. Moreover, fluorescence emission of etheno-analogs also allows the experimenter to access the nucleic structure in the complex. This is because the emission of etheno-adenosine, εA, has little dependence upon the nature of the environment but dramatically quenched (8–12-fold) in etheno-oligomers and poly(εA), as compared to free εAMP (Baker *et al.*, 1978; Jones and Nakai, 1999, 2001; Ledneva *et al.*, 1978; Lee and Marians, 1987; Nurse *et al.*, 1990; Tolman *et al.*, 1974). Stacking interactions between neighboring εA bases is similar to stacking interactions in unmodified adenosine polymers or oligomers (Baker *et al.*, 1978). A dynamic model, in which the motion of εA leads to quenching via intramolecular collisions, has been proposed as a predominant mechanism of the observed strong quenching (Baker *et al.*, 1978). Thus, increased viscosity of the solvent or immobilization and separation of the bases in the binding site partially eliminates the fluorescence quenching.

An example of fluorescence titrations of the ssDNA 24-mer, dεA(pεA)$_{23}$, with the PriA helicase at two different nucleic acid concentrations, in 10 m*M* sodium cacodylate/HCl (pH 7.0, 10 °C), containing 100 m*M* NaCl, 0.1 m*M* EDTA, 1 m*M* DTT, and 25% glycerol (buffer C), is shown in Fig. 13.2A. Binding of the protein to the oligomer induces a large ~280% increase of the nucleic acid fluorescence. The selected DNA concentrations provide a significant separation of the titration curves, up to the relative fluorescence increase, $\Delta F_{obs} \sim 2.3$. The shift of the titration curve at a higher oligomer concentration results from the fact that more protein is required to obtain the same total average degree of binding, $\sum\Theta_i$ (Fig. 13.1A). The fluorescence titration curves in Fig. 13.2A have been analyzed, using the quantitative approach outlined above. A typical dependence of the observed relative fluorescence increase of the ssDNA 24-mer, ΔF_{obs}, as a function of the total average degree of binding, $\sum\Theta_i$, of the PriA helicase is shown in Fig. 13.2B. The values of $\sum\Theta_i$ could reliably be

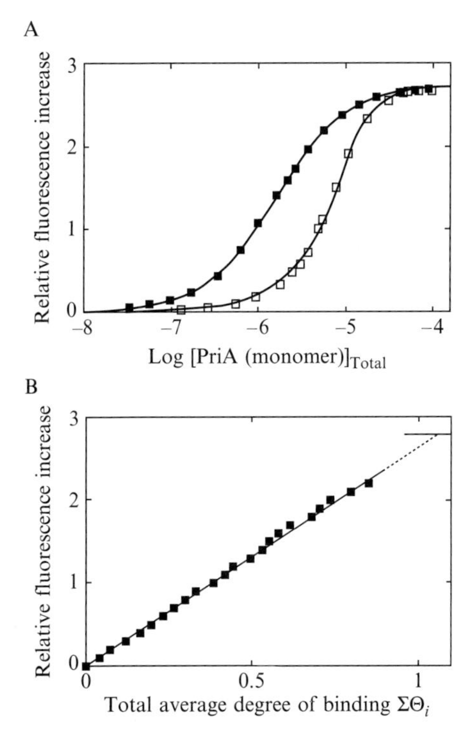

Figure 13.2 (A) Fluorescence titrations of the 24-mer, dεA(pεA)$_{23}$, with the PriA protein (λ_{ex} = 325 nm, λ_{em} = 410 nm) in buffer C (10 mM sodium cacodylate adjusted to pH 7.0 with HCl, 0.1 mM EDTA, 1 mM DTT, 25% glycerol, 10 °C), containing 100 mM NaCl, at two different nucleic acid concentrations: (■) 4.7×10^{-7} M; (□) 1.2×10^{-5} M (oligomer). The solid lines are nonlinear least-squares fits of the titration curves, using a single-site binding isotherm (Eq. 13.11) with the macroscopic binding constant $K_{24} = 7.5 \times 10^5$ M^{-1} and the relative fluorescence change $\Delta F_{max} = 2.8$. (B) Dependence of the relative fluorescence of the 24-mer, ΔF_{obs}, upon the total average degree of binding of the PriA protein, $\sum\Theta_i$ (■). The solid line follows the experimental points and has no theoretical basis. The dashed line is the extrapolation of ΔF_{obs} to the maximum value of $\Delta F_{max} = 2.8$.

determined up to ΔF_{obs} ~2.3. Short extrapolation to the maximum fluorescence change, $\Delta F_{max} = 2.8 \pm 0.2$ shows that only one molecule of the PriA helicase binds to the ssDNA 24-mer.

The maximum stoichiometry of the enzyme–ssDNA oligomer is different in the case of the 30-mer, dεA(pεA)$_{29}$, although this oligomer is only 6 nucleotides longer than the 24-mer. An example of fluorescence titrations of dεA(pεA)$_{29}$ with the PriA helicase at two different nucleic acid concentrations is shown in Fig. 13.3A. Separation of the titration curves allowed us to determine the total average degree of binding, $\sum\Theta_i$, up to ~1.8 at the value of ΔF_{obs} ~1.8. The dependence of ΔF_{obs}, as a function of the total

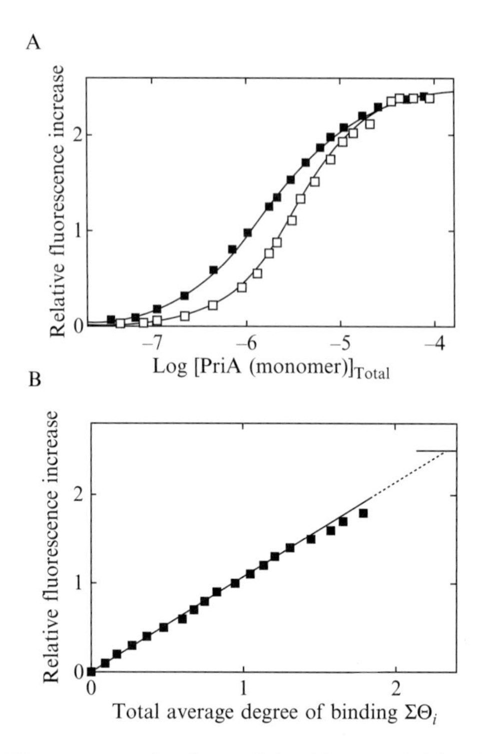

Figure 13.3 (A) Fluorescence titrations of the 30-mer, $d\varepsilon A(p\varepsilon A)_{29}$, with the PriA protein ($\lambda_{ex} = 325$ nm, $\lambda_{em} = 410$ nm) in buffer C (pH 7.0, 10 °C), containing 100 mM NaCl, at two different nucleic acid concentrations: (■) 4.7×10^{-7} M; (□) 2.1×10^{-6} M. The solid lines are nonlinear least-squares fits of the titration curves, using the statistical thermodynamic model for the binding of two PriA molecules to the 30-mer, described by Eqs. (13.14)–(13.17). The intrinsic binding constant $K_p = 1.3 \times 10^5$ M^{-1}, cooperativity parameter $\omega = 11$, and relative fluorescence change $\Delta F_1 = 1.25$, and $\Delta F_{max} = 2.5$. (B) Dependence of the relative fluorescence of the 30-mer, ΔF_{obs}, upon the total average degree of binding of the PriA protein, $\sum\Theta_i$ (■). The solid line follows the experimental points and has no theoretical basis. The dashed line is the extrapolation of ΔF_{obs} to the maximum value of $\Delta F_{max} = 2.5$.

average degree of binding of the PriA helicase on the 30-mer, is shown in Fig. 13.3B. The plot is also linear. Extrapolation to the maximum fluorescence increase $\Delta F_{max} = 2.5 \pm 0.2$ provides $\sum\Theta_i = 2.3 \pm 0.2$. Thus, the presence of an extra 6 nucleotides in the 30-mer, as compared to the 24-mer, provides enough interaction space for the binding of the second PriA protein molecule.

Analogous quantitative analysis of the maximum stoichiometry of the PriA–ssDNA complexes has been performed for a series of ssDNA oligomers. Recall, the discussed method of determining the maximum stoichiometry does not depend upon any binding model. The dependence of the parameters that characterize the examined complex is only possible when such a relationship is available and is achieved by comparing the experimental isotherms to theoretical predictions that incorporate known molecular aspects of the examined system (Bujalowski, 2006; Bujalowski and Jezewska, 2000; Bujalowski and Lohman, 1987; Bujalowski *et al.*, 1989; Epstein, 1978; Jezewska and Bujalowski, 1996; Lohman and Bujalowski, 1991; McGhee and von Hippel, 1974).

Fluorescence, or any spectroscopic titration method, is an indirect method of determining the binding isotherm, as the association of the ligand with a macromolecule is examined by monitoring changes in a spectroscopic parameter of the system accompanying the formation of the complex (Bujalowski, 2006; Lohman and Bujalowski, 1991). These changes are then correlated with the concentration of the free and bound ligand or with the fractional saturation of the macromolecule. Nevertheless, the functional relationship between the observed spectroscopic signal and the degree of binding or binding density is never *a priori* known, with exception of the systems where, at saturation, only a single ligand molecule binds to the macromolecule. But then, one has to establish that only one ligand molecule associates with the macromolecule, in the first place. On the other hand, in much more general cases where multiple ligand molecules can participate in the binding process, the observed fractional change of the spectroscopic signal and the extent of binding may not and, frequently will not, have such a simple linear relationship (Bujalowski, 2006; Bujalowski and Jezewska, 2000; Bujalowski and Lohman, 1987; Jezewska and Bujalowski, 1996; Lohman and Bujalowski, 1991).

Analysis of the binding of a protein to a nucleic acid can be performed using two different types of equilibrium spectroscopic titrations, "normal" or "reverse" titration approach (Bujalowski, 2006; Bujalowski and Jezewska, 2000; Bujalowski and Lohman, 1987; Jezewska and Bujalowski, 1996; Lohman and Bujalowski, 1991). Generally, the type of titration that is performed will depend on whether or not the monitored signal is from the macromolecule (normal) or the ligand (reverse). In the normal titration, a fluorescing macromolecule is titrated with a nonfluorescing ligand and the total average degree of binding, $\sum\Theta_i$ (average number of moles of ligand bound, L_B, per mole of the macromolecule, N_T), $\sum\Theta_i = L_B/N_T$, increases as the titration progresses (Bujalowski, 2006; Bujalowski and Jezewska, 2000; Bujalowski and Lohman, 1987; Jezewska and Bujalowski, 1996; Lohman and Bujalowski, 1991). In the reverse titration approach, the fluorescing ligand is titrated with the nonfluorescing macromolecule and the total average degree of binding decreases throughout the titration (Bujalowski, 2006; Bujalowski and Jezewska, 2000; Bujalowski and Lohman, 1987; Jezewska and Bujalowski, 1996; Lohman and Bujalowski, 1991). In our studies of the PriA helicase binding to the ssDNA, we used the normal titration approach. The interactions have been examined using the fluorescent derivative of the homoadenosine oligomers and the binding was

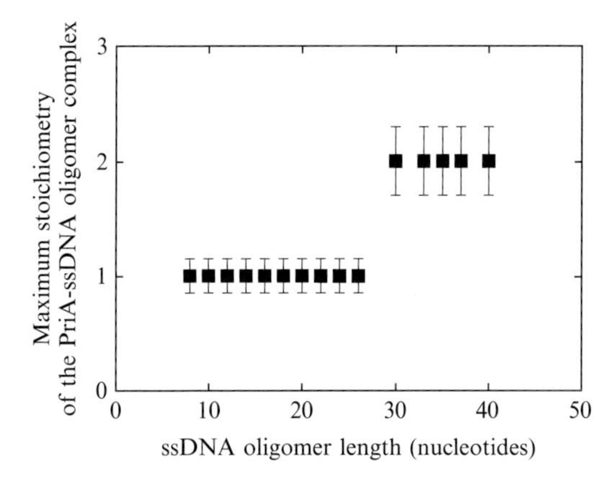

Figure 13.4 The maximum number of bound PriA molecules, as a function of the length of the ssDNA oligomer in buffer C (pH, 10 °C), containing 100 m*M* NaCl. The solid lines follow the experimental points and have no theoretical bases.

maximum number of bound PriA molecules per selected ssDNA oligomer is shown in Fig. 13.4. Oligomers from 8 to 26 nucleotide residues bind a single PriA molecule. On the other hand, transition from a single enzyme bound per ssDNA oligomer to two bound protein molecules occurs between 26- and 30-mers (Fig. 13.4). Further, an increase in the length of the oligomer, up to 40 nucleotides, does not lead to an increased number of bound PriA molecules per ssDNA oligomer. These data provide the first indication that the total site-size of the PriA–ssDNA complex is definitely less than 24 nucleotides, but that it must contain at least 14–15 nucleotides per bound protein molecule to be compatible with the observed maximum stoichiometries of the examined complexes.

3.2. Macroscopic affinities of PriA–ssDNA complexes containing a single PriA protein molecule bound

The next step in the analysis comprises examination of macroscopic affinities of the PriA helicase for different ssDNA oligomers. The striking feature of the data in Figs. 13.2–13.4 is that only a single PriA molecule binds to ssDNA oligomers of very different length. Thus, both 8- and 26-mer bind only a single enzyme molecule, yet the length of the 26-mer is more than three times longer than the length of the 8-mer. These data already indicate that only a part of the protein total DNA-binding site is engaged in direct interactions with the DNA, that is, the enzyme must possess an ssDNA-binding site proper within the total site, which predominantly interacts with the nucleic acid. However, the decisive evidence for the presence of the ssDNA-binding site proper comes from the determination of the macroscopic binding constant for enzyme binding to the ssDNA oligomers, which can accommodate only one protein molecule, as a function of the length of the nucleic acid (Galletto *et al.*, 2004c; Jezewska *et al.*, 2000a,b). Moreover, this analysis also allows us to determine the site-size of the ssDNA-binding site proper.

Binding of a single PriA molecule to 8-, 10-, 12-, 14-, 16-, 18-, 20-, 22-, 24-, and 26-mers can be analyzed using a single-site-binding isotherm described by

$$\Delta F_{\text{obs}} = \Delta F_{\text{max}} \left[\frac{K_N P_F}{1 + K_N P_F} \right] \tag{13.11}$$

where K_N is the macroscopic binding constant characterizing the affinity for a given ssDNA oligomer containing N nucleotides, and ΔF_{max} is the maximum relative fluorescence increase. The solid lines in Figs. 13.2A and 13.3A are nonlinear least-squares fits of the experimental titration curves, using Eq. (13.11). The values of K_N and ΔF_{max} for the studied ssDNA oligomers are included in Table 13.1. The value of K_N increases as the length of the ssDNA oligomers increase. The simplest explanation of this result is that the values of K_N contain a statistical factor resulting from the fact that the number of nucleotides engaged in interactions with the ssDNA-binding site proper of the protein is lower than the length of the ssDNA oligomers. In other words, the enzyme engages a small number of the nucleotides in direct interactions, characterized by the intrinsic binding constant, K_p, while the macroscopic binding constant, K_N, reflects the presence of extra potential binding sites on the ssDNA oligomers (Galletto *et al.*, 2004c; Jezewska *et al.*, 2000a,b). In terms of potential binding sites and the intrinsic binding constant, K_p, characterizing the interactions with the ssDNA-binding site proper, the macroscopic binding constant, K_N, for the PriA helicase binding to the ssDNA oligomer containing N nucleotides, is analytically defined as (Bujalowski, 2006; Bujalowski and Jezewska, 2000; Bujalowski and Lohman, 1987; Galletto *et al.*, 2004c; Jezewska and Bujalowski, 1996; Jezewska *et al.*, 2000a; Lohman and Bujalowski, 1991):

$$K_N = (N - m + 1)K_p \tag{13.12a}$$

and

$$K_N = NK_p - (m - 1)K_p \tag{13.12b}$$

Thus, the plot of K_N as a function of N should be linear with respect to N. This is clearly evident in Fig. 13.5, which shows the overall equilibrium constant, K_N, for PriA binding to ssDNA oligomers with a different number of nucleotides, as a function of the length of the ssDNA oligomer. Within experimental accuracy, the plot is linear. Moreover, the plot extrapolates to a zero value of the macroscopic binding constant, while intercepting the

Table 13.1 Macroscopic and intrinsic binding constants characterizing the association of the *E. coli* PriA helicase with ssDNA oligomers, which binds only one enzyme molecule

	8-mer	10-mer	12-mer	14-mer	16-mer	18-mer	20-mer	22-mer	24-mer
	$d\varepsilon A(p\varepsilon A)_7$	$d\varepsilon A(p\varepsilon A)_9$	$d\varepsilon A(p\varepsilon A)_{11}$	$d\varepsilon A(p\varepsilon A)_{13}$	$d\varepsilon A(p\varepsilon A)_{15}$	$d\varepsilon A(p\varepsilon A)_{17}$	$d\varepsilon A(p\varepsilon A)_{19}$	$d\varepsilon A(p\varepsilon A)_{21}$	$d\varepsilon A(p\varepsilon A)_{23}$
K_N (M^{-1})	$(8.5 \pm 03) \times 10^4$	$(1.7 \pm 03) \times 10^5$	$(2.4 \pm 03) \times 10^5$	$(3.6 \pm 0.4) \times 10^5$	$(3.9 \pm 0.5) \times 10^5$	$(4 \pm 0.5) \times 10^5$	$(5.6 \pm 0.6) \times 10^5$	$(6.4 \pm 03) \times 10^5$	$(7.5 \pm 03) \times 10^5$
K_p (M^{-1})	$(43 \pm 0.3) \times 10^4$	$(4.3 \pm 0.4) \times 10^4$	$(4.0 \pm 0.4) \times 10^4$	$(4.5 \pm 0.4) \times 10^4$	$(3.9 \pm 0.3) \times 10^4$	$(33 \pm 03) \times 10^4$	$(4.0 \pm 0.5) \times 10^4$	$(4.0 \pm 0.4) \times 10^4$	$(4.2 \pm 0.5) \times 10^4$

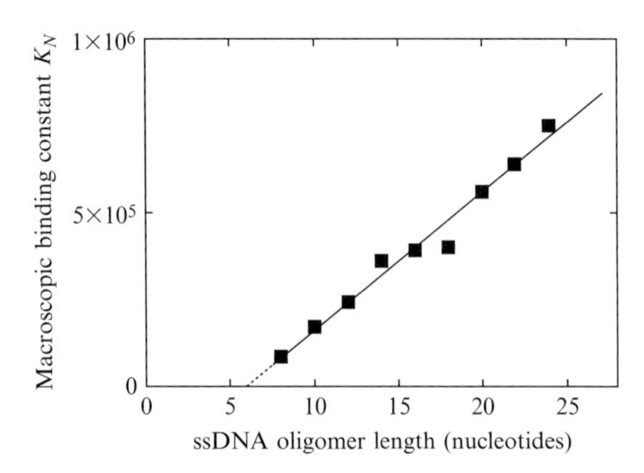

Figure 13.5 The dependence of the macroscopic binding constant, K_N, for the PriA helicase binding to the etheno-derivatives of the ssDNA oligomers upon the length of the oligomer (nucleotides). The solid line is the linear least-squares fit of the plot to Eq. (13.12b). The dashed line is an extrapolation of the plot to the value of $K_N = 0$. The plot intersects the DNA length axis at the value, $m - 1$ (details in text).

DNA length axis at a specific length. Such behavior of K_N as a function of the length of ssDNA oligomers provides strong experimental evidence of the presence of the ssDNA-binding proper within the total site of the helicase, which explores several potential binding sites on the ssDNA oligomers (Galletto *et al.*, 2004c; Jezewska *et al.*, 2000a,b). The plot intercepts the nucleic acid length axis at $N = m - 1$ (Eq. (13.12b)). Extrapolation of the plot in Fig. 13.5 to $K_N = 0$ provides the site-size of the ssDNA-binding site proper of the PriA helicase, $m = 7.1 \pm 1$ (Jezewska *et al.*, 2000a). This value is, within experimental accuracy, the same as the more conservative value of 8 ± 1, which we have estimated before (Jezewska *et al.*, 2000a). The slight difference does not affect any structure–function conclusions of the previous studies and results discussed in this work. Additional crucial information, which often escapes in such analyses, is that the obtained data on the binding of different oligomers to the enzyme clearly show that the protein possesses only one ssDNA-binding site proper, in spite of the fact that the total site-size is at least 15 nucleotides in length. If there was more than one ssDNA-binding site proper on the enzyme, then the PriA helicase would bind two 8-mer molecules, which is not experimentally observed.

3.3. The site-size of the total DNA-binding site of the PriA helicase

3.3.1. Model of PriA protein–ssDNA interactions

As discussed above, the transition from a single PriA molecule bound per ssDNA oligomer to two molecules bound per the oligomer occurs between 26- and 30-mers, indicating that the minimum, total site-size of the

PriA–ssDNA complex is at least 14–15 nucleotides per bound protein (Fig. 13.4). On the other hand, binding studies with oligomers, which can accommodate only one enzyme molecule, indicate that the protein engages only 7.1 ± 1 nucleotides in interactions with its ssDNA-binding site proper (see above). These results indicate that a significant part of the total site-size of the PriA helicase (at least ~7–8 nucleotides) results from the protruding of the large protein molecule over the nucleotides adjacent to the ssDNA-binding site proper.

Inspection of the data on the binding of the PriA helicase to the ssDNA oligomers of different lengths allows the experimenter to deduce the site-size of the total DNA-binding site of the enzyme, as well as the location of the ssDNA-binding site proper within the global structure of the protein matrix, using two, limiting binding models (Galletto *et al.*, 2004c; Jezewska *et al.*, 2000a,b). The first model of the structure of the total DNA-binding site of the PriA helicase is depicted in Fig. 13.6. The ssDNA-binding site proper, which encompasses ~7 nucleotides, is located on one side of the protein molecule (Fig. 13.6A). We are still assuming that the total

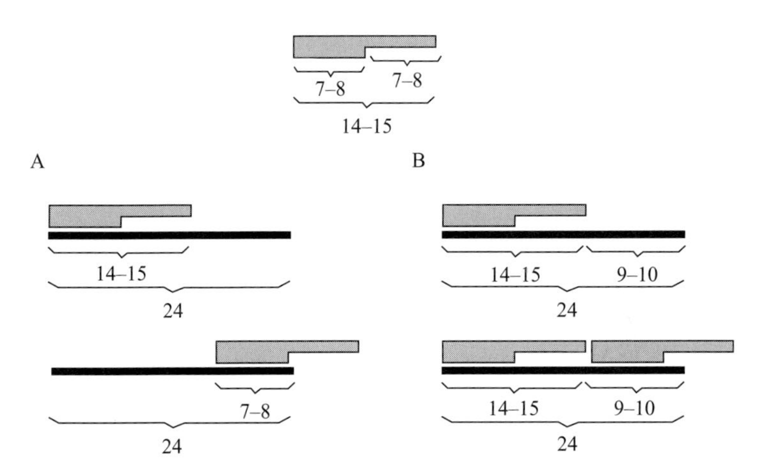

Figure 13.6 Schematic model for the binding of the PriA helicase to the ssDNA, based on the minimum site-size of the protein–nucleic acid complex, $n = 15$, and the size of the binding site engaged in protein ssDNA interactions, $m = 7$. The helicase binds the ssDNA in a single orientation with respect to the polarity of the sugar-phosphate backbone of the ssDNA. The ssDNA-binding site proper, which encompasses 7–8 nucleotides, is located on one side of the enzyme molecule, with the rest of the protein matrix protruding over the extra 7–8 nucleotides, without engaging in thermodynami cally significant interactions with the DNA (black ribbon) (A). When bound at the ends of the nucleic acid, or in its center, the protein can occlude 7 or 15 nucleotides (B). Therefore, this model would allow the binding of two molecules of the PriA helicase to the ssDNA 24-mer and three molecules of the enzyme to the 40-mer (C). However, this is not experimentally observed.

DNA-binding site size is 14–15 nucleotides, that is, the protein matrix protrudes over 7–8 nucleotides outside of the ssDNA-binding site proper. However, if the ssDNA-binding site of the protein is located on one side of the molecule, with a part of the enzyme protruding over the extra 7–8 nucleotides, then the 24-, 26-, and 40-mers would be able to accommodate two and three PriA molecules, respectively (Fig. 13.6B). These are not experimentally observed maximum stoichiometries (Fig. 13.4). Only one PriA molecule binds to the 24- and 26-mers, and only two enzyme molecules bind to the 40-mer (Fig. 13.4). Therefore, the model shown in Fig. 13.6 cannot represent the PriA–ssDNA complex.

The second model where the ssDNA-binding site proper, which engages 7–8 nucleotides, is located in the central part of the protein and is depicted in Fig. 13.7. The protein molecule now has two parts that are protruding over 7–8 nucleotide residues on both sides of the ssDNA-binding site proper. In this model, only one PriA molecule can bind to the 24- and 26-mers (Fig. 13.7A). This is because the first bound molecule

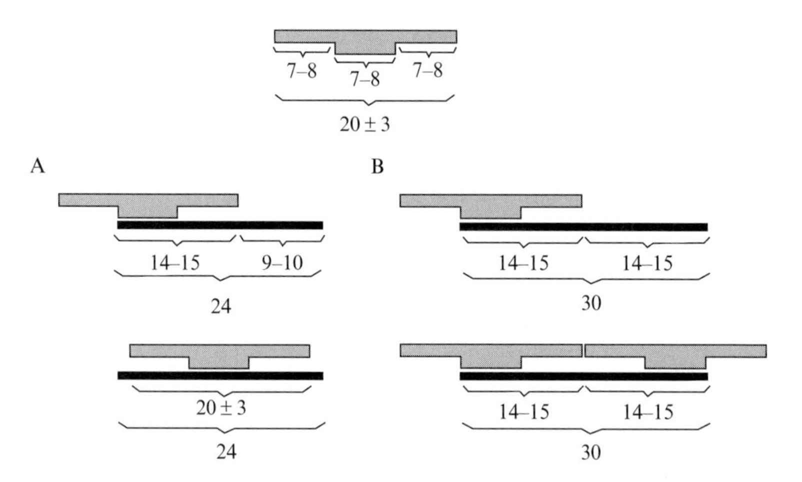

Figure 13.7 Schematic model for the binding of the PriA helicase to the ssDNA, based on the total site-size of the protein–nucleic acid complex, $n = 20$, and the size of the ssDNA-binding site engaged in protein–ssDNA interactions, $m = 7$. The helicase binds the ssDNA in a single orientation with respect to the polarity of the sugar-phosphate backbone of the ssDNA. The ssDNA-binding site of the PriA helicase, which encompasses only 7–8 nucleotides, is located in the center of the enzyme molecule. The protein matrix protrudes over 7–8 nucleotides on both sides of the ssDNA-binding site without engaging in interactions with the nucleic acid (black ribbon). When bound at the 5′ or the 3′ end of the nucleic acid, the protein always occludes 14–15 nucleotides, while in the complex in the center of the ssDNA oligomer, 20–23 nucleotides are occluded (A). This model would allow the binding of only one molecule of the PriA helicase to the 24-mer and only two molecules of the enzyme to the 30-, 33-, 35-, 37-, and 40-mers (B). Such maximum stoichiometries are in complete agreement with the experimental data.

can now block at least 14–15 nucleotides. For efficient binding, the second protein molecule also needs a fragment of at least 14 nucleotides, which is larger than the remaining 11 and 12 residues of the 24- and 26-mers, respectively. On the other hand, such a location of the ssDNA-binding site proper allows two molecules of the enzyme to bind to the 30-, 33-, 35-, 37-, and 40-mers (Fig. 13.7B). In the case of the 40-mer, the remaining fragment of 4–5 nucleotides is too short, that is, it does not provide efficient interacting space to allow the third PriA protein to associate with the oligomer. This is exactly what is experimentally observed (Fig. 13.4). Therefore, the model of the protein–ssDNA, presented in Fig. 13.7, adequately describes all experimentally determined stoichiometries of the PriA with the series of ssDNA oligomers (Fig. 13.4). Moreover, these data and analyses indicate that the actual total site-size of the PriA–ssDNA complex is 20 ± 3 nucleotide residues and include the 7.1 ± 1 nucleotides encompassed by the ssDNA-binding site of the enzyme, as well as the 14–16 nucleotides occluded by the protruding protein matrix (Fig. 13.7B).

3.3.2. Intrinsic affinities of PriA–ssDNA interactions

Examination of the intrinsic affinities of the PriA helicase for different ssDNA oligomers, that is, the affinity of the ssDNA-binding site proper characterized by the binding constant, K_p, provides additional support for the proposed model of the PriA–ssDNA complex (Galletto *et al.*, 2004c; Jezewska *et al.*, 2000a,b; Lucius *et al.*, 2006). The macroscopic binding constant, K_N, for the binding of a single PriA molecule to 8- 10-, 12-, 14-, 16-, 18-, 20-, 22-, 24-, and 26-mers is defined in terms of m and K_p by Eq. (13.12a). Using the lattice model for the nucleic acid and taking integer value of $m \approx 7$, one can recalculate the values of K_p, which are included in Table 13.1 (Galletto *et al.*, 2004c; Jezewska *et al.*, 2000a,b). As expected (Fig. 13.5), the values of K_p are very similar for all examined oligomers, indicating that very similar interactions are present in all examined complexes. However, if the model in Fig. 13.7B is correct, then the intrinsic affinity of the PriA helicase in the complexes with the ssDNA oligomers, which can accommodate two enzyme molecules, should also be similar to the value determined for the oligomers, which accommodate only a single protein molecule.

Quantitative determination of the value of K_p for the PriA helicase to the ssDNA 30-, 33-, 35-, 37-, and 40-mers, which can accommodate two enzyme molecules, requires a more complex statistical thermodynamic approach. In general, the binding process includes intrinsic affinity of the ssDNA-binding site proper, possible cooperative interactions between bound protein molecules, and the overlap between potential binding sites on the nucleic acid lattice (Bujalowski, 2006; Bujalowski and Jezewska, 2000; Bujalowski and Lohman, 1987; Galletto *et al.*, 2004c; Jezewska and Bujalowski, 1996; Jezewska *et al.*, 2000a,b; Lohman and Bujalowski, 1991).

We know that the total site-size of the PriA–ssDNA complex is $n = 20 \pm 3$. However, the number of nucleotides engaged in interactions with the ssDNA-binding site proper of the enzyme is only $m = 7.1 \pm 1$, and that the protein protrudes over a distance of 7–8 nucleotides on both sides of the binding site proper (Fig. 13.7B). Therefore, the partial degree of binding that involves only the first PriA molecule bound to the ssDNA 30-, 35-, and 40-mers, is described by

$$\sum \Theta_i = \frac{(N - m + 1)K_p L_F}{1 + (N - m + 1)K_p L_F} \tag{13.13}$$

where $N = 30$, 33, 35, 37, or 40, $m = 7$, and K_p is the intrinsic binding constant for the given N-mer. The factor, $N + m + 1$, indicates that the first single PriA molecule experiences the presence of a multitude of potential binding sites on each examined N-mer. However, as the protein concentration increases, this complex is replaced by the complex, in which two PriA molecules are bound to the oligomer and there are several different possible configurations of the two proteins on each N-mer (Fig. 13.7B). In order to derive the part of the partition function corresponding to the binding of two PriA molecules, we apply an exact combinatorial theory for the cooperative binding of a large ligand to a finite one-dimensional lattice (Epstein, 1978; Jezewska *et al.*, 2000b). The complete partition function for the PriA–N-mer system, Z_N, is then

$$Z_N = 1 + (N - m + 1)K_p L_F + \sum_{j=0}^{k-1} S_N(k,j)(K_p L_F)^k \omega^j \tag{13.14}$$

where $k = 2$ and j is the number of cooperative contacts between the bound PriA molecules in a particular configuration on the lattice, and ω is the parameter characterizing the cooperative interactions. The factor $S_N(k, j)$ is the number of distinct ways that two protein ligands bind to a lattice with j cooperative contacts and is defined by (Jezewska *et al.*, 2000a)

$$S_N(k,j) = \frac{(N - (m+7)k + 1)!(k-1)!}{(N - (m+7)k - k + j + 1)!(k-j)!j!(k-j-1)!} \tag{13.15}$$

The factor, $m + 7$, arises from the fact that, with two protein molecules bound, each bound protein occludes the nucleic acid with its ssDNA-binding site proper (m nucleotides) and the protruding protein matrix on one side of the protein (in the considered case, 7 nucleotides). The total average degree of binding, $\sum \Theta_i$, is defined as

$$\sum \Theta_i = \frac{(N - m + 1)K_p L_F + \sum_{j=0}^{k-1} S_N(k,j)k(K_p L_F)^k \omega^j}{Z_N} \tag{13.16}$$

The observed relative fluorescence increase of the nucleic acid, ΔF_{obs}, is then

$$\Delta F_{obs} = \Delta F_1 \left[\frac{(N - m + 1) K_p L_F}{Z_N} \right] + \Delta F_{max} \left[\frac{\sum_{j=0}^{k=1} S_N(k,j) (K_p L_F)^k \omega^j}{Z_N} \right] \tag{13.17}$$

where ΔF_1 and ΔF_{max} are relative molar fluorescence parameters that characterize the complexes with one and two PriA molecules bound to the N-mer, respectively.

An example of fluorescence titrations of the 40-mer dA(pεA)$_{39}$ with the PriA helicase at two different nucleic acid concentrations is shown in Fig. 13.8A. Separation of the titration curves allowed us to determine the total average degree of binding, $\sum\Theta_i$, up to ~1.5 at the value of ΔF_{obs} ~2.4. The dependence of ΔF_{obs}, as a function of the total average degree of binding of the PriA helicase on the 40-mer, is shown in Fig. 13.8B. The plot is linear and extrapolation to the maximum fluorescence increase $\Delta F_{max} = 3.5 \pm 0.2$ provides $\sum\Theta_i = 2.0 \pm 0.2$. Also, using the plot in Fig. 13.8B, one can estimate the values of $\Delta F_1 = 1.7 \pm 0.05$. Therefore, there are only two unknown parameters, K_p and ω, in Eqs. (13.14)–(13.17). The solid lines in Fig. 13.8A are nonlinear least-squares fits using Eqs. (13.14)–(13.17) with only two fitted parameters K_p and ω. The obtained values are $K_p = (5.6 \pm 0.3) \times 10^4\ M^{-1}$ and $\omega = 0.8 \pm 0.3$. Analogously, the solid lines in Fig. 13.3A are nonlinear least-squares fits using Eqs. (13.14)–(13.17) for the PriA binding to the 30-mer, dεA(pεA)$_{29}$, with only two fitted parameters, K_p and ω, which provides $K_p = (1.3 \pm 0.3) \times 10^5\ M^{-1}$ and $\omega = 11 \pm 3$. The same analysis for the 35-mer, dεA(pεA)$_{29}$, provides $K_p = (1.3 \pm 0.3) \times 10^5\ M^{-1}$ and $\omega = 2 \pm 0.6$ (Jezewska *et al.*, 2000a).

The values of the intrinsic binding constant, K_p, for all examined oligomers fall between ~$3.3 \times 10^4\ M^{-1}$ and $1.3 \times 10^5\ M^{-1}$. Such similar values of K_p, obtained with the ssDNA oligomers of different lengths, indicate similar types of intrinsic interactions between the PriA protein and the nucleic acid in the studied complexes, corroborating the proposed model of the PriA protein–ssDNA complex, which takes into account the total site-size of the protein–nucleic acid complex, n, the number of the nucleotide residues engaged in interactions with the ssDNA-binding site, m, and the central location of the ssDNA-binding site within the protein matrix. Moreover, the obtained value of ω ranging between 11 ± 3 and 0.8 ± 0.3 also indicates that the binding of the PriA helicase to the ssDNA is not accompanied by any significant cooperative interactions.

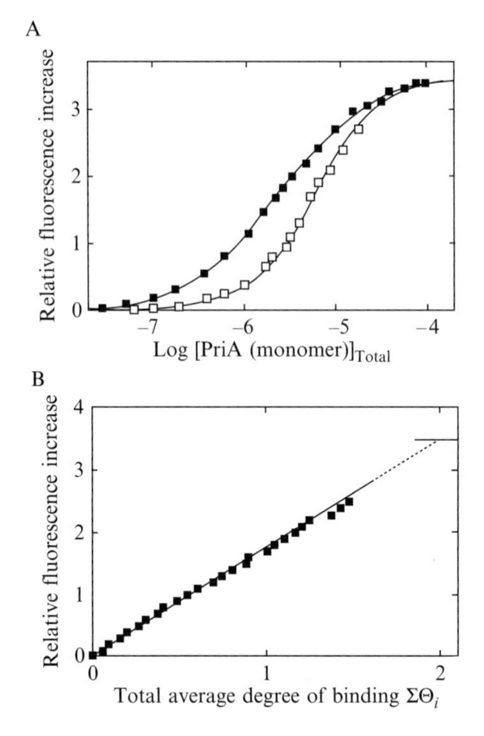

Figure 13.8 (A) Fluorescence titrations of the 40-mer, dεA(pεA)$_{39}$, with the PriA protein ($\lambda_{ex} = 325$ nm, $\lambda_{em} = 410$ nm) in buffer C (pH 7.0, 10 °C), containing 100 m*M* NaCl, at two different nucleic acid concentrations: (■) 4.6×10^{-7} *M*; (□) 4×10^{-6} *M*. The solid lines are nonlinear least-squares fits of the titration curves, using the statistical thermodynamic model for the binding of two PriA molecules to the 40-mer, described by Eqs. (13.14)–(13.17). The intrinsic binding constant $K_p = 5.6 \times 10^4\ M^{-1}$, cooperativity parameter $\omega = 0.8$, and relative fluorescence change $\Delta F_1 = 1.7$, and $\Delta F_{max} = 3.5$, respectively. (B) Dependence of the relative fluorescence of the 40-mer, ΔF_{obs}, upon the total average degree of binding of the PriA proteins, $\sum\Theta_i$ (■). The solid line follows the experimental points and has no theoretical basis. The dashed line is the extrapolation of ΔF_{obs} to the maximum value of $\Delta F_{max} = 3.5$.

3.3.3. Binding modes in PriA–ssDNA interactions

It should be pointed out that the binding of the PriA protein to the ssDNA constitutes an example, albeit a very specific one, of a protein binding to the nucleic acid in two binding modes, differing by the number of the occluded nucleotides in the complex (Bujalowski and Lohman, 1986; Chabbert *et al.*, 1987; Jezewska *et al.*, 1998b, 2003, 2006; Lohman and Ferrari, 1994; Menetski and Kowalczykowski, 1985; Rajendran et al., 1998, 2001). In one mode, the PriA protein forms a complex with the site size of 20 ± 3, that is, the total ssDNA-binding site is occluding the nucleic acid lattice and in the other binding mode, the protein can associate with the DNA using only its ssDNA-binding site proper, with the site size of ~7 nucleotides, with an additional ~7–8 nucleotides occluded by the protruding protein matrix (Fig. 13.7B). However, on the polymer ssDNA lattices only a single binding mode with the site size of 20 ± 3 nucleotides would be detectable, as in the predominant fraction of the bound protein molecules the PriA helicase occludes 20 ± 3 nucleotides. In other words, the detection of the PriA binding modes was only possible through the experimental strategy of examining the stoichiometry of the protein–ssDNA complex using an extensive series of ssDNA oligomers (Bujalowski and Lohman, 1986; Galletto *et al.*, 2004c; Jezewska *et al.*, 2000a,b, 2004; Lohman and Ferrari, 1994; Lucius *et al.*, 2006).

4. Structure–Function Relationship in the Total ssDNA-Binding Site of the DNA Repair Pol X From ASFV

4.1. Model of the ASFV Pol X–ssDNA interactions

The DNA pol X from the ASFV provides an analogous, although more complex, system of the protein–ssDNA complex, as compared to the PriA–ssDNA complex. We will limit our discussion to studies of the structure–function relationship within the total ssDNA-binding site of the enzyme. The reader is advised to consult the original paper on the full analysis of the system (Jezewska *et al.*, 2006). The DNA genome of the ASFV encodes a DNA polymerase, a member of the pol X family referred to as ASFV pol X, which is engaged in the repair processes of the viral DNA (Garcia-Escudero *et al.*, 2003; Oliveros *et al.*, 1997; Yanez *et al.*, 1995). With a molecular weight of ~20,000, ASFV pol X is currently the smallest known DNA polymerase, whose structure has been solved by NMR (Maciejewski *et al.*, 2001; Showwalter *et al.*, 2003). The enzyme is built only of the palm domain, which includes the first 105 amino acids from the N-terminus of the protein and the C-terminal domain, which is built of the remaining 69 amino acid residues.

The strategy of addressing the function–structure relationship in the total DNA-binding site of the ASFV pol X has been analogous to the strategy discussed for the *E. coli* PriA helicase. Binding of the enzyme to the nucleic acid was examined using a series of etheno-homo-adenosine ssDNA oligomers (Jezewska *et al.*, 2006). The first tactical step is to determine the maximum stoichiometry of the enzyme–ssDNA oligomer complex, as a function of the nucleic acid length. The stoichiometry of the complexes was determined using the quantitative fluorescence titrations method. The dependence of the maximum stoichiometry of the ASFV pol X–ssDNA

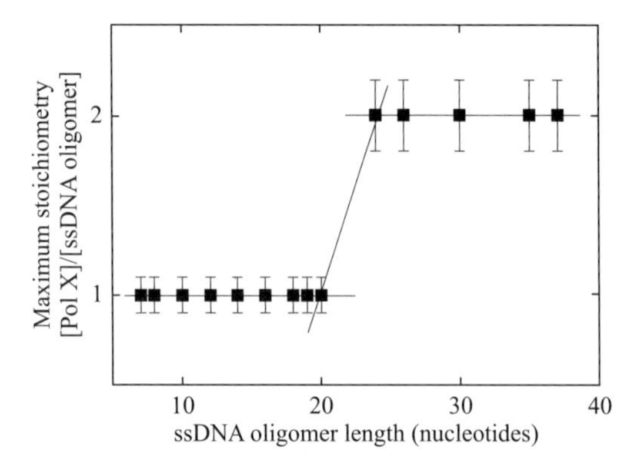

Figure 13.9 The dependence of the total average degree of binding, $\sum\Theta_i$, of the ASFV pol X–ssDNA complex upon the length of the ssDNA oligomer in buffer C (10 m*M* sodium cacodylate adjusted to pH 7.0 with HCl, 1 m*M* $MgCl_2$, and 10% glycerol (10 °C), containing 100 m*M* NaCl and 1 m*M* $MgCl_2$. The solid lines follow the experimental points and have no theoretical basis (Jezewska *et al.*, 2006).

oligomer for a series of ssDNA oligomers, $d\varepsilon A(p\varepsilon A)_{N-1}$, as a function of the length of the ssDNA oligomer is shown in Fig. 13.9. Only a single ASFV pol X molecule binds to ssDNA oligomers, ranging from 7 to 20 nucleotides. A jump in the maximum stoichiometry of the complex, from one to two, occurs between the 20- and 24-mer. The maximum stoichiometry does not increase for the oligomer containing 37 nucleotides, which is 13 nucleotides longer than the 24-mer (Fig. 13.9). These data provide the first indication that a total site-size of the ASFV pol X–ssDNA complex is less than 19 nucleotides, but it must contain at least 12 nucleotides per bound protein molecule.

The dependence of the macroscopic binding constant, K_N, for the ASFV pol X binding to the ssDNA oligomers, containing from 7 to 20 residues, that is, oligomers which can accommodate only a single enzyme molecule, upon the length of the ssDNA oligomer, is shown in Fig. 13.10. The plot has an unusual feature, when compared to the behavior of the PriA helicase. It is built of two linear regions separated by an intermediate plateau. For the ssDNA oligomers containing 7–12 nucleotides, the values of K_N increase linearly with the length of the oligomers. As we discussed earlier, the simplest explanation of such empirical linear behavior of K_N, as a function of the length of the ssDNA, is that there is a small, discrete binding region within the total DNA-binding site of the ASFV pol X that experiences the presence of several potential binding sites on the ssDNA oligomers, that is, there is the DNA-binding site proper within the total DNA-binding site. The DNA-binding site proper engages in interactions a number of p nucleotides, which must be smaller than the length of the

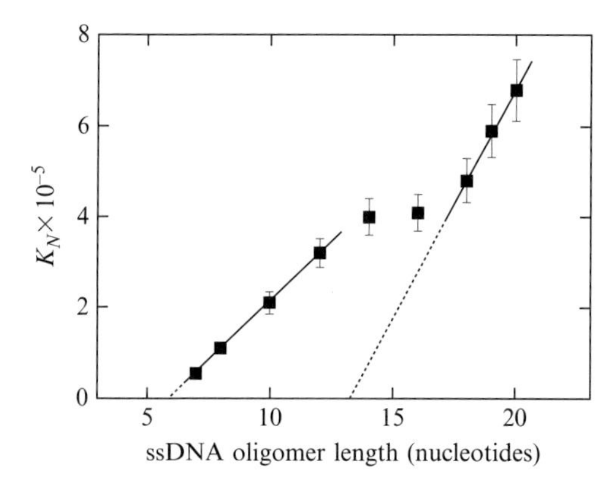

Figure 13.10 (A) The dependence of the macroscopic binding constant K_N, upon the length of the ssDNA for oligomers containing 7–20 nucleotides, that is, the oligomers that can accommodate only one ASFV pol X molecule. The solid line for the part of the plot corresponding to oligomers from 7 to 12 nucleotide residues is a linear least-squares fit to Eq. (13.12b). Extrapolation of the line to $K_N = 0$ intercepts the DNA length axis at 6.0 ± 0.5. The solid line for the part of the plot corresponding to oligomers from 18 to 20 nucleotide residues is a linear least-squares fit to Eq. (13.18). Extrapolation of the line intercepts the DNA length axis at 13.2 ± 1.0 (Jezewska *et al.*, 2006).

examined ssDNA oligomers. Therefore, the values of K_N contain a statistical factor that can be analytically defined in terms of the site-size, p, and the intrinsic binding constant, K_p, of the DNA-binding site proper, as defined by Eqs. (13.12a) and (13.12b). Extrapolation of the plot in Fig. 13.10 to the zero value of the macroscopic equilibrium constant, K_N, intercepts the abscissa at the DNA length corresponding to the length of the ssDNA oligomer, which is too short to be able to form a complex with the enzyme. The plot in Fig. 13.10 gives that length as 6 ± 0.5. Therefore, the DNA-binding site proper of the ASFV pol X requires $p = 7 \pm 1$ nucleotide residues of the ssDNA to engage in energetically efficient interactions with the nucleic acid. The intrinsic binding constant, K_p, can be determined from the slope of the first linear region in Fig. 13.10 (Eq. (13.12b)). For the ssDNA oligomers with the length between 7 and 12 nucleotides, the average value of $K_p = (5.4 \pm 0.8) \times 10^4\ M^{-1}$.

For the ssDNA oligomers, with the length exceeding 12 nucleotide residues, the plot in Fig. 13.7A becomes nonlinear and passes through an intermediate plateau into the second linear region for the oligomers, with the length longer than ~16–17 nucleotides. Such a transition between two linear regions of the plot of the macroscopic binding constant as a function of the length of the ssDNA oligomers indicates that different areas of the total DNA-binding site of the polymerase, beyond the DNA-binding site proper, become involved in interactions with the longer nucleic acid

(Jezewska *et al.*, 2006). The second linear region of the plot in Fig. 13.7A can be described by expressions analogous to Eqs. (13.12a) and (13.12b), as

$$K_N = NK_q - (q+1)K_q \tag{13.18}$$

However, the parameters in Eq. 13.18 now have different physical meanings. The quantity, q, is the minimum length of the ssDNA oligomer that can engage the total DNA-binding site of the polymerase, and K_q is the intrinsic binding constant for the total DNA-binding site. Extrapolation of this region to the $K_N = 0$ provides $q = 14.3 \pm 1.0$. Thus, the data indicate that in order to engage the total DNA-binding site of the ASFV pol X, the ssDNA must have at least ~14–15 nucleotides. Notice, this is a minimum estimate, not the actual site-size of the total DNA-binding site, which can be even larger (Jezewska *et al.*, 2006). The slope of the linear region, for the ssDNA oligomers with the length between 18 and 20 nucleotides, provides $K_q = (9.3 \pm 2.1) \times 10^4\ M^{-1}$. This value is higher than the corresponding value of the intrinsic binding constant characterizing the interactions of the short ssDNA oligomers with the DNA-binding site proper of the polymerase. Such a difference reflects the fact that one observed different intrinsic interactions when the total DNA-binding site is engaged in interactions with the enzyme, as compared to the ssDNA-binding site proper (Jezewska *et al.*, 2006).

4.2. Model of the ASFV Pol X–ssDNA complex

The next step is to address all obtained stoichiometry data to deduce the structural features of the total DNA-binding site. As discussed earlier, the transition of the maximum stoichiometry of the ASFV pol X–ssDNA oligomer complexes, from a single pol X molecule bound per an ssDNA oligomer to two molecules bound per oligomer, occurs between 20- and 24-mers (Fig. 13.9). However, binding studies with short ssDNA oligomers indicate that the polymerase engages only 7 ± 1 nucleotides in interactions with its DNA-binding site proper and the total DNA-binding site engages at least ~14–15 nucleotides. On the other hand, the 37-mer can still accommodate only two pol X molecules, that is, it does not allow the enzyme to associate with the nucleic acid using its ssDNA-binding site proper. These data indicate that the total site-size of the pol X–ssDNA complex is 16 ± 1 nucleotides per bound protein.

To address the structure of the total DNA-binding site of pol X, we consider the following two limiting models of the ASFV pol X–ssDNA complexes, depicted schematically in Fig. 13.11A and B (Jezewska *et al.*, 2006). In Fig. 13.11A, the ssDNA-binding site proper, which engages 7 nucleotides, is located in the central part of the protein molecule. If this model of the total DNA-binding site applies, it would require that the

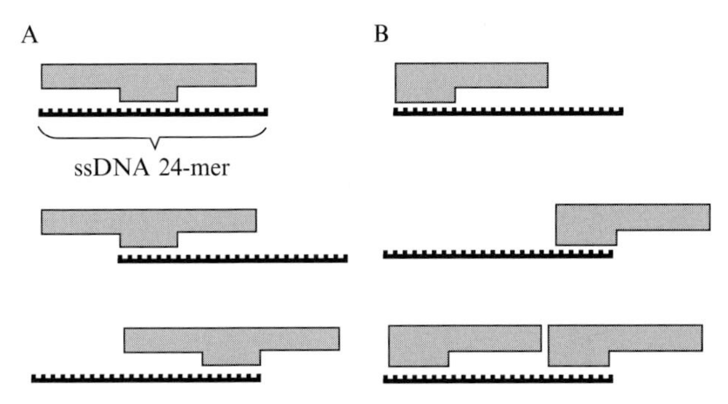

Figure 13.11 (A) Schematic model for the binding of the ASFV pol X to the ssDNA 24-mer, based on the site-size of the DNA-binding site proper of the complex $p = 7$ nucleotides and the site-size of the total DNA-binding site of the polymerase, $q = 16$ (Jezewska *et al.*, 2006). The DNA-binding site proper is located in the center of the enzyme molecule. In this model the protein matrix would have to protrude on both sides of the proper DNA-binding site over ~7–8 nucleotides in order to provide the minimum total DNA-binding site-size of ~15 nucleotides, independent of the location of the proper DNA-binding site. However, this model would allow the binding of only one molecule of pol X to the ssDNA 24-mer, which is not experimentally observed (Fig. 13.9). (A) Schematic model for the binding of the ASFV pol X to the ssDNA 24-mer, based on the site-size of the proper DNA-binding site of the complex $p = 7$ nucleotides and the site-size of the total DNA-binding site of the polymerase, $q = 16$. The enzyme binds the ssDNA in a single orientation with respect to the polarity of the sugar-phosphate backbone of the ssDNA. The DNA-binding site proper is located on one side of the enzyme molecule, with the rest of the protein matrix protruding over the extra 9 nucleotides. When bound at the ends of the nucleic acid, or in its center, the enzyme can occlude from 7 or 16 nucleotides. This model would allow the binding of two molecules of pol X to the ssDNA 24-mer (Jezewska *et al.*, 2006).

protein molecule possesses two parts, each part protruding over ~8 nucleotides on both sides of the DNA-binding site proper giving a total DNA-site-size of ~23 nucleotides. This is because the minimum number of nucleotides engaged in interactions with the total DNA-binding site must be, at least, ~14–15, independent of the location of the DNA-binding site proper on the nucleic acid (Fig. 13.10). First, such a total site-size would be much larger than estimated from the fluorescence titration data. Second, the first bound pol X molecule would block ~15 nucleotides and the next bound molecule would also require an additional ~15 residues, giving the total requirement of 30 nucleotides. As a result, the ssDNA 24-mer would be able to accommodate only one pol X molecule. This is not experimentally observed (Fig. 13.9).

Next, we consider a limiting model where the DNA-binding site proper of the enzyme engages 7 nucleotides and is asymmetrically located on one side of the polymerase molecule, as depicted in Fig. 13.11B. Part of the enzyme protrudes over the nucleic acid and engages an additional area in interactions at a distance from the proper binding site corresponding to ~8–9 nucleotides of the ssDNA. In this model, the total site-size of the complex is 16 nucleotides. The ssDNA oligomers with 20 or less nucleotides would bind only one ASFV pol X molecule, while oligomers from 24 to 37 residues would be able to accommodate two pol X molecules. This is exactly what is experimentally observed (Fig. 13.9). Therefore, the model of the ASFV pol X–ssDNA complex, depicted in Fig. 13.11B, adequately describes all experimentally determined stoichiometries of the ASFV pol X complexes with the examined series of ssDNA oligomers.

Notice, if the total DNA-binding site was only 15 nucleotides then the 37-mer would be able to accommodate three pol X molecules, which is not experimentally observed (Fig. 13.9). This is because two bound pol X molecules would block only ~30 nucleotides, leaving 7 nucleotides to accommodate an additional enzyme molecule, bound through its DNA-binding site proper. In other words, these data and analyses of the two alternative limiting models indicate that the actual total site-size of the pol X–ssDNA complex is 16 ± 1 nucleotides and includes 7 nucleotides encompassed by the DNA-binding site proper of the polymerase, as well as 9 nucleotides occluded by the protein only on one side of the DNA-binding site proper (Fig. 13.11B).

ACKNOWLEDGMENTS

We thank Gloria Drennan Bellard for her help in preparing the manuscript. This work was supported by NIH Grants GM46679 and GM58565 (to W. B.).

REFERENCES

Ali, J. A., and Lohman, T. M. (1997). *Science* **275,** 377.
Amaratunga, M, and Lohman, T. M. (1993). *Biochemistry* **32,** 6815.
Baker, B. M., Vanderkooi, J., and Kallenbach, N. R. (1978). *Biopolymers* **17,** 1361.
Baker, T. A., Funnell, B. E., and Kornberg, A. (1987). *J. Biol. Chem.* **262,** 6877.
Bjorson, K. P., Moore, K. J., and Lohman, T. M. (1996). *Biochemistry* **35,** 2268.
Bujalowski, W. (2003). *Trends Biochem. Sci.* **28,** 116.
Bujalowski, W. (2006). *Chem. Rev.* **106,** 556.
Bujalowski, W., and Jezewska, M. J. (2000). *In* "Spectrophotometry and Spectrofluorimetry. A Practical Approach," (M. G. Gore, ed.), pp. 141–165. Oxford University Press, Oxford.
Bujalowski, W., and Lohman, T. M. (1986). *Biochemistry* **25,** 7779.
Bujalowski, W., and Lohman, T. M. (1987). *Biochemistry* **26,** 3099.
Bujalowski, W., Lohman, T. M., and Anderson, C. F. (1989). *Biopolymers* **28,** 1637.
Chabbert, M., Cazenave, C., and Hélène, C. (1987). *Biochemistry* **26,** 2218.
Enemark, E. J., and Joshua-Tor, L. (2008). *Curr. Opin. Struct. Biol.* **18,** 243.
Epstein, I. R. (1978). *Biophys. Chem.* **8,** 327.
Galletto, R., Jezewska, M. J., and Bujalowski, W. (2004a). *J. Mol. Biol.* **343,** 83.

Galletto, R., Jezewska, M. J., and Bujalowski, W. (2004b). *J. Mol. Biol.* **343,** 101.
Galletto, R., Jezewska, M. J., and Bujalowski, W. (2004c). *Biochemistry* **43,** 11002.
Garcia-Escudero, R., Garcia-Diaz, M., Salas, M. L., Blanco, L., and Salas, J. (2003). *J. Mol. Biol.* **326,** 1403.
Heller, R. C., and Marians, K. J. (2007). *DNA Repair* **6,** 945.
Hubscher, U., Nasheuer, H.-P., and Syvaoja, J. E. (2000). *Trends Biochem. Sci.* **25,** 143.
Jankowsky, E., Gross, C. H., Shuman, S., and Pyle, A. M. (2002). *Nature* **403,** 447.
Jezewska, M. J., and Bujalowski, W. (1996). *Biochemistry* **35,** 2117.
Jezewska, M. J., Kim, U.-S., and Bujalowski, W. (1996a). *Biochemistry* **35,** 2129.
Jezewska, M. J., Kim, U.-S., and Bujalowski, W. (1996b). *Biochemistry* **35,** 2129.
Jezewska, M. J., Rajendran, S., and Bujalowski, W. (1998a). *Biochemistry* **37,** 3116.
Jezewska, M. J., Rajendran, S., and Bujalowski, W. (1998b). *J. Biol. Chem.* **273,** 9058.
Jezewska, M. J., Rajendran, S., and Bujalowski, W. (1998c). *Biochemistry* **37,** 3116.
Jezewska, M. J., Rajendran, S., and Bujalowski, W. (1998d). *J. Mol. Biol.* **284,** 1113.
Jezewska, M., Rajendran, S., and Bujalowski, W. (2000a). *J. Biol. Chem.* **275,** 27865.
Jezewska, M., Rajendran, S., and Bujalowski, W. (2000b). *Biochemistry* **39,** 10454.
Jezewska, M. J., Rajendran, S., and Bujalowski, W. (2001). *Biochemistry* **40,** 3295.
Jezewska, M. J., Galletto, R., and Bujalowski, W. (2003). *Biochemistry* **42,** 5955.
Jezewska, M. J., Galletto, R., and Bujalowski, W. (2004). *J. Mol. Biol.* **343,** 115.
Jezewska, M. J., Marcinowicz, A., Lucius, A. L., and Bujalowski, W. (2006). *J. Mol. Biol.* **356,** 121.
Jones, J. M., and Nakai, H. (1999). *J. Mol. Biol.* **289,** 503.
Jones, J. M., and Nakai, H. (2001). *J. Mol. Biol.* **312,** 935.
Joyce, K. M., and Benkovic, S. J. (2004). *Biochemistry* **43,** 14317.
Kaplan, D. L., and O'Donnell, M. (2002). *Mol. Cell* **10,** 647.
Ledneva, R. K., Razjivin, A. P., Kost, A. A., and Bogdanov, A. A. (1978). *Nucleic Acids Res.* **5,** 4225.
Lee, M. S., and Marians, K. J. (1987). *Proc. Natl. Acad. Sci. USA* **84,** 8345.
Lohman, T. M., and Bjornson, K. P. (1996). *Annu. Rev. Biochem.* **65,** 169.
Lohman, T. M., and Bujalowski, W. (1991). *Methods Enzymol.* **208,** 258.
Lohman, T.M, and Ferrari, M. E. (1994). *Annu. Rev. Biochem.* **63,** 527.
Lucius, A. L., Vindigni, A., Gregorian, R., Ali, J. A., Taylor, A. F., Smith, G. R., and Lohman, T. M. (2002). *J. Mol. Biol.* **324,** 409.
Lucius, A. L., Jezewska, M. J., and Bujalowski, W. (2006). *Biochemistry* **45,** 7217.
Maciejewski, M., Shin, R., Pan, B., Marintchev, A., Denninger, A., Mullen, M. A., Chen, K., Gryk, M. R., and Mullen, G. P. (2001). *Nat. Struct. Biol.* **8,** 936.
Marcinowicz, A., Jezewska, M. J., Bujalowski, P. J., and Bujalowski, W. (2007). *Biochemistry* **46,** 13279.
Marians, K. J. (1999). *Prog. Nucleic Acid Res. Mol. Biol.* **63,** 39.
McGhee, J. D., and von Hippel, P. H. (1974). *J. Mol. Biol.* **86,** 469.
Menetski, J. P., and Kowalczykowski, S. C. (1985). *J. Mol. Biol.* **181,** 281.
Morales, J. C., and Kool, E. T. (2000). *Biochemistry* **39,** 12979.
Nanduri, B., Byrd, A. K., Eoff, R. L., Tackett, A. J., and Raney, K. D. (2002). *Proc. Natl. Acad. Sci. USA* **99,** 14722.
Nurse, P., DiGate, R. J., Zavitz, K. H., and Marians, K. J. (1990). *Proc. Natl. Acad. Sci. USA* **87,** 4615.
Oliveros, M., Yanez, R. R., Salas, M. L., Slas, J., Vinuela, E., and Blanco, L. (1997). *J. Biol. Chem.* **272,** 30899.
Rajendran, S., Jezewska, M. J., and Bujalowski, W. (1998). *J. Biol. Chem.* **273,** 31021.
Rajendran, S., Jezewska, M. J., and Bujalowski, W. (2001). *J. Mol. Biol.* **308,** 477.
Sangler, S. J., and Marians, K. J. (2000). *J. Bacteriol.* **182,** 9.

Showwalter, A. K., Byeon, I-J., Su, M- I., and Tsai, M- D. (2003). *Nat. Struct. Biol.* **8,** 942.
Tolman, G. L., Barrio, J. R., and Leonard, N. J. (1974). *Biochemistry* **13,** 4869.
von Hippel, P. H., and Delagoutte, E. (2002). *Q. Rev. Biophys.* **35,** 431.
von Hippel, P. H., and Delagoutte, E. (2003). *Q. Rev. Biophys.* **36,** 1.
West, S. C. (1996). *Cell* **86,** 177.
Yanez, R. J., Rodriguez, J. M., Nogal, M. L., Yuste, L., Enriquez, C., Rodriguez, J. F., and Vinuela, E. (1995). *Virology* **208,** 249.

CHAPTER FOURTEEN

Equilibrium and Kinetic Approaches for Studying Oligomeric Protein Folding

Lisa M. Gloss

Contents

Abstract

This chapter describes the approaches and considerations necessary for extension of current protein folding methods to the equilibrium and kinetic reactions of oligomeric proteins, using dimers as the primary example. Spectroscopic and transport methods to monitor folding and unfolding transitions are summarized. The data collection and analyses to determine protein stability and kinetic folding mechanisms are discussed in the context of the additional dimension of complexity that arises in higher order folding processes, compared to first order monomeric proteins. As a case study to illustrate the data analysis process, equilibrium, and kinetic data are presented for SmtB, a homodimeric DNA-binding protein from *Synechococcus* PCC7942.

School of Molecular Biosciences, Washington State University, Pullman, Washington, USA

Methods in Enzymology, Volume 466
ISSN 0076-6879, DOI: 10.1016/S0076-6879(09)66014-6

1. Introduction

The seminal studies of Christian B. Anfinsen and colleagues on the refolding of RNase A (Sela *et al.*, 1957; White, 1961) established the paradigm that a protein's native, functional, three-dimensional structure is encoded in the linear sequence of amino acids dictated by its genetic sequence. Ever since, scientists have been trying to decipher this second half of the genetic code, how the directions for the native, folded structure are written in the primary structure; even more elusive has been the deciphering of instructions that dictate the pathway to reach the native state. Examples of early protein folding model systems include nucleases, lysozymes, dihydrofolate reductase, apomyoglobin, cytochrome *c*, and bovine pancreatic trypsin inhibitor (for review, Kim and Baldwin, 1990; Matthews, 1993). Studies of these proteins provided insights into the kinetic mechanisms of protein folding at the level of secondary and tertiary structural units and then residue-specific information. These model systems shared common features: a modest size (<20 kDa), monomeric structure and a lack of dissociable cofactors. The key characteristic is that the refolding reactions of these model systems were first order processes.

In the 1990s, advances in recombinant protein expression and computational methods spawned a reductionist revolution, leading to a focus on smaller, single domain proteins. Some model proteins, such as chymotrypsin inhibitor 2 (CI2), ubiquitin, cold shock protein B (CspB) were full-length polypeptides, but others represented isolated domains of larger proteins (e.g., SH3 domains, monomeric λ repressor and coiled coil leucine zipper peptides). These reductionist folding models, still mostly monomeric, often fold by two-state kinetic processes (Jackson, 1998). This simplicity made their folding mechanisms amenable to synergistic dissection by experimental and computational approaches, providing novel, atomic-level insights into the sequence determinants of protein stability and folding mechanisms (Onuchic and Wolynes, 2004; Schaeffer *et al.*, 2008; Vendruscolo and Dobson, 2005).

However, monomeric systems cannot elucidate how the development of quaternary structure and interchain interactions are coordinated with intrachain formation of secondary and tertiary structure. For example, partially folded monomers likely have exposed hydrophobic surfaces, and these surfaces must oligomerize correctly, while avoiding inappropriate intermolecular interactions that could lead to aggregation. Understanding the stability and folding mechanisms of oligomeric proteins, wherein the protein folding code is written on multiple polypeptide chains, is a necessary evolution of protein folding studies with high biological significance. It has been estimated that ~80% of *Escherichia coli* proteins are oligomeric, and the percentage may be even higher in eukaryotic organisms (Goodsell and Olson, 2000). There is a substantial body of literature on the equilibrium stability of oligomeric proteins, predominantly homodimers, but there are significantly fewer studies describing the kinetic folding mechanisms of oligomers. The key complexity that distinguishes the study of oligomers from monomers is the impact of a change in molecularity as the protein folding landscape is traversed. These higher order reactions add a new dimension to the perturbations necessary to characterize the folding reaction, namely measurements must be conducted as a function of protein concentration. A comprehensive review of the folding mechanisms of dimeric proteins is provided by Rumfeldt *et al.* (2008).

Two recent methodological papers (Street *et al.*, 2008; Walters *et al.*, 2009) provide extensive detail on the design and execution of general protein folding studies. This chapter will expand on this framework and discuss experiments and analyses for studying oligomeric protein folding, with an emphasis on obligatory dimers, that is sub-μM dissociation constants for the native state. However, many of the concepts are also applicable to transient and/or weaker protein–protein interactions which are prevalent in biological processes. To illustrate various points in Sections 3 and 4, data are shown for folding studies on SmtB, a homodimeric, Zn^{2+} responsive, helix-turn-helix DNA-binding transcription repressor from *Synechococcus* PCC7942 with 120 residues per monomer (for review of structure and function, see Robinson *et al.*, 2001).

2. Methods to Monitor Folding and Association

2.1. Perturbation methods

Perturbation methods are the fundamental experimental approach for the study of protein folding, that is altering some environmental condition (temperature, pH, denaturant concentration) to alter the relative populations of the native and unfolded species. The practical reason for this approach is readily apparent when one considers the relative concentrations of native and unfolded protein ([N] and [U], respectively) under nondenaturing conditions. Typical monomer stabilities exhibit a range of free energies of unfolding from 3 to 15 kcal mol^{-1} (Jackson, 1998). At room temperature, given the relationship:

$$\Delta G^\circ = -RT\cdot\ln(K_{eq}); \quad \text{where } K_{eq} = \frac{[U]}{[N]} \tag{14.1}$$

the ratio of [U]:[N] for a protein stability of 3 kcal mol^{-1} is ~1:150. To determine K_{eq}, [N] and [U] must be detected simultaneously. Similarly, to monitor folding kinetics, there must be a significant change in signal as U is converted to N. Few biophysical or biochemical methods are capable of distinguishing one unfolded protein per 150 molecules, or per 10^{11} molecules at the other extreme of ~15 kcal mol^{-1}. A noteworthy exception is the native-state hydrogen/deuterium exchange of amide protons (for review, Englander, 2006; Krishna *et al.*, 2004). Thus, it is necessary to use perturbations to systematically vary [N] and [U] over a range of conditions where both species can be quantified. Of course, one wants to know the stability at a standard condition comparable between proteins, namely the absence of denaturant, $\Delta G^\circ(H_2O)$. A well-established linear extrapolation (Santoro and Bolen, 1992; Street *et al.*, 2008) is used to relate K_{eq} values in the presence of denaturant to the stability in the absence of denaturant:

$$\Delta G^\circ = \Delta G^\circ(H_2O) - m[\mathrm{Denaturant}] \tag{14.2}$$

where the slope of the linear dependence is the *m* value, describing the steepness of the transition. In two-state equilibrium reactions, there is a strong correlation between the *m* value and the change in solvent-accessible surface area between the folded and unfolded species (Myers *et al.*, 1995).

For an oligomeric protein, the $[x\mathrm{U}]:[\mathrm{N}_x]$ ratio (where x is the number of subunits in the folded oligomer) is protein concentration dependent. The range of $\Delta G^\circ(H_2O)$ values reported for dimeric proteins is 8–40 kcal mol^{-1} at a standard state of 1 *M* dimer (Rumfeldt *et al.*, 2008). Note: this is a different standard state than for the monomeric stabilities given above; see Section 3.3. To emphasize the need for perturbation methods, consider the relative concentrations of native dimer, N_2, and unfolded monomers, 2U, at protein concentrations typically used in the methods discussed in Sections 2.2 and 2.3. The apparent fraction of unfolded monomers, F_{app}, can be defined as:

$$F_{app} = \frac{[U]}{[P_{total}]}; \quad \text{where } P_{total} = [U] + 2[N_2] \tag{14.3}$$

For a dimeric system, the F_{app} value is related to the equilibrium constant (and thus the $\Delta G^\circ(H_2O)$ value) as follows:

$$F_{app} = \frac{-K_{eq} + \sqrt{K_{eq}^2 + 8K_{eq}P_{total}}}{4P_{total}} \tag{14.4}$$

For a relatively unstable dimer, $\Delta G^\circ(H_2O) = 8$ kcal $\mathrm{mol}^{\circ 1}$ (reported for some leucine zipper peptides of ~30 residues per monomer), the expected F_{app} values are 0.4 and 0.17 at 1 and 10 μM dimer, respectively. For a moderately stable dimer, $\Delta G^\circ(H_2O) = 12$ kcal mol^{-1} (typical of dimers with 50 or more residues per monomer), the F_{app} values are 0.02 and 0.006 at 1 and 10 μM dimer, respectively.

The perturbation method must induce reversible unfolding, that is the unfolded protein will refold when the perturbation is removed. Common reversible perturbations include temperature, pressure, pH, and chemical denaturants. Thermal denaturation, especially with differential scanning calorimetry, is particularly useful because it allows the dissection of the free energy of unfolding into enthalpic and entropic contributions (for review, Privalov and Dragan, 2007). One concern with thermal denaturation is the tendency of proteins to aggregate irreversibly at high temperature as well as undergo chemical modifications, such as deamidation or Cys oxidation. Another complication of thermal denaturation is the difficulty of complementing equilibrium and kinetic experiments. Protein unfolding kinetics can be measured with rapid heating in T-jump experiments, but extracting folding kinetic information is complicated.

Many proteins unfold at acidic or alkaline pH extremes, and pH-jump experiments are excellent for inducing kinetic folding and unfolding reactions. However, determination of ΔG° values typically employ an orthogonal perturbation, such as chemical or thermal denaturation at various pH values (for example, Eftink *et al.*, 1994; Trevino *et al.*, 2005).

This chapter will focus on unfolding induced by chemical denaturants, specifically urea and guanidinium chloride (GdmCl). Chemical denaturation has three notable advantages. First, the perturbations are amenable to both equilibrium and kinetic studies. Second, denaturation occurs at room temperature, so a variety of methods can be used to detect the signal change between native and unfolded species. The transport methods described in Section 2.3 are not easily adapted to the elevated temperatures needed to unfold many proteins. Third, because denaturants are effective at solubilizing partially or fully unfolded proteins, denaturant-induced unfolding is often more reversible than thermal or pH, where aggregation can be problematic. Detailed descriptions of the preparation and use of urea and GdmCl in protein folding studies are provided in (Shaw *et al.*, 2009; Street *et al.*, 2008; Walters *et al.*, 2009).

The choice between urea and GdmCl often seems arbitrary, although there are certain aspects that should be considered. GdmCl is a more potent denaturant, usually ~two-fold more effective than urea (Pace, 1986); this feature can be important for studying particularly stable proteins. Conversely, urea has the advantage of being nonionic, and thus does not alter the ionic strength which impacts electrostatic interactions. Finally, some proteins unfold reversibly in one denaturant but not the other. As an example, the H2A–H2B histone heterodimer unfolds with high reversibility in urea and GdmCl, whereas the $(H3–H4)_2$ histone heterotetramer only exhibits high reversibility in GdmCl (Banks and Gloss, 2003; Gloss and Placek, 2002).

2.2. Spectroscopy

Many types of spectroscopy can assay various levels of structure in a protein. These can be grouped based on the use of intrinsic probes, such as nuclear magnetic resonance spectroscopy (NMR), Fourier-transformed infrared spectroscopy (FTIR), Raman and resonance Raman spectroscopy, or the use of extrinsic probes, such as electron paramagnetic resonance spectroscopy (EPR) or Förster resonance energy transfer (FRET). This chapter will focus on two spectroscopic methods that are the most accessible to the general audience, namely circular dichroism (CD) and various fluorescence (FL)-based approaches. Furthermore, both CD and FL methods are applicable to equilibrium and kinetic studies of protein folding.

2.2.1. Circular dichroism

Only the basics regarding application of CD to protein folding will be summarized here. A comprehensive description of the biophysics and application of CD methods to macromolecules is provided by Fasman (1996) and Berova *et al.* (2000). The fundamental premise of CD is that chromophores in a chiral environment, such as the peptide bond in secondary structure or aromatic residues in tertiary structure, will preferentially absorb either left or right circularly polarized light. Far-UV CD (230–190 nm) monitors secondary structure in polypeptides. Proteins with predominantly α-helical structure will exhibit spectra with minima at 222 and 208 nm. The signature of β-sheet structure is a minimum around 218 nm. Spectral features below 200 nm are quite useful when employing deconvolution software to estimate the relative amount of α-helix, β-sheet, and β-turns. However, in practice, the signal-to-noise ratio becomes problematic below 200 nm even in the best conditions (avoiding buffer components with significant absorbance in the far-UV range). In the presence of urea or GdmCl, it is difficult to obtain useful data below ~215 nm. Near-UV CD (270–300 nm) can report on the chiral packing around aromatic side chains, namely Phe, Tyr, and Trp residues. The spectra of folded proteins often exhibit several maxima and minima between 270 and 300 nm, while the spectra of unfolded proteins are relatively featureless (e.g., see Figure 2B of Gloss and Matthews, 1997). The difficulty with near-UV CD spectra is the relatively low signal. The mean residue ellipticity (MRE) difference between an entirely α-helical folded and unfolded polypeptide is ~35,000 deg cm^2 $dmol^{-1}$ at 222 nm (Berova *et al.*, 2000). In contrast, the change in ellipticity in the near-UV CD region is at least one to two orders of magnitude lower (Woody and Dunker, 1996). Therefore, far-UV CD experiments are carried out in cuvettes with 0.1–1 cm path lengths, but near-UV CD experiments may require path lengths up to 10 cm and/or much higher protein concentrations.

2.2.2. Fluorescence

There are several methods for applying fluorescence (FL) in protein chemistry; comprehensive reviews are provided by Royer (1995) and Lakowicz (2006). This chapter will focus on FL intensity and anisotropy of intrinsic fluorophores and FRET, which generally requires extrinsic fluorophores. The intrinsic fluorophores, Tyr and Trp, report on tertiary structure as they are typically buried in a protein's hydrophobic core; in oligomers, these residues are often buried in intermonomer interfaces, and thus are sensitive to quaternary structure as well.

To monitor FL intensity, Tyr is excited at 280 nm, with an emission maximum (λ_{max}) between 305 and 310 nm. Upon unfolding, Tyr FL intensity generally decreases as the residue moves from a nonpolar environment to the more polar aqueous solvent, with little change in λ_{max}. The more fluorescent Trp can also be excited at 280 nm, the absorbance maxima. However, for proteins that contain both Tyr and Trp, excitation at 290–295 nm is often chosen to specifically excite Trp, as Tyr has very little absorbance at these longer wavelengths. Note: it is easier to relate spectral effects to specific conformational changes if one is monitoring a unique fluorophore, and there are generally many fewer Trp residues than Tyr residues in a protein. On the other hand, multiple FL probes can provide a more global picture of the unfolding of tertiary and quaternary structure.

When buried in a hydrophobic environment, Trp emission λ_{max} is between 320 and 340 nm; upon unfolding and exposure to the aqueous solvent, the λ_{max} exhibits a significant red shift to 355–360 nm. It can be difficult to predict, *a priori,* how the FL intensity of the unfolded protein will change relative to that of the folded species because Trp residues are sensitive to quenching by polar moieties buried in the protein's hydrophobic core, such as backbone carbonyls or amides. The maximum FL intensities of N and U can be similar; the λ_{max} is simply red shifted in the unfolded state. In other cases, the FL intensity of U, at ~360 nm, may be less than or greater than the FL intensity of the native state at its λ_{max}. Spectra reflecting these two scenarios are shown in Fig. 14.1 for two mutants that introduce unique Trp residues into the hydrophobic dimer interfaces of closely related homodimeric archaeal histones, hMfB and hPyA1 (Stump and Gloss, 2008b). For hPyA1-M32W, FL intensity at the relevant λ_{max} decreases upon unfolding. Conversely, the FL intensity at the λ_{max} is greater for unfolded hMfB-M35W than for the native dimer, suggesting that there is a specific interaction in N_2 that quenches the FL of the M35W Trp residue. This example highlights the structural subtleties that complicate interpretation of changes in FL intensity and demonstrate the need to monitor Trp FL intensities at multiple wavelengths, as discussed further in Section 3.2.

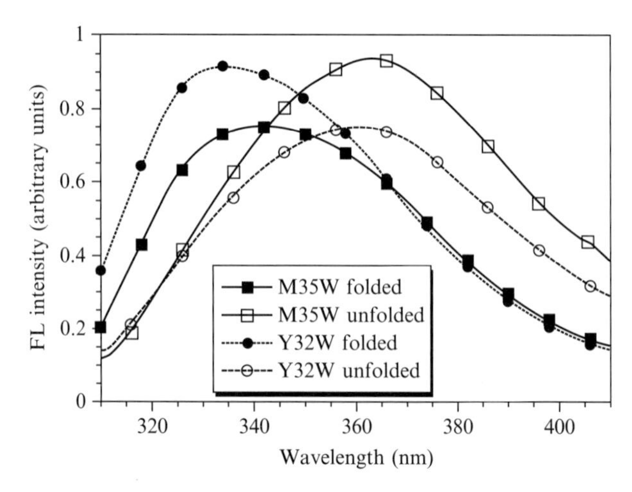

Figure 14.1 Representative fluorescence spectra to monitor unfolding of the homodimeric archaeal histones hMfB from *Methanothermus fervidus* and hPyA1 from *Pyrococcus strain* GB-3A (described by Stump and Gloss, 2008b). Data are shown for unique Trp residues in hMfB-M35W (dark, solid lines with square symbols) and hPyA1-Y32W (gray, dotted lines with circle symbols). Spectra for the folded and GdmCl-denatured species are indicated by solid and open symbols, respectively.

The participation of Tyr and Trp residues in tertiary and quaternary structure can also be interrogated by FL anisotropy, which monitors the rotational freedom of the fluorophore, using plane polarized excitation light and assessing the degree to which this polarization is scrambled (by rotation of the fluorophore) prior to fluorescence emission. If the fluorophore is relatively rigid within the macromolecule (e.g., a buried Tyr or Trp) so that it reorients slowly relative to the lifetime of the excited state, then the polarization of the incoming excitation light is largely maintained in the fluorescence emission. Conversely, if the fluorophore has relatively free rotation, such as a Tyr or Trp side chain in an unfolded random coil or on the surface of a protein with few tertiary structural constraints, then the polarization of the excitation light is scrambled during the lifetime of the excited state, and anisotropy is significantly diminished in the emitted fluorescence. The anisotropy of a buried, rigid fluorophore will reflect the rate of tumbling of the protein, which has an additional benefit for oligomeric protein folding studies. As a protein assembles from a monomeric species into a higher order oligomer that tumbles more slowly, there will be a significant increase in the anisotropic emission of the fluorophore.

2.2.3. Förster resonance energy transfer

FRET is a powerful tool in various fields for the study of protein–protein and protein–nucleic acid interactions, using with either purified components or intact cells. This fluorescence-based method reports on the distance between two fluorophores, a donor and an acceptor, and is suitable for both equilibrium and kinetic approaches. Many protein folding FRET studies have characterized the intramolecular collapse and condensation of a single polypeptide (for review, Haas, 2005; Schuler and Eaton, 2008). However, FRET is particularly well suited for studying oligomeric protein folding and the assembly of macromolecular complexes, where the donor and acceptor moieties reside on different molecules, and thus FRET is sensitive to association/dissociation reactions.

The FRET donor and acceptor must be appropriately matched, and the most important criteria are: (1) the fluorescence quantum yield of the donor; (2) the spectral overlap between the emission of the donor and the absorption of the acceptor; and (3) the distance between the donor and acceptor, usually denoted r. Donor–acceptor (D–A) pairs have a characteristic Förster radius, R_o, the distance at which the efficiency of energy transfer is 50%. The R_o values range from 10 to 60 Å for D–A pairs commonly used in biochemistry (Lakowicz, 2006). FRET efficiency, E, depends on distance to the sixth power:

$$E = \frac{R_o^6}{R_o^6 + r^6} \quad (14.5)$$

making FRET a sensitive "molecular ruler." For an oligomeric system, the dynamic signal range for FRET measurements is quite high as r essentially goes to infinity upon dissociation.

FRET can be monitored by the fluorescence emission of the donor, which is quenched upon association, and by fluorescence emission of the acceptor, which should increase concomitantly. Ideally, both signals should yield similar responses. To accurately determine FRET efficiency and thus distance, the fluorescence of the donor and acceptor in isolation must be insensitive to the conformational change or association reaction, that is if one monitors the folding of the protein modified with only the donor or with only the acceptor, there should be no fluorescence change of the individual fluorophore. This point is less critical if the goal is simply to engineer a spectral probe that is sensitive to association and monitoring relative distances is sufficient.

A FRET system generally requires the introduction of extrinsic fluorophores, usually incorporated on the surface of the protein to minimally perturb structure and stability. There are no natural amino acids which can serve as D–A pairs, although Trp can serve as a donor for several extrinsic acceptors. In general, it is desirable to have a unique donor if one wants to monitor the distances between two specific sites on a protein; conveniently, Trp is a relatively rare amino acid, not present in many small proteins or present as only a few nonessential residues in larger proteins. Unfortunately, many native Trp residues are excluded from solvent in the native state, so that their FL is sensitive to folding/unfolding reactions, a consideration noted above. There are three common strategies for introducing a D–A FRET pair:

1. Introduce a unique Trp by site-directed mutagenesis to serve as a donor. This may require additional mutagenesis to replace native Trp residues.
2. Introduce a Cys residue that can be uniquely modified to introduce an acceptor and/or a donor. Like Trp, Cys is a less common amino acid, and if present, native Cys residues are often partially excluded from solvent. Thus, a Cys introduced onto a protein's surface may be more reactive to alkylation or disulfide exchange than wild-type Cys residues. Alternatively, there are numerous examples of proteins where mutagenesis of multiple Cys residues yielded a Cys-free protein with wild-type like properties. For an intermolecular FRET system, a unique Cys residue can be incorporated on each subunit and modified to introduce both the donor and acceptor. For hetero-oligomers, this should be straightforward; for homo-oligomers, protein engineering will be more complicated to maximize specific labeling and/or minimize intramolecular FRET. To complement a Trp donor, Cys modification with IAEDANS (5-((((2-iodoacetyl)amino)ethyl)-amino)-naphthalene-1-sulfonic acid) generates a well-characterized D–A pair (e.g., Dalbey *et al.*, 1983; Soumillion *et al.*, 1998; Hoch *et al.*, 2007). Alternative acceptors for a Trp donor are described by Lakowicz (2006), with most pairs exhibiting R_o values of 22–30 Å. Common extrinsic donors and acceptors include dansyl, coumarin, and fluorescein derivatives as well as the Cy3–Cy5 pair and a host of Alexa Fluor derivatives. Compared to a Trp/Cys-AEDANS pair, the advantages of these D–A pairs are a higher donor quantum yield and a longer R_o, 40–60 Å. The disadvantage is that these fluorophores are bulkier and more hydrophobic, with a tendency to sequester themselves from solvent, leading to potentially destabilizing interactions with the protein or oligomer interface.
3. Introduce an unnatural amino acid as a donor and/or acceptor using the engineered tRNA/amino-acyl-tRNA synthetase approach developed by the Schultz lab (Wang *et al.*, 2006). A recent report describes the incorporation of L-4-cyano-Phe and 4-ethynyl-Phe as donors for a Trp acceptor (Miyake-Stoner *et al.*, 2009).

Regardless of the strategy and chosen D–A pair, one must always be cognizant of the potential effects of the FRET pair on protein stability and folding pathway, even if the position of the fluorophores is designed to be largely solvent accessible. Key controls are measuring the FL anisotropy in the folded and unfolded species of the donor and acceptor (in constructs with only one of the fluorophores) as well as determining the equilibrium stability of the modified protein.

2.3. Monitoring radius of gyration and molecular weight

The radius of gyration of a polypeptide changes upon unfolding, and this can provide an important probe of the equilibrium unfolding reaction. This is particularly true for oligomeric proteins to monitor both subunit dissociation (a decrease in radius of gyration) as well as expansion of the monomers upon unfolding to a random coil. Determining the radius of gyration, at various denaturant concentrations, can be instrumental in characterizing partially folded intermediates populated at equilibrium. Two of the most common methods are analytical ultracentrifugation (AUC) and size-exclusion chromatography (SEC).

The past decade has seen a resurgence of the utility of AUC as new instrumentation has been developed. AUC can provide information on the hydrodynamic size and shape of macromolecules as well as the thermodynamics of association. There are two modes of AUC experiments, sedimentation velocity (SV) and sedimentation equilibrium (SE). In SV experiments, the rate of sedimentation is measured, providing a sedimentation coefficient, the *s* value, which reflects the particle's molecular mass, partial specific volume and shape. Dynamic oligomers and macromolecular assemblies that undergo association and dissociation on the time scale of sedimentation can be characterized by SE experiments, and the equilibrium distribution of the particles can be analyzed to determine dissociation constants. A more comprehensive description of the fundamentals and applications of AUC is provided by Lebowitz *et al.* (2002) and Howlett *et al.* (2006).

While not as versatile and informative as AUC, high-resolution HPLC SEC is a fairly accessible method for characterizing the hydrodynamic properties of oligomers. As with centrifugation, a protein's movement is dictated by its molecular mass, shape, and degree of compactness, that is a fully folded polypeptide will elute slower than an expanded, random coil, unfolded state. The utility of SEC is enhanced if the column is coupled to an in-line static light scattering detector, such as those made by Wyatt Technology Corporation (Santa Barbara, CA). The light scattering profile provides an independent measurement of the molecular mass of the eluting species that can be compared to the mass predicted based on elution time relative to protein standards. An example is the study of the rapid dissociation and refolding of the FIS (Factor for Inversion Stimulation) homodimer from *E. coli*. The SEC elution time of FIS is protein concentration dependent, with a systematic increase in elution time as dimer concentrations decrease (Hobart *et al.*, 2002a; Topping *et al.*, 2004). Initially, this observation was interpreted as evidence that FIS dissociated to monomers between 1 and 10 μM (Hobart *et al.*, 2002a). However, static light scattering confirmed that the eluting species remained dimeric at these concentrations (Topping *et al.*, 2004). In fact, the elution profile reflected the population-weighted average of the predominant native dimer and a transiently populated monomeric species. The fraction of this transient monomer increased as the protein concentration applied to the column was decreased, resulting in the protein concentration dependent change in the elution profile. This example also highlights how SEC can provide information on kinetic aspects of the association/dissociation of oligomers; further details are provided by Shalongo *et al.* (1993a,b).

3. Equilibrium Studies

3.1. General approach

The goal of equilibrium unfolding studies is two-fold: to determine the stability of a protein and identify any stably populated partially folded species. The general approach is to measure some signal (like those covered in Section 2) over a range of denaturant concentrations where the folded and unfolded species are populated. Details of the practical aspects of setting up an equilibrium experiment are provided by Street *et al.* (2008) and Walters *et al.* (2009); only a brief summary is presented here.

Individual samples can be made for each denaturant concentration, or automated titrators can be interfaced with many spectrometers. Making individual samples is more laborious, often requires more protein and may result in more scatter in the dataset. Automated titrations are set up with folded protein in a cuvette (with stirring), and small aliquots are repeatedly removed and replaced with aliquots from a stock of protein unfolded in a high concentration of denaturant, resulting in incremental increases in the denaturant concentration. Of course, refolding titrations can also be done, starting with unfolded protein in the cuvette and titrated with aliquots from a stock of folded protein. Titrations are certainly less laborious, but there are some caveats to consider. First, there is autocorrelation between the data points since each sample is derived from the previous. Thus, one must be aware of the potential for systematic errors. It is important to determine the final denaturant concentration by measuring the refractive index (Street *et al.*, 2008). Second, titrations become impractical when long incubations are required for the sample to reach equilibrium. Even for titrations that only take a few hours, there can be complications from evaporation and "creeping" of dried denaturant up the sides of the cuvette.

Whether preparing a titration or individual samples, one must ensure that the sample has reached equilibrium before the measurement is made. The simple approach is to equilibrate for a certain period of time, measure the signal, wait an additional period of time and remeasure the signal to verify that no further reaction has occurred. The complication is that folding and unfolding rates are dependent on the final denaturant concentration, being slowest in the middle of the transition region where both folded and unfolded species are populated. Therefore, it is advisable to start by doing rudimentary manual mixing kinetics to monitor the approach to equilibrium at low, moderate, and high denaturant concentrations to estimate a relaxation time, τ, which can vary by an order of magnitude over the range of denaturant concentrations. A general guideline is to use an equilibration time that corresponds to 4–5 x τ. When performing a titration, relatively short incubation times should be used in the folded and unfolded baselines (defined below) and longer incubation times used in the transition region.

3.2. Specific considerations for data collection

The experimental design for an equilibrium study needs to address three general folding questions as well as a fourth specific to oligomeric systems: (1) Is the unfolding reaction reversible, without hysteresis? (2) Can multiple detection methods be employed, and for spectroscopic methods, should data be collected at one or multiple wavelengths? (3) Are there sufficient data points to define the native and unfolded baselines as well as the transition region? and (4) Does the dataset span a sufficiently broad range of protein concentrations?

To obtain meaningful thermodynamic parameters, the equilibrium unfolding reaction must be highly reversible. Furthermore, transitions initiated from the folded and unfolded species should be coincident, that is there is no hysteresis. Therefore, to demonstrate reversibility, it is not sufficient to show that the denaturant-unfolded protein adopts the same native structure (as detected by the methods described in Sections 2.2 and 2.3) after dilution to a single low denaturant concentration in the folded baseline region. The equilibrium signals for each assay method should be determined at several denaturant concentrations spanning the transition region; the signals should be compared for samples made by: (1) adding denaturant to folded protein; and (2) diluting denaturant-unfolded protein stocks. An example of the overlay of unfolding and refolding titrations monitored by far-UV CD is shown in Fig. 14.2A.

To determine if the equilibrium reaction is two-state or a multistate process with populated equilibrium intermediates, it is important to monitor unfolding by different methods. The methods described in Section 2 probe different aspects of protein structure. For example, far-UV CD and FL intensities provide complementary information on secondary structure and tertiary and quaternary structures, respectively. A second consideration is what wavelength(s) to monitor for spectroscopic assays. It may be sufficient to collect data at a single wavelength that represents a maximal difference in signal between the folded and unfolded states. However, there are advantages to collecting full spectra at each denaturant

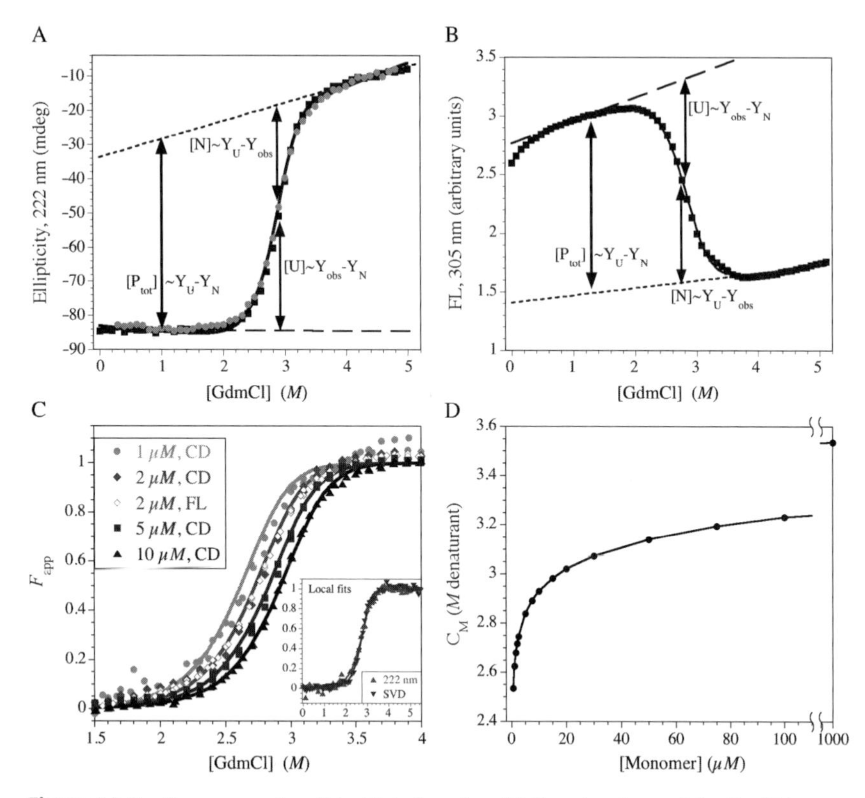

Figure 14.2 Representative GdmCl-induced unfolding titrations of the SmtB homodimer. In all panels, the solid lines represent the global fit of the equilibrium dataset of 11 titrations including far-UV CD and Tyr FL data spanning 1–10 μM monomer (SmtB has no Trp residues). Native baselines are indicated by dashed lines; unfolded baselines, by dotted lines. (A) Unfolding (dark squares) and refolding (gray circles) titrations at 5 μM monomer monitored by far-UV CD at 222 nm. (B) Unfolding titration at 2 μM monomer monitored by Tyr FL at 305 nm. (C) F_{app} curves for representative CD and FL data at monomer concentrations of 1 μM (circles), 2 μM (diamonds), 5 μM (squares), and 10 μM (triangles). Data were collected between 0 and 6 *M* GdmCl, but for clarity, only the transition region is shown. Inset: F_{app} curves for local fits of the 1 μM monomer data. (D) Dependence of C_M on monomer concentration. Calculated using Eq. (14.7) and the fitted parameters for SmtB: $\Delta G°(H_2O) = 20$ kcal mol^{-1}; $m = 4.5$ kcal mol^{-1} M^{-1}. Conditions: 200 m*M* KCl, 1 m*M* EDTA, 20 m*M* KPi, pH 7.5, 25 °C.

concentration and analyzing by singular value decomposition (SVD). The details of the application of SVD to spectroscopic data are described by Henry and Hofrichter (1992). In brief, SVD analyzes the matrix of spectra versus denaturant concentration for systematic changes and extracts a basis spectrum and associated amplitude vector for each populated species. An example application is the FL spectra shown in Fig. 14.1, where there are significant differences in Trp FL around 330 nm (N_2 disappearing) and 360 nm (2U appearing). SVD can distinguish if these differences are concerted changes, reflecting a single component, consistent with two-state equilibrium unfolding, or if there are multiple components, indicating the population of an intermediate species with spectral properties distinct from both N_2 and 2U. SVD can also be valuable for analyzing spectroscopic signals where there is a change in intensity, but no significant change in the wavelength of the maxima or minima, such as Tyr FL or far-UV CD. Because SVD focuses on systematic changes, it can minimize scatter in the data across a titration or between individually prepared samples.

For most spectral probes, there is a denaturant concentration dependence of the signal, even at denaturant concentrations above and below the transition region where only one species is present, that is N_2 in the native baseline region and 2U in the unfolded baseline region. These baseline regions are indicated on the CD and FL data in Fig. 14.2A and B. To determine the concentrations of the native and unfolded species in the transition region, this denaturant dependence of their spectral signal must be known. This requirement is explained by the definition of F_{app} cast in terms of the observed signal, Y_{obs}, and the difference between the signals for the native species, Y_N, and the unfolded species, Y_U, as a function of denaturant:

$$F_{app} = \frac{(Y_{obs} - Y_N)}{(Y_U - Y_N)} \quad (14.6)$$

This relationship is depicted graphically in Fig. 14.2A and B. The accuracy of determining [N_2] and [2U], and thus the equilibrium constants, is clearly dependent on the selection of the appropriate baselines. Thus, it is important to have enough data points, covering a range of denaturant concentrations, to unambiguously define the baselines. This generally is not an issue on the unfolded side, except for very stable proteins. For unstable proteins, having a sufficiently long native baseline can be a problem. Decreased temperature or addition of cosolutes can be used to stabilize the protein and lengthen the native baseline. Trimethylamine-*N*-oxide (TMAO) is commonly used for this purpose; for an example, see Topping and Gloss (2004). For reviews regarding TMAO and other protein stabilizers, see Timasheff (1998) and Bolen (2001).

Nonlinear baselines are also a complication, for example the native baseline for the Tyr FL of SmtB below ~0.5 *M* GdmCl (Fig. 14.2B). Curvature in the native baseline may reflect an additional unfolding step. However, Tyr FL (in the absence of any Trp residues) tends to exhibit an ionic strength effect. Like SmtB, the eukaryotic histones H2A–H2B and H3–H4 exhibit curvature in the native baseline, but only in GdmCl, not urea (Banks and Gloss, 2003). Furthermore, this nonlinear change in FL can be mimicked by titrating in similar concentrations of KCl. Therefore, in analyzing the SmtB data, the FL native baseline was defined using the linear region between 0.6 and 1.5 *M* GdmCl.

Figure 14.2C shows F_{app} curves for SmtB unfolding at different monomer concentrations. For an oligomeric protein, the unfolding transition moves to higher denaturant concentrations with higher monomer concentrations because of mass action. This has the effect of extending the native baseline, which as noted above, can be important for unstable proteins. Measuring unfolding transitions at multiple oligomer concentrations is essential for determination of stability and whether the equilibrium process is two-state. For a dimer, the protein concentration dependence of the C_M, the midpoint of the transition where $F_{app} = 0.5$, can be calculated from Eq. (14.7):

$$C_M = \frac{\Delta G^\circ(H_2O) + RT\ln[P_{total}]}{m} \tag{14.7}$$

Figure 14.2D shows the C_M values expected for the SmtB dimer between 0.5 μM and 1 mM monomer, given the fitted $\Delta G^\circ(H_2O)$ and m values described in Section 3.3. The magnitude of $\Delta G^\circ(H_2O)$ shifts the line vertically, but for a given stability, the m value has little effect on the shape of the curve. The C_M increases significantly at low monomer concentrations, but approaches an asymptote. For example, the difference in C_M for 1 and 5 μM monomer is 0.2 *M*, but to increase the C_M an additional 0.2 *M*, a monomer concentration of 25 μM is required. This example should illustrate the importance of performing titrations across a range of monomer concentrations, with a focus on the lower concentration range.

3.3. Data analysis for two-state unfolding

The basic function to describe the equilibrium unfolding data is the linear extrapolation of Eq. (14.2). The general outline for deriving an analytical expression relating the observed signals, Y_{obs}, to the F_{app} values, K_{eq}, $\Delta G^\circ(H_2O)$ and m values is provided by Street *et al.* (2008). Specific derivations are described elsewhere for dimers (Rumfeldt *et al.*, 2008; Walters *et al.*, 2009), trimers (Backmann *et al.*, 1998), and tetramers (Fairman *et al.*, 1995; Mateu and Fersht, 1998). To fit the unfolding transitions, seven parameters are necessary. The monomer concentration represents one fixed parameter (see Eq. (14.4)); the six adjustable parameters are the slopes and intercepts for the native and unfolded baselines plus the $\Delta G^\circ(H_2O)$ and m values.

There are two approaches to nonlinear least squares fitting for a series of titrations, such as that in Fig. 14.2C: locally, where each titration is fit in isolation to yield its own six adjustable parameters; or globally, where all the titrations are fit simultaneously to yield global $\Delta G^\circ(H_2O)$ and m values as well as local parameters for the baseline slopes and intercepts. If the data are scaled by protein concentration (such as converting CD ellipticity in mdeg to mean residue ellipticity or molar ellipticity), then it is possible to link the parameters for the baselines between different titrations. This semiglobal definition of the baselines can be particularly useful in fitting titrations with limited native baselines.

Global fitting provides a more rigorous analysis of the data for evaluating two features that are consistent with a two-state equilibrium mechanism: do different methods measure coincident transitions, and do the $\Delta G°(H_2O)$ and m values correctly describe the observed protein concentration dependence? However, it is generally informative to do local fits (particularly to select appropriate starting guesses for the global fitting) and compare the average and standard deviation for the $\Delta G°(H_2O)$ and m values as well as the reduced χ^2 value to those obtained for the global fits. For example, for 11 titrations monitoring far-UV CD and Tyr FL at monomer concentrations from 1 to 10 μM SmtB monomer, the average $\Delta G°(H_2O)$ and m values from local fits were 19.3 ± 1.4 kcal mol^{-1} and 4.2 ± 0.5 kcal mol^{-1} M^{-1}, respectively. These averages are in good agreement with the globally fitted values of 19.9 ± 1.0 kcal mol^{-1} and 4.5 ± 0.3 kcal mol^{-1} M^{-1}, consistent with the graphical impression from Fig. 14.2 that the lines for the global fits adequately describe all of the data. For comparison, the inset of Fig. 14.2C shows the overlay of F_{app} curves obtained from local fits for the 1 μM CD data. While the local fits better describe the individual titrations, the improvement is not significant when one considers the increase in the number of fitted parameters used in the local fits.

One disadvantage that has limited the use of global fitting approaches is the accessibility of appropriate software. However, there are a number of options. Commercial packages include SAS (SAS Institute, Cary, NC), Pro-K from Applied Photophysics (Surrey, England) and the free software R statistics package (http://www.r-project.org/). There are also programs developed in-house by academic researchers such as Fitsim/Kinsim (Dang and Frieden, 1997) and Savuka (http://www.osmanbilsel.net/software/savuka).

For monomers, the free energy for unfolding is protein concentration independent, so a standard state of 1 M monomer is easy to understand. Furthermore, the fitted $\Delta G°(H_2O)$ value should be the same whether unfolding data were collected at 1 μM or 100 μM monomer. In contrast, the $\Delta G°(H_2O)$ values reported for most oligomers are at a standard state of 1 M oligomer. These different standard states complicate direct comparison of stability for monomeric and oligomeric systems. When one compares the range of stabilities reported for monomers versus dimers as in Section 2.1, dimers seem much more stable, which is true at 1 M concentrations. It would be more meaningful to compare the stability of different oligomeric states at more experimentally and physiologically relevant conditions.

There are two approaches to this issue in the literature. A rather simplistic approach that has been used by the author (for example, Banks and Gloss, 2003; Stump and Gloss, 2008a) is to report not only the $\Delta G^{\circ}(H_2O)$ for 1 M dimer but also to calculate the $\Delta G(H_2O)$ at experimental conditions, such as 10 μM monomer, using the following equation:

$$\Delta G(H_2O) = \Delta G^{\circ}(H_2O) + RT\ln[P_{total}] \qquad (14.8)$$

An alternative approach, generalized for any type of oligomer, is to define a $\Delta G_{eff}(H_2O)$:

$$\Delta G_{eff}(H_2O) = \frac{\Delta G^{\circ}(H_2O)}{n} + \frac{RT}{n}\ln(n[P_{total}]^{n-1}) \qquad (14.9)$$

where n is the number of subunits in the oligomer (Park and Marqusee, 2004). The first term corresponds to the ΔG° per monomer for the specific protein; however, the second term contains no protein-specific information, but is rather a protein-concentration dependent scaling function for the type of oligomer. The standard state free energy of unfolding of SmtB, 19.9 kcal mol^{-1}, converts to a $\Delta G_{10\ \mu M}(H_2O)$ of 13.1 kcal mol^{-1} or $\Delta G_{1\ \mu M}(H_2O)$ of 11.7 kcal mol^{-1}. These values are still at the high end of reported monomer stabilities, consistent with the relatively high C_M of $\sim$3 M GdmCl. The $\Delta G_{eff}(H_2O)$ at 10 μM monomer is 6.7 kcal mol^{-1}, and a monomer with comparable $\Delta G^{\circ}(H_2O)$ and m values would have a C_M of 1.5 M GdmCl. One might conclude that ΔG_{eff} is more useful for comparing oligomers of different subunit composition, but $\Delta G_{x\ \mu M}$ determined from Eq. (14.8) may be of greater functional utility when comparing stabilities under experimental conditions.

3.4. Identifying equilibrium intermediates

What type of data would indicate that the equilibrium mechanism is not two-state, and that there may be populated equilibrium intermediates? The classic test for a two-state equilibrium is the coincidence of F_{app} curves from probes that monitor different aspects of structure. The overlap of F_{app} curves for far-UV CD and Tyr FL at 2 μM monomer in Fig. 14.2C are an example. The stringency of the test is increased if one probe monitors global structure, such as far-UV CD, and another probe monitors a specific region of the structure, such as the FL of a single Trp or a specific FRET pair (Street *et al.*, 2008).

For oligomeric systems, the quality of the global fit over a range of monomer concentrations is another criterion for two-state equilibrium. Monomeric unfolding transitions are quite symmetric, but the transitions for oligomeric proteins are somewhat asymmetric, with a shallower departure from the native baseline (evident in the FL data in Fig. 14.2B and the

F_{app} plot of Fig. 14.2C). With increasing protein concentrations, the transitions move to higher denaturant concentrations, but the shape should stay the same, that is the *m* value is protein concentration independent. Mutants of the FIS homodimer are an example of an intermediate species revealed by changes in the apparent ΔG and *m* values as a function of protein concentration (Hobart *et al.*, 2002b).

For oligomers that unfold via equilibrium intermediates, it is essential to determine the molecularity of the intermediate species to define the appropriate mechanism. For example, in dimers, there are proteins that populate either dimeric or monomeric intermediates. Many tetrameric proteins are organized structurally as dimers of dimers, and unfold with dimeric equilibrium intermediates. Determining the protein concentration dependence and/or methods such as AUC and SEC (Section 2.3) can be used to deduce the molecularity of the intermediates. For dimeric systems, detailed discussion, simulation of unfolding transitions and the relevant equations are covered by Rumfeldt *et al.* (2008) and Walters *et al.* (2009).

4. Kinetic Studies

4.1. General approach

The goal of kinetic studies is to monitor the time-dependent change of a signal after shifting a protein rapidly from conditions favoring one state to those favoring a different state, for example from unfolded species in 6 *M* urea or GdmCl to the native state by rapid dilution into buffer without denaturant. Stopped-flow (SF) methods, often coupled to CD or FL detection, are commonly employed to initiate folding. The dead times of SF mixing are 1–10 ms, limited by the time it takes for the solution to move from the mixer to the observation cuvette. Even with ms resolution, significant reactions can occur before the first observable time point, often called burst-phase (BP) reactions (discussed in Section 4.2.1). Faster mixing, with dead times of $\sim$100 μs, can be accomplished with continuous flow instruments, again usually coupled to spectroscopic detection (for example, Maki *et al.*, 2007; Wu *et al.*, 2008). Stopped and continuous flow mixing are necessary for rapid reactions, but both methods are subject to solution drift and remixing at longer time periods. Therefore, manual mixing methods, with standard cuvettes in a steady-state spectrophotometer, are recommended for reactions on timescales greater than 60 s. A detailed description of experimental design and SF instrument set up is provided by Walters *et al.* (2009). The remainder of this section will focus on the rationale for the systematic variation of experimental conditions, namely denaturant and protein concentrations.

For both unfolding and refolding kinetics, it is necessary to "jump" to different final denaturant concentrations. The basic rationale is because determining the dependence of the observed rates on the denaturant concentration allows extrapolation to the absence of denaturant, akin to the extrapolation made in equilibrium studies. The benefit of having rates extrapolated to the absence of denaturant is discussed further in Section 4.2.2. The need for extrapolation is obvious for unfolding kinetics, because unfolding rates are only measurable at high denaturant concentrations. However, the limits of dilution also necessitate extrapolation for folding kinetics. The highest ratio for effective mixing is 1:10 in most SF instruments, though the viscosity and density of solutions with high denaturant concentrations often necessitate smaller ratios, like 1:6, to achieve reproducible, stable mixing. Consider refolding of SmtB; the unfolded protein stock must be in at least 4 *M* GdmCl, so the lowest possible denaturant concentration for refolding would be 0.4 *M* GdmCl. This is in the folded baseline region, but still a long way from the absence of denaturant.

On a practical note, one must consider the protein concentration dependence of the unfolding transition for oligomers when choosing the denaturant concentration for the unfolded stock. An unfolded protein solution of 100 μM monomer is needed to achieve a 1:10 dilution to final folding conditions of 10 μM monomer. The difference between 10 and 100 μM monomer shifts the C_M (and the approach to the unfolded baseline) by $\sim$0.3 *M* GdmCl for SmtB. For dimers with a lower *m* value, the shift in C_M will be even greater (Eq. (14.7)). Therefore, the initial denaturant concentration in a refolding experiment should be sufficient to fully unfold the protein at the monomer concentrations of the starting conditions (e.g., 100 μM), not the final conditions (e.g., 10 μM).

A second rationale for varying the final concentration of denaturant in folding and unfolding kinetic studies is because the denaturant dependence of the rates provides parameters that describe the nature of the transition state(s), such as the $m^{\ddagger}$ value, analogous to the equilibrium *m* value. This will be discussed further in Section 4.2.2.

A third rationale to vary denaturant concentrations is to measure folding and unfolding rates at overlapping denaturant concentrations to assess microscopic reversibility. In the transition region, the observed rates for the folding and unfolding of a given species should converge to similar values, providing a strong indication that the refolding kinetic phase is the reverse reaction of unfolding phase, initiated from the native state. Thus, one can conclude that the kinetic folding reaction leads to the native oligomer. If the denaturant dependence of a kinetic folding step does not converge with an unfolding phase, it is likely that this folding reaction does not lead directly to the native state. Such information is important for constructing the appropriate mechanism to describe the folding pathway for proteins that fold via kinetic intermediates.

In general, the final conditions dictate the observed rate of a folding or unfolding reaction. However, varying the initial conditions can elucidate the relative populations of species present in the ensemble from which the kinetic reaction is initiated. Consider a native-state ensemble that contains N_2 and $N_2^\star$, with N_2 being slightly more stable. The two native species may not be sufficiently different in stability or spectroscopic properties to be resolvable in equilibrium studies; however, they unfold by different pathways (i.e., over different transition states). As the initial denaturant concentration is varied from 0 M to higher concentrations in the native baseline, the fraction of the population will shift toward the more stable N_2. If unfolding proceeds by two observable kinetic phases, one for N_2 and another for $N_2^\star$, the amplitude for the $N_2^\star$-associated phase, which is proportional to the initial concentration of $N_2^\star$, will diminish with increasing initial denaturant concentration. An example of this application is described by Banks and Gloss (2003) for the preequilibrium between the eukaryotic histone $(H3–H4)_2$ tetramer and the dissociated H3–H4 heterodimers. Alternatively, if unfolding rates for N_2 and $N_2^\star$ do not differ sufficiently to be resolved into different kinetic phases, the observed rate will reflect the fractional population of the two species, $f(N_2)$ and $f(N_2^\star)$, and their individual rates, such that $k_{obs} = \{f(N_2) \times k(N_2)\} + \{f(N_2^\star) \times k(N_2^\star)\}$. As the fraction of $N_2^\star$ is altered by changing the initial conditions, k_{obs} will change accordingly.

For an oligomeric system, a protein concentration dependent kinetic refolding phase is expected. To extract an accurate rate of association, kinetic responses must be determined for a range of protein concentrations; this is particularly important in the application of global fitting methods which are very helpful in determining the most reasonable folding mechanism. Another complexity is that the folding mechanisms of many oligomeric proteins involve first order and higher order reactions. For example, there are dimers that fold via a monomeric intermediate prior to dimerization ($2U \rightarrow 2I \rightarrow N_2$) or fold to a dimeric intermediate that then folds to the native dimer ($2U \rightarrow I_2 \rightarrow N_2$) (reviewed by Rumfeldt *et al.*, 2008). Protein concentration dependent studies are required to distinguish between these possible mechanisms. Ideally, the range of concentrations studied should be similar to that employed in equilibrium studies as described above.

4.2. Data analysis

A complete analysis of any kinetic process must consider two components: (1) the signal change (amplitude) associated with the observed kinetic phase, particularly with respect to the signal change expected from equilibrium studies; and (2) the time dependence of the observed signal change. Folding and unfolding data for SmtB are presented as a case study to demonstrate various scenarios for analyzing kinetic data. Representative refolding SF-CD and SF-FL kinetic traces are shown in Fig. 14.3. The GdmCl dependence of the signal changes and rates are presented in Fig. 14.4.

4.2.1. Burst-phase reactions

The initial and final signals associated with a kinetic process, illustrated for the SF-CD trace in Fig. 14.3A, can be very informative. The CD ellipticity at the end of the refolding reaction matches that expected from the

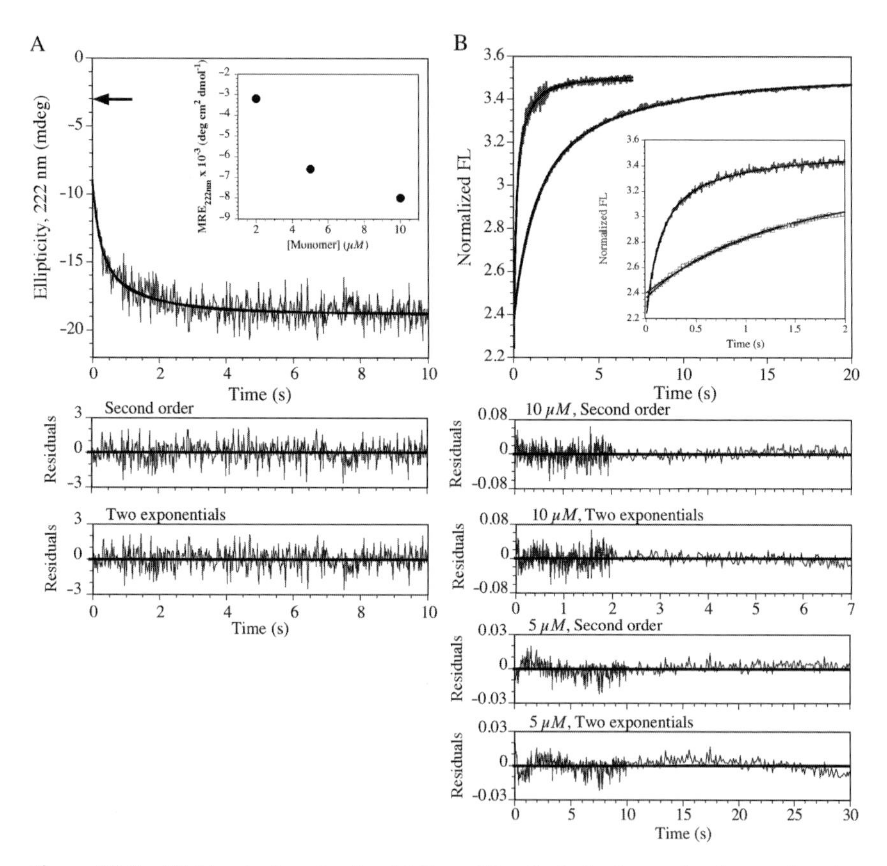

Figure 14.3 SF kinetic data for the folding of SmtB. In both panels, the overlapping dark lines represent local fits to a single second-order equation (Eq. (14.12)) and two first-order exponentials (Eq. (14.11)). Residuals for the different fits are shown. (A) SF-CD refolding kinetic trace to final concentration of 10 μM monomer. The arrow denotes the expected ellipticity of the unfolded monomers. Inset: SF-CD burst-phase signal, normalized to mean residue ellipticity, as a function of monomer concentration. (B) SF-FL refolding kinetic traces to final concentrations of 5 and 10 μM monomer (bottom and top traces, respectively). Inset: Expansion of the initial portion of the kinetic traces. Conditions: Final GdmCl concentration of 2.2 M, 200 mM KCl, 1 mM EDTA, 20 mM KPi, pH 7.5, 25 °C.

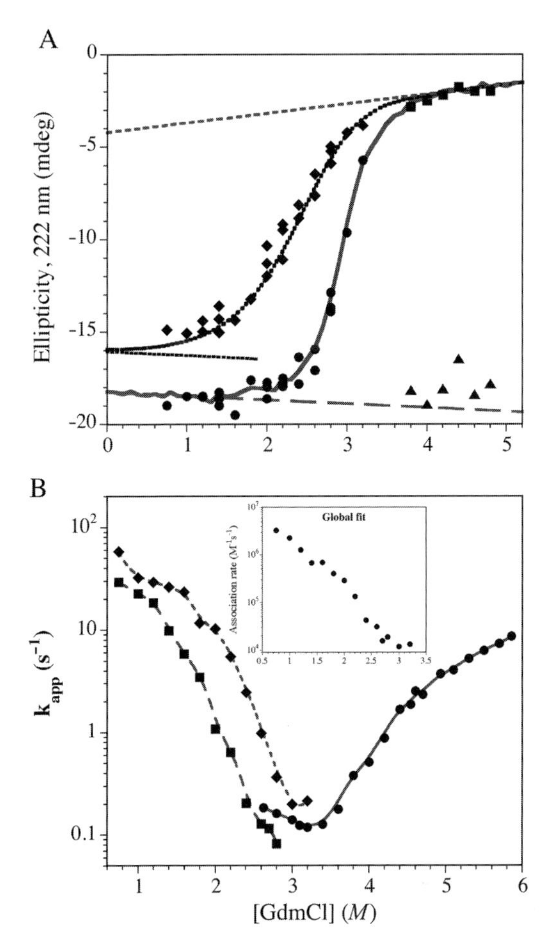

Figure 14.4 GdmCl dependence of SmtB folding and unfolding data. (A) SF-CD ellipticity as a function of final [GdmCl] at 10 μM monomer. Initial ellipticity represented by diamonds, folding; triangles, unfolding; ellipticity at $t = \infty$ shown as circles, folding; squares, unfolding. The corresponding equilibrium transition and baselines are shown in gray lines. The dark, dotted lines represent the fit of the burst-phase amplitude to a two-state dimeric unfolding model with the Y_U baseline and the slope of the Y_N baseline fixed to that observed for the equilibrium data. (B) GdmCl dependence of the folding and unfolding rates. The average k_{app} values were determined from local fits of SF-CD and SF-FL folding traces at 5 and 10 μM monomer (squares and diamonds, respectively) to Eq. (14.12). Inset: second order rates of association determined from global fits of multiple folding traces (both SF-CD and SF-FL as well as 5 and 10 μM monomer) at a given GdmCl concentration. Buffer conditions are described in the legend of Fig. 14.2.

equilibrium transitions. However, the initial signal observed in folding kinetics, -9 mdeg, is much greater than that expected from the unfolding baseline, -4 mdeg (Fig. 14.4A). This difference indicates that some burst-phase reaction, characterized by an increase in secondary structure, has occurred during the SF mixing dead time. Studying the dependence of this BP signal on protein and denaturant concentrations is required to identify the nature of the BP reaction product.

The inset of Fig. 14.3A shows the CD BP signal, normalized to mean residue ellipticity (MRE), for different final monomer concentrations at one GdmCl concentration. If the partially folded BP species were monomeric, the MRE should not change significantly with increased protein concentration. The marked increase in MRE indicates that an association reaction has occurred within the 5 ms SF dead time. A lower limit for the BP association reaction can be estimated, assuming a maximum relaxation time for a reaction that is largely complete in 5 ms, $\tau \sim 2$ ms, the lowest monomer concentration employed (2 μM), and the following equation for a homodimer:

$$\frac{1}{\tau} = k_{\text{assoc}}[\text{Monomer}_{\text{total}}] \tag{14.10}$$

The estimated association rate is $\geq 10^8\ M^{-1}\text{s}^{-1}$, approaching the diffusion limit. Similarly rapid, efficient BP association reactions have been reported for other dimeric systems (Gloss and Matthews, 1998; Placek and Gloss, 2005; Topping *et al.*, 2004).

The denaturant dependencies of the BP and final ellipticity are shown in Fig. 14.4A. The data points represent values determined from fits of kinetic traces to the appropriate equation (Section 4.2.2 below). The use of fitted values, rather than the initial and final observed signals, allows extrapolation forward to $t = \infty$, as well as back to $t = 0$ to account for any signal change from the observed kinetic phase that may have occurred in the 5 ms dead time.

The equilibrium transition can be reconstructed from the final signals of the refolding and unfolding reactions (circles and squares in Fig. 14.4A). For CD, direct comparison of the kinetic signals and the equilibrium data (gray lines in Fig. 14.4A) is relatively straightforward, simply requiring normalization of the data for differences in cuvette path length and buffer baseline subtraction. This comparison is useful for determining the appropriate baselines (gray dotted and dashed lines in Fig 14.4A). For example, this comparison shows that the initial signal from fits of the unfolding kinetics (triangles, Fig. 14.4A) agrees reasonably well with the extrapolated native baseline (gray dashed line); this agreement suggests that there is little to no BP amplitude associated with the unfolding reaction.

For FL, comparison of kinetic and equilibrium data may be complicated for two reasons. Firstly, unlike equilibrium measurements made on steady-state fluorometers, many SF-FL measurements are made without an emission monochromator. Instead, total FL emission is detected with a PMT (photomultiplier tube) positioned at 90° to the excitation beam with an appropriate cutoff filter, that is a piece of quartz or glass that blocks all light below a certain wavelength, such as a 295 nm cutoff filter for Tyr FL centered around 305 nm. Secondly, it is necessary to apply different voltages to the PMT when collecting data using different path lengths and slit widths, and these instrument settings make normalization nonlinear. However, if the range of final denaturant concentrations is extensive enough, with sufficient overlap in the transition region between folding and unfolding kinetics, the FL equilibrium transition and baselines can still be reconstructed from the kinetic data in order to identify BP reactions.

The denaturant dependence of the BP signal describes the stability of the species formed by that reaction, much like the final signal in a refolding reaction describes the equilibrium stability of the native state. A relatively linear denaturant dependence indicates that folding during the dead time is a noncooperative process, suggesting a nonspecific collapse, rather than development of specific structure. Conversely, if the BP signal displays a sigmoidal, cooperative denaturant dependence, it may be possible to fit the data to a two-state unfolding model such as described above for equilibrium transitions (indicated by the black, dotted line in Fig. 14.4A). Of course, the baselines may not be well determined, so it is reasonable to fix the parameters for the unfolded baseline to that observed for the equilibrium transition or determined from the final signal observed in unfolding kinetics, as well as link the slope of the native baseline between the BP species and final equilibrium data. Ideally, the BP signal should be determined at multiple protein concentrations across the range of denaturant concentrations, and then this matrix of data is used in global analysis to fit a $\Delta G°(H_2O)$ and *m* value for the BP ensemble (for examples, see Gloss and Matthews, 1998; Placek and Gloss, 2005; Topping *et al.*, 2004). One complication in analyzing data for an oligomeric BP species is determining the appropriate monomer concentration to use (e.g., Eq. (14.4)). If the BP intermediate is obligatory, that is all molecules pass through this state as they traverse the folding landscape, then the monomer concentration is simply the final concentration after dilution. However, if the BP ensemble is transiently populated by only a fraction of the folding molecules, it may be unclear what the appropriate monomer concentration to use in fitting the data is. One means to distinguish between obligatory and non-obligatory intermediates is a double-jump experiment as described in Section 4.3.

4.2.2. Analyzing kinetic traces

The first step in analyzing the data for kinetic reactions observable after the mixing dead time is to determine the appropriate number of kinetic phases and the order of the associated reaction(s). First order processes can be fit to a sum of n exponentials:

$$Y(t) = Y_\infty + \sum_{i=1}^{n} \Delta Y_i \exp(-k_i t) \tag{14.11}$$

where $Y(t)$ is the signal observed at time t, Y_∞ is the signal as the reaction reaches equilibrium, and ΔY_i is the amplitude associated with the apparent rate, k_i. Note, this generic equation describes both increasing and decreasing exponential processes, with negative and positive signs for ΔY, respectively. The kinetic trace for a second order process is described by the following equation:

$$Y(t) = Y_\infty + \Delta Y \left(\frac{k_{app}}{1 + k_{app}} \right) \tag{14.12}$$

Given the different time dependence, there is a clear distinction between the kinetic response for a first-order reaction ($n = 1$ in Eq. (14.11)) and a second order reaction.

Distinguishing between a second order process and the sum of two or more exponentials can be ambiguous. For example, the SF-CD kinetic trace in Fig. 14.3A is described equally well by a sum of two exponentials or a single second order process. Differentiating between the quality of the fits is difficult even with the improved signal-to-noise of the SF-FL data in Fig. 14.3B. The fit to Eq. (14.12) is marginally better, most notably at the shortest and longest times; however, these are the data collection times most subject to mixing and remixing artifacts. Analysis requires data at various protein concentrations, ideally at least spanning a 10-fold range. The refolding reaction monitored by SF-FL at 10 μM monomer is clearly faster than at 5 μM (Fig. 14.3B). When kinetic traces over a range of GdmCl concentrations are fit to Eq. (14.12), the 10 μM monomer k_{app} values are consistently greater than for 5 μM (Fig. 14.4B), indicative of a second order process. However, it should be noted that a first order process coupled to a preceding second order reaction (i.e., the second step in $2U \rightarrow I_2 \rightarrow N_2$) will exhibit moderate protein concentration dependence because the final equilibrium stability of N_2 as well as the stability and thus population of I_2 is protein concentration dependent. Examples are described by Topping *et al.* (2004) and Placek and Gloss (2005).

The appropriate mechanism and order of the reaction can be further delineated from local and global fitting of the data. The k_{app} values from local fits of kinetic traces to Eq. (14.12) should increase linearly with monomer concentration; the slope of this line represents the second order rate constant for the association reaction (Eq. (14.10), where $k_{app} = 1/\tau$). Global fitting provides a rigorous examination of the consistency of the protein concentration dependence in determining the second order rate constant across the range of denaturant concentrations (inset of Fig. 14.4B).

As mentioned in Section 4.1, the denaturant dependence provides additional information about the folding and unfolding kinetics. Generally, in a "chevron" plot of observed rate as a function of final denaturant concentration, there is a log-linear dependence of the rates on final denaturant concentration under strongly folding and unfolding conditions:

$$k_{obs} = k(H_2O)\exp\left(\frac{m^{\ddagger}[\text{Denaturant}]}{RT}\right) \quad (14.13)$$

where $k(H_2O)$ is the rate in the absence of denaturant, and $m^{\ddagger}$ is the slope, which like the equilibrium m value (Eq. (14.2)) is correlated with the change in solvent accessible surface area in conversion of the ground state to the transition state. Equation (14.13) can be integrated into the global analyses of folding kinetics so that a set of SF-FL and SF-CD traces at multiple denaturant and monomer concentrations can be fit simultaneously with the $k(H_2O)$ and $m^{\ddagger}$ values linked across all data (with locally determined values for Y_∞ and ΔY (Eq. (14.12)). For first order processes coupled to an overall second order process, such as unfolding kinetics or the second step in $2U \rightarrow I_2 \rightarrow N_2$, where there is moderate (less than second order) protein concentration dependence as noted above, Eq. (14.13) can be used for global analysis of kinetic traces across the range of denaturant concentrations for a given monomer concentration.

The SmtB chevron in Fig. 14.4B is an example of nonlinearity and "roll-over" at low and high GdmCl concentrations. There are multiple explanations for roll-over, but most generally involve a change in the rate determining step such as movement of the transition state, population of an additional kinetic intermediate or a preceding step becoming partially rate-determining (for further discussion, see references in Maxwell *et al.* (2005)). The kinetic responses of many proteins also exhibit roll-over in the transition region, where the folding and unfolding rates converge; this roll-over is also apparent in the SmtB chevron around 3 *M* GdmCl. The significance of this convergence is the demonstration that the folding and unfolding phases exhibit microscopic reversibility and share a common transition state.

For converging folding and unfolding reactions, the ratio of $k_{unfold}(H_2O)$ and $k_{fold}(H_2O)$ can be used to determine the change in free energy, $\Delta G^\circ(H_2O)$, associated with the kinetic reaction. Similarly, the sum of the absolute values of $m^{\ddagger}_{unf}$ and $m^{\ddagger}_{fold}$ should reflect the difference in solvent accessible surface area between the initial and final states of the kinetic process. For systems that appear to fold by a two-state mechanism, without transient kinetic intermediates, comparison of the $\Delta G°(H_2O)$ and m values from kinetic and equilibrium experiments can test the two state hypothesis and suggest the presence of hidden intermediates.

The $m^{\ddagger}$ values can also describe the relative position of the transition state in terms of similarity to the native state. The α-value is such a parameter:

$$\alpha = \frac{m^{\ddagger}_{\text{fold}}}{m_{\text{equil}}} \tag{14.14}$$

which ranges from 0, an unfolded-like transition state, to 1, a native-like transition state. A complementary term that is also used is the β-value, where the numerator in Eq. (14.14) is $m^{\ddagger}_{\text{unf}}$.

4.3. Transient kinetic intermediates

There are four types of data that indicate that a protein does not fold by a single two-state kinetic process without transiently populated, partially folded intermediates. First, there is development of a significant burst-phase signal, especially if this signal exhibits a sigmoidal dependence on denaturant concentration (Fig. 14.4A). Second, the kinetic traces exhibit multiple phases, requiring fits to a combination of rates (first and/or second order). Third, different kinetic responses are observed by methods that probe different types of structure (such as SF-FL vs. SF-CD). Fourth, the folding rate(s) do not exhibit a log-linear dependence on the denaturant concentration (Fig 14.4B). Two questions then arise: does the mechanism involve parallel or sequential steps, and are the intermediates obligatory (i.e., do all molecules populate the kinetic intermediate while folding?). A review of various approaches to answering these questions is provided by Wallace and Matthews (2002). This chapter will focus on one method, folding double-jump experiments, that is well suited to address whether intermediates are obligatory and can also discriminate between parallel and sequential mechanisms.

The goal of a folding double-jump (DJ) experiment is to assay the formation of the native state by virtue of its slower unfolding rate compared to partially folded intermediates; the amplitude associated with the unfolding rate is proportional to the amount of native species formed. The basic assay starts with unfolded protein, and folding is initiated by jumping to low denaturant concentration, allowing folding to proceed for various delay times, then jumping to high denaturant concentrations and monitoring the unfolding reaction. The amount of native species should increase with longer delay times, and the time dependence of the increase in observed amplitude indicates the rate that leads to the N_2 species.

Example data for SmtB are shown in Fig. 14.5. Protein unfolded in 4 *M* GdmCl was diluted to 2 *M* GdmCl and allowed to refold for various delay times, followed by a second jump to 4 *M* GdmCl where unfolding was monitored by SF-FL. The unfolding rate determined from the double-jump experiment agreed well with that measured for direct unfolding from N_2 to 4 *M* GdmCl, confirming that the amplitudes measured during the DJ did reflect the formation of N_2. With longer delay times, more unfolding amplitude was observed. Importantly, the time dependence of the increasing amplitude was consistent with a second order process (indicated by the fitted line in Fig. 14.5). This argument would be stronger if additional DJ experiments were performed at different monomer concentrations.

Consider if two rates were observed in refolding kinetics, k_{fast} and k_{slow}. Do both kinetic phases lead to the native dimer in parallel reactions, or are they sequential steps in a single pathway? The rate(s) associated with the growing amplitude in the DJ assay will indicate which observed refolding rate constrains the formation of N_2. For a parallel mechanism, the kinetics for the increase in unfolding amplitude should be described by two exponentials with rates similar to k_{fast} and k_{slow}. In a sequential mechanism, with k_{slow} being the rate-determining step leading to N_2, then the increase in DJ amplitude should fit to a single kinetic phase with the rate of k_{slow}. (Note, if the reaction described by k_{fast} occurred after k_{slow}, then the fast reaction would not be observed in direct refolding experiments.) An example of this application of double-jump analysis is given by Gloss *et al.* (2001).

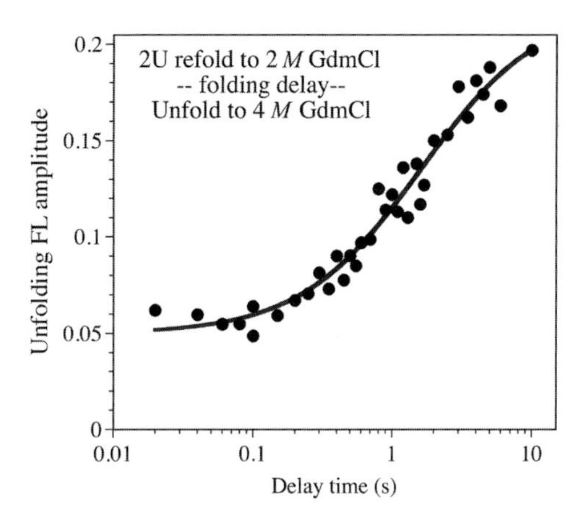

Figure 14.5 SF-FL refolding double-jump experiment for SmtB. The data points are the results from global fitting of the data at all delay times with the unfolding rate linked across all kinetic traces. The unfolding rate thus determined was 0.58 s^{-1}, in good agreement with that observed in direct unfolding jumps, 0.51 s^{-1}. The solid line is the fit of the unfolding amplitude to Eq. (14.12) for a second order reaction. Buffer conditions are described in the legend of Fig. 14.2.

Refolding double-jump experiments can also assess whether an intermediate is obligatory. For the case of SmtB, the question arises whether the burst-phase species is an obligatory intermediate (sequential) or whether there is a parallel, rapid folding track to N_2 within 5 ms. The amount of amplitude present after the shortest delay time addresses these questions. If the BP species were an obligatory intermediate, then the unfolding amplitude should disappear as the delay time approaches zero. Such a result for obligatory BP intermediates are presented by Gloss and Matthews (1998), Topping *et al.* (2004), and Placek and Gloss (2005). However, for SmtB, the limiting amplitude as the delay time approaches zero is $\sim$30% of the total amplitude expected after infinite refolding delays (Fig. 14.5). Therefore, it appears that SmtB can fold to N_2 within the SF dead time.

In conclusion, the case study here supports a parallel folding mechanism for SmtB, with a fraction of unfolded monomers folding to N_2 within 5 ms (association rate $> 10^8\ M^{-1}\ s^{-1}$), and the majority folding via a slower dimerization step on the order of $10^6\ M^{-1}\ s^{-1}$, and perhaps approaching $10^7\ M^{-1}\ s^{-1}$ in the absence of denaturant. It is unclear whether there is folding in the monomers (in less than 5 ms) before association for either of the parallel channels leading to native SmtB.

ACKNOWLEDGMENTS

Helpful discussions and critical reading of the manuscript by Paul Guyett, Traci Topping, and Andy Galbraith are gratefully acknowledged. The equilibrium and kinetic data for SmtB were collected by Traci Topping. This research was supported by an NIGMS grant (GM073787).

REFERENCES

Backmann, J., Schafer, G., Wyns, L., and Bonisch, H. (1998). Thermodynamics and kinetics of unfolding of the thermostable trimeric adenylate kinase from the archaeon Sulfolobus acidocaldarius. *J. Mol. Biol.* **284,** 817–833.

Banks, D. D., and Gloss, L. M. (2003). Equilibrium folding of the core histones: The H3–H4 tetramer is less stable than the H2A–H2B dimer. *Biochemistry* **42,** 6827–6839.

Berova, N., Nakanishi, K., and Woody, R. W. (eds.) (2000). *In* "Circular dichroism: Principles and applications". Wiley-VCH, New York.

Bolen, D. W. (2001). Protein stabilization by naturally occurring osmolytes. *Methods Mol. Biol.* **168,** 17–36.

Dalbey, R. E., Weiel, J., and Yount, R. G. (1983). Förster energy transfer measurements of thiol 1 to thiol 2 distances in myosin subfragment 1. *Biochemistry* **22,** 4696–4706.

Dang, Q., and Frieden, C. (1997). New PC versions of the kinetic-simulation and fitting programs. KINSIM and FITSIM. *Trends Biochem. Sci.* **22,** 317.

Eftink, M. R., Helton, K. J., Beavers, A., and Ramsay, G. D. (1994). The unfolding of trp aporepressor as a function of pH: Evidence for an unfolding intermediate. *Biochemistry* **33,** 10220–10228.

Englander, S. W. (2006). Hydrogen exchange and mass spectrometry: A historical perspective. *J. Am. Soc. Mass Spectrom.* **17,** 1481–1489.
Fairman, R., Chao, H. G., Mueller, L., Lavoie, T. B., Shen, L., Novotny, J., and Matsueda, G. R. (1995). Characterization of a new four-chain coiled-coil: Influence of chain length on stability. *Protein Sci.* **4,** 1457–1469.
Fasman, G. D. (ed.) (1996). *In* "Circular dichroism and the conformational analysis of biomolecules". Springer, New York.
Gloss, L. M., and Matthews, C. R. (1997). Urea and thermal equilibrium denaturation studies on the dimerization domain of *Escherichia coli Trp* repressor. *Biochemistry* **36,** 5612–5623.
Gloss, L. M., and Matthews, C. R. (1998). Mechanism of folding of the dimeric core domain of *Escherichia coli Trp* repressor: A nearly diffusion-limited reaction leads to the formation of an on-pathway dimeric intermediate. *Biochemistry* **37,** 15990–15999.
Gloss, L. M., and Placek, B. J. (2002). The effect of salts on the stability of the H2A–H2B histone dimer. *Biochemistry* **41,** 14951–14959.
Gloss, L. M., Simler, B. R., and Matthews, C. R. (2001). Rough energy landscapes in protein folding: Dimeric *E. coli* Trp repressor folds through three parallel channels. *J. Mol. Biol.* **312,** 1121–1134.
Goodsell, D. S., and Olson, A. J. (2000). Structural symmetry and protein function. *Annu. Rev. Biophys. Biomol. Struct.* **29,** 105–153.
Haas, E. (2005). The study of protein folding and dynamics by determination of intramolecular distance distributions and their fluctuations using ensemble and single-molecule FRET measurements. *Chemphyschem.* **6,** 858–870.
Henry, E. R., and Hofrichter, J. (1992). Singular value decomposition: Application of analysis of experimental data. *Methods Enzymol.* **210,** 129–192.
Hobart, S. A., Ilin, S., Moriarty, D. F., Osuna, R., and Colon, W. (2002a). Equilibrium denaturation studies of the *Escherichia coli* factor for inversion stimulation: Implications for in vivo function. *Protein Sci.* **11,** 1671–1680.
Hobart, S. A., Meinhold, D. W., Osuna, R., and Colon, W. (2002b). From two-state to three-state: The effect of the P61A mutation on the dynamics and stability of the Factor for Inversion Stimulation in an altered equilibrium denaturation mechanism. *Biochemistry* **41,** 13744–13754.
Hoch, D. A., Stratton, J. J., and Gloss, L. M. (2007). Protein-protein Förster resonance energy transfer analysis of nucleosome core particles containing H2A and H2A.Z. *J. Mol. Biol.* **371,** 971–988.
Howlett, G. J., Minton, A. P., and Rivas, G. (2006). Analytical ultracentrifugation for the study of protein association and assembly. *Curr. Opin. Chem. Biol.* **10,** 430–436.
Jackson, S. E. (1998). How do small single-domain proteins fold? *Fold Des.* **3,** R81–R91.
Kim, P. S., and Baldwin, R. L. (1990). Intermediates in the folding reactions of small proteins. *Annu. Rev. Biochem.* **59,** 631–660.
Krishna, M. M., Hoang, L., Lin, Y., and Englander, S. W. (2004). Hydrogen exchange methods to study protein folding. *Methods* **34,** 51–64.
Lakowicz, J. R. (2006). Principles of Fluorescence Spectroscopy. 3rd edn. Springer, New York.
Lebowitz, J., Lewis, M. S., and Schuck, P. (2002). Modern analytical ultracentrifugation in protein science: A tutorial review. *Protein Sci.* **11,** 2067–2079.
Maki, K., Cheng, H., Dolgikh, D. A., and Roder, H. (2007). Folding kinetics of staphylococcal nuclease studied by tryptophan engineering and rapid mixing methods. *J. Mol. Biol.* **368,** 244–255.
Mateu, M. G., and Fersht, A. R. (1998). Nine hydrophobic side chains are key determinants of the thermodynamic stability and oligomerization status of tumour suppressor p53 tetramerization domain. *EMBO J.* **17,** 2748–2758.

Matthews, C. R. (1993). Pathways of protein folding. *Annu. Rev. Biochem.* **62,** 653–683.
Maxwell, K. L., *et al.* (2005). Protein folding: Defining a "standard" set of experimental conditions and a preliminary kinetic data set of two-state proteins. *Protein Sci.* **14,** 602–616.
Miyake-Stoner, S. J., Miller, A. M., Hammill, J. T., Peeler, J. C., Hess, K. R., Mehl, R. A., and Brewer, S. H. (2009). Probing protein folding using site-specifically encoded unnatural amino acids as FRET donors with tryptophan. *Biochemistry* **48,** 5953–5962.
Myers, J. K., Pace, C. N., and Scholtz, J. M. (1995). Denaturant *m* values and heat capacity changes: Relation to changes in accessble surface areas of protein folding. *Protein Sci.* **4,** 2138–2148.
Onuchic, J. N., and Wolynes, P. G. (2004). Theory of protein folding. *Curr. Opin. Struct. Biol.* **14,** 70–75.
Pace, C. N. (1986). Determination and analysis of urea and guanidine hydrochloride denaturation curves. *Methods Enzymol.* **131,** 266–280.
Park, C., and Marqusee, S. (2004). Analysis of the stability of multimeric proteins by effective DeltaG and effective m-values. *Protein Sci.* **13,** 2553–2558.
Placek, B. J., and Gloss, L. M. (2005). Three-state kinetic folding mechanism of the H2A/H2B histone heterodimer: The N-terminal tails affect the transition state between a dimeric intermediate and the native dimer. *J. Mol. Biol.* **345,** 827–836.
Privalov, P. L., and Dragan, A. I. (2007). Microcalorimetry of biological macromolecules. *Biophys. Chem.* **126,** 16–24.
Robinson, N. J., Whitehall, S. K., and Cavet, J. S. (2001). Microbial metallothioneins. *Adv. Microb. Physiol.* **44,** 183–213.
Royer, C. A. (1995). Approaches to teaching fluorescence spectroscopy. *Biophys. J.* **68,** 1191–1195.
Rumfeldt, J. A., Galvagnion, C., Vassall, K. A., and Meiering, E. M. (2008). Conformational stability and folding mechanisms of dimeric proteins. *Prog. Biophys. Mol. Biol.* **98,** 61–84.
Santoro, M. M., and Bolen, D. W. (1992). A test of the linear extrapolation of unfolding free energy changes over an extended denaturant concentration range. *Biochemistry* **31,** 4901–4907.
Schaeffer, R. D., Fersht, A., and Daggett, V. (2008). Combining experiment and simulation in protein folding: Closing the gap for small model systems. *Curr. Opin. Struct. Biol.* **18,** 4–9.
Schuler, B., and Eaton, W. A. (2008). Protein folding studied by single-molecule FRET. *Curr. Opin. Struct. Biol.* **18,** 16–26.
Sela, M., White, F. H. Jr., and Anfinsen, C. B. (1957). Reductive cleavage of disulfide bridges in ribonuclease. *Science* **125,** 691–692.
Shalongo, W., Heid, P., and Stellwagen, E. (1993a). Kinetic analysis of the hydrodynamic transition accompanying protein folding using size exclusion chromatography. 1. Denaturant dependent baseline changes. *Biopolymers* **33,** 127–134.
Shalongo, W., Jagannadham, M., and Stellwagen, E. (1993b). Kinetic analysis of the hydrodynamic transition accompanying protein folding using size exclusion chromatography. 2. Comparison of spectral and chromatographic kinetic analyses. *Biopolymers* **33,** 135–145.
Shaw, K. L., Scholtz, J. M., Pace, C. N., and Grimsley, G. R. (2009). Determining the conformational stability of a protein using urea denaturation curves. *Methods Mol. Biol.* **490,** 41–55.
Soumillion, P., Sexton, D. J., and Benkovic, S. J. (1998). Clamp subunit dissociation dictates bacteriophage T4 DNA polymerase holoenzyme disassembly. *Biochemistry* **37,** 1819–1827.
Street, T. O., Courtemanche, N., and Barrick, D. (2008). Protein folding and stability using denaturants. *Methods Cell Biol.* **84,** 295–325.

Stump, M. R., and Gloss, L. M. (2008a). Mutational analysis of the stability of the H2A and H2B histone monomers. *J. Mol. Biol.* **384,** 1369–1383.
Stump, M. R., and Gloss, L. M. (2008b). Unique fluorophores in the dimeric archaeal histones hMfB and hPyA1 reveal the impact of nonnative structure in a monomeric kinetic intermediate. *Protein Sci.* **17,** 322–332.
Timasheff, S. N. (1998). Control of protein stability and reactions by weakly interacting cosolvents: The simplicity of the complicated. *Adv. Protein Chem.* **51,** 356–432.
Topping, T. B., and Gloss, L. M. (2004). Stability and folding mechanism of mesophilic, thermophilic and hyperthermophilic archael histones: The importance of folding intermediates. *J. Mol. Biol.* **342,** 247–260.
Topping, T. B., Hoch, D. A., and Gloss, L. M. (2004). Folding mechanism of FIS, the intertwined, dimeric factor for inversion stimulation. *J. Mol. Biol.* **335,** 1065–1081.
Trevino, S. R., Gokulan, K., Newsom, S., Thurlkill, R. L., Shaw, K. L., Mitkevich, V. A., Makarov, A. A., Sacchettini, J. C., Scholtz, J. M., and Pace, C. N. (2005). Asp79 makes a large, unfavorable contribution to the stability of RNase Sa. *J. Mol. Biol.* **354,** 967–978.
Vendruscolo, M., and Dobson, C. M. (2005). Towards complete descriptions of the free-energy landscapes of proteins. *Phil. Trans. A Math. Phys. Eng. Sci.* **363,** 433–450, discussion 450–452.
Wallace, L. A., and Matthews, C. R. (2002). Sequential vs. parallel protein-folding mechanisms: Experimental tests for complex folding reactions. *Biophys. Chem.* **101–102,** 113–131.
Walters, J., Milam, S. L., and Clark, A. C. (2009). Practical approaches to protein folding and assembly: spectroscopic strategies in thermodynamics and kinetics. *Methods Enzymol.* **455,** 1–39.
Wang, L., Xie, J., and Schultz, P. G. (2006). Expanding the genetic code. *Annu. Rev. Biophys. Biomol. Struct.* **35,** 225–249.
White, F. H. Jr. (1961). Regeneration of native secondary and tertiary structures by air oxidation of reduced ribonuclease. *J. Biol. Chem.* **236,** 1353–1360.
Woody, R. W., and Dunker, A. K. (1996). Aromatic and cystine side-chain circular dichroism in proteins. *In* "Circular Dichroism and the Conformational Analysis of Biomolecules," (G. D. Fasman, ed.), pp. 109–158. Springer, New York.
Wu, Y., Kondrashkina, E., Kayatekin, C., Matthews, C. R., and Bilsel, O. (2008). Microsecond acquisition of heterogeneous structure in the folding of a TIM barrel protein. *Proc. Natl. Acad. Sci. USA* **105,** 13367–13372.

CHAPTER FIFTEEN

Methods for Quantifying T cell Receptor Binding Affinities and Thermodynamics

Kurt H. Piepenbrink,* Brian E. Gloor,* Kathryn M. Armstrong,* *and* Brian M. Baker*,†

Contents

Abstract

$\alpha\beta$ T cell receptors (TCRs) recognize peptide antigens bound and presented by class I or class II major histocompatibility complex (MHC) proteins. Recognition of a peptide/MHC complex is required for initiation and propagation of a cellular immune response, as well as the development and maintenance of the T cell

* Department of Chemistry and Biochemistry, University of Notre Dame, Notre Dame, Indiana, USA
† Walther Cancer Research Center, University of Notre Dame, Notre Dame, Indiana, USA

Methods in Enzymology, Volume 466
ISSN 0076-6879, DOI: 10.1016/S0076-6879(09)66015-8

repertoire. Here, we discuss methods to quantify the affinities and thermodynamics of interactions between soluble ectodomains of TCRs and their peptide/MHC ligands, focusing on titration calorimetry, surface plasmon resonance, and fluorescence anisotropy. As TCRs typically bind ligand with weak-to-moderate affinities, we focus the discussion on means to enhance the accuracy and precision of low-affinity measurements. In addition to further elucidating the biology of the T cell mediated immune response, more reliable low-affinity measurements will aid with more probing studies with mutants or altered peptides that can help illuminate the physical underpinnings of how TCRs achieve their remarkable recognition properties.

1. Introduction

$\alpha\beta$ T cell receptors (TCRs) are clonotypic membrane proteins on the surface of T lymphocytes responsible for recognizing peptide antigens bound and "presented" by class I or class II major histocompatibility complex (MHC) proteins. TCR recognition of a peptide/MHC complex is necessary for the initiation and propagation of a cellular immune response, as well as the development and maintenance of the T cell repertoire. TCR recognition of peptide/MHC is also involved in pathological conditions such as autoimmunity and transplant rejection. Given the central role these interactions play in health and disease, there has been intense interest in the physical mechanisms underlying TCR recognition of peptide/MHC as well as the physical correlates with immunological function.

TCRs are similar in some respects to antibodies, consisting of four immunoglobulin domains and an antigen-binding site with multiple CDR (complementarity determining region) loops. However, TCRs and antibodies differ strikingly in the nature of the antigen that is recognized. Whereas antibodies can be elicited to molecules of nearly unlimited structural or chemical diversity, TCRs recognize a composite surface consisting of the antigenic peptide in an extended form flanked by the α-helices that form the walls of the MHC peptide-binding groove (Fig.15.1). The peptide typically contributes approximately 30% of the recognized solvent accessible surface area (Rudolph *et al.*, 2006), meaning that the MHC contributes significantly to the interface. This combined recognition of nonself (the peptide) in the context of self (the MHC) is a fundamental facet of cellular immunity. TCRs are also cross-reactive, capable of binding and initiating responses to multiple peptide/MHC antigens (Wucherpfennig *et al.*, 2007). The properties of dual recognition of a composite surface together with extensive cross-reactivity have further stimulated interest in the physical underpinnings of TCR recognition of ligand.

Figure 15.1 Structural overview of a complex between a T cell receptor and a peptide/MHC molecule. The receptor is positioned at the top in dark gray. The peptide/MHC complex is underneath in light gray, with the peptide in black rendered in stick format. The structure is that of the B7 TCR bound to the Tax_{11-19} peptide presented by the class I MHC HLA-A2 (Ding *et al.*, 1998).

Here, we discuss approaches that are useful in characterizing the affinities and thermodynamics for interactions between soluble ectodomains of TCRs and their peptide/MHC ligands, focusing primarily on isothermal titration calorimetry (ITC), surface plasmon resonance (SPR), and fluorescence anisotropy, and highlighting advantages, disadvantages, and potential pitfalls of each. As TCR–peptide/MHC interactions are typically of low-to-moderate affinity (K_D values for soluble constructs are typically in the single-to-double digit micromolar range) (Davis *et al.*, 1998), we also discuss approaches beneficial in obtaining more accurate binding data useful in the analysis of mutants for alanine scanning studies or more complex experiments such as double-mutant cycles. The latter approach might be expected to shed light on the distribution of binding energy within TCR–peptide/MHC interfaces, addressing questions such as the "basal" level of affinity

TCRs maintain towards MHC and the extent to which various loops are directed energetically towards the peptide vs. the MHC α-helices (Collins and Riddle, 2008; Garcia *et al.*, 2009).

2. Isothermal Titration Calorimetry of TCR–Peptide/MHC Interactions

2.1. Introduction to titration calorimetry

ITC is an ideal method for characterizing receptor–ligand interactions, as it does not require the addition of a potentially interfering label nor does it require attachment of a binding partner to a surface. Further, as the signal reports directly on the binding enthalpy change ($\Delta H°$) as well as the equilibrium constant and thus the free energy change ($\Delta G°$), it is possible to obtain a nearly complete thermodynamic profile ($\Delta G°$, $\Delta H°$, and $\Delta S°$) from a single experiment (ITC has recently been reviewed several times, most recently by Freyer *et al.*, 2008). A fourth thermodynamic parameter, the binding heat capacity change (ΔC_p) is available through an analysis of enthalpy changes measured as a function of temperature. Knowledge of binding thermodynamics is becoming increasingly desirable when examining receptor–ligand interactions, as it can aid in deconvoluting the forces driving binding. Thermodynamic information is particularly useful when interpreting the physical consequences of mutations or in efforts to guide the design of interactions with stronger affinity. Interactions between TCRs and peptide/MHC complexes are no exception to these questions, and since the first report in 1999 (Willcox *et al.*, 1999), ITC has been used several times to probe TCR–peptide/MHC interactions (Armstrong and Baker, 2007; Colf *et al.*, 2007; Davis-Harrison *et al.*, 2005; Jones *et al.*, 2008; Krogsgaard *et al.*, 2003; Miller *et al.*, 2007; Piepenbrink *et al.*, 2009), providing information about the forces driving individual interactions as well as general information about the role of flexibility in receptor specificity and cross-reactivity.

2.2. Concentration requirements and data quality

Although calorimetry remains one of the foremost techniques for characterizing macromolecular interactions, a downside of the technique is its relative insensitivity, requiring high concentrations and large volumes of protein. Although this has been mitigated somewhat with the introduction of new instrumentation with greater sensitivity and smaller cell volumes, it is still a problem for TCRs and peptide/MHC complexes given that these proteins require considerable effort to produce recombinantly. Although other expression systems are occasionally used, most recombinant TCR and class I peptide/MHC is produced by refolding from bacterially expressed inclusion bodies (Garboczi *et al.*, 1992, 1996b). Refolding yields can vary dramatically, particularly with TCRs, for which no single stabilizing strategy has proved consistently successful. The situation is further compounded by the weak-to-moderate affinity of most TCR–peptide/MHC interactions.

Importantly, in the absence of experimental constraints, obtaining a full set of accurate binding data from an ITC experiment normally requires the concentrations to be matched to both the affinity and the enthalpy change. The issue of the enthalpy change is obvious, as the smaller in magnitude this becomes, the weaker the ITC signal will be and correspondingly, the more difficult binding will be to detect. Weak enthalpies can be overcome by injecting more protein. Yet this brings up the more complex issue of concentrations and affinities: measuring both accurate affinities and enthalpy changes by ITC requires an optimal degree curvature to the data that also allows estimation of pre- and postsaturation baselines. Practically, this is expressed in the concept of the *c* value: the product of the binding constant ($1/K_D$) and the concentration of the protein in the cell is recommended to lie in the range of 1–1000 in order to obtain an accurate fit to the data (Wiseman *et al.*, 1989), although in practice a range of 10–500 is more realistic given the noise and experimental error present in most experiments. For TCR–peptide/MHC interactions, with a typical affinity (K_D) of 20 μM, the concentration of protein in the cell should thus lie somewhere in the range of 200 μM–10 mM. With the need for much higher concentrations in the syringe, the practical challenges with calorimetry are understandable.

The challenge most likely to be encountered in ITC of TCRs and peptide/MHC interactions is very low c (<1), stemming from the weak binding affinities and low availability of protein. In low c experiments, the enthalpy change is usually poorly defined, leading to accuracy errors in $\Delta H°$ (and thus $\Delta S°$) and precision errors in K_D. Fixing the stoichiometry (n value) can to some extent mitigate this problem (Turnbull and Daranas, 2003). However, this is likely to be a poor solution when working with TCRs or MHC proteins. Although only one peptide/MHC binds to a TCR, in practice the stoichiometry is often used as a correction factor accounting for inaccuracies in protein concentration. As the activity of refolded TCR or peptide/MHC is rarely 100%, fitting for stoichiometry is necessary to account for the level of inactive protein. In our experience, whether TCRs or peptide/MHC complexes are in the calorimeter cell, despite exhaustive purification and using fresh protein, the fitted stoichiometry parameter is often in the range of 0.8–0.9, but occasionally much lower.

In some cases, accurate equilibrium constants can be determined with ITC performed at very low c values, even if enthalpy changes cannot (Tellinghuisen, 2008; Turnbull and Daranas, 2003). Affinities determined this way still provide the opportunity to determine enthalpy changes (and thus entropy and heat capacity changes) via van't Hoff analysis. Yet such experiments require reaching full saturation in a binding experiment. If a low *c* experiment is being performed due to a low-affinity interaction, the protein requirements for achieving saturation may be prohibitive.

2.3. Linkage effects in calorimetric experiments

Although the high sample requirements can make calorimetry with T cell receptors difficult, this challenge clearly can be overcome, and as noted earlier new instrumentation with greater sensitivity and lower sample needs is becoming increasingly available. Furthermore, there is interest in engineering high affinity TCR variants (Li *et al.*, 2005; Shusta *et al.*, 2000), and an ideal use of calorimetry is to examine the thermodynamic basis for improvements in binding affinity. For those interactions that can be characterized by ITC, what other opportunities and challenges can ITC provide?

A key issue when interpreting thermodynamic data for protein binding reactions is the influence of other equilibria that are linked to binding. Calorimetry, with its ability to measure a nearly complete suite of thermodynamic parameters in a single experiment, is especially useful for examining linked equilibria (Baker and Murphy, 1996; Fisher and Singh, 1995; Horn *et al.*, 2002). The most commonly encountered form of linked equilibria is linkage of binding to changes in protonation, which occurs when the pK_a of an ionizable group changes upon binding. This can occur, for instance, when a charged group is placed in a less polar environment or becomes involved in a hydrogen bond or other electrostatic interaction (Fitch *et al.*, 2002). When the pK_a shifts, protons are released to or taken up from solution. As binding reactions are invariably carried out in a pH-buffered solution, release of a proton into or removal of a proton from solution is countered by absorption or release of a proton from the buffer. Importantly, most biological buffers have very large enthalpies of proton absorption/release (HEPES, e.g., has an ionization enthalpy of 5 kcal/mol). Thus, even a fractional pK_a change occurring upon binding will have enthalpic consequences, contributing significantly to the ΔH° that is measured in an ITC experiment. As the proton exchange does not influence the affinity, there will be a compensatory change in ΔS° (Baker and Murphy, 1996). In some cases, linkage to protonation can dramatically influence the binding heat capacity change (Guinto and Di Cera, 1996).

If the reasons for performing a calorimetric experiment are to obtain thermodynamic data for comparison with structural information, or if different interactions are to be compared (e.g., two different TCR–peptide/MHC interactions), the potential influence of linked protonation should be examined for the comparisons to be most meaningful. An example of linked protonation occurring in TCR–peptide/MHC interactions is provided by recognition of the Tax_{11-19}/HLA-A2 ligand by the

A6 TCR (Armstrong and Baker, 2007), where a pK_a change from 7.5 to 6.9 occurring upon binding imparts such an influence that the binding ΔH° and ΔS° varies by as much as fourfold in different buffers (Fig. 15.2).

The diagnostic for the influence of linked protonation is easy, if expensive from a protein requirement standpoint: perform multiple titrations at the same pH in buffers with different ionization enthalpies (e.g., HEPES, phosphate, and imidazole). Ionization enthalpies are known for all common biological buffers (Christensen *et al.*, 1976; Fukada and Takahashi, 1998). A plot of the measured ΔH° of binding versus the ionization enthalpy of the buffer will reveal the extent of linkage present; this plot will be linear according to the following equation:

$$\Delta H^\circ_{\text{obs}} = \Delta H^\circ_{\text{O}} + n\text{H}^+ \Delta H^{\text{ion}}_{\text{buff}} \tag{15.1}$$

where $\Delta H^\circ_{\text{obs}}$ is the measured binding enthalpy, $n\text{H}^+$ is the number of protons released at that pH, and $\Delta H^{\text{ion}}_{\text{buff}}$ is the ionization enthalpy of the buffer. $\Delta H^\circ_{\text{O}}$ is the intercept of the line, and can be interpreted as the protein binding enthalpy removed from the influence of buffer effects. If the

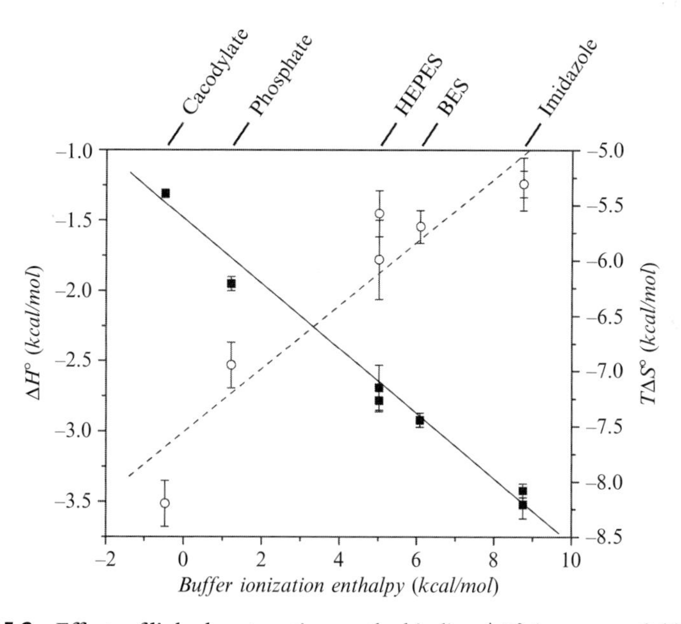

Figure 15.2 Effects of linked protonation on the binding ΔH° (squares, solid line) and ΔS° (circles, dashed line) for recognition of Tax_{11-19}/HLA-A2 by the A6 TCR. The linkage results from a pK_a shift from 7.5 to 6.9 that occurs upon binding. The slope of the binding ΔH° versus buffer ionization enthalpy yields the number of protons released at the experimental pH (6.4), and the intercept yields the buffer-independent binding enthalpy change at the experimental pH. The various buffers used are listed across the top according to ionization enthalpy. Data are from Armstrong and Baker (2007).

slope of this analysis is zero within error, then no proton linkage is present and no further decisions are necessary. However, a nonzero slope, as shown in Fig.15.2 for the binding of A6 to Tax_{11-19}/HLA-A2, requires further consideration, particularly when interpreting the intercept (ΔH°_{O}). This value, the binding enthalpy no longer influenced by the buffer, still contains a contribution from the magnitude of the pK_a shift and its enthalpic component (Baker and Murphy, 1996). If the goal of the ITC experiments are to compare binding thermodynamics with those estimated from structure using methods that do not account for the energetics of the pK_a shift, such as commonly applied empirical surface area-based algorithms (Baker and Murphy, 1998; Spolar and Record, 1994), then further measurements also varying pH are necessary to extract the "intrinsic" binding thermodynamics (Armstrong and Baker, 2007; Baker and Murphy, 1997; Barbieri and Pilch, 2006). However, for comparison of different interactions, values of ΔH°_{O} (and entropy changes determined from them) are often sufficient, provided the comparisons are performed at the same pH.

Calorimetry can also be used to characterize other forms of equilibria linked to binding, such as ion binding or release, or even conformational changes. To characterize these, the linked equilibria must have significant enthalpic consequences, as well as be present at sufficient levels to tease out during data analysis. Although such effects have yet to be explored calorimetrically in the study of TCR–peptide/MHC interactions, they remain a promising avenue of investigation, particularly with new, higher sensitivity instrumentation (Armstrong *et al.*, 2008).

2.4. Other practical concerns for titration calorimetry

Beyond the issues outlined above, what other concerns are manifest in calorimetry on TCR–peptide/MHC interactions? One issue frequently encountered is the need to perform blank (baseline) titrations of titrant into sample buffer. These are usually described as being necessary to counter the heats of dilution and mixing that always exist when a titrant is titrated into a calorimeter cell, regardless of whether binding occurs. Yet performing blank titrations requires twice the titrant, further increasing the protein cost of an ITC experiment. One way this can be avoided is if the experiment clearly exceeds saturation, the postsaturation heats can be used to determine the dilution/mixing heat. Yet as noted earlier, exceeding saturation may not always be possible in TCR binding experiments.

A more practical method for accounting for the heat of dilution/mixing is to simply include it as a baseline offset during curve fitting. Baseline offsets are included in many other analysis techniques, and there is no fundamental reason they cannot be included in the analysis of ITC data. Indeed, this is routinely done in our laboratory (e.g., Armstrong and Baker, 2007; Davis-Harrison *et al.*, 2005; Piepenbrink *et al.*, 2009). The approach requires a simple modification to the fitting function, adding an adjustable baseline parameter to the penultimate equation describing the heat that is released at each injection. Unfortunately, this is not easily achievable with the software distributed with current commercial calorimetric instrumentation, requiring the user to use other software tools for data analysis. However, if the integrated heats are available from the instrument software, writing a fitting function in any number of data analysis packages is straightforward and a good exercise for investigators wishing to gain insight into the equations describing calorimetric data and the process of nonlinear least squares analysis. Note that adding a baseline offset does add another adjustable parameter to the fit. In our experience, at very low c this can negatively impact the fitting such that fits cannot converge. However, for data with a modicum of pre- and postsaturation baselines, it provides no disadvantages to the quality of the fit. In a detailed, global analysis of multiple ITC experiments, inclusion of a baseline offset did not result in suboptimal parameter correlation or negatively impact the precision of the other fitted parameters (Armstrong and Baker, 2007).

3. Surface Plasmon Resonance Studies of TCR–Peptide/MHC Interactions

3.1. Introduction to SPR

TCR–peptide/MHC interactions are notable in that, beginning with some of the first studies in 1994, they helped popularize the use of SPR spectroscopy in characterizing macromolecular interactions (Corr *et al.*, 1994; Matsui *et al.*, 1994). The sample requirements for SPR are much lower than that of ITC, and the technique is more amenable for measuring weak-to-moderate affinities. Since the late 1990s, numerous studies have used SPR to characterize TCR–peptide/MHC interactions (we regret not having sufficient space to reference all published studies using SPR with TCRs; the number of publications as of this writing exceeds 200).

As a technique for measuring biomolecular interactions, SPR is now well established. Briefly, a binding partner is tethered (either covalently via cross-linking or noncovalently via an affinity tag) to a sensor surface. A second binding partner is flowed over the surface, and the signal increases as mass accumulates on the sensor surface due to binding. The technique can be used to obtain binding kinetics and affinities, and via van't Hoff analysis, underlying binding thermodynamics ($\Delta H°$, $\Delta S°$, and ΔC_p).

Since its introduction, many investigators have discussed SPR experimental design, data acquisition, and data analysis. The technique's versatility and ease of use naturally lends itself to wide applicability, but this same versatility and ease has led to concerns about the way in which SPR is sometimes applied. Commonly discussed issues include immobilization chemistry, flow rates, blank corrections, replicate injections, model choice for analysis, and curve fitting strategies. These concerns have been reviewed several times (e.g., Myszka, 1999; Rich and Myszka, 2008). This literature is worth consulting to ensure the acquisition of high-quality data and its proper analysis.

In addition to such concerns, TCRs and peptide/MHC complexes provide some additional challenges. One unique aspect is the noncovalent nature of the peptide/MHC complex. At the concentrations used for calorimetry, peptide dissociation is usually not an issue, as for most peptides the equilibrium will be shifted far toward the complexed state. For example, the Tax_{11-19} peptide binds HLA-A2 with an affinity near 20 nM, well below the peptide/MHC concentrations needed to characterize TCR binding (Binz *et al.*, 2003). However, in SPR, if very low peptide/MHC concentrations are used, or if the peptide/MHC is tethered to the surface, peptide dissociation could be problematic. In our laboratory, we typically couple TCRs to the sensor surface. Peptide/MHC complexes are stabilized by maintaining the sample storage chamber at low temperature, which reduces the peptide-MHC dissociation rate and thus limits accumulation of any peptide-free MHC. However, measurements with peptides that bind weakly to MHC molecules may necessitate additional safeguards, such as ensuring all samples are diluted with buffer containing a constant concentration of excess peptide (Jones *et al.*, 2008). If the peptide/MHC complex is tethered to the surface, use of model that accounts for a decaying surface may be needed (Joss *et al.*, 1998).

3.2. Use of SPR to measure low-affinity TCR–peptide/MHC interactions

Although SPR may be more amenable for low-affinity interactions than ITC, accurate measurements of low-affinity interactions will usually be difficult when protein is limiting due to the inability to generate a full titration curve. In some cases, if the low affinity is due to a slow association rate with a reasonably long dissociation rate, then measuring affinity via kinetic methods may circumvent this problem (e.g., the recognition of the Tax_{11-19}-IBA ligand by the A6 TCR occurs with an affinity near 160 μM (Gagnon *et al.*, 2006), a measurement that was obtained via kinetic rather than equilibrium methods). Yet often, low affinities arise from very rapid dissociation rates, which can preclude the use of kinetics in determining affinities by SPR.

However, SPR provides a means for greatly increasing the accuracy of binding constants that is particularly useful for low-affinity TCR–peptide/MHC interactions. The primary problem in low-affinity titration curves lies in knowing where saturation is. Very simply, if the K_D is the free ligand concentration where 50% binding occurs, how can one determine the concentration that gives 50% bound if the concentration that gives 100% bound cannot be determined or reliably estimated?

The general hyperbolic equation that is fit to in a 1:1 equilibrium binding experiment is

$$RU = RU_{max} \frac{K[L]}{1 + K[L]} \tag{15.2}$$

where RU is the instrument response, RU_{max} is the activity of the sensor surface, K is the binding equilibrium constant (equal to $1/K_D$), and [L] is the concentration of injected ligand. Typically, if 100% saturation is not reached, RU_{max} is estimated during the fitting process from the curvature of the binding response. However, how much saturation is required to accurately determine RU_{max} (and thus K_D) will be determined by many variables, including the number of data points, instrument noise, dilution errors, etc. Clearly, the greater the degree of saturation the more reliable the fit will be. Yet if a high degree of saturation cannot be reached, what options are available to ensure an accurate measurement?

One method available in SPR is to independently determine the activity of the sensor surface and fix this value in subsequent analyses. For example, if a peptide/MHC complex is on the sensor surface and a high affinity TCR variant is available, RU_{max} can be determined with the high-affinity receptor before or after experiments are performed with the weaker binding wild-type molecule. Alternatively, if peptide/MHC or TCR variants are of interest, RU_{max} can be determined independently with the wild-type molecule.

The advantages of this approach are demonstrated in Fig.15.3, which highlights results from the analysis of 100 simulated, noisy datasets that reach only 33% saturation. The data are for an interaction proceeding with a 1-mM K_D and for a sensor surface with a RU_{max} of 1000. By most standards, give noise and experimental error, reaching only 33% saturation in a binding experiment would lead to suspicions about the accuracy of the fitted K_D. This is easily demonstrated, as highlighted in Fig.15.3B. A 1-mM K_D corresponds to a dissociation ΔG° of 4092 cal/mol at 25 °C. Analysis of the 100 datasets by the traditional means in which both RU_{max} and K_D are fitted parameters leads to an average of ΔG° 4098 $\pm$ 613 cal/mol. While at first glance the agreement with the actual value (4092 cal/mol) is impressive, this agreement only demonstrates the power of repeating experiments—one is likely not to repeat a binding experiment 100 times, and the standard deviation of 613 cal/mol indicates that any one ΔG° measurement is likely to be inaccurate. The situation is much worse if one examines the precision in the experiments: the average fitting error is 1208 cal/mol, or 30% of the actual binding free energy. This result indicates that any particular fit, even if it converged on an affinity close to the actual value, will have a large

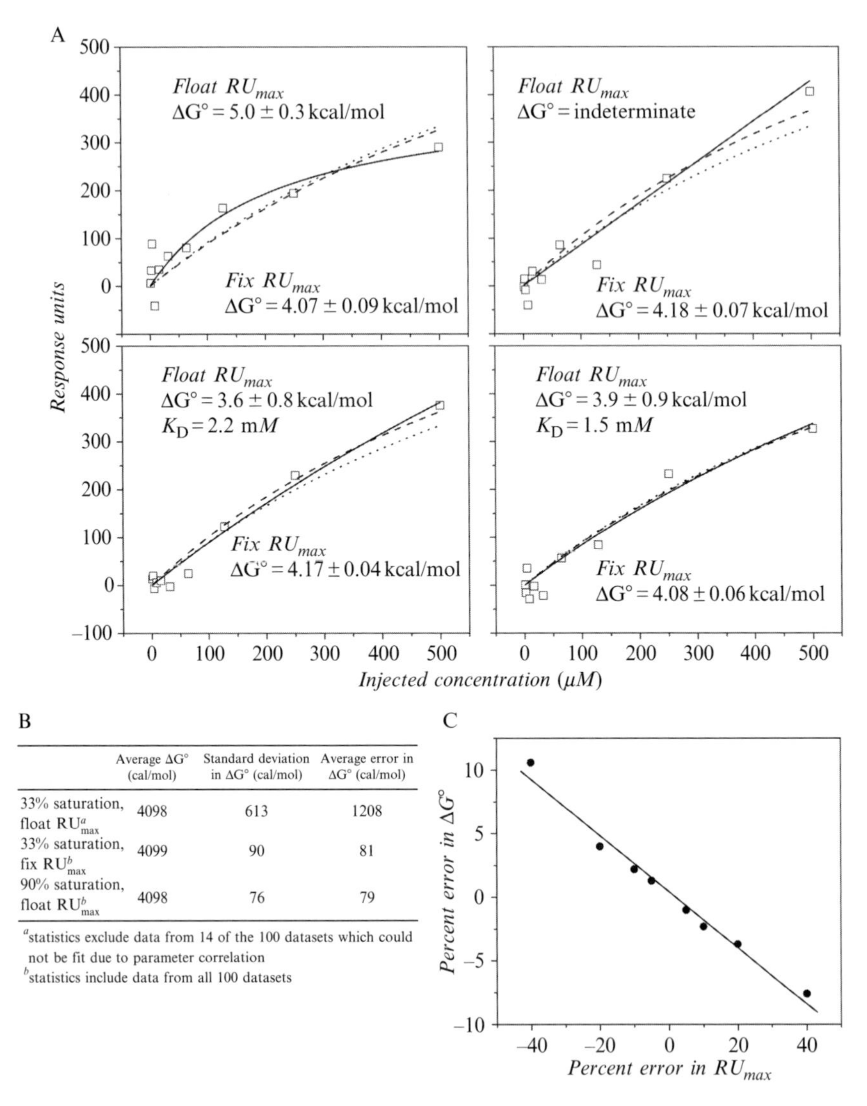

	Average $\Delta G°$ (cal/mol)	Standard deviation in $\Delta G°$ (cal/mol)	Average error in $\Delta G°$ (cal/mol)
33% saturation, float RU^a_{max}	4098	613	1208
33% saturation, fix RU^b_{max}	4099	90	81
90% saturation, float RU^b_{max}	4098	76	79

[a]statistics exclude data from 14 of the 100 datasets which could not be fit due to parameter correlation
[b]statistics include data from all 100 datasets

Figure 15.3 Fixing the activity of a SPR sensor surface dramatically enhances the ability to recover affinities from noisy, incomplete binding data. Binding data to 33% saturation were simulated for an interaction proceeding with a 1-mM K_D ($\Delta G°$ of 4092 cal/mol at 25 °C) and an RU_{max} of 1000, and 100 noisy datasets were generated by adding Gaussian-distributed random noise with a Gaussian width of 40 RU. The 100 noisy datasets were then fit with either RU_{max} fixed or floated as a fitting parameter. (A) Representative results from four of the noisy datasets. Solid lines represent the fits with both RU_{max} and K_D floated, dashed lines represent theoretically correct, perfect curves, and dotted lines represent fits with RU_{max} fixed at 1000. With RU_{max} floated, the $\Delta G°$ values recovered from the fits vary in their accuracy and have substantial errors in precision. The sample on the top right could not be fit due to extensive parameter uncertainty. Moreover, of the 100 datasets simulated for Fig.15.3, 14 could not be fit due to near-perfect parameter correlation. Clearly, the odds of obtaining an accurate and precise affinity from one of these datasets are low if one is fitting for both surface activity and K_D.

Yet when the same data are fit with the RU_{max} fixed at 1000, the situation is vastly improved. The average ΔG° is again very close to the real value, as expected (4099 cal/mol vs. 4092 cal/mol). However, the standard deviation in this value is only 90 cal/mol, a nearly sevenfold improvement in accuracy over the case when RU_{max} is floated. Furthermore, the average error in any one experiment is only 81 cal/mol, a 15-fold improvement in precision. Every dataset could be fit, and for those datasets which could not be fit when RU_{max} was floated, the average ΔG° is still almost exactly correct: 4084 ± 95 cal/mol, with an average error of 82 cal/mol. For comparison, knowing the RU_{max} in advance when only 33% saturation is achieved results in the same level of accuracy and precision as would be achieved in a traditional analysis in which 90% saturation is reached. However, with a 1-m*M* affinity, reaching 33% saturation requires injecting a sample at a concentration of 500 μM, difficult but achievable for TCRs or peptide/MHC complexes. Reaching 90% saturation, on the other hand, would require injecting a 9-m*M* sample, well above what is reasonably achievable for these molecules.

Obviously though, any independent measurement of RU_{max} includes error: when fitting 100 simulated noisy datasets that reach 90% saturation, RU_{max} was determined with an average error of 4%. What is the effect of error in a predetermined RU_{max} in this analysis? Surprisingly, the effect is small. In Fig.15.3C, errors in RU_{max} were introduced into the analysis of the 33% saturation data. Every 1% error in RU_{max} led to an error in ΔG° of approximately 0.2%. Thus, even a very large error of 20% in RU_{max} leads to a tolerable error in ΔG° of only 4%.

What else is needed to accurately fix the maximum response in an SPR experiment? Given that the signal is proportional to the amount of active

---

correlation between RU_{max} and K_D. With RU_{max} fixed at 1000, however, the ΔG° values are much more accurate and the standard errors are substantially lower. (B) Summary statistics from the analysis of the 100 datasets. Floating RU_{max} results in a large standard deviation in the recovered ΔG° values, and 14% of the datasets could not be fit. The average error in the recovered ΔG° values, 1208 cal/mol, is 30% of the actual ΔG°. Fixing RU_{max} brings the standard deviation in the recovered ΔG° down sevenfold, and results in a 15-fold improvement in the average error. As shown in the third entry in the table, fitting the noisy datasets that only go to 33% saturation with RU_{max} fixed is equivalent to traditional analyses (both RU_{max} and K_D floated) of datasets that reach 90% saturation. (C) Errors in RU_{max} have a small effect on the error in the recovered ΔG°, with an approximately 0.2% error in ΔG° for every 1% error in RU_{max}.

material on the sensor surface, the more the better. This advice is counter to common recommendations to limit the amount of material on a surface. However, low surface activities are usually recommended for kinetic experiments in order to reduce mass transport and rebinding effects, neither of which are issues when performing equilibrium experiments. Note that for these reasons flow rates can be reduced to a minimum when performing equilibrium studies with SPR, reducing the sample requirements for the injected ligand.

While of great utility, this method of enhancing accuracy and precision of binding affinities in SPR is not amenable in all cases. If a TCR or peptide/MHC surface decays significantly over the course of an experiment (due, perhaps to peptide dissociation from MHC or the detrimental effects of any necessary regeneration steps performed between injection cycles), RU_{max} will not be constant. More practically, if one is studying an interaction without the availability of a high affinity variant or not studying mutants of what is otherwise a reasonably affinity interaction, fixing RU_{max} is simply not an option. Yet in the study of T cell receptor recognition, many lines of investigation have turned toward examining the effects of mutations in the TCR or MHC or substitutions in the peptide, and the approach outlined above will be of considerable utility in such cases. One promising avenue of investigation is double-mutant cycle experiments, which involve measurements with two single amino acid mutants and one double mutant (Schreiber and Fersht, 1995). Depending upon the strength and type of interaction between the two sites that are mutated, affinities could drop substantially in these experiments. Yet double mutant cycle experiments have potential to address outstanding questions in TCR–peptide/MHC interactions, such as the "basal" level of affinity TCRs maintain towards MHC and the extent to which various loops are directed energetically towards the peptide versus the MHC α-helices (Collins and Riddle, 2008; Garcia *et al.*, 2009). The method outlined above may thus prove particularly useful in addressing immunologically relevant questions about how TCRs engage peptide/MHC.

3.3. Underlying binding thermodynamics from SPR experiments

Binding thermodynamics beyond $\Delta G°$ are available from SPR via van't Hoff-style analyses. Because interactions for protein–protein interactions are usually associated with large negative heat capacity changes (Stites, 1997), a direct fit to the temperature dependence of the free energy change is likely to be more preferable than a traditional natural log of K versus inverse temperature van't Hoff analysis (provided the traditional analysis incorporates a ΔC_p the approaches will be numerically similar, but fitting to $\Delta G°$ versus temperature will more effectively illustrate the heat capacity change). This approach has been used numerous times with T cell receptors, yielding data about the relationships between entropy, heat capacity, and conformational changes that may be occurring upon receptor binding (reviewed by Armstrong *et al.*, 2008). It is important though to appreciate that the accuracy of the $\Delta H°$, $\Delta S°$, and ΔC_p values determined in this manner depends on the accuracy of the individual $\Delta G°$ measurements. Ideally, realistic errors in $\Delta G°$ should be incorporated into the fit, which can easily be performed in most nonlinear curve fitting packages. In this regard, the approach outlined above for increasing the accuracy for low-affinity measurements by SPR may be useful for determine underlying binding thermodynamics by SPR. Finally, note that the effects of linked equilibria on measured binding thermodynamics that were outlined for titration calorimetry will also be manifest in thermodynamics determined by van't Hoff methods (Horn *et al.*, 2001, 2002). Although a linkage analysis by van't Hoff methods may be impractical, in its absence caution may be warranted when interpreting binding thermodynamics for very different protein–protein interfaces.

4. Fluorescence Anisotropy as a Tool for Characterizing TCR–Peptide/MHC Interactions

4.1. Introduction to fluorescence anisotropy

Fluorescence has not routinely been used to characterize TCR–peptide/MHC interactions, primarily because there is little or no change in intrinsic fluorescence upon binding and the introduction of a probe within the protein–protein interface would likely interfere with recognition. FRET has recently been used in kinetic studies, with a donor on the MHC protein and an acceptor on the TCR, bypassing some of the limitations inherent to SPR (Gakamsky *et al.*, 2007). Yet beyond these studies, the vast majority of quantitative studies on TCRs and peptide/MHC interactions has been performed with SPR, and to a lesser extent, ITC.

Fluorescence anisotropy (equivalent to fluorescence polarization via a simple mathematical relationship and recently reviewed by Jameson and Mocz, 2005) reports on molecular motion that occurs over the lifetime of a fluorescent probe. A mode of motion occurring in every molecule in solution is molecular tumbling, the rate of which to a first approximation is proportional to a molecule's size: the larger the size, the slower the tumbling, and the less tumbling contributes to the loss of anisotropy over the course of the fluorescence lifetime. For measuring binding, one simply measures the anisotropy of a fluorescent molecule in the presence of increasing concentrations of a binding partner. As the fluorescent molecule forms a larger complex, the slower tumbling of the complex is reflected as an increase in the anisotropy. Fluorescence anisotropy has a long history in the analysis of macromolecular interactions, and has seen particular use in investigating interactions between proteins and small molecules or nucleic acids (Heyduk *et al.*, 1996; Sportsman, 2003), including the interactions of peptides with MHC proteins (Baxter *et al.*, 2004; Binz *et al.*, 2003; Buchli *et al.*, 2006; Chen and Bouvier, 2007; Dedier *et al.*, 2001). Fluorescence anisotropy is less commonly used to monitor protein–protein interactions, largely due to the availability and success of other techniques, as well as limitations placed by the relationship between the size and tumbling rate of the fluorescent molecule and the magnitude of the anisotropy change that occurs upon binding.

Yet fluorescence anisotropy deserves special mention here due to its potential for efficiently characterizing low-affinity interactions, and TCR–peptide/MHC interactions in particular. This is because the advantages provided by SPR described in Section 3.2 also apply to fluorescence anisotropy, but as discussed below, because the maximum shift in anisotropy is an intrinsic property of a molecular complex rather than a property unique to each experiment, fluorescence anisotropy provides further advantages over SPR.

For a 1:1 binding interaction monitored by changes in fluorescence anisotropy, one way to represent the binding response is via a traditional hyperbolic binding curve of the form:

$$\Delta A = \Delta A_{\max} \frac{K[\mathrm{L}]}{1 + K[\mathrm{L}]} \tag{15.3}$$

where ΔA is the measured change in anisotropy that occurs upon binding, [L] is the concentration of free ligand, and ΔA_{max} is the maximum change in anisotropy, equal to the anisotropy of the complex minus the anisotropy of the free labeled ligand ($A_{complex} - A_{ligand}$). The anisotropy of the free ligand is easily measured. The anisotropy of the complex can be determined by either fitting for it in a binding experiment that achieves complete (or near-complete) saturation. Once the anisotropy of the complex is measured though, this value can be fixed for the analysis of lower affinity interactions involving similar proteins, such as MHC proteins presenting different peptides, or MHC or TCR mutants. All the advantages described in Section 3.2 apply to such an experimental approach, the advantage over SPR being that the anisotropy of the complex is unchanging from experiment to experiment (provided the mutations or different peptides do not fundamentally alter the shape and thus rotational properties of the ternary complex). In SPR, on the other hand, the maximum surface capacity changes with the preparation of each new surface and can degrade over the course of an experiment. As in SPR, the anisotropy of the complex can be measured through the use of high affinity TCR variants or determined with a wild-type TCR that binds with reasonably tight affinity.

Obviously, one of the proteins must be fluorescently labeled in order to use fluorescence anisotropy. The site of labeling must be distal enough from the binding interface so as not to influence binding, and if a peptide/MHC complex is labeled, not interfere with the peptide–MHC interaction. The site should not possess a high level of intrinsic mobility (e.g., a disordered loop), as this could result in such rapid depolarization of fluorescence that there is insufficient signal to report on changes in molecular tumbling (although this can be offset with the use of a longer lifetime fluorescent probe). A final requirement is tolerance to a cysteine mutation, as in most cases a cysteine reactive probe will be needed for site-specific labeling. An ideal position for labeling is thus an amino acid in a rigid unit of secondary structure with a solvent exposed side chain, preferably polar, as most commonly used fluorescence probes are highly polar themselves.

4.2. An example TCR–peptide/MHC interaction characterized by fluorescence anisotropy

To demonstrate the utility of using fluorescence anisotropy to monitor TCR binding to a peptide/MHC complex, we produced variants of the HLA-A2 class I MHC complex separately labeled with fluorescein at position 145 and position 195 of the HLA-A2 heavy chain. Position 145 is near the edge of the $\alpha 2$ helix in the HLA-A2 peptide binding domain, and position 195 is in a loop in the $\alpha 3$ domain distal to the peptide binding domain. The position 145 label required mutating a histidine to cysteine, whereas the position 195 label required mutating a serine. The side chains of both positions extend away from the protein surface and in available crystallographic structures do not appear to interact with other protein atoms. Despite being in the peptide binding domain, position 145 is fully solvent exposed and atoms of the imidazole ring are least 11 Å away from any TCR atoms in known ternary complexes with HLA-A2 (Buslepp *et al.*, 2003; Chen *et al.*, 2005; Ding *et al.*, 1998; Gagnon *et al.*, 2006; Garboczi *et al.*, 1996a; Ishizuka *et al.*, 2008; Miller *et al.*, 2007; Stewart-Jones *et al.*, 2003). The primary reason for choosing these two positions is that 195 appears to be highly flexible as it is in an exposed loop that is occasionally disordered in structures with HLA-A2, and position 145 is in a relatively rigid unit of secondary structure. These two positions should thus highlight the extent intrinsic flexibility has on signal and data quality: compared to the more rigid location at position 145, the more flexible location at position 195 should result in a lower overall change in anisotropy, as the greater flexibility will diminish the contribution of molecular tumbling to the depolarization of fluorescence, degrading the signal that should be produced upon TCR binding.

The two mutant HLA-A2 complexes were expressed and refolded in the presence of the HTLV-1 $Tax_{11-19;}$ peptide and purified according to standard procedures (Davis-Harrison *et al.*, 2005; Garboczi *et al.*, 1992), and labeled with a cysteine-reactive fluorescein derivative (fluorescein-5-maleimide). Extensive postlabeling dialysis and chromatographic purification was performed to ensure the samples were free of unreacted label, as the presence of free label, with its very low anisotropy would negatively influence the assay. Control labeling reactions performed with wild-type Tax_{11-19}/HLA-A2 indicated that this procedure resulted in no nonspecific labeling and no residual free label.

Aliquots (120 μL) of 100 n*M* fluorescently labeled Tax_{11-19}/HLA-A2 were then incubated with increasing amounts of purified A6 TCR (assembled with the aid of a leucine zipper, see Ding *et al.*, 1999), and the anisotropy measured using a Beacon 2000 fluorescence spectrometer. Final volumes for each sample were 200 μL. As shown in Fig.15.4, for both samples, increasing the TCR concentration resulted in an increase in anisotropy. However, as anticipated, the more flexible position 195-labeled sample generated substantially poorer data than the position 145-labeled

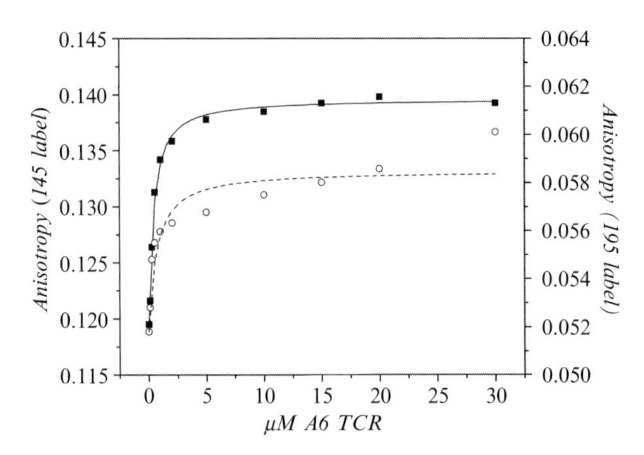

Figure 15.4 Binding of the A6 TCR to the Tax_{11-19}/HLA-A2 complex monitored by fluorescence anisotropy. Data are shown for Tax_{11-19}/HLA-A2 labeled at position 145 (squares, solid line) and position 195 (circles, dashed line). The starting anisotropy, dynamic range, and overall data quality data with the position 145-labeled sample are substantially greater than with the position 195-labeled sample, reflecting the influence of site-specific flexibility on the anisotropy data. Fitting the 145-labled sample to a single site binding model yielded a K_D of 0.37 $\pm$ 0.04 μM, or a ΔG° of 8.77 $\pm$ 0.06 kcal/mol, in close agreement with the value previously determined by ITC (Armstrong and Baker, 2007).

sample, with greater scatter and an overall change in anisotropy only 1/3 that of the position 145-labeled sample. Indeed, the starting anisotropy for the position 195-labeled sample was much less than that of the position 145-labeled sample (0.05 for position 195, 0.12 for position 145), reflecting the greater flexibility of the loop at position 195 compared to the more rigid helix at position 145. As the overall change in anisotropy reflects the dynamic range available to the assay, measurements with HLA-A2 labeled at position 145 will be substantially more reliable than those with protein labeled at position 195. On this note, a longer lifetime fluorescent probe is likely to provide even greater dynamic range, as the molecules will experience greater tumbling over the lifetime of the probe.

Fitting the position 145-labeled data to a single-site binding model yielded an affinity of 0.4 μM, identical within error to the 0.3 μM affinity determined by ITC (Armstrong and Baker, 2007). Note that the values of 0.3–0.4 μM are slightly tighter than the 1–2 μM affinity measured by SPR (Davis-Harrison *et al.*, 2005). Yet unlike SPR the ITC and fluorescence anisotropy measurements are pure solution measurements and slight variations between SPR and pure solution measurements are not unusual. These results demonstrate the potential for using fluorescence anisotropy in monitoring TCR–peptide/MHC interactions. The small sample volumes, low protein requirements, and ability to fix the maximum shift in anisotropy make fluorescence anisotropy particularly attractive for assaying mutants, perhaps even more so than SPR. The technique is also easily adaptable for measuring binding thermodynamics via van't Hoff analysis: sealing the samples and simply measuring them at multiple temperatures will obviate the need to create new samples for each temperature point, as is necessary with SPR. For such an experiment, fluorimeters that can measure anisotropy in plate format will be particularly useful. Note that as with SPR, incubating the samples with excess peptide may be needed to ensure peptide/MHC stability over the course of the measurements. A final caution is that as changes in the anisotropy signal are closely linked to the size of the fluorescently labeled molecule and its binding partner, different TCR constructs may yield data of differing quality. As noted above, the A6 TCR construct used in Fig.15.4 was assembled with the aid of a leucine zipper (Ding *et al.*, 1999; O'Shea *et al.*, 1993). It is becoming more common to stabilize soluble TCRs with an engineered disulfide bond linking the α and β chain (Laugel *et al.*, 2005); the resulting molecules are smaller than the equivalent zippered constructs and will likely yield smaller anisotropy changes than those shown in Fig.15.4. This could potentially be offset by the use of a longer lifetime fluorophore as discussed above.

5. Concluding Remarks

Measurements of the interactions between T cell receptors and their peptide/MHC ligands continue to provide valuable insight into the biology of the cellular immune system. Although challenges in producing the proteins recombinantly and their weak-to-moderate affinities can make detailed physical studies difficult, the extraordinary molecular recognition properties displayed by TCRs, such as dual recognition of self and nonself and the simultaneous display of both cross-reactivity and specificity, make such studies well worth the effort. Comparisons of binding thermodynamics with structural and functional properties are likely to shed considerable light on how these properties are achieved, and new instrumentation will facilitate these experiments. SPR has played and will continue to play a dominant role in characterizing TCR–peptide/MHC interactions, and its usefulness in characterizing low-affinity interactions can be extended with simple experimental approaches. The approaches described here are particularly attractive for the study of mutants, and make informative double-mutant cycle experiments an attractive line of experimentation. Less commonly used techniques such as fluorescence anisotropy may be of even greater utility for such experiments.

Acknowledgments

We thank Alison Wojnarowicz and Emily Doan for technical assistance, and all the members of the Baker lab for helpful discussion. Supported by grant GM067079 from the National Institutes of General Medical Sciences, National Institutes of Health and MCB0448298 from the National Science Foundation. KHP was supported by the Notre Dame Chemistry-Biochemistry-Biology Interface training program, funded by GM075762 from the National Institutes of General Medical Sciences.

References

Armstrong, K. M., and Baker, B. M. (2007). A comprehensive calorimetric investigation of an entropically driven T cell receptor-peptide/major histocompatibility complex interaction. *Biophys. J.* **93,** 597–609.

Armstrong, K. M., *et al.* (2008). Thermodynamics of T-cell receptor-peptide/MHC interactions: Progress and opportunities. *J. Mol. Recognit.* **21,** 275–287.

Baker, B. M., and Murphy, K. P. (1996). Evaluation of linked protonation effects in protein binding reactions using isothermal titration calorimetry. *Biophys. J.* **71,** 2049–2055.

Baker, B. M., and Murphy, K. P. (1997). Dissecting the energetics of a protein-protein interaction: The binding of ovomucoid third domain to elastase. *J. Mol. Biol.* **268,** 557–569.

Baker, B. M., and Murphy, K. P. (1998). Prediction of binding energetics from structure using empirical parameterization. *Methods Enzymol.* **295,** 294–315.

Barbieri, C. M., and Pilch, D. S. (2006). Complete thermodynamic characterization of the multiple protonation equilibria of the aminoglycoside antibiotic paromomycin: A calorimetric and natural abundance ^{15}N NMR study. *Biophys. J.* **90,** 1338–1349.
Baxter, T. K., *et al.* (2004). Strategic mutations in the class I MHC HLA-A2 independently affect both peptide binding and T cell receptor recognition. *J. Biol. Chem.* **279,** 29175–29184.
Binz, A. K., *et al.* (2003). Thermodynamic and kinetic analysis of a peptide-class I MHC interaction highlights the noncovalent nature and conformational dynamics of the class I heterotrimer. *Biochemistry* **42,** 4954–4961.
Buchli, R., *et al.* (2006). Critical factors in the development of fluorescence polarization-based peptide binding assays: An equilibrium study monitoring specific peptide binding to soluble HLA-A*0201. *J. Immunol. Methods* **314,** 38–53.
Buslepp, J., *et al.* (2003). A Correlation between TCR Va docking on MHC and CD8 dependence: Implications for T cell selection. *Immunity* **19,** 595–606.
Chen, M. N., and Bouvier, M. (2007). Analysis of interactions in a tapasin/class I complex provides a mechanism for peptide selection. *EMBO J.* **26,** 1681–1690.
Chen, J.-L., *et al.* (2005). Structural and kinetic basis for heightened immunogenicity of T cell vaccines. *J. Exp. Med.* **201,** 1243–1255.
Christensen, J. J., *et al.* (1976). *Handbook of proton ionization heats.* Wiley, New York.
Collins, E., and Riddle, D. (2008). TCR-MHC docking orientation: Natural selection or thymic selection? *Immunol. Res.* **41,** 267–294.
Colf *et al.* (2007) How a single T cell receptor recognizes both self and foreign MHC. *Cell* **129,** 135–146.
Corr, M., *et al.* (1994). T cell receptor-MHC class I peptide interactions: Affinity, kinetics, and specificity [see comments]. *Science* **265,** 946–949.
Davis, M. M., *et al.* (1998). Ligand recognition by alpha beta T cell receptors. *Annu. Rev. Immunol.* **16,** 523–544.
Davis-Harrison, R. L., *et al.* (2005). Two different T cell receptors use different thermodynamic strategies to recognize the same peptide/MHC ligand. *J. Mol. Biol.* **346,** 533–550.
Dedier, S., *et al.* (2001). Use of fluorescence polarization to monitor MHC-peptide interactions in solution. *J. Immunol. Methods* **255,** 57–66.
Ding, Y. H., *et al.* (1998). Two human T cell receptors bind in a similar diagonal mode to the HLA- A2/Tax peptide complex using different TCR amino acids. *Immunity* **8,** 403–411.
Ding, Y. H., *et al.* (1999). Four A6-TCR/peptide/HLA-A2 structures that generate very different T cell signals are nearly identical. *Immunity* **11,** 45–56.
Fisher, H. F., and Singh, N. (1995). Calorimetric methods for interpreting protein-ligand interactions. *Methods Enzymol.* **259,** 194–221.
Fitch, C. A., *et al.* (2002). Experimental pKa values of buried residues: Analysis with continuum methods and role of water penetration. *Biophys. J.* **82,** 3289–3304.
Freyer, M. W., *et al.* (2008). Isothermal titration calorimetry: Experimental design, data analysis, and probing macromolecule/ligand binding and kinetic interactions. *Methods Cell Biol.* **84,** 79–113; Academic Press.
Fukada, H., and Takahashi, K. (1998). Enthalpy and heat capacity changes for the proton dissociation of various buffer components in 0.1 M potassium chloride. *Proteins* **33,** 159–166.
Gagnon, S. J., *et al.* (2006). T cell receptor recognition via cooperative conformational plasticity. *J. Mol. Biol.* **363,** 228–243.
Gakamsky, D. M., *et al.* (2007). Kinetic evidence for a ligand-binding-induced conformational transition in the T cell receptor. *Proc. Natl. Acad. Sci. USA* **104,** 16639–16644.
Garboczi, D. N., *et al.* (1992). HLA-A2-peptide complexes: Refolding and crystallization of molecules expressed in *Escherichia coli* and complexed with single antigenic peptides. *Proc. Natl. Acad. Sci. USA* **89,** 3429–3433.

Garboczi, D. N., *et al.* (1996a). Structure of the complex between human T-cell receptor, viral peptide and HLA-A2. *Nature* **384,** 134–141.
Garboczi, D. N., *et al.* (1996b). Assembly, specific binding, and crystallization of a human TCR-alphabeta with an antigenic Tax peptide from human T lymphotropic virus type 1 and the class I MHC molecule HLA-A2. *J. Immunol.* **157,** 5403–5410.
Garcia, K. C., *et al.* (2009). The molecular basis of TCR germline bias for MHC is surprisingly simple. *Nat. Immunol.* **10,** 143–147.
Guinto, E. R., and Di Cera, E. (1996). Large heat capacity change in a protein-monovalent cation interaction. *Biochemistry* **35,** 8800–8804.
Heyduk, T., *et al.* (1996). Fluorescence anisotropy: Rapid, quantitative assay for protein-DNA and protein-protein interaction. *Methods Enzymol.* **274,** 492–503.
Horn, J. R., *et al.* (2001). Van't Hoff and calorimetric enthalpies from isothermal titration calorimetry: Are there significant discrepancies? *Biochemistry* **40,** 1774–1778.
Horn, J. R., *et al.* (2002). van't Hoff and calorimetric enthalpies II: Effects of linked equilibria. *Biochemistry* **41,** 7501–7507.
Ishizuka, J., *et al.* (2008). The structural dynamics and energetics of an immunodominant T cell receptor are programmed by its V[beta] domain. *Immunity* **28,** 171–182.
Jameson, D. M., and Mocz, G. (2005). Fluorescence polarization/anisotropy approaches to study protein-ligand interactions: Effects of errors and uncertainties. *Methods Mol. Biol.* **305,** 301–322.
Jones, L. L., *et al.* (2008). Different thermodynamic binding mechanisms and peptide fine specificities associated with a panel of structurally similar high-affinity T cell receptors. *Biochemistry* **47,** 12398–12408.
Joss, L., *et al.* (1998). Interpreting kinetic rate constants from optical biosensor data recorded on a decaying surface. *Anal. Biochem.* **261,** 203–210.
Krogsgaard, M., *et al.* (2003). Evidence that structural rearrangements and/or flexibility during TCR binding can contribute to T cell activation. *Mol. Cell.* **12,** 1367–1378.
Laugel, B., *et al.* (2005). Design of soluble recombinant T cell receptors for antigen targeting and T cell inhibition. *J. Biol. Chem.* **280,** 1882–1892.
Li, Y., *et al.* (2005). Directed evolution of human T-cell receptors with picomolar affinities by phage display. *Nat. Biotech.* **23,** 349–354.
Matsui, K., *et al.* (1994). Kinetics of T-cell receptor binding to peptide/I-Ek complexes: Correlation of the dissociation rate with T-cell responsiveness. *Proc. Natl. Acad. Sci. USA* **91,** 12862–12866.
Miller, P. J., *et al.* (2007). Single MHC mutation eliminates enthalpy associated with T cell receptor binding. *J. Mol. Biol.* **373,** 315–327.
Myszka, D. G. (1999). Improving biosensor analysis. *J. Mol. Recognit.* **12,** 279–284.
O'Shea, E. K., *et al.* (1993). Peptide 'velcro': Design of a heterodimeric coiled coil. *Curr. Biol.* **3,** 658–667.
Piepenbrink *et al.* (2009). Fluorine substitutions in an antigenic peptide selectively modulate T-cell receptor binding in a minimally perturbing manner. *Biochem. J.* **423,** 353–361.
Rich, R. L., and Myszka, D. G. (2008). Survey of the year 2007 commercial optical biosensor literature. *J. Mol. Recognit.* **21,** 355–400.
Rudolph, M. G., *et al.* (2006). How TCRs bind MHCs, peptides, and coreceptors. *Annu Rev Immunol.* **24,** 419–466.
Schreiber, G., and Fersht, A. R. (1995). Energetics of protein-protein interactions: Analysis of the Barnase-Barstar interface by single mutations and double mutant cycles. *J. Mol. Biol.* **248,** 478–486.
Shusta, E. V., *et al.* (2000). Directed evolution of a stable scaffold for T-cell receptor engineering. *Nat. Biotechnol.* **18,** 754–759.
Spolar, R. S., and Record, M. T. Jr. (1994). Coupling of local folding to site-specific binding of proteins to DNA. *Science* **263,** 777–784.

Sportsman, J. R. (2003). Fluorescence anisotropy in pharmacologic screening. *Methods Enzymol.* **361,** 505–529.

Stewart-Jones, G. B., *et al.* (2003). A structural basis for immunodominant human T cell receptor recognition. *Nat Immunol.* **4,** 657–663.

Stites, W. E. (1997). Protein-protein interactions: Interface structure, binding thermodynamics, and mutational analysis. *Chem. Rev.* **97,** 1233–1250.

Tellinghuisen, J. (2008). Isothermal titration calorimetry at very low c. *Anal. Biochem.* **373,** 395–397.

Turnbull, W. B., and Daranas, A. H. (2003). On the value of *c*: Can low affinity systems be studied by isothermal titration calorimetry? *J. Am. Chem. Soc.* **125,** 14859–14866.

Willcox, B. E., *et al.* (1999). TCR binding to peptide-MHC stabilizes a flexible recognition interface. *Immunity* **10,** 357–365.

Wiseman, T., *et al.* (1989). Rapid measurement of binding constants and heats of binding using a new titration calorimeter. *Anal. Biochem.* **179,** 131–137.

Wucherpfennig, K. W., *et al.* (2007). Polyspecificity of T cell and B cell receptor recognition. *Semin. Immunol.* **19,** 216–224.

CHAPTER SIXTEEN

Thermodynamic and Kinetic Analysis of Bromodomain–Histone Interactions

Martin Thompson

Contents

Abstract

Multiple factors are involved when selecting a technique and designing experiments to investigate emerging questions in biochemistry. Success depends on a conceptual understanding of a given spectroscopic approach and the ability to design a system to optimize the quality of the acquired data. In this chapter, we discuss fluorescence anisotropy and its application in characterizing the factors that drive the acetylation-dependent interactions between histone and bromodomain proteins. The steady-state and pre-steady-state binding events associated with biomolecular assemblies can be quantified with this technique, so long as the proper assays and binding models are developed. To accomplish this, the continuum of experimental considerations from instrumental setup and choice of fluorophore, to experimental procedures, and data analysis is

Department of Chemistry, Michigan Technological University, Houghton, Michigan, USA

Methods in Enzymology, Volume 466
ISSN 0076-6879, DOI: 10.1016/S0076-6879(09)66016-X

described. The methodology is discussed in sufficient detail such that this chapter is a complete guide to setting up and performing fluorescence anisotropy measurements to study biomolecular interactions. A thermodynamic and kinetic analysis is performed to determine the factors that drive molecular recognition and binding affinity, resulting in the identification of an induced-folding mechanism.

1. Introduction

This chapter describes the application of fluorescence anisotropy to quantify the thermodynamic and kinetic parameters that define biomacromolecular assembly dynamics. Fluorescence anisotropy is a powerful method to characterize steady-state and time-resolved binding events as, it is rapid, quantitative, and adaptable to a broad range of experimental conditions. The high sensitivity makes the approach ideal for using the low sample concentrations and short integration times that are oftentimes required for equilibrium titrations and time-based polarization measurements. We start by reviewing the conceptual basis of fluorescence anisotropy that makes it so well-suited to study solution-phase protein interactions. This permits an examination into practical matters and technical issues that need to be addressed to successfully apply the methodology. Such insights will be followed by an exploration of the models used to correlate the observed fluorescence anisotropy data with the physical changes occurring upon protein association. Next, a series of thermodynamic and kinetic binding studies are performed to illustrate data acquisition and analysis. Finally, the molecular level driving forces that describe bromodomain–histone binding events are discussed.

2. Fluorescence Anisotropy Theory and Concepts

Successful application of fluorescence anisotropy in the study of protein interactions requires an understanding of the conceptual framework that defines the approach. Such insights give meaning to the anisotropy changes observed under different experimental conditions for successful interpretation of the acquired data and the troubleshooting that inevitably occurs when developing novel assays to approach new problems in biochemistry. Simply stated, fluorescence anisotropy reveals changes in the rotational diffusion of a fluorescent molecule occurring while that molecule is in the excited state. Because the rate of rotational diffusion is faster than the rate of emission for most small molecules, events occurring on a given timescale (usually nanoseconds) can be determined if a means to monitor these changes can be employed. This is accomplished by measuring the change in the angular displacement of a molecule while in the lowest energy singlet state (S_1).

Consider a homogeneous population of randomly oriented fluorophores in the ground state (S_0). Upon exposure to polarized light, only those molecules that have their absorption transition moments aligned with the incident light are excited. This initial selection of a population of specifically oriented fluorophores from the bulk ensemble of randomly oriented fluorophores means that only events occurring during the fluorescence lifetime of this excited-state population is observed. It is the presence of absorption and emission transition moments oriented in specific directions within the fluorophore structure that give rise to the concept of anisotropy (Lakowicz, 2006). Rotational diffusion of the molecule changes the direction of its transition moments, giving rise to changes in polarization. The central notion of measuring the rotation of biomolecules with a fluorescent probe relies upon the change in emission polarization associated with the unbound to bound transition. Because angular displacement is proportional to the spherical volume, the sizes of these complexes can be correlated to changes (loss or gain) in polarization between the exciting and emitting light (Fig. 16.1). Molecules that experience changes in angular displacement at a slower rate (e.g., tumble slowly) will retain more polarization, whereas molecules that tumble rapidly relative to the fluorescence lifetime will lose polarization (become depolarized) (Lakowicz, 1991).

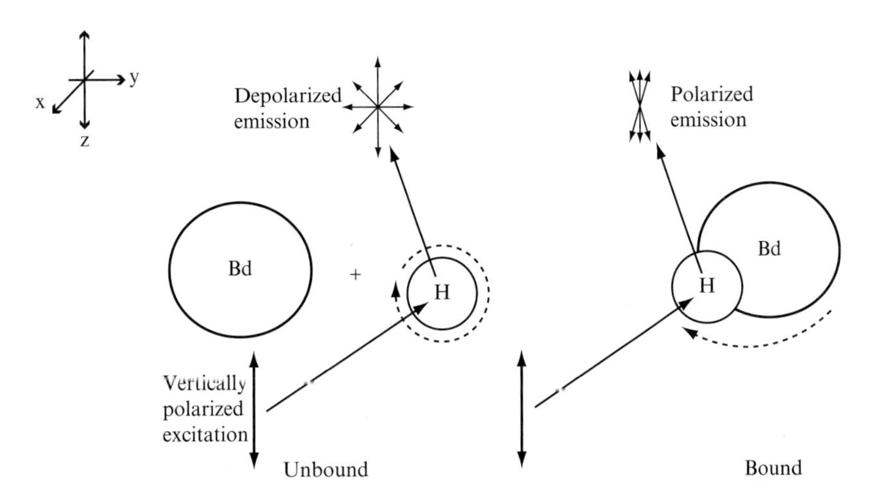

Figure 16.1 Schematic representation of the changes in polarized emission between unbound and bound states.

Fluorescence anisotropy measurements are performed with the incident light vertically polarized such that the electric vector of the excitation light is oriented parallel to the z-axis (Fig. 16.1). The measured quantities are the components of fluorescence intensity which are polarized in the vertical direction (parallel) or the horizontal direction (perpendicular) offset by 90°. Fluorescence anisotropy using vertically polarized excitation is defined as

$$A = \frac{(I_{\mathrm{vv}} - gI_{\mathrm{vh}})}{(I_{\mathrm{vv}} + 2gI_{\mathrm{vh}})}. \tag{16.1}$$

Here, I_{vv} is the fluorescence intensity with vertically polarized excitation and vertically polarized emission, I_{vh} is the fluorescence intensity with vertically polarized excitation and horizontally polarized emission, and $g = (I_{\mathrm{hv}}/I_{\mathrm{hh}})$ is a factor correcting for the polarization dependence of the spectrofluorometer (Lundblad *et al.*, 1996; Valeur, 2001). The anisotropy is a dimensionless quantity independent of the total fluorescence intensity and fluorophore concentration. Fluorescence emission anisotropy (A) is preferred over polarization because the difference in emission intensity between the two polarizations (numerator) is normalized by the total fluorescence (denominator) simplifying the equations (Lakowicz, 2006).

3. Developing Binding Models for the Analysis of Fluorescence Anisotropy Data

To understand the development of protein-binding models used to interpret anisotropy data, take the case where a protein interacts with a fluorophore-labeled biomolecule at a single site. Here, the change in anisotropy is measured as a function of time or binding partner concentration. Because the labeled biomolecule will exist in more than one state, the observed anisotropy is given by

$$A = \sum_i A_i f_i, \tag{16.2}$$

where A_i and f_i are the emission anisotropies and the fraction of the total population for species i, respectively. This means that for a system with population heterogeneity, the steady-state anisotropy of the emission will be the weighted average of the constituent values, where the sum of all values for f equals 1. For a system undergoing a single transition, the value for A has two components; one for the unbound fluorophore-labeled biomolecule and the other for the protein-bound complex, where f describes the fractional contribution of each species as they change throughout the titration.

The relative populations described by Eq. (16.2) can be determined as the anisotropy changes throughout a titration where the minimum and maximum anisotropy reflects the fully unbound and bound populations. In this case, the "fractional occupancy" refers to the percent of the labeled biomolecule complexed with its binding partner. The value of f can be determined throughout the course of the titration using the mass action expression:

$$[\mathrm{Bd}] + [\mathrm{H}] \overset{K_{\mathrm{eq}}}{\leftrightarrow} [\mathrm{Bd–H}]. \tag{16.3}$$

For the purposes of this work, we will assign an identity to the biomolecules under study. The above equilibrium describes the concentration of bromodomain protein [Bd] interacting with histone [H] to form the bromodomain–histone complex [Bd–H]. It has been suggested that the bound complex should be represented as [Bd′–H′], to draw attention to the structural changes that occur upon binding and the preferential interaction between biomolecules over protein–solvent interactions present in the unbound state (Chowdhry, 2001). For the purposes of developing a binding model, several important assumptions are made: (1) all acetylhistones are either bound or unbound (alternate binding modes are not distinguishable), (2) all binding partners are equally accessible, and (3) binding is reversible. When at equilibrium, the rate of complex association (k_{on}) and dissociation (k_{off}) are the same and can be represented by

$$k_{\mathrm{on}}[\mathrm{Bd}][\mathrm{H}] = k_{\mathrm{off}}[\mathrm{Bd–H}]. \tag{16.4}$$

Based on the law of mass action, the corresponding equilibrium dissociation constant can be expressed as

$$\frac{[\mathrm{Bd}][\mathrm{H}]}{[\mathrm{Bd–H}]} = \frac{k_{\mathrm{off}}}{k_{\mathrm{on}}} = K_{\mathrm{D}}. \tag{16.5}$$

The equilibrium molar dissociation constant or binding constant (K_{D}) is the more conventional and intuitive representation because it is expressed in units of molar concentration (M). It should be noted that the equilibrium association and dissociation constants are inversely related ($K_{\mathrm{D}} = 1/K_{\mathrm{A}}$). K_{D} is defined by the bromodomain concentration at which 50% of the labeled histones are bound ($f = 0.50$). Therefore, smaller K_{D} values mean a lower bromodomain concentration is required to attain 50% bound histone, indicating a high-affinity complex. Conversely, the larger K_{D} values observed for low-affinity complexes indicate a requirement for high concentrations to attain comparable fractional occupancies.

The law of mass action predicts the fractional occupancy of labeled histone at equilibrium can be expressed at each point of the titration as the ratio of bromodomain–histone complex [Bd–H] to the total histone concentration $[\mathrm{H}]_{\mathrm{T}}$:

$$f = \frac{[\text{Bd–H}]}{[\text{H}]_\text{T}} = \frac{[\text{Bd–H}]}{[\text{H}] + [\text{Bd–H}]}. \tag{16.6}$$

The total amount of labeled histone does not change over the course of the titration, whereas the relative populations of bound ([Bd–H]) and unbound histone ([H]) change as a function of bromodomain concentration. This means that $[\text{H}]_\text{T} = [\text{H}] + [\text{Bd–H}]$ at any point throughout the titration. Therefore, when $[\text{Bd}] = 0$ at the start of the titration, $[\text{Bd–H}] = 0$, $[\text{H}]_\text{T} = [\text{H}]$, and the fractional occupancy equals 0. At saturation, when $[\text{Bd}] \gg K_\text{D}$, the fractional occupancy approaches 100% (the maximum anisotropy value), unbound $[\text{H}] = 0$ and $[\text{H}]_\text{T} = [\text{Bd–H}]$.

To determine the concentration of bromodomain corresponding to the K_D, a few simple algebraic rearrangements need to be performed. First, multiply the numerator and denominator in the rightmost equation given in Eq. (16.6) by [Bd] and second, divide both by [B–H]. The K_D, as given in Eq. (16.5), appears in the denominator, and the equation takes the form

$$f = \frac{[\text{Bd}]}{K_\text{D} + [\text{Bd}]}. \tag{16.7}$$

It becomes clear that the equation defines the point at which $f = 0.5$, $[\text{H}] = [\text{Bd–H}]$, and $[\text{Bd}] = K_\text{D}$. Now, the fractional occupancy equation is described in the form of a rectangular hyperbola, where the horizontal asymptote is defined as 100% bound. The characteristic hyperbolic shape of the binding curve can also be described as a rapid binding below K_D and slower association at higher [Bd] where fewer unbound histones are available. Because of this phenomenon, saturation-binding experiments are required to determine the maximum anisotropy value corresponding to 100% bound histone.

To solve for fractional occupancy continuously throughout the titration, the concentration of unbound [H] (and inherently [Bd–H]) must be determined explicitly. This is accomplished by using the known values for the concentration of $[\text{H}]_\text{T}$, which is constant throughout the titration and $[\text{Bd}]_\text{T}$ for each data point throughout the titration. By substituting the concentration of unbound bromodomain in the form of total bromodomain added, minus the bound complex ($[\text{Bd}] - [\text{Bd–H}]$) and solving for $[\text{Bd–H}]/[\text{H}]_\text{T}$ using the quadratic equation (Eftink *et al.*, 1997), fractional occupancy can be expressed as

$$f = \frac{(K_\text{D} + [\text{H}]_\text{T} + [\text{Bd}]_\text{T}) - \sqrt{(K_\text{D} + [\text{H}]_\text{T} + [\text{Bd}]_\text{T})^2 - (4[\text{H}]_\text{T}[\text{Bd}]_\text{T})}}{2[\text{H}]_\text{T}}. \tag{16.8}$$

Here, f corresponds to the fraction of the total histone concentration $[\mathrm{H}]_\mathrm{T}$ bound by bromodomain, where $[\mathrm{Bd}]_\mathrm{T}$ is the total concentration at any given point. In this work, histones are labeled with the fluorophore and the concentration will not change for the complete titration. The parameter of the fit is the equilibrium dissociation constant. This analysis is based on the assumption that site-specific binding can be described as a two-state system (also referred to as a single site-occupancy model). Considering that there is only one acetyllysine per histone and bromodomain proteins shows negligible binding to unmodified lysines, the two-state model is an appropriate description of the unbound to bound transition (Chandrasekaran and Thompson, 2006, 2007). If a histone with multiple binding sites or a protein with multiple bromodomains were used, as opposed to the system we describe here, then complicated binding behavior or cooperativity between domains must be considered (Eftink *et al.*, 1997).

These complicated equations are used to evaluate experimental results to understand the observed anisotropy changes as a function of bromodomain concentration. Anisotropy values for each data point are plotted versus bromodomain protein concentration and fit to the two-state binding model to attain the equilibrium dissociation constant for bromodomain binding to the acetylated histone (Kohler and Schepartz, 2001):

$$A = A_{\mathrm{min}} + (A_{\mathrm{max}} - A_{\mathrm{min}}) \left[\frac{(K_\mathrm{D} + [\mathrm{H}]_\mathrm{T} + [\mathrm{Bd}]_\mathrm{T}) - \sqrt{(K_\mathrm{D} + [\mathrm{H}]_\mathrm{T} + [\mathrm{Bd}]_\mathrm{T})^2 - (4[\mathrm{H}]_\mathrm{T}[\mathrm{Bd}]_\mathrm{T})}}{2[\mathrm{H}]_\mathrm{T}} \right]. \tag{16.9}$$

Here, A_{min} and A_{max} are the minimum and maximum anisotropies observed under conditions where the fluorophore-labeled histone is 100% unbound (when $[\mathrm{Bd}] = 0, f = 0$) and 100% bound (when $[\mathrm{H}]_\mathrm{T} = [\mathrm{Bd–H}]$, $f = 1$), respectively. Notice the equation in brackets is simply the fraction bound given in Eq. (16.8) and has values ranging from 0 to 1. The term $(A_{\mathrm{max}} - A_{\mathrm{min}})$ is the experimental range of anisotropy values spanning the complete titration. When plotting fluorescence anisotropy data, the y-axis can be presented as (and fit to) fractional occupancy or raw anisotropy using Eq. (16.8) or (16.9), respectively. When plotting as fraction of histone bound, the raw anisotropy data (y-axis) are converted to the form $(A_i - A_{\mathrm{min}})/(A_{\mathrm{max}} - A_{\mathrm{min}})$, where A_i is the anisotropy at $[\mathrm{Bd}]_i$ or time i (x-axis) for steady-state and pre-steady-state analysis, respectively. The anisotropy data are rescaled along the y-axis from 0 (experimentally measured lower baseline) to 1 (measured upper baseline).

4. Experimental Considerations in Designing Fluorescence Anisotropy Assays

The successful development of a highly sensitive fluorescence-based assay to examine the acetylation dependence of bromodomain–histone interactions requires the careful evaluation of multiple parameters and experimental issues to optimize the system under examination. Examples include the fluorescence lifetime of the fluorophore, the attachment of the fluorophore to the biomolecule, and the relative size of the biomolecules in the free and bound states. The ideal fluorescent probe will have a high extinction coefficient and high fluorescence quantum yield to minimize background scattering and maximize detection sensitivity. The fluorescence lifetime (τ_f) may be the single most important photophysical parameter simply because all measured information is based on what occurs within the time the fluorophore is in the excited state. If τ_f is too short or too long, binding events may not be observable. Furthermore, because polarization can be lost through other photophysical mechanisms including, intersystem crossing into triplet states or excitation into high-energy excited states (S_n, where $n > 1$), dyes susceptible to these mechanisms should be avoided. Molecules that undergo changes in fluorescence quantum yield (Φ_f) or fluorescence lifetime upon conjugation to a biomolecule are acceptable, so long as the photophysical parameters are quantifiable and, most importantly, remain unchanged throughout the course of the experiment. Photophysical changes occurring for any reason during the titration complicates interpretation of the data, because the anisotropy change cannot be solely attributed to the binding event. Furthermore, changes in Φ_f, τ_f, and/or fluorescence intensity that accompany depolarization are indicators of quenching or competing processes that increase error and make the data difficult, if not impossible, to interpret accurately.

In designing an assay, an extrinsic probe, such as fluorescein or rhodamine is tethered to the biomolecule of interest to reflect the motion of the protein. This motion includes the overall rotational mode of the biomolecule(s) and any localized motion of the attached fluorescent probe. Because the biomolecule and the probe rotate as a unit, care must be taken in the choice of, and linkage to the fluorophore to ensure anisotropy provides useful information about changes in the size and shape of the biomolecule upon binding. Typically, the fluorophore is attached to the smaller of the two binding partners, as this increases the sensitivity of measurement (Fig. 16.1). In effect, the relative change in angular displacement upon binding is proportional to the change in the size of the complex. A larger change in anisotropy is observed when the smaller biomolecule binds the larger one in forming the complex.

Solution conditions experienced by the fluorophore (or probe) need to be carefully controlled. For example, solvent viscosity will impact the rate of rotational diffusion and therefore the measured anisotropy of a molecule in solution, so care must be taken that throughout the course of the titration, that the solvent (or buffer solution) is unchanged. The probe should experience little or no change in solution conditions that might affect its quantum yield or fluorescence lifetime throughout the titration. Changes in percent glycerol, pH, or salt concentration can change the photophysical parameters of many fluorophores, which will alter the apparent volume of the complex. Such complications are readily avoided by dialyzing the two samples in the same buffer solution prior to measurement and quantification. This is to ensure the environment of the binding partners and the fluorophore do not change as a result of mixing. The acquired data will reflect all processes and therefore it is critical to minimize or eliminate these contributions so that the observed changes in anisotropy can be exclusively attributed to the binding event.

5. Preparation of Histone and Bromodomain Samples

Peptides were synthesized on Rink amide resin by solid-phase peptide synthesis using standard F-moc chemistry. The peptides used here are based on amino acids 2–26 of histone H3 (GenBank Accession No. P68431) and have the sequence: ARTKQTAR**K**STGG**K**APRKQLATKAA. Acetylated lysines were incorporated at either the lysine 9 (K9) or lysine 14 (K14) position (shown in bold). Fluorescein was coupled to the α-amine of the N-terminal end using HATU activated coupling chemistry. The peptides were removed from the resin and deprotected by cleavage using a 90% trifluoroacetic acid (TFA) solution and precipitated with ice-cold diethyl ether as previously described (Chandrasekaran and Thompson, 2006). The fluorescein-labeled peptides were purified using HPLC by running water (0.1% TFA): acetonitrile (0.1% TFA) gradient. Identities and purities of the mutant histone, tail peptide, and full-length histone were confirmed by MALDI–TOF mass spectrometry.

Alternatively, full-length proteins displaying a specific chemical modification at a single position can be readily synthesized using an adaptation of the native chemical ligation (NCL) strategy. The histone N-terminal "tail," consisting of the first 25 amino acids, is replaced with a "prosthetic tail" displaying a specific chemical modification (He *et al.*, 2003; Michael and Shrogen-Knaak, 2004). NCL is a method in which two unprotected peptides; one bearing a reactive C-terminal thioester, and the other with an N-terminal cysteine residue can be selectively condensed in aqueous solution at neutral pH to form an amide bond (Futaki *et al.*, 2004). This is accomplished in three steps. First, a "tailless" histone is prepared by deleting the N-terminal 25 amino acids and introducing a cysteine at the first amino acid position of the mutant. The mutant histone is expressed in *Escherichia coli* and purified using affinity chromatography. Next, a peptide with a C-terminal thioester is generated as described above with a few adaptations. A 2-chlorotrityl resin is used so the peptide can be removed from the solid supports using acetic acid/methanol/trifluoroethanol (1:1:8, v/v), without removing the side chain protecting groups. The C-terminal end of the peptide is functionalized with a thioester using a threefold excess of benzyl mercaptan in DMF. After lyophilization, the side chains are deprotected, and the peptide is precipitated and purified using HPLC as described above. Finally, the fluorescein-labeled peptide is ligated to the tailless histone protein using a 2.5-fold molar excess of the peptide in 10 m*M* phosphate (pH 7.5), 6 *M* GdnHCl, and 2% thiophenol (v/v) for 16 h at room temperature. The modified histones are purified using nickel-affinity chromatography and refolded by stepwise dialysis in 10 m*M* Tris (pH 7.5), 1 m*M* EDTA, and 1 m*M* DTT where the concentration of KCl is varied at each step from 2 to 0.25 *M*. Identities and purities of the mutant histone, tail peptide, and full-length histone were confirmed by MALDI–TOF mass spectrometry. Concentrations of fluorescein-labeled peptide and fluorescein-labeled histones were determined spectrophotometrically based on the extinction coefficient of $\varepsilon_{491\text{nm}} = 68{,}000\ M^{-1}\ \text{cm}^{-1}$ for protein conjugated fluorescein.

The bromodomain (Bd) used for this study corresponds to amino acids 315–473 from the human polybromo-1 protein (Thompson, 2009). Sequence alignments using known bromodomain sequences were used to determine the coding region expected to represent fully functional bromodomain. Cloning, expression, and purification of a bromodomain from the *PB1* gene (GenBank Accession No. AF225871) were performed as previously described (Chandrasekaran and Thompson, 2006). The high purity of the bromodomain was achieved by one-step nickel-affinity chromatography (Ni^{2+}–NTA) and confirmed by MALDI–TOF. Bromodomain samples, whether purified by native or denaturing protocols, are dialyzed in reaction buffer containing 20 m*M* Tris (pH 7.8), 100 m*M* NaCl. The concentration of purified Bd was determined by Bradford assay using a bovine serum albumin (BSA) standard.

6. Fluorescence Anisotropy Measurements

All fluorescence anisotropy measurements are performed on a QuantaMaster QM-6 spectrofluorometer in T-format with dual-photomultiplier tubes a 75 W Xenon lamp and analyzed with Felix32 Fluorescence Analysis

Software (Photon Technology International, Inc., Canada). Fluorescence anisotropy measurements used vertically polarized excitation at 488 nm and emission detection in the vertical and horizontal planes at 520 nm, corresponding to the spectral properties of fluorescein. In general, 10–50 n*M* of a given fluorescein-labeled histone is used to have the histone concentration at least one to two orders of magnitude lower than the K_D estimated from published work (Jacobson *et al.*, 2000; Pizzitutti *et al.*, 2006). Using slit widths of 0.25–1.0 mm yields signal intensity in the range of 20,000–200,000 fluorescence counts per channel. The slit widths can be increased or decreased to control the amount of light passing through as needed.

Before bromodomain titrations are performed several controls and determination of the *g*-factor need to be performed. The background fluorescence intensity of the buffer, or "working solution" is measured ($I_{vv(bg)}$ and $I_{vh(bg)}$) and subtracted from each measurement made throughout the titration. This means that the value given in Eq. (16.1) as I_{vv} is actually $I_{vv} - I_{vv(bg)}$ and similarly for horizontal emission. This correction removes contributions from the buffer, which scatters a small amount of the incident light into the detector. If the titration is performed by the progressive addition of bromodomain into a given amount of labeled histone, then fluorescence intensities need to be corrected for dilution. Alternatively, if multiple preincubated samples are prepared, where volumes of decreasing buffer and increasing bromodomain sample are added to maintain a constant concentration of fluorophore, no dilution corrections are necessary, although random error needs to be carefully controlled. The *g*-factor (I_{hv}/I_{hh}) corrects for variations in detection efficiencies of the instrumentation for different emission polarizations. This correction factor is determined separately using only fluorophore or prior to taking the first sample measurement by switching the excitation polarization to horizontal and measuring the I_{hv} and I_{hh} for each channel. Although the *g*-factor does not change much over time, it should be measured regularly, especially if changes are made in the instrumental setup (slit widths, new lamp, etc.).

Steady-state fluorescence anisotropy measurements were performed as forward titrations, in 20 m*M* Tris (pH 7.8) and 100 m*M* NaCl. Acetylhistone and bromodomain samples were prepared fresh and preincubated for at least 15 min prior to measurement to ensure equilibrium was reached. The time required for complexes to reach equilibrium was determined by monitoring the fluorescence anisotropy until no further anisotropy change was observed, indicating no net change in complex formation. Each data point is measured for approximately 30–60 s to ensure the anisotropy value is unchanged. Fluorescein-labeled histone is titrated with 0–0.5 m*M* (final concentration) of Bd or until the maximum anisotropy was attained indicating saturation of the histone-binding site. Anisotropy values for each data point are the average of at least three independent titrations. The change in anisotropy was plotted versus protein concentration and fit to the two-state model developed in Eq. (16.9) using a nonlinear least-squares algorithm to attain the K_D values (Fig. 16.2). The bromodomain shows a strong preference for the histone acetylated at K9 over K14, which have binding constants of 1.1 and 30.6 μM, respectively.

Careful controls should be performed to ensure any changes in anisotropy are exclusively a function of complex formation, rather than dilution, aggregation, solvent conditions, time, etc. Titrations of unmodified histone H3 with Bd show negligible anisotropy change in the concentration range and solution conditions indicating that Bd does not interact with a protein in the absence of acetyllysine. Parallel titrations substituting Bd with BSA were performed to rule out nonspecific binding to the fluorescein-labeled histone, as BSA is not expected to interact with histones. BSA concentrations up to 10 mM caused no change in anisotropy (data not shown) confirming that anisotropy changes result from the formation of the Bd–histone complex. Titrations without bromodomain using the buffer itself resulted in no anisotropy change.

No discussion of binding analysis would be complete without a few remarks on error analysis. To measure the variability of the data set, standard deviation is used. In general, the lower the standard deviation, the closer are the individual data points to the mean. Standard deviation describes not only the variability of the data, but as a description of the margin of error, the measure of confidence in the statistical conclusion. The primary source

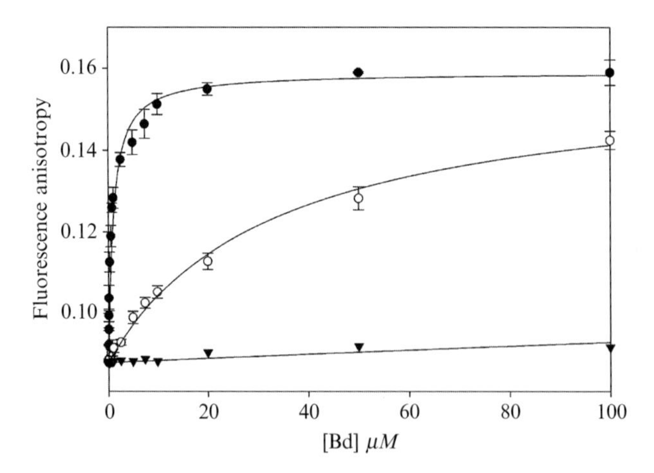

Figure 16.2 Plot of the change in fluorescence anisotropy as a function of bromodomain concentration. A 20 nM fluorescein-labeled histone was titrated with 0–200 μM bromodomain. The anisotropy is plotted as a function of the Bd concentration for acetylhistone K9 (●) and K14 (○). The titration of an unmodified histone (▼) with Bd is shown as a reference. Data were fit to Eq. (16.9).

of error is random in nature (e.g., sample handling, pipetting, etc.) and such variations in measurement should not be confused with causal variations. For an identical measurement that is made multiple times, error analysis is determined by squaring the difference of each data point (x_i) with the mean of the data set (x). The sum of these values is divided by the total number of measurements (N) to get the average, of which the square root is taken to give the standard deviation. Using summation notation the standard deviation (σ) is described as (Harnett, 1982)

$$\sigma = \sqrt{\frac{1}{N}\sum_{i=1}^{N}(x_i - \bar{x})^2}. \tag{16.10}$$

This information is applied to each data point using, for example, error bars spanning one standard deviation unit to give a sense of the variation of each data point. This is important because the standard deviation tells us how far from the mean the data points are and has the same units. Perhaps more importantly standard deviation is a measure of uncertainty and if, for example, the data points are fit with an expression derived based on assumptions, as discussed here, and the fit does not fall within the standard deviation ranges for multiple data points, then the assumptions (and the model) need revision.

7. Kinetic Analysis

Pre-steady-state fluorescence anisotropy was performed to compare the time-dependent binding features for Bd–histone interactions. To accomplish this, the anisotropy was measured for both acetylation sites at different Bd concentrations as a function of time. Time-resolved fluorescence anisotropy measurements are performed using a dual-syringe system. The two syringes of the stopped-flow apparatus are loaded with (1) Bd in reaction buffer consisting of 20 mM Tris (pH 7.8) and 100 mM NaCl and (2) labeled histone in the same reaction buffer. Equal volumes of Bd and the indicated acetylhistone in reaction buffer were mixed and fluorescence anisotropy measured until no further change in anisotropy was observed (~60 min). Final concentrations of histone ranged from 50 to 200 nM and Bd from 1 to 50 μM to achieve a large molar excess. Association reactions were initiated by mixing equal volumes of bromodomain and histone solution. Control reactions were performed by mixing equal volumes of histone and reaction buffer in the absence of bromodomain to measure the anisotropy of the fluorescein-labeled histone alone. Data points were taken every 10 s with an integration time of 1 s to collect fluorescence signal for each data point over a full-time course. If low signal strength is a problem due to short-time intervals, a stronger signal can be attained by increasing the concentration of fluorescein-labeled histone, changing the excitation and emission slit widths, or averaging multiple runs to increase the signal to noise ratio.

Analogous to the section on equilibrium binding, the pre-steady-state analysis described here must relate the raw data to binding models that represent the interaction between bromodomain and acetylhistone proteins. The complete solution to Eq. (16.4) for time-dependent changes in unbound and bound populations is discussed in a treatise by Eccleston *et al.* for stopped-flow analysis (Harding and Chowdhry, 2001). Here, we provide a distilled series of equations and solutions as it pertains to kinetic analysis of data acquired using time-based fluorescence anisotropy. The rate for most bimolecular interactions is second order, meaning the complex association rate is controlled by the concentration of two species (see Eq. 16.4). Second-order reactions can be very complicated and are best studied by using a great excess of one species ([Bd] $\gg$ [H]), such that the binding reaction mimics a first-order process. This pseudo-first-order reaction assumes the concentration of the species in excess does not change over the course of the measurement. Based on this assumption, k_{on}[Bd] is effectively a constant (k'), permitting us to set k' equivalent to k_{on}[Bd]. Therefore, substituting k' into Eq. (16.4) at equilibrium, we get an expression for the rate of complex formation:

$$\frac{d[\text{Bd–H}]_i}{dt} = k'[\text{H}]_i - k_{\text{off}}[\text{Bd–H}]_i \tag{16.11}$$

where $[\text{H}]_i$ and $[\text{Bd–H}]_i$ are the histone and bromodomain–histone concentration at time i. If $[\text{H}]_i = [\text{H}]_{eq} - [\text{Bd–H}]_i$ is substituted into Eq. (16.11) and rate and concentration terms are isolated, we get an expression that takes on the form of a first-order reaction (Harding and Chowdhry, 2001):

$$\frac{d[\text{Bd–H}]_i}{dt} = (k' + k_{\text{off}})([\text{Bd–H}]_{eq} - [\text{Bd–H}]_i). \tag{16.12}$$

The value for $-k_{obs}t$ is proportional to the natural log of $([\text{Bd–H}]_{eq} - [\text{Bd–H}]_i)$ and therefore, a plot of $\ln([\text{Bd–H}]_{eq} - [\text{Bd–H}]_i)$ versus time will yield the first-order rate constant k_{obs}. Note the subscript "eq" replaces "T" to draw attention to the population of a given species when the association reaction reaches equilibrium. Solving the above differential equation and fitting to the experimentally determined fluorescence anisotropy data yields

$$A_{(\text{calculated})} = A_{\min} + (A_{\max} - A_{\min})[1 - e^{-(k_{obs}t)}]. \tag{16.13}$$

Careful examination of Eq. (16.13) indicates that like Eq. (16.9), the first term sets the minimum and maximum anisotropies at the measured values, and the term in brackets is the exponential that correlates the observed complex association rate with fraction of bound histone over the time course and has a value between 0 and 1. The time-dependent association reaction can be expressed conveniently by the anisotropy function given above. Because the term ($k_{obs}t$) is unitless and the measured anisotropy at each time (t) is in seconds, the units for the pseudo-first-order rate constant are k_{obs} is s^{-1}.

Plots of the change in fluorescence anisotropy as a function of time are shown for the association of 50 n*M* acetylhistone mixed with molar excess of Bd (Fig. 16.3). Data reveal the time-dependent increase in fluorescence anisotropy. The kinetic data were fit to a single-exponential growth model (Eq. 16.13) to calculate the observed association rate (k_{obs}) (Hoopes *et al.*, 1992). The observed association rates for 1–20 μM Bd ranged from 1.02×10^{-3} to 6.58×10^{-3} s^{-1} for acetylated K9 and 0.89×10^{-3} to 1.33×10^{-3} s^{-1} for acetylated K14, respectively. From these data, the half-time of binding ($t_{1/2}$) can be determined using the following relationship:

$$t_{1/2} = \frac{\ln([H]_T/[H]_i)}{k_{obs}} = \frac{0.69}{k_{obs}}. \tag{16.14}$$

Because $k_{obs}t$ is proportional to the natural log of the ratio of $[H]_T/[H]_i$, the ratio is equivalent to two when 50% of the total histone proteins are bound by bromodomain. Comparison of the values for the pseudo-first-order rate constants at 20 μM Bd corresponds to a half-time of binding ($t_{1/2}$) 1.8 and 8.6 min for histones acetylated at K9 and K14, respectively.

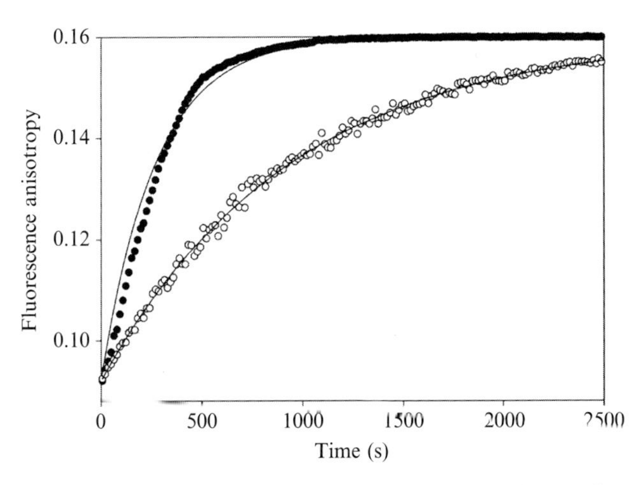

Figure 16.3 Time-dependent association of bromodomain with acetylhistones. Representative curves are shown corresponding to the titration of 50 n*M* of acetylhistone K9 (●) and K14 (○) histone with 10 μM Bd. Data were fit to Eq. (16.13).

Determination of the on- and off-rates that describe Eq. (16.4) requires dissecting the first-order rate constant k_{obs} derived from the previous plot. Because k_{obs} was set equal to $k' + k_{off}$ and $k' = k_{on}[Bd]$, the rate of complex formation (Eq. 16.11) is represented by the following equation (Fierke and Hammes, 1996; Kupitz *et al.*, 2008):

$$k_{obs} = k_{on}[Bd] + k_{off}. \tag{16.15}$$

If the reaction is performed for a range of bromodomain concentrations, the dependence of k_{obs} on [Bd] is a line that can be fit to Eq. (16.15). Here, the slope of a plot of k_{obs} versus [Bd] is the bimolecular association rate constant, k_{on}, and the y-intercept is the dissociation rate constant, k_{off} (Fig. 16.4). The k_{on} for the Bd high-affinity target site acetylated K9 was determined to be 312 M^{-1} s^{-1}. The value obtained from the y-intercept, 0.44×10^{-3} s^{-1}, suggests a slow dissociation, k_{off}. The k_{on} for the Bd nonspecific or low-affinity acetylated K14 site was determined to be 29.9 M^{-1} s^{-1}. The value obtained from the y-intercept, 1.02×10^{-3} s^{-1}, suggests a comparable dissociation rate. These data represent a greater than 10-fold difference in the on-rate and roughly a twofold difference in the off-rate.

The off-rate has proven to be sufficiently accurate using the aforementioned approach, but it can be determined directly by loading a syringe with preincubated bromodomain–histone complex in reaction buffer and rapidly mixing with an excess of nonlabeled histone. The idea is that as complexes dissociate and reassociate in an attempt to reach equilibrium, there will be a greater probability of the bromodomain finding a nonlabeled histone due to the large excess over labeled histones. These "unseen" complexes will make

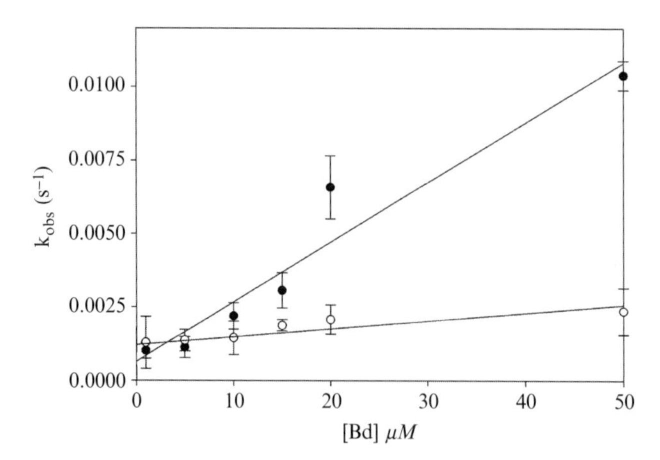

Figure 16.4 Plot of the observed association rate at different bromodomain concentrations is shown for acetylhistone K9 (●) and K14 (○). Data were fit to Eq. (16.15).

the increasing population of labeled histones appear unbound (and therefore a decreasing anisotropy with time). The amount of Bd–H complex decreases at a rate proportional to its concentration and is fit using an exponential decay (Patrick and Turchi, 2001). Control reactions can be performed in an analogous manner to the association experiments to measure any inherent changes in anisotropy that may be attributable to dilution.

The k_{on} and k_{off} values were used to calculate the equilibrium constants (Eq. 16.5) for comparison with steady-state determinations at 298 K. The K_D values calculated from k_{on} and k_{off} values for each acetylation site are 1.4 and 34.3 μM for Bd binding to histone H3 acetylated at K9 and 14, respectively. Comparing these values to steady-state measurements shown in Fig. 16.2, the K_D values were determined to be 1.1 and 30.6 μM for Bd binding to K9 and 14. Comparing the two approaches indicates a 10–20% difference, which is within experimental error.

8. Determination of Thermodynamic Parameters

Thermodynamic analysis is a well-established means for dissecting the molecular driving forces that give rise to stability and specificity attributes of protein complexes. The extent to which histone binding by the bromodomain will be enthalpically or entropically driven, is a function of solution conditions, such as temperature, salt concentration, and pH, giving information about the hydrophobic and electrostatic effects that drive complex formation. Studies of interactions ranging from specific to nonspecific show large variations in the enthalpic and entropic contributions for a given protein (Frank *et al.*, 1997; Oda *et al.*, 1998), suggesting that the energetics of the binding event is unique to the particular protein–ligand interface. This section describes a detailed series of fluorescence measurements to determine the dependence of the K_D values with temperature. Subsequent analysis permits the specific quantification of the driving forces contributing to complex formation and specificity in terms of the thermodynamic parameters (i.e., ΔG, ΔH, ΔS, ΔC_p) that describe the system.

The overall affinity for the interaction of any two biomolecules is determined by the free energy difference between any two states, which is proportional to the experimentally determined equilibrium constant by the well-established relationship:

$$\Delta G^\circ = -RT\ln\left(\frac{1}{K_{obs}}\right) = \Delta H^\circ - T\Delta S^\circ, \qquad (16.16)$$

where the free energy (ΔG°) is a function of the natural log of ($1/K_{obs}$) with temperature (T) and R is the gas constant (8.315 J/mol K). It should be noted that rather than K_D, the observed equilibrium constant is given as

K_{obs} due to the dependence of binding equilibria on solution conditions. The stability of the bound complex is determined by the differences in the noncovalent interactions between the Bd and each histone as solution conditions are varied. By convention, many biologically oriented researchers will want to convert these values to calories by using the appropriate value for R (1.987 cal/mol K).

The contribution of the hydrophobic effect to the stability of each Bd–histone complex was determined by analyzing the temperature dependence of K_{obs}. It should be noted that values are converted to the corresponding association constant using the relationship $K_A = 1/K_D$ and fit to Eq. (16.17) using a van't Hoff plot to determine the standard heat capacity change (ΔC_p) for the complexes (Ha *et al.*, 1989; Jen-Jacobson *et al.*, 2000; Thompson and Chandrasekaran, 2008):

$$\ln\left(\frac{1}{K_{calc}}\right) = \ln\left(\frac{1}{K_{obs}}\right) - \frac{\Delta H^{\circ}_{obs}}{R}\left(\frac{1}{T_i} - \frac{1}{T^*}\right) + \frac{\Delta C_p}{R}\left[\left(\ln\frac{T_i}{T^*}\right) + \frac{T^*}{T_i} - 1\right]. \quad (16.17)$$

Here, $\ln(1/K_{obs})$ and enthalpy (ΔH°) are values at $T^* = 298$ K. The choice of reference temperature is arbitrary and does not affect the results of the fit. Equation (16.17) assumes that ΔC_p is independent of temperature within experimental uncertainty (Frank *et al.*, 1997). It has been shown that the thermodynamics of processes where ΔC_p is large and temperature independent can be completely characterized by the heat capacity and two temperatures (Baldwin, 1986; Record *et al.*, 1991). Therefore, the determination of the change of enthalpy and entropy with temperature are given as

$$\Delta H^{\circ} = \Delta C_p(T - T_H), \quad (16.18)$$

$$\Delta S^{\circ} = \Delta C_p \ln(T/T_S), \quad (16.19)$$

where T_H is the temperature at which $\Delta H^{\circ} = 0$ and occurs at the maxima of the van't Hoff plot ($1/K_{obs}$ is maximum) and T_S is the temperature at which entropy (ΔS°) equals 0 and occurs at the minima of a plot of ΔG° versus temperature (ΔG° is minimum).

9. Thermodynamic Measurements

The measurements are performed using the same protocol as described for the steady-state K_D determinations above, with the exception that measurements are performed at specific temperatures in the range from 5 to 40 °C. The fluorometer has a thermal cell fitting to maintain a constant temperature for pre-equilibration and sample measurement. To minimize sample evaporation during experiments, cuvettes are capped during titrations at higher temperatures. The change in the total fluorescence signal, after being adjusted at each titration point for dilution, varied by less than 2% over the course of these measurements indicating evaporation did not adversely affect these measurements.

Binding affinity was determined over a range of temperatures by monitoring the anisotropy of the fluorescein-labeled histones as a function of the Bd concentration. The bromodomain interacts with the histone displaying an acetylated K9 with high affinity ($K_{obs} \sim 1\ \mu M$) over the range of 293–303 K (Fig. 16.5). However, binding at the temperature extremes are weaker ($K_{obs} \sim 4\ \mu M$), representing nearly a fourfold difference in the dissociation constant for the temperature range examined here. Changes in binding strength are observed for both complexes as the temperature is increased from 278 to 313 K. For the interaction of Bd with acetylated K9, binding affinity increased from 283 to 300 K, followed by a decrease in the binding strength as the temperature increases beyond 300 K. This is due to an unfavorable binding enthalpy at lower temperature that becomes favorable at higher temperature, as described in more detail below. A plot of the natural log of $1/K_{obs}$ as a function of temperature is used to attain the change in heat capacity and the observed enthalpy upon complex formation (given in Eq. 16.17) (Fig. 16.5). The temperature dependence of K_{obs} for the interaction of Bd with histone acetylated at K9 and 14 yields ΔC_p values of -1.33 and -0.46 kcal/K mol, respectively. The nonspecific value is roughly a third of that for the specific complex.

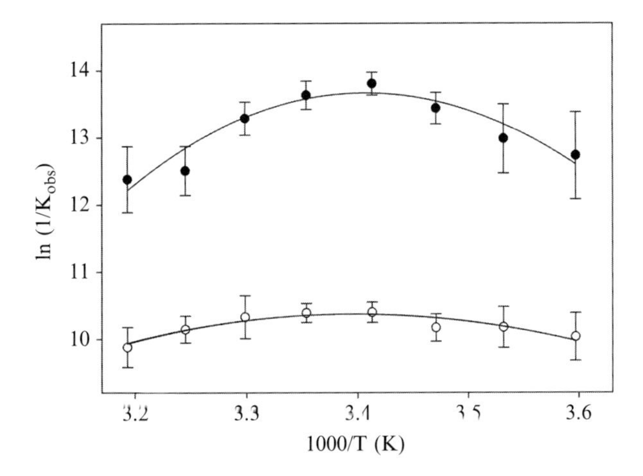

Figure 16.5 Dependence of the observed binding constant on temperature for the interaction of Bd with acetylhistone K9 (●) and K14 (○). $\ln(1/K_{obs})$ versus reciprocal temperature ($1000/T$) is plotted over a range of temperatures (278–313 K) and fit to Eq. (16.17).

The temperature dependence of the enthalpic and entropic contributions to the binding free energy was determined for both complexes using Eqs. (16.18) and (16.19). The resulting thermodynamic profiles of both Bd–histone complexes are shown in Fig. 16.6, which plots the free energy, entropy, and enthalpy as functions of temperature. Two interesting revelations from this type of analysis are the offsetting effects of the entropy and enthalpy across the temperature range and the relative steepness of the slopes. The high-affinity K9 site shows a stronger transition between enthalpy and entropy as the primary driving force across the temperature range. Interestingly, binding is both enthalpically and entropically favorable for Bd binding to the histone acetylated at K9 at physiological temperature (298 K). Site-specific binding for Bd becomes enthalpically favorable and

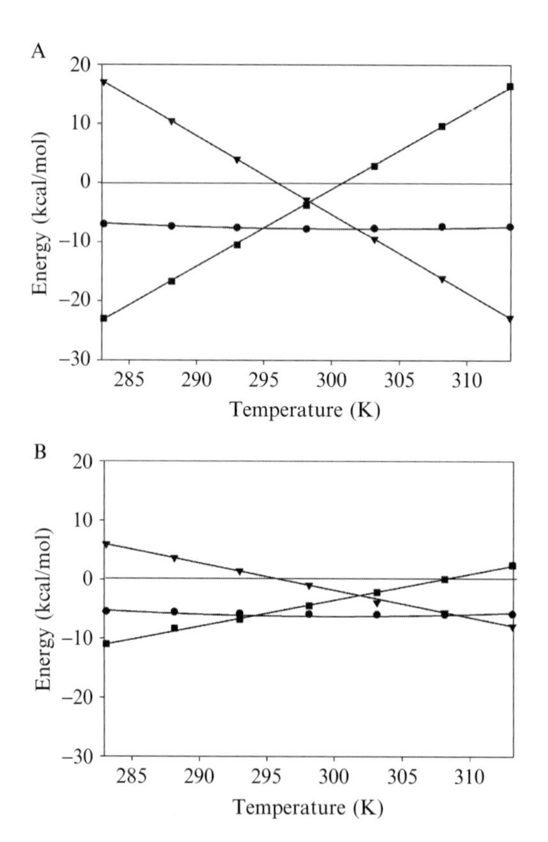

Figure 16.6 Thermodynamic profiles of the interactions of Bd with acetylhistones. Plot of free energy (●), enthalpy (▼), and entropy (■) of binding versus temperature are given for (A) acetylhistone K9 and (B) acetylhistone K14. It should be noted that the entropy is given as $-T\Delta S$ to present the data with enthalpy and free energy in the same units (Thompson and Woodbury, 2001).

entropically unfavorable as temperature increases, such that, at 308 K, association is entirely enthalpically driven. Often, the enthalpy term (ΔH) is dominated by a ligand-induced-folding mechanism that accompanies observable structural changes upon binding (Spolar and Record, 1994). The entropic contribution to the binding free energy has been shown to drive both specific and nonspecific binding (Bulinsk *et al.*, 1985; Takeda *et al.*, 1992). The large positive change in entropy (ΔS), which seems contrary to formation of a higher order complex, is partially derived from the net release of ions and water molecules from the complementary macromolecular surfaces upon protein–ligand association.

10. Developing a Binding Model

The threefold difference in ΔC_p values is a strong indication of the variation in the structures of the two Bd–histone complexes. In fact, the magnitude of the heat capacity change is proportional to the change in water exposed hydrophobic surface area (Eq. 16.20), yielding important information about protein folding and conformational changes that occur upon binding (Brenowitz *et al.*, 1990; Privalov and Gill, 1988). The ΔC_p values determined experimentally were compared to the values calculated based on the change in solvent exposed surface area (Spolar and Record, 1994) using

$$\Delta C_p = 0.32\Delta SA_{np} - 0.14\Delta SA_p, \tag{16.20}$$

where ΔSA_{np} and ΔSA_p are the changes in nonpolar and polar surface area, respectively.

To accomplish this, structural data for several bromodomain–histone complexes derived from the Protein Database (PDB IDs: 2RNY, 2DVQ, 2ROY, 1E6I, 1JSP, 1JM4) were examined. The total surface area for the aforementioned complexes was estimated using Surface Racer (Tsodikov *et al.*, 2002). The average surface area of the bromodomain–histone interface determined from the structures is 300 ± 40 Å^2. This corresponds to a calculated change in heat capacity (ΔC_{calc}) of 0.593 ± 0.079 kcal/K mol. The surface area formed by the Bd-binding pocket with the acetyllysine is 178 ± 16 Å^2 and between other regions of the histone is 151 ± 30 Å^2. Using Eq. (16.20), the estimated ΔC_{calc} contribution is 0.352 ± 0.032 and 0.299 ± 0.059 kcal/K mol for Bd contacts with the acetyllysine and flanking side chains, respectively. If the individual contributions are compared to the experimental results, 80% of the ΔC_p value for binding of Bd to the histone acetylated at K14 is from the interface formed between the hydrophobic-binding pocket and the acetyllysine. The remaining contribution comes from all other contacts between the two proteins and any minor structural perturbations, suggesting that nonspecific binding may be limited to the interface formed with the acetyllysine. Comparing this to the K9 site indicates roughly half of the experimentally determined ΔC_p value is derived from the complete binding interface, as observed in crystal structures. The additional 730 cal/K mol observed in the specific complex is a significantly more dehydrated interface formed by optimal contacts from the flanking side chains of the histone and optimized folding by the Bd. Previous studies concluded that when regions of the protein fold upon binding, significant contributions to the change in heat capacity are made due to the removal of nonpolar regions form solvent exposure (Spolar and Record, 1994). When such variation between different bromodomain complexes are considered, the contact interface alone does not account for approximately 370 Å^2 of water exposed hydrophobic surface area.

Taken together, the results from thermodynamic and kinetic studies are consistent with a position-dependent binding-induced-folding mechanism. The mechanism we propose occurs in the following sequence: (1) the Bd must first discriminate the acetylation state of the peptide, (2) the complementary binding interface forms between Bd and acetylhistone, and (3) certain sites induce folding of the bromodomain to form a thermodynamically stable complex (Fig. 16.7). This stable complex has a higher energetic barrier to dissociation and consequently, has a slower dissociation rate. In this case, the interaction between proteins is more favorable than between unbound proteins and solvent. If the complex is a nontarget or low-affinity site, the many water bridges that are integrated into the binding interface minimize structural alteration of the bromodomain. In the absence of

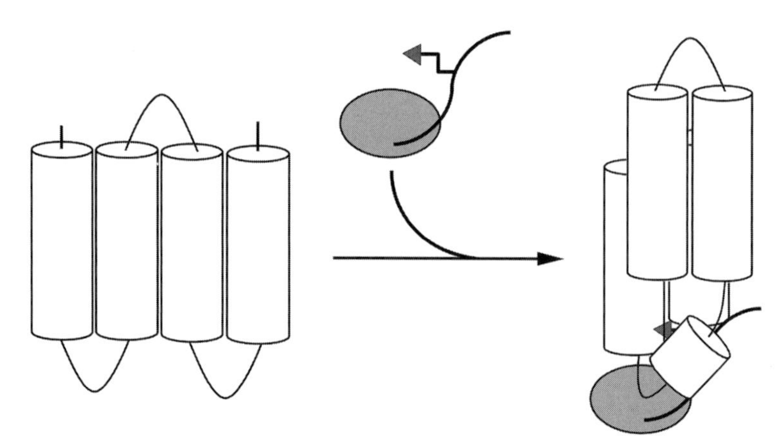

Figure 16.7 Schematic of the proposed bromodomain–acetylhistone binding mechanism. The helical regions (cylinders) of the bromodomain adopt native-like structure upon binding the histone (shaded) protein in which the histone tail is acetylated (triangle).

forming this stable complex, the faster off-rate is attributable to a lower energetic barrier to dissociation. In other words, the interactions between biomolecules are not preferred over solvent molecules simply because many solvent molecules bridge the two proteins. Ultimately, Bd structural changes required to adopt a low-energy configuration are only initiated by certain acetylation sites.

11. Concluding Remarks

Fluorescence anisotropy is an analytical tool well suited to study solution-phase protein interactions due to its convenience, quantitative results, and high sensitivity. Using the methodology discussed herein, we performed a series of kinetic and thermodynamic analysis to characterize the acetylation-dependent binding interactions of histone and bromodomain proteins. The results support an induced folding mechanism, which explains both the specificity and stability in terms of acetylation site preferences among bromodomains.

Acknowledgments

We thank the Thompson lab for helpful discussions. This research was supported by National Institutes of Health Grant GM076055.

References

Baldwin, R. L. (1986). Temperature-dependence of the hydrophobic interaction in protein folding. *Proc. Natl. Acad. Sci. USA* **83,** 8069–8072.

Brenowitz, M., Jamison, E., Majumdar, E., and Adhya, S. (1990). Interaction of the *Escherichia coli* Gal repressor protein with its DNA operators *in vitro*. *Biochemistry* **29,** 3374–3383.

Bulinsk, H., Harmsen, B., and Hilbers, C. (1985). Specificity of the binding of bacteriophage-M13 encoded gene-5 protein to DNA and RNA studied by means of fluorescence titrations. *J. Biomol. Struct. Dyn.* **3,** 227–247.

Chandrasekaran, R., and Thompson, M. (2006). Expression, purification and characterization of individual bromodomains from human polybromo-1. *Protein Expr. Purif.* **50,** 111–117.

Chandrasekaran, R., and Thompson, M. (2007). Polybromo-1 bromodomains bind histone H3 at specific acetyl-lysine positions. *Biochem. Biophys. Res. Commun.* **355,** 661–666.

Chowdhry, S. E. H. A. B. (2001). Protein–Ligand Interactions: A Practical Approach. Oxford University Press, Oxford.

Eccleston, J. F., Martin, S. R., *et al.* (2008). Rapid kinetic techniques. *In* "Biophysical Tools for Biologists: Vol 1 *In Vitro* Techniques," Vol. 84, pp. 445–477. Elsevier Academic Press Inc., San Diego.

Eftink, M., Brand, L., and Johnson, M. (1997). Fluorescence methods for studying equilibrium macromolecule–ligand interactions. *Methods Enzymol.* **278,** 221–257.

Fierke, C. A., and Hammes, G. G. (1996). Contemporary Enzyme Kinetics and Mechanism. Academic Press, New York, NY.

Frank, D. E., Saeker, R. M., Bond, J. P., Capp, M. W., Tsoidkov, O. V., Melcher, S. E., Levandoski, M. M., and Record, M. T. (1997). Thermodynamics of the interactions of lac repressor with variants of the symmetric lac operator: Effects of converting a consensus site to a non-specific site. *J. Mol. Biol.* **267,** 1186–1206.

Futaki, S., Tatsuto, K., Shiraishi, Y., and Sugiura, Y. (2004). Total synthesis of artificial zinc-finger proteins: Problems and perspectives. *Biopolymers* **76,** 98–109.

Ha, J.-H., Spolar, R. S., and Record, M. T. (1989). Role of the hydrophobic effect in stability of site-specific protein–DNA complexes. *J. Mol. Biol.* **209,** 801–816.

Harding, S. E., and Chowdhry, B. (eds.) (2001). *In* "Protein–Ligand Interactions: A Practical Approach, Structure and Spectroscopy," Oxford University Press, Oxford.

Harnett, D. L. (1982). Statistical Methods. Addison-Wesley, Reading, MA.

He, S., Bauman, D., Davis, J. S., Loyola, A., Nishioka, K., Gronlund, J. L., Reinberg, D., Meng, F., Kelleher, N., and McCafferty, D. G. (2003). Facile synthesis of site-specifically acetylated and methylated histone proteins: Reagents for evaluation of the histone code hypothesis. *Proc. Natl. Acad. Sci. USA* **100,** 12033–12038.

Hoopes, B. C., Leblanc, J. F., and Hawley, D. K. (1992). Kinetic analysis of yeast TFIID–TATA box complex formation suggests a multistep pathway. *J. Biol. Chem.* **267,** 11539–11547.

Jacobson, R. H., Ladurner, A. G., King, D. S., and Tjian, R. (2000). Structure and function of a human TAF(II)250 double bromodomain module. *Science* **288,** 1422–1425.

Jen-Jacobson, L., Engler, L. E., and Jacobson, L. A. (2000). Structural and thermodynamic strategies for site-specific DNA binding proteins. *Structure* **8,** 1015–1023.

Kohler, J. J., and Schepartz, A. (2001). Kinetic studies of Fos Jun DNA complex formation: DNA binding prior to dimerization. *Biochemistry* **40,** 130–142.

Kupitz, C., Chandrasekaran, R., and Thompson, M. (2008). Kinetic analysis of acetylation-dependent Pb1 bromodomain–histone interactions. *Biophys. Chem.* **136,** 7–12.

Lakowicz, J. R. (ed.) (1991). *In* "Topics in Fluorescence Spectroscopy," Plenum Press, New York, NY.

Lakowicz, J. R. (ed.) (2006). *In* "Principles of Fluorescence Spectroscopy," Plenum Press, New York, NY.

Lundblad, J. R., Laurance, M., and Goodman, R. H. (1996). Fluorescence polarization analysis of protein–DNA and protein–protein interactions. *Mol. Endocrinol.* **10,** 607–612.

Michael, A., and Shrogen-Knaak, C. L. P. (2004). Creating designer histones by native chemical ligation. *Methods Enzymol.* **375,** 62–77.

Oda, M., Furukawa, K., Ogata, K., Sarai, A., and Nakamura, H. (1998). Thermodynamics of specific and non-specific DNA binding by the c-Myb DNA-binding domain. *J. Mol. Biol.* **276,** 571–590.

Patrick, S. M., and Turchi, J. J. (2001). Stopped-flow kinetic analysis of replication protein A-binding DNA. *J. Biol. Chem.* **276,** 22630–22637.

Pizzitutti, F., Giansanti, A., Ballario, P., Ornaghi, P., Torreri, P., Ciccotti, G., and Filetici, P. (2006). The role of loop ZA and Pro371 in the function of yeast Gcn5p bromodomain revealed through molecular dynamics and experiment. *J. Mol. Recognit.* **19,** 1–9.

Privalov, P. L., and Gill, S. J. (1988). Stability of protein–structure and hydrophobic interaction. *Adv. Protein Chem.* **39,** 191–234.

Record, M. T., Ha, J. H., and Fisher, M. A. (1991). Analysis of equilibrium and kinetic measurements to determine thermodynamic origins of stability and specificity and mechanism of formation of site-specific complexes between proteins and helical DNA. *Methods Enzymol.* **208,** 291–343.

Spolar, R. S., and Record, M. T. (1994). Coupling of local folding to site-specific binding of proteins to DNA. *Science* **263,** 777–784.

Takeda, Y., Ross, P. D., and Mudd, C. P. (1992). Thermodynamics of Cro protein–DNA interactions. *Proc. Natl. Acad. Sci. USA* **89,** 8180–8184.

Thompson, M. (2009). Polybromo-1: The chromatin targeting subunit of the PBAF complex. *Biochimie* **91,** 309–319.

Thompson, M., and Chandrasekaran, R. (2008). Thermodynamic analysis of the acetylation dependence of bromodomain–histone interactions. *Anal. Biochem.* **374,** 304–312.

Thompson, M., and Woodbury, N. (2001). Thermodynamics of specific and non-specific DNA-binding by cyanine dye labeled DNA-binding domains. *Biophys. J.* **81,** 1793–1804.

Tsodikov, O. V., Record, M. T., and Sergeev, Y. V. (2002). Novel computer program for fast exact calculation of accessible and molecular surface areas and average surface curvature. *J. Comput. Chem.* **23,** 600–609.

Valeur, B. (2001). Molecular Fluorescence: Principles and Applications. Wiley-VCH, Weinheim, Germany.

CHAPTER SEVENTEEN

Thermodynamics of 2-Cys Peroxiredoxin Assembly Determined by Isothermal Titration Calorimetry

Sergio Barranco-Medina *and* Karl-Josef Dietz

Contents

Biochemistry and Physiology of Plants, W5-134, Bielefeld University, Bielefeld, Germany

Methods in Enzymology, Volume 466
ISSN 0076-6879, DOI: 10.1016/S0076-6879(09)66017-1

Abstract

Oligomerization is a frequently encountered physical characteristic of biological molecules that occurs for a wide number of transcription factors, ion channels, oxygen-carrying macromolecules such as hemocyanin and enzymes. On the other hand, unwanted protein oligomerization can lead to the formation of pathogenic structures related with Alzheimer and other diseases. Self-assembly is also a well-described phenomenon within peroxiredoxins, a family of thiol peroxidases. Peroxiredoxin hyperaggregate formation is the key mechanism that triggers the switch between Prx activity as peroxidase and chaperone. The oligomerization process is fundamental for understanding the multiple peroxiredoxin function. The chapter gives a detailed description of typical 2-Cys Peroxiredoxin oligomerization using isothermal titration calorimetry (ITC) and provides a recipe for studying the thermodynamic parameters of peroxiredoxin assembly, that is, association and dissociation constant, enthalpy, entropy, and the Gibbs free energy of the process.

Abbreviations

2-CysPrx	2-cysteine peroxiredoxin
CC	critical concentration
CTC	critical transition concentration
DTT	dithiothreitol
HMW	high molecular weight
ITC	isothermal titration calorimetry

1. Introduction

Protein oligomerization can be defined as the process of generating protein assemblies composed of a small number of identical component monomers. Oligomers are formed by association of monomers or by depolymerization of large protein polymers. Homooligomerization is found in 50–70% of known complexes (Levy *et al.*, 2006). This high incidence in nature implies high functionality. First, there is a scale in terms of economy. Binding of monomers to build an oligomeric protein minimizes the genome sizes and the multiple subunits can fold and unfold more readily than a single large protein. Second, the generation of intermolecular interfaces increases the sites for regulation by providing combinatorial specificity, as well as allosteric mechanisms in activation and inhibition.

Third, formation of oligomers can enhance protein stability but also enables disassembly (Meng *et al.*, 2009). For these reasons, evolution has favored homooligomers versus single chain multidomain proteins.

In many oligomerization processes, the degree of aggregation depends on subunit concentration and solvent activity. The effect of water is critical for the dissociation of oligomers whereas it seems to be less crucial in the case of nondissociating macromolecules. Furthermore, the degree of aggregation of some macromolecules, for example, hemoglobin, is sensitive to ligand binding (Wyman and Stanely, 1990). The solubility of oligomer subunits is also an important aspect of oligomerization. About two-thirds of the amino acid residues located in the intersubunit interfaces within oligomers is nonpolar while one-fifth is polar. The latter are essential to facilitate solubilization of the unassociated state of the subunits (D'Alessio, 1999).

Peroxiredoxins (Prx) are thiol peroxidases of ubiquitous occurrence which catalyze the reduction of a broad range of peroxide substrates. Prxs have been grouped into 1-CysPrx and 2-CysPrx, the latter being divided into atypical and typical 2-CysPrx according to the number of cysteinyl residues implicated in catalysis as well as the formation of either inter- or intramolecular disulfide bonds during the reaction mechanism (Barranco-Medina *et al.*, 2009; Dietz, 2003). Initially, Prx function was exclusively seen in terms of turnover number moderate to low peroxide detoxification activity partly compensated by its high cellular concentration. Subsequently, the floodgate theory linked 2-CysPrx to cell signaling. While under normal conditions 2-CysPrx reduces H_2O_2 and other peroxide substrates including peroxinitrite, a transient, for example, stress-dependent overproduction of H_2O_2 facilitates hyperoxidation and inhibition of Prx. Inhibited 2-CysPrx enables local spreading of H_2O_2 as signal (Wood *et al.*, 2003). Thus, 2-CysPrx appears to play a role as redox sensor in signaling process. Different Prx types are known that occur in a variety of aggregation states in solutions, for example, dimers, tetramers, hexamers, octamers, decamers, dodecamers, and dodecahedron species (Aran *et al.*, 2009; Barranco-Medina *et al.*, 2009; Hall *et al.*, 2009). Typical 2-CysPrx oligomerization has extensively been characterized by diverse methods such as gel filtration, analytical ultracentrifugation, and microthermocalorimetry. Also the crystal structure has been resolved under different redox states (Karplus and Hall, 2007). During homooligomerization, four, five, or six dimers form a donut-like shaped oligomer with a central hole. Decamer formation seems to occur most frequently. The possibility that oligomerization is an artifact due to high Prx concentrations used for crystal formation may be excluded since decamers and other higher mass oligomers have also been observed under low Prx concentration. The interconversion of 2-CysPrx dimers to oligomer is a dynamic process controlled by different chemical and physical parameters. With some exceptions, high ionic strength, high pH, phosphorylation, and reducing conditions favor decamerization (Barranco-Medina *et al.*, 2009). In 2004, Jang *et al.* described the physiological function of Prx oligomers as chaperone. The change in quaternary 2-CysPrx structure enables a switch from peroxidase when present as decamers to chaperone-like activity as high molecular weight form (HMWe.g., dodecamers). 2-CysPrxs, chaperones, and chaperone-assisted proteases seem to share similar multioligomerization properties in which the association of mono- or dimers leads to the formation of toroid- or ring-like structures (Jang *et al.*, 2004; Jorgensen et al., 2003; Ogura and Wilkinson, 2001; Yonehara *et al.*, 1996). Heat and other physical and chemical factors stimulate the formation of these assemblies that function as molecular machines. Furthermore, mono-/dimer–oligomer transitions offer a mechanism to sense state-specific protein concentration as observed for reduced-Prxs at the threshold of the critical transition concentration (CTC) (Barranco-Medina *et al.*, 2008).

The energetics of dissociation of Prxs oligomers can be estimated from ITC dilution experiments. Such measurements involve sequential injection of a concentrated protein solution of about 50–100 μmol/L into the buffer contained in the stirred calorimeter cuvette. Any exo- or endothermic reaction results in a compensatory temperature readjustment of the sample cell resulting in a transient down- or upward deflection of the baseline signal. Consecutive injections within defined time intervals produce series of exo- or endothermic heat peaks. With increasing number further injections give progressively smaller peaks as the protein concentration of Prx builds up in the cuvette. From the data association parameters such as K_d can be calculated if the transition between the heat change during dissociation and the dilution heat after saturation resolves in at least two measuring points. To this end, an appropriate model is fitted to the absolute heat changes per injection during titration. A buffer-mixing control by injection of equal volumes of buffer solution is used as a blank to subtract the heat produced by dilution.

This chapter offers a protocol for characterizing Prx oligomerization/dissociation using calorimetric dilution experiments with a MicrocalVP calorimeter (Northampton, MA) which is adaptable to the measurement of other oligomerization processes. In the following, first a simple mathematical model will be derived for dimer–decamer transition, in the second part the method for the isothermal titration calorimetry (ITC) will be given in detail, as a general guideline and for two experimental protocols.

2. Dimer–Decamer Equilibrium

As depicted above, 2-CysPrx switches between a decameric and dimeric state. A dimer–decamer model can be derived as follows. The developed mathematical model can be used as template for other oligomerization states of Prxs by simply following the equations and considering the degree of oligomerization, that is, hexamer, octamer, dodecamer, etc. Throughout the experiments, the total concentration of reactant $[M_T]$ is the known independent variable. Due to the absence of monomers within 2-CysPrx, it can be assumed that a decamer $[M_5]$ is constituted by five dimers $[M_1]$ following the equilibrium:

$$5M_1 \leftrightarrow M_5 \tag{17.1}$$

The association constant L_5 is defined as

$$L_5 = \frac{[M_5]}{[M_1]^5} \tag{17.2}$$

$$[M_5] = L_5[M_1]^5 \tag{17.2'}$$

The total macromolecule concentration $[M_T]$ is defined as the sum of concentrations of the different species present in the equilibrium

$$[M_T] = [M_1] + 5[M_5] \tag{17.3}$$

and substituting $[M_5]$ as depicted in Eq. (17.2)

$$[M_T] = [M_1] + 5L_5[M_1]^5 \tag{17.4}$$

α defines the fraction of monomers contained within a j-mer $\alpha = j[M_j]/[M_T]$

$$\alpha_1 = [M_1]/[M_T] \tag{17.5}$$

Therefore, $[M_T]$ can also be represented as function of α_1

$$[M_T] = \frac{[M_1]}{\alpha_1} = \alpha_1^{-1}[M_1] \tag{17.6}$$

To express $[M_T]$ as function of $[M_1]$ and L_5, we need some additional equations

$$\alpha_1 = \frac{[M_1]}{[M_T]} = \frac{[M_1]}{([M_1] + 5L_5[M_1]^5)} = \frac{[M_1]}{([M_1](1 + 5L_5[M_1]^4))} \tag{17.7}$$

$$\alpha_1 = (1 + 5L_5[M_1]^4)^{-1} = \frac{1}{(1 + 5L_5[M_1]^4)} \tag{17.7'}$$

$$1 + 5L_5[M_1]^4 = \frac{1}{\alpha_1} \tag{17.8}$$

$$5L_5[M_1]^4 = (1 - \alpha_1)/\alpha_1 \tag{17.8'}$$

$$[M_1] = \left((1 - \alpha_1)/5L_5\alpha_1\right)^{1/4} \tag{17.8''}$$

and substituting $[M_1]$ and α_1 in Eq. (17.6)

$$[M_T] = \alpha_1^{-1}\left(\frac{(1-\alpha_1)}{5L_5\alpha_1}\right)^{1/4} \tag{17.9}$$

To obtain the association constant $L_5\alpha_1$ we must develop some more formulation

$$\alpha_1 + \alpha_5 = 1 \tag{17.10}$$

and expressing $\alpha_1 + \alpha_5$ by its values according to $\alpha = j[M_j]/[M_T]$

$$\frac{[M_1]}{[M_T]} + \frac{5[M_5]}{[M_T]} = 1 \tag{17.11}$$

$$[M_5] = \frac{([M_T] - [M_1])}{5} \tag{17.3'}$$

The equilibrium constant is defined as

$$L_5 = \frac{[M_5]}{[M_1]^5} \tag{17.12}$$

By substituting $[M_5]$ using Eq. (17.3′) gives

$$L_5 = \frac{([M_T] - [M_1])}{5[M_1]^5} = \frac{[M_T]}{5[M_1]^5} - \frac{[M_1]}{5[M_1]^5} = \frac{[M_T]}{5[M_1]^5} - \frac{1}{5[M_1]^4} \tag{17.13}$$

By multiplication and division of the first term for $[M_T]^4$ gives

$$\begin{aligned} L_5 &= \left(\frac{[M_T][M_T]^4}{5[M_1]^5[M_T]^4}\right) - \frac{1}{5[M_1]^4} = \left(\frac{[M_T]^5}{5[M_1]^5[M_T]^4}\right) - \frac{1}{5[M_1]^4} \\ &= \frac{1}{5[M_T]^4}\frac{1}{\alpha_1^5} - \frac{1}{5[M_1]^4} \end{aligned} \tag{17.14}$$

Substituting the value of $[M_1] = ((1 - \alpha_1)/(5L_5\,\alpha_1))^{1/4}$ in the previous equation

$$L_5 = \frac{1}{5[M_T]^4}\frac{1}{\alpha_1^5} - \frac{5L_5\alpha_1}{5(1-\alpha_1)} \tag{17.15}$$

$$L_5 + \frac{(L_5\alpha_1)}{(1-\alpha_1)} = \frac{1}{5[M_T]^4\alpha_1^5} \tag{17.16}$$

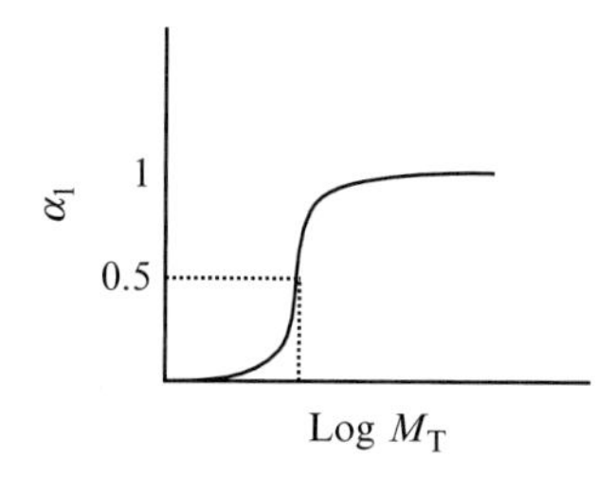

Figure 17.1 Dependency of the oligomerization fraction α_1 on total concentration of 2-CysPrx dimers M_T.

Rearranging:

$$L_5\left(1+\frac{\alpha_1}{(1-\alpha_1)}\right)=\frac{1}{5[M_T]^4\alpha_1^5} \tag{17.17}$$

Hence, the dimer–decamer equilibrium constant can be expressed as

$$L_5=\frac{(1-\alpha_1)}{(5[M_T]^4\alpha_1^5)} \tag{17.18}$$

Considering that in the midpoint transition of the decamer–dimer equilibrium $\alpha = 0.5$, L_5 can be defined as

$$L_5=\frac{(1-0.5)}{(5[M_T]^4(0.5)^5)}=\frac{2^4}{5[M_T]^4} \tag{17.19}$$

Finally, L_5 is defined by the equation

$$L_5=\frac{16}{5[M_T]^4} \tag{17.20}$$

As denoted in Eq. (17.20), the defined dimer–decamer equilibrium constant L_5 depends only on the known independent variable $[M_T]$.

The $[M_T]$ concentration can be related to α_1. The extrapolation of $[M_T]$ to α_1 equal to 0.5 as shown in Fig. 17.1 allows us the determination of L_5. A high cooperativity of association is seen. As will be shown below, the transition from complete dissociation of 2-CysPrx decamer to dimers to stable decamer occurs in an even more narrow transition zone.

3. Isothermal Titration Calorimetry—General Concepts

ITC is a highly advanced and commercially available technique that can be used to quantify protein–protein, protein–ligand, or other molecular interactions. ITC is the only quantitative biochemical method that directly provides the thermodynamic parameters of a reaction, that is, enthalpy ΔH, entropy ΔS, free energy ΔG, and the calorimetric capacity change ΔC_p. A single ITC experiment does not only address the thermodynamics but also the stoichoimetry and equilibrium properties of the reaction such as association constant K_a and dissociation constant K_d, respectively. ITC is based on the heat absorbed or released during binding and is useful for any reaction within the sensitivity range of the instrument.

A typical ITC instrument has two cells, the reference cell filled with water and the sample cell containing the macromolecule, both of which are surrounded by an adiabatic jacket. A sensitive thermopile circuit detects changes in temperature between the reference cell and the sample cell containing the reactants. The temperature differences between both cuvettes are monitored. The feedback circuit activates a variable power to the sample cuvette in order to keep both cells at the same temperature. In the absence of any reaction, the feedback power maintains the initial baseline level. During the experiment, computer controlled aliquots of reactant, for example, ligand, are injected from the syringe into the sample cell containing, for example, the respective receptor. The heat change during interaction can be negative or positive depending on the nature of the reaction. For an endothermic reaction the feedback power circuit administrates heat to the sample cell to keep it at the same temperature as the reference cell. In case of an exothermic reaction the opposite occurs, the heat released in the sample cell is detected and the feedback power is administrated to keep both cells at the same temperature. In a normal ligand–macromolecule interaction, the ligand concentration in the syringe is higher than the macromolecule concentration in the cell to allow for saturation of binding sites on the macromolecule with ligand during the titration experiment. Initially, the injected ligand binds to the free macromolecule and heat is released or absorbed in direct proportion to the amount of binding. Upon ligand titration, macromolecule binding sites become saturated and heat signal diminishes until only heat of dilution is observed.

4. ITC Dilution Experiments

Although mostly used for thermodynamic characterization of ligand–macromolecule interactions, ITC has been shown to be an effective tool to study protein oligomerization and micelle formation (Burrows *et al.*, 1994; Heerklotz and Seelig, 2000; Lovatt *et al.*, 1996; Luke *et al.*, 2005). The injection of reduced-Prx solution into the calorimeter cell containing buffer leads to approximately 1000-fold dilution of Prx. Thus, the Prx concentration in the cell is significantly below the dissociation constant or the CTC. The injected Prx solution presented as decamers disintegrate into dimers in an exothermic manner.

The heat released during the decamer breakdown is measured by the microcalorimeter using the power-compensation feedback. Upon each injection, the Prx concentration in the cell increases, and the magnitude of the heat released diminishes because the extent of dissociation decreases at high Prx concentrations. Thereby the system passes through a sequence of equilibrium states. As the protein concentration approaches the CTC, injected decamers dissociate only partly or cease to dissociate and the heat pulses fall to control levels corresponding to the heat of dilution. This gives rise to small heat dilution responses. In the case of a gradual transition, the data can be fitted as a monomer–dimer dissociation using the Microcal Origin ITC software yielding K_d and ΔH the dissociation constant. Our experience with 2-CysPrx indicates that in a very narrow concentration range near the CTC a rather sudden transition occurs from dissociation heat release to dilution heat (Fig. 17.2).

The abrupt transition with less than the minimal two points needed to establish the equilibrium impeded to deduce the K_d and also the indirect calculation of the entropy change. The lack of decamer–dimer dissociation

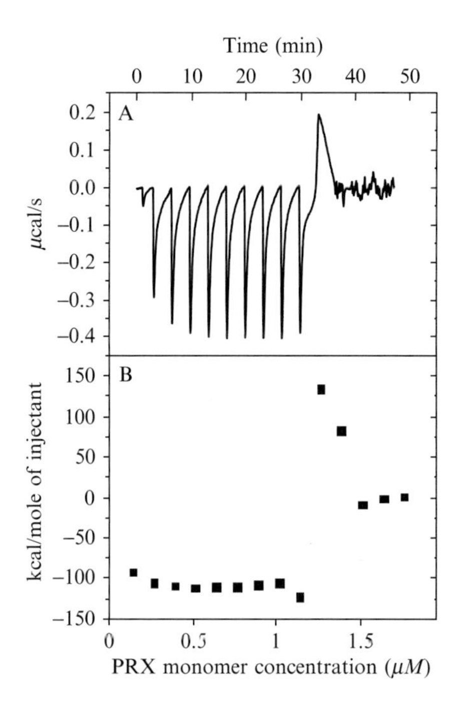

Figure 17.2 ITC dilution experiment performed with 2-CysPrx. (A) Raw titration data, plotted as heat (microcalories per second) versus time (minutes) obtained after injecting reduced typical 2-CysPrx into buffer. (B) Integrated heat responses from the experiment in (A) normalized to the moles of the injected Prx.

program in the ITC software necessitated the recalculation of the heat released for decamer–dimer.

5. Material and Instruments

Protein of interest: for example, purified recombinant 2-CysPrx
Dialysis membrane of ≤15.000 molecular weight cutoff
Sterile filters (2 μm)
Buffer: 40 m*M* K-phosphate, pH 8.0, and 1 m*M* DTT
Isothermal titration calorimeter (Microcal, Calorimetry Sciences, GE Healthcare)
Software (provided by the instrument manufacturer)
Degassing apparatus: Thermo Vac system
UV–visible spectrophotometer
Hamilton gas-tight syringe (2.5-ml) with blunt-end needle long enough to reach the sample cell

6. Experimental Procedure

6.1. Expression of 2-CysPrx protein

The cDNA encoding *Arabidopsis thaliana* or human 2-CysPrx (Barranco-Medina *et al.*, 2008) was cloned into TOPO vector to generate a 6×histidine-tagged fusion protein. After verification of correctness by nucleotide sequencing, plasmid DNA was transformed into *E. coli* BL21 cells. A preculture of 10 mL was grown over night and used to inoculate 1 L of Luria-Bertani medium supplemented with selective antibiotic. Cells were grown at 37 °C until OD reached the value of 0.6. At this point, 0.4 m*M* isopropyl-β-D-thiogalactopyranoside (IPTG) was added to induce the expression for 5 h. Afterward, cells were harvested by centrifugation and stored at − 80 °C until use. It should be noted that terminal tags ease protein purification but may alter aggregation properties (Cao *et al.*, 2007). Thus, either purified native protein or protein processed by enzymatic cleaving of the tag are preferable or need to be included for comparative reasons.

6.2. 2-CysPrx purification

Frozen bacteria were resuspended in 50 m*M* potassium phosphate buffer, pH 8.0, 100 m*M* NaCl, and 1 m*M* dithiothreitol (DTT). Cell lysis was achieved by adding 0.1 mg/mL lysozyme at 4 °C for 1 h, followed by sonication of the suspension. After centrifugation at 12,000 rpm for 60 min, the supernatant containing soluble Prx was loaded onto Ni-nitrilotriacetic acid sepharose preequilibrated with the same buffer at 4 °C. Column was extensively washed with potassium phosphate buffer, pH 8.0, 200 m*M* NaCl, 50 m*M* imidazole, and 1 m*M* DTT. Finally, Prx protein was recovered from the column with elution buffer consisting of 50 m*M* potassium phosphate, pH 8.0, 200 m*M* NaCl, 250 m*M* imidazole, and 1 m*M* DTT. Protein purity was checked by SDS–PAGE separation and subsequent staining with Coomassie brilliant blue.

6.3. Selection of appropriate buffer

Firstly, a buffer must be chosen in which the protein maintains its structure and activity. Prx is known to be quite stable which allows us to work with a broad range of buffers. However, the buffer must not only stabilize the 2-CysPrx but also enable its oligomerization. Based on known biochemical properties, 2-CysPrx is present as a stable decamer, for example, in 40 m*M* potassium phosphate, pH 8.0, and 1 m*M* DTT. For other proteins, buffer type, salts, and protein concentration need to be selected to establish proper conditions in which the protein is present as oligomer before starting the ITC dilution experiments. To ensure oligomeric state of the injected protein, the aggregation state should be checked by gel filtration, analytical ultracentrifugation, SDS–PAGE analysis, or other appropriate methods. High imidazole amounts in the buffer should be avoided because they may induce Prx precipitation during overnight dialysis (Bystrova *et al.*, 2007). Buffers with low proton ionization heat such as phosphate (0.9 kcal/mol) are preferable for ITC to minimize background level.

6.4. Selection of Prx concentration in the syringe

In ITC experiments aimed at analyzing intermolecular interactions, the concentration of cell reactant is an important parameter to consider since it must be about 10–20 times lower than the protein concentration in the syringe. Furthermore, the cell reactant concentration must be between 1 and 500 times the K_a of interaction as pointed out by Wisemann *et al.* (1989). In ITC dilution experiments focused on quantifying dissociation features, only buffer is present in the ITC cell. In this case, the syringe concentration of oligomers is a critical factor. On the one hand, the oligomer concentration in the syringe must be significantly higher than the K_d in order to maintain the oligomeric state in the syringe. On the other hand, it must be low enough for sufficient dilution and disaggregation, to enable repeated determinations of maximal dissociation heat during the first set of injections. For example, 2 μL injections of 100 μM protein into the 1.4 mL cuvette will result in 700-fold dilution and a first concentration of 0.143 μM protein. Complete dissociation should occur under these conditions and an endothermic or exothermic heat pulse will be detected in the microcalorimeter. It is recommended to test protein solubility at high concentrations prior to starting the experiment since precipitation may occur in the syringe. In the case of 2-CysPrx, a syringe concentration of 50 μM was used. For proteins where the K_d cannot be determined beforehand, a high concentration of about 100 μM in the syringe is recommended. The concentration must be increased if no heat is observed due to very high K_d or decreased if K_d is very low. The needed syringe volume depends on the ITC instrument model. The syringe volume available for injection amounts to 330 μL in case of the MicrocalVP. This corresponds to about 400 μL filling volume due to the dead volume of the syringe.

6.5. Protein dialysis

The chemical environments in the syringe and cuvette solutions must match except for the reactant(s). Small differences in ionic strength, reductants, or pH between both solutions result in high heat dilution background. Therefore, the Prx should be dialyzed extensively against the desired buffer 40 mM K-phosphate, pH 8.0, and 1 mM DTT at 4 °C using 15,000 Da molecular weight cutoff dialysis tubing. Buffer used in the last dialysis step may be used to fill the ITC cell.

6.6. Filtration of solutions

Buffer and protein solutions should be filtered using 0.2 μm filter to eliminate undissolved particles and contaminants such as bacteria. Microorganic metabolism interferes with the measurement. Since the protein may not fully be recovered during filtration the protein concentration needs to be redetermined after filtration.

6.7. Determination of protein concentration

The accuracy of protein concentration adjustment in the syringe is absolutely critical for quantification of thermodynamic binding parameters. Prx concentration after filtration can be determined by reading absorbance at 280 nm and calculating the concentration from its molar extinction coefficient. Since this coefficient changes depending on the amino acid composition, the precise amino acid sequence including, for example, tags must be taken into consideration. The recording of a UV–vis spectrum between 380 and 230 nm is recommended to get information on purity and solubility of the sample. The 320–380 nm region of the spectrum is flat if absorbing cofactors are lacking while sloping in comparison to baseline reveals light scattering indicating aggregation of part of the protein.

6.8. Degassing of solutions

Formation of gas bubbles during the experiment will generate heat pulses detected by the microcalorimeter. Gas bubbles commonly occur if the samples were previously kept at 4 °C, for example, during overnight dialysis. To ensure bubble-free loading of protein solution and buffer, samples are degassed for 5–10 min while stirring in the Thermo Vac station provided with the VP ITC. Cuvette loading with buffer normally involves pressure application resulting in buffer warming. Due to the low cooling capacity of the ITC and in order to decrease equilbration times after filling the cuvette with buffer, the temperature of the Thermo Vac station may be set below the desired experimental temperature. Here, buffer and 2-CysPrx were degassed at 21 °C (4° below the intended experimental temperature). Also a pre-cooled syringe may be used to minimize the warming of the samples.

6.9. Cleaning of the instrument

Before filling the cell and the syringe with the experimental solutions the ITC must be (i) cleaned with 5 *M* NaOH, (ii) washed with water, and (iii) rinsed with the experimental buffer. The ITC is provided with a guideline describing all cleaning steps.

To check that the ITC cuvette has been cleaned properly, a water–water experiment can be done. Sample cell and loading syringe are filled with degassed water (for technical operations see below). After equilibration of the microcalorimeter, 10 μL serial injections of water into water should release heat pulses $\leq$0.2 μCal/s.

6.10. Loading buffer into the cuvette

Buffer is filled into the sample cell using a precooled 2.5 ml Hamilton syringe with blunt-end needle. Remove bubbles from the filling syringe. Slide the needle gently downward until reaching the bottom of the cuvette. Lift the needle about 1–2 mm above the bottom of the cell. The syringe will remain in this position during the filling process. Inject the buffer slowly into the cell. When the water spills out at the top of the cuvette, strongly press the plunger in order to displace bubbles possibly present in the cell. Finally, remove the overflow solution with the syringe. Close the cell and start the first ITC equilibration.

6.11. Equilibration period

After filling the cuvette with buffer, temperatures between the measuring and reference cells and the adiabatic jacket must be equilibrated. Equilibration is reached when ΔT, that is, the difference between the adiabatic jacket and both cells, is less than 0.001°, which takes up to 15 min.

6.12. Selection of experimental setup parameters

(a) *Injection volume:* Depending on the protein concentration in the cell and the expected K_d the number and volume of injections can be estimated. Prx at 50–100 μM was injected in series of 1.2 μL each. A good experiment must contain several measuring points in the transition region.
(b) *Duration of each injection:* 2.4 s which is twice the injected volume.
(c) *Time between consecutive injections:* Intervals of 180 s are long enough to reach equilibrium between each heat pulse. If needed, time intervals are increased until peaks return to the baseline.
(d) *Filter period:* It is defined as the time taken to calculate a single data point. Set to 2 s.
(e) *Cell temperature:* 2-CysPrx ITC dilution experiments were carried out at 25 °C.
(f) *Reference power:* Is the power applied to the heater of the reference cell? Set to 10 μCal/s.
(g) *Initial delay:* Time from the start of the data collection until the first injection. Set to 60 s.
(h) *Stirring speed:* Set to 310 rpm.

6.13. Loading the Prx solution into the syringe

To load the syringe, click on the Open Fill Port button (in ITC control window). The pipette retracts the plunger until it is positioned just above the fill port allowing the filling of the syringe. Prx solution can be sucked up into the syringe using a clean syringe (see instructions). When the syringe has been filled, click on the Close Fill Port button. At this point, the plunger moves down slightly, the system is closed and ready for injections. To remove any bubbles present in the syringe one might purge the syringe by clicking on the Purge-ReFill button.

6.14. Loading the syringe into the cell

Once the system is in its first equilibrium one can proceed to load the syringe into the ITC cell. Insert vertically the syringe into the cuvette through the central hole on the top of the cell and lower the pipette to the end position. Avoid bending the needle or touching any surface with the needle tip. The insertion of the syringe into the cell generates a temperature shift in the cell necessitating a second equilibration period. This period should take no longer than 5 min to prevent diffusion of Prx from the syringe into the buffer in the cell. Once the differential power and ΔT in the VPViewer stabilize the start button is pressed to determine the baseline for the experiment. When the VPViewer determines that the final baseline has equilibrated properly (close to 10 μCal/s in our experiments), the real analysis starts automatically with an initial 60 s delay to record the baseline prior to the first injection of Prx. Consecutive injections will be carried out according to the desired injection program (see Section 6.11).

6.15. Blank titration for background heat subtraction

Once the Prx dilution experiment has concluded, the ITC compartments are emptied and cleaned. The absolute heat corresponding to the 2-CysPrx dissociation effect must be corrected for heat dilution background which corresponds to the heat released when injecting buffer into buffer. Heat dilution background is determined under the same conditions as used in the titration but using buffer in the syringe. Software provided by manufacturers will allow the subtraction of the heat dilution. At the end of the experiment, cell and syringe are cleaned manually as described in the manufacturer's instructions. Vacuum is applied to dry both compartments.

7. Results, Data Analysis, and Discussion

In this chapter, two protocol variants are presented to characterize the oligomerization process of typical 2-CysPrx based on two different ITC dilution experiments. 2-CysPrx decamerization was maximized by adjusting reducing conditions with DTT in the buffer (Bernier-Villamor *et al.*, 2004; König *et al.*, 2002; Wood *et al.*, 2002). In both protocols, concentrated decameric Prx is injected into the cell containing either buffer or Prx at varying concentrations near the CTC. Oligomerization and redox state of the protein were tested by size exclusion chromatography and nonreducing SDS–PAGE, respectively. During electrophoretic separation oxidized typical 2-CysPrx separates as dimer due to the presence of intermolecular disulfide bonds whereas the hyperoxidized Prx migrates as monomer (around 20 kDa) and is unable to form disulfide bridges under oxidizing conditions. In a converse manner, the reduced form separates as monomer and can be converted to dimers under oxidizing conditions.

7.1. Protocol one

The first type of experiment consisted of injections of reduced decameric 2-CysPrx at high concentrations into the buffer-containing cell. Under these conditions, the injected 2-CysPrx decamers dissociate into the dimeric subunits in an exothermic manner. The released heat is detected by the microcalorimeter reporting a negative heat pulse in the thermogram. In typical ITC dilution experiments, further injections should result in a stepwise decrease of the released/absorbed heat until the low heat level of dilution indicates that no further oligomers dissociate. This smooth transition can be analyzed allowing the calculation of the thermodynamic parameters of the assembly/disassembly process. However, for typical 2-CysPrx, the transition between full disassembly and background as indicated by dilution heat occurred abruptly (Fig. 17.2), and points in the transition region were not observed. As a consequence, values for K_d, ΔS, and ΔG could not be derived from the experimental data. For this reason instead of K_d the terms critical concentration (CC) and CTC were chosen to describe the sharp transition.

The term CC comes from micelle formation observed in detergent solutions once a critical concentration is surpassed (Heerklotz and Seelig, 2000). CTC defines the critical concentration of 2-CysPrx which triggers the dimer–decamer transition. The oligomerization midpoint or CTC defines the transition point where half of the injected decamers dissociate. Due to the high cooperativity of the Prx oligomerization process, the CTC term is the closest experimental value that can be estimated with ITC and it approximates the K_d parameter calculated for the decamer–dimer dissociation model described in the dimer–decamer equilibrium section. Above the CTC the injected Prx decamers do not dissociate and the injection peaks approach the level of heat dilution. ITC dilution experiments using typical 2-CysPrx from different species revealed a highly similar behavior with CTC values between 1 and 2 μM monomer equivalents. In some experiments, a single endothermic peak was observed in the abrupt transition between the dissociation region and the heat of dilution as depicted in Fig. 17.2. This endothermic pulse near the CTC may indicate the reversibility of the disassembly process. The endothermic peak may reflect a rapid dissociation followed by a small oligomerization. Taking into account that typical 2-CysPrx concentrations at the site of their subcellular localization amount to >60 μM (König *et al.*, 2002) and that the *in vitro* value of the decamer–dimer CTC is around 1.6 μM monomer equivalents, one can conclude that the reduced form of typical 2-CysPrx *in vivo* predominantly exists as decamer. The very sharp transition witnesses high cooperativity in line with the hypothesis of a critical nucleation process. Cooperativity in oligomerization has also been described for proteins such as actin and tubulin (Popodi *et al.*, 2008) and typically takes place when a multitude of weak interactions act at once. The highly cooperative process implies that intermediate species between decamers and dimers in the $M_{(2)5} \leftrightarrow 5M_2$ equilibrium do not exist. The maximum Hill coefficient that might be expected is five. The high cooperativity observed indicates that only dimers and decamers exist in this equilibrium. Figure 17.3 gives the interpretation of the results in a schematic representation.

The areas underneath heat peaks can be integrated and normalized with respect to the molar amount of 2-CysPrx injected and plotted versus the

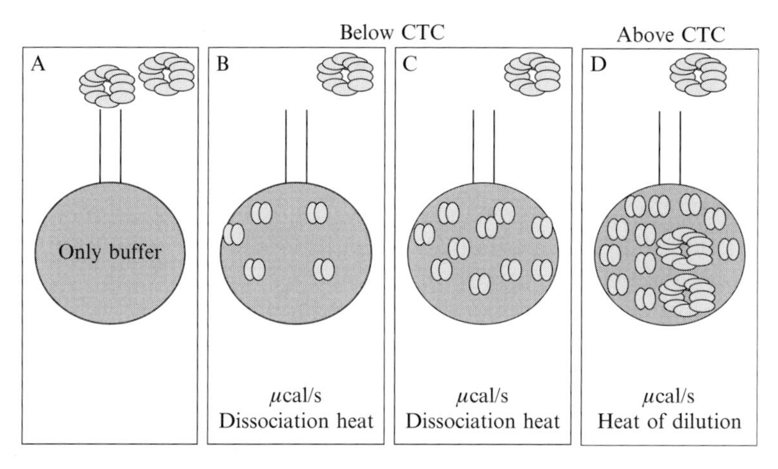

Figure 17.3 Representation of ITC dilution experiment with decameric 2-CysPrx. Firstly, Prx oligomers are injected into the cell containing buffer only and dissociate into dimers with heat release. Injection of Prx decamers above the CTC produced only a background heat indicating the end of the titration.

monomer 2-CysPrx concentration in the cell. Disassembly heat is determined using the dimer–monomer model provided by Microcal software. The reaction enthalpy of the 2-CysPrx decamer breakdown was calculated as the average of the integrated heat (kcal/mol) of all points before the sharp transition after subtracting the heat dilution of the blank experiment. Only the very first injection was excluded. The value was 142 ± 24 kcal/mol dimer calculated according to equation (Barranco-Medina *et al.*, 2008).

$$\Delta H_{\mathrm{d}} = \Delta H_{\mathrm{m}} \frac{c_{\mathrm{m}}}{c_{\mathrm{d}}}$$

where ΔH_{d} is the enthalpy per mol dimer, ΔH_{m}, the measured value per mol monomer, and $c_{\mathrm{m}}/c_{\mathrm{d}}$ represents the molar conversion factor. The favorable exothermic enthalpy in the decamer breakdown offers valuable information regarding the nature of the oligomerization process. Firstly, breakdown of the decamers is favored at concentrations below the CTC. Naturally, ΔH of oligomerization has the same value as ΔH of disassembly with opposite sign, that is, -142 ± 24 kcal/mol dimer. Furthermore, the negative (favorable) enthalpy of dissociation indicates that Prx oligomerization is endothermic and driven entropically. The dissection of enthalpy in binding must address three different components, namely ΔH_{s}, ΔH_{i}, and ΔH_{c}, and can be mathematically represented as

$$\Delta H = \Delta H_{\mathrm{s}} + \Delta H_{\mathrm{i}} + \Delta H_{\mathrm{c}}$$

The solvation enthalpy ΔH_s corresponds to hydration or dehydration phenomena at the protein–protein interface. The release or uptake of highly ordered water molecules in the protein–protein interface contributes to the total enthalpy change. The reorientation of ordered water molecules affects not only the reaction enthalpy but is also related to changes in entropy. ΔH_i reflects formation or breakage of noncovalent bonds that contribute to the binding process such as hydrogen bonding, electrostatic forces, and van der Waals interactions. The conformational enthalpy ΔH_c quantifies the contribution of conformation changes to ΔH of oligomerization. As stated above 2-CysPrx oligomerization seems to be enthalpically unfavorable due to the positive value of oligomerization ΔH. Although ΔH is positive, the two components of the equation $[\Delta H_c + \Delta H_i]$ must be exothermic (negative value) since protein–protein interaction and protein folding involves favorable (negative) enthalpy (Lakshminarayanan *et al.*, 2007). Hence, overall endothermic enthalpy of 2-CysPrx oligomerization is attributed to the first component of the equation, ΔH_s which must comprise a large unfavorable and thus, positive solvation enthalpy value. Protein folding and ligand binding to protein through hydrophobic interactions are often accompanied by the burial of nonpolar surfaces from water (Eftink and Biltonen, 1980).

Barranco-Medina *et al.* (2008) carried out similar experiments with nondissociating Prx, that is, AtPrxQ, as well as oxidized and overoxidized At-2-CysPrx. In these cases, injection of protein resulted in heat dilution background. Oxidized 2-CysPrx is present as dimer impeding the dissociation after injection into the cell. The overoxidized 2-CysPrx mimics the reduced decameric form. However, some structural changes trap the enzyme in the stable decameric state and block the oligomer dissociation (Barranco-Medina *et al.*, 2009; Wood *et al.*, 2002). This observation indicates that hyperoxidized and reduced decamers are structurally different and fulfill different functions in the cell. Indeed, recently Jang *et al.* (2004) reported that only hyperoxidized HMW species of typical Prx possess a protective chaperone-like activity. Under physiological conditions 2-CysPrx exists mostly as decamer favored by reducing conditions and concentrations higher than the CTC. 2-CysPrx acts as redox sensor and its oligomers as a messenger in redox signaling (Barranco-Medina *et al.*, 2009; Dietz, 2008). When the peroxide concentration in the living cell increases, oxidation of cysteinyl residues leads to the disassembly of the decamers releasing free dimers with intermolecular disulfide bridges. The dimers can react with their physiologically regenerating redox partner such as thioredoxins, glutaredoxins, or glutathione. Several oxidative conditions will produce an excess of H_2O_2 yielding an irreversible hyperoxidation of the cysteinyl groups to sulfinic acid enabling 2-CysPrx function as protective chaperone.

7.2. Protocol two

This protocol allows to study Prx re-reduction and to investigate oligomerization complementation by Prx of different origin. To confirm CTC and enthalpic parameters obtained with *protocol one* a variation of ITC dilution experiments is proposed where reduced typical 2-CysPrx (50 μM) is injected in the cell preloaded with the same protein at different concentrations (0–5 μM monomer concentration). If the Prx concentration inside the cell is below the CTC an exothermic reaction is observed whereas above the CTC the heat peak approaches the background level. Injection of 50 μM reduced At-2-CysPrx into the cell filled with the same solution only showed heat dilution indicating the absence of dissociation processes while identical injections into buffer produced the typical exothermic answer described in *Protocol one*. Injection of 50 μM reduced At-2-CysPrx into the cell filled with the same enzyme at concentrations slightly below the CTC = 1.58 μM, for example, 1.55 μM enabled an almost complete disassembly whereas a slight increase in the 2-CysPrx concentration to 1.6 μM inhibited dissociation. The obtained CTC is similar to that estimated with *protocol one* validating that both methods yield accurate CTC values. Reversibility of hyperoxidized Prx can also be assessed with ITC dilution experiments. Hyperoxidized Prx treated with high concentration of H_2O_2 can be incubated with DTT or sulfiredoxins in an attempt to re-reduce the enzyme. Injection of 50 μM reduced Prx as a probe into the cell containing formerly treated hyperoxidized Prx should not produce exothermic peaks, if the 2-CysPrx was successfully regenerated and added at concentrations above the CTC. The advantage is that a much lower concentration is needed, although the total protein amount is still high, for example, 2 μM $\times$1400 μl = 2.8 nmol. Conversely, an exothermic dissociation will occur if the 2-CysPrx has not been re-reduced. Accordingly, reduced At-2-CysPrx dissociated when injected into DTT-treated hyperoxidized enzyme under reducing conditions with a CTC of 1.61 μM, suggesting that the overoxidized At-2-CysPrx does not interact with the reduced form (Barranco-Medina *et al.*, 2008). The experiment confirmed the hypothesis that hyperoxidized Prx adopts a structure distinct from reduced decamer.

Complementation assays can be designed to investigate whether dimer–dimer interfaces are conserved among 2-CysPrx isoforms from a single or different species. To this end, reduced Prx from one species, for example, At-2CysPrxA is injected into the cell containing other Prx, for example, Hs-2-CysPrx. If the dimer–dimer interface is conserved, the number of injections required to arrive at the CTC of the injected Prx should be smaller than in a normal ITC dilution experiments. Injection of At-2-CysPrx into human PRDX1 and Ps-2-CysPrx revealed that, despite the similarity in the CTC among the Prx of different species, the dissociation process of At-2-CysPrx was not affected by the presence of any other Prx indicating differentiation of interfaces during evolution. However, the similarity of the CTC value among the different Prx species suggests that the dimer–decamer transition is essential for 2-CysPrx function.

8. Conclusions

Oligomerization of reduced 2-CysPrx is a highly cooperative process. Below the CTC, all injected decameric 2-CysPrx disassemble. Above the CTC, additionally added protein remains in the decameric form. This behavior resembles that of detergent micelle formation (Heerklotz and Seelig, 2000). The extremely sharp transition from exclusive presence of dimers to that of decamers ranged between 1.55 and 1.6 μmol monomer equivalent/L for At-2-CysPrx (Barranco-Medina *et al.*, 2008) and suppresses formation of intermediate aggregates. In this way, the assembly and disassembly process dissembles that of other oligo-/polymerizing proteins such as actin where also low concentrations of low mass aggregates and a wide range of different oligo-/polymers occur (Terada *et al.*, 2007). The presented method can be adapted to describe quantitatively other reversible *n*-mer formation processes by ITC.

ACKNOWLEDGMENTS

The work was supported by the SFB613 and by the DFG (Di346/14). SBM is indebted to Christine Allison for her support.

REFERENCES

Aran, M., Ferrero, D. S., Pagano, E., and Wolosiuk, R. A. (2009). Typical 2-Cys peroxiredoxins—Modulation by covalent transformations and noncovalent interactions. *FEBS J.* **276,** 2478–2493.

Barranco-Medina, S., Kakorin, S., Lázaro, J. J., and Dietz, K. J. (2008). Thermodynamics of the dimer–decamer transition of reduced human and plant 2-cys peroxiredoxin. *Biochemistry* **47,** 7196–7204.

Barranco-Medina, S., Lazaro, J. J., and Dietz, K. F. (2009). The oligomeric conformation of peroxiredoxins links redox state to function. *FEBS Lett.* **583,** 1809–1816.

Bernier-Villamor, L., Navarro, E., Sevilla, F., and Lazaro, J. J. (2004). Cloning and characterization of a 2-Cys peroxiredoxin from *Pisum sativum*. *J. Exp. Bot.* **55,** 2191–2199.

Burrows, S. D., Doyle, M. L., Murphy, K. P., Franklin, S. G., White, J. R., Brooks, I., McNulty, D. E., Scott, M. O., Knutson, J. R., Porter, D., Young, P. R., and Hensley, P. (1994). Determination of the monomer–dimer equilibrium of interleukin-8 reveals it is a monomer at physiological concentrations. *Biochemistry* **33,** 12741–12745.

Bystrova, M. F., Budanova, E. N., Novoselov, V. I., and Fesenko, E. E. (2007). Study of the quarternary structure of rat 1 Cys peroxiredoxin. *Biofzika* **52,** 436–442.

Cao, Z., Bhella, D., and Lindsay, J. G. (2007). Reconstitution of the mitochondrial Prx3 antioxidant defence pathway: General properties and factors affecting Prx3 activity and oligomeric state. *J. Mol. Biol.* **372,** 1022–1033.

D'Alessio, G. (1999). The evolutionary transition from monomeric to oligomeric proteins. Tools, the environment, hypotheses. *Prog. Biophys. Mol. Biol.* **72,** 271–298.

Dietz, K. J. (2003). Plant peroxiredoxins. *Annu. Rev. Plant Biol.* **54,** 93–107.

Dietz, K. J. (2008). Redox signal integration: From stimulus to networks and genes. *Physiol. Plant.* **133,** 459–468.

Eftink, M., and Biltonen, R. (1980). Thermodynamics of interacting biological systems. *In* "Biological Microcalorimetry," (A.E Beezer, ed.), pp. 343–412. Academic Press, London.

Hall, A., Karplus, P. A., and Poole, L. B. (2009). Typical 2-Cys peroxiredoxins—Structures, mechanisms and functions. *FEBS J.* **276,** 2469–2477.

Heerklotz, H., and Seelig, J. (2000). Titration calorimetry of surfactant-membrane partitioning and membrane solubilization. *Biochim. Biophys. Acta* **1508,** 69–85.

Jang, H. H., Lee, K. O., Chi, Y. H., Jung, B. G., Park, S. K., Park, J. H., Lee, J. R., Lee, S. S., Moon, J. C., Yun, J. W., Choi, Y. O., Kim, W. Y., *et al.* (2004). Two enzymes in one: Two yeast peroxiredoxin display oxidative stress-dependent switching from a peroxidase to a molecular chaperone function. *Cell* **117,** 625–635.

Jorgensen, C. S., Ryder, L. R., Steino, A., Hojrup, P., Hansen, J., Beyer, N. H., Heegaard, N. H. H., and Houen, G. (2003). Dimerization and oligomerization of the chaperone calreticulin. *Eur. J. Biochem.* **270,** 4140–4148.

Karplus, P. A., and Hall, A. (2007). Structural survey of the peroxiredoxins. *Subcell. Biochem.* **44,** 41–60.

König, J., Baier, M., Horling, F., Kahmann, U., Harris, G., Schürmann, P., and Dietz, K. J. (2002). The plant-specific function of 2-Cys peroxiredoxin-mediated detoxification of peroxides in the redox-hierarchy of photosynthetic electron flux. *Proc. Natl. Acad. Sci. USA* **99,** 5738–5743.

Lakshminarayanan, R., Fan, D., Du, C., and Moradian-Oldak, J. (2007). The role of secondary structure in the entropically driven amelogenin self-assembly. *Biophys. J.* **107,** 3664–3674.

Levy, E. D., Pereira-Leal, J. B., Chothia, C., and Teichmann, S. A. (2006). 3D complex. A Structural classification of protein complexes. *PLoS Comput. Biol.* **2,** e155.

Lovatt, M., Cooper, A., and Camilleri, P. (1996). Energetics of cyclodextrin-induced dissociation of insulin. *Eur. Biophys. J.* **24,** 354–357.

Luke, K., Apiyo, D., and Wittung-Stafshede, P. (2005). Dissecting homo-heptamer thermodynamics by isothermal titration calorimetry: Entropy-driven assembly of co-chaperonin protein 10. *Biophys. J.* **89,** 3332–3336.

Meng, G., Fronzes, R., Chandran, V., Remaut, H., and Waksman, G. (2009). Protein oligomerization in the bacterial outer membrane. *Mol. Membr. Biol.* **26,** 136–145.

Ogura, T., and Wilkinson, A. J. (2001). AAA+ superfamily ATPases. Common structure-diverse function. *Genes Cells* **6,** 575–597.

Popodi, E. M., Hoyle, H. D., Turner, F. R., and Raff, E. C. (2008). Cooperativity between the beta-tubulin carboxy tail and the body of the molecule is required for microtubule function. *Cell Motil. Cytoskeleton* **65,** 955–963.

Terada, N., Shimozawa, T., Ishiwata, S., and Funatsu, T. (2007) Size distribution of linear and helical polymers in actin solution analyzed by photon counting histogram. *Biophys. J.* **92,** 2162–2171.

Wisemann, T., Willston, S., Brandts, J. F., and Lung-Nan, L. (1989). Rapid measurement of binding constant and heats of binding using a new titration calorimeter. *Biochemistry* **179,** 131–137.

Wood, Z. A., Poole, L. B., Hantgan, R. R., and Karplus, A. (2002). Dimers to doughnuts: Redox-sensitive oligomerization of 2-Cysteine peroxiredoxins. *Biochemistry* **41,** 493–5504.

Wood, Z. A., Poole, L. B., and Karplus, P. A. (2003). Peroxiredoxin evolution and the regulation of hydrogen peroxide signaling. *Science* **300,** 650–653.

Wyman, J., and Stanely, J. G. (1990). Binding and Linkage. Functional Chemistry of Biological Macromolecules. University Science Books, California.

Yonehara, M., Minami, Y., Kawata, Y., Nagai, J., and Yahara, I. (1996). Heat-induced chaperone activity of HSP90. *J. Biol. Chem.* **271,** 2641–2645.

CHAPTER EIGHTEEN

Protein–Lipid Interactions: Role of Membrane Plasticity and Lipid Specificity on Peripheral Protein Interactions

Jesse Murphy, Kristofer Knutson, *and* Anne Hinderliter

Contents

Abstract

Lipid mixtures are inherently nonrandom as each lipid species differs slightly in its chemical structure. A protein associates not with a lipid but with a membrane comprised of lipids where the chemical activities of each lipid is determined by the composition of the mixture. There can be selectivity in this association

Department of Chemistry and Biochemistry, University of Minnesota Duluth, Duluth, Minnesota, USA

Methods in Enzymology, Volume 466
ISSN 0076-6879, DOI: 10.1016/S0076-6879(09)66018-3

because a protein can enhance the underlying tendency of lipids to be heterogeneously distributed. This is dependent on the protein having a preferential association of sufficient magnitude with some of the lipids within the membrane. To measure and model protein–lipid interactions, an understanding of the underlying lipid behavior is necessary to interpret their association constants. Methods to measure protein–lipid interactions are discussed within the context of using these techniques in modeling and a general framework is presented for the use of a signal arising from these interactions. The use of binding partition functions is presented as this allows the modeling of cooperative or independent (noncooperative) interactions of protein with lipids and of proteins with additional ligands as well as lipids. A model is also provided using the binding partition function formalism where protein dimerization, and by extension, oligomerization is enhanced at the membrane compared to in solution.

1. Introduction

The cellular membrane is nonuniform in its distribution of proteins and lipids (Baumgart *et al.*, 2007; Bernardino de la Serna *et al.*, 2004; Field *et al.*, 1997; Rogasevskaia and Coorssen, 2006; Sheets *et al.*, 1999; Simons and van Meer, 1988). There is an intense interest in the basis for the heterogeneous distribution of these signaling components at the membrane. A myriad of names such as rafts, complexes, domains, and lipid shells has been proposed in attempt to capture the fleeting organization of these membrane localized components (Brown and London, 2000; Edidin, 2003; McConnell and Vrljic, 2003; Simons and Ikonen, 1997; Simons and Vaz, 2004; Thompson and Tillack, 1985). In comparison, older and ongoing bodies of work have established a strong understanding, both experimental and theoretical, of the nonideal mixing behavior of lipids (Bloom and Thewalt, 1995; Cao *et al.*, 2005; Garidel *et al.*, 1997; Heimburg *et al.*, 1992; Huang and Feigenson, 1999; Ipsen *et al.*, 1987; Keller and McConnell, 1999; Mendelsohn *et al.*, 1995; Mountcastle *et al.*, 1978; Radhakrishnan and McConnell, 1999; Sankaram and Thompson, 1990; Silvius and Gagné, 1984; Suurkuusk *et al.*, 1976). These findings have long suggested that both the small, cooperative interactions between lipids as well as the dramatic changes in lipid domain size, as lipid mixtures approach a phase transition boundary, provide the organizational underpinning to signal complexes (Kinnunen, 1991). Clustering of signaling components, such as in a lipid domain is a means to enhance the probability of the components interacting (Thompson *et al.*, 1992, 1995), and these interactions are necessary to convey a triggered signal in the cell. In an analogous manner, signaling components sequestered into separate regions of the membrane would decrease the probability of interactions occurring

(Melo *et al.*, 1992). Signal transduction complexes are comprised of lipids and proteins where individually the interactions are often small and cooperative so coalescence of the components into a signaling complex becomes very sensitive to the composition of the complex. Furthermore, proteins can redistribute lipids, leading to clustering of lipids and proteins alike. This is dependent on whether proteins are selective in the lipids they interact with and whether their interactions are of sufficient magnitude to overcome the entropy of mixing. Here, we present a thermodynamic perspective on how the inherent nonrandom nature of lipid distributions influences the interaction of protein with the membrane surface and how protein–protein interactions can be more probable at the membrane surface than in solution.

2. Defining Protein–Lipid Interactions

Proteins and lipids have plasticity in their interactions because the surface of the membrane can be altered by such interactions. When a protein partitions onto the membrane surface, the initial distribution of lipids underneath the protein can differ from the final distribution upon reaching equilibrium. To quantitatively define the system, association constants may be measured between proteins and lipids. This does not, however, convey the sensitivity of such interactions to lipid phase behavior and how this coupling leads to lipid and protein redistribution. To understand the system, the dynamic behavior of the interactions must be conceptualized as the system approaches equilibrium. While the cell does not exist in a state of equilibrium, as this would correspond to death, evaluation of a system at equilibrium does allow the tendencies of the interactions to be ascertained.

The terms binding and partitioning are not used interchangeably as elements of both are needed to convey this model of protein–lipid interactions with lipid selectivity. Binding requires specificity, implies identifiable binding sites and is more robust than partitioning. Lipid-binding sites with both specificity and high affinity have been identified. For example, human prothrombinase complex requires the acidic phospholipid phosphatidylserine (PS) but not a membrane surface to activate the complex (Majumder *et al.*, 2005a) and high-affinity binding sites for PS have been identified within the complex (Majumder *et al.*, 2005b).

Partitioning is the nonspecific interaction of a protein with the membrane and as it is nonspecific (White *et al.*, 1998), the interactions are weak relative to binding. For example, the interaction energy of a charged amino acid (such as lysine) within a peptide with an oppositely charged individual lipid (such as phosphatidylserine) is $\sim$1.4 kcal/mol (Kim *et al.*, 1991).

3. Selective Partitioning and Lipid Activities

We define a protein–lipid interaction that resides between partitioning and binding as selective partitioning. Rather than using the criteria of an identifiable binding site within the protein for a particular lipid for specificity, selectivity is instead based upon relative affinity within the protein for each lipid in the membrane. If this relative selectivity is large enough, the process by which proteins partitioning onto and off of the membrane surface is not a passive phenomena; it has the capacity to redistribute the underlying lipids.

The initial association of a protein with a membrane surface we described as partitioning, and the extent by which the protein partitions from solution onto the membrane is dictated by lipid chemical activities. The chemical activity of each lipid species in a mixture is determined by the mixing behavior of that lipid, and it is this activity that dictates the strength of association. Each lipid's activity is its effective concentration, as it can be greater or less than the physical concentration of that lipid within the mixture. Activities of lipids vary with their distribution and can be visualized by $a = (c)(\gamma)$, where a is the activity, c is the concentration of the lipid within the membrane suspension, and γ is the activity coefficient. An activity coefficient is the measure of the deviation of a mixture from a random distribution of components. An activity coefficient greater than one leads to a greater effective than physical concentration; less than one the inverse holds and if the activity coefficient is one, the effective and physical concentrations are equal. Defining the strength of a specific lipid–protein interaction must then be placed within the context of the overall lipid mixture and the effective concentration of the lipids.

Lipid mixtures are inherently nonuniform as each lipid species differs slightly in its chemical structure (Almeida *et al.*, 2005a). As the differences in lipid chemical structure are small, the unlike nearest neighbor interaction Gibbs free energy, or interaction, energy between like and unlike lipids are also small. Each lipid is surrounded by six nearest neighbors and how weakly attractive or weakly repulsive to one another these lipids are defined by the interaction energy, ω (Almeida, 2009). An interaction energy is a measure of, like an activity coefficient, the deviation of the mixture from a random distribution of components in a lattice such as in a membrane. From the partitioning or macroscopic framework, lipids do not act individually, but a microscopic view must also be incorporated to gain understanding of protein influence on lipid reorganization. Interaction energies extend to those lipids underneath the protein as well as the leading edge of lipids beyond the edge of the proteins due to the weakly cooperative nearest neighbor interactions between lipids.

Upon initiation of a favorable protein–lipid contact, additional protein–lipid and lipid–lipid interactions may be established. Favorable lipid contacts that lie beyond the edge of the protein can be switched underneath the protein resulting in the expulsion of an unfavorable lipid and the gain of a protein–lipid interaction and a favorable nearest neighbor lipid contact. The magnitude of the unlike nearest neighbor interaction energy is indicative of the nearness of the lipid mixture to being separated into phases of distinct lipid composition. As these interaction energies are weakly cooperative and therefore not additive, even small changes in these energies lead to dramatic changes in the distribution of lipids. For a binary mixture of lipids, interaction energies approaching 320 cal/mol are nearly phase separated into clusters enriched in each lipid. A positive ω indicates a repulsive interaction between unlike lipids, subsequently; like lipids have a greater tendency to be segregated together. Each lipid in such a scenario has a greater probability to have a like nearest neighbor beyond the edge of an associated protein. As ω decreases, the lipids appear as a nearly random pattern (Fig. 18.1). In a binding framework, especially if lipids are systematically substituted in a mixture to establish a trend, but, the binding behavior is not known, indeed, a change in binding affinity will be observed. This change in affinity may be attributed to an increase in the effective concentration of the lipid with clustering of the preferred lipids and perhaps not to a change in either specificity of the protein for the lipid or presence of a specific lipid-binding site. Conversely, a protein could conceivably have a distribution of contact points across its membrane-binding surface that favors interacting with an ordered array of unlike lipids. Such a pattern correlates with a negative interaction energy as unlike lipid contacts are now favored. The mixing behaviors of the lipid mixture can dramatically impact the strength of the interaction of a protein with a membrane surface.

4. Protein–Protein Interactions at the Membrane Surface

The membrane is more than a surface which provides a plasticity to potentially be optimized by the interacting proteins; it is a surface which may enhance protein–protein interactions. To illustrate this concept, we define a scenario in which there is monomeric protein and dimeric protein. Each form of the protein can interact with lipids. If the dimeric and oligomerized form of the protein had greater affinity for lipids, or binds more lipids, than the monomeric form, the dimer would be depleted from solution as it binds lipids. To repopulate the dimer and reestablish

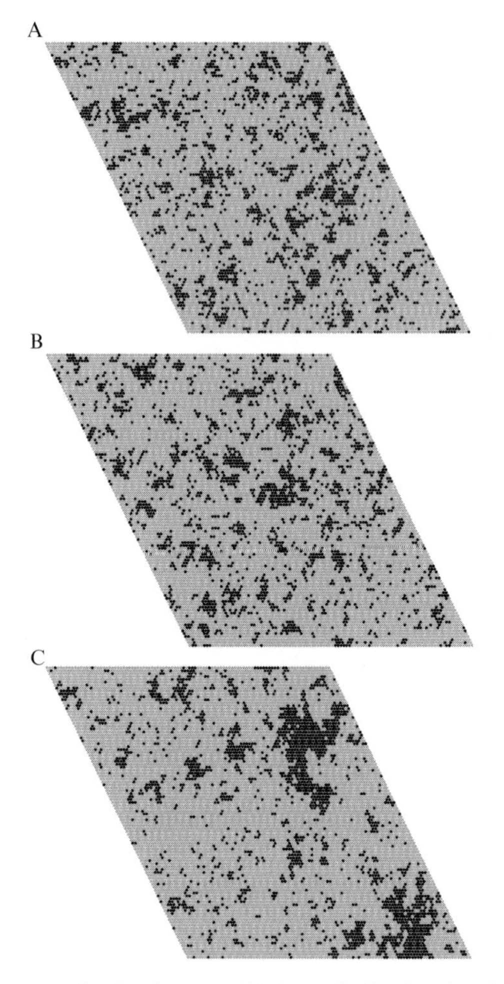

Figure 18.1 Monte Carlo simulations of 20% acidic lipids, phosphatidylserine (PS), shown with black dots in a background of 80% zwitterionic lipids, phosphatidylcholine (PC) shown with gray dots, where the interaction energies (ω) increased from 240, 280 to 320 cal/mol. This range of interaction energies correlate, for example, to mixtures of the acidic phospholipid 16:0,18:1PS in a background of different PCs. Observed is the composition of the lipid mixture can concentrate or disperse the PC. This behavior is consistent within a partitioning framework but could lead to erroneous conclusions if a binding framework that assumes ideal mixing of the ligands is assumed. This is as the interaction of the protein with the membrane surface would be improved by the clustering of like lipids.

equilibrium, the monomer would dimerize. As the dimer has a greater affinity for lipid than the monomer, the overall equilibrium will shift and the other states are depleted with the membrane-bound dimer predominating.

5. Measuring Protein–Lipid Interactions

Although a variety of techniques are used to measure the affinity of a protein for a membrane surface, we limit our methods discourse to the uses of fluorescence spectroscopy and isothermal titration calorimetry.

5.1. Fluorescence spectroscopy

Fluorescence spectroscopy is a commonly used technique to detect an interaction between a protein and ligand. Either the ligand or the protein may be the source of the signal. If there is a change of fluorescence signal upon a binding event, this change may be transformed via the Lever rule into fractional saturation of the macromolecule as a function of ligand concentration. To interpret the association event, the relationship of saturation of macromolecule to ligand concentration must be related to the equilibration constant.

5.2. The general use of a signal: Lever rule derivation

$S = f_b S_b + f_f S_f$, where $f_b + f_f = 1$ and S is any signal comprised of f_f is the fraction free ligand with the corresponding associated signal of S_f and f_b is fraction bound with the corresponding associated signal of S_b, with rearrangement $f_b = S - S_f/S_b - S_f$ and $f_f = S - S_b/S_f - S_b$. Therefore, when a system exhibits a change of signal upon an interaction (binding event) as a function of ligand concentration, this change in signal can be plotted as fraction bound versus free ligand concentration. A relationship that relates these variables to an association constant must now be derived to allow evaluation of the interaction in the context of the proposed models. The interaction [P] + [L] = [PL] at equilibrium is represented by K = [PL]/[P][L]. The protein, or any macromolecule, in this scenario exists in two forms, bound ([PL]) and free ([P]). The fraction of the protein in the bound state, f_b, is then equal to [PL]/([P] + [PL]), where ([P] + [PL]) is the distribution of states the protein can exist in the system. In its present state, this equation is not of use. However, by recognizing that with rearrangement, K[L] = [PL]/[P], the equation may be modified. Each form of the macromolecule is normalized against the unbound protein ([P]) by multiplying through by 1/[P]. The fractional saturation (f_b) of the protein is now equal to K[L]/(1 + K[L]) and is a function of an equilibrium constant and the free ligand. When compared to $f_b = S - S_f/S_b - S_f$, the connection is apparent, this is the change of signal with each addition of ligand as well as the dependent variable of the overall binding isotherm, $f_b = K[L]/(1 + K[L])$. The independent variable [L] when representing unbound lipids and more specifically lipids within single lamellar liposomes must be transformed to calculate the change in concentration with interaction with protein. Within a thermodynamic system, the ligand must be accessible to the macromolecule to interact. Therefore, for a protein that interacts with the surface of a liposome, only the outer monolayer is considered. Furthermore, each protein has a greater footprint than an individual lipid, so the number of lipids bound/protein must be determined. This can be problematic if the orientation of the protein on the membrane is not known, but a reasonable estimate can be made. Five to ten lipids could be estimated to be occupied by each protein. If the available lipid concentration is not corrected, the affinity will be underestimated. In this context, correcting multilamellar liposome (MLV) systems is difficult as only a fraction of the lipid comprises the outer layer and small unilamellar vesicles (SUVs) have approximately 60% of their lipids on the outer surface. Large unilamellar vesicles (LUVs) are assumed to have 50% of their lipids on the outer surface. It is important to note that concentrations are used and not a count of individual lipids. Chemical potential (or normalized free energy) of the ligand is directly proportional to the logarithm of its chemical activity. Here, chemical activity is represented for systems that are a dilute aqueous solution as a concentration. Liposomes, (such as LUVs), comprised of individual lipids, are a colloid of a solid phase suspended in a liquid phase and are a homogeneous suspension.

Finally, the independent variable is free lipid, [L]. To calculate free lipid, the previously determined fraction bound (f_b) term that was generated from the change in signal must be used. For each protein bound to lipid (or to any ligand), there are some number of ligands per protein, this number of ligands multiplied by fraction bound of protein with each addition of ligand is the number of bound ligands. By subtracting the bound ligand from the total ligand at each step of the titration, the free ligand, (lipid), is determined.

5.3. Extension beyond fluorescence spectroscopy

Transformation of a signal is the same regardless of the source of the signal. A few caveats exist, such as the signal in question must have a linear response (Ladokhin *et al.*, 2000). We most often use fluorescence steady-state intensities at a constant wavelength. The system must also be in equilibrium and hence, reversible necessitating that there is not a vast excess of both macromolecule and ligand. A benefit of the use of fluorescence spectroscopy is the relative limited amount of macromolecule necessary to generate signal. This is worth considering when the association reaction involves membrane suspensions. Membrane suspensions strongly scatter light. This can impede the use of optical techniques over a broad range of membrane concentrations.

5.4. Isothermal titration calorimetry

Isothermal titration calorimetry is another commonly used technique to detect an interaction between protein and ligand. The use of isothermal titration calorimetry to characterize protein–lipid interactions requires orders of magnitude more material (μM) in contrast to fluorescence spectroscopy (nM). The signal in isothermal titration calorimetry is heat and the source of this signal may not be as readily apparent as with fluorescence spectroscopy. Buffer and pH mismatch between syringe and cell can be a robust source of heat (Freire *et al.*, 1990). Protein–membrane interactions can be weak (Seelig, 1997). This necessitates the use of enough macromolecule for a detectable signal and makes characterizing these interactions more susceptible to the contribution of unanticipated heats.

6. Modeling of Protein–Lipid Interactions

6.1. Binding partition functions

An example of a binding partition function is where the interaction of $P + L = PL$ is represented by the binding isotherm of $f_b = K[L]/(1 + K[L])$. A binding partition function is the distribution of states that the macromolecule exits in and the relative probability of the macromolecule within each state (Dill and Bromberg, 2002; Wyman and Gill, 1990). We will illustrate this by examining the normalized probability of the total distribution of states which is commonly represented by a capital Q where $Q = [PL] + [P]$, the denominator of the fractional saturation relationship for f_b. To convert this relationship to a relative probability, Q is divided through by a reference state which is arbitrary in choice but often the free macromolecule is selected. The two states are now $[P]/[P] = 1$ and $[PL]/[P] = K[L]$, we can now compare the probability of being bound (interacting) to the normalized reference free state (not bound) of one. The normalized probability $[PL]/[P]$ of being bound scales with affinity of macromolecule for ligand (the association constant, K) and with the concentration of ligand, [L] (as expected by Le Chatelier's principle). When the overall fractional saturation relationship, $f_b = [PL]/([P] + [PL])$ is normalized by the reference state [P] and the same substitutions are used; the binding isotherm, $f_b = K[L]/(1 + K[L])$ results.

6.2. Interpretation of binding partition functions

The binding isotherm, $f_b = K[L]/(1 + K[L])$ is an independent site model as being bound is independent of being free. If multiple sites existed of the same (or within 10-fold similarity) affinity, the probability of any one site on the macromolecule interacting with ligand is independent of the other sites.

The liposome surface is comprised of lipids that during the course of reaching equilibrium may redistribute, but, as a suspension of liposomes is titrated in as the ligand form, binding of protein to each liposome remains independent of binding to another protein to a separate liposome. The functional form of this isotherm is hyperbolic and all such variations on an independent model will likewise be hyperbolic (see Fig. 18.2).

6.3. Comparison to Langmuir model

This may be contrasted to a system where the macromolecule is titrated into a fixed concentration of lipids. This interaction is classically modeled by a Langmuir model. Development of the Langmuir model follows a similar formalism as where liposomes are titrated into a fixed concentration of macromolecule. In the Langmuir model there are as in the previous model, two states but these two states are of the liposome surface which is occupied and unoccupied by the macromolecule. We may represent this fractional saturation of the exposed liposome surface by θ, and the unoccupied surface by $1 - \theta$. The mole fraction of the protein occupied liposome surface (χ_{PL}) is θ and the unoccupied surface is $1 - \theta = \chi_L$ as $\chi_{PL} + \chi_L = 1$. P (protein in solution) + L (unoccupied surface lipids) = PL (occupied lipids). $K_{as} = [PL]/[P][L]$ may then be rewritten as $K_{as} = \theta/[P](1 - \theta)$ which upon rearrangement yields $\theta = (K_{as})[P]/1 + (K_{as})[P]$. We will focus our discussion on systems where the concentration of lipids varies as these systems are amenable to evaluating the impact of multiple ligands on lipid binding.

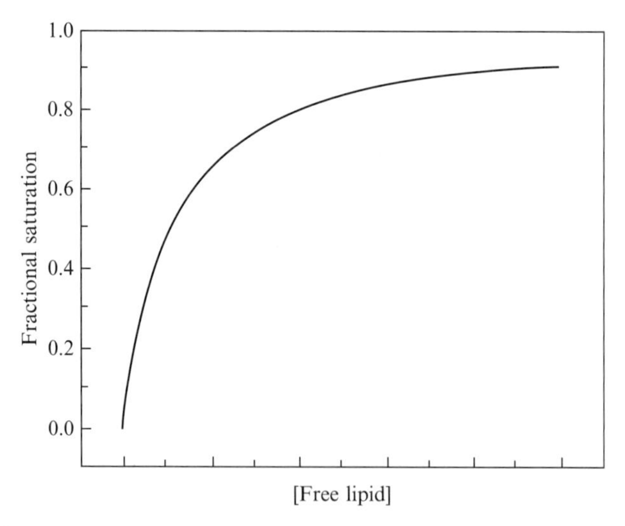

Figure 18.2 Hyperbolic binding isotherm generated by titrating liposomes into a macromolecule in solution as a ligand. This figure depicts the macromolecule having a liposome-binding affinity of $K_L = 2.50 \times 10^{-3}\ \mu M^{-1}$.

6.4. Binding partition functions: Linked binding equilibrium

The single-site binding model where the liposome is the ligand will be expanded to include an additional ligand, [X] where [X] can be bound by protein independent of the liposome or where the binding of lipid influences the binding of the other ligand (coupled or linked equilibrium). The benefit of using a binding partition function approach is that the distribution of states of the macromolecule may be directly calculated whether the binding equilibrium is linked or independent. If the binding sites are linked, advantage may be taken of how the binding of one ligand by macromolecule will alter the probability of the macromolecule binding the second ligand. This can allow the determination of the association constant for a ligand, such as to a liposome system that could be difficult to measure directly.

To identify all possible states of the macromolecule that interacts with two ligands, a thermodynamic cycle is constructed. In these scenarios, it will first be assumed that the two ligands interact with the macromolecule independent of one another (Fig. 18.3).

The total distribution of states is then $Q = [P] + [PL] + [XP] + [XPL]$, using the same approach as with a single binding site. After normalizing against the reference state [P], substitutions are made using the association constants for each reaction. The total distribution becomes $Q = 1 + K_L[L] + K_X[X] + K_L[L]K_X[X]$ and the probability of being in the unbound state is $1/Q$, of being only bound to liposome is $K_L[L]/Q$, of being only bound to X is $K_X[X]/Q$, and of being double bound to X and L is $K_L[L]K_X[X]/Q$ (see Fig. 18.4), where θ (fractional saturation) represents the binding of X over all of the possible states the protein can exist.

$$\theta = \frac{[X]}{Q}\frac{dQ}{d[X]} = \frac{K_X[X] + K_X[X]K_L[L]}{1 + K_X[X] + K_L[L] + K_X[X]K_L[L]}. \quad (18.1)$$

If the binding of one ligand now influences the binding of the second ligand, the distribution of states of the macromolecule, and the consequent probabilities will differ compared to the independent site model. This depends on the extent of the coupling which is reflected in the term σ (Fig. 18.5) and the greater σ, the greater the linkage between the binding

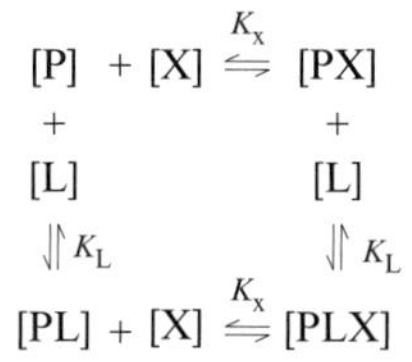

$$\begin{array}{ccc} [P] + [X] & \overset{K_X}{\rightleftharpoons} & [PX] \\ + & & + \\ [L] & & [L] \\ \downarrow\uparrow K_L & & \downarrow\uparrow K_L \\ [PL] + [X] & \overset{K_X}{\rightleftharpoons} & [PLX] \end{array}$$

Figure 18.3 Thermodynamic cycle showing the independent binding relationship the protein [P] has for ligand [X] and liposome [L].

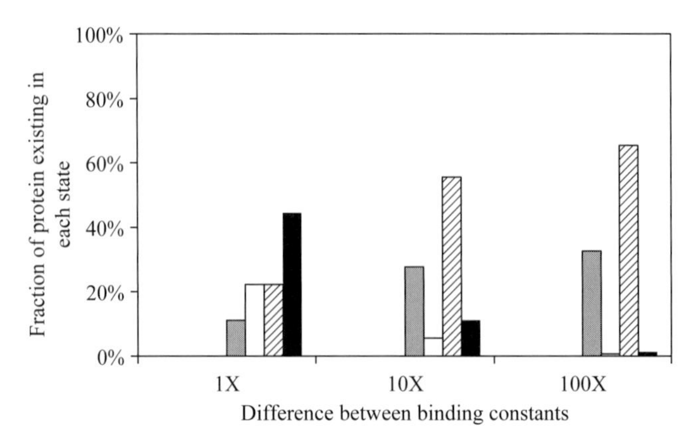

Figure 18.4 Graphical interpretation of the probability that the protein exists in any of the possible bound states. Here the four possible states are: unbound (shown in gray), protein bound to liposomes (shown in white), protein bound to ligand [X] (shown in stripes), and protein bound to both ligand [X] and liposomes (shown in black). We show that when the binding constants the protein has for liposomes and ligand [X] are equal ($K_X = 0.05\ \mu M^{-1}$ and $K_L = 0.05\ \mu M^{-1}$), there is an equal probability of the protein existing in either state. By changing the K_L to $0.005\ \mu M^{-1}$ and keeping K_X unchanged, the shift in probability towards being bound to ligand [X] alone is clear (~10x). A greater difference between binding constants, $K_X = 0.05\ \mu M^{-1}$, $K_L = 0.0005\ \mu M^{-1}$, has an even more dramatic shift (~100x).

sites (Fig. 18.6). This term is incorporated into the thermodynamic cycle as well as in the calculation of probabilities of each state of the protein (Fig. 18.7). The protein binds ligand [X] in the absence of the liposome with an affinity of K_X. However, the protein if already associated with the liposome now binds ligand [X] with affinity σK_X. The difference in free energies of association is $\Delta G = -RT\ \ln K_X$ compared to $\Delta G = -RT\ \ln \sigma K_X$.

With linked sites, a change of signal upon binding one ligand [X] may be used to evaluate the binding to liposomes; even if interacting with liposomes does not exhibit a change in signal upon binding. Shown is an example where there is an enhanced affinity for ligand [X] by the protein when the protein is bound to liposome compared to the affinity in the absence of liposome (Fig. 18.8, leftmost panel). Advantage may be taken of this coupling to evaluate the liposome affinity by protein. A suspension of liposomes is titrated into a solution of protein and subsaturating concentrations of [X], binding of liposome increases the affinity of protein for [X] and thereby leading to a change in signal (Fig. 18.8, rightmost panel). A maximum change in signal is achieved upon titrating in liposomes by selecting the concentration of [X] that corresponds to the greatest difference in affinities of protein for ligand [X] in the absence and presence of liposome.

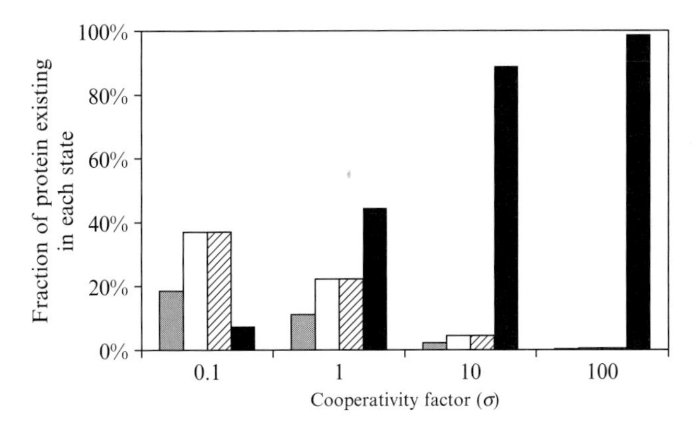

Figure 18.5 Graphical interpretation of the shift in probability toward doubly bound (bound to both liposomes and ligand [X]) by increasing the cooperativity factor σ. Unbound shown in gray, singly bound to liposomes shown in white, singly bound to ligand [X] shown in stripes, and doubly bound shown in black. The plot is based on the following binding polynomial: $Q = 1 + K_X[X] + K_L[L] = \sigma K_X[X]K_L[L]$, where $K_X = 0.05\ \mu M^{-1}$, $K_L = 0.05\ \mu M^{-1}$, $[X] = 40.0\ \mu M$ and [L] is 40 μM. σ was allowed to vary, and the effects are obvious. σ less than 1 indicates negative cooperativity and manifests itself as increasing probability of being completely unbound protein. Unit cooperativity is identical to independent binding. Finally, the more positive the cooperativity factor, the higher the probability of being doubly bound.

The representative equations for these binding isotherms are

$$\theta = \frac{[X]}{Q}\frac{dQ}{d[X]} = \frac{K_X[X] + \sigma K_X[X]K_L[L]}{1 + K_X[X] + K_L[L] + \sigma K_X[X]K_L[L]}, \quad (18.2)$$

$$\theta = \frac{[L]}{Q}\frac{dQ}{d[L]} = \frac{K_L[L] + \sigma K_X[X]K_L[L]}{1 + K_X[X] + K_L[L] + \sigma K_X[X]K_L[L]}. \quad (18.3)$$

It is important to note that for the binding isotherms where one ligand is titrated into the protein suspension and in the presence of a constant concentration of the other ligand, that ligand is also accounted for in the binding isotherm relationship. For the liposomes, once again, only the outer leaflet lipids will be accessible to a peripheral membrane-binding protein.

6.5. Isothermal titration calorimetry to measure protein–lipid interactions

When an interaction does not have a representative signal change, the use of linkage relationships is advantageous. A similar approach with isothermal titration calorimetry would complicate interpretation of the signal. The signal for isothermal titration calorimetry is heat. To measure the

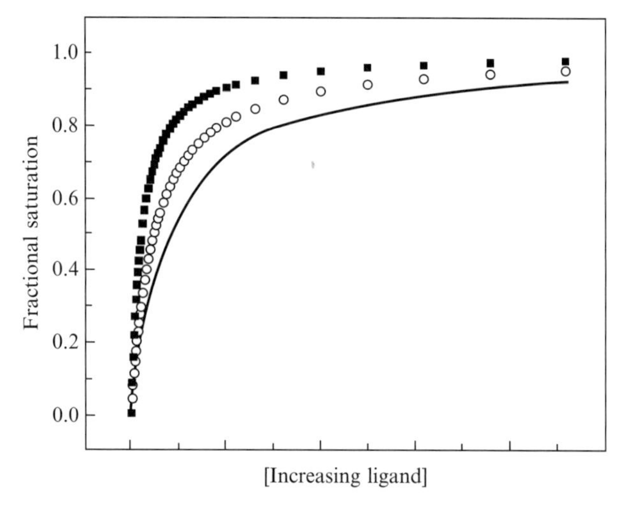

Figure 18.6 The effects of increasing the cooperativity factor on binding isotherms. The three isotherms were created by plotting Eq. (18.2), where K_X was held at $0.05\ \mu M^{-1}$, K_L was held at $2.50 \times 10^{-3}\ \mu M^{-1}$ and [L] was held at $20.0\ \mu M$. Unit cooperativity ($\sigma = 1$) is shown in black. A cooperativity factor of 2 is shown in the open circle isotherm and a cooperativity factor of 5 is shown in the black square isotherm. It is clear that an increase in σ increases the affinity, thus becoming fully saturated with ligand at a lower and lower ligand concentration.

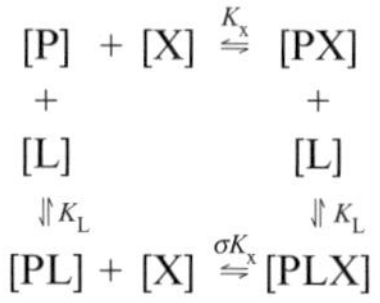

$$
\begin{array}{ccc}
[P] + [X] & \overset{K_X}{\rightleftharpoons} & [PX] \\
+ & & + \\
[L] & & [L] \\
\Updownarrow K_L & & \Updownarrow K_L \\
[PL] + [X] & \overset{\sigma K_X}{\rightleftharpoons} & [PLX]
\end{array}
$$

Figure 18.7 Thermodynamic cycle showing linked binding between ligand [X] and liposome [L].

heats of association of protein for lipid in the presence of a second ligand via the ITC, the second ligand needs to be at saturating levels as attributing the resultant cumulative heats to separate binding processes is difficult (e.g., see Fig. 18.9, the associated thermodynamic cycle is similar to Fig. 18.7). For instance, subsaturating levels of X would increase the affinity of the protein for lipid, leading to a heat of lipid binding; lipid binding in turn would increase the affinity of protein for X leading to an additional heat as the protein binds X. These heats could be both endothermic, both exothermic or endothermic and exothermic. The binding experiment is much easier to interpret if it reflects a single binding event. Essentially for this approach to

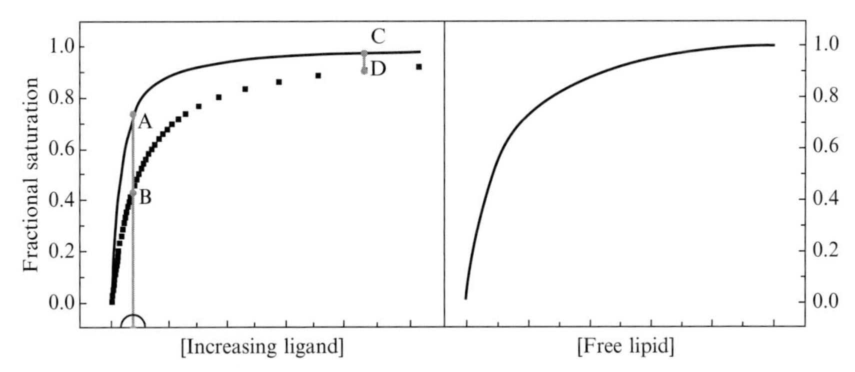

Figure 18.8 Leftmost panel: binding isotherms showing the binding of protein to ligand [X] in the absence (dotted line) and presence (solid line) of liposomes. The shift in affinity in the presence of liposomes is indicative of linkage between the two sites. To exploit this linkage relationship for the determination of liposome-binding affinity, a lipid titration must be done in the presence of subsaturating ligand [X]. The optimum ligand [X] concentration corresponds to the greatest difference between isotherms (depicted by points *A* and *B*, and the ligand [X] concentration is found half circled on the *x*-axis). If a ligand [X] concentration corresponding to a smaller difference between isotherms is used (points *C* and *D*), the resulting lipid titration will be greatly truncated as the protein is already much closer to fully saturated with lipid. In this example, using a ligand [X] concentration corresponding to points *C* and *D* would lead to approximately 10% of the signal that would have been seen using the ideal concentration. Rightmost panel: the resulting liposome titration in the presence of subsaturating ligand [X], the concentration of which what determined in Fig. 18.6.

be effective, a signal needs to arise from just one of the binding events (Table 18.1).

6.6. Modeling protein–protein interactions

Signal transduction complexes are based upon protein–lipid and lipid–lipid interactions, a model for how protein–protein interactions are enhanced at the membrane is presented. Such interactions would lead to further heterogeneous distributions of membrane components. A thermodynamic cycle and derivation of the model illustrates how differences in affinities between monomeric and dimeric proteins toward lipid will drive protein interactions at the membrane (Figs. 18.8 and 18.10). [P] is the monomeric form and [PP] is the dimer. If [PP] has greater affinity for a lipid than [P], the addition of this ligand will induce the protein to bind the membrane as a dimer. How cooperatively this occurs will depend on the difference in ligand affinity between the two forms of the protein and the concentration of [P] and [PP].

The fraction of protein existing as a dimer is simply all the dimerized states divided by all the possible states (Eq. 18.4):

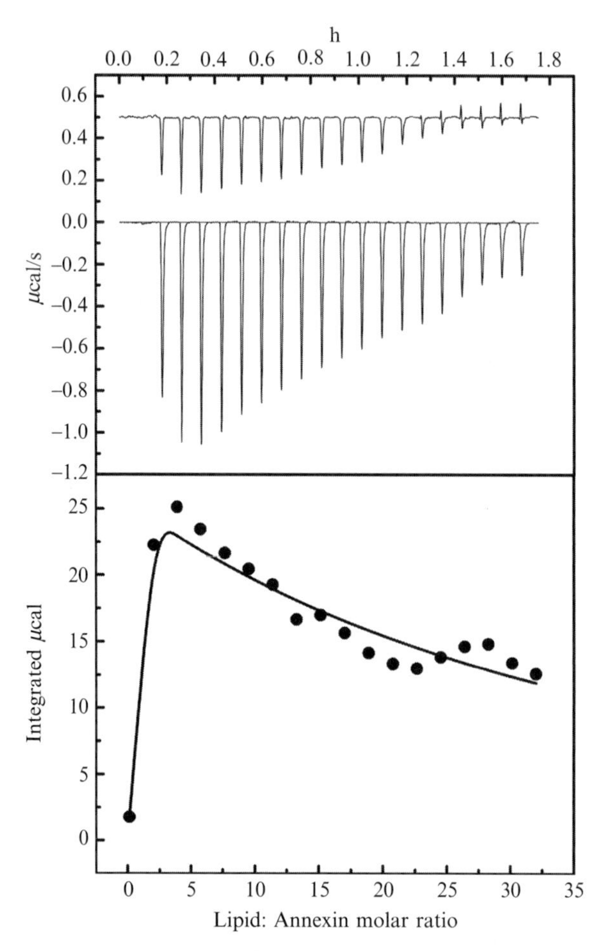

Figure 18.9 A representative plot is shown for a protein–lipid interaction where the lipids in the titration are in the gel state to minimize lipid–lipid rearrangement. The conditions were a binary lipid mixture comprised of 14:0,14:0PS:14L0,14:0PC (40%:60%) LUVs in the presence of Ca^{2+} titrated into a solution of 80 μM annexin a5 in the presence of the same concentration of Ca^{2+} at 15 °C with heat of dilution offset. The lipids in this titration are in the gel state to minimize lipid–lipid rearrangement. The upper panel shows the raw data for the titration with 30 mM large unilamellar vesicles (LUVs) in the presence of 0.75 mM constant [Ca^{2+}]. The Ca^{2+} concentration present corresponds to 88% saturation of the protein, where the association reaction $[PX] + [L] = [PXL]$ is used to calculate the Ca^{2+} saturation in the absence of lipid with the assumption that each protein binds nine Ca^{2+} ions independently (Almeida *et al.*, 2005b). The lower panel shows the integrated data from the previous raw data following the subtraction of the heat of dilution. The solid line represents the best fit of the data using the independent model and 15 mM total outer leaflet lipid concentration as the ligand concentration. The titration was conducted in decalcified 20 mM Mops, 100 mM KCl pH 7.5. The thermodynamic parameters for this titration are listed in Table 18.1. Both the Ca^{2+} and lipid titrant solutions were prepared using the dialysate of the protein to maintain a perfectly matched buffer system.

Table 18.1 Thermodynamic parameters of 14:0,14:0PS:14:0,14:0PC (40%:60%) LUVs binding to annexin a5 in the presence of saturating Ca^{2+} at 15 °C

Number of bound ligand molecules (*N*)	K_d (μM)	ΔG (kcal/mol)	ΔH (kcal/mol)	$T\Delta S$ (kcal/mol)
15 ± 4	6000 ± 1000	−2.9 ± 0.2	0.8 ± 0.2	3.7 ± 0.2

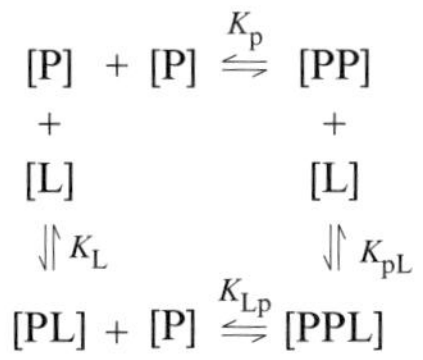

$$
\begin{array}{ccc}
[\mathrm{P}] + [\mathrm{P}] & \overset{K_\mathrm{p}}{\rightleftharpoons} & [\mathrm{PP}] \\
+ & & + \\
[\mathrm{L}] & & [\mathrm{L}] \\
\Downarrow\!\Uparrow K_\mathrm{L} & & \Downarrow\!\Uparrow K_\mathrm{pL} \\
[\mathrm{PL}] + [\mathrm{P}] & \overset{K_\mathrm{Lp}}{\rightleftharpoons} & [\mathrm{PPL}]
\end{array}
$$

Figure 18.10 Thermodynamic cycle showing the dimerization of proteins, [P] and interaction with lipid [L] of the monomers and dimers.

$$\theta_{\text{dimerized}} = \frac{[\mathrm{PP}] + [\mathrm{PPL}]}{[\mathrm{P}] + [\mathrm{PP}] + [\mathrm{PL}] + [\mathrm{PPL}]}. \tag{18.4}$$

To obtain a more useful form of this equation, each term must be normalized against a reference state which we select as the monomeric form ([P]). Then, substituting each normalized term with the respective equilibrium constant from the thermodynamic cycle, (Eq. 18.5) is generated:

$$\theta_{\text{dimerized}} = \frac{K_\mathrm{P}[\mathrm{P}] + K_\mathrm{P}[\mathrm{P}]K_\mathrm{PL}[\mathrm{L}]}{1 + K_\mathrm{P}[\mathrm{P}] + K_\mathrm{L}[\mathrm{L}] + K_\mathrm{P}[\mathrm{P}]K_\mathrm{PL}[\mathrm{L}]}. \tag{18.5}$$

This equation can be simplified by factoring ($K_\mathrm{P}[\mathrm{P}]$) from the numerator and rearranging the denominator:

$$\theta_{\text{dimerized}} = \frac{K_\mathrm{P}[\mathrm{P}](1 + K_\mathrm{PL}[\mathrm{L}])}{1 + K_\mathrm{L}[\mathrm{L}] + K_\mathrm{P}[\mathrm{P}](1 + K_\mathrm{PL}[\mathrm{L}])}. \tag{18.6}$$

A similar approach can be used to generate the fraction in the monomeric state:

$$\theta_{\text{monomer}} = \frac{1 + K_\mathrm{L}[\mathrm{L}]}{1 + K_\mathrm{L}[\mathrm{L}] + K_\mathrm{P}[\mathrm{P}](1 + K_\mathrm{PL}[\mathrm{L}])}. \tag{18.7}$$

The ratio of the fraction dimerized to the fraction in the monomeric state may now be calculated, we define this as K_{apparent}:

$$K_{\mathrm{apparent}} = \frac{\theta_{\mathrm{dimer}}}{\theta_{\mathrm{monomer}}} = K_{\mathrm{P}}[\mathrm{P}]\frac{(1 + K_{\mathrm{PL}}[\mathrm{L}])}{(1 + K_{\mathrm{L}}[\mathrm{L}])}. \tag{18.8}$$

In this final form, it is obvious that the greater K_{PL} is than K_{L}, the higher the probability that the protein will exist as a dimer at the membrane. Conversely, if K_{L} is greater than K_{PL}, there would be a higher probability of the protein existing as a monomer. It is also important to note that if K_{PL} is greater than K_{L} then the dimeric form has a higher affinity for lipid than the monomer. This will also further drive the protein into the dimeric form.

We can also introduce a scenario where the dimer and monomeric form can interact with difference numbers of lipids, where n or m are the number of lipids. The representative equation is

$$K_{\mathrm{apparent}} = \frac{\theta_{\mathrm{dimer}}}{\theta_{\mathrm{monomer}}} = K_{\mathrm{P}}[\mathrm{P}]\frac{(1 + K_{\mathrm{PL}}[\mathrm{L}])^{n}}{(1 + K_{\mathrm{L}}[\mathrm{L}])^{m}}. \tag{18.9}$$

If $n > m$, or the dimer binds more lipid than the monomer, even if $K_{\mathrm{PL}} \sim K_{\mathrm{L}}$, the protein will have a greater probability to be a dimer when membrane associated.

7. Synopsis

Proteins partition onto the membrane surface from the surrounding solution and this extent of partitioning will vary with lipid composition. How partitioning varies with lipid composition and lipid concentration can be difficult to determine as partitioning and binding is dependent on the chemical activity of the unbound lipid within the mixture. As relatively few lipid mixtures have been characterized on a rigorous thermodynamic level of the almost infinite possible combinations, this becomes a limiting factor. To gain predictability of which lipids that proteins interact with as well as which lipid mixtures proteins can redistribute requires such characterization.

Depending on the strength of the lipid–protein interaction and how specifically a protein interacts with a particular lipid, this interaction can redistribute the lipids within the mixture. It is the coupling of the inherent nonrandom mixing behavior of lipids with protein selectivity in association which leads to lipid reorganization and nonrandom distributions of proteins on the membrane surface. If the lipid mixture was randomly distributed, or, the protein binds each of the lipids within the mixture with equal affinity, the protein–lipid interaction would not redistribute lipids (Hinderliter *et al.*, 2001, 2004). Defining whether a protein has the capacity to reorder a

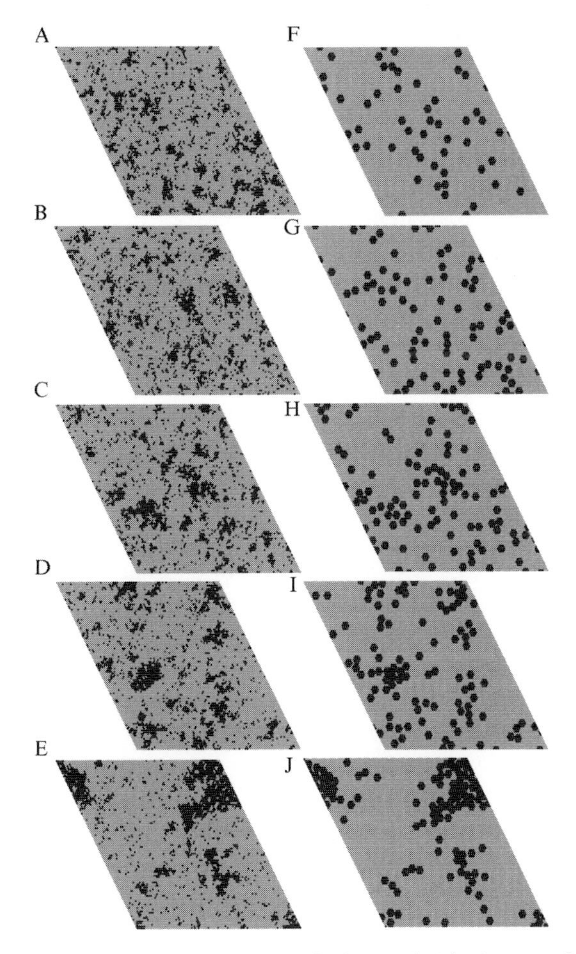

Figure 18.11 The system was comprised of a binary lipid mixture of 20%PS (black dots) in a background of 80%PC (gray dots) with a unlike nearest neighbor interactions, ω, of 280 cal/mol. There is a base protein–lipid interaction strength of 5 kcal/mol for either PS or PC and the protein interacts with 19 lipids. To this base protein–lipid interaction, specificity is added to the interaction with an extra PS–protein interaction. This extra interaction for each PS with protein was increased in increments of 200 from (A) 200 cal/mol, (B) 400 cal/mol, (C) 600 cal/mol, (D) 800 cal/mol, and (E) 1000 cal/mol. 100 proteins were added to the system for each simulation (shown as hexagons in rightmost panels), the numbers of proteins associated with the membrane surface increased as extra interaction energy increased where (F) 57 proteins were associated with the surface at 200 cal/mol, (G) 80 proteins were associated at 400 cal/mol, (H) 91 proteins were associated at 600 cal/mol, (I) 97 proteins were associated at 800 cal/mol, and (J) 100 proteins were associated at 1000 cal/mol. If the protein–lipid interaction strength is increased to 8 kcal/mol, all of the proteins were bound for each of the extra interaction energies but the protein and lipid simulations appeared the same as with 5 kcal/mol.

membrane surface is not necessarily straightforward. To explore how much difference in affinity that is necessary for one lipid over another in a binary lipid mixture by a protein for the protein to redistribute the lipids, a series of Monte Carlo simulations were performed (kindly provided by Paulo Almeida). The system was comprised of a binary lipid mixture of 20%PS in a background of 80%PC (phosphatidylcholine) with a lipid–lipid interaction parameter of 280 cal/mol. There is a base protein–lipid interaction strength of 5 kcal/mol for the membrane comprised of PS and PC and the protein interacts with 19 lipids. To this base protein–lipid interaction, specificity is added with an extra PS–protein interaction. This extra interaction for each PS with protein was increased in increments of 200 from 200 to 1000 cal/mol (Fig. 18.11). Observed is with additional interaction energy of 600 cal/mol for each PS contact with the protein, clustering of PS as well as proteins occurs in tandem. In comparison, the extra PS-protein interaction energy an extra interaction energy of 200–400 cal/mol did not lead to a pronounced redistribution of proteins and lipids while an additional PS-protein interaction energy interaction energy of 800–1000 cal/mol lead to an increasing to almost complete segregation of lipid and proteins. The sensitivity of the system to small differences in specificity to which lipid a protein preferentially interacts with is illustrated.

In quantitating protein–lipid interactions, the concept of specificity becomes relevant although specificity must be carefully defined. Specificity in this case, we define as an interaction of a protein for a particular lipid in the mixture over the other lipids present. The question is then raised, how much binding affinity signifies specificity? Specificity is enough affinity that the protein can induce reorganization of the membrane surface upon interaction with the surface. In this sense, the protein is forming not a lipid-binding pocket, but the optimal lipid surface to maximize protein–lipid contacts. In other words, the protein forms the lipids into its binding surface. This malleability is a manifestation of the small interaction energies between lipids. If lipids strongly adhered to one another, a protein-induced optimization of its membrane-binding surface would not be possible. The surface would be static rather than plastic and correspondingly less responsive to protein interactions.

ACKNOWLEDGMENTS

This material is based upon work supported by the National Science Foundation under CAREER MCB 0747339 (AH) and the Office of the Vice President for Research at the University of Minnesota Duluth (JM). The project described was also supported by NIH GM64443 (AH), from the National Institutes of Health. The content is solely the responsibility of the authors. We would also like to kindly thank Paulo Almeida for providing all of the Monte Carlo computer simulations as an invaluable asset to illustrate the concepts. Additionally, the authors thank Jacob Gauer and Emily Frey for critical reading and aid in preparation of figures for the chapter.

REFERENCES

Almeida, P. F. F. (2009). Thermodynamics of lipid interactions in complex bilayers. *Biochim. Biophys. Acta* **1788,** 72–85.

Almeida, P. F. F., Pokorny, A., and Hinderliter, A. (2005a). Thermodynamics of membrane domains. *Biochim. Biophys. Acta* **1720,** 1–13.

Almeida, P. F. F., Sohma, H., Rasch, K., Wieser, C. M., and Hinderliter, A. (2005b). Allosterism in membrane binding: A common motif of the annexins? *Biochemistry* **44,** 10905–10913.

Baumgart, T., Hunt, G., Farkas, E. R., Webb, W. W., and Feigenson, G. W. (2007). Fluorescence probe partitioning between L_o/L_d phases in lipid membranes. *Biochim. Biophys. Acta* **1768,** 2182–2194.

Bernardino de la Serna, J., Perez-Gil, J., Simonsen, A. C., and Bagatolli, L. A. (2004). Cholesterol rules: Direct observation of the coexistence of two fluid phases in native pulmonary surfactant membranes at physiological temperatures. *J. Biol. Chem.* **279,** 40715–40722.

Bloom, M., and Thewalt, J. L. (1995). Time and distance scales of membrane domain organization. *Mol. Membr. Biol.* **12,** 9–13.

Brown, D. A., and London, E. (2000). Structure and function of sphingolipid- and cholesterol-rich membrane rafts. *J. Biol. Chem.* **275,** 17221–17224.

Cao, H., Zhang, J., Jing, B., and Regen, S. L. (2005). A chemical sensor for the liquid-ordered phase. *J. Am. Chem. Soc.* **127,** 8813–8816.

Dill, K. A., and Bromberg, S. (2002). *Molecular Driving Forces.* Garland Science, New York, NY.

Edidin, M. (2003). The state of lipid rafts: From model membranes to cells. *Annu. Rev. Biophys. Biomol. Struct.* **32,** 257–283.

Field, K. A., Holowka, D., and Baird, B. (1997). Compartmentalized activation of the high affinity immunoglobulin E receptor within membrane domains. *J. Biol. Chem.* **272,** 4276–4280.

Freire, E., Mayorga, O., and Straume, M. (1990). Isothermal titration calorimetry. *Anal. Chem.* **62,** 950A–959A.

Garidel, P., Johann, C., and Blume, A. (1997). Nonideal mixing and phase separation in phosphatidylcholine-phosphatidic acid mixtures as a function of acyl chain length and pH. *Biophys. J.* **72,** 2196–2210.

Heimburg, T., Würz, U., and Marsh, D. (1992). Binary phase diagram of hydrated dimyristoylglycerol-dimyristoylphosphatidylcholine mixtures. *Biophys. J.* **63,** 1369–1378.

Hinderliter, A. K., Almeida, P. F. F., Creutz, C. E., and Biltonen, R. L. (2001). Domain formation in a fluid mixed bilayer modulated through binding of the C2 protein motif. *Biochemistry* **40,** 4181–4191.

Hinderliter, A., Biltonen, R. L., and Almeida, P. F. F. (2004). Lipid modulation of protein induced membrane domains as a mechanism for controlling signal transduction. *Biochemistry* **43,** 7102–7110.

Huang, J., and Feigenson, G. (1999). A microscopic interaction model of maximum solubility of cholesterol in lipid bilayers. *Biophys. J.* **4,** 2142–2157.

Ipsen, J. H., Karlström, G., Mouritsen, O. G., Wennerström, H., and Zuckermann, M. J. (1987). Phase equilibria in the phosphatidylcholine-cholesterol system. *Biochim. Biophys. Acta* **1,** 162–172.

Keller, S. L., and McConnell, H. M. (1999). Stripe phases in lipid monolayers near a miscibility critical point. *Phys. Rev. Lett.* **82,** 1602–1605.

Kim, J., Mosior, M., Chung, L. A., Wu, H., and McLaughlin, S. (1991). Binding of peptides with basic residues to membranes containing acidic phospholipids. *Biophys. J.* **60,** 135–148.

Kinnunen, P. K. J. (1991). On the principles of functional ordering in biological membranes. *Chem. Phys. Lipids* **57,** 375–399.

Ladokhin, A. S., Jayasinghe, S., and White, S. H. (2000). How to measure and analyze tryptophan fluorescence in membranes properly, and why bother? *Anal. Biochem.* **285,** 235–245.

Majumder, R., Quinn-Allen, M. A., Kane, W. H., and Lentz, B. R. (2005a). The phosphatidylserine binding site of the factor V(a) C2 domain accounts for membrane binding but does not contribute to the assembly or activity of a human factor X(a)–factor V(a) complex. *Biochemistry* **44,** 711–718.

Majumder, R., Weinreb, G. E., and Lentz, B. R. (2005b). Efficient thrombin generation requires molecular phosphatidylserine, not a membrane surface. *Biochemistry* **51,** 16998–17006.

McConnell, H. M., and Vrljic, M. (2003). Liquid–liquid immiscibility in membranes. *Annu. Rev. Biophys. Biomol. Struct.* **32,** 469–492.

Melo, E. C. C., Lourtie, I. M. G., Sankaram, M. B., Thompson, T. E., and Vaz, W. L. C. (1992). Effects of domain connection and disconnection on the yields of in-plane bimolecular reaction in membranes. *Biophys. J.* **63,** 1506–1512.

Mendelsohn, R., Liang, G. L., Strauss, H. L., and Snyder, R. G. (1995). IR spectroscopic determination of gel state miscibility in long-chain phosphatidyl choline mixtures. *Biophys. J.* **69,** 1987–1998.

Mountcastle, D. B., Biltonen, R. L., and Halsey, M. J. (1978). Effect of anesthetics and pressure on the thermotropic behavior of multilamellar dipalmitoylphosphatidylcholine liposomes. *Proc. Natl. Acad. Sci. USA* **10,** 4906–4910.

Radhakrishnan, A., and McConnell, H. M. (1999). Condensed complexes of cholesterol and phospholipids. *Biophys. J.* **77,** 1507–1517.

Rogasevskaia, T., and Coorssen, J. R. (2006). Sphingomyelin-enriched microdomains define the efficiency of native Ca^{2+}-triggered membrane fusion. *J. Cell Sci.* **119,** 2688–2694.

Sankaram, M. B., and Thompson, T. E. (1990). Modulation of phospholipid acyl chain order by cholesterol. A solid state ^{2}H nuclear magnetic resonance study. *Biochemistry* **29,** 10676–10684.

Seelig, J. (1997). Titration calorimetry of lipid–peptide interactions. *Biochim. Biophys. Acta* **1331,** 103–116.

Sheets, E. D., Holowka, D., and Baird, B. (1999). Membrane organization in immunoglobulin E receptor signaling. *Curr. Opin. Chem. Biol.* **3,** 95–99.

Silvius, J. R., and Gagné, J. (1984). Calcium-induced fusion of lateral phase separations in phosphatidylcholine-phosphatidylserine vesicles. Correlation by calorimetric and fusion measurements. *Biochemistry* **23,** 3232–3240.

Simons, K., and Ikonen, E. (1997). Functional rafts in cell membranes. *Nature* **387,** 569–572.

Simons, K., and van Meer, G. (1988). Lipid sorting in epithelial cells. *Biochemistry* **17,** 6197–6202.

Simons, K., and Vaz, W. L. (2004). Model systems, lipid rafts, and cell membranes. *Annu. Rev. Biophys. Biomol. Struct.* **33,** 269–295.

Suurkuusk, J., Lentz, B. R., Barenholz, Y., Biltonen, R. L., and Thomson, T. E. (1976). A calorimetric and fluorescent probe study of the gel-liquid crystalline phase transition in small, single-lamellar dipalmitoylphosphatidylcholine vesicles. *Biochemistry* **15,** 1393–1401.

Thompson, T. E., and Tillack, T. W. (1985). Organization of glycosphingolipids in bilayers and plasma membranes of mammalian cells. *Annu. Rev. Biophys. Biophys. Chem.* **14,** 361–386.

Thompson, T. E., Sankaram, M. B., and Biltonen, R. L. (1992). Biological membrane domains: Functional significance. *Comments Mol. Cell. Biophys.* **8,** 1–15.

Thompson, T. E., Sankaram, M. B., Biltonen, R. L., Marsh, D., and Vaz, W. L. C. (1995). Effects of domain structure of in-plane reactions and interactions. *Mol. Membr. Biol.* **12,** 157–162.

White, S. H., Wimley, W. C., Ladokhin, A. S., and Hristova, K. (1998). Protein folding in membranes: Determining energetics of peptide–bilayer interactions. *Methods Enzymol.* **295,** 62–87.

Wyman, J., and Gill, S. J. (1990). *Binding and Linkage.* University Science Books, Mill Valley, CA.

CHAPTER NINETEEN

Predicting pK_a Values with Continuous Constant pH Molecular Dynamics

Jason A. Wallace *and* Jana K. Shen

Contents

Abstract

Knowledge of pK_a values is important for understanding structure and function relationships in proteins. Over the past two decades, theoretical methods for pK_a calculations have been mainly based on macroscopic models, in which the protein is considered as a low-dielectric cavity embedded in a high-dielectric continuum. In recent years, constant pH molecular dynamics methods have been developed based on a microscopic description of the protein. We describe here the methodology of continuous constant pH molecular dynamics (CPHMD), which has emerged as one of the most robust and accurate tools for predicting protein pK_as and for the study of pH-modulated conformational dynamics. We illustrate the utility of CPHMD by the calculation of pK_as for surface residues in ribonuclease A, buried residues in staphylococcal nuclease, and titratable groups in the intrinsically flexible protein α-lactalbumin. We will compare the CPHMD results with experimental data as well as calculations from PB-based and empirical methods. These examples demonstrate the accuracy and

Department of Chemistry and Biochemistry, University of Oklahoma, Norman, Oklahoma, USA

Methods in Enzymology, Volume 466
ISSN 0076-6879, DOI: 10.1016/S0076-6879(09)66019-5

robustness of the CPHMD method and its ability to capture the correlation between ionization equilibria and conformational dynamics as well as the local dielectric response to structural rearrangement. Finally, we discuss future improvement of the CPHMD method.

1. Introduction

Proteins can gain or lose protons in response to changes in solution acidity or pH. Many important biological processes such as ligand binding (Warshel, 1981b), enzyme reactions (Warshel, 1981a), protein–protein recognition (Sheinerman *et al.*, 2000), and protein folding (Bierzynski *et al.*, 1982) are modulated by pH. For example, NADH binding to human alcohol dehydrogenase is dependent upon a single ionization that occurs at pH 8.1 such that the binding is substantially diminished at elevated pH (Stone *et al.*, 1999). The ability to predetermine the ionization states of proteins is useful not only for understanding pH-mediated processes, but also for protein design and engineering. Electrostatic complementarity is a common feature in protein–protein (Lee and Tidor, 2001; Schreiber and Fersht, 1996) and protein–substrate recognition (Getzoff *et al.*, 1983). Knowledge of, and the ability to tune, protonation states at the binding interface can lend much to protein design by allowing engineered proteins to have optimal electrostatic complementarity with their binding partners. Using this strategy the binding affinity between β-lactamase and inhibitor BLIP was increased by 200-fold (Selzer *et al.*, 2000).

Protonation or deprotonation of proteins often occurs at ionizable, also known as titratable, side chains. The ionization equilibrium of a titratable side chain, $HA \rightleftharpoons A^- + H^+$, is described by a dissociation constant:

$$K_a = \frac{[H^+][A^-]}{[HA]}. \tag{19.1}$$

Rearrangement of this relationship gives the Henderson–Hasselbach (HH) equation:

$$pK_a = pH + \log\left(\frac{[A^-]}{[HA]}\right), \tag{19.2}$$

where pK_a is related to K_a by $pK_a = -\log K_a$. Thus, the value of pK_a equals pH if the populations of protonated and deprotonated states are identical. The HH equation can be rearranged and written in a generalized form in the presence of multiple sites that titrate in a similar pH range:

$$S^{\text{unprot}} = \frac{1}{1 + 10^{n(\text{p}K_\text{a} - \text{pH})}}, \tag{19.3}$$

where S^{unprot} is the unprotonated fraction for the titration site of interest and n is the Hill coefficient. The deviation of the Hill coefficient from 1 reflects the extent of cooperative ($n > 1$) or anticooperative ($n < 1$) interactions between multiple titrating groups. Fitting the titration data, S^{unprot} versus pH, to the generalized HH equation (Eq. 19.3) offers the desired pK_a value and Hill coefficient.

An important concept in protein pK_a calculations is model pK_a which is defined as the pK_a of an isolated side chain fully exposed to water. Model pK_a values can be experimentally determined using model compounds containing a single amino acid side chain terminated by hydrogens or blocking groups (Nozaki and Tanford, 1967; Thurlkill *et al.*, 2006). Using model pK_as as reference, pK_a shifts (ΔpK_a) can be obtained for titratable residues in proteins, which can be positive (up), negative (down), or zero depending on the local electrostatic environment. ΔpK_a of a side chain is caused by a shift in the deprotonation equilibrium, which gives rise to a change in the relative free energy between the protonated and deprotonated forms:

$$\Delta\Delta G^{\text{deprot}} = \ln(10) RT \Delta \text{p}K_\text{a}. \tag{19.4}$$

Thus, pK_a shifts are determined by factors stabilizing protonated or deprotonated forms. These are typically desolvation, hydrogen bonding, charge–charge, and charge–dipole interactions with nearby groups.

2. Theoretical Methods for pK_a Predictions

The absolute pK_a can in principle be calculated quantum mechanically for small molecules by making use of the following thermodynamic cycle (Lim *et al.*, 1991; Richardson *et al.*, 1997):

$$\begin{array}{lcccc} & \text{HA}_{(\text{gas})} & \xrightarrow{\Delta G_{gas}} \text{H}^+_{(\text{gas})} & + & \text{A}^-_{(\text{gas})} \\ -\Delta G_{\text{solv.}}(\text{HA}) \uparrow & & \downarrow \Delta G_{\text{solv.}}(\text{H}^+) & & \downarrow \Delta G_{\text{solv.}}(\text{A}^-) \\ & \text{HA}_{(\text{aq.})} & \xrightarrow{\Delta G_{aq.}} \text{H}^+_{(\text{aq.})} & + & \text{A}^-_{(\text{aq.})} \end{array}$$

which gives the deprotonation free energy as

$$\Delta G_{\text{aq}} = \Delta G_{\text{gas}} + \Delta G_{\text{solv}}(\text{A}^-) + \Delta G_{\text{solv}}(\text{H}^+) - \Delta G_{\text{solv}}(\text{HA}). \tag{19.5}$$

However, there are several obstacles to this approach. First, the solvation free energies in Eq. (19.5) are on the order of hundreds of kcal/mol. Small relative errors in the calculation can result in large error in the predicted pK_a considering one pH unit is equivalent to a free energy change of 1.2 kcal/mol. Second, the accuracy for the calculation of gas-phase proton affinity is 1–2 kcal/mol (Chipman, 2002). Nevertheless, recent work shows that the pK_a values for a heterogeneous group of organic molecules can be determined with an RMSD of 0.6 pH units using density functional calculations (Schmidt am Busch and Knapp, 2004). To circumvent the problems encountered in the calculation of absolute pK_as, various methods have been developed to determine relative pK_as, also known as pK_a shifts in proteins. pK_a shifts can be calculated with high accuracy by considering the energetic difference between a titratable residue in the protein environment and in solution.

2.1. Methods based on macroscopic description of proteins

Since the major contributor to the pK_a shift from solution to the protein environment is electrostatics, much effort has focused on calculating the energy difference in transferring a residue from the high-dielectric water environment to the low-dielectric protein interior. This has led to the development of methods based on the model that protein solvated in water can be considered as a low-dielectric cavity embedded in a high-dielectric medium. In this case, the electrostatics is fully described by the Poisson–Boltzmann (PB) equation from classical electrostatics theory (Bashford and Karplus, 1990):

$$\nabla\cdot[\varepsilon(r)\nabla\phi(r)] - \kappa^2(r)\varepsilon(r)\phi(r) = -4\pi\rho_0(r) \tag{19.6}$$

where $\varepsilon(r)$ is the dielectric constant, $\phi(r)$ is the electrostatic potential, $\rho_0(r)$ is the fixed solute charge density, and $\kappa(r)$ is related to the ionic strength, I, by $\kappa(r) = 8\pi eI/\varepsilon(r)k_BT$. The linearized PB equation can be solved analytically for a spherical protein as in the Tanford–Kirkwood model (Tanford and Kirkwood, 1957). Over the past two decades efficient algorithms based on finite-difference approach have been developed for solving the linearized PB equation with an arbitrarily shaped dielectric boundary. These algorithms are implemented in several computer programs such as DELPHI (Nicholls and Honig, 1991), MEAD (Bashford, 1997), and UHBD (Madura *et al.*, 1995) that have been widely used for solving the PB equation and computing pK_a values for proteins (Antosiewicz *et al.*, 1994; Bashford and Karplus, 1990; Yang *et al.*, 1993).

In the PB-based methods, the pK_a for a protein ionizable site is given by (Bashford, 2004)

$$pK_a^{prot} = pK_{intr} - \frac{1}{\ln(10)RT}\Delta G_{Coul}, \qquad (19.7)$$

where pK_{intr} is the intrinsic pK_a according to Tanford, which is the pK_a an ionizable group would have if all other groups were held in the neutral state (Bashford, 2004). ΔG_{Coul} represents the Coulomb interaction between other charged sites in the protein. The intrinsic pK_a is given by

$$pK_{intr} = pK_a^{mod} - \frac{1}{\ln(10)RT}(\Delta G_{Born} + \Delta G_{bg}), \qquad (19.8)$$

where ΔG_{Born} is the Born or reaction field energy of the charge at the titratable site, and ΔG_{bg} is the interaction energy between the titratable site and nontitratable "background" charges in the protein. This division of energy terms allows the analysis of molecular determinants of pK_a shifts (García-Moreno and Fitch, 2004). The same method as in Eqs. (19.7) and (19.8) can be applied with generalized Born (GB) implicit-solvent models (Still *et al.*, 1990) to reduce computational cost of pK_a calculations (Kuhn *et al.*, 2004).

The PB- or GB-based continuum methods assume a fixed dielectric constant for the protein, neglecting the dielectric heterogeneity as well as response to structural rearrangement. To compensate for the lack of explicit treatment of dielectric response, the use of an effective dielectric constant has been adopted, which is typically between 4 and 20. However, as noted recently, no single dielectric constant can be used to correctly predict pK_as for both surface and buried residues (García-Moreno and Fitch, 2004). This is due to the fact that the dielectric constant of proteins switches from 2–3 in the interior (Bone and Pethig, 1985) to 14–25 in the outer region (Simonson and Brooks, 1996). To correct for this change in protein dielectrics, sigmoidal-shaped dielectric functions have been suggested (Mehler and Guarnieri, 1999; Mehler *et al.*, 2002).

Another way to compensate for the lack of dielectric response is to approximately include conformational flexibility. Several approaches have been suggested, which include scanning side-chain torsion angles (You and Bashford, 1995), combining short molecular dynamics (MD) trajectories with PB calculations (Koumanov *et al.*, 2001; van Vlijmen *et al.*, 1998), and sampling positions of hydroxyl and other polar protons using a Monte Carlo (MC) protocol (Alexov and Gunner, 1997; Georgescu *et al.*, 2002). Although these developments have proven useful in bringing the predicted pK_as closer to experiment, they do not address the fundamental limitation of the macroscopic approaches, namely, the description of protein as a dielectric continuum. This limitation can only be overcome by introducing microscopic treatment of the protein such that the ionization states of titratable residues are determined by the microscopic environment (Schutz and Warshel, 2001).

2.2. Methods based on microscopic description of proteins

The coupling between local conformational rearrangement and ionization of titratable residues can be accounted for in pK_a calculations based on free energy simulations (Merz, 1991; Riccardi *et al.*, 2005; Simonson *et al.*, 2004; Warshel *et al.*, 1986). In this approach, the free energy of charging a side chain in the protein environment is obtained using molecular dynamics in either explicit or implicit solvent. However, this approach is computationally demanding and cannot be used to simultaneously obtain pK_as of multiple sites. To enable the calculation of pK_as with one molecular dynamics simulation run, the Gaussian fluctuation method based on linear response theory (Levy *et al.*, 1991) was developed (Del Buono *et al.*, 1994). Another method that accounts for the local electrostatic environment is based on the protein dipoles/Langevin dipoles (PDLD) model, which describes the protein explicitly while treating solvent molecules as point dipoles placed on a lattice (Lee *et al.*, 1993; Sham *et al.*, 1997). The PDLD model is able to reproduce experimental pK_a shifts of interior groups for which the macroscopic models assuming high protein dielectric constant cannot (Schutz and Warshel, 2001). The above-mentioned methods cannot be applied to study pH-dependent conformational dynamics of proteins or to predict pK_as where ionization is coupled to a large conformational transition, such as folding or unfolding of proteins (Bashford, 2004).

Below, we will outline recent progress in the development of a class of methods that enables simultaneous description of conformational dynamics and the titration processes of ionizable residues in the protein under a specified pH condition. These methods assume an infinite proton bath and are thus referred to as constant pH molecular dynamics (PHMD) (Chen and Khandogin, 2008; Khandogin and Brooks, 2007b; Mongan and Case, 2005). Our particular emphasis will be on the λ-dynamics (Kong and Brooks, 1996) based continuous constant pH molecular dynamics (CPHMD) method (Khandogin and Brooks, 2005; Lee *et al.*, 2004).

2.2.1. Methods based on discrete protonation states

Since a titratable group can either be protonated or deprotonated, the most intuitive way to address the side-chain titration during the course of MD simulation is to periodically interrupt the simulation by attempting a protonation/deprotonation event using MC sampling. Methods based on discrete protonation states mainly differ in the solvent representation for the sampling of conformational and protonation states and the method for evaluating the free energy of deprotonation necessary for the MC step. Baptista *et al.* (2002) developed a mixed solvent scheme based on explicit-solvent MD and MC sampling using the free energy obtained from PB calculations. After a switch in protonation states, explicit-solvent molecules are allowed to relax at the fixed solute conformation, a step necessary to alleviate a problem due to the abrupt change in the charge states, which is a potential pitfall for all methods based on discrete protonation states. An instantaneous switch in the charge states may lead to a large increase in energy, resulting in a low acceptance ratio in the MC move (Stern, 2007). It also presents a problem for the accurate treatment of long-range electrostatics in explicit-solvent molecular dynamics simulations. Baptista and coworker tested two approaches, the generalized reaction field (GRF) method and the Particle mesh Ewald (PME) method with inclusion of an approximate number of counter ions (Machuqueiro and Baptista, 2006, 2008). Bürgi *et al.* (2002) developed a method based on the explicit-solvent model for sampling both conformational and protonation states, where the deprotonation free energy used in MC trial moves is estimated with free energy simulations. This method is computationally demanding and convergence is a major issue.

Employing implicit-solvent models in both MD and MC steps can significantly reduce the computational cost. Also, the potential pitfall associated with the large energy change may be circumvented by the instantaneous adjustment of solvent to the new protonation states of the protein. Dlugosz and Antosiewicz (2004) demonstrated a method that combines the analytical continuum electrostatics model (Schaefer and Karplus, 1996) for conformational sampling with the PB calculation for MC moves. Mongan *et al.* (2004) developed a protocol employing the GB implicit-solvent model in both molecular dynamics and protonation states sampling. Because of the extra computational time spent on the MC evaluation, simulations using discrete protonation states are much slower than standard molecular dynamics. Nevertheless, recent results of the pK_a calculation for hen egg-white lysozyme (HEWL) using the GB model and the mixed PB/explicit solvent scheme are encouraging (Machuqueiro and Baptista, 2008; Mongan *et al.*, 2004). Future benchmark studies are needed to assess the accuracy of the discrete states approaches and their capability in modeling pH-dependent conformational dynamics.

2.2.2. Methods based on continuous protonation states

In methods based on continuous protonation states, the ionization state of a titratable site is gradually switched between protonated and deprotonated forms. Mertz and Pettitt (1994) devised a grand canonical molecular dynamics approach based on an extended Hamiltonian to allow protonation and deprotonation, and tested the method using acetic acid in water. Börjesson and Hünenberger (2001) developed the acidostat procedure, which allows the titration coordinates, λ, bound between 0 and 1, to relax toward the equilibrium values in analogy to Berendsen's thermostat (Berendsen *et al.*, 1984) in explicit-solvent molecular dynamics simulations. The acidostat method has not been tested on protein titrations. Recently, Lee, Khandogin, and Brooks developed a method based on the λ-dynamics approach to free energy calculations (Kong and Brooks, 1996) and GB implicit-solvent models (Im *et al.*, 2003; Lee *et al.*, 2002, 2003). This method, referred to as CPHMD (Khandogin and Brooks, 2005; Lee *et al.*, 2004), has been extensively tested on protein titrations (Khandogin and Brooks, 2006) and applied to study pH-dependent mechanisms of protein folding (Khandogin and Brooks, 2007b; Khandogin *et al.*, 2006, 2007). We will discuss the methodology here in more detail.

In the CPHMD method, protonation/deprotonation of a titrating residue is described by a continuous coordinate, λ_j, bound between 0 and 1, which is a function of an unbounded variable θ_j:

$$\lambda_j = \sin^2(\theta_j), \quad j = 1, \ldots, n. \tag{19.9}$$

The deprotonated and protonated states are defined as $\lambda > \text{cut}_{\text{high}}$ and $\lambda < \text{cut}_{\text{low}}$, respectively, where the two cutoffs are typically, 0.9 and 0.1. The deprotonated fraction is then calculated from the relative occupancy of the deprotonated state. The titration coordinates, λ_j, are propagated simultaneously with the spatial coordinates, r_i, using the extended Hamiltonian given below (Khandogin and Brooks, 2005):

$$H^{\text{ext}}(\{r_i\}, \{\theta_j\}) = H^{\text{hyb}}(\{\mathbf{r}_i\}, \{\theta_j\}) + \sum_j \frac{m_j}{2}\dot{\theta}_j^2 + U^*(\{\theta_j\}). \tag{19.10}$$

Here, the second term represents the kinetic energy of θ_j, which carries a fictitious mass similar to that of a heavy atom. The last term represents the biasing potential defined as

$$U^*(\{\theta_j\}) = \sum_j [-U^{\text{barr}}(\theta_j) - U^{\text{mod}}(\theta_j) + U^{\text{pH}}(\theta_j)], \tag{19.11}$$

where U^{barr} is a barrier potential:

$$U^{\text{barr}}(\lambda_j) = 4\beta_i\left(\lambda_j - \frac{1}{2}\right)^2, \tag{19.12}$$

which serves to suppress the population of the mixed states ($\text{cut}_{\text{low}} < \lambda < \text{cut}_{\text{high}}$), which are unphysical. Previous studies (Khandogin and Brooks, 2005, 2006, 2007a; Khandogin *et al.*, 2006, 2007) show that a barrier height (β_j) of 1.5 kcal/mol (or 2.5 kcal/mol for double-site titration, see later discussions) offers a good tradeoff between the transition rate for protonation/deprotonation and a low population (below 15%) of mixed states. $U^{\text{mod}}(\theta_j)$ is an analytic function of the potential of mean force (PMF) for the deprotonation of a model compound along the titration coordinate, λ_j. Thus, the titration simulation of a model compound in solution at $\text{pH} = \text{p}K_{\text{a}}^{\text{mod}}$ yields approximately 50% protonated and 50%

deprotonated states. Due to the pairwise form of the GB energy function, U^{mod} is a quadratic function of λ for single-site titrations (see later discussions):

$$U^{\mathrm{mod}}(\lambda_j) = A_j(\lambda_j - B_j)^2, \tag{19.13}$$

where parameters A_j and B_j can be obtained via thermodynamic integration using titration simulations of the model compound at different θ-values. Finally, $U^{\mathrm{pH}}(\lambda_j)$ models the free energy dependence on the external pH by

$$U^{\mathrm{pH}}(\lambda_j) = \ln 10 \cdot \kappa_{\mathrm{B}} T \cdot \lambda_j(\mathrm{p}K_{\mathrm{a}} - \mathrm{pH}), \tag{19.14}$$

where $\mathrm{p}K_{\mathrm{a}}$ refers to the model $\mathrm{p}K_{\mathrm{a}}$. The coupling between conformational dynamics and proton titration is enabled through the hybrid Hamiltonian in Eq. (19.10), which is a sum of van der Waals, Coulomb, and GB electrostatic energies that are attenuated by the values of λ. Thus, the protonation equilibrium of a protein titratable site is controlled by the difference between the simulation pH and the model $\mathrm{p}K_{\mathrm{a}}$ as well as between the environment in the protein and that in solution.

The formalism introduced so far considers ionization at different titratable sites independently. We will refer to it as the single-site model. This is, however, not realistic for side chains such as histidine and carboxylates. Histidine can lose or gain a proton from either Nδ or Nε atom, while protonation/deprotonation can occur at either of the carboxylate oxygen atoms in Asp and Glu residues. Another way of viewing competitive titration is to consider the titration reaction products as tautomers, although in the case of a carboxylate side chain, the two protonated forms are chemically equivalent. Nevertheless, we can treat them as tautomers because rotation of the C–O bond in a carboxylate group is slow on a MD time scale. To couple the protonation equilibria of tautomers, a double-site model was developed, which utilizes an extended variable, x, to describe the tautomeric interconversion. Consequently, the hybrid Hamiltonian is reformulated to include the dependence of x. Also, the PMF for titrating a model compound is no longer a simple quadratic function of λ, but rather, a bivariate polynomial, quadratic in either λ or x. The coefficients in the polynomial can be obtained through thermodynamic integration at different λ- and x-values. The development of double-site titration model has significantly improved the accuracy of the CPHMD method as demonstrated by the qualitative agreement between predicted and experimental $\mathrm{p}K_{\mathrm{a}}$s (Khandogin and Brooks, 2005). However, several deficiencies remained to be addressed, which we discuss in Section 2.2.3.

2.2.3. Addressing issues in the microscopic methods

One major bottleneck in methods based on the microscopic description of proteins is convergence (Mongan and Case, 2005). As mentioned earlier, employment of GB implicit-solvent models in constant pH molecular dynamics greatly speeds up simulation convergence. The current CPHMD implementation utilizes the implicit-solvent GB model, GBSW (Im *et al.*, 2003) with a simple surface area-dependent term for modeling the nonpolar solvation effects. The GBSW simulations offer high accuracy (<1% error with respect to PB solvation energies for a large set of proteins; Feig *et al.*, 2004b) but are significantly less computationally demanding relative to PB calculations or explicit-solvent simulations. Nevertheless, random errors in the CPHMD titration of model compounds were as large as 0.5 pH units (Khandogin and Brooks, 2005). Considering the tight coupling between fluctuations in protonation states and local conformational rearrangement, one way to improve convergence is to enhance sampling of conformational states using the replica-exchange (REX) technique (Sugita and Okamoto, 1999). In a REX–CPHMD scheme, multiple independent copies of the protein are simulated in parallel at different temperatures. Conformational as well as protonation states at adjacent temperatures are allowed to swap periodically based on the Monte Carlo criterion. By incorporating the REX algorithm, the magnitude of random errors was reduced to below 0.16 pH units in REX–CPHMD titration simulations of model compounds using the same amount of simulation time (Khandogin and Brooks, 2006).

Although GB implicit-solvent models greatly accelerate simulations, it comes with a price. Overstabilization of attractive electrostatic interactions was a common problem (Im *et al.*, 2006) and led to a systematic underestimation of carboxylate pK_as due to overstabilized salt-bridge interactions (Khandogin and Brooks, 2005). By employing an optimized set of atomic input radii for the GBSW model and the improved torsion energetics for the implicit-solvent force field (Chen *et al.*, 2006), this problem was largely alleviated (Khandogin and Brooks, 2006). In the PB-based methods, the effects due to mobile ions are naturally accounted for in solving the PB equation. In the GB-based methods, however, one has to resort to some approximation. A Debye–Hückel screening function, $e^{\kappa r}$, was applied to the solvent dielectric constant ε, where κ depends on the ionic strength of the solution and was defined earlier (Srinivasan *et al.*, 1999). Although the extent of salt dependence recovered by this crude treatment may be somewhat larger than that in the PB calculation (Srinivasan *et al.*, 1999), REX–CPHMD simulations using the approximate screening function gave pK_as in better agreement with experiments conducted in 100–300 m*M* salt solution (Khandogin and Brooks, 2006).

Having addressed the issues in coupled ionization, sampling convergence, implicit-solvent model, and salt screening of electrostatics, the REX–CPHMD method is able to offer quantitative prediction of protein pK_as. A benchmark study demonstrated that 1-ns REX–CPHMD simulations gave pK_a predictions with an RMSD of 1 pH unit or below for a test set of proteins, in which anomalously large pK_a shifts are observed (Khandogin and Brooks, 2006). The CPHMD method is currently implemented in the CHARMM program package (version c33a and higher) (Brooks *et al.*, 2009). REX–CPHMD can be accessed through a Perl interface package, MMTSB Toolset (Feig *et al.*, 2004a). Below, we will examine the performance of REX–CPHMD titration simulations in predicting pK_as for residues in various chemical environments.

2.3. Empirical methods

With the growth of the number of experimentally determined pK_a values, it has become possible to develop very simple structure-based empirical methods to calculate pK_a shifts. One of such methods, PROPKA, utilizes a simple model parameterized to account for the effects on pK_a shifts due to hydrogen bonding, desolvation, and charge–charge interactions based on a large training set of known pK_as (Bas *et al.*, 2008; Li *et al.*, 2005). Empirical methods offer an attractive alternative to PB- or MD-based approaches because of the computational efficiency. Since a recent benchmark study noted PROPKA as more accurate than PB-based methods (Davies *et al.*, 2006), we will compare the CPHMD results with those from PROPKA calculations in the context of the examples that will be discussed.

3. Predicting Protein pK_as with REX–CPHMD Simulations

3.1. Surface residues in RNase A

Ribonuclease A (RNase A) catalyzes the hydrolysis of single stranded RNA in the absence of metal ions or cofactors. This enzyme has several surface residues with strongly shifted pK_a values, which provide excellent benchmarks for testing the theoretical description of local electrostatic interactions in the solvent medium. We compare calculated pK_a values from REX–CPHMD titration simulations with the experimental data (Baker and Kintanar, 1996) (Table 19.1). The overall RMSD is 0.83 pH unit. Figure 19.1 illustrates the calculation of pK_a from the REX–CPHMD titration data. One of the most drastically shifted pK_a is that of Asp-38. This residue is located in a loop region that connects an α-helix with a β-strand, and is flanked by Lys-37 and Arg-39. Asp-38 also interacts with Lys-41 as well as Lys-1 and Arg-10. Thus, interactions with these positively charged residues lead to stabilization of the deprotonated form of Asp-38. In the CPHMD method, the pK_a value of a side chain is obtained at a pH, where the protonated and deprotonated conformational states are of equal free energy. Thus, CPHMD directly couples the ionization equilibrium with the equilibrium for conformational populations. This point can be illustrated by the correlation between the relative positions of the nearby

Table 19.1 Calculated and experimental pK_a values for ribonuclease A

	CPHMD		Expt	
Residue	pK_a	ΔpK_a	pK_a	ΔpK_a
Glu-2	3.6	−0.4	2.6	−1.8
Glu-9	3.8	−0.6	–	–
His-12	5.8	−0.8	6.0	−0.6
Asp-14	3.4	−0.6	1.8	−2.2
Asp-38	3.0	−1.0	2.1	−1.9
His-48	4.9	−1.7	6.1	−0.5
Glu-49	3.3	−1.1	4.3	−0.1
Asp-53	4.0	0.0	3.7	−0.3
Asp-83	3.2	−0.7	3.3	−0.7
Glu-86	4.5	0.1	4.0	−0.4
His-105	6.4	−0.2	6.5	−0.1
Glu-111	3.4	−1.0	–	–
His-119	5.6	−1.0	6.5	−0.1
Asp-121	2.7	−1.3	3.0	−1.0
RMSD	0.8			

REX–CPHMD simulations were conducted using the crystal structure (PDB ID: 7RSA) at pH 2, 3, 4, 5, 6, and 7 with an ionic strength of 60 m*M*. Sixteen replicas were used that occupied exponentially spaced temperature windows ranging from 298 to 400 K. Simulations were run for a total of 500 replica-exchange steps or 1 ns. The deprotonated fractions were extracted from the replica at 298 K. Other simulation details are given in Khandogin and Brooks (2006). pK_a shifts were calculated relative to the model values: Asp-4.0, Glu-4.4, and His-6.5. Experimental data were obtained with an ionic strength ≤60 m*M* (Baker and Kintanar, 1996).

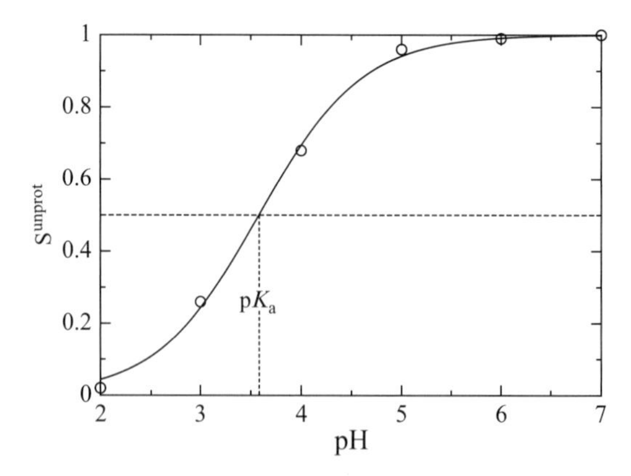

Figure 19.1 Illustration of curve fitting to calculate pK_a from REX–CPHMD simulations. Unprotonated fractions are shown in circles. Solid line represents curve fitting of data to the HH equation. Data are taken from the titration simulations of Glu-2 in RNase A.

positively charged residues and the protonation state of Asp-38 (Fig. 19.2). When Asp-38 is deprotonated, the probability shifts toward these positively charged residues being nearby Asp-38.

3.2. Buried groups in staphylococcal nuclease

Knowledge of the protonation states of ionizable residues buried in the protein interior is important for understanding the functions of many enzymes. However, calculation of pK_a values for buried groups is challenging because pK_a shifts of these groups tend to be very large and require accurate calculation of desolvation energy, and electrostatic interactions in a low-dielectric environment. While the former is sensitive to errors in the underlying solvent model, the latter is sensitive to errors in the force field and the structure used for the calculation. Thus, pK_as of buried groups provide a sensitive probe of the local electrostatic environment and hence they are excellent targets for testing theoretical methods for pK_a prediction. We use REX–CPHMD titration simulations to calculate pK_as for two deeply buried sites in the wild type (WT) and two variants of staphylococcal nuclease (SNase), PHS, and Δ+PHS. PHS is a hyperstable variant of SNase with three point mutations, P117G, H124L, and S128A, while Δ+PHS is PHS with two additional substitutions, G50F and V51N and deletion of residues 44–49. In the first site Leu-38 is replaced by Asp, Glu, or Lys. NMR titration data show that while the pK_a values for Asp-38 and Glu-38

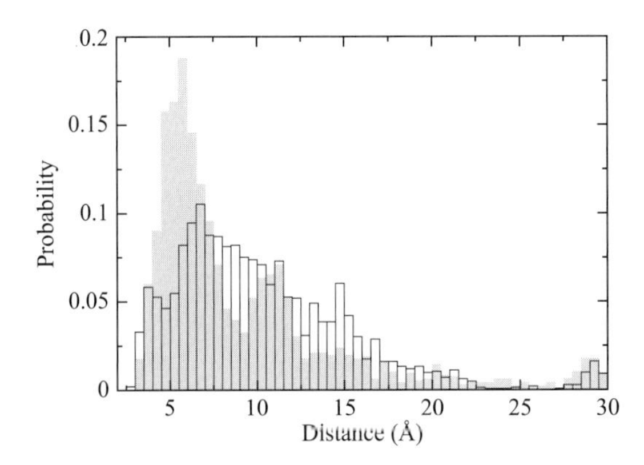

Figure 19.2 Coupling between protonation equilibria and conformational dynamics in RNase A. Histogram for the distance between the protonated (black) or deprotonated (gray) form of Asp-38 and nearby positively charged groups, Lys-1, Arg-10, and Lys-41. When Asp-38 is deprotonated, there is an increased likelihood of finding positively charged residues nearby. Data are collected from REX–CPHMD simulations of RNase A at pH 3, where the protonated and deprotonated populations of Asp-38 coexist.

are shifted up relative to the model value, the pK_a for Lys-38 is not shifted, probably due to the balance between desolvation and favorable electrostatic interaction with Glu-122 (Harms *et al.*, 2009). Compared to experiment, the calculated pK_as for Asp-38, Glu-38, and Lys-38 are lower by 0.6, 0.1, and 1.1 pH units, respectively (Table 19.2). We compare our results with calculations using the multiple-conformation continuum electrostatics (MCCE) method with the protein dielectric constant set to 8 (Alexov and Gunner, 1997; Georgescu *et al.*, 2002). MCCE uses finite-difference PB calculations with MC sampling of side-chain rotamers and tautomer states. The MCCE method overestimates the pK_as for Asp-38 and Glu-38 by 1.5 pH units, and underestimates the pK_a for Lys-38 by 3.1 pH units (Table 19.2). Overestimation of pK_a shifts is typical for PB-based approaches and can be attributed to the lack of explicit treatment of structural rearrangement. Although the MCCE method accounts for side-chain reorientations, it is obviously not sufficient to fully capture the dielectric response due to conformational dynamics of the local environment. In order for the PB-based approach to reproduce experimental pK_a values, the effective dielectric constant for protein needs to be increased further (Harms *et al.*, 2008). By contrast, dielectric response due to structural rearrangement is explicitly taken into account through conformational sampling in the REX–CPHMD method, in which the dielectric constant of protein is set to 1. For comparison, Table 19.2 also lists the published pK_a values computed with the empirical method PROPKA, which are lower than experimental values by 2.2, 1.7, and 0.7 pH units for Asp-38, Glu-38, and Lys-38, respectively. Thus, the CPHMD method gives the most accurate and consistent performance for predicting buried pK_as at this site.

Table 19.2 Calculated and experimental pK_a values for two buried sites in staphylococcal nuclease

PDB	Protein	Residue	CPHMD	Expt	MCCE	PROPKA
–	L38D/Δ+PHS	Asp-38	6.6	7.2	8.7	5.0
3D6C	L38E/Δ+PHS	Glu-38	6.9	7.0	8.5	5.3
2RKS	L38K/Δ+PHS	Lys-38	9.3	10.4	7.3	9.7
2SNM	V66K/WT	Lys-66	7.5	≤6.4	–	–
–	V66K/PHS	Lys-66	6.9	6.35	–	–
–	V66K/Δ+PHS	Lys-66	7.0	5.8	–	–

The structures of proteins without PDB identification were generated by computational mutation. The model pK_a for Lys is 10.4. The REX–CPHMD simulations were performed with an ionic strength of 100 m*M* as in experiments (Harms *et al.*, 2008, 2009). Other details are given in Table 19.1. Experimental data for Asp-38, Glu-38, and Lys-38 are taken from Harms *et al.* (2008, 2009).
The PROPKA and MCCE results (using an effective dielectric constant of 8) for the L38 mutants are taken from the same references. Experimental data for Lys-66 are taken from Fitch *et al.* (2002), Garcia-Moreno *et al.* (1997), and Stites *et al.* (1991).

Errors in the REX–CPHMD method can be best probed by evaluating the pK_as for ionizable residues at the second buried site, Val-66, in SNase. Substitution of Val-66 by Lys in WT and variants of SNase shifts the pK_a from 10.4 to about 6 (Fitch *et al.*, 2002; Garcia-Moreno *et al.*, 1997; Stites *et al.*, 1991). The pK_as obtained from REX–CPHMD simulations are about 1 pH unit too high (Table 19.2). In view of the absence of charged residues within 7 Å of Lys-66, the pK_a shift is mainly caused by the desolvation effect, which stabilizes the neutral form, thereby lowering the pK_a value for Lys-66. Thus, underestimation of the pK_a shift in the REX–CPHMD method can be attributed to the underestimation of desolvation energy, which is likely caused by the underestimation of Born radii in the underlying GBSW model (Khandogin and Brooks, 2006). The GBSW model uses van der Waals surface to represent solvent-solute dielectric boundary (Im *et al.*, 2003), which treats the small crevices between the van der Waals spheres as being filled with water, thus resulting in an underestimation of Born radii for buried atoms and an overestimation of the magnitude of the self-solvation energy (Onufriev *et al.*, 2002). Future improvement of the CPHMD method will need to address this issue.

3.3. Conformational fluctuations in α-lactalbumin

Bovine α-lactalbumin (BLA) is a genetic and structural homolog to HEWL. However, while having almost identical tertiary and secondary structure, the titration property and stability of BLA differs drastically from HEWL (Halskau *et al.*, 2004). A recent NMR study shows that the pK_a values of carboxylic groups are depressed relative to the model values in HEWL (Table 19.3) but not in BLA (Table 19.4). The folding free energy barrier of HEWL is about 35 kJ/mol, while that of BLA is on the order of kT (Halskau *et al.*, 2004). Thus, it was proposed that the marginal folding barrier of BLA leads to large conformational fluctuations causing the electrostatic environment around titratable residues to be similar to a disordered structure (Halskau *et al.*, 2004). Since CPHMD explicitly accounts for structural fluctuations, we expect that it should outperform methods based on static structures, such as the empirical or PB-based methods, for predicting pK_as of BLA. For the same reason, we also expect that the static structure-based methods should perform better for HEWL as compared to BLA. Indeed, PROPKA predicts the pK_as for HEWL and BLA with RMSDs of 0.8 and 1.5, respectively, while the RMSDs using CPHMD are 0.6 and 0.9, respectively. Although to a lesser extent than PROPKA, the pK_as from CPHMD are generally lower than those from experiment, which are close to the model values. Future study will address the question as to whether the small pK_a shifts (relative to the models) obtained from the NMR titration data are the result of local unfolding or the coexistence of folded and denatured states.

Table 19.3 Calculated and experimental pK_a values for hen egg-white lysozyme

Residue	CPHMD	PROPKA	Expt
Glu-7	3.2	3.1	2.9
His-15	6.2	6.4	5.7
Asp-18	3.3	3.7	2.7
Glu-35	5.5	5.8	6.2
Asp-48	3.5	1.4	2.5
Asp-52	4.7	4.8	3.7
Asp-66	1.9	0.5	2.0
Asp-87	2.7	2.2	2.1
Asp-101	4.0	4.0	4.1
Asp-119	2.5	3.4	3.2
RMSD	0.6	0.8	

Both REX–CPHMD simulations (taken from Khandogin and Brooks, 2006) and PROPKA calculations were based on the crystal structure (PDB ID: 1LSA). Experimental pK_a values were obtained with 100 mM NaCl (Bartik *et al.*, 1994; Takahashi *et al.*, 1992).

4. Conclusions

Accurate prediction of amino acid pK_a values in the protein environment is challenging. The main obstacles arise from the small energy differences that cause the pK_a shifts and the need to sample different conformational substates accessible to the native structure. Empirical and methods based on a macroscopic description of proteins do not explicitly account for dielectric heterogeneity and response to conformational rearrangement, nor do they take into account the conformational rearrangement due to charging. The CPHMD method addresses these issues by enabling the direct coupling between ionization processes and conformational dynamics. Another advantage of the CPHMD method is that its accuracy can be systematically improved through further development of the underlying solvent model and force field. REX–CPHMD simulations offer a robust and accurate tool for pK_a calculations and studies of pH-dependent conformational processes. The examples of RNase A, SNase, and BLA demonstrate that REX–CPHMD titration simulations can offer not only the accurate pK_a predictions, but also the microscopic factors governing pK_a shifts, which are not readily accessible by using empirical or macroscopic approaches. Finally, we note that while the CPHMD method is currently implemented with the GB implicit-solvent models, extension to explicit-solvent simulations is possible. The development of the latter technique will hinge upon advances to accelerate sampling convergence.

Table 19.4 Calculated and experimental pK_as of carboxylic groups in bovine α-lactalbumin

	CPHMD			
Residue	25 °C	43 °C	PROPKA	Expt
Glu-1	3.8	3.8	4.6	–
Glu-7	4.8	4.6	3.6	4.9
Glu-11	4.6	4.6	4.8	4.7
Asp-14	2.9	2.9	3.8	3.5
Glu-25	3.9	4.0	2.3	4.9
Asp-37	3.7	3.8	2.3	4.2
Asp-46	2.7	2.6	2.5	3.8
Glu-49	5.0	4.9	4.9	4.0
Asp-63	4.4	4.3	1.8	4.5
Asp-64	3.0	2.9	3.8	4.1
Asp-78	2.3	2.7	2.9	–
Asp-82	4.6	4.7	3.5	–
Asp-83	3.3	3.1	4.0	4.5
Asp-84	2.5	2.4	2.4	4.1
Asp-87	3.7	3.8	2.0	4.4
Asp-88	5.9	5.9	9.1	–
Asp-97	2.6	2.5	3.4	3.5
Glu-113	3.9	3.7	4.5	4.1
Asp-116	2.6	2.5	3.8	3.5
RMSD	0.9	0.9	1.5	

Both REX–CPHMD simulations and PROPKA calculations were performed using the crystal structure (PDB ID: 1F6S) of *holo*-α-lactalbumin with Ca^{2+} removed. REX–CPHMD simulations were conducted with an ionic strength of 150 m*M* using 16 replicas at pH 2, 3, 4, 5, and 6. pK_as computed from the temperature windows of 25 and 43 °C are reported here. Experimental pK_a values were estimated from semicomplete NMR titration curves of *holo*-α-lactalbumin obtained at 40 °C over the pH range of 3.5–7.5 (Halskau *et al.*, 2004). Not all pK_as could be determined because the protein underwent partial unfolding at pH $\leq$ 3.5.

ACKNOWLEDGMENT

Financial support from the University of Oklahoma is greatly appreciated.

REFERENCES

Alexov, E. G., and Gunner, M. R. (1997). Incorporating protein conformational flexibility into the calculation of pH-dependent protein properties. *Biophys. J.* **72,** 2075–2093.

Antosiewicz, J., McCammon, J. A., and Gilson, M. K. (1994). Prediction of pH-dependent properties of proteins. *J. Mol. Biol.* **238,** 415–436.

Baker, W. R., and Kintanar, A. (1996). Characterization of the pH titration shifts of ribonuclease A by one- and two-dimensional nuclear magnetic resonance spectroscopy. *Arch. Biochem. Biophys.* **327,** 189–199.

Baptista, A. M., Teixeira, V. H., and Soares, C. M. (2002). Constant-pH molecular dynamics using stochastic titration. *J. Chem. Phys.* **117,** 4184–4200.

Bartik, K., Redfield, C., and Dobson, C. M. (1994). Measurement of the individual p*Ka* values of acidic residues of hen and turkey lysozymes by two-dimensional ^{1}H NMR. *Biophys. J.* **66,** 1180–1184.

Bas, D. C., Rogers, D. M., and Jensen, J. H. (2008). Very fast prediction and rationalization of pK_a values for protein–ligand complexes. *Proteins* **73,** 765–783.

Bashford, D. (1997). An object-oriented programming suite for electrostatic effects in biological molecules. *In* "Scientific Computing in Object-Oriented Parallel Environments." *Lecture Notes in Computer Science*. Vol. 1343, pp. 233–240. Springer, Berlin.

Bashford, D. (2004). Macroscopic electrostatic models for protonation states in proteins. *Front. Biosci.* **9,** 1082–1099.

Bashford, D., and Karplus, M. (1990). pK_as of ionizable groups in proteins: Atomic detail from a continuum electrostatic model. *Biochemistry* **29,** 10219–10225.

Berendsen, H. J. C., Postma, J. P. M., van Gunsteren, W. F., Dinola, A., and Haak, J. R. (1984). Molecular dynamics with coupling to an external bath. *J. Chem. Phys.* **81,** 3684–3690.

Bierzynski, A., Kim, P. S., and Baldwin, R. L. (1982). A salt bridge stabilizes the helix formed by isolated C-peptide of RNase A. *Proc. Natl. Acad. Sci. USA* **79,** 2470–2474.

Bone, S., and Pethig, R. (1985). Dielectric studies of protein hydration and hydration-induced flexibility. *J. Mol. Biol.* **181,** 323–326.

Börjesson, U., and Hünenberger, P. H. (2001). Explicit-solvent molecular dynamics simulation at constant pH: Methodology and application to small amines. *J. Chem. Phys.* **114**(22), 9706–9719.

Brooks, B. R., Brooks, C. L., III., Mackerell, A. D., Jr., Nilsson, L., Petrella, R. J., Roux, B., Won, Y., Archontis, G., Bartles, C., Boresch, S., Caflisch, A., Caves, L., *et al.* (2009). CHARMM: The biomolecular simulation program. *J. Comput. Chem.* **30,** 1545–1614.

Bürgi, R., Kollman, P. A., and van Gunsteren, W. F. (2002). Simulating proteins at constant pH: An approach combining molecular dynamics and Monte Carlo simulation. *Proteins* **47,** 469–480.

Chen, J. C. L. B., III., and Khandogin, J. (2008). Recent advances in implicit solvent-based methods for biomolecular simulations. *Curr. Opin. Struct. Biol.* **18,** 140–148.

Chen, J., Im, W., and Brooks, C. L., III. (2006). Balancing solvation and intramolecular interactions: Toward a consistent generalized Born force field. *J. Am. Chem. Soc.* **128,** 3728–3736.

Chipman, D. M. (2002). Computation of p*Ka* from dielectric continuum theory. *J. Phys. Chem. A* **106,** 7413–7422.

Davies, M. N., Toseland, C. R., Moss, D. S., and Flower, D. R. (2006). Benchmarking p*Ka* prediction. *BMC Biochem.* **7,** 18.

Del Buono, G. S., Figueirido, F. E., and Levy, R. M. (1994). Intrinsic p*Ka*s of ionizable residues in proteins: An explicit solvent calculation for lysozyme. *Proteins* **20,** 85–97.

Dlugosz, M., and Antosiewicz, J. M. (2004). Constant-pH molecular dynamics simulations: A test case of succinic acid. *Chem. Phys.* **302,** 161–170.

Feig, M., Karanicolas, J., and Brooks, C. L., III. (2004a). MMTSB tool set: Enhanced sampling and multiscale modeling methods for applications in structure biology. *J. Mol. Graph. Model.* **22,** 377–395.

Feig, M., Onufriev, A., Lee, M. S., Im, W., Case, D. A., and Brooks, C. L., III. (2004b). Performance comparison of generalized born and Poisson methods in the calculation of electrostatic solvation energies for protein structures. *J. Comput. Chem.* **25,** 265–284.

Fitch, C. A., Karp, D. A., Lee, K. K., Stites, W. E., Lattman, E. E., and García-Moreno, E. B. (2002). Experimental p*Ka* values of buried residues: Analysis with continuum methods and role of water penetration. *Biophys. J.* **82,** 3289–3304.

García-Moreno, E. B., and Fitch, C. A. (2004). Structural interpretation of pH and salt-dependent processes in proteins with computational methods. *Methods Enzymol.* **380,** 20–51.

Garcia-Moreno, E. B., Dwyer, J. J., Gittis, A. G., Lattman, E. E., Spencer, D. S., and Stites, W. E. (1997). Experimental measurement of the effective dielectric in the hydrophobic core of a protein. *Biophys. Chem.* **64,** 211–224.

Georgescu, R. E., Alexov, E. G., and Gunner, M. R. (2002). Combining conformational flexibility and continuum electrostatics for calculating p*K*as in proteins. *Biophys. J.* **83,** 1731–1748.

Getzoff, E. D., Tainer, J. A., Weiner, P. K., Kollman, P. A., Richardson, J. S., and Richardson, D. C. (1983). Electrostatic recognition between superoxide and copper, zinc superoxide dismutase. *Nature* **306,** 287–290.

Halskau, Ø., Jr., Perez-Jimenez, R., Ibarra-Molero, B., Underhaug, J., Muñoz, V., Martinez, A., and Sanchez-Ruiz, J. M. (2004). Large-scale modulation of thermodynamic protein folding barriers linked to electrostatics. *Proc. Natl. Acad. Sci. USA* **105,** 8625–8630.

Harms, M. J., Schlessman, J. L., Chimentl, M. S., Sue, G. R., Damjanovic, A., and Garcia-Moreno, E. B. (2008). A buried lysine that titrates with a normal pK_a: Role of conformational flexibility at the protein water interface as a determinant of pK_a values. *Protein Sci.* **17,** 833–845.

Harms, M. J., Castañeda, C. A., Schlessman, J. L., Sue, G. R., Isom, D. G., Cannon, B. R., and García-Moreno, E. B. (2009). The pK_a values of acidic and basic residues buried at the same internal location in a protein are governed by different factors. *J. Mol. Biol.* **389,** 34–47.

Im, W., Lee, M. S., and Brooks, C. L., III. (2003). Generalized Born model with a simple smoothing function. *J. Comput. Chem.* **24,** 1691–1702.

Im, W., Chen, J., and Brooks, C. L., III. (2006). Peptide and protein folding and conformational equilibria: Theoretical treatment of electrostatics and hydrogen bonding with implicit solvent models. *Adv. Protein Chem.* **72,** 173–198.

Khandogin, J., and Brooks, C. L., III. (2005). Constant pH molecular dynamics with proton tautomerism. *Biophys. J.* **89,** 141–157.

Khandogin, J., and Brooks, C. L., III. (2006). Toward the accurate first-principles prediction of ionization equilibria in proteins. *Biochemistry* **45,** 9363–9373.

Khandogin, J., and Brooks, C. L., III. (2007a). Linking folding with aggregation in Alzheimer's beta amyloid peptides. *Proc. Natl. Acad. Sci. USA* **104,** 16880–16885.

Khandogin, J., and Brooks, C. L., III. (2007b). Molecular simulations of pH-mediated biological processes. *Annu. Rep. Comput. Chem.* **3,** 3–13.

Khandogin, J., Chen, J., and Brooks, C. L., III. (2006). Exploring atomistic details of pH-dependent peptide folding. *Proc. Natl. Acad. Sci. USA* **103,** 18546–18550.

Khandogin, J., Raleigh, D. P., and Brooks, C. L., III. (2007). Folding intermediate in the villin headpiece domain arises from disruption of a N-terminal hydrogen-bonded network. *J. Am. Chem. Soc.* **129,** 3056–3057.

Kong, X., and Brooks, C. L., III. (1996). λ-Dynamics: A new approach to free energy calculations. *J. Chem. Phys.* **105,** 2414–2423.

Koumanov, A., Karshikoff, A., Friis, E. P., and Borchert, T. V. (2001). Conformational averaging in p*K* calculations: Improvement and limitations in prediction of ionization properties of proteins. *J. Phys. Chem. B* **105,** 9339–9344.

Kuhn, B., Kollman, P. A., and Stahl, M. (2004). Prediction of pK_a shifts in proteins using a combination of molecular mechanical and continuum solvent calculations. *J. Comput. Chem.* **25,** 1865–1872.

Lee, L.-P., and Tidor, B. (2001). Barstar is electrostatically optimized for tight binding to barnase. *Nat. Struct. Biol.* **8,** 73–76.

Lee, F. S., Chu, Z. T., and Warshel, A. (1993). Microscopic and semimicroscopic calculations of electrostatic energies in proteins by the POLARIS and ENZYMIX programs. *J. Comput. Chem.* **14,** 161–185.

Lee, M. S., Salsbury, F. R., Jr., and Brooks, C. L., III. (2002). Novel generalized Born methods. *J. Chem. Phys.* **116,** 10606–10614.

Lee, M. S., Feig, M., Salsbury, F. R., Jr., and Brooks, C. L., III. (2003). New analytic approximation to the standard molecular volume definition and its application to generalized Born calculations. *J. Comput. Chem.* **24,** 1348–1356.

Lee, M. S., Salsbury, F. R., Jr., and Brooks, C. L., III. (2004). Constant-pH molecular dynamics using continuous titration coordinates. *Proteins* **56,** 738–752.

Levy, R. M., Belhadj, M., and Kitchen, D. B. (1991). Gaussian fluctuation formula for electrostatic free-energy changes in solution. *J. Chem. Phys.* **95,** 3627–3633.

Li, H., Robertson, A. D., and Jensen, J. H. (2005). Very fast empirical prediction and rationalization of protein pK_a values. *Proteins* **61,** 704–721.

Lim, C., Bashford, D., and Karplus, M. (1991). Absolute pK_a calculations with continuum dielectric methods. *J. Phys. Chem.* **95,** 5610–5620.

Machuqueiro, M., and Baptista, A. M. (2006). Constant-pH molecular dynamics with ionic strength effects: Protonation–conformation coupling in decalysine. *J. Phys. Chem. B* **110,** 2927–2933.

Machuqueiro, M., and Baptista, A. M. (2008). Acidic range titration of HEWL using a constant-pH molecular dynamics method. *Proteins* **72,** 289–298.

Madura, J. D., Briggs, J. M., Wade, R. C., Davis, M. E., Luty, B. A., Ilin, A., Antosiewicz, J., Gilson, M. K., Bagheri, B., Scott, L. R., and McCammon, J. A. (1995). Electrostatics and diffusion of molecules in solution: Simulations with the University of Houston Brownian Dynamics program. *Comput. Phys. Commun.* **91,** 57–95.

Mehler, E. L., and Guarnieri, F. (1999). A self-consistent, microenvironment modulated screened coulomb potential approximation to calculate pH-dependent electrostatic effects in proteins. *Biophys. J.* **75,** 3–22.

Mehler, E. L., Fuxreiter, M., Simon, I., and Garcia-Moreno, B. (2002). The role of hydrophobic microenvironments in modulating pK_a shifts in proteins. *Proteins* **48,** 283–292.

Mertz, J. E., and Pettitt, B. M. (1994). Molecular dynamics at a constant pH. *Int. J. Supercomput. Appl. High Perform. Comput.* **8,** 47–53.

Merz, K. M., Jr. (1991). Determination of pK_a's of ionizable groups in proteins: The pK_a of Glu 7 and 35 in hen egg white lysozyme and Glu 106 in human carbonic anhydrase II. *J. Am. Chem. Soc.* **113,** 3572–3575.

Mongan, J., and Case, D. A. (2005). Biomolecular simulations at constant pH. *Curr. Opin. Struct. Biol.* **15,** 157–163.

Mongan, J., Case, D. A., and McCammon, J. A. (2004). Constant pH molecular dynamics in generalized Born implicit solvent. *J. Comput. Chem.* **25,** 2038–2048.

Nicholls, A., and Honig, B. (1991). A rapid finite difference algorithm, utilizing successive over-relaxation to solve the Poisson–Boltzmann equation. *J. Comput. Chem.* **12,** 435–445.

Nozaki, Y., and Tanford, C. (1967). Examination of titration behavior. *Methods Enzymol.* **11,** 715–734.

Onufriev, A., Case, D. A., and Bashford, D. (2002). Effective Born radii in the generalized Born approximation: The importance of being perfect. *J. Comput. Chem.* **23,** 1297–1304.

Riccardi, D., Schaefer, P., and Cui, Q. (2005). pK_a Calculations in solution and proteins with QM/MM free energy perturbation simulations: A quantitative test of QM/MM protocols. *J. Phys. Chem. B* **109,** 17715–17733.

Richardson, W. H., Peng, C., Bashford, D., Noodleman, L., and Case, D. A. (1997). Incorporating solvation effects into Density Functional theory: Calculation of absolute acidities. *Int. J. Quantum Chem.* **61,** 207–217.

Schaefer, M., and Karplus, M. (1996). A comprehensive analytical treatment of continuum electrostatics. *J. Phys. Chem.* **100**(5), 1578–1600.

Schmidt am Busch, M., and Knapp, E.-W. (2004). Accurate pK_a determination for a heterogeneous group of organic molecules. *ChemPhysChem* **5,** 1513–1522.

Schreiber, G., and Fersht, A. R. (1996). Rapid, electrostatically assisted association of proteins. *Nat. Struct. Biol.* **3,** 427–431.

Schutz, C. N., and Warshel, A. (2001). What are the dielectric constants of proteins and how to validate electrostatic models? *Proteins* **44,** 400–417.

Selzer, T., Albeck, S., and Schreiber, G. (2000). Rational design of faster associating and tighter binding protein complexes. *Nat. Struct. Biol.* **7,** 537–541.

Sham, Y. Y., Chu, Z. T., and Warshel, A. (1997). Consistent calculations of pK_as of ionizable residues in proteins: Semi-microscopic and microscopic approaches. *J. Phys. Chem. B* **101,** 4458–4472.

Sheinerman, F. B., Norel, R., and Honig, B. (2000). Electrostatic aspects of protein–protein interactions. *Curr. Opin. Struct. Biol.* **10,** 153–159.

Simonson, T., and Brooks, C. L., III. (1996). Charge screening and the dielectric constant of proteins: Insights from molecular dynamics. *J. Am. Chem. Soc.* **118,** 8452–8458.

Simonson, T., Carlsson, J., and Case, D. A. (2004). Proton binding to proteins: pK_a Calculations with explicit and implicit solvent models. *J. Am. Chem. Soc.* **126,** 4167–4180.

Srinivasan, J., Trevathan, M. W., Beroza, P., and Case, D. A. (1999). Application of a pairwise generalized Born model to proteins and nucleic acids: Inclusion of salt effects. *Theor. Chem. Acc.* **101,** 426–434.

Stern, H. A. (2007). Molecular simulation with variable protonation states at constant pH. *J. Chem. Phys.* **126,** 164112.

Still, W. C., Tempczyk, A., Hawley, R. C., and Hendrickson, T. (1990). Semianalytical treatment of solvation for molecular mechanics and dynamics. *J. Am. Chem. Soc.* **112,** 6127–6129.

Stites, W. E., Gittis, A. G., Lattman, E. E., and Shortle, D. (1991). In a staphylococcal nuclease mutant the side-chain of a lysine replacing valine 66 is fully buried in the hydrophobic core. *J. Mol. Biol.* **221,** 7–14.

Stone, C. L., Jipping, M. B., Owuso-Dekyi, K., Hurley, T. D., Li, T.-K., and Bosron, W. F. (1999). The pH-dependent binding of NADH and subsequent enzyme isomerization of human liver $\beta_3\beta_3$ alcohol dehydrogenase. *Biochemistry* **38,** 5829–5835.

Sugita, Y., and Okamoto, Y. (1999). Replica-exchange molecular dynamics method for protein folding. *Chem. Phys. Lett.* **314,** 141–151.

Takahashi, T., Nakamura, H., and Wada, A. (1992). Electrostatic forces in two lysozymes: Calculations and measurements of histidine pK_a values. *Biopolymers* **32,** 897–909.

Tanford, C., and Kirkwood, J. G. (1957). Theory of protein titration curves. I. General equations for impenetrable spheres. *J. Am. Chem. Soc.* **79,** 5333–5339.

Thurlkill, R. L., Grimsley, G. R., Scholtz, J. M., and Pace, C. N. (2006). pK Values of the ionizable groups of proteins. *Protein Sci.* **15,** 1214–1218.

van Vlijmen, H. W., Schaefer, M., and Karplus, M. (1998). Improving the accuracy of protein pK_a calculations: Conformational averaging versus the average structure. *Proteins* **33,** 145–158.

Warshel, A. (1981a). Calculations of enzymatic reactions: Calculations of pK_a, proton transfer reactions, and general acid catalysis reactions in enzymes. *Biochemistry* **20,** 3167–3177.

Warshel, A. (1981b). Electrostatic basis of structure–function correlation in proteins. *Acc. Chem. Res.* **14,** 284–290.

Warshel, A., Sussman, F., and King, G. (1986). Free energy of charges in solvated proteins: Microscopic calculations using a reversible charging process. *Biochemistry* **25,** 8368–8372.

Yang, A.-S., Gunner, M. R., Sampogna, R., Sharp, K., and Honig, B. (1993). On the calculation of pK_a's in proteins. *Proteins* **15,** 252–265.

You, T. J., and Bashford, D. (1995). Conformation and hydrogen ion titration of proteins: A continuum electrostatic model with conformational flexibility. *Biophys. J.* **69,** 1721–1733.

CHAPTER TWENTY

Unfolding Thermodynamics of DNA Intramolecular Complexes Involving Joined Triple- and Double-Helical Motifs

Irine Khutsishvili, Sarah Johnson, Hui-Ting Lee, *and* Luis A. Marky

Contents

Department of Pharmaceutical Sciences, College of Pharmacy, University of Nebraska Medical Center, Omaha, Nebraska, USA

Methods in Enzymology, Volume 466
ISSN 0076-6879, DOI: 10.1016/S0076-6879(09)66020-1

Abstract

Our laboratory is interested in predicting the thermal stability and melting behavior of nucleic acids from knowledge of their sequence. One focus is to understand how sequence, duplex and triplex stabilities, and solution conditions affect the melting behavior of complex DNA structures, such as intramolecular DNA complexes containing triplex and duplex motifs. For these reasons, in this chapter, we used a combination of UV and circular dichroism (CD) spectroscopies and differential scanning calorimetry (DSC) techniques to obtain a full thermodynamic description of the melting behavior of six intramolecular DNA complexes with joined triplex and duplex motifs. The CD spectra at low temperatures indicated that these complexes maintained the "B" conformation. UV and DSC melting curves of each complex show biphasic or triphasic transitions. However, their corresponding transition temperatures (T_ms) remained constant with increasing strand concentration, confirming their intramolecular formation. Deconvolution of the DSC thermograms allowed us to determine standard thermodynamic profiles for the transitions of each complex. For each transition, the favorable free energy terms result from the characteristic compensation of a favorable enthalpy and unfavorable entropy contributions. The magnitude of these thermodynamic parameters (and associated T_ms) indicate that the overall folding of each complex depends on several factors: (a) the extent of the favorable heat contributions (formation of base-pair and base-triplet stacks) that are compensated with both the ordering of the oligonucleotide and the putative uptake of protons and ions; (b) inclusion of the more stable C^+GC base triplets; (c) stabilizing the duplex stem of the complex; and (d) solution conditions, such as pH and salt concentration. Overall, the temperature-induced unfolding of each complex corresponds to the initial disruption of the triplex motif (removal of the third strand) followed by the partial or full unfolding of the duplex stem.

1. Introduction

There is considerable interest to study a variety of noncanonical DNA structures, including triplexes, Holliday junctions, H-DNA, G- and C-quadruplexes (i-motifs). To understand how these structures carry out their biological roles, it is essential to have a complete physical description of the molecular forces that are involved in their folding. In principle, the folding of nucleic acids is controlled by the local base sequence in a precise and potentially predictable way; however, this knowledge alone cannot provide an understanding of the forces responsible for maintaining the distinct structures in nucleic acids. The overall physical properties of a nucleic acid molecule depend on its chemical architecture, but globally on contributions from base pairing, base stacking, polyelectrolyte behavior (ion binding), and hydration (water binding).

It has been long recognized that control over gene expression may open the possibility of a number of therapeutic as well as biotechnological applications. In the cell, regulatory proteins modulate gene expression through specific binding of selected DNA sequences (Bloomfield *et al.*, 2000). The high affinity of proteins for their nucleic acid targets allows the formation of tight complexes that effectively inhibit and/or promote downstream interactions (Bloomfield *et al.*, 2000). To mimic the affinity and specificity of these regulatory proteins, synthetic ligands should be able to recognize nucleic acid targets differing by a single base pair. Nucleic acids and their exquisite selectivity offer a rational way of recognizing DNA or RNA sequences with high affinity and specificity (Thuong and Hélène, 1993). The use of nucleic acids as modulators of gene expression has been exploited through two main approaches: the antisense and the antigene strategies (Helene, 1991, 1994). In the antisense strategy an oligonucleotide (ODN) binds messenger RNA and inhibits the translation of the corresponding protein (Dove, 2002). In the antigene strategy the ODN binds to the major groove of a DNA double helix, resulting in the formation of a triple helix, and inhibiting the transcription of its target gene (Duval-Valentin *et al.*, 1992). The formation of triple helices is sequence specific and may interfere with transcription by competing with the binding of proteins that activate the transcription machinery (Helene, 1994; Maher *et al.*, 1989). The specificity of triplexes relies in the formation of Hoogsteen hydrogen bonds between the third-strand bases and the purine bases of the duplex (Rajagopal and Feigon, 1989b; Soyfer and Potaman, 1996; Thuong and Hélène, 1993). Triplexes can be classified based on the composition and orientation of the third strand. The most common triplex motif involves the asymmetric binding of a homopyrimidine third strand to the major groove of a homopurine–homopyrimidine duplex, in a parallel orientation with respect to the homopurine strand. In this structure, high specificity is achieved by Hoogsteen pairing of dT with dA•dT and of protonated cytosine (dC^{+}) with dG·dC Watson–Crick base pairs (De los Santos *et al.*, 1989; Moser and Dervan, 1987; Rajagopal and Feigon, 1989a,b), thymines and cytosines in the third strand can be arranged to recognize the dA•dT and dG•dC base pairs of many gene sequences (Helene and Thuong, 1989; Moser and Dervan, 1987; Perrouault *et al.*, 1990; Strobel *et al.*, 1988; Sun *et al.*, 1989), while tight affinity is achieved by the formation of at least 10 base triplets, assuming an average free energy contribution of a base-triplet stack is -3.3 kcal mol^{-1} (Soto *et al.*, 2002). Furthermore, oligonucleotide-directed triplex formation has been implicated as a possible means of controlling gene expression, both by an endogenous and exogenous mechanism (Cooney *et al.*, 1988; Duval-Valentin *et al.*, 1992; Francois *et al.*, 1989; Helene, 1990, 1991; Maher *et al.*, 1992). The inhibition of gene expression via triple-helix formation or antigene strategy is feasible due to the natural abundance of homopurine–homopyrimidine tracts in genomes (Behe, 1987, 1995; Kato *et al.*, 1990; Larsen and Weintraub, 1982; Yavachev *et al.*, 1986). It is also reasonable to think that their presence in biological systems may be exerting a variety of effects, justifying their complete physical characterization in terms of stability as a function of sequence and solution conditions.

It has been demonstrated that the presence of hairpin loops in DNA may play an important role in biological processes, such as the palindromic sequences of plasmids forming cruciform structures in response to a topological stress which are cleaved by specific endonucleases (Lilley, 1980, 1981; Panayotatos and Wells, 1981; Pollack *et al.*, 1980), providing considerable interest in both the structure and overall physical properties of nucleic acid hairpin loops. Oligonucleotide hairpin loops form intramolecularly and their unfolding resembles the pseudomonomolecular melting of a nucleic acid polymer. Their stable formation is accompanied by a lower entropy penalty and their unfolding takes place at convenient transition temperatures (Lee *et al.*, 2008a; Marky *et al.*, 2007; Rentzeperis *et al.*, 1993; Soto *et al.*, 2002).

Our laboratory has investigated a variety of intramolecular DNA triplexes (Lee *et al.*, 2008a; Marky *et al.*, 1983; Rentzeperis and Marky, 1995; Soto *et al.*, 2002); their design involves the inclusion of two hairpin loops that will render them slightly more hydrophobic and compact. These double-hairpin triplexes can be used to target duplex DNA, via triplex formation. Hence, to develop oligonucleotide-forming triplexes as therapeutic agents, it is necessary to have a clear understanding on how the sequence, base composition, and solution conditions affect their stability. To this end, we have substituted their duplex closing loop with a hairpin loop containing 4 bp in their stem generating intramolecular DNA complexes containing triplex and duplex motifs with an extended duplex stem. Other applications of the use of intramolecular DNA molecules complexes includes a better understanding of the melting behavior of DNA, the construction of nanoscale two-dimensional shapes (DNA Origami) (Rothemund, 2006), and the mimicking of the putative structures of mRNA that can be targeted with partially complementary strands (Lee *et al.*, 2008b).

In this chapter, we used a combination of UV and circular dichroism (CD) spectroscopies and calorimetric techniques to obtain a full thermodynamic description of the melting behavior of six intramolecular DNA complexes containing joined triplex–duplex motifs with two loops. The major focus of this investigation is to report the relative thermodynamic contributions of base stacking and ionic effects to the stability and melting behavior of single-stranded DNA secondary structures. The overall results should improve our current understanding of how sequence, loops, ion, and proton binding control the stability and melting behavior of a nucleic acid molecule.

2. Materials and Methods

2.1. Materials

All oligonucleotides were synthesized in the Eppley Institute Molecular Biology Core Facility at UNMC, reverse-phase HPLC purified, desalted on a G-10 Sephadex column, and lyophilized to dryness prior to experiments. The sequences of the six complexes, and control molecules of Scheme 20.1, and their designation are reported in Table 20.1. The concentration of each oligomer solution was determined from absorbance measurements at 260 nm and 90 °C using the molar absorptivities, in mM^{-1} cm^{-1} of strands, reported in the last column of Table 20.1. These values were calculated by extrapolation of the tabulated values of the dimers and monomer bases from 25 °C to high temperatures, using procedures reported earlier (Cantor and Schimmel, 1980; Marky *et al.*, 1983). Buffer solutions consisted of 10 mM sodium phosphate buffer, 0.2 M NaCl, adjusted to pH of 6.2 or 7.0. All other chemicals used in this study were reagent grade.

2.2. Temperature-dependent UV spectroscopy (UV melts)

Absorbance versus temperature profiles (UV melting curves) were measured at 260 nm with a thermoelectrically controlled Aviv 14 DS UV/Vis spectrophotometer (Lakewood, NJ). The absorbance was scanned with a temperature ramp of approximately 0.4 °C min^{-1}. The analysis of the shape of the melting curves yielded transition temperatures, T_m, which correspond to the inflection point of the order–disorder transitions. To determine the

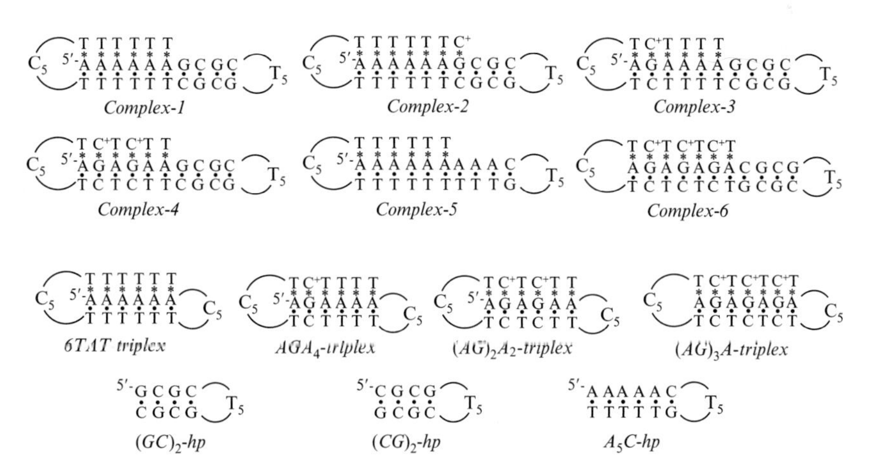

Scheme 20.1 Cartoon of the structures of DNA complexes, control triplexes and hairpins, and their designations.

Table 20.1 Deoxyoligonucleotide sequences and their molar extinction coefficients (ε)

5′ to 3′ Deoxyoligonucleotide sequence	Designation	ε (mM^{-1} cm^{-1})
$A_6GCGCT_5GCGCT_6C_5T_6$	*Complex-1*	317
$A_6GCGCT_5GCGCT_6C_5T_6C$	*Complex-2*	324
$AGA_4GCGCT_5GCGCT_4CTC_5TCT_4$	*Complex-3*	319
$AGAGA_2GCGCT_5GCGCT_2CTCTC_5TCTCT_2$	*Complex-4*	316
$A_9CT_5GT_9C_5T_6$	*Complex-5*	331
$AGAGAGACGCGT_5CGCGTCTCTCTC_5TCTCTCT$	*Complex-6*	339
$A_6C_5T_6C_5T_6$	*6TAT-Triplex*	246
$AGA_4C_5T_4CTC_5TCT_4$	AGA_4*-Triplex*	240
$AGAGA_2C_5T_2CTCTC_5TCTCT_2$	$(AG)_2A_2$*-Triplex*	239
$AGAGAGAC_5TCTCTCTC_5TCTCTCT$	$(AG)_3A$*-Triplex*	259
$GCGCT_5GCGC$	$(GC)_2$*-hp*	108
$CGCGT_5CGCG$	$(CG)_2$*-hp*	115
$A_5CT_5GT_5$	A_5C*-hp*	183

molecularity of the transition(s) of each DNA complex, we investigated the dependence of T_m over at least a 10-fold range of total strand concentration, 4–50 μM. If the T_m remains constant in this range of strand concentration, this indicates a monomolecular or intramolecular transition (Marky and Breslauer, 1987). Additional UV melting curves were obtained as a function of salt concentration and pH, to determine the differential binding of ions and protons, respectively, as described in a later section.

2.3. Circular dichroism spectroscopy

We use a thermoelectrically controlled Aviv circular dichroism spectrometer Model 202SF (Lakewood, NJ) to measure the CD spectrum of each oligonucleotide. The analysis of these spectra yielded the conformation adopted by the helical state of each oligonucleotide. Typically, we prepared an oligonucleotide solution with an absorbance of 1 (total strand concentration of $\sim 4\ \mu M$) in appropriate buffered solutions, and the CD spectrum is measured from 320 to 200 nm every 1 nm, using a strained free quartz cuvette with a path length of 1 cm, and at temperatures that the oligonucleotide is 100% in the helical state. The reported spectra correspond to an average of at least two scans.

2.4. Differential scanning calorimetry

Heat capacity functions of the helix–coil transition of each oligonucleotide were measured with a Microcal VP-DSC (Northampton, MA) instrument. Two cells, the sample cell containing 0.5 ml of a DNA solution (~ 0.15 mM in total strands) and the reference cell filled with the same volume of buffer solution, were heated from 0 to 110 °C at a heating rate of 0.75 °C min^{-1}. Analysis of the resulting thermographs yielded T_ms and standard thermodynamic unfolding profiles: ΔH_{cal}, ΔS_{cal}, and $\Delta G^{\circ}_{(T)}$ (Marky and Breslauer, 1987). These parameters are measured from DSC experiments using the following relationships: $\Delta H_{cal} = \int \Delta C_p^a dT$ and $\Delta S_{cal} = \int (\Delta C_p^a/T) dT$, where ΔC_p^a represents the anomalous heat capacity during the unfolding process (Marky and Breslauer, 1987). The free energy, $\Delta G^{\circ}_{(T)}$, is obtained at any temperature with the Gibbs relationship: $\Delta G^{\circ}_{(T)} = \Delta H_{cal} - T\Delta S_{cal}$. Alternatively, $\Delta G^{\circ}_{(T)}$ can be obtained using the following equation: $\Delta G^{\circ}_{(T)} = \Delta H_{cal}\,(1 - T/T_m)$, which can be applied rigorously for intramolecular transitions. For multiphasic unfolding curves, the heat associated with each transition are obtained using a non-two-state transitions deconvolution procedure of the Origin software, provided with the VP-DSC instrument.

2.5. Differential binding of protons and counterions

The helical and coil states of an oligonucleotide are associated with a different number of bound protons and ions (Manning, 1978; Rentzeperis, 1995; Soto *et al.*, 2002). For example, DNA triplexes with C^+GC base triplets need to protonate the third-strand cytosines, yielding stronger base triplets, that is, two Hoogsteen hydrogen bonds are formed between C^+ and G. Therefore, their Helix → Coil transition is accompanied by a differential release (or uptake) of protons and counterions. The differential release (or uptake) of each of these species can be measured experimentally using the following reaction (Kaushik *et al.*, 2007):

$$\text{Helix}(a\text{H}^+, b\text{Na}^+) \rightarrow \text{Coil}(x\text{H}^+, y\text{Na}^+) + \Delta n_{\text{H}^+}\text{H}^+ + \Delta n_{\text{Na}^+}\text{Na}^+$$

where $\Delta n_{\text{H}^+} = x - a$ and $\Delta n_{\text{Na}^+} = y - b$; each of these terms correspond to the differential binding of protons and counterions, respectively, and are written on the right-hand side of the reaction to indicate "releases"; however, if there is an "uptake" of any of these species, they should be written on the left-hand side.

The corresponding reaction constant can be written as

$$K = \left\{\frac{(\text{Coil})}{(\text{Helix})}\right\}(\text{H}^+)^{\Delta n_{\text{H}^+}}(\text{Na}^+)^{\Delta n_{\text{Na}^+}}. \quad (20.1)$$

Using a logarithmic function for simplicity: $\ln K = \ln K\{T, P, \ln(\text{H}^+), \ln(\text{Na}^+)\}$; its total differential is $\text{d}\ln K = (\partial \ln K/\partial T)\text{d}T + (\partial \ln K/\partial P)\text{d}P + (\partial \ln K/\partial \ln(\text{H}^+))\text{d}\ln(\text{H}^+) + (\partial \ln K/\partial \ln(\text{Na}^+))\text{d}\ln(\text{Na}^+)$; the last two partial differentials correspond to the desired quantities or linking numbers: $\Delta n_{\text{H}^+} = (\partial \ln K/\partial \ln(\text{H}^+))_{(\text{Na}^+)}$ and $\Delta n_{\text{Na}^+} = (\partial \ln K/\partial \ln(\text{Na}^+))_{(\text{H}^+)}$. These two linking numbers are measured experimentally with the assumption that proton or counterion binding to the helical and coil states of the oligonucleotide takes place with a similar type of binding. Applying the chain rule to each partial differential and converting ionic activities to concentrations and natural logarithms to decimal logarithms, we obtain (Kaushik *et al.*, 2007)

$$\Delta n_{\text{H}^+} = (\partial \ln K/\partial T_{\text{m}})(\partial T_{\text{m}}/\partial \ln(\text{H}^+)) = -0.434[\Delta H_{\text{cal}}/RT_{\text{m}}^2](\partial T_{\text{m}}/\partial \text{pH}), \quad (20.2)$$

$$\begin{aligned}\Delta n_{\text{Na}^+} &= (\partial \ln K/\partial T_{\text{m}})(\partial T_{\text{m}}/\partial \ln(\text{Na}^+)) \\ &= 0.483[\Delta H_{\text{cal}}/RT_{\text{m}}^2](\partial T_{\text{m}}/\partial \log[\text{Na}^+]).\end{aligned} \quad (20.3)$$

The first term in brackets on the right-hand side of Eqs. (20.2) and (20.3), $[\Delta H_{\text{cal}}/RT_{\text{m}}^2]$, is a constant determined directly in differential scanning calorimetric experiments, R is the gas constant; while the second term in parenthesis is also determined experimentally from UV or DSC melting curves, from the T_{m}-dependences on the concentration of protons and counterions, respectively.

In the determination of Δn_{H^+}, UV melts were carried out at the pH of 5.8 and 7.2, and at 200 mM NaCl; while for Δn_{Na^+}, UV melts were measured in the salt range of 10–200 mM NaCl at pH 6.2 or 7.

3. Results and Discussion

3.1. Design of DNA intramolecular complexes with joint triplex and duplex helical stems

A cartoon of the putative structure of all six complexes and control molecules are shown in Scheme 20.1. For simplicity, our starting molecule is the control triplex, *6TAT-Triplex*. What we have done to obtain *Complex-1* is to substitute the C_5 loop of the duplex domain of this triplex with the hairpin loop, GCGCT_5GCGC, *(GC)$_2$-hp*, see Scheme 20.1, yielding a duplex stem of 10 bp, A_6GCGC/GCGCT_6, closed with a T_5 end loop. *Complex-2* is similar to *Complex-1* but has an additional cytosine at the 3′-end forming an additional C^+GC base triplet in the helical stem; *Complex-3* and *Complex-4* are similar to *Complex-1* but have one and two TAT $\rightarrow$ C^+GC substitutions in their helical stems, respectively. *Complex-5* is like *Complex-1* but with a different hairpin loop (AAACT_5GTTT), while *Complex-6* has a different stem sequence with 11 bp, $(AG)_3A(CG)_2/(CG)_2T(CT)_3$, and a triplex domain consistent of three C^+GC base triplets and four TAT base triplets.

3.2. Overall experimental protocol

Initially, we use temperature-dependent UV spectroscopy to measure the UV unfolding of each complex. Analysis of the resulting UV melting curves yields the optical transitions for each complex and associated T_ms, indicating the thermal stability of the melting domains. We then test if the folding of each DNA complex takes place intramolecularly, by following the dependence of the T_m on strand concentration, if the T_m remains constant then complex formation is intramolecular. The next step is to use CD spectroscopy to check for the particular conformation that each complex adopts, and to confirm the formation of specific helical motifs using the CD spectral characteristics or physical signatures of each complex component. Then, we use DSC to determine complete thermodynamic profiles for the temperature unfolding of each DNA complex. Finally, the measurement of the differential binding of protons (Δn_{H^+}) and counterions (Δn_{Na^+}), for the unfolding of each complex is done experimentally using UV and/or DSC melting profiles.

Therefore, in the following sections, we first describe in general, the melting behavior of these triplex–duplex complexes followed by a discussion of the specific thermodynamic contributions for the inclusion of a C^+GC base triplet, TAT $\rightarrow$ C^+GC substitutions, and overall changes in the sequence of the duplex stem.

3.3. UV unfolding of complexes

The helix–coil transition of each complex is initially characterized by UV melting techniques. Figure 20.1 shows typical melting curves for all complexes at pH 7 (*Complex-1 and Complex-5*) and at pH 7 and 6.2 for the other four complexes. Clearly, the unfolding of each complex at these pHs is biphasic. Each transition in these curves follows the characteristic sigmoidal behavior for the unfolding of a nucleic acid helix. Their hyperchromicities at 260 nm ranged from 0.79 (*Complex-2* and *Complex-5*) to 0.85 (all other complexes), which are proportional to the number of base triplets and/or sequence of the duplex stem; however, the contributions from GC base pairs and $C^{+}GC$ base pairs is minimal at this wavelength. T_ms were obtained from analysis of the differential curves (data not shown), the T_ms of the first transition of each complex ranged from 24.7 °C (*Complex-6* at pH 7) to 52.7 °C (*Complex-2* at pH 6), while the T_ms of the second transition of each complex ranged from 62.3 °C (*Complex-5* at pH 7) to 83.9 °C (*Complex-6* at pH 6.2). The results indicate that the transitions of the complexes with $C^{+}GC$ base triplets at pH 6.2 are more clearly resolved. Furthermore, the stabilization of the duplex domain of each complex, by the incorporation of additional base pairs and/or incorporation of AT → GC base pairs, induces the dissociation of the third strand of the complexes. To determine

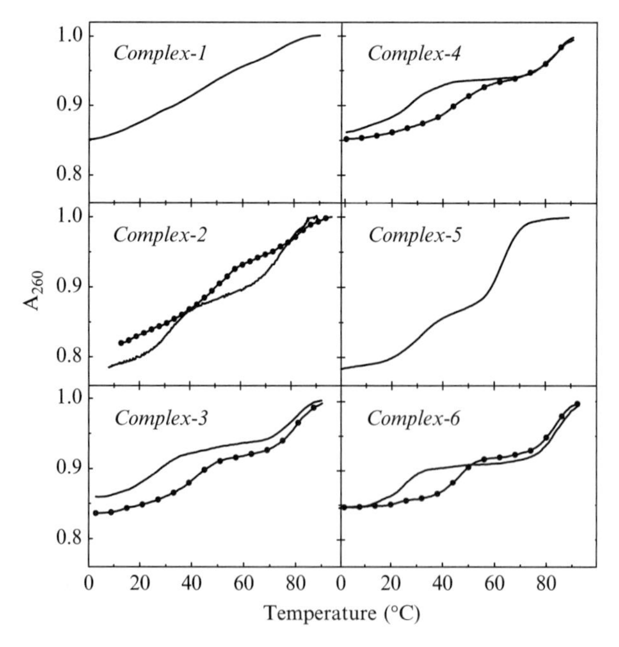

Figure 20.1 UV melting curves of complexes in 10 m*M* sodium phosphate buffer, 0.2 *M* NaCl at pH 7 (solid lines) and pH 6.2 (circles).

transition molecularity, we follow the T_m as a function of strand concentration for the transition of each complex. These T_m-dependences for some complexes are shown in Fig. 20.2; we obtained similar T_ms over a 50-fold increase in strand concentration, confirming their intramolecular formation. The use of intramolecular complexes has the main advantage of forming at low sodium concentrations, allowing us to study their physical properties in a wide variety of solution conditions (Lee *et al.*, 2008a; Soto *et al.*, 2002).

3.4. CD spectroscopy and complex conformation

Analysis of the CD spectra (Fig. 20.3) indicates that all complexes exhibit the typical CD spectra of a nucleic acid helix in the "B" conformation, that is, the positive band centered at 280 nm has an area/magnitude comparable to that of the negative band centered at approximately 246 nm. This indicates that binding of the third strand does not impose major distortions in the geometry of the duplex. All complexes exhibit a negative band at 210 nm, which has been observed in previous reports (Escude *et al.*, 1993; Shikiya and Marky, 2005; Soto *et al.*, 2002), and it has been considered as a characteristic feature of the formation of a triplex. Furthermore, the small split of the positive band

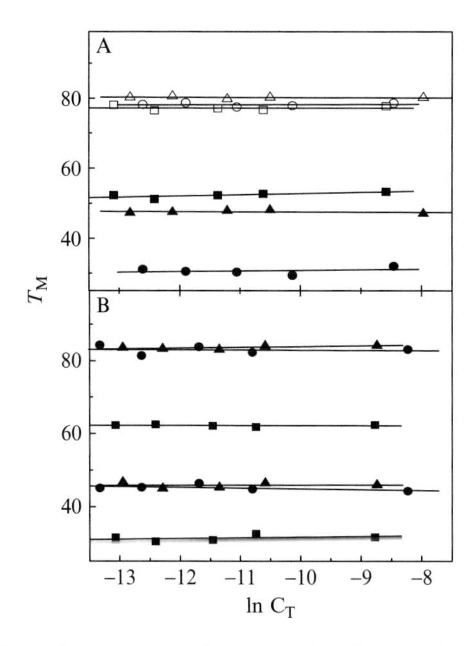

Figure 20.2 T_m-dependences on strand concentration for complexes in 10 m*M* sodium phosphate buffer, 0.2 *M* NaCl, at the indicated pH. (A) *Complex-1* at pH 7 (circles), *Complex-2 at* pH 6.2 (squares), and *Complex-3 at* pH 6.2 (triangles). (B) *Complex-4* at pH 6.2 (circles), *Complex-5 at* pH 7 (squares), and *Complex-6 at* pH 6.2 (triangles).

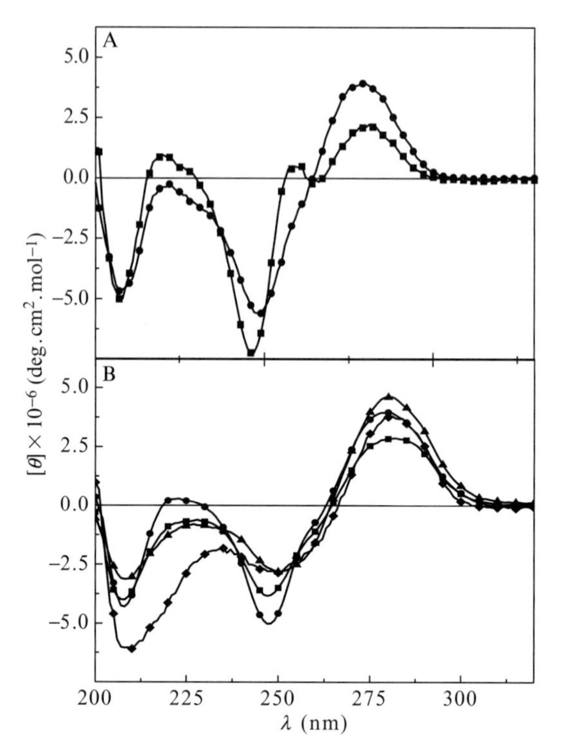

Figure 20.3 Circular dichroism spectra of complexes in 10 m*M* sodium phosphate buffer, 0.2 *M* NaCl at the indicated pH. (A) *Complex-1* (circles) and *Complex-5* (squares) at pH 7. (B) *Complex-2* (circles), *Complex-3* (squares), *Complex-4* (triangles), and *Complex-6* (diamonds) at pH 6.2.

(255–275 nm) of *Complex-5* is characteristic of the homopurine/homopyrimidine sequence (A_9/T_9) of its duplex stem, while the lower ellipticity values of the band at 246 nm (*Complex-1*, *Complex-2*, and *Complex-5*) indicate additional base stacking interactions of the thymidine third strand.

3.5. Calorimetric unfolding of DNA complexes

In the DSC experiments, only pH 7 was used for *Complex-1* and *Complex-5*, while for the other four complexes pH of 6.2 and 7 were used. Typical DSC profiles are shown in Fig. 20.4; the transitions of each complex are clearly defined because of their differential nature. At pH 7, the unfolding of complexes 1–3 has three transitions and complexes 4–6 show two transitions, while at pH 6.2 complexes 2–6 show two transitions, the one exception is *Complex-6* showing an additional small transition at low temperatures. Table 20.2 lists the T_ms for the transitions of each complex, the T_ms of the first transition of each complex range from 26.1 °C (*Complex-4* at

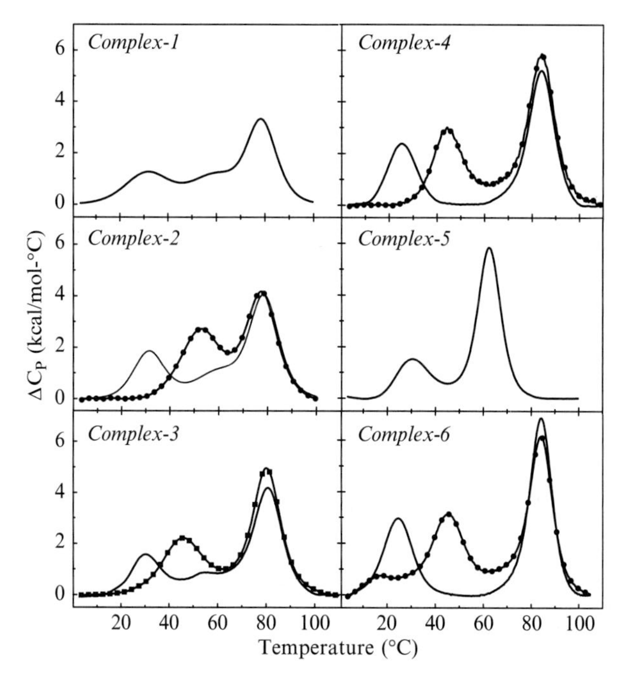

Figure 20.4 Differential scanning calorimetry curves for each complex in 10 m*M* sodium phosphate buffer, 0.2 *M* NaCl at pH 7 (solid lines) or pH 6.2 (circles).

pH 7) to 47.1 °C (*Complex-3* at pH 6.2), while the T_ms of the last transition of each complex range from 62.4 °C (*Complex-5* at pH 7) to 84 °C (*Complex-6* at pH 7). Overall, the observed transitions of each complex correspond to the temperature removal of the third strand followed by the opening of its whole duplex stem. The lowering of the pH yielded a stabilization of the triplex transitions for complexes with C^+GC base triplets. We did not find heat capacity effects between the initial and final states of these curves, that is, $\Delta C_p = 0$. However, small heat capacity effects may be present but the sensitivity of the VP-DSC calorimeter does not allow their direct determination, as these are within the experimental noise of the DSC baselines. Standard thermodynamic profiles at 5 °C for each transition in the folding of each complex are summarized in Table 20.2. Inspection of Table 20.2 indicates that the folding of each complex is accompanied by favorable free energy terms resulting from the characteristic compensation of a favorable enthalpy and unfavorable entropy terms. In general, these favorable enthalpy terms correspond mainly to the formation of base-pair, and base-triplet stacks, while the unfavorable entropy terms arise from contributions of the unfavorable ordering of strands and the uptake of counterions, protons, and water molecules. In terms of the overall free energy contribution for the folding of each complex at 5 °C, and at the

Table 20.2 Folding thermodynamic profiles of DNA complexes[a]

pH		T_m (°C)	$\Delta G°$ (kcal mol^{-1})	ΔH_{cal} (kcal mol^{-1})	$T\Delta S_{cal}$ (kcal mol^{-1})	Δn_{H+} (mol$_{Na+}$/mol)	Δn_{Na+} (mol$_{H+}$/mol)
Complex-1							
7.0	First	32.0	−2.5	−27.7	−25.2		−1.0
	Second	60.2	−5.2	−31.6	−26.4		−1.0
	Third	78.6	−10.7	−50.9	−40.2		−1.4
	Total	–	−18.4	−110.2	−91.8		−3.4
Complex-2							
7.0	First	31.4	−2.7	−31.5	−28.8		
	Second	60.5	−4.2	−25.0	−20.8		
	Third	78.8	−13.1	−62.4	−49.3		
	Total	–	−20.0	−118.9	−98.9		
6.2	First	53.3	−9.1	−61.4	−52.3	−1.9	−1.9
	Second	77.7	−13.8	−66.7	−52.9	0	−1.8
	Total	–	−22.9	−128.1	−105.2	−1.9	−3.7
Complex-3							
7.0	First	30.4	−1.9	−23.2	−21.3		
	Second	59.7	−4.8	−29.2	−24.4		
	Third	80.1	−12.6	−59.4	−46.8		
	Total	–	−19.3	−111.8	−92.5		
6.2	First	47.1	−7.3	−55.9	−48.6	−2.1	−1.1
	Second	80.2	−16.6	−77.8	−61.2	0	−2.0
	Total	–	−23.9	−133.7	−109.8	−2.1	−3.1

Complex-4							
7.0	First	26.1	−2.4	−34.0	−31.6		
	Second	83.8	−16.9	−76.5	−59.6		
	Total	–	−19.3	−110.5	−91.2		
6.2	First	44.3	−6.9	−55.9	−49.0	−2.4	−0.7
	Second	83.7	−20.3	−92.2	−71.9	0	−2.2
	Total	–	−27.2	−148.1	−120.9	−2.4	−2.9
Complex-5							
7.0	First	31.8	−2.6	−29.5	−26.9		−1.4
	Second	62.4	−13.1	−76.4	−63.3		−3.0
	Total		−15.7	−105.9	−90.2		−4.4
Complex-6							
7.0	First	24.7	−3.1	−46.2	−43.1		
	Second	84.0	−18.9	−85.5	−66.6		
	Total		−22.0	−131.7	−117.0		
6.2	First	17.4	−0.5	−10.6	−10.1		
	Second	46.0	−9.0	−69.9	−60.9	−3.1	0.9
	Third	83.8	−18.9	−85.5	−66.6	0	−2.4
	Total		−28.4	−166.0	−137.6	−3.1	−1.5

[a] All parameters are measured in 10 m*M* sodium phosphate buffer, 0.2 *M* NaCl at pH 7 or 6.2. Experimental uncertainties are as follows: T_m (±0.5 °C), $\Delta G°$ (±5%), ΔH_{cal} (±5%), $T\Delta S$ (±7%), Δn_{H^+} (15%), and Δn_{Na^+} (15%).

corresponding pH values (Table 20.2), we obtained $\Delta G^{\circ}_{(5)}$ of -18.4 kcal mol^{-1} (*Complex-1*), -20.0 to -22.9 kcal mol^{-1} (*Complex-2*), -19.3 to -23.9 kcal mol^{-1} (*Complex-3*), -19.3 to -27.2 kcal mol^{-1} (*Complex-4*), -15.7 kcal mol^{-1} (*Complex-5*), and -22.0 to -28.2 kcal mol^{-1} (*Complex-6*). These values are robust, indicating that each complex folds with their particular triplex and duplex domains. Furthermore, the thermodynamic profiles for each of their transitions are consistent with a sequential unfolding of their triplex and duplex domains. All other thermodynamic parameters of Table 20.2 will be discussed in the following sections.

3.6. Melting behavior of complexes relative to its control triplex and hairpin loops

Figure 20.5 shows the calorimetric unfolding of four complexes and corresponding control triplex and hairpin loop molecules. This comparison is made at pH 6.2 for the complexes with $C^{+}GC$ base triplets and pH 7 for complexes with exclusive TAT base triplets, as indicated in Fig. 20.5. Standard thermodynamic profiles for the folding of each complex are presented in Table 20.2 while those of the control molecules are presented in Table 20.3. *Complex-1* show a triphasic transition while the transition

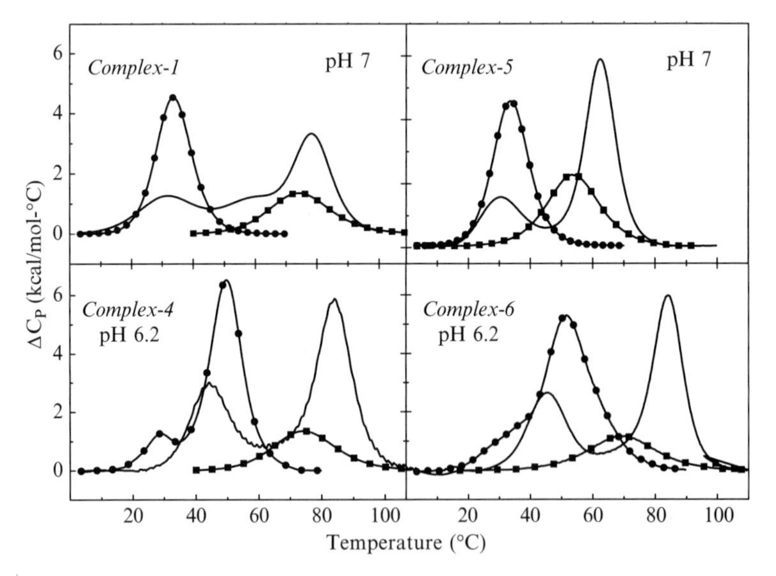

Figure 20.5 Differential scanning calorimetry curves for complexes and control molecules in 10 m*M* NaPi buffer and 0.2 *M* NaCl at the indicated pH. Top panels: *Complex-1* (solid line), *6TAT-Triplex* (circles), and *(GC)$_2$-hp* (squares) and *Complex-5* (solid line), *6TAT-Triplex* (circles), *A$_5$C-hp* (squares). Bottom panels: *Complex-4* (solid line), *AGA$_4$-Triplex* (circles), *(GC)$_2$-hp* (squares) and *Complex-6* (solid line), *AG7-Triplex* (circles), *(CG)$_2$-hp* (squares).

Table 20.3 Folding thermodynamic profiles of control triplexes and hairpins[a]

pH		T_m (°C)	$\Delta G°$ (kcal mol^{-1})	ΔH_{cal} (kcal mol^{-1})	$T\Delta S_{cal}$ (kcal mol^{-1})	Δn_{H+} (mol$_{Na+}$/mol)	Δn_{Na+} (mol$_{H+}$/mol)
6TAT-Triplex							
7.0		33.7	−6.6	−70.9	−64.3		−2.1
AGA_4-Triplex							
6.2		46.4	−11.6	−89.8	−78.2	−1.4	−2.2
$(AG)_2A_2$-Triplex							
6.2	First	29.4	−1.5	−18.0	−16.5		
	Second	49.7	−12.4	−89.3	−76.9	−2.7	−1.7
	Total		−13.9	−107.3	−93.4	−2.7	−1.7
$(AG)_3A$-Triplex							
6.2	First	31.2	−1.2	−13.6	−12.4		
	Second	52.5	−16.0	−110.0	−94.0	−3.1	−1.0
	Total		−17.2	−124.6	−106.4	−3.1	−1.0
$(GC)_2$-hp							
7.0		73.9	−7.2	−36.3	−29.1		−0.4
A_5C-hp							
7.0		54.0	−5.2	−49.6	−44.4		−0.8
$(CG)_2$-hp							
7.0		69.6	−6.2	−33.1	−26.9		−0.4

[a] All parameters are measured in 10 m*M* sodium phosphate buffer, 0.2 *M* NaCl at pH 7 or 6.2. Experimental uncertainties are as follow: T_m (±0.5 °C), $\Delta G°$ (±5%), ΔH_{cal} (±5%), $T\Delta S$ (±7%), Δn_{H^+} (15%), and Δn_{Na^+} (15%).

of the other three complexes are biphasic, all control molecules show monophasic transitions with the exception of two control triplexes, *(AG)$_2$A$_2$-Triplex* and *(AG)$_3$A-Triplex*, showing a small peak and shoulder, respectively. The first transition of each complex has T_ms that corresponds more or less (2–6 °C) to the T_ms of their control triplexes; however, their unfolding enthalpies are much lower, by 41–55 kcal mol^{-1}. This indicates that the elongation of the duplex domain of each complex, by replacing the right-hand side loop of the control triplex with its control hairpin loop, forces the removal of the third strand. On the other hand, the last transition of each complex shows higher T_ms, by 5–14 °C, than the T_ms of their corresponding control hairpin loops; however, their unfolding enthalpies are much higher and ranging from 15 kcal mol^{-1} (*Complex-1*) to 56 kcal mol^{-1} (*Complex-4*), the enthalpy of the middle transition of *Complex-1* has not been included in this exercise. These results indicate that the upper temperature transitions of each complex correspond to the unfolding of their duplex domain. In summary, the intramolecular unfolding of each complex takes place in sequential transitions, corresponding to the removal of the third strand followed by the unfolding of their duplex domain, which unfolds in two sequential transitions (*Complex-1*) or in single transitions (*Complex-4*, *Complex-5*, and *Complex-6*).

The differences in the overall unfolding energies between each complex and the sum of its control molecules are small, in kcal mol^{-1}: 4.6 (*Complex-1*), 3.9 (*Complex-5*), −6.1 (*Complex-4*), and −5.1 (*Complex-6*), while the comparison of the unfolding enthalpies yielded closer values: −3 (*Complex-1*), 14.6 (*Complex-5*), 4.5 (*Complex-4*), and 1.2 (*Complex-6*). These small differences in the thermodynamic parameters may be explained by an enthalpy compensation in the formation of the triplex–duplex junction, that is, removal of the stacked bases in the triplex loop with the formation of an additional base-pair stack upon the inclusion of the hairpin loop. Furthermore, the complexes with positive differential free energy changes are the ones with exclusive TAT base triplets while the complexes with negative differential free energy changes are the complexes with 2–3 C^+GC base triplets, this may be due to a pH effect, which may be associated with the protonation of cytosines for effective C^+GC base triplets (Soto *et al.*, 2002). In Section 3.7, we discuss the effects of proton and counterion binding, which contribute to the entropy contribution.

3.7. Differential binding of protons and counterions in the folding of complexes

Typical UV melts for the unfolding of *Complex-3* as a function of pH at the total Na^+ concentration of 216 m*M* are shown in Fig. 20.6A. The first transition, which corresponds to the removal of the third strand, is shifted to

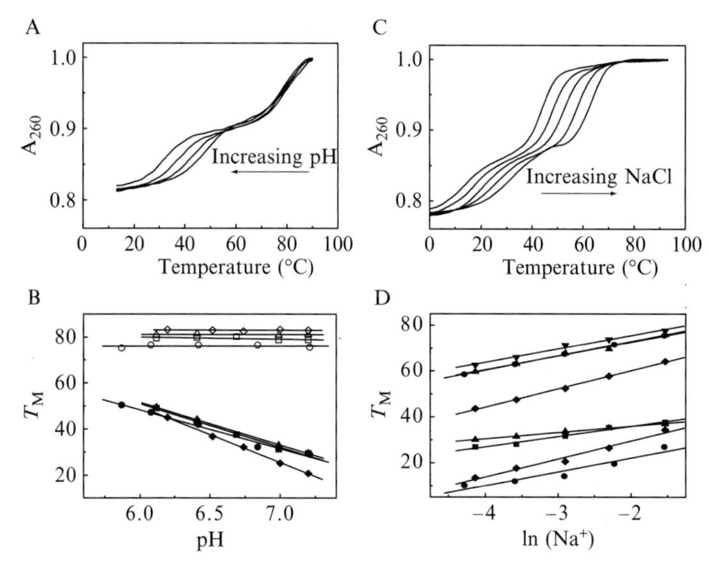

Figure 20.6 (A) Typical UV melting curves of *Complex-3* in 10 m*M* sodium phosphate buffer, 0.2 *M* NaCl, as a function of pH. (B) T_m-dependences on pH, first transition (solid symbols) and second transition (open symbols): *Complex-2* (circles), *Complex-3* (squares), *Complex-4* (triangles), and *Complex-6* (diamonds). (C) Typical UV melting curves of *Complex-5* in 10 m*M* sodium phosphate buffer as a function of salt concentration. (D) T_m-dependences on Na^+ concentration, first transition (solid symbols) and second transition (open symbols): *Complex-1* at pH 7 (circles), *Complex-2* at pH 6.2 (squares), *Complex-3* at pH 6.2 (triangles), and *Complex-5* at pH 7 (diamonds).

higher temperatures when the pH is decreased, while its second transition remains the same. This is consistent with protonation of the cytosines in the third strand. Similar melting behavior is observed with complexes containing one to three C^+GC base triplets in their triplex domain, while the first transition of the complexes with exclusive TAT base triplets (*Complex-1* and *Complex-5*) remains the same as the pH is decreased (data not shown). We obtained UV melts for the complexes with C^+GC base triplets, and their control triplexes (shown in Scheme 20.1), as a function of pH to determine the slope of the lines of the T_m versus pH plots. These T_m-dependences on pH are shown in Fig. 20.6B. The slopes of these lines were averaged with the ones obtained from the DSC experiments, and together with the enthalpy term of Eq. (20.2) allowed us to measure the release/uptake of protons, Δn_{H^+}. The complexes yielded Δn_{H^+} values ranging from -1.9 ± 0.2 (*Complex-2*) to -3.1 ± 0.2 (*Complex-6*), while the control triplexes yielded values of -1.4 ± 0.2 (*AGA_4-Triplex*) to -3.1 ± 0.2 (*$(AG)_3A$-Triplex*). However, the transitions of the duplex component of these complexes (control hairpins) yielded negligible Δn_{H^+} values. This clearly shows that the differential binding of protons is associated with the protonation of the third-strand cytosines, and to a smaller extent to cytosine protonation of the loops (Soto, 2002; Zimmer and Venner, 1966; Zimmer *et al.*, 1968). These Δn_{H^+} values, in general, indicate how strong or compact the triplex motif is formed; moreover, our results are more or less consistent with the expectation of Δn_{H^+} values equal to -1 for every C^+GC base triplet that is included.

Typical UV melting curves as a function of sodium concentration of *Complex-5* are shown in Fig. 20.6C. As the concentration of sodium increases, both transitions of this complex are shifted to higher temperatures. This is consistent with the stabilizing effect of cations on nucleic acid helices. However, this effect is not seen with the first transition of the complexes containing C^+GC base triplets (data not shown) because of the exclusion of counterions competing with the protonated cytosines (Soto *et al.*, 2002). We obtained UV melts for the complexes and their control molecules (Scheme 20.1) with Na^+ concentration in the range of 16–216 m*M* at pH 7, or 6.2, to determine the slope of the T_m-dependences on salt concentration. The slope of these T_m-dependences (Fig. 20.6D) together with the enthalpy term obtained from the DSC experiments, and Eq. (20.3), allowed us to measure the release of counterions, Δn_{Na^+}, for each of the two transitions of each complex. The resulting Δn_{Na^+} values for the complexes and control molecules are summarized in the last column of Tables 20.2 and 20.3, respectively. It should be noted that the average of the T_m-dependences of the first and third transition of the triphasic unfolding of *Complex-1* was used to estimate the Δn_{Na^+} value for the second transition of this complex, and the Δn_{Na^+} value of *6TAT-Triplex* was used for the control triplex of *Complex-2*. We obtained total Δn_{Na^+} values (the second value corresponds to the sum of their control triplex and hairpin loop molecules), in mol Na^+ per mol complex, of 3.4 and 2.5 (*Complex*-1 at pH 7), 3.7 and 2.5 (*Complex-2* at pH 6.2), 3.1 and 2.6 (*Complex-3* at pH 6.2), 2.9 and 2.1 (*Complex-4* at pH 6.2), 4.4 and 2.9 (*Complex-5* at pH 7), and 1.5 and 1.4 (*Complex-6* at pH 6.2). In general, the Δn_{Na^+} values for each complex are higher than the sum of their control molecules; however, the differences are lower for the complexes with a higher number of C^+GC base triplets. This is explained in terms of the longer duplex stem of each complex, one helical turn, their response to salt is approaching polymer behavior. For instance, the Δn_{Na^+} value of *Complex-5* corresponds to 0.163 mol Na^+ per phosphate (4.4/27). In this calculation, the two loop phosphates adjacent to the duplex stem of the two loops are considered. However, the lower Δn_{Na^+} values of the complexes with C^+GC base triplets are due to the effective exclusion of counterions by the protonated cytosines. This counterion–proton competition is in excellent agreement with previous reports on the unfolding of intramolecular triplexes with C^+GC base triplets (Soto, 2002).

3.8. Unfolding of complexes: Enthalpy contributions of their triplex and duplex motifs

The comparison of the melting behavior of *Complex-1* and *Complex-2* is shown in Fig. 20.7A. The T_ms of the first transition of each complex are 32–53.3 °C, respectively, while the T_m for the single transition of *6TAT-Triplex* is 33.7; however, their enthalpies of 27.7 kcal mol^{-1} (*Complex-1*) and 61.4 kcal mol^{-1} (*Complex-2*) are lower than the enthalpy of 70.9 kcal mol^{-1} of the *6TAT-Triplex*. These results clearly demonstrate that the stabilization of the duplex domain of each complex, by the incorporation of additional GC base pairs, induce the dissociation of the third strand of the complexes. Further inspection of Table 20.2 and Fig. 20.7A shows that the T_ms of the second transition of *Complex-1* (60.2 °C) are

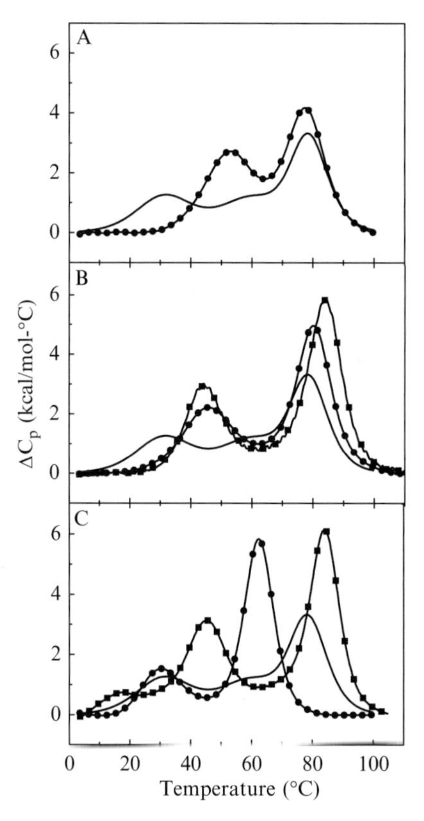

Figure 20.7 Calorimetric unfolding of complexes in 10 m*M* NaPi buffer, 0.2 *M* NaCl, at the indicated pH. (A) *Complex-1* at pH 7 (solid line) and *Complex-2* at pH 6.2 (circles). (B) *Complex-1* at pH 7 (solid line), *Complex-3* at pH 6.2 (circles), and *Complex-4* at pH 6.2 (squares). (C) *Complex-1* at pH 7 (solid line), *Complex-5* at pH 7 (circles), and *Complex-6* at pH 6.2 (squares).

much higher that the T_m of *6TAT-Triplex* with lower enthalpy contributions; on the other hand, the T_ms of the last transition (78.7 °C) are close to the T_m of 73.0 °C of *(GC)₂-hp*, while their unfolding enthalpies are much higher, by 14.6and 30.4 kcal mol^{-1}, respectively. This indicates that these transitions correspond to the sequential melting of their duplex stem, that is, initial disruption of 3–4 AA/TT base-pair stacks followed by the cooperative disruption of the remaining 6–7 bp stacks (2–3 AA/TT, 1 AG/CT, 2 GC/GC, and 1 CG/CG). Furthermore, the total unfolding enthalpy of *Complex-2* is 17.9 kcal mol^{-1} higher than *Complex-1* (Table 20.2), which corresponds to the disruption of the additional TAT/C^+GC base-triplet stack. This is in good agreement with previous findings of base-triplet stacking contributions of intramolecular triplexes (Soto *et al.*, 2002).

Figure 20.7B shows the DSC melting curves of *Complex-1*, *Complex-3*, and *Complex-4* and their thermodynamic profiles are listed in Table 20.2. *Complex-3* and *Complex-4* unfold in biphasic transitions, their first transition have T_ms of 47.1 and 44.3 °C higher than the T_m of 32 °C for the first transition of *Complex-1*. The enthalpies (in kcal mol^{-1}) associated with these transitions are 27.7, 55.9, and 55.9, respectively. The incorporation of one or two TAT → C^+GC substitutions in *Complex-1* eliminates the second transition observed in *Complex-1*; therefore, the observed two transitions of *Complex-3* and *Complex-4* correspond to disruption of the third strand followed by the unfolding of their duplex stem. Relative to the total enthalpy of *Complex-1*, the unfolding enthalpies of *Complex-3* and *Complex-4* are higher, by 23.5 and 37.9 kcal mol^{-1}, respectively. This indicates that the higher stability of these complexes is due to favorable enthalpy contributions. The inclusion of C^+GC base triplets renders more stable complexes (more favorable $\Delta G°$) which are consistent with what has been observed with similar intramolecular triplexes (Soto *et al.*, 2002). This suggests that these complexes are more compact.

The calorimetric unfolding of *Complex-1*, *Complex-5*, and *Complex-6* is compared in Fig. 20.7C, and the associated thermodynamic profiles are listed in Table 20.2. The comparison of the melting behavior of *Complex-1* with that of *Complex-5* (with similar triplex motif) allow us to discuss the effect of duplex sequence on complex formation, $A_6(GC)_2/(GC)_2T_6$ versus A_9C/GT_9. *Complex-1* shows three transitions while *Complex-5* shows only two transitions with similar total enthalpies of 110.2 and 105.9 kcal mol^{-1}. Their first transition have similar T_ms of 32 °C and similar unfolding enthalpies of 28.6 kcal mol^{-1}, whereas the second transition of *Complex-5* has a T_m slightly higher, and larger enthalpy (by 44.8 kcal mol^{-1}) than the second transition of *Complex-1*. This triphasic to biphasic transition behavior (*Complex-1* → *Complex-5*) may be explained by the different unfolding of their component decamer duplexes. Their first transition corresponds to the removal of the third strand. The second transition of *Complex-5* is due to the cooperative melting of all 10 bp while the second and third transitions of *Complex-1* correspond to the sequential melting of 3–4 AA/TT base-pair stacks followed by the cooperative disruption of the remaining 5–6 bp stacks. Furthermore, the total folding free energy of *Complex-1* (-18.4 kcal mol^{-1}) is more favorable than *Complex-5* (-15.7 kcal mol^{-1}) due to more favorable enthalpy contributions (-2.7).

Complex-6 has an entirely different duplex stem sequence, $(AG)_3A(CG)_2/(GC)_2T(CT)_2$, consistent with a triplex motif of 7 base triplets and a duplex motif of 4 bp (Scheme 20.1). This complex unfolds in a triphasic transition with T_ms of 17.4, 46, and 83.8 °C, and unfolding enthalpies of 10.6, 60.9, and 85.5 kcal mol^{-1}, respectively. The two main transitions take place at higher temperatures than the corresponding transitions of the other two complexes, and its overall enfolding enthalpy (166 kcal mol^{-1}) is in excellent agreement with the sum of the enthalpies of its control molecules (157.7 kcal mol^{-1}). This suggests that the first two transitions correspond to the removal of the third strand while the third transition corresponds to the cooperative melting of all 11 bp. Furthermore, the total folding free energy of *Complex-6* (-28.4 kcal mol^{-1}) is more favorable than *Complex-1* (-18.4 kcal mol^{-1}) due to more favorable enthalpy contributions.

4. Conclusions

We have investigated the unfolding of six intramolecular DNA complexes containing joined triplex and double-helical motifs. Complete thermodynamic profiles are reported for the observed transitions of each complex, and their overall thermodynamic profiles are compared with those of their control molecules. Overall, the temperature-induced unfolding of each complex corresponds to the initial disruption of the triplex motif (removal of the third strand) followed by the partial or full unfolding of the duplex stem.

The resulting data should improve our current picture of how sequence, loops, proton, and ion binding control the stability and melting behavior of nucleic acid molecules, and would supplement existing nearest-neighbor thermodynamic parameters to help in the prediction of secondary structure from a given sequence, as demonstrated earlier (Breslauer *et al.*, 1986; SantaLucia *et al.*, 1996; Sugimoto *et al.*, 1995; Xia *et al.*, 1998). Furthermore, the design of these complexes mimic portions of the complex structures presented by RNA molecules; therefore, the data presented should help in the optimization of oligonucleotide reagents for the targeting of specific transient structures that form *in vivo*.

Acknowledgments

This chapter was supported by grants MCB-0315746 and MCB-0616005 from the National Science Foundation.

REFERENCES

Behe, M. J. (1987). The DNA sequence of the human β-globin region is strongly biased in favor of long strings of contiguous purine or pyrimidine residues. *Biochemistry* **26,** 7870–7875.

Behe, M. J. (1995). An overabundance of long oligopurine tracts occurs in the genome of simple and complex eukaryotes. *Nucleic Acids Res.* **23,** 689–695.

Bloomfield, V. A., Crothers, D. M., and Tinoco, I. (2000). *Nucleic Acids: Structures, Properties, and Functions.* University Science Books, Sausalito, CA.

Breslauer, K. J., Frank, R., Blocker, H., and Marky, L. A. (1986). Predicting DNA duplex stability from the base sequence. *Proc. Natl. Acad. Sci. USA* **83,** 3746–3750.

Cantor, C. R., and Schimmel, P. R. (1980). Biophysical Chemistry Part III: The Behavior of Biological Macromolecules. W.H. Freeman, New York, NY.

Cooney, M., Czernuszewicz, G., Postel, E. H., Flint, S. J., and Hogan, M. E. (1988). Site-specific oligonucleotide binding represses transcription of the human c-myc gene *in vitro.* *Science* **241,** 456–459.

De los Santos, C., Rosen, M., and Patel, D. (1989). NMR studies of DNA (R+)n·(Y−)n·(Y+)n triple helixes in solution: Imino and amino proton markers of T·A·T and C·G·C+ base-triple formation. *Biochemistry* **28,** 7282–7289.

Dove, A. (2002). Antisense and sensibility. *Nat. Biotechnol.* **20,** 121–124.

Duval-Valentin, G., Thuong, N. T., and Helene, C. (1992). Specific inhibition of transcription by triple helix-forming oligonucleotides. *Proc. Natl. Acad. Sci. USA* **89,** 504–508.

Escude, C., Francois, J.-C., Sun, J.-S., Ott, G., Sprinzl, M., Garestier, T., and Helene, C. (1993). Stability of triple helixes containing RNA and DNA strands: Experimental and molecular modeling studies. *Nucleic Acids Res.* **21,** 5547–5553.

Francois, J. C., Saison-Behmoaras, T., Nguyen, T. T., and Helene, C. (1989). Inhibition of restriction endonuclease cleavage via triple helix formation by homopyrimidine oligonucleotides. *Biochemistry* **28,** 9617–9619.

Helene, C. (1990). Specific regulation of gene expression by antisense, sense and antigene nucleic acids. *Biochim. Biophys. Acta* **1049,** 99–125.

Helene, C. (1991). Rational design of sequence-specific oncogene inhibitors based on antisense and antigene oligonucleotides. *Eur. J. Cancer* **27,** 1466–1471.

Helene, C. (1994). Control of oncogene expression by antisense nucleic acids. *Eur. J. Cancer* **30A,** 1721–1726.

Helene, C., and Thuong, N. T. (1989). Control of gene expression by oligonucleotides covalently linked to intercalating agents. *Genome* **31,** 413–421.

Kato, M., Kudoh, J., and Shimizu, N. (1990). The pyrimidine/purine-biased region of the epidermal growth factor receptor gene is sensitive to S1 nuclease and may form an intramolecular triplex. *Biochem. J.* **268,** 175–180.

Kaushik, M., Suehl, N., and Marky, L. A. (2007). Calorimetric unfolding of the bimolecular and i-motif complexes of the human telomere complementary strand, d(C3TA2)4. *Biophys. Chem.* **126,** 154–164.

Larsen, A., and Weintraub, H. (1982). An altered DNA conformation detected by S1 nuclease occurs at specific regions active chick globine chromatin. *Cell* **29,** 609–622.

Lee, H.-T., Arciniegas, S., and Marky, L. A. (2008a). Unfolding thermodynamics of DNA pyrimidine triplex with different molecularities. *J. Phys. Chem.* **112,** 4833–4840.

Lee, H.-T., Olsen, C. M., Waters, L., Sukup, H., and Marky, L. A. (2008b). Thermodynamic contributions of the reactions of DNA intramolecular structures with their complementary strands. *Biochimie* **90,** 1052–1063.

Lilley, D. M. J. (1980). The inverted repeat as a recognizable structural feature in supercoiled DNA molecules. *Proc. Natl. Acad. Sci. USA* **77,** 6468–6472.

Lilley, D. M. J. (1981). Hairpin-loop formation by inverted repeats in supercoiled DNA is a local and transmissible property. *Nucleic Acids Res.* **9,** 1271–1290.

Maher, L. J. III, Wold, B., and Dervan, P. B. (1989). Inhibition of DNA binding proteins by oligonucleotide-directed triple helix formation. *Science* **245,** 725–730.

Maher, L. J., Dervan, P. B., and Wold, B. (1992). Analysis of promoter-specific repression by triple helical DNA complexes eukaryotic cell-free transcription system. *Biochemistry* **31,** 70–81.

Manning, G. S. (1978). The molecular theory of polyelectrolyte solutions with applications to the electrostatic properties of polynucleotides. *Q. Rev. Biophys.* **11,** 179–246.

Marky, L. A., and Breslauer, K. J. (1987). Calculating thermodynamic data for transitions of any molecularity from equilibrium melting curves. *Biopolymers* **26,** 1601–1620.

Marky, L. A., Blumenfeld, K. S., Kozlowski, S., and Breslauer, K. J. (1983). Salt-dependent conformational transitions in the self-complementary deoxydodecanucleotide d (CGCGAATTCGCG): Evidence for hairpin formation. *Biopolymers* **22,** 1247–1257.

Marky, L. A., Maiti, S., Olsen, C. M., Shikiya, R., Johnson, S. E., Kaushik, M., and Khutsishvili, I. (2007). Building blocks of nucleic acid nanostructures: Unfolding thermodynamics of intramolecular DNA complexes. *In* "Biomedical Applications of Nanotechnology," (V. Labhasetwar and D. Leslie-Pelecky, eds.), pp. 191–225. John Wiley & Sons, New York, NY.

Moser, H. E., and Dervan, P. B. (1987). Sequence-specific cleavage of double helical DNA by triple helix formation. *Science* **238,** 645–650.

Panayotatos, N., and Wells, R. D. (1981). Cruciform structures in supercoiled DNA. *Nature (London)* **289,** 466–470.

Perrouault, L., Asseline, U., Rivalle, C., Thuong, N. T. B. E., Giovannangeli, C., Le Doan, T., and Helene, C. (1990). Sequence-specific artificial photoinduced endonucleases based on triple helix-forming oligonucleotides. *Nature (London)* **344,** 358.

Pollack, Y., Stein, R., Razin, A., and Cedar, H. (1980). Methylation of foreign DNA sequences in eukaryotic cells. *Proc. Natl. Acad. Sci. USA* **77,** 6463–6467.

Rajagopal, P., and Feigon, J. (1989a). NMR studies of triple-strand formation from the homopurine–homopyrimidine deoxyribonucleotides d(GA)4 and d(TC)4. *Biochemistry* **28,** 7859–7870.

Rajagopal, P., and Feigon, J. (1989b). Triple-strand formation in the homopurine:homopyrimidine DNA oligonucleotides d(G-A)4 and d(T-C)4. *Nature (London)* **339,** 637–640.

Rentzeperis, D. (1995). *Thermodynamics and Ligand Interactions of DNA Hairpins.* New York University, New York. Doctoral Dissertation, p. 246.

Rentzeperis, D., and Marky, L. A. (1995). Ligand binding to the Hoogsteen-WC groove of TAT base-triplets: Thermodynamic contribution of the thymine methyl groups. *J. Am. Chem. Soc.* **117,** 5423.

Rentzeperis, D., Alessi, K., and Marky, L. A. (1993). Thermodynamics of DNA hairpins: Contribution of loop size to hairpin stability and ethidium binding. *Nucleic Acids Res.* **21,** 2683–2689.

Rothemund, P. W. K. (2006). Folding DNA to create nanoscale shapes and patterns. *Nature (London)* **440,** 297–302.

SantaLucia, J. Jr., Allawi, H. T., and Seneviratne, P. A. (1996). Improved nearest-neighbor parameters for predicting DNA duplex stability. *Biochemistry* **35,** 3555–3562.

Shikiya, R., and Marky, L. A. (2005). Calorimetric unfolding of intramolecular triplexes: Length dependence and incorporation of single AT → TA substitutions in the duplex domain. *J. Phys. Chem. B* **109,** 18177–18183.

Soto, A. M. (2002). Thermodynamics for the Unfolding of Non-Canonical Nucleic Acids: Bent Duplexes, Okazaki Fragments, and DNA Triplexes. University of Nebraska Medical Center, Omaha, NE. Doctoral Dissertation.

Soto, A. M., Loo, J., and Marky, L. A. (2002). Energetic contributions for the formation of TAT/TAT, TAT/CGC+, and CGC+/CGC+ base triplet stacks. *J. Am. Chem. Soc.* **124,** 14355–14363.

Soyfer, V. N., and Potaman, V. N. (1996). *Triple-Helical Nucleic Acids.* Springer-Verlag, New York, NY.

Strobel, S. A., Moser, H. E., and Dervan, P. B. (1988). Double strand cleavage of genomic DNA at a single site by triple helix formation. *J. Am. Chem. Soc.* **110,** 7927–7929.

Sugimoto, N., Nakano, S.-I., Katoh, M., Matsumura, A., Nakamuta, H., Ohmichi, T., Yoneyama, M., and Sasaki, M. (1995). Thermodynamic parameters to predict stability of RNA/DNA hybrid duplexes. *Biochemistry* **34,** 11211–11216.

Sun, J. S., Francois, J. C., Montenay-Garestier, T., Saison-Behmoaras, T., Roig, V., Thuong, N. T., and Helene, C. (1989). Sequence-specific intercalating agents: Intercalation at specific sequences on duplex DNA via major groove recognition by oligonucleotide-intercalator conjugates. *Proc. Natl. Acad. Sci. USA* **86,** 9198–9202.

Thuong, N. T., and Hélène, C. (1993). Sequence-specific recognition and modification of double-helical DNA by oligonucleotides. *Angew. Chem. Int. Ed. Engl.* **32,** 666–690.

Xia, T., SantaLucia, J. Jr., Burkard, M. E., Kierzek, R., Schroeder, S. J., Jiao, X., Cox, C., and Turner, D. H. (1998). Thermodynamic parameters for an expanded nearest-neighbor model for formation of RNA duplexes with Watson–Crick base pairs. *Biochemistry* **37,** 14719–14735.

Yavachev, L. P., Georgiev, O. I., Braga, E. A., Avdonina, T. A., Bogomolova, A. E., Zhurkin, V. B., Nosikov, V. V., and Hadjiolov, A. A. (1986). Nucleotide sequence analysis of the spacer regions flanking the rat rRNA transcription unit and identification of repetitive elements. *Nucleic Acids Res.* **14,** 2799–2810.

Zimmer, C., and Venner, H. (1966). Protonation of cytosine in DNA. *Biopolymers* **4,** 1073–1079.

Zimmer, C., Luck, G., Venner, H., and Fric, J. (1968). Studies on the conformation of protonated DNA. *Biopolymers* **6,** 563–574.

CHAPTER TWENTY-ONE

Thermodynamics and Conformational Change Governing Domain–Domain Interactions of Calmodulin

Susan E. O'Donnell, Rhonda A. Newman, Travis J. Witt, Rainbo Hultman, John R. Froehlig, Adam P. Christensen, *and* Madeline A. Shea

Contents

Abstract

Calmodulin (CaM) is a small (148 amino acid), ubiquitously expressed eukaryotic protein essential for Ca^{2+} regulation and signaling. This highly acidic polypeptide (p*I* < 4) has two homologous domains (N and C), each consisting

Department of Biochemistry, Roy J. and Lucille A. Carver College of Medicine, University of Iowa, Iowa City, Iowa, USA

Methods in Enzymology, Volume 466
ISSN 0076-6879, DOI: 10.1016/S0076-6879(09)66021-3

of two EF-hand Ca^{2+}-binding sites. Despite significant homology, the domains have intrinsic differences in their Ca^{2+}-binding properties and separable roles in regulating physiological targets such as kinases and ion channels. In mammalian full-length CaM, sites III and IV in the C-domain bind Ca^{2+} cooperatively with ~10-fold higher affinity than sites I and II in the N-domain. However, the difference is only twofold when CaM is severed at residue 75, indicating that anticooperative interactions occur in full-length CaM. The Ca^{2+}-binding properties of sites I and II are regulated by several factors including the interplay of interdomain linker residues far from the binding sites. Our prior thermodynamic studies showed that these residues inhibit thermal denaturation and decrease calcium affinity. Based on high-resolution structures and NMR spectra, there appear to be interactions between charged residues in the sequence 75–80 and those near the amino terminus of CaM. To explore electrostatic contributions to interdomain interactions in CaM, KCl was used to perturb the Ca^{2+}-binding affinity, thermal stability, and hydrodynamic size of a nested set of recombinant mammalian CaM (rCaM) fragments terminating at residues 75, 80, 85, or 90. Potassium chloride is known to decrease Ca^{2+}-binding affinity of full-length CaM. It may act directly by competition with acidic side chains that chelate Ca^{2+} in the binding sites, and indirectly elsewhere in the molecule by changing tertiary constraints and conformation. In all proteins studied, KCl decreased Ca^{2+}-affinity, decreased Stokes radius, and increased thermal stability, but not monotonically. Crystallographic structures of Ca^{2+}-saturated $rCaM_{1-75}$ (3B32.pdb) and $rCaM_{1-90}$ (3IFK.pdb) were determined, offering cautionary notes about the effect of packing interactions on flexible linkers. This chapter describes an array of methods for characterizing system-specific thermodynamic properties that in concert govern structure and function.

1. Introduction

Calmodulin (CaM, Fig. 21.1A) is an essential intracellular Ca^{2+} sensor that plays important roles in Ca^{2+}-mediated signaling pathways. It is comprised of two homologous "4-helix bundle" domains that each bind two Ca^{2+} ions via EF-hand (helix-loop-helix) motifs. A short flexible tether known as the interdomain linker region (spanning residues 76–80) connects the N- (residues 1–80) and C-domains (76–148) (Fig. 21.1A). NMR structures of CaM indicate that in the absence of Ca^{2+}, the linker region is flexible (Kuboniwa *et al.*, 1995; Zhang *et al.*, 1995). Upon binding Ca^{2+}, a portion of the linker remains flexible in solution structures (Ikura *et al.*, 1991). However, early crystallographic structures of Ca^{2+}-saturated CaM showed this region adopting a rigid, helical structure (Babu *et al.*, 1988), where the fourth helix of the N-domain (helix D of CaM) and the first helix of the C-domain (helix E of CaM) were aligned. Residues in this so-called "central helix" of CaM are known to play

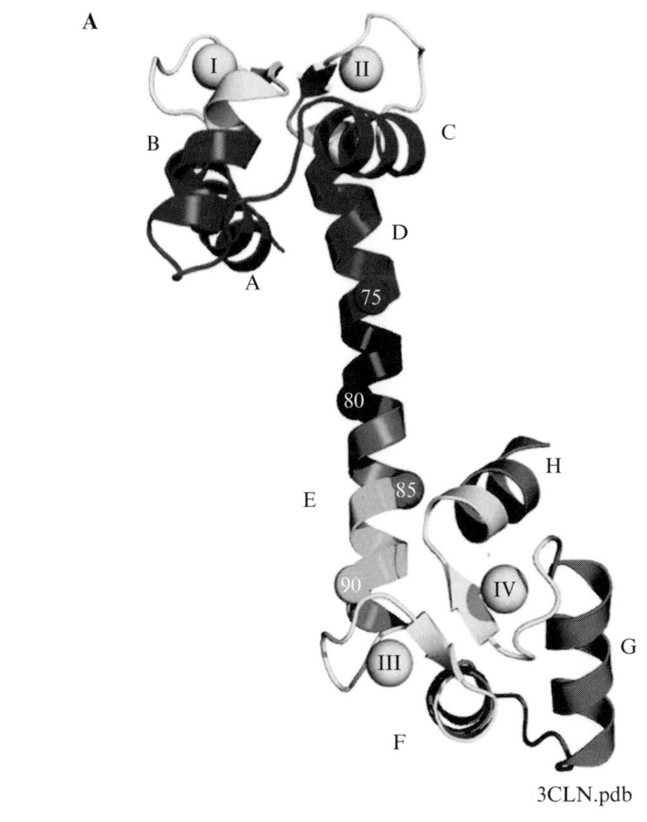

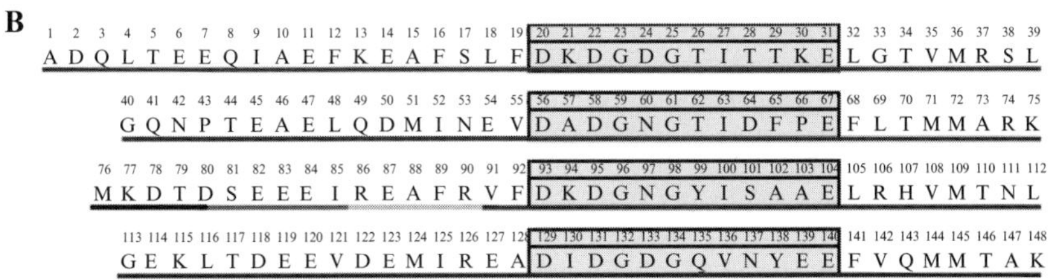

Figure 21.1 *Ribbon diagram of CaM and sequence information.* (A) Ribbon diagram of $(Ca^{2+})_4$-CaM (3CLN.pdb). Ca^{2+}-binding sites (yellow) I and II are in the N-domain (blue), and sites III and IV are in the C-domain (red). Residues 76–80 (black), 81–85 (purple), and 86–90 (green) are highlighted. (B) Amino acid sequence for CaM with corresponding color schematic described for (A). (See Color Insert.)

important roles in mediating conformational changes in the structure of CaM (Houdusse *et al.*, 1996; Ikura *et al.*, 1992; Meador *et al.*, 1993).

Residues at the boundary between the N- and C-domains were shown to contribute to energetic properties of Ca^{2+}-binding and stability of sites I and II in the N-domain, as well as mediating anticooperative energetic interactions between the two domains of CaM (Faga *et al.*, 2003; Sorensen and Shea, 1998; Sorensen *et al.*, 2002; Sun *et al.*, 2001). To further determine how helix E affects the thermodynamic behavior of the N-domain, it was of interest to explore residues 81–90 that extend almost to the beginning of site III in the C-domain. Because the sequence of CaM from 81 to 90 (SEEEIREAFR, Fig. 21.1B) is highly charged at neutral pH (Fig. 21.2A and B), and potassium levels reach 300 m*M* in the interior of a neuron, KCl-induced effects in thermodynamic properties and conformation of CaM were studied.

We have previously described KCl-dependent changes in interdomain cooperativity of full-length CaM using 1D NMR (Pedigo and Shea, 1995a).

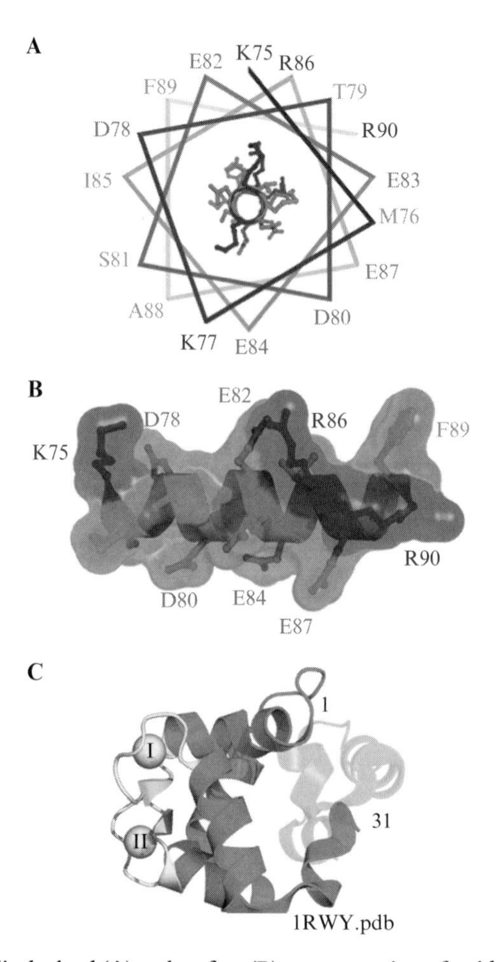

Figure 21.2 Helical wheel (A) and surface (B) representation of residues 75–90 of CaM with basic residues shown in blue, acidic residues shown in red, and all other residues shown in green. (C) Ribbon diagram of $(Ca^{2+})_2$-parvalbumin (1RWY.pdb). Ca^{2+}-binding sites are shown in yellow (residues 1–31 are transparent). (See Color Insert.)

The chapter will discuss the application of several quantitative biophysical methods used to study how residues near a domain boundary contribute to domain–domain interactions. Using KCl as a perturbant, we report the Ca^{2+}-binding affinity, thermal stability, and Stokes radii values of a series of N-domain fragments ($rCaM_{1-75}$, $rCaM_{1-80}$, $rCaM_{1-85}$, and $rCaM_{1-90}$) and compare them to full-length CaM ($rCaM_{1-148}$). Results from these studies prompted us to further explore the structural features of $rCaM_{1-75}$ and $rCaM_{1-90}$ using X-ray crystallography. This study yielded unexpected results that illustrate the propensity of aromatic residues to bind to the FLMM pocket (Ataman *et al.*, 2007) in the hydrophobic clefts of Ca^{2+}-saturated CaM, as well as the challenge of correlating structures observed in crystals to those found in aqueous solution.

2. Overexpression and Purification of rCaM Fragments

Reagents were of the highest grade commercially available. IPTG-induced CaM overexpression was performed using transformed BL21(DE3) cells containing the recombinant pT7-7 vector of full-length mammalian CaM ($rCaM_{1-148}$) and the N-domain fragments ($rCaM_{1-75}$, $rCaM_{1-80}$, $rCaM_{1-85}$, and $rCaM_{1-90}$) using standard methods (Sorensen and Shea, 1998). Cells were lysed using a French Press. The proteins were then purified using phenyl-sepharose CL-4B (hydrophobic interaction chromatography) as previously described (Putkey *et al.*, 1985); some samples required further purification on a DEAE Sephacel anion exchange column using a 100–700 m*M* NaCl gradient. The recombinant proteins were 97–99% pure as judged by silver stained SDS–PAGE gels and RP-HPLC. Protein concentrations were determined by UV spectroscopy of protein under native conditions at pH 7.4 (Crouch and Klee, 1980) or denatured with NaOH (Beaven and Holiday, 1952).

3. Calcium-Binding Properties of N-Domain CaM Fragments

3.1. Experimental conditions

Equilibrium Ca^{2+} titrations of $rCaM_{1-75}$, $rCaM_{1-80}$, $rCaM_{1-85}$, $rCaM_{1-90}$, and $rCaM_{1-148}$ (6 μM) (Fig. 21.3) were monitored using a PTI-QM4 fluorimeter (Photon Technology International) and an SLM 4800™ fluorimeter (SLM Instruments, Inc.), with 4 nm excitation and 6 nm emission band passes. Calcium titrations were conducted in the presence of Oregon

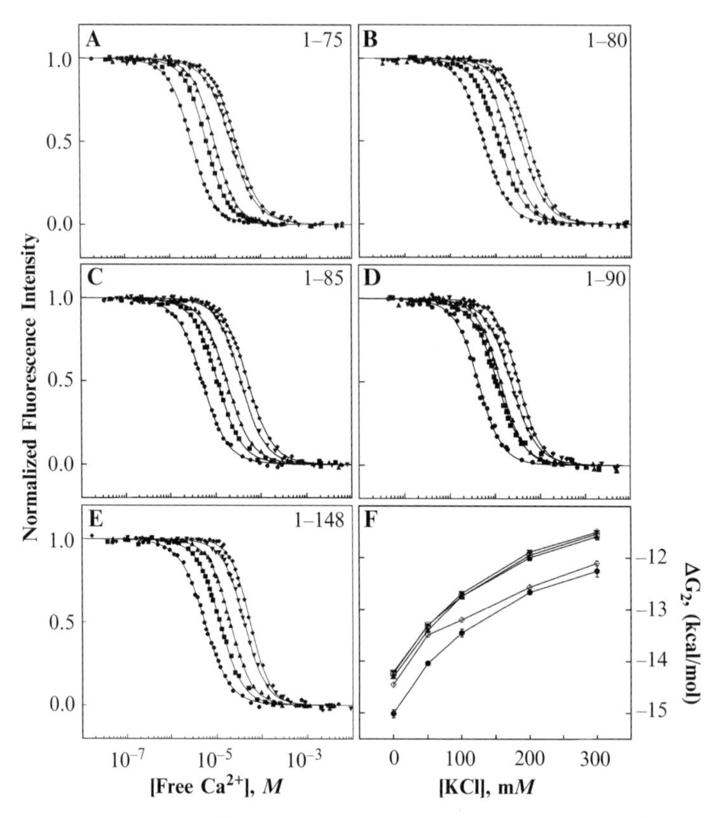

Figure 21.3 *Equilibrium Ca^{2+}-titrations of CaM fragments.* Equilibrium Ca^{2+} titrations of (A) CaM_{1-75}, (B) CaM_{1-80}, (C) CaM_{1-85}, (D) CaM_{1-90}, and (E) CaM_{1-148} at various KCl levels. Titrations of CaM at 0 m*M* KCl (●), 50 m*M* KCl (■), 100 m*M* KCl (▲), 200 m*M* KCl (▼), and 300 m*M* KCl (◆) were monitored by changes in Phe intensity. Curves were simulated using the free energies resolved for Ca^{2+} binding to CaM at each KCl level using Eq. (21.2). Comparison of ΔG_2 values as a function of KCl concentration (F) for CaM_{1-75} (●), CaM_{1-80} (□), CaM_{1-85} (▲), CaM_{1-90} (◇), and CaM_{1-148} (▼) is shown. Results are summarized in Table 21.1.

Green 488 BAPTA-5N (Molecular Probes, Eugene, OR; $\lambda_{ex} = 494$ nm, $\lambda_{em} = 521$ nm) to monitor the free calcium concentration at each point of the titration as determined using Eq. (21.1) (Pedigo and Shea, 1995b):

$$[Ca^{2+}]_{free} = K_d \frac{[Indicator : Ca^{2+}]}{[Indicator]_{free}} \tag{21.1}$$

The K_d values for Oregon Green were determined in 50 m*M* HEPES and 1 m*M* $MgCl_2$, pH 7.4 at 22 °C; they were 9.11 μM in the presence of no added KCl, 20.8 μM in 50 m*M* KCl, 32.24 μM in 100 m*M* KCl, 70.2 μM in 200 m*M* KCl, and 98.4 μM in 300 m*M* KCl. Samples of

CaM (6 μM) in 50 mM HEPES, with "0" (no added) KCl, 50, 100, 200, or 300 mM KCl in 0.05 mM EGTA, 5 mM NTA, and 1 mM $MgCl_2$, pH 7.4 at 22 °C were titrated with 5, 50, and 500 mM $CaCl_2$ stocks in matching buffer delivered with a microburette fitted with a 250 μl Hamilton syringe. Due to KCl contamination in the buffer of the original protein sample and the Ca^{2+}-titrant, the titration measured in buffer with no added KCl had a final concentration of 1–4 mM KCl.

Calcium binding to sites I and II was monitored by phenylalanine fluorescence ($\lambda_{ex} = 250$ nm, $\lambda_{em} = 280$ nm) as reported previously (VanScyoc and Shea, 2001). Three independent titrations of each protein were conducted. Representative normalized $((F - F_{min})/(F_{max} - F_{min}))$ titrations of each sample at each salt condition tested are shown in Fig. 21.3.

3.2. Analysis of free energy of calcium binding

The diagram below describes calcium binding to two sites in a domain of CaM.

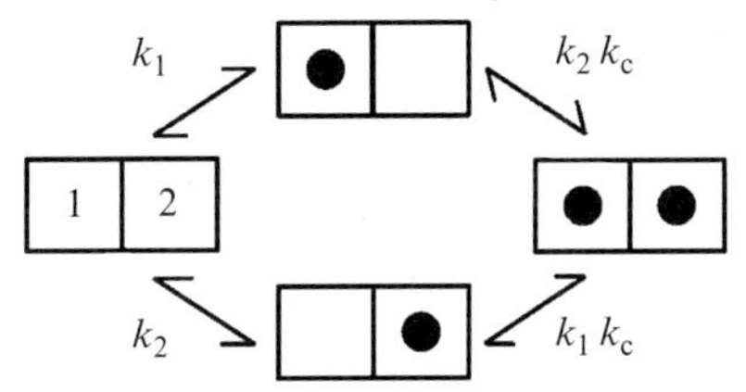

Gibbs free energies for Ca^{2+} binding to sites I and II of $rCaM_{1-75}$, $rCaM_{1-80}$, $rCaM_{1-85}$, $rCaM_{1-90}$, and $rCaM_{1-148}$ were determined by nonlinear least squares analysis, using a model-independent Adair function for two sites as shown in Eq. (21.2):

$$\bar{Y}_2 = \frac{K_1[X] + 2K_2[X]^2}{2(1 + K_1[X] + K_2[X]^2)} \tag{21.2}$$

where $\bar{Y}_2$ is the average degree of saturation for two sites, K_1 describes binding of the first ligand to that domain and is the sum of two intrinsic microscopic binding constants (k_1 and k_2) that may or may not be equal, K_2 is a macroscopic binding constant that describes the total free energy of calcium binding to that domain and is the product of k_1, k_2, and k_c (cooperativity constant), and [X] indicates the concentration of free calcium. This equation allows for fitting of the extent of calcium binding to two potentially nonequivalent, cooperative sites within a domain of CaM (Shea *et al.*, 2000). The parameters ΔG_1 and ΔG_2 are macroscopic binding free energies, with ΔG_i equal to $-RT \ln K_i$. Thus, ΔG_2 represents the total free energy of saturating both Ca^{2+}-binding sites in either domain, while ΔG_1 corresponds to the total free energy of saturating one Ca^{2+}-binding site in either domain.

The equilibrium calcium titrations of $rCaM_{1-75}$, $rCaM_{1-80}$, $rCaM_{1-85}$, $rCaM_{1-90}$, and $rCaM_{1-148}$ were fit to a function ($f(X)$) given in Eq. (21.3):

$$f(X) = Y_{[X]low} + \bar{Y}_2 \times \text{Span} \qquad (21.3)$$

where the fluorescence signal ($f(X)$) is related to the Adair equation (Eq. 21.2), but accounts for experimental variations in the asymptotes of the titration profiles. As described in Eq. (21.2), $\bar{Y}_2$ represents the average fractional saturation of the Ca^{2+}-binding sites. $Y_{[X]low}$ corresponds to the value of the fluorescence intensity of protein in the absence of Ca^{2+}, and Span accounts for the magnitude and direction of the signal change upon titration. In titrations of rCaM, Phe intensity always decreases; therefore, Span is negative.

Values for all parameters were fit simultaneously using nonlinear least squares analysis (Johnson and Frasier, 1985). The quality of each fit was evaluated as described (Pedigo and Shea, 1995b). Free energies of Ca^{2+} binding were determined from three or more independent titrations of each rCaM sample; averages and standard deviations of those values are summarized in Fig. 21.3 and are reported in Table 21.1.

3.3. Experimental considerations

3.3.1. Fluorimeter and cuvettes

The choice of fluorimeter should include considerations of the fluorophores to be monitored. The sequence of $rCaM_{1-148}$ does not contain Trp. Although it does contain Tyr and Phe, Tyr is found only in sites III and IV in the C-domain (Fig. 21.1B). Thus, Phe is the only aromatic amino acid in the sequence of the N-domain fragments. In calcium titrations, it is essential to detect Ca^{2+}-dependent changes in the intensity of phenylalanine; however, it is a weak fluorophore. It was imperative that the instrument lamp provide sufficient intensity at 250 nm (this required avoiding a "nonozone producing" lamp), and that the intrinsic fluorophores of CaM not be bleached by lamp intensity (some lamps are too bright). Many instruments were tested; those giving acceptable signal-to-noise ratios included fluorimeters made by SLM, PTI, Varian, and Jasco.

Calcium titrations are conducted in 3 ml capacity Suprasil Quartz cuvettes from Hellma. Prior to experiments, cuvettes are cleaned thoroughly with a detergent such as RBS 35 or a mixture of ethanol and detergent and washed extensively with water. For stoichiometric titration controls that begin with apo-CaM, the cuvettes are soaked in dilute nitric acid for several hours or overnight to minimize calcium contamination. We assiduously avoid using cleaning agents such as Chromerge™ that contain metals.

Table 21.1 Effect of KCl on free energies of calcium binding

Protein	[KCl] (m*M*)	ΔG_1 (kcal/mol)	ΔG_2 (kcal/mol)
CaM_{1-75}	0 added	-6.94 ± 0.86	-15.01 ± 0.08
	50	-6.29 ± 0.08	-14.04 ± 0.04
	100	-6.17 ± 0.20	-13.53 ± 0.08
	200	-5.72 ± 0.13	-12.66 ± 0.03
	300	-5.68 ± 0.20	-12.24 ± 0.11
CaM_{1-80}	0 added	-6.77 ± 0.08	-14.23 ± 0.05
	50	-6.33 ± 0.08	-13.30 ± 0.02
	100	-5.83 ± 0.15	-12.74 ± 0.04
	200	-5.50 ± 0.13	-11.94 ± 0.04
	300	-4.95 ± 0.25	-11.53 ± 0.06
CaM_{1-85}	0 added	-6.75 ± 0.10	-14.30 ± 0.03
	50	-6.23 ± 0.11	-13.40 ± 0.03
	100	-5.90 ± 0.13	-12.74 ± 0.05
	200	-5.49 ± 0.15	-11.99 ± 0.05
	300	-5.26 ± 0.15	-11.58 ± 0.06
CaM_{1-90}	0 added	-6.77 ± 0.15	-14.45 ± 0.04
	50	-6.48 ± 0.10	-13.49 ± 0.04
	100	-6.04 ± 0.18	-13.20 ± 0.05
	200	-5.86 ± 0.18	-12.56 ± 0.04
	300	-5.34 ± 0.22	-12.10 ± 0.05
CaM_{1-148}	0 added	-6.80 ± 0.14	-14.21 ± 0.01
	50	-6.02 ± 0.03	-13.31 ± 0.01
	100	-5.52 ± 0.08	-12.68 ± 0.02
	200	-4.77 ± 0.33	-11.88 ± 0.03
	300	-4.09 ± 0.66	-11.50 ± 0.07

Gibbs free energies (in kcal/mol: 1 kcal = 4.184 kJ) resolved from fitting titrations to a two-site Adair function for ligand binding (sites may be nonequivalent and cooperative; Eq. 21.2). Reported free energies and errors represent averages and standard deviations for three or more trials.

3.3.2. Ca^{2+}-indicator dye

We have used several commercially available Ca^{2+}-indicators for monitoring titrations as described above. Depending upon the protein and ligand under investigation, care should be taken in selecting an indicator that allows for the acquisition of 30 or more points over a wide range of Ca^{2+} concentrations, thus allowing the experimental determination of both low and high plateaus. For example, in the absence of a target peptide, the median ligand activity for the domains of CaM is in the micromolar range; for these studies, we routinely use Oregon Green 488 BAPTA-5N ($K_d = 32.32\ \mu M$), an indicator dye that binds a single calcium ion.

However, in the presence of a peptide (such as CaMKIIp) that matches the sequence of a CaM-binding domain in a target protein, the Ca^{2+}-binding affinity of both domains of CaM increases significantly, shifting the binding curves to a median ligand activity below μM (Evans and Shea, 2009). Thus, accurate analysis of these titrations required a Ca^{2+}-indicator such as XRhod5F that has a higher affinity ($K_d = 1.78\ \mu M$). These dyes are sensitive to pH and salt; therefore, values tabulated in the literature may not be pertinent and can dramatically affect inferred energetic properties. Thus, the K_d of each Ca^{2+}-indicator was determined experimentally in our laboratory for each buffer condition. Levels of contaminating calcium in buffer components were determined using atomic absorption analysis, calibrated with serial dilutions of commercially NIST-certified $CaCl_2$ solutions. To avoid perturbations caused by high ionic strength when determining the dissociation constant for each dye, it is best to establish the steady-state fluorescence intensity at low calcium (i.e., after addition of EGTA), and subsequently saturate with calcium to establish the plateau intensity of the saturated indicator dye. Depending on the indicator, the plateau at low calcium may be the minimum or maximum value.

4. Tertiary Constraints of N-Domain CaM Fragments

We have previously shown that basic residues in the flexible linker of CaM (75–80) decrease the Ca^{2+}-binding affinity of sites I and II of CaM_{1-80} and increase melting temperature (Sorensen *et al.*, 2002) of the 4-helix bundle that comprises the N-domain. A study of incremental additions showed that individual residues beyond position 74 decreased calcium-binding affinity and increased the thermostability of the domain until position 77 was reached. Fragments terminating at domain boundary residues 77–80 recapitulated the calcium-binding energetics of sites I and II and thermal denaturation of the N-domain in context of CaM_{1-148} (Faga *et al.*, 2003). Here, we further explore the role of electrostatic contributions to the interdomain boundary by studying the effect of KCl on calcium binding and thermal stability of N-domain CaM fragments carrying extensions into helix E (Fig. 21.1), which is highly charged (Fig. 21.2).

4.1. Experimental conditions

The thermal stability of apo- (calcium-depleted) rCaM was monitored using CD spectroscopy (Fig. 21.4). Thermal unfolding experiments presented here were performed on an AVIV 62DS CD spectrometer equipped with a thermoelectric temperature controller and an immersible

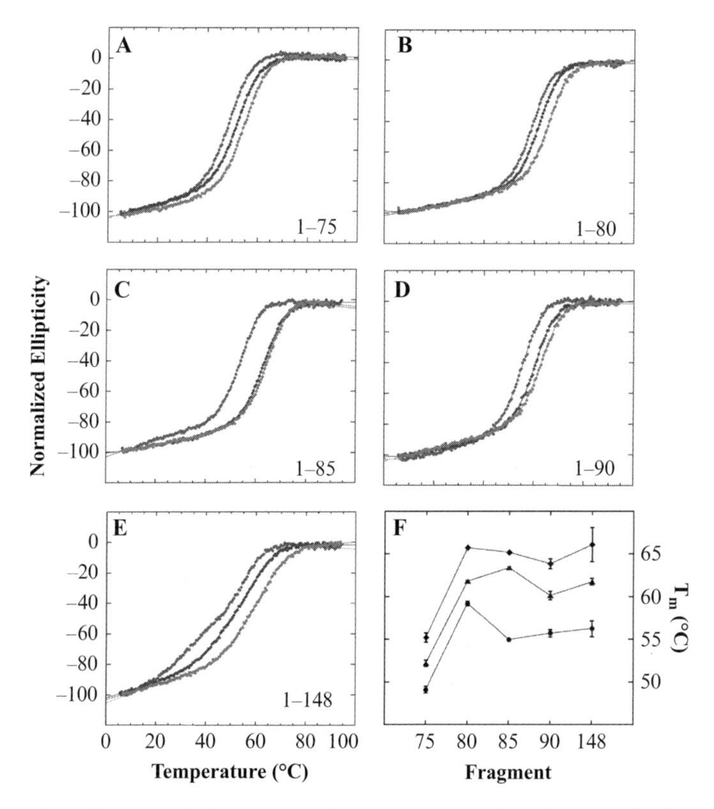

Figure 21.4 *Thermal unfolding of CaM fragments.* Thermal unfolding of (A) CaM_{1-75}, (B) CaM_{1-80}, (C) CaM_{1-85}, (D) CaM_{1-90}, and (E) CaM_{1-148} in the presence of 0 m*M* added KCl (red), 100 m*M* KCl (green), or 300 m*M* KCl (purple) as monitored by circular dichroism at 222 nm. Curves were simulated using parameters resolved for unfolding of CaM at each KCl level using a two-state model for unfolding (A–D, Eq. 21.4) or a three-model (E, Eq. 21.5). Comparison of T_m (°C) for each fragment tested (F) at 0 m*M* added KCl (●), 100 m*M* KCl (▲), and 300 m*M* KCl (◆) is shown. Results are summarized in Table 21.2. (See Color Insert.)

thermocouple, accurate to ±0.4 °C as described previously (Sorensen and Shea, 1998). Proteins ($rCaM_{1-75}$, $rCaM_{1-80}$, $rCaM_{1-85}$, $rCaM_{1-90}$, and $rCaM_{1-148}$) were diluted to 10 μM in a total volume of 3 ml of 2 m*M* HEPES, 5 m*M* NTA, 0.05 m*M* EGTA, and 1 m*M* $MgCl_2$ with either no added KCL, 100 or 300 m*M* KCl; pH 7.4. Samples were equilibrated at 5 °C and denatured with increasing temperature from 5 to 95 °C at a rate of 1 °C/min. Ellipticity at 222 nm was averaged for 20 s every 30 s and stored with experimental measurements of temperature. At the end of the heating ramp, samples were rapidly cooled to 5 °C; the final ellipticity reading was compared to that at the beginning of the experiment.

4.2. Analysis of thermal denaturation

The normalized ellipticity values were fit to two models of unfolding: two-state (native-to-unfolded, Eq. 21.4a; Santoro and Bolen, 1988) and/or three-state (native-to-intermediate-to-unfolded, Eq. 21.4b; Carra *et al.*, 1994; Eftink *et al.*, 1996) using nonlinear least squares analysis (Johnson and Frasier, 1985) as described previously (Sorensen and Shea, 1998):

$$Y_{obs} = f_N(y_N + m_N T) + f_U(y_U + m_U T) \qquad (21.4a)$$

$$Y_{obs} = f_N(y_N + m_N T) + f_I(y_I + m_I T) + f_U(y_U + m_U T) \qquad (21.4b)$$

where the contribution of native (N), intermediate (I), and unfolded (U) proteins to the overall observed spectral signal was estimated based on its fractional population f_N, f_I, and f_U at each temperature, T, and their corresponding linear baselines (fit to the equation $y_i + m_i T$, where y_i is the intercept and m_i is the slope of the baseline for species N, I, or U).

The only denaturation profile that fit well to a three-state model was that of full-length $rCaM_{1-148}$ (Fig. 21.4E). In that case, a baseline for the intermediate species was barely apparent with no added KCl, and not visible by inspection at 100 and 300 mM KCl. Therefore, the slope of the intermediate baseline was fixed to the value of the average of the baselines for the native (m_N) and unfolded (m_U) species. The equilibrium constant (K_i) for unfolding each transition i [native-to-unfolded (N–U) in the two-state model and native-to-intermediate (N–I), intermediate-to-unfolded (I–U) in the three-state model] was calculated from the fractional populations and converted to the free energy of unfolding (ΔG_i) for each transition ($K_i = \exp(-\Delta G_i RT)$).

The free energy relates to the van't Hoff enthalpy ($\Delta H^\circ_{VH,i}$), the heat capacity (ΔC_{pi}), and the melting temperature (T_{mi}) according to the modified Gibbs–Helmholtz equation:

$$\Delta G_i = \Delta H_i\left(1 - \frac{T}{T_{mi}}\right) + \Delta C_{pi}\left[(T - T_{mi}) - T\ln\frac{T}{T_{mi}}\right] \qquad (21.5)$$

In fitting the heat-induced changes in ellipticity to a two-state model, all parameters were allowed to vary unless stated otherwise. In contrast, in the three-state analysis, ΔC_p was fixed to 0. Criteria for "goodness-of-fit" included (a) the value of the square root of variance, (b) the values of asymmetric 65% confidence intervals, (c) examination for any trends in the distribution of residuals, (d) the magnitude of the span of the residuals, and (e) the absolute value of elements of the correlation matrix (Johnson and Frasier, 1985). Parameters were determined from two to five independent trials of each protein at the indicated salt concentrations (Fig. 21.4); averages and standard deviations of those values are reported in Table 21.2.

Table 21.2 Thermal denaturation

Protein	[KCl] (mM)	T_m (°C)	ΔH (kcal/mol)	ΔC_p (cal/K mol)
CaM_{1-75}[a]	0 added	49.10 ± 0.39	45.76 ± 1.85	524 ± 938
	100	52.17 ± 0.37	44.26 ± 1.02	717 ± 412
	300	55.21 ± 0.54	45.04 ± 0.12	948 ± 201
CaM_{1-80}[a]	0 added	59.23 ± 0.26	50.99 ± 1.48	789 ± 473
	100	61.83 ± 0.06	52.06 ± 0.49	1099 ± 57
	300	65.72 ± 0.06	54.47 ± 1.80	1337 ± 181
CaM_{1-85}[a]	0 added	54.95 ± 0.03	53.59 ± 1.03	850 ± 1068
	100	63.40 ± 0.10	49.60 ± 0.10	820 ± 435
	300	65.19 ± 0.17	51.89 ± 1.05	1020 ± 333
CaM_{1-90}[a]	0 added	55.71 ± 0.42	48.17 ± 0.41	608 ± 608
	100	60.36 ± 0.49	51.52 ± 1.20	899 ± 779
	300	63.87 ± 0.58	52.82 ± 1.80	1050 ± 193
		$T_{m(N-I)}$ (°C)	ΔH_{N-I} (kcal/mol)	ΔC_p (cal/K mol)
CaM_{1-148}[b] "N"	0 added	56.25 ± 0.95	48.22 ± 2.37	0
	100	61.78 ± 0.39	47.24 ± 1.17	
	300	66.11 ± 1.98	51.68 ± 0.11	
		$T_{m(I-U)}$ (°C)	ΔH_{I-U} (kcal/mol)	
CaM_{1-148}[b] "C"	0 added	35.33 ± 0.19	25.74 ± 2.45	0
	100	48.06 ± 0.46	30.82 ± 0.02	
	300	54.63 ± 1.01	37.19 ± 0.56	

[a] Values were resolved from fitting to a two-state model for unfolding (Eq. 21.4a) as described in Section 4.2.

[b] Values were resolved from fitting to a three-state model for unfolding (Eq. 21.4b) as described in Section 4.2. ΔC_p was fixed at zero.

4.3. Experimental considerations

4.3.1. Buffer selection

For the best signal-to-noise ratio, the protein should account for the majority of the spectroscopic signal and the buffer preferably does not contribute to the signal at the wavelength(s) being monitored. For example, titrations of CaM with calcium monitored using steady-state fluorescence intensity, and measurements of Stokes radii monitored using UV absorbance were conducted with HEPES as a pH-buffering component at a concentration of 50 mM. However, spectra of buffer tested for CD spectroscopy in which ellipticity was monitored at 222 nm identified a contaminant in commercially purchased HEPES such that a concentration of

50 m*M* HEPES was not optically clear (as evidenced by an intolerably high dyna voltage detected by the instrument). Thus, concentrated samples of CaM that had been dialyzed against buffer at pH 7.4 containing 50 m*M* HEPES were then diluted into 3 ml of buffer of the same pH with a HEPES concentration of 2 m*M*. As a positive control, standard Ca^{2+}-binding properties were determined to be identical in buffers with 2 and 50 m*M* HEPES. Based on this experience, it is recommended to test any buffer intended for CD measurements before testing a protein solution in that buffer. Other buffer systems that are often used for CD spectra include phosphate, which tends to be optically clear. However, because we are interested in properties of CaM that are dependent on calcium, and calcium phosphate forms insoluble flakes, our studies were not conducted in phosphate buffers. Tris-based buffers were excluded because of their heat of protonation.

4.3.2. Renaturation

To accurately calculate the melting temperature, enthalpy, and heat capacity for a protein, thermal denaturation must be reversible. This is tested by determining whether the protein sample that was fully unfolded refolds completely after rapid cooling to 5 °C. For CaM and its fragments discussed here, the percent renaturation was 97 ± 1% for all samples, consistent with purity of the samples.

4.3.3. Ellipticity signal

The secondary structure of CaM is primarily α-helical with a short stretch of β-sheet between paired Ca^{2+}-binding sites (see Fig. 21.1). It displays characteristic minima at 208 and 222 nm. Loss of ellipticity is observed as CaM unfolds. The concentration of protein was chosen to give sufficient signal-to-noise to monitor unfolding. This is dependent on the sensitivity of the instrument, extent of ordered secondary structure at low temperature, and the path length of the cuvette; therefore, it must be determined empirically.

5. Tertiary Conformation of N-Domain CaM Fragments

To establish the effect of KCl on the hydrodynamic behavior of N-domain rCaM fragments, the Stokes radius of $rCaM_{1-75}$, $rCaM_{1-80}$, $rCaM_{1-85}$, $rCaM_{1-90}$, and $rCaM_{1-148}$ was determined in the absence of calcium or presence of saturating calcium.

5.1. Experimental conditions

Analytical gel chromatography studies were performed on an ÄKTA FPLC (model UPC-900; Amersham Pharmacia Biotech) using a 10 × 300 mm, 24 ml Superdex-75 column (GE Healthcare), with a flow rate of 0.4 ml/min at room temperature (23 ± 1 °C). The column was pre-equilibrated with at least three column volumes of running buffer (50 m*M* HEPES, ±10 m*M* $CaCl_2$, 50 μM EGTA, with either no added KCl, 50, 100, 200, or 300 m*M* KCl; pH 7.42). Samples of Ca^{2+}-saturated or Ca^{2+}-depleted (apo) $rCaM_{1-75}$, $rCaM_{1-80}$, $rCaM_{1-85}$, and $rCaM_{1-90}$ were diluted to 245 μM, $rCaM_{1-148}$ was diluted to 10 μM using running buffer for each salt and Ca^{2+} condition studied. Samples were filtered using a disposable syringe equipped with a 4 mm diameter, 0.2 μm filter (Nalgene, 0.4 μM filters were used for Blue Dextran samples) prior to injection (200 μl) on the sizing column. Elution of protein samples was monitored at 280 nM. The Stokes radius (R_s) was calculated based on the elution volumes for Blue Dextran, acetone (0.1%), and the following standard globular proteins: bovine serum albumin (BSA), ovalbumin, chymotrypsin, and ribonuclease A used at 4 mg/ml each, as described previously (Sorensen and Shea, 1996). The average R_s from at least three determinations is reported in Fig. 21.5 and Table 21.3. A discussion of protein separation by analytical gel chromatography and shape analysis is provided in (Sorensen and Shea, 1996), based on the analytical theory and methods developed by Gary K. Ackers and coworkers. Note

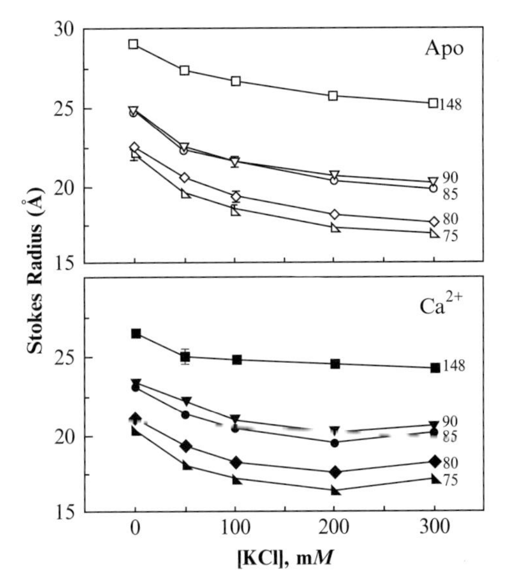

Figure 21.5 *Stokes radii values for CaM fragments*. Stokes radii values for CaM_{1-75} (◣), CaM_{1-80} (◆), CaM_{1-85} (●), CaM_{1-90} (▼), and CaM_{1-148} (■) in the absence (open symbols) or presence (closed symbols) of saturating levels of Ca^{2+} at 0 m*M* added KCl, 50, 100, 200, and 300 m*M* KCl levels. Results are summarized in Table 21.3.

Table 21.3 KCl dependence of hydrodynamic radius

Protein	[KCl] (m*M*)	R_s^{apo}	$R_s^{calcium}$
CaM_{1-75}	0[a]	22.02 ± 0.42	20.02 ± 0.06
	50	19.53 ± 0.13	17.84 ± 0.10
	100	18.52 ± 0.27	16.92 ± 0.02
	200	17.36 ± 0.07	16.11 ± 0.03
	300	16.95 ± 0.14	16.98 ± 0.00
CaM_{1-80}	0[a]	22.55 ± 0.24	20.80 ± 0.11
	50	20.56 ± 0.00	19.08 ± 0.00
	100	19.34 ± 0.35	18.04 ± 0.07
	200	18.15 ± 0.06	17.29 ± 0.10
	300	17.66 ± 0.03	17.98 ± 0.00
CaM_{1-85}	0[a]	24.73 ± 0.28	22.79 ± 0.18
	50	22.28 ± 0.09	21.14 ± 0.22
	100	21.60 ± 0.32	20.15 ± 0.02
	200	20.35 ± 0.25	19.25 ± 0.02
	300	19.78 ± 0.01	19.82 ± 0.24
CaM_{1-90}	0[a]	24.80 ± 0.06	23.11 ± 0.12
	50	22.51 ± 0.07	21.93 ± 0.00
	100	21.55 ± 0.30	20.65 ± 0.02
	200	20.71 ± 0.23	19.89 ± 0.02
	300	20.16 ± 0.21	20.32 ± 0.24
CaM_{1-148}	0[a]	29.11 ± 0.16	26.34 ± 0.05
	50	27.45 ± 0.03	24.78 ± 0.39
	100	26.78 ± 0.18	24.62 ± 0.00
	200	25.76 ± 0.24	24.27 ± 0.06
	300	25.26 ± 0.21	24.01 ± 0.00

[a] 0 KCl denotes no added KCl in buffer. Protein samples carry counterions based on purification procedures; the final concentration was estimated to be less than 4 m*M* KCl.
Stokes radii were determined as described in Section 5.1. Apo refers to calcium-depleted CaM (dialyzed against EGTA-containing buffer); calcium refers to buffer containing 10 m*M* $CaCl_2$.

that the expression for elution volume (V_e) is the sum of the void volume (V_o) and the product of the partition coefficient (σ) and the internal volume of the column beads (V_i) (i.e., the corrected form of Eq. 2 in Sorensen and Shea, 1996 is $V_e = V_o + \sigma V_i$).

5.2. Experimental considerations

5.2.1. Ca^{2+}-independent behavior of calibration standards

Stokes radius values reported here were determined in the presence and absence of saturating calcium levels. To ensure accurate calculation of R_s, it was necessary to ensure there was no Ca^{2+}-dependent change in the elution profile of the calibrations standards. Column characteristics should be determined prior to use by observing consistent elution of Blue Dextran (to determine V_o) and acetone (to determine V_i).

5.2.2. Acetone interactions with proteins

As described above, acetone was used at 0.1% concentration as a marker for internal volume, and was added to each CaM sample (200 μl) immediately before loading onto the column (the protein was not stored with acetone). This provided a measure of the reproducibility of a column and consistency of the chromatography pumps from day to day, over the life of the column. A shift in the elution volume for acetone is usually an indication that the column matrix is compromised (cleaning and repacking may restore a column) or that the volume measurement or chart paper readings are wrong or inconsistent (either of which needs to be corrected before proceeding further). Extensive studies were performed to ensure that acetone did not interact with CaM, and did not change its size, or its Ca^{2+}-binding properties at the level used. Because acetone is small, it enters the interstices of the resin and separates quickly from CaM after the sample is loaded onto the column.

5.2.3. Interaction of proteins with column matrix

Care should be taken to ensure that calibration standards and proteins of interest do not interact with the column matrix. Each protein used as a size standard and CaM fragment tested eluted with consistently symmetrical peaks. Evidence for nonspecific interactions with matrix would include trailing elution profile, or delay relative to expected elution time (i.e., larger elution volume) for a given protein size. In general, if a protein is unduly retarded on a size-exclusion column, the manufacturer recommends raising the level of salt in the buffer. Note that the opposite effect was observed for KCl acting on CaM; higher [KCl] led to increased retention (attributed to more compact size of CaM rather than any interaction with the column).

6. High-Resolution Studies of N-Domain CaM Fragments

To shed light on unique similarities between $rCaM_{1-75}$ and $rCaM_{1-90}$ (discussed in Section 21.7), we determined crystal structures for these fragments shown in Fig. 21.6. We have not included an exhaustive report regarding selection of crystallization conditions, cryoprotectants, and other details of structure determination. Instead, the reader is directed to standard crystallography texts widely available such as "Crystallography Made

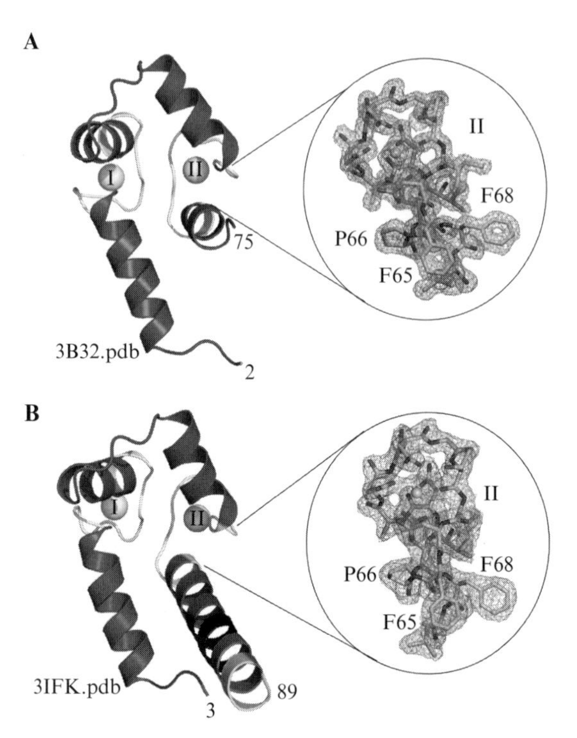

Figure 21.6 *Ribbon diagram of CaM_{1-75} and CaM_{1-90}.* Ribbon diagram of (A) $(Ca^{2+})_2$-CaM_{1-75}. (3B32.pdb) and (B) $(Ca^{2+})_2$-CaM_{1-90} (3IFK.pdb) displayed with electron density for site II. Ca^{2+}-binding sites I and II are shown in yellow. Residues 1–75 (blue, A and B) and residues 76–80 (black), 81–85 (purple), and 86–90 (green) (B) are shown. (See Color Insert.)

Crystal Clear" by Gale Rhodes, resources available from the Protein Data Bank (http://www.rcsb.org) and commercial Web sites such as Hampton (http://hamptonresearch.com).

6.1. Optimization of crystallization conditions

Crystals of $rCaM_{1-75}$ grew at 4 °C with reservoir solution of 7 m*M* $CaCl_2$, 100 m*M* Na acetate (pH 5.5), 0.01% Na azide, 21.25% PEG8000. Crystals of $rCaM_{1-90}$ grew at 4 °C with reservoir solution of 5 m*M* $CaCl_2$, 100 m*M* citrate (pH 5.0), 20% PEG8000. Both used the hanging-drop vapor-diffusion technique. A 2 μl of a solution containing 8.5 mg/ml $rCaM_{1-75}$ or 7.4 mg/ml $rCaM_{1-90}$ (in 50 m*M* HEPES, 100 m*M* KCl, 50 μ*M* EGTA, pH 7.4) were mixed with an equal volume of the reservoir solutions. Crystals were cryoprotected in 50% PEG400 and 50% mineral oil and flash cooled in liquid nitrogen. The final salt concentrations were 100 m*M* total for $rCaM_{1-75}$ (50 m*M* potassium and 50 m*M* sodium) and 50 m*M* (potassium only) for $rCaM_{1-90}$.

6.2. Diffraction data collection and processing

Crystals of $rCaM_{1-75}$ (Fig. 21.6A) diffracted to 1.60 Å resolution. Data for $rCaM_{1-75}$ were collected at the University of Iowa (Carver College of Medicine Protein Crystallography Facility). Crystals of $rCaM_{1-90}$ (Fig. 21.6B) crystals diffracted to 2.03 Å resolution. Data for $rCaM_{1-90}$ were collected at the Advanced Photon Source (APS: Argonne National Laboratory) at beamline 17-ID, Industrial Macromolecular Crystallography Association Collaborative Access Team (IMCA-CAT). All data were processed by using d*TREK(X) from Rigaku/MSC and REFMAC5 (Murshudov *et al.*, 1997); the starting model was 3CLN.pdb. Water was added to the structural model by using Arp/Warp (part of CCP4 suite; Collaborative Computational Project, No. 4, 1994) and was verified in O. Table 21.4 summarizes crystallographic statistics. Note that the unit cell of the crystal of $rCaM_{1-90}$ contained two molecules of $rCaM_{1-90}$.

Table 21.4 Crystallographic data collection and refinement statistics

	$rCaM_{1-75}$	$rCaM_{1-90}$
Data collection statistics		
Cell dimensions (*a*, *b*, *c*) (Å)	34.4, 66.1, 57.8, 90, 90, 90	55.3, 58.9, 67.1, 90, 90, 90
Space group	$C222_1$	$P2_12_12_1$
Resolution (Å)	15.24–1.60	67.12–2.03
Highest resolution shell (Å)	1.64–1.60	2.08–2.03
Completeness (%)	98.6 (96.8)	99.9 (99.9)
Refinement statistics		
Number of protein atoms	575	1368
Number of solvent atoms	110	58
Number of calcium ions	2	4
R–factor	0.1975	0.2397
Free *R*–factor	0.2692	0.2985
Overall weighted *R*-factor	0.2094	0.2428
Average *B*-factors ($Å^2$)	13.91	31.00

Values in parenthesis refer to the highest resolution shell.

6.3. Stoichiometry and biological units

Although we chose to pursue crystallographic studies of rCaM fragments, studies of CaM interacting with peptides representing CaM-binding domains have highlighted the importance of assessing protein size and structure in solution, rather than relying on crystallographic data. We have shown that the complex of CaM_{1-148} bound to the CaM-binding domain of CaN (calcineurin)-β (CaNp) forms exclusively a 1:1 complex, as determined using size-exclusion chromatography, analytical ultracentrifugation and ^{15}N-T_2 relaxation (O'Donnell, 2009). This result contrasted with three crystal structures of CaM_{1-148} bound to CaN-α which is identical in amino acid sequence at all but one position where an Ile/Val substitution occurs. They all displayed a 2:2 binding stoichiometry (Majava and Kursula, 2009; Ye *et al.*, 2006, 2008).

7. Conclusions

Several biophysical techniques were used to assess the effects of Ca^{2+} and KCl on the thermodynamic behavior and structures of a nested series of CaM fragments all containing the core of the N-domain (i.e., helices A–D). For each protein studied, increasing the concentration of KCl decreased the Ca^{2+}-binding affinity of sites I and II (Fig. 21.3). At all levels, $rCaM_{1-75}$ had the most favorable total free energy of calcium binding. All proteins except for $rCaM_{1-90}$ displayed identical changes in total free energy of calcium binding (to lower affinity) in response to increasing KCl (Fig. 21.3F). Unexpectedly, despite their initial separation at low [KCl], $rCaM_{1-75}$ and $rCaM_{1-90}$ displayed remarkably similar Ca^{2+}-affinities at KCl concentrations of 100, 200, and 300 mM KCl. This suggested that higher KCl resulted in release of tertiary constraints in $rCaM_{1-90}$ that otherwise occur in $rCaM_{1-80}$, $rCaM_{1-85}$, and $rCaM_{1-148}$ so that Ca^{2+} binds to sites I and II of $rCaM_{1-90}$ as it does to sites I and II of $rCaM_{1-75}$.

Thermal denaturation studies of apo-CaM (Fig. 21.4) indicated that addition of residues in helix E increased thermal stability and all fragments displayed KCl-dependent increases in thermal stability. The estimated charge of the CaM sequence from 75 to 90 is -2.1 at pH 7.4. Despite their similar Ca^{2+}-binding properties, $rCaM_{1-75}$ and $rCaM_{1-90}$ differed in T_m by ~8 °C at 100 and 300 mM KCl. The fragment with an exceptional response to KCl was $rCaM_{1-85}$ which differed more in T_m and ΔH between no added KCl and 100 mM KCl than any other protein tested. We hypothesize that this may related to the SXXE "helix-capping" sequence in this fragment, and the sequence containing three contiguous glutamate residues. Further residue-specific experiments such as NMR will be required to determine the nature of this stabilization.

Calcium binding decreased the Stokes radius of all CaM fragments studied (Fig. 21.5) as has been observed previously for a subset of these proteins at 100 m*M* KCl (Sorensen and Shea, 1996). For each protein, increasing [KCl] decreased the R_s with the exception of that determined for calcium-saturated proteins at 300 m*M* KCl, where a very slight increase was noted relative to the properties at 200 m*M* KCl. If residues beyond 80 rotated independently from the 4-helix bundle core of the N-domain, it was expected that the Stokes radius would increase incrementally as additional residues of helix E were added. However, fragments $rCaM_{1-85}$ and $rCaM_{1-90}$ were surprisingly similar, especially under apo conditions where they were virtually indistinguishable. Coupled with the results from the Ca^{2+}-binding studies where $rCaM_{1-90}$ behaved more like the smaller $rCaM_{1-75}$ fragment, it is conceivable that a portion of the E-helix may wrap back to make contacts with residues in the 4-helix bundle (helices A–D), thus accounting for a smaller size and release of tertiary constraints. An example of a similar compact tertiary structure for a calcium-binding protein with more than four helices is found in parvalbumin, the original protein to be identified as having EF-hand motifs, and containing two Ca^{2+}-binding sites (Fig. 21.2C).

Because the Ca^{2+}-binding properties of the $rCaM_{1-90}$ fragment demonstrated a unique response to [KCl] at levels $\geq$100 m*M*, and behaved more similarly to $rCaM_{1-75}$ than to either $rCaM_{1-80}$ or $rCaM_{1-85}$ (Fig. 21.3F), we embarked on a crystallographic study to test the hypothesis that it resembled parvalbumin. Although the relative orientation of helices A–D is identical in $rCaM_{1-75}$ (Fig. 21.6A, 3B32.pdb) and $rCaM_{1-90}$ (Fig. 21.6B, 3IFK.pdb), $rCaM_{1-90}$ was found to be a dimer in the unit cell, with helix E extended in an orientation resembling its position in the structure of CaM_{1-148} shown in Fig. 21.1A. This was a surprise, and not consistent with the hydrodynamic radius measured for $rCaM_{1-90}$.

Relative to the other fragments containing sites I and II, a special feature of the sequence of $rCaM_{1-90}$ is the addition of a Phe residue at position 89. In the homodimer observed in the unit cell, F89 of one $rCaM_{1-90}$ molecule is buried in the hydrophobic pocket of a second $rCaM_{1-90}$ molecule. We believe that dimerization may have occurred because of the high concentration of protein in the crystallization drop, and the high affinity of the N-domain cleft of calcium-saturated CaM for an aromatic residue in close proximity (Fig. 21.7A). Upon further analysis, it is apparent that F89 is behaving like polycyclic drugs or like the aromatic groups in target recognition motifs of CaM-regulated proteins (Fig. 21.7B and C). In particular, F89 is interacting with a set of four highly conserved residues in each domain of CaM (F19/92, L32/105, M51/124, M71/144) that form a cavity in which bulky, hydrophobic (usually aromatic) residues of CaM-target sequences are easily accommodated and frequently utilized (Ataman *et al.*, 2007). In the structure of $rCaM_{1-90}$ dimer, the cavity created by the FLMM

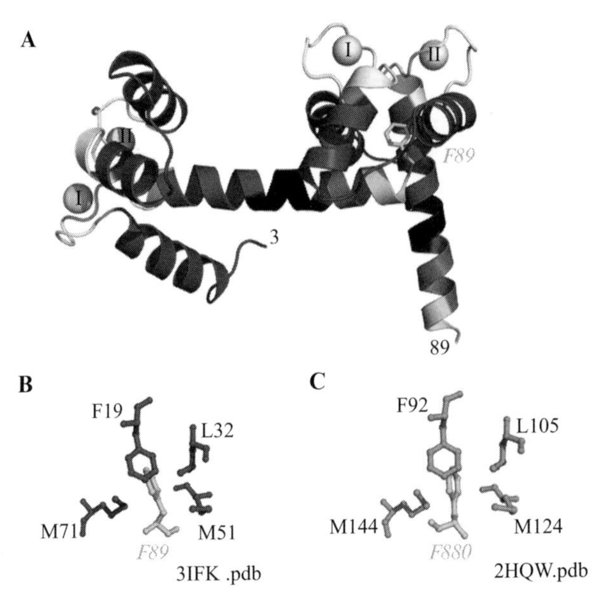

Figure 21.7 *Crystal packing-induced dimerization of* $rCaM_{1-90}$ *(3IFK).* (A) Ca^{2+}-binding sites I and II (yellow), residues 1–75 (blue), 76–80 (black), 81–85 (purple), and 86–90 (green) are shown. (B) Occupancy of the N-domain FLMM pocket (blue) of $rCaM_{1-90}$ by F89 (green). (C) Occupancy of the C-domain FLMM pocket (red) of CaM_{1-148} by F880 (green) of the CaM-binding domain of NR1C0 (2HQW.pdb). (See Color Insert.)

residues of the N-domain houses F89 (Fig. 21.7B). Although interesting, it contradicts the hydrodynamic data. It is likely that this is an anomaly of crystallization where high concentrations of CaM resulted in this interaction. Future experiments will include the use of NMR to explore how interactions between helices D and E of $rCaM_{1-90}$ behave in solution, which the crystal structure does not explain.

ACKNOWLEDGMENTS

These studies were supported by an NIH Training Grant in Biotechnology (T32 GM08365) to R.A.N.; University of Iowa Center for Biocatalysis and Bioprocessing Fellowship and an American Heart Association Predoctoral Fellowship to S. E. O.; a University of Iowa Center for Research by Undergraduates Research Fellowship to J. R. F., and a grant from the National Institutes of Health (RO1 GM 57001) to M. A. S. Orijit Kaar prepared the tray of hanging drops that yielded the crystal of $rCaM_{1-75}$; the Carver College of Medicine Crystallography Core Research Facility, Arthur Arnone, Jeff Kavanough, and Paul Rogers assisted with data collection and interpretation.

REFERENCES

Ackers, G. K. (1970). Analytical gel chromatography of proteins. *Adv. Protein Chem.* **24,** 343–446.

Ataman, Z. A., Gakhar, L., Sorensen, B. R., Hell, J. W., and Shea, M. A. (2007). The NMDA receptor NR1 C1 region bound to calmodulin: Structural insights into functional differences between homologous domains. *Structure* **15,** 1603–1617.

Babu, Y. S., Bugg, C. E., and Cook, W. J. (1988). Structure of calmodulin refined at 2.2 Å resolution. *J. Mol. Biol.* **204,** 191–204.

Beaven, G. H., and Holiday, E. R. (1952). Ultraviolet absorption spectra of proteins and amino acids. *Adv. Protein Chem.* **7,** 319–386.

Carra, J. H., Anderson, E. A., and Privalov, P. L. (1994). Three-state thermodynamic analysis of the denaturation of staphylococcal nuclease mutants. *Biochemistry* **33,** 10842–10850.

Collaborative Computational Project, No. 4 (1994). The CCP4 suite: Programs for protein crystallography. *Acta Crystallogr. D Biol. Crystallogr.* **D50,** 760–763.

Crouch, T. H., and Klee, C. B. (1980). Positive cooperative binding of calcium to bovine brain calmodulin. *Biochemistry* **19,** 3692–3698.

Eftink, M. R., Ionescu, R., Ramsay, G. D., Wong, C., Wu, J. Q., and Maki, A. H. (1996). Thermodynamics of the unfolding and spectroscopic properties of the V66W mutant of *staphylococcal* nuclease and its 1–136 fragment. *Biochemistry* **35,** 8084–8094.

Evans, T. I., and Shea, M. A. (2009). Energetics of calmodulin domain interactions with the calmodulin binding domain of CaMKII. *Proteins* **76,** 47–61.

Faga, L. A., Sorensen, B. R., VanScyoc, W. S., and Shea, M. A. (2003). Basic interdomain boundary residues in calmodulin decrease calcium affinity of sites I and II by stabilizing helix–helix interactions. *Proteins Struct. Funct. Genetics* **50,** 381–391.

Houdusse, A., Silver, M., and Cohen, C. (1996). A model of Ca(2+)-free calmodulin binding to unconventional myosins reveals how calmodulin acts as a regulatory switch. *Structure* **4,** 1475–1490.

Ikura, M., Kay, L. E., Krinks, M., and Bax, A. (1991). Triple-resonance multidimensional NMR study of calmodulin complexed with the binding domain of skeletal muscle myosin light-chain kinase: Indication of a conformational change in the central helix. *Biochemistry* **30,** 5498–5504.

Ikura, M., Barbato, G., Klee, C. B., and Bax, A. (1992). Solution structure of calmodulin and its complex with a myosin light chain kinase fragment. *Cell Calcium* **13,** 391–400.

Johnson, M. L., and Frasier, S. G. (1985). Nonlinear least-squares analysis. *Methods Enzymol.* **117,** 301–342.

Kuboniwa, H., Tjandra, N., Grzesiek, S., Ren, H., Klee, C. B., and Bax, A. (1995). Solution structure of calcium-free calmodulin. *Nat. Struct. Biol.* **2,** 768–776.

Majava, V., and Kursula, P. (2009). Domain swapping and different oligomeric states for the complex between calmodulin and the calmodulin-binding domain of calcineurin a. *PLoS One* **4,** e5402.

Meador, W. E., Means, A. R., and Quiocho, F. A. (1993). Modulation of calmodulin plasticity in molecular recognition on the basis of X-ray structures. *Science* **262,** 1718–1721.

Murshudov, G. N., Vagin, A. A., and Dodson, E. J. (1997). Refinement of macromolecular structures by the maximum-likelihood method. *Acta Crystallogr. D Biol. Crystallogr.* **53,** 240–255.

O'Donnell, S. E. (2009). Recognition of Calcineurin by the Domains of Calmodulin: Structural and Thermodynamic Determinants, Ph.D. Dissertation. University of Iowa, Iowa.

Pedigo, S., and Shea, M. A. (1995a). Discontinuous equilibrium titrations of cooperative calcium binding to calmodulin monitored by 1-D ^{1}H-nuclear magnetic resonance spectroscopy. *Biochemistry* **34,** 10676–10689.

Pedigo, S., and Shea, M. A. (1995b). Quantitative endoproteinase GluC footprinting of cooperative Ca^{2+} binding to calmodulin: Proteolytic susceptibility of E31 and E87 indicates interdomain interactions. *Biochemistry* **34,** 1179–1196.

Putkey, J. A., Slaughter, G. R., and Means, A. R. (1985). Bacterial expression and characterization of proteins derived from the chicken calmodulin cDNA and a calmodulin processed gene. *J. Biol. Chem.* **260,** 4704–4712.

Santoro, M. M., and Bolen, D. W. (1988). Unfolding free energy changes determined by the linear extrapolation method. 1. Unfolding of phenylmethanesulfonyl alpha-chymotrypsin using different denaturants. *Biochemistry* **27,** 8063–8068.

Shea, M. A., Sorensen, B. R., Pedigo, S., and Verhoeven, A. (2000). Proteolytic footprinting titrations for estimating ligand-binding constants and detecting pathways of conformational switching of calmodulin. *Methods Enzymol.* **323,** 254–301.

Sorensen, B. R., and Shea, M. A. (1996). Calcium binding decreases the stokes radius of calmodulin and mutants R74A, R90A, and R90G. *Biophys. J.* **71,** 3407–3420.

Sorensen, B. R., and Shea, M. A. (1998). Interactions between domains of apo calmodulin alter calcium binding and stability. *Biochemistry* **37,** 4244–4253.

Sorensen, B. R., Faga, L. A., Hultman, R., and Shea, M. A. (2002). Interdomain linker increases thermostability and decreases calcium affinity of calmodulin N-domain. *Biochemistry* **41,** 15–20.

Sun, H., Yin, D., Coffeen, L. A., Shea, M. A., and Squier, T. C. (2001). Mutation of Tyr138 disrupts the structural coupling between the opposing domains in vertebrate calmodulin. *Biochemistry* **40,** 9605–9617.

VanScyoc, W. S., and Shea, M. A. (2001). Phenylalanine fluorescence studies of calcium binding to N-domain fragments of *Paramecium* calmodulin mutants show increased calcium affinity correlates with increased disorder. *Protein Sci.* **10,** 1758–1768.

Ye, Q., Li, X., Wong, A., Wei, Q., and Jia, Z. (2006). Structure of calmodulin bound to a calcineurin peptide: A new way of making an old binding mode. *Biochemistry* **45,** 738–745.

Ye, Q., Wang, H., Zheng, J., Wei, Q., and Jia, Z. (2008). The complex structure of calmodulin bound to a calcineurin peptide. *Proteins* **73,** 19–27.

Zhang, M., Tanaka, T., and Ikura, M. (1995). Calcium-induced conformational transition revealed by the solution structure of apo calmodulin. *Nat. Struct. Biol.* **2,** 758–767.

CHAPTER TWENTY-TWO

Use of Pressure Perturbation Calorimetry to Characterize the Volumetric Properties of Proteins

Katrina L. Schweiker*,† *and* George I. Makhatadze*,†

Contents

Abstract

Pressure perturbation calorimetry (PPC) is a new technique that makes possible to study the volumetric changes that occur upon protein unfolding. Here, we summarize the thermodynamic foundation of the method and introduce a two-state model for the analysis of the unfolding data monitored by PPC. Several examples of data analysis illustrating potential pitfalls and solutions are discussed.

* Center for Biotechnology and Interdisciplinary Studies, Rensselaer Polytechnic Institute, Troy, New York, USA
† Department of Biochemistry and Molecular Biology, Penn State University College of Medicine, Hershey, Pennsylvania, USA

Methods in Enzymology, Volume 466
ISSN 0076-6879, DOI: 10.1016/S0076-6879(09)66022-5

1. Introduction

Understanding the forces that govern protein folding and stability is one of the central problems of protein biophysics. Through the use of thermal denaturation or chemical (i.e., urea or GuHCl) denaturation, much has been learned about how intramolecular interactions such as charge–charge interactions, hydrogen bonding, van der Waals interactions, or the hydrophobic effect, contribute to protein stability. However, there are still some aspects regarding the thermodynamic characterization of proteins that have yet to be extensively studied, such as the response of proteins to pressure and the volumetric changes that occur upon unfolding. A more complete understanding of how intramolecular interactions govern protein folding and stability can be obtained only after these areas are as well studied as other biophysical responses, such as thermal denaturation.

The study of the transfer free energies of model compounds from liquid hydrocarbons to water has been very successful in helping to understand the relative contributions of intramolecular interactions, such as the hydrophobic effect, to thermal and chemical denaturation (i.e., urea- or GuHCl-induced) of proteins (see, e.g., Baldwin, 1986; Kauzmann, 1959; Khechinashvili, 1990; Livingstone *et al.*, 1991; Makhatadze and Privalov, 1992, 1993, 1995; Ooi and Oobatake, 1988; Pace *et al.*, 1996; Privalov and Makhatadze, 1990, 1993; Tanford, 1968). Unfortunately, these studies largely fail to explain the volumetric changes that should occur upon isothermal pressure-induced denaturation (Kauzmann, 1987). Based on model compound data, the solvation of polar groups and the transfer of nonpolar groups from a hydrophobic to an aqueous environment were both expected to contribute negatively to the changes in the specific volume of a protein upon unfolding (Brandts, 1969; Gross and Jaenicke, 1994; Kauzmann, 1959; Mozhaev *et al.*, 1996). The change in the intrinsic void volume (volume of cavities) of proteins was also expected to have a negative contribution to the volumetric changes upon unfolding. As a result of these measurements, it was believed that the unfolding of proteins should be accompanied by a large decrease in their specific volumes. However, in most cases, only small decreases, or even small increases, in the partial specific volume of proteins upon unfolding were observed (Chalikian and Breslauer, 1996; Gross and Jaenicke, 1994; Lin *et al.*, 2002; Mitra *et al.*, 2008; Royer, 2002).

Chalikian and Breslauer (1996) were the first to try to resolve this issue by introducing the concept of thermal volume, such that the specific volume of a protein (V_{Pr}) is actually made up of three components:

$$V_{Pr} = v_{int} + v_{hyd} + v_t, \tag{22.1}$$

where ν_{int} is the intrinsic volume of the protein, which is a sum of the van der Waals volumes of all atoms in the protein and the internal cavities and voids; ν_{hyd} is the volume change in the solvent due to the hydration of the solvent accessible surface of the protein; and ν_t is the thermal volume that results from the thermally induced molecular vibrations of the protein and solvent. The effect of the thermal volume is to expand the solvent away from the surface of the protein, such that solvent-free volume element forms around the protein. It is possible then that the negative contribution of ν_{hyd} for hydrophobic residues measured in the model compound studies is simply a reflection of the lower thermal volume of water compared to nonpolar solvent (Chalikian and Breslauer, 1996). In addition, the protein interior is most likely denser and more heterogeneous than a nonpolar solvent, providing another explanation for why the model compound studies were unable to accurately describe the volumetric changes in proteins upon unfolding (Chalikian and Breslauer, 1996; Royer, 2002). Chalikian and Breslauer (1996) also demonstrated how the thermal volume of the solvent can compensate for the negative changes in ν_{int} and ν_{hyd} in such a way that the overall V_{Pr} is only slightly negative. Furthermore, if these three contributions to specific volume respond differently to changes in temperature or pressure, then it is possible that the protein could react such that V_{Pr} can also be positive (Chalikian and Breslauer, 1996). Indeed, such behavior has been observed for a number of proteins (Chalikian and Breslauer, 1996; Garcia-Moreno and Fitch, 2004; Lin *et al.*, 2002; Rosgen and Hinz, 2000).

The idea that V_{Pr} can change with response to pressure, as well as temperature, led to the development of methods to study various aspects of proteins at high pressure, which makes it possible to stabilize conformational states that are not usually populated enough to be studied under standard conditions (i.e., atmospheric pressure, ~14.7 psi). These methods have been successfully used to study the structures of kinetic intermediates and protein aggregation pathways (for a review, see Silva *et al.*, 2001). However, one of the aspects of the pressure–volume relationship that is difficult to measure using the high-pressure techniques of densitometry, FITR, or SAXS is the thermal expansivity coefficient, $\alpha(T)$. This parameter is the temperature derivative of the $V_{Pr}(T)$ function, so methods that measure only the volume of a protein (usually indirectly) will inherently have large errors in $\alpha(T)$

Pressure perturbation calorimetry (PPC) is a relatively new experimental method that overcomes the problems associated with indirect measurements of α (Lin *et al.*, 2002). In a PPC experiment, α is measured directly as the difference between the heats produced by a calorimetric cell containing dilute protein solution and that of a cell containing only buffer as they are subjected to rapid changes in pressure (~80 psi) under isothermal conditions. By performing PPC at a series of different temperatures, it is possible to measure α as a function of temperature (Lin *et al.*, 2002). A few PPC experiments on several different proteins have demonstrated how this valuable biophysical technique can be used, not only to measure α and $\Delta V_{Pr}/V_{Pr}$, but also to give information about the interactions between the solvent and proteins in their native and unfolded states.

The first PPC studies on proteins focused on developing a framework for understanding pressure-induced denaturation. Studying the pressure responses of small molecules (Lin *et al.*, 2002; Mitra *et al.*, 2006), single amino acids (Lin *et al.*, 2002; Mitra *et al.*, 2006), and tripeptides (Mitra *et al.*, 2006) in water showed how sensitive $\alpha(T)$ is to the hydrophobicity of the solute. For example, the polar amino acids tend to have a large, positive value of α at lower temperatures, which decreases as a function of T, eventually leveling off at higher temperatures. In contrast, the hydrophobic amino acids tend to have large, negative values of α at low T, which increase as a function of temperature, and also level off at high T (<100 °C).

PPC experiments have also been performed on several model protein systems (Lin *et al.*, 2002; Mitra *et al.*, 2006, 2008; Ravindra and Winter, 2004). In addition to measuring the volumetric changes of the proteins upon unfolding, these studies provided insight into what defines the expansivity, $\alpha(T)$ of the native and unfolded states of proteins. The general observations from these studies are that proteins with large numbers of hydrophilic residues on the surface have larger $\alpha(T)$ values and a steeper temperature dependence of $\alpha(T)$ at lower temperatures (Mitra *et al.*, 2006, 2008). It also appears that the absolute value and temperature dependence of $\alpha(T)$, and resulting ΔV, are highly dependent on the nature of the cosolvent (Lin *et al.*, 2002; Mitra *et al.*, 2006; Ravindra and Winter, 2004). For example, in the presence of denaturant, the absolute value of $\alpha(T)$ at low temperatures is smaller than in water, and the temperature dependence is shallower. The measured α also changes sign under these conditions (Lin *et al.*, 2002).

It has been discussed that one of the advantages of PPC compared to other methods (i.e., Rosgen and Hinz, 2000) is that it does not rely on the validity of a two-state model of unfolding for interpreting the data, and as such, can be used to measure volumetric changes upon unfolding in a model-independent manner (Lin *et al.*, 2002). In the current analysis of PPC data, the user defines the beginning and the end of the transition, and a progress baseline is extrapolated between them using a polynomial function. The volumetric change upon unfolding ($\Delta V/V$) is then determined by calculating the area between the baseline and the experimental $\alpha(T)$ profile. While a model-independent method might produce similar results to those obtained by other high-pressure methods that use a two-state model of unfolding to analyze data (Lin *et al.*, 2002; Rosgen and Hinz, 2000), there are instances, such as broad unfolding transitions or cold denaturation, where the beginning and the end of transition are impossible to define given the experimental data. Furthermore, the model-independent method could make it difficult to reproduce results for proteins with small $\Delta V/V$, and therefore more sensitive to the definition of the baselines. In this chapter, a two-state model for analyzing PPC experiments is described, and it is shown that this approach can alleviate many the complications described above if the PPC results are analyzed in conjunction with those from a DSC experiment. By analyzing PPC data in the context of a two-state model of unfolding, it becomes possible to analyze effects in cases where baselines are difficult to define, and to directly fit the PPC data to get a more complete thermodynamic description of unfolding, including the ability to construct P–T diagrams.

2. Determination of the Coefficient of Thermal Expansion (α_{PR}) Using PPC

PPC experiments are performed using a DSC instrument (e.g., MicroCal VP-DSC) operating in isothermal mode. Two cells, sample and reference, are equilibrated at a given temperature using heaters 1 and 2 (Fig. 22.1). After equilibration is achieved, a small amount of pressure is applied to both cells. The change in pressure leads to a change in the temperature difference between the two cells, ΔT, which is detected by a computer. Additional power, ΔQ, is applied to one of the cells to re-equilibrate the temperatures. Once equilibrium is reestablished, the pressure is released, and

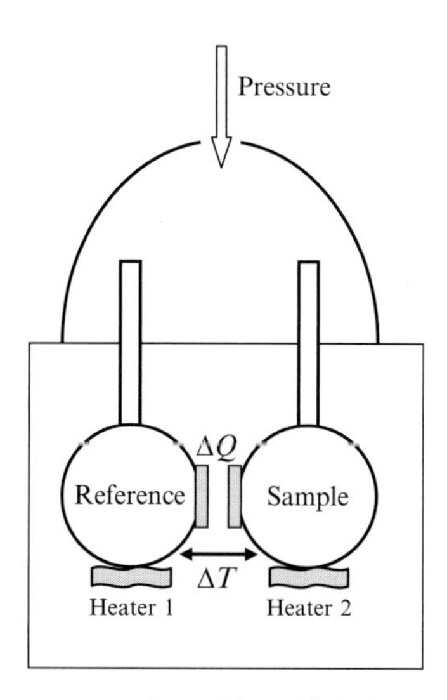

Figure 22.1 Schematic representation of the cell design used in a PPC experiment.

this also leads to a change in the temperature difference between cells that are again re-equilibrated by applying additional power. This process can be repeated several times at a given temperature. The temperature is then raised and the procedure is repeated. As a result of this method, a set of at least two measurements of ΔQ at each temperature is obtained. The ΔQ between the sample and reference cells is directly proportional to the linear expansion coefficient of the solute.

To calculate the coefficient of thermal expansion (α_{Pr}) for a protein in solution, the expansion effects of water and buffer salts must be also taken into account. Therefore, control runs of water/water, buffer/water, and buffer/buffer are performed before each experiment. The temperature range for the control experiments is 5–110 °C, with data collected every 5 °C. The protein/buffer experiments are also performed in the temperature range of 5–110 °C, with data collected every 5 °C or every 2 °C in the transition region, which is determined by corresponding DSC experiments.

Although the foundations for analyzing data from a PPC experiment have already been described in great detail (Lin *et al.*, 2002), a brief overview of the thermodynamic relationship between changes in pressure, volume, and heat will help make the analysis presented here more clear. A PPC experiment measures small changes in the heat absorbed/released (ΔQ) by the solution in the calorimetric cell as small perturbations in pressure ($\Delta P \approx 0.55$ MPa) are applied. Starting from the second law of thermodynamics, and differentiating with respect to pressure, this change in heat can be represented as (Lin *et al.*, 2002)

$$\mathrm{d}Q = \mathrm{d}S\,T \Rightarrow \left(\frac{\partial Q}{\partial P}\right)_T = T\left(\frac{\partial S}{\partial P}\right)_T \tag{22.2}$$

where $\mathrm{d}S$ is the change in entropy at temperature, T, for a reversible change in heat, $\mathrm{d}Q$. The Maxwell relationship $(\partial S/\partial P)_T = -(\partial V/\partial T)_P$ makes it possible to express the pressure-induced changes in heat to the change in the volume (V) of the system:

$$\left(\frac{\partial Q}{\partial P}\right)_T = -T\left(\frac{\partial V}{\partial T}\right)_P = -TV\alpha, \tag{22.3}$$

where α is the coefficient of thermal expansion:

$$\alpha = \frac{1}{V}\left(\frac{\partial V}{\partial T}\right)_P. \tag{22.4}$$

Integrating Eq. (22.3) yields an expression for Q for the reference cell:

$$Q_{\text{reference}} = -TV_{\text{cell}}\alpha_S\Delta P. \tag{22.5}$$

For the sample cell, with a solution containing m_{Pr} grams of protein, and m_S grams of solvent, the total volume of the cell, V_{cell}, can be represented by

$$V_{cell} = m_{Pr} V_{Pr} + m_S V_S, \tag{22.6}$$

where V_{Pr} is the partial specific volume of protein in the cell and V_S is the specific volume of buffer. Differentiating Eq. (22.6) with respect to temperature and substituting into Eq. (22.3) yields

$$\left(\frac{\partial Q_{sample}}{\partial P} \right)_T = -T(m_{Pr} V_{Pr} \alpha_{Pr} + m_S V_S \alpha_S). \tag{22.7}$$

By integrating Eq. (22.7) over a small pressure range, we can obtain an expression for the thermal expansivity coefficient of the protein, α_{Pr}, at temperature, T, in terms of the thermal expansion of the buffer, α_S, the change in the heat ($\Delta Q_{Pr/buf} = Q_{sample} - Q_{reference}$) and the change in the pressure (ΔP) of the system:

$$\alpha_{Pr} = \alpha_S - \frac{\Delta Q_{Pr/buf}}{T \Delta P m_{Pr} V_{Pr}}. \tag{22.8}$$

The thermal expansion coefficient of the buffer is determined from the buffer/water scans in a similar manner:

$$\alpha_S = \alpha_{H_2O} - \frac{\Delta Q_{buf/H_2O}}{T \Delta P V_{cell}}, \tag{22.9}$$

where V_{cell} is the volume of the calorimetric cell. The water/water scans, mentioned above, are used to determine the value of the thermal expansion coefficient of water (α_{H_2O}) and also to account for the difference in volumes between the two cells:

$$\alpha_{Pr} = \alpha_{H_2O} - \frac{\Delta Q_{buf/H_2O}}{T \Delta P V_{cell}} - \frac{\Delta Q_{Pr/buf}}{T \Delta P m_{Pr} V_{Pr}}. \tag{22.10}$$

In this equation, all parameters but V_{Pr} are determined from PPC experiments. The values of V_{Pr} are usually calculated using an additivity assumption (see, e.g., Lee *et al.*, 2008; Makhatadze *et al.*, 1990). The errors of calculating V_{Pr} using additivity are on the order of 5%, but its variation has only a very small (<1% effect on the value of α_{Pr}). The raw data from the PPC experiments are usually processed using the scripts in the Origin PPC data analysis software supplied by MicroCal (Northampton, MA) to obtain values for α_{Pr} as a function of temperature.

3. Sample Preparation

Experiments should be performed in buffers with an intrinsic α as close as possible to that of water. Therefore, glycine- and phosphate-based buffers are better choices than acetate or Tris-based buffers. Other requirements of sample preparation should follow the same general guidelines as for DSC experiments (Lopez and Makhatadze, 2002; Makhatadze, 1998; Streicher and Makhatadze, 2007). Protein samples need to be extensively dialyzed against several changes of desired buffer. All insoluble material should be removed by centrifugation at $\sim$13,000*g*. Knowledge of exact protein concentration is very important and can be measured spectrophotometrically using extinction coefficients either experimentally determined or calculated from the amino acid composition (Pace *et al.*, 1995). For some proteins, correction for light scattering must be made (Lopez and Makhatadze, 2002). Finally, a PPC experiment is several times slower than a DSC experiment. Thus, the protein reversibility should be established from independent experiments that mimic the time that a protein is exposed to high temperatures during the PPC run. The reversibility of proteins under these conditions is a major limitation of PPC experiments.

Figure 22.2 shows the data for PPC experiments performed on ribonuclease A (RNase A) at four different concentrations. This figure clearly demonstrates that in the concentration range of 1–4 mg/ml, protein concentration has very little effect on the experimental results. It has been previously argued that increasing concentration will have an effect on the shape of the temperature dependence of the α_{Pr} (and potentially the measurement of $\Delta V_{Pr}/V_{Pr}$) because increased intermolecular interactions will

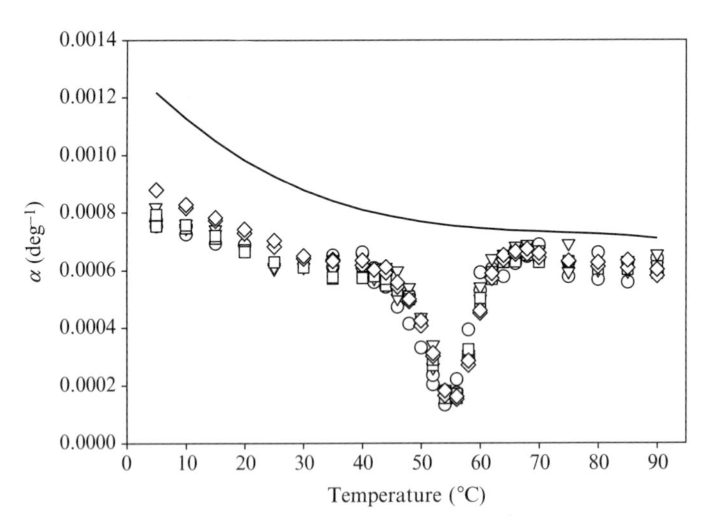

Figure 22.2 PPC profiles for RNase A at different concentrations: ○—1.3 mg/ml ($\Delta V_{Pr}/V_{Pr} = 4.8 \times 10^{-3}$), □—1.6 mg/ml ($\Delta V_{Pr}/V_{Pr} = 4.9 \times 10^{-3}$), ◇—4.0 mg/ml ($\Delta V_{Pr}/V_{Pr} = 4.6 \times 10^{-3}$), and ∇—4.8 mg/ml ($\Delta V_{Pr}/V_{Pr} = 4.4 \times 10^{-3}$). All experiments were performed in 10 m*M* glycine buffer, pH 3.2. Solid line, $\alpha(T)$ for the unfolded RNAse A calculated from the amino acid composition as described in the text (see Eqs. 22.24, 22.28, and 22.29).

affect the hydration properties of the native state (Mitra *et al.*, 2006). However, these effects were only observed for high protein concentrations. It seems that in relatively dilute protein solutions (the protein concentration <5 mg/ml), there should be very little dependence of $\alpha_{exp}(T)$ or $\Delta V_{Pr}/V_{Pr}$ on protein concentration. This is an important observation because it means that it is possible to perform accurate PPC experiments with lower protein concentrations than has been previously used or recommended.

4. Derivation of a Two-State Model for Analysis of PPC Data

The concepts used to analyze the data from PPC experiments are analogous in many ways to those used to analyze the experimental data from DSC. In other words, the relationship of $\alpha(T)$ to $\Delta V_{Pr}/V_{Pr}$ is akin to the relationship between $C_p(T)$ and ΔH. Figure 22.3 highlights some of the similarities and differences between PPC and DSC experiments, and

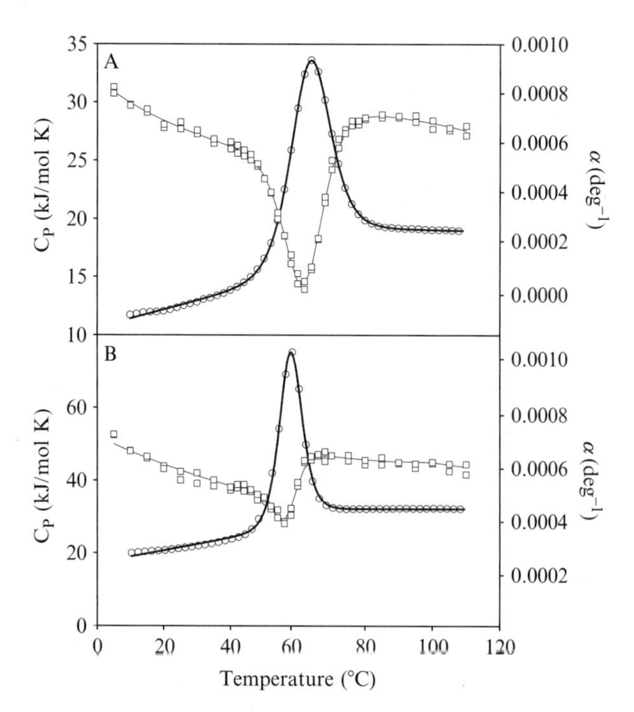

Figure 22.3 Comparison of thermal denaturation curves obtained from DSC and PPC experiments for ubiquitin in 50 m*M* glycine buffer, pH 2.6 (A) and lysozyme in 50 m*M* glycine buffer, pH 2.2 (B). In both panels, the symbols represent the experimental data (○—DSC data, shown every 5 °C for clarity; □—PPC data). The solid lines represent the fit of the experimental data to a two-state model of unfolding.

demonstrates some of the considerations that need to be addressed in the analysis of the experimental data. In this figure, the experimental PPC and DSC data for ubiquitin and lysozyme are shown. For both proteins, the $\alpha(T)$ profiles suggest negative volumetric changes, although the magnitudes of $\Delta V_{Pr}/V_{Pr}$ are markedly different. Notice that the peak of the $C_p(T)$ profile measured by DSC for ubiquitin and lysozyme occurs at a similar temperature as the minimum of the $\alpha(T)$ profile measured by PPC, suggesting that the T_m values measured by each technique are, as expected, similar. The area under the $C_p(T)$ profile is the enthalpy of unfolding (ΔH), while the area under the $\alpha(T)$ profile represents the change in volume upon unfolding ($\Delta V_{Pr}/V_{Pr}$).

Since the results of PPC experiments are analogous to those obtained from DSC, and since the analysis of DSC data is well established, it should be relatively straightforward to analyze the PPC data in the context of a two-state model of unfolding, which has two major advantages. First, it provides a standard method to analyze experimental results, which could make it possible to decrease the current errors in the measurement of $\Delta V_{Pr}/V_{Pr}$ that most likely stem from how the experimental baselines are defined (see Lin *et al.*, 2002; Rosgen and Hinz, 2000). Since the temperature-dependent behaviors of the native and unfolded state baselines are vital for determining the area under the $\alpha(T)$ profile, and hence the volume changes upon denaturation, it is important to develop an analysis that will decrease the errors associated with user-defined baselines. The second advantage of developing a two-state model for analyzing experimental data is that such a model will make it possible to fit data in circumstances where current methods fail, such as small changes in $\Delta V_{Pr}/V_{Pr}$, broad unfolding transitions, or the presence of cold denaturation.

The following is a derivation of analytical form for PPC data fit to a two-state transition. The experimental $\alpha_{exp}(T)$ profile consists of two terms:

$$\alpha_P^{exp}(T) = \alpha_P^{progress}(T) + \alpha_P^{excess}(T). \tag{22.11}$$

The $\alpha_P^{progress}(T)$ (Fig. 22.4B) is defined by the fraction of native, $F_N(T)$, and unfolded, $F_U(T)$, protein as

$$\alpha_P^{progress}(T) = F_N(T)\alpha_N(T) + F_U(T)\alpha_U(T), \tag{22.12}$$

where $\alpha_N(T)$ and $\alpha_U(T)$ are the temperature dependencies of the expansion coefficient for the native and unfolded proteins, respectively. Subtracting $\alpha_P^{progress}(T)$ from $\alpha_P^{exp}(T)$ gives the $\alpha_P^{excess}(T)$ profile (Fig. 22.4C). The area under this curve is equal to $\Delta V_{Pr}/V_{Pr}$, and we will now derive the relationship between $\alpha_P^{excess}(T)$ and $\Delta V_{Pr}/V_{Pr}$ for a protein that undergoes two-state unfolding.

For a two-state system, the Gibbs free energy (ΔG) of unfolding is equal to

$$\Delta G(T) = -RT\ln(K_{eq}(T)) = -RT\ln\left(\frac{F_U(T)}{F_N(T)}\right), \tag{22.13}$$

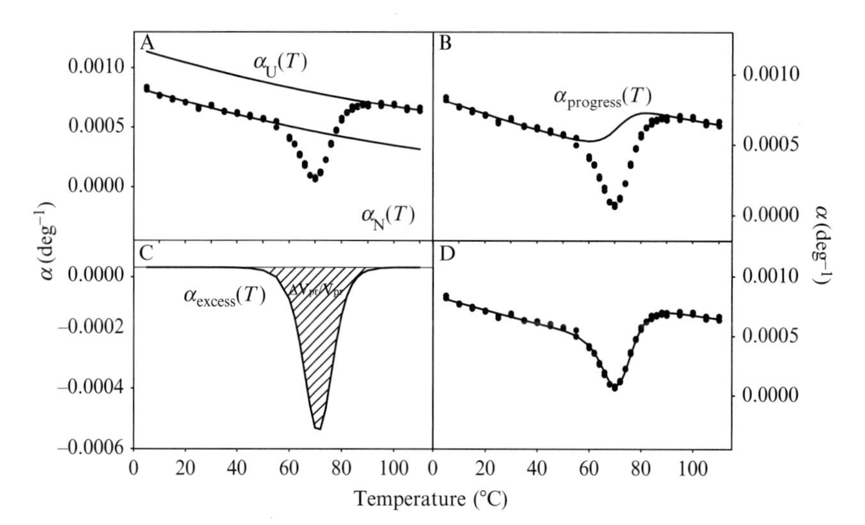

Figure 22.4 Example of PPC data fit to a two-state model of unfolding. In all parts of the figure, the symbols represent the experimental data. (A) Native ($\alpha_N(T)$) and unfolded ($\alpha_U(T)$) baselines. (B) Definition of the $\alpha_{progress}(T)$. (C) Relationship between $\alpha_{excess}(T)$ and $\Delta V_{Pr}/V_{Pr}$. (D) The fit of the experimental data to the two-state model of unfolding using Eq. (22.22). Note that all parameters are fitted simultaneously.

where R is the universal gas constant, $K_{eq}(T)$ is the unfolding equilibrium constant, $F_N(T)$ is the fraction of folded protein, and $F_U(T)$ is the fraction of unfolded protein in the population. The Gibbs free energy can also be related to the changes in enthalpy (ΔH) and entropy (ΔS) upon unfolding via the Gibbs–Helmholtz relationship:

$$\Delta G(T) = \Delta H(T) - T\Delta S(T), \tag{22.14}$$

where the temperature dependencies of $\Delta H(T)$ and $\Delta S(T)$ are defined by the heat capacity change upon unfolding, ΔC_p, which is assumed to be temperature independent:

$$\Delta H(T) = \Delta H(T_m) + \Delta C_p(T - T_m), \tag{22.15}$$

$$\Delta S(T) = \Delta H(T_m)/T_m + \Delta C_p \ln(T/T_m). \tag{22.16}$$

Knowing that $K_{eq}(T) = F_U(T)/F_N(T)$ and that $F_U(T) + F_N(T) = 1$, we can solve for $F_U(T)$:

$$F_U(T) = \frac{\exp\left(\frac{-\Delta G(T)}{RT}\right)}{1 + \exp\left(\frac{-\Delta G(T)}{RT}\right)} = \frac{\exp\left(\frac{-\Delta H(T)}{RT}\right)\exp\left(\frac{\Delta S(T)}{R}\right)}{1 + \exp\left(\frac{-\Delta H(T)}{RT}\right)\exp\left(\frac{\Delta S(T)}{R}\right)}. \tag{22.17}$$

If we assume that ΔV_{Pr} is temperature independent, then $\alpha_{\mathrm{P}}^{\mathrm{excess}}(T)$ is related to the fraction of unfolded protein in the sample (Fig. 22.4C), in such a way that Eq. (22.3) becomes

$$\alpha_{\mathrm{P}}^{\mathrm{excess}}(T) = \frac{1}{V_{\mathrm{P}}}\frac{\partial}{\partial T}(F_{\mathrm{U}}(T)\Delta V_{\mathrm{P}}) = \frac{\Delta V_{\mathrm{Pr}}}{V_{\mathrm{Pr}}}\frac{\partial}{\partial T}(F_{\mathrm{U}}(T)). \tag{22.18}$$

To obtain the complete relationship between $\alpha_{\mathrm{P}}^{\mathrm{excess}}(T)$ and $\Delta V_{\mathrm{Pr}}/V_{\mathrm{Pr}}$, we take the derivative of Eq. (22.18) with respect to temperature:

$$\frac{\partial}{\partial T}(F_{\mathrm{U}}(T)) = \frac{\partial}{\partial T}\left(\frac{\exp\left(\frac{-\Delta H(T)}{RT}\right)\exp\left(\frac{\Delta S(T)}{R}\right)}{1+\exp\left(\frac{-\Delta H(T)}{RT}\right)\exp\left(\frac{\Delta S(T)}{R}\right)}\right)$$
$$= \frac{\exp\left(\frac{\Delta S(T)}{R}\right)\exp\left(\frac{-\Delta H(T)}{RT}\right)\frac{\Delta H(T)}{RT^2}}{\left(1+\exp\left(\frac{\Delta S(T)}{R}\right)\exp\left(\frac{-\Delta H(T)}{RT}\right)\right)^2}, \tag{22.19}$$

$$\frac{\partial}{\partial T}(F_{\mathrm{U}}(T)) = \frac{\exp\left(\frac{-\Delta G(T)}{RT}\right)\frac{\Delta H(T)}{RT^2}}{\left(1+\exp\left(\frac{-\Delta G(T)}{RT}\right)\right)^2} = \frac{K_{\mathrm{eq}}(T)}{(1+K_{\mathrm{eq}}(T))^2}\frac{\Delta H(T)}{RT^2}. \tag{22.20}$$

Substituting Eq. (22.20) into Eq. (22.18) gives the relationship between $\alpha_{\mathrm{P}}^{\mathrm{excess}}(T)$ and $\Delta V_{\mathrm{Pr}}/V_{\mathrm{Pr}}$ (Heerklotz, 2007):

$$\alpha_{\mathrm{P}}^{\mathrm{excess}}(T) = \frac{K_{\mathrm{eq}}(T)}{(1+K_{\mathrm{eq}}(T))^2}\frac{\Delta H(T)}{RT^2}\frac{\Delta V_{\mathrm{Pr}}}{V_{\mathrm{Pr}}}. \tag{22.21}$$

Finally, by incorporating Eqs. (22.12) and (22.21) into Eq. (22.11), we obtain the following representation for $\alpha_{\mathrm{P}}^{\mathrm{exp}}(T)$, which fits the PPC data to a two-state model of unfolding (Fig. 22.4D):

$$\alpha_{\mathrm{P}}^{\mathrm{exp}}(T) = F_{\mathrm{N}}(T)\alpha_{\mathrm{N}}(T) + F_{\mathrm{U}}(T)\alpha_{\mathrm{U}}(T) + \frac{K_{\mathrm{eq}}(T)}{(1+K_{\mathrm{eq}}(T))^2}\frac{\Delta H(T)}{RT^2}\frac{\Delta V_{\mathrm{Pr}}}{V_{\mathrm{Pr}}}. \tag{22.22}$$

From Eq. (22.22), we can see that placing PPC data in the context of a two-state model of unfolding makes it possible to directly fit the experimental data for $\Delta V_{\mathrm{Pr}}/V_{\mathrm{Pr}}$.

5. Practical Considerations

5.1. Absolute values of α for native and unfolded proteins and their temperature dependencies

The α_{Pr} values for the native state are strongly temperature dependent (assuming a linear dependence the slope of $d\alpha_{Pr}(T)/dT \approx (-1.0 \pm 0.5) \times 10^{-5}\,\text{deg}^{-2}$; Ravindra and Winter, 2004). There is significant variability in the α_{Pr} (20 °C) values for different proteins. Experimentally measured values for a number of globular proteins are $(7.0 \pm 0.2) \times 10^{-4}\,\text{deg}^{-1}$ for ribonuclease A (Ravindra and Winter, 2004; Schweiker *et al.*, 2009), $(8.5 \pm 0.2) \times 10^{-4}\,\text{deg}^{-1}$ for staphylococcal nuclease (Ravindra and Winter, 2004), $(6.0 \pm 0.2) \times 10^{-4}\,\text{deg}^{-1}$ for lysozyme (Schweiker *et al.*, 2009), $(6.2 \pm 0.3) \times 10^{-4}\,\text{deg}^{-1}$ for cytochrome *c* (Schweiker *et al.*, 2009), $(7.4 \pm 0.2) \times 10^{-4}\,\text{deg}^{-1}$ for eglin c (Schweiker *et al.*, 2009), and $(7.0 \pm 0.2) \times 10^{-4}\,\text{deg}^{-1}$ for ubiquitin (Schweiker *et al.*, 2009). The values of α for the unfolded state are also for different proteins (see, e.g., Fig. 22.3).

Overall, the dependence of α on temperature for both native and unfolded states should not be linear. This follows from the experimental data (see, e.g., Ravindra and Winter, 2004; Schweiker *et al.*, 2009 and also Figs. 22.2 and 22.3) as well from the data on model compounds (Lin *et al.*, 2002; Mitra *et al.*, 2006). The model compound data of Lin *et al.* (2002) are particularly complete and allows estimates of the expected shape of the α for unfolded state of proteins based on their amino acid composition. The major assumption here is that individual amino acids behave independently in the unfolded polypeptide chain.

To model the unfolded state baselines based on the amino acid composition of proteins, it is important to note that there is no evidence that molar expansivity coefficient is additive (i.e., $\alpha_{Pr} = \sum \alpha_i$.). It is known, however, that to a rather good approximation, the partial volume of a protein can be described by the sum of the partial volumes of amino acids (i.e., $V_{Pr} = \sum v_i$) (Makhatadze *et al.*, 1990). This additivity of partial molar volumes can be used to calculate α from the amino acid composition. Indeed, if $\alpha_{Pr} = (1/V_{Pr})(\partial V_{Pr}/\partial T)$, and the partial volume of the protein is described by the sum of the partial volumes of its amino acids (i.e., $V_{Pr} = \sum v_i$) (Makhatadze *et al.*, 1990), then

$$\begin{aligned}\alpha_{Pr} &= \frac{1}{\sum v_i}\left(\frac{\partial \sum v_i}{\partial T}\right) = \frac{1}{\sum v_i}\left(\frac{\partial v_1}{\partial T} + \frac{\partial v_2}{\partial T} + \ldots\right) \\ &= \frac{1}{\sum v_i}\left(\frac{v_1}{v_1}\frac{\partial v_1}{\partial T} + \frac{v_2}{v_2}\frac{\partial v_2}{\partial T} + \ldots\right) = \frac{1}{\sum v_i}(v_1\alpha_1 + v_2\alpha_2 + \ldots),\end{aligned} \tag{22.23}$$

$$\alpha_{\mathrm{Pr}} = \frac{\sum \nu_i \alpha_i}{\sum \nu_i}. \tag{22.24}$$

Since it is known that the specific volumes of proteins also have temperature-dependent behavior (Makhatadze *et al.*, 1990), an accurate representation of $\alpha_{\mathrm{U}}(T)$ will also take the temperature dependence of $\bar{\nu}_{\mathrm{p}}$ into account. Once again, we will utilize the relationship $\alpha_i = (1/V_i)(\partial V_i/\partial T)$ to derive a temperature-dependent function for $V_i(V_i(T))$:

$$\int (\alpha_i) \mathrm{d}T = \int \frac{\mathrm{d}V_i}{V_i}. \tag{22.25}$$

Brandts and coworkers demonstrated that the $\alpha(T)$ profiles for individual amino acids could be represented by a cubic function $\alpha(T) = a + bT + cT^2 + dT^3$ (Lin *et al.*, 2002). It follows then that

$$\int (\alpha_i) \mathrm{d}T = a_i T + \frac{b_i}{2} T^2 + \frac{c_i}{3} T^3 + \frac{d_i}{4} T^4 + \alpha_{i,o}, \tag{22.26}$$

where a_i, b_i, c_i, and d_i are the coefficients for a given amino acid, given in Lin *et al.* (2002), and $\alpha_{i,o}$ is an arbitrary constant of integration. The other half of Eq. (22.25) is given by

$$\int \frac{\mathrm{d}V_i}{V_i} = \ln(V_i) + V_{i,o}. \tag{22.27}$$

By combining Eqs. (22.26) and (22.27), we can solve for V_i as a function of temperature:

$$V_i(T) = \exp\left(a_i T + \frac{b_i}{2} T^2 + \frac{c_i}{3} T^3 + \frac{d_i}{4} T^4 + F_i \right), \tag{22.28}$$

where F_i is the combination of the two arbitrary integration constants (i.e., $F_i = \alpha_{i,o} + V_{i,o}$). The partial molar volumes of the amino acids at 25 °C (Lin *et al.*, 2002) are used to define F_i:

$$F_i = \ln(\bar{V}_i^{25\,^\circ\mathrm{C}}) - a_i(25) - \frac{b_i}{2}(25)^2 - \frac{c_i}{3}(25)^3 - \frac{d_i}{4}(25)^4. \tag{22.29}$$

By incorporating Eqs. (22.28) and (22.29) into Eq. (22.24), we can develop a model for the temperature dependence of $\alpha_{\mathrm{U}}(T)$.

Figure 22.2 shows the $\alpha_{\mathrm{U}}(T)$ profile for RNase A, calculated based on the amino acid composition, as described above. From this figure, it can be seen that the calculated unfolded state baseline is remarkably similar to the experimental values at high temperatures. However, at low temperatures the slope is steeper than expected based on the experimental data. One of the possible explanations is that individual amino acids are not the best model compounds to represent proteins in the unfolded state. Indeed, comparison of the data on amino acids (Lin *et al.*, 2002) with the data of Gly-X-Gly tripeptides (Mitra *et al.*, 2006) suggests that the calculated $\alpha(T)$ per amino acid side chain depends on the model compound used. Furthermore, this difference is larger particularly at low temperatures. (It must be noted, however, that Mitra *et al.*, 2006 calculated their $\alpha(T)$ per amino acid as a simple difference of $\alpha(T)$ functions for Gly-X-Gly and Gly-Gly-Gly instead of using volume-weighted difference as suggested by Eq. (22.23).) Nevertheless, these calculations based on the individual amino acid model compounds, establish that the shape of the temperature dependence of α for the native and unfolded states should be represented in a first approximation by at least third order polynomial of temperature.

5.2. Importance of ΔH for quantitative estimates of $\Delta V/V$

Equation (22.22) provides the analytical form for a two-state analysis of the experimentally determined linear expansion coefficient of a protein as a function of temperature. The fitted parameters are $\Delta H(T_m)$, ΔC_p, T_m, $\Delta V_{Pr}/V_{Pr}$, and temperature dependencies of the linear expansion coefficient of the native and unfolded states of the protein. Alternatively, $\Delta H(T_m)$, ΔC_p, and T_m can be experimentally determined from the simultaneous DSC experiments performed on each protein under given solvent conditions. A comparison of the direct fit of $\alpha(T)$ to Eq. (22.22) with the fit in which ΔH and T_m values are based on the DSC data is shown in Table 22.1. Indeed, it is possible to fit the PPC data with both ΔH and T_m as parameters. This will typically result in a fitted PPC T_m that is within 2–3 °C of the DSC T_m. However, the fitted ΔH values based solely on PPC data can differ from the DSC measurements by as much as 15%. This occurs mostly because of the lower information content of PPC data: fewer experimental points are collected in a PPC experiment, where data points are collected every 5 °C outside the transition range and every 2 °C in the transition range, than in a DSC experiment, where data points are collected every ~0.1 °C. As a result, the temperature dependence of the equilibrium constant, and consequently the enthalpy of unfolding, cannot be defined with the same accuracy in PPC analysis as in DSC analysis. The combinations of these errors could result in erroneous estimates of the volumetric changes ($\Delta V_{Pr}/V_{Pr}$) upon unfolding (Table 22.1), which might then result in an incorrect interpretation of experimental data. This was especially evident for proteins like lysozyme (Fig. 22.3B), where $\Delta V_{Pr}/V_{Pr}$ is smaller than $\Delta V_{Pr}/V_{Pr}$ of ubiquitin (Fig. 22.3A). For these reasons, it is recommended to always perform a corresponding DSC experiment so that the T_m and ΔH can be properly constrained in the PPC data analysis.

Table 22.1 Comparison of fitted parameters for ubiquitin PPC data

pH	T_m (°C)		$\Delta H(T_m)$ (kJ/mol)		$\Delta V_{Pr}/V_{Pr}$		
	DSC fit	PPC fit	DSC fit	PPC fit	ΔH and T_m from DSC	ΔH and T_m fitted	Difference (%)
2.4	61.5	60.2	231	245	-9.9×10^{-3}	-9.1×10^{-3}	−9
2.6	63.1	61.8	244	240	-9.5×10^{-3}	-10.0×10^{-3}	6
2.8	66.2	64.5	261	247	-8.7×10^{-3}	-9.3×10^{-3}	6
3.0	71.1	70.3	266	252	-8.1×10^{-3}	-8.9×10^{-3}	9
3.2	73.2	72.1	277	239	-8.1×10^{-3}	-11.0×10^{-3}	25

5.3. Example of global analysis of PPC data and *P–T* diagram

Figure 22.5 shows the results of PPC experiments performed on ubiquitin. The temperature dependence of the thermal expansion coefficient was measured at different pH values, and all profiles were analyzed simultaneously with the common dependencies of the $\alpha_N(T)$ and $\alpha_U(T)$, and with $\Delta H(T)$ and ΔC_p obtained from the DSC experiments. The resulting fit described the experimental values of $\alpha(T)$ very well. The fitted values of $\Delta V_{Pr}/V_{Pr}$, when plotted as a function of transition temperature, show that $\Delta V_{Pr}/V_{Pr}$ depends on temperature. The values of $\Delta V_{Pr}/V_{Pr}$ are negative, suggesting that ubiquitin unfolding is accompanied by a negative volume changes, as has been found for other proteins (see, e.g., Mitra *et al.*, 2008; Royer, 2002; Schweiker *et al.*, 2009). Furthermore, the slope of $\Delta V_{Pr}/V_{Pr}$ versus T_m is positive, which is in agreement with the observation that $\alpha_U(T)$ is typically larger than $\alpha_N(T)$. Using thermodynamic values obtained from

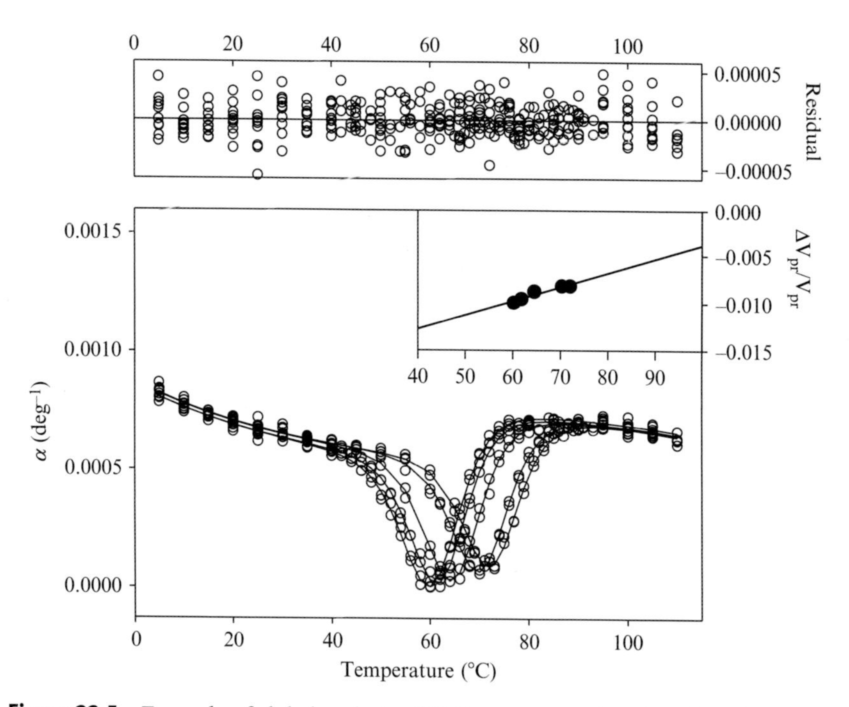

Figure 22.5 Example of global analysis of PPC experiments for ubiquitin using a two-state model. pH dependence of $\alpha_{exp}(T)$ for ubiquitin. The solid lines represent the fits of the data to a two-state model of unfolding. The symbols represent the experimental data for pH 2.4 (○), pH 2.6 (▽), pH 2.8 (□), pH 3.0 (◇), and pH 3.2 (△). Inset: dependence of the specific change in volume $\Delta V_{Pr}/V_{Pr}$ on the transition temperature T_m, using values given in columns 3 and 6 of Table 22.1. The solid line represents the linear fit of the data with a slope ($\Delta\alpha$) of 1.5×10^{-4} deg^{-1}.

the DSC and PPC experiments, one can construct the pressure–temperature diagram for ubiquitin. This can be done using the following relationship (Paschek *et al.*, 2008; Smeller, 2002):

$$\Delta G(P,T) = \Delta\beta/2(P-P_0)^2 + \Delta\alpha(P-P_0)(T-T_0)$$
$$-\Delta C_p[T(\ln(T/T_0)) + T_0] + \Delta V_0(P-P_0) - \Delta S_0(T-T_0) + \Delta G_0, \quad (22.30)$$

where $\Delta\beta$ is the change in compressibility upon unfolding, and ΔV_0, ΔS_0, and ΔG_0 are the unfolding volume, entropy, and free energy at the reference temperature T_0 and reference pressure P_0. If the reference temperature is taken to be the transition temperature and the reference pressure is 1 atm, then $\Delta G_0(T_m, 1\,\mathrm{atm}) = 0$ and $\Delta S_0(T_m) = \Delta H(T_m)/T_m$. This makes it possible to solve for $\Delta G(T,P) = 0$, and to compare the calculated result to the experimental values shown in Fig. 22.6. As expected, $\Delta G(T,P)$ calculated using parameters from PPC and DSC follows an ellipsoidal dependence, and is very similar to that obtained experimentally from pressure-induced denaturation experiments (Herberhold and Winter, 2002). However, this only required two experiments, one DSC and one PPC, at low pressure.

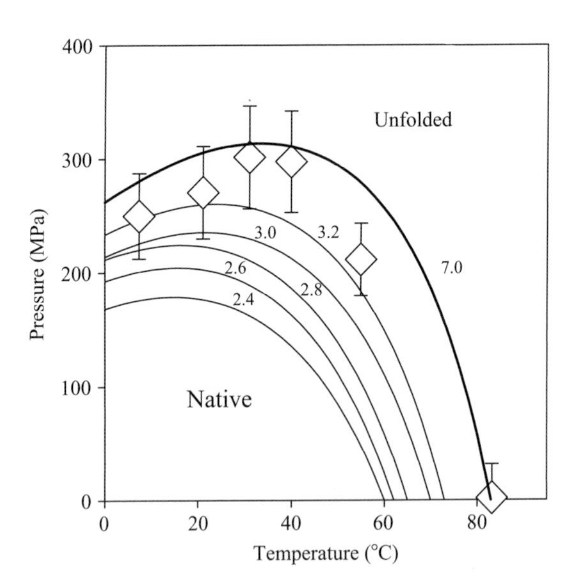

Figure 22.6 A pressure–temperature diagram for ubiquitin calculated using the data from Table 22.1 and Eq. (22.30). pH values are indicated above each curve. A *P–T* diagram for ubiquitin at pH 7.0 was constructed using extrapolated thermodynamic parameters for comparison with the experimental data of Herberhold and Winter (2002) (shown as symbols).

6. Implications of Two-State Model for Future PPC Experiments

This chapter has discussed a method for directly analyzing the data from a PPC experiment. The main advantage of using a two-state model of unfolding to calculate the volumetric changes upon unfolding is that placing the results in this context allows one to understand volumetric changes when the native and unfolded state baselines are not easily defined.

In the future, it will be interesting to extend the use of PPC to examine the role of solvent in protein unfolding. Since $\alpha(T)$ seems to be sensitive to solvent conditions, especially at low T (Barrett *et al.*, 2006; Lin *et al.*, 2002; Mitra *et al.*, 2006; Ravindra and Winter, 2004), PPC provides a sensitive method for determining the extent of interactions of the protein surface with bulk solvent and will aid the development of a better understanding of how intramolecular interactions contribute to stability and solubility.

REFERENCES

Baldwin, R. L. (1986). Temperature dependence of the hydrophobic interaction in protein folding. *Proc. Natl. Acad. Sci. USA* **83,** 8069–8072.

Barrett, D. G., Minder, C. M., Mian, M. U., Whittington, S. J., Cooper, W. J., Fuchs, K. M., Tripathy, A., Waters, M. L., Creamer, T. P., and Pielak, G. J. (2006). Pressure perturbation calorimetry of helical peptides. *Proteins* **63,** 322–326.

Brandts, J.F (1969). *In* "Structure and Stability of Biological Macromolecules," (S. N. Timasheff and G. D. Fasman, eds.), pp. 213–290. Marcel Dekker, New York, NY.

Chalikian, T., and Breslauer, K. (1996). On volume changes accompanying conformational transitions of biopolymers. *Biopolymers* **39,** 619–626.

Garcia-Moreno, E. B., and Fitch, C. A. (2004). Structural interpretation of pH and salt-dependent processes in proteins with computational methods. *Methods Enzymol.* **380,** 20–51.

Gross, M., and Jaenicke, R. (1994). Proteins under pressure. The influence of high hydrostatic pressure on structure, function and assembly of proteins and protein complexes. *Eur. J. Biochem.* **221,** 617–630.

Heerklotz, P. D. (2007). Pressure perturbation calorimetry. *Methods Mol. Biol.* **400,** 197–206.

Herberhold, H., and Winter, R. (2002). Temperature- and pressure-induced unfolding and refolding of ubiquitin: A static and kinetic Fourier transform infrared spectroscopy study. *Biochemistry* **41,** 2396–2401.

Kauzmann, W. (1959). Some factors in the interpretation of protein denaturation. *Adv. Protein Chem.* **14,** 1–63.

Kauzmann, W. (1987). Thermodynamics of unfolding. *Nature* **325,** 763–764.

Khechinashvili, N. N. (1990). Thermodynamic properties of globular proteins and the principle of stabilization of their native structure. *Biochim. Biophys. Acta* **1040,** 346–354.

Lee, S., Tikhomirova, A., Shalvardjian, N., and Chalikian, T. V. (2008). Partial molar volumes and adiabatic compressibilities of unfolded protein states. *Biophys. Chem.* **134,** 185–199.

Lin, L. N., Brandts, J. F., Brandts, J. M., and Plotnikov, V. (2002). Determination of the volumetric properties of proteins and other solutes using pressure perturbation calorimetry. *Anal. Biochem.* **302,** 144–160.

Livingstone, J. R., Spolar, R. S., and Record, M. T. Jr. (1991). Contribution to the thermodynamics of protein folding from the reduction in water-accessible nonpolar surface area. *Biochemistry* **30,** 4237–4244.

Lopez, M. M., and Makhatadze, G. I. (2002). Differential scanning calorimetry. *Methods Mol. Biol.* **173,** 113–119.

Makhatadze, G. I. (1998). Measuring protein thermostability by differential scanning calorimetry. *In* "Current Protocols in Protein Chemistry," (T. J. Wiley, ed.)Vol. 2. John Wiley & Sons, New York, NY.

Makhatadze, G. I., and Privalov, P. L. (1992). Protein interactions with urea and guanidinium chloride: A calorimetric study. *J. Mol. Biol.* **226,** 491–505.

Makhatadze, G. I., and Privalov, P. L. (1993). Contribution of hydration to protein folding thermodynamics. I. The enthalpy of hydration. *J. Mol. Biol.* **232,** 639–659.

Makhatadze, G. I., and Privalov, P. L. (1995). Energetics of protein structure. *Adv. Protein Chem.* **47,** 307–425.

Makhatadze, G. I., Medvedkin, V. N., and Privalov, P. L. (1990). Partial molar volumes of polypeptides and their constituent groups in aqueous solution over a broad temperature range. *Biopolymers* **30,** 1001–1010.

Mitra, L., Smolin, N., Ravindra, R., Royer, C., and Winter, R. (2006). Pressure perturbation calorimetric studies of the solvation properties and the thermal unfolding of proteins in solution—Experiments and theoretical interpretation. *Phys. Chem. Chem. Phys.* **8,** 1249–1265.

Mitra, L., Rouget, J. B., Garcia-Moreno, B., Royer, C. A., and Winter, R. (2008). Towards a quantitative understanding of protein hydration and volumetric properties. *Chemphyschem* **9,** 2715–2721.

Mozhaev, V. V., Heremans, K., Frank, J., Masson, P., and Balny, C. (1996). High pressure effects on protein structure and function. *Proteins* **24,** 81–91.

Ooi, T., and Oobatake, M. (1988). Effects of hydrated water on protein unfolding. *J. Biochem.* **103,** 114–120.

Pace, C. N., Vajdos, F., Fee, L., Grimsley, G., and Gray, T. (1995). How to measure and predict the molar absorption coefficient of a protein. *Protein Sci.* **4,** 2411–2423.

Pace, C. N., Shirley, B. A., McNutt, M., and Gajiwala, K. (1996). Forces contributing to the conformational stability of proteins. *FASEB J.* **10,** 75–83.

Paschek, D., Hempel, S., and Garcia, A. E. (2008). Computing the stability diagram of the Trp-cage miniprotein. *Proc. Natl. Acad. Sci. USA* **105,** 17754–17759.

Privalov, P. L., and Makhatadze, G. I. (1990). Heat capacity of proteins. II. Partial molar heat capacity of the unfolded polypeptide chain of proteins: Protein unfolding effects. *J. Mol. Biol.* **213,** 385–391.

Privalov, P. L., and Makhatadze, G. I. (1993). Contribution of hydration to protein folding thermodynamics. II. The entropy and Gibbs energy of hydration. *J. Mol. Biol.* **232,** 660–679.

Ravindra, R., and Winter, R. (2004). Pressure perturbation calorimetry: A new technique provides surprising results on the effects of co-solvents on protein solvation and unfolding behaviour. *Chemphyschem* **5,** 566–571.

Rosgen, J., and Hinz, H. J. (2000). Response functions of proteins. *Biophys. Chem.* **83,** 61–71.

Royer, C. A. (2002). Revisiting volume changes in pressure-induced protein unfolding. *Biochim. Biophys. Acta* **1595,** 201–209.

Schweiker, K. A., Fitz, V., and Makhatadze, G. I. (2009). Biochemistry, In press.

Silva, J. L., Foguel, D., and Royer, C. A. (2001). Pressure provides new insights into protein folding, dynamics and structure. *Trends Biochem. Sci.* **26,** 612–618.

Smeller, L. (2002). Pressure-temperature phase diagrams of biomolecules. *Biochim. Biophys. Acta* **1595,** 11–29.

Streicher, W. W., and Makhatadze, G. I. (2007). Advances in the analysis of conformational transitions in peptides using differential scanning calorimetry. *Methods Mol. Biol.* **350,** 105–113.

Tanford, C. (1968). Protein denaturation. *Adv. Protein Chem.* **23,** 121–282.

CHAPTER TWENTY-THREE

Solvent Denaturation of Proteins and Interpretations of the *m* Value

J. Martin Scholtz,[*,†] Gerald R. Grimsley,[*] *and* C. Nick Pace[*,†]

Contents

Abstract

The stability of globular proteins is important in medicine, proteomics, and basic research. The conformational stability of the folded state can be determined experimentally by analyzing urea, guanidinium chloride, and thermal denaturation curves. Solvent denaturation curves in particular may give useful information about a protein such as the existence of domains or the presence of stable folding intermediates. The linear extrapolation method (LEM) for analyzing solvent denaturation curves gives the parameter *m*, which is a measure of the dependence of ΔG on denaturant concentration. There is much recent interest in the *m* value as it relates to the change in accessible surface area of a protein when it unfolds and what it may reveal about the denatured states of proteins.

1. Introduction

Urea has been known to act as a protein unfolding agent since 1900 (Spiro, 1900). The even greater effectiveness of guanidine hydrochloride to unfold proteins was first reported by Greenstein (1938). Guanidine

[*] Department of Molecular and Cellular Medicine, Texas A&M Health Science Center, College Station, Texas, USA
[†] Department Biochemistry and Biophysics, Texas A&M University, College Station, Texas, USA

Methods in Enzymology, Volume 466
ISSN 0076-6879, DOI: 10.1016/S0076-6879(09)66023-7

hydrochloride is sometimes called guanidinium chloride (GdmCl) to emphasize the fact that it is a salt. Together these two agents, and compounds related to them, are called protein denaturants and their action on proteins has proved to be quite valuable to protein chemists. Here, we will discuss how to perform and analyze solvent denaturation curves and how to use them to learn about protein stability, folding, and structure.

Tanford (1964) was the first to discuss in quantitative terms the unfolding of proteins by urea. He also provided the first simple framework for understanding how and why denaturant solutions promote the unfolding of proteins. The schematic in Fig. 23.1 is from Tanford's paper; it shows that when a protein unfolds many nonpolar side chains and peptide groups that were buried in the folded protein become exposed to solvent in the unfolded protein. Conceptually, this diagram shows that the unfolding process can be modeled by considering a series of transfer reactions of the side chains and polypeptide backbone from the buried environment of the folded protein to a solvent-exposed state in the unfolded chain. In the lower half of the figure are the propensities for the free energy of transfer, ΔG_{tr}, for a leucine side chain and a peptide group from water to the solvents shown. This construct shows why proteins unfold in urea and GdmCl solutions: the free energy of both leucine side chains and peptide groups is lower in the presence of urea and GdmCl than they are in water. It turns out that this is true for most of the constituent groups of a protein. Thus, it is easy to see why urea and GdmCl are protein denaturants and why GdmCl is more potent than urea. It is also clear why both folded and unfolded proteins are generally more soluble in aqueous urea and GdmCl solutions than in pure water. The transfer data in Fig. 23.1 also illustrate why other cosolvents act as protein stabilizers (glycerol, sucrose, sacrosine, and TMAO) while ethanol and trifluoroethanol often act to unfold proteins, yet stabilize hydrogen-bonded structures like α-helices (Buck, 1998).

2. Protein Unfolding or Denaturation

The basic approach here will be to describe how to determine and analyze urea and GdmCl denaturation curves using optical spectroscopic techniques to follow unfolding. These are traditional methods involving relatively simple experiments that can be done in almost any laboratory. The information obtained can be used to estimate $\Delta G(H_2O)$, to determine the stability curve for a protein, and to measure differences in stability among proteins, $\Delta(\Delta G)$. These experiments sometimes reveal additional features of a protein such as the existence of domains or the presence of stable folding intermediates. Details for performing these experiments are explained to a greater degree in Pace *et al.* (2005). Finally, the analysis of a solvent

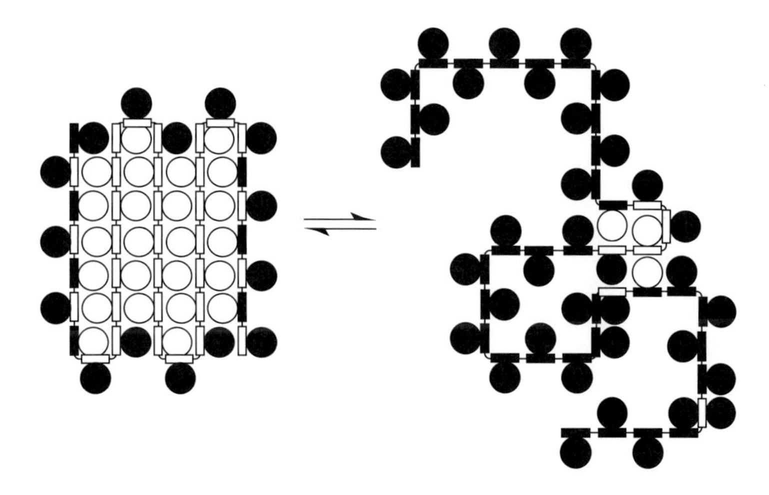

	Propensity to transfer from H_2O to solvent	
Solvent	Peptide group	Leu side chain
Urea (2 *M*)	+	+
GdmCl (2 *M*)	++	++
Ethanol (20%)	−−−	+++
Glycerol (20%)	−	−
Sucrose (2 *M*)	−	−
Sarcosine (2 *M*)	−	−
TMAO (2 *M*)	−−	−

Figure 23.1 Schematic from Tanford (1964) illustrating the changes in solvent accessibility of individual amino acid residues upon protein unfolding. The closed symbols represent groups that are accessible to solvent and the open symbols represent groups that are not accessible to solvent. The table below shows the propensity for the free energy of transfer, ΔG_{tr}, (−, unfavorable; +, favorable) for the leucine side chain and the peptide group illustrating the importance of constituents of proteins to solutions of the indicated cosolvent (protein denaturant or osmolytes). For the quantified ΔG_{tr} values, see Liu and Bolen (1995), Auton and Bolen (2005) for osmolyte data, and Pace *et al.* (1990) for GdmCl data.

denaturation provides, in addition to ΔG, a parameter called the *m* value which provides a link to the transfer models like those in Fig. 23.1 and also provides some insight into the mechanism and extent of protein folding.

The optical techniques used most often for solvent denaturation studies are fluorescence, circular dichroism (CD), and UV absorption. In all cases, the spectra of the folded (no denaturant) and unfolded protein (high concentration of denaturant) should be determined to help select the most useful technique. The best technique is generally the one that shows the biggest change between the folded and unfolded forms, although sometimes there are other concerns. The spectral changes observed upon unfolding often depend upon different features of protein structure. Fluorescence and near-UV spectroscopy respond to changes in the environment of the tryptophan and tyrosine residues, and hence to changes in tertiary structure, while CD measurements below 250 nm depend mainly on changes in the secondary structure. This may also be a consideration in determining the technique you use to follow unfolding. Usually, UV absorbance measurements are less convenient for following urea and GdmCl unfolding because the pre- and posttransition baselines are steeper than with most of the other techniques. In general, unfolding curves with steeper pre- and posttransition baselines lead to larger errors in the parameters determined in the analysis.

Figure 23.2A shows a typical urea unfolding curve for RNase Sa using CD to follow unfolding. To remain completely general, the physical parameter used to follow unfolding will be called y. The curves can be conveniently divided into three regions: (1) The pretransition region, which shows how y for the folded protein, y_F, depends upon the concentration of the denaturant. (2) The transition region, which shows how y varies as unfolding occurs. (3) The posttransition region, which shows how y for the unfolded protein, y_U, varies with denaturant concentration. All of these regions are important for analyzing unfolding curves. As a minimum, four points need to be determined in the pre- and posttransition regions, and five points in the transition region. Of course, the more points determined the better defined the curve and often times these can be automated and it is usual to collect 30–40 data points per curve.

As with all thermodynamic measurements, it is essential that equilibrium is reached before measurements are made, and that the unfolding reaction is reversible. The time required to reach equilibrium can vary from seconds to days, depending upon the protein and the conditions. For example, the unfolding of RNase T1 requires hours to reach equilibrium at 25 °C, as compared to only minutes for RNase Sa (Pace *et al.*, 1998; Thomson *et al.*, 1989). For any given protein, the time to reach equilibrium is longest at the midpoint of the transition and decreases in both the pre- and posttransition regions. To ensure that equilibrium is reached, y is measured as a function of time to establish the time required to reach equilibrium.

To test the reversibility of unfolding, allow a solution to reach equilibrium in the posttransition region and then, by dilution, return the solution to the pretransition region, and measure y. If the reaction is reversible, the value of y measured after complete unfolding should not be less than 90% of that determined directly. For most proteins, unfolding in urea and GdmCl is

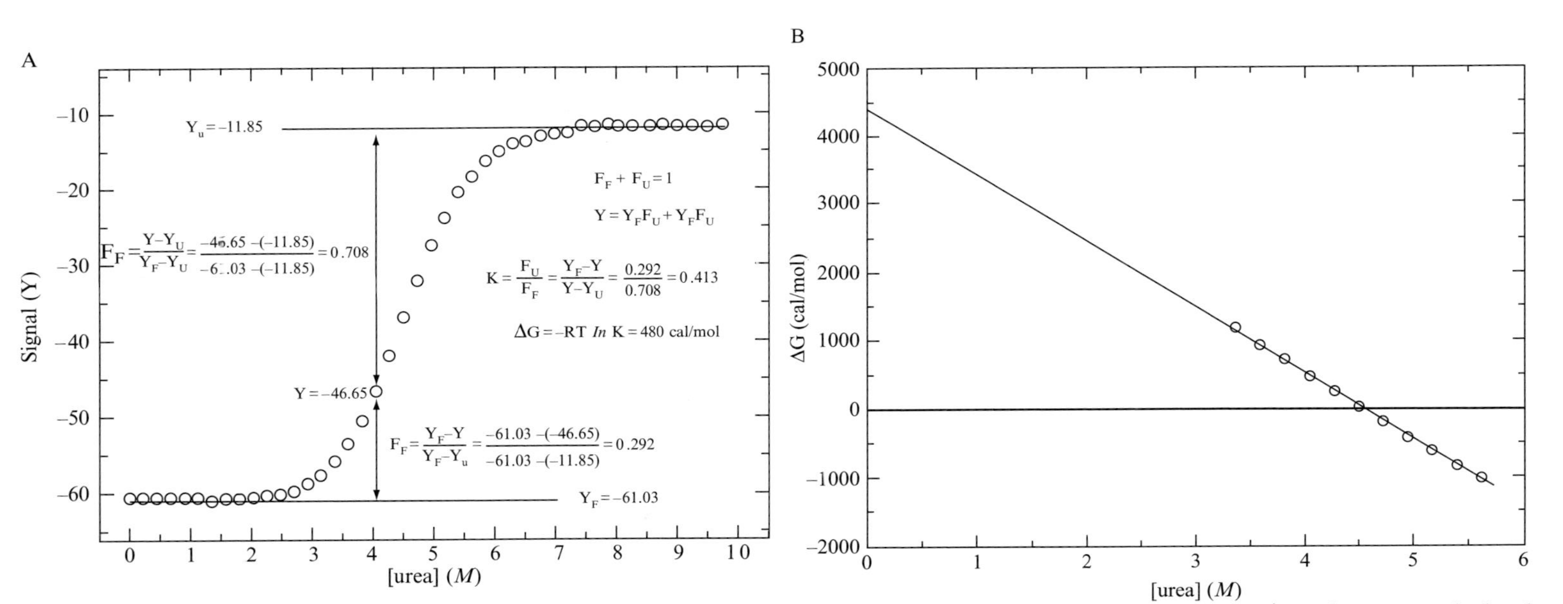

Figure 23.2 (A) Urea denaturation curve for RNase Sa, a typical small protein. (B) ΔG as a function of urea molarity. The ΔG values were calculated from the data in the transition region as described in the text. The $\Delta G(H_2O)$ and *m* values can be determined by fitting the data in (B) to Eq. (23.4), or by analyzing the data in (A) with Eq. (23.5).

completely reversible. For example, RNase T1 has been left in 6 M GdmCl for 3 months and found to refold completely on dilution.

Both urea and GdmCl can be purchased commercially in highly purified forms (e.g., from United States Biochemical or AMRESCO). However, some lots of these denaturants may contain fluorescent or metallic impurities. Methods are available for checking the purity of GdmCl and for recrystallization if necessary (Nozaki, 1972). A procedure for purifying urea has also been described (Prakash *et al.*, 1981). GdmCl solutions are stable for months, but urea solutions slowly decompose to form cyanate and ammonium ions in a process accelerated at high pH (Hagel *et al.*, 1971). The cyanate ions can react with amino groups on proteins (Stark, 1965). Consequently, a fresh urea stock solution should be prepared for each unfolding curve and used within the day.

Table 23.1 summarizes useful information for preparing urea and GdmCl stock solutions. Urea stock solutions should be prepared by weight, and the concentration verified by refractive index measurements using the equation given in Table 23.1. If the concentrations do not vary by more than 3%, the solution can be used for determining an unfolding curve. Since GdmCl is quite hygroscopic, it is more difficult to prepare stock solutions by weight. Consequently, the molarity of GdmCl stock solutions is generally based on refractive index measurements using the equation given in Table 23.1.

Table 23.1 Information for preparing urea and GdmCl stock solutions

Property	Urea	GdmCl
Molecular weight	60.056	95.533
Solubility (25.0 °C)	10.49 *M*	8.54 *M*
d/d_0[a]	$1 + 0.2658W + 0.0330W^2$	$1 + 0.2710W + 0.0330W^2$
Molarity[b]	$117.66(\Delta N) + 29.753(\Delta N)^2 + 185.56(\Delta N)^3$	$57.147(\Delta N) + 38.68(\Delta N)^2 - 91.60(\Delta N)^3$
Grams of denaturant per gram of water to prepare		
6 *M*	0.495	1.009
8 *M*	0.755	1.816
10 *M*	1.103	–

[a] *W* is the weight fraction denaturant in the solution, *d* is the density of the solution, and d_0 is density of water (see Kawahara and Tanford, 1966).

[b] ΔN is the difference in refractive index between the denaturant solution and water (or buffer) at the sodium D line. The equation for urea solutions is based on data from Warren and Gordon (1966), and the equation for GdmCl solutions is from Nozaki (1972).

3. Linear Extrapolation Method

When a protein unfolds by a two-state mechanism, the equilibrium constant, K, and the standard free energy change, ΔG°, can be calculated from the experimental data using:

$$K = \frac{y_F - y_{obs}}{y_{obs} - y_U} \tag{23.1}$$

$$\Delta G^{\circ} = -RT \ln K \tag{23.2}$$

where y_{obs} is the observed value of the parameter used to follow unfolding, and y_F and y_U are the values y would have for the folded state and the unfolded state under the same conditions where y_{obs} was measured. In the original analyses of urea denaturation curves (Alexander and Pace, 1971; Pace and Tanford, 1968), log K was found to vary linearly as a function of log [urea] and the slope of the plot was denoted by n, and the midpoint of the curve by $(\text{urea})_{1/2}$ (where $\log K = 0$). These parameters could then be used to calculate the dependence of the standard free energy of denaturation, on urea concentration (Tanford, 1964) with:

$$\partial(\Delta G^{\circ})/\partial(\text{urea}) = RTn/(\text{urea})_{1/2} \tag{23.3}$$

This equation was used by Alexander and Pace (1971) to estimate the differences in stability among three genetic variants of β-lactoglobulin for the first time and was the forerunner of the linear extrapolation method (LEM).

An investigation of urea and GdmCl denaturation curves was reported again by the Pace laboratory in 1974 (Greene and Pace, 1974). When ΔG is calculated at several urea concentrations using data such as those shown in Fig. 23.2A and Eqs. (23.1) and (23.2), ΔG is found to vary linearly with urea concentration as shown in Fig. 23.2B. Thus, it appears that a simple linear equation could be used for analyzing the data:

$$\Delta G = \Delta G(H_2O) - m[\text{urea}] \tag{23.4}$$

where $\Delta G(H_2O)$ is an estimate of the conformational stability of a protein in the absence of denaturant and m is a measure of the dependence of ΔG on urea concentration or the slope of the plot shown in Fig. 23.2B. The general method and relationship in Eq. (23.4) assumes that the linear dependence continues to 0 M denaturant. This analysis has become known as the LEM and is now the accepted way to analyze solvent denaturation curves for proteins.

The general analysis of a solvent denaturation curve like that in Fig. 23.2A requires estimates of the pre- and posttransition baselines to determine y_F and y_U in the transition region so Eqs. (23.1) and (23.2) can be used to calculate K and $\Delta G°$. Instead of the stepwise analysis illustrated in Fig. 23.2B, Santoro and Bolen (1992) had a better idea and proposed that nonlinear least squares be used to directly fit data such as those shown in Fig. 23.2A. With their approach, six parameters are used to fit the data: a slope and an intercept each for the pre- and posttransition baselines, and $\Delta G(H_2O)$ and the m value leading to:

$$y=\left\{\left(y_F^\circ+m_F[\text{urea}]\right)+\left(y_U^\circ+m_U[\text{urea}]\right) \times \exp-\left(\left(\Delta G(H_2O)-m[\text{urea}]\right)/RT\right)\right\} /\left(1+\exp-(\Delta G(H_2O)-m[\text{urea}])/RT\right) \quad (23.5)$$

where y_F° and y_U° are the intercepts and m_F and m_U the slopes of the pre- and posttransition baselines, and $\Delta G(H_2O)$ and m are defined by Eq. (23.4).

The LEM is now widely used for estimating the conformational stability of proteins and for measuring the difference in stability between proteins differing slightly in structure. In addition, the interpretation of the m value has provided insight into the properties of the unfolded ensemble of proteins and the denatured state (Shortle, 1995). These topics will be discussed further below.

4. ΔG(H₂O): Conformational Stability

The analysis of solvent denaturation curves provides estimates of two key parameters: $\Delta G(H_2O)$ and the m value. The measurement of the conformational stability of a protein and an understanding of the forces involved in protein folding are longstanding questions in protein science. The measurement of the stability for even the simplest proteins that exhibit two-state folding under normal laboratory conditions is nontrivial. For example, the conformational stability of a model protein that has been extensively studied, ribonuclease (RNase) Sa is 7.0 kcal mol^{-1} at 25 °C and pH 5. A simple two-state analysis shows that there is one unfolded molecule for each 135,000 folded molecules ($K = 7.4 \times 10^{-6}$). The equilibrium between the folded and unfolded conformations can only be easily determined when both conformations are present at concentrations that can be measured and quantified. With urea or thermal denaturation this is done by increasing the urea concentration or the temperature, respectively. It is clear from Fig. 23.2A that the unfolding equilibrium can be studied only near 4 M urea, and that a long extrapolation is needed to estimate $\Delta G°$ in the absence of urea, $\Delta G(H_2O)$. The same is true for thermal denaturation. The unfolding equilibrium can only be studied near the thermal unfolding or melting temperature (55 °C for RNase Sa at pH 5). Thermal denaturation curves can be analyzed to obtain the midpoint of the thermal denaturation curve, T_m, and the enthalpy change at T_m, ΔH_m. These two parameters and the heat capacity change for folding, ΔC_p, can then be used with the Gibbs–Helmholtz equation

$$\Delta G(T) = \Delta H_m(1 - T/T_m) + \Delta C_p[T - T_m - T\ln(T/T_m)] \quad (23.6)$$

to estimate the conformational stability at any temperature, denoted by $\Delta G(T)$. Thermal and solvent denaturation experiments thus provide complementary methods for estimating ΔG, although each requires extrapolation to a common reference temperature and solvent. Because of the fundamental differences between the methods, there has been much discussion about the validity and suitability of both to provide reliable estimates for $\Delta G(H_2O)$.

Previously, RNase T1 was used to show that estimates of $\Delta G(T)$ based on Differential Scanning Calorimetry (DSC) experiments were in good agreement with estimates of $\Delta G(H_2O)$ from urea denaturation (Hu *et al.*, 1992; Yu *et al.*, 1994). A few years ago, a carefully controlled study was performed and concluded that $\Delta G(T)$ values from DSC agree with $\Delta G(H_2O)$ for RNase A unfolding using urea denaturation and the LEM (Pace *et al.*, 1999). Table 23.2 shows some of the key results of this study and illustrates the excellent agreement between $\Delta G(T)$ and $\Delta G(H_2O)$. As a further test of the complementarily of thermal and urea denaturation,

Table 23.2 Comparison of $\Delta G(T)$ values from DSC with $\Delta G(H_2O)$ values from urea denaturation curves for RNase A

pH	T (°C)	T_m[a] (°C)	ΔH_m[a] (kcal mol^{-1})	$\Delta G(T)$[a] (kcal mol^{-1})	$\Delta G(H_2O)$[b] (kcal mol^{-1})
2.8	17.1	44.9	79.4	5.5	5.4
2.8	21.1	44.9	79.4	4.9	4.9
2.8	24.9	44.9	79.4	4.3	4.3
2.8	27.8	44.9	79.4	3.7	3.5
2.8	25	44.9	79.4	4.5	4.3
3	25	49.1	82.7	5.1	5.2
3.55	25	54.5	91.5	6.7	6.4
4	25	56.1	94.2	7.2	7.3
5	25	58.6	99.1	8.1	7.9
6	25	60.3	100.7	8.5	8.6
7	25	61.8	102.3	8.9	9.1

[a] The T_m, ΔH_m, and $\Delta C_p = 1.15$ kcal mol^{-1} K^{-1} values are from Pace *et al.* (1999). They were used in Eq. (23.6) to calculate the $\Delta G(T)$ values.

[b] The first four $\Delta G(H_2O)$ values are from Pace and Laurents (1989), and the last seven $\Delta G(H_2O)$ values are interpolated from Fig. 6 in Pace *et al.* (1990).

Nicholson and Scholtz (1996) combined both types of denaturation experiments to determine the conformation stability of a small protein as functions of both temperature and urea concentration. In all these cases, the agreement between $\Delta G(T)$ and $\Delta G(H_2O)$ is remarkably good, suggesting that both are providing reliable and thermodynamic estimates of the conformational stability of proteins. Finally, in a completely different study, conformational stabilities estimated from hydrogen exchange rates measured under native state conditions with those from thermal or solvent denaturation were compared (Huyghues-Despointes *et al.*, 1999). Again, the conformational stabilities were in good agreement when compared under identical conditions.

The excellent agreement between conformational stabilities based on DSC and other thermal unfolding methods, the hydrogen exchange results and the LEM suggest that urea denaturation curves and the LEM is a reliable method for estimating the conformational stability of a protein. In contrast, the analysis of GdmCl denaturation curves do not always provide $\Delta G(H_2O)$ values that are as reliable, as pointed out by Makhatadze (1999). There are several reasons why urea is preferred to GdmCl as a denaturant as summarized a few years ago by Schellman and Gassner (1996): "Finally this study as well as a number of others indicates that urea has a number of advantages over guanidinium chloride as far as thermodynamic interactions are concerned. The free energy and enthalpy functions of both proteins and model compounds are more linear; the solutions more ideal; extrapolations to zero concentration are more certain; and least-square analysis of the data is more stable." Perhaps the largest problem is that the ionic strength cannot be controlled with GdmCl, and this has been shown to effect the results in several studies (see, e.g., Ibarra-Molero *et al.*, 1999; Monera *et al.*, 1994; Santoro and Bolen, 1992).

In summary, the agreement between $\Delta G(T)$ values from DSC and thermal denaturation studies and $\Delta G(H_2O)$ values from urea denaturation studies analyzed by the LEM suggests that either method can be used to reliably measure the conformational stability of a protein.

5. The *m* Value

The form and interpretation of the relationship between the protein unfolding conformational change and the concentration of denaturant has been a puzzle for as long as it was realized that proteins could be reversibly unfolded (denatured) with certain chemicals in aqueous solution. To our knowledge, the first attempt to describe the relationship used Eq. (23.3) to provide n as the slope of the plot of log K as a function of log [denaturant]. Several years later (1974), it was realized that a simpler relationship could be used and the Gibbs energy change (or log K) for the conformational transition varied in a linear manner with the molar concentration of denaturant as expressed in Eq. (23.4). Therefore, this simple LEM provides the m value as the empirical parameter linking conformational stability and denaturant concentration. The basic treatment of the m value and the LEM has recently been expanded to include the detergent-induced destabilization of membrane proteins (Lau and Bowie, 1997; Hong *et al.*, 2009) and even to the effects of stabilizing compounds, like osmolytes, on increasing the conformational stability of globular proteins (Holthauzen and Bolen, 2007; Mello and Barrick, 2003). The simplicity of the LEM makes the application to other systems especially attractive.

Many attempts to provide a thermodynamic or physical explanation for the ability of denaturants to unfold proteins have been explored over the years and two main models have emerged: the denaturant binding model and a model based on the transfer free energies, the basic framework of which is provided in Fig. 23.1. The denaturant binding model as originally formulated treats the interaction of urea or GdmCl with the protein as a "binding" event, with the binding constant (k) and the number of new binding sites uncovered upon unfolding of the protein (Δn) as the main parameters:

$$\Delta G = \Delta G(H_2O) - \Delta n RT \ln(1 + ka) \tag{23.7}$$

In this model, the Gibbs free energy change is predicted to show a logarithmic dependence on the activity (a) of the denaturant concentration. A difficulty with the application of the binding model, besides the need to determine the activity and not the molar concentration of the denaturant solution, is the empirical nature of the analysis of the number of new sites (Δn) with binding constants (k) for the denaturant, since these two parameters cannot be determined independently in the analysis. Related thermodynamic approaches to general protein–cosolvent interactions, and particularly those involving protein denaturation, have been summarized by both Record (Courtenay *et al.*, 2000) and Schellman (2003). In these models, the direct exchange between water and urea at the protein surface is considered and the basic results are consistent with the results of the LEM. Because of some of the difficulties in applying the thermodynamic models and the simplicity of the LEM, the binding model and its derivatives are no longer in widespread use as a tool to analyze solvent denaturation curves.

The other major method proposed to analyze solvent denaturation curves is the transfer free energy model as originally described by Tanford (1964). The basic form is

$$\Delta G = \Delta G(H_2O) - \Sigma \alpha_i n_i \Delta g_{tr,i} \approx \Delta G(H_2O) - (\Delta\alpha) \Sigma n_i \Delta g_{tr,i} \tag{23.8}$$

where n_i is the number of groups of type i, α_i is the fraction of type i groups exposed to solvent in the unfolded protein but not in the folded form and $\Delta g_{tr,i}$ is the free energy of transfer of groups of type i from water to a given concentration of denaturant. In the approximation, $\Delta\alpha$ is the average change in exposure of all peptide groups and side chains in the protein when it unfolds.

While the LEM and denaturant binding model are empirical, the transfer free energy model has a thermodynamic basis ($\Delta g_{tr,i}$) and furthermore can provide a molecular description and explanation for solvent-induced protein denaturation (n_i and α_i). However, the model has limitations and uncertainties, specifically with regard to the component free transfer energy terms for all the side chains and backbone. Auton and Bolen (2004) have recently revisited the transfer free energy model including the component transfer free energy terms of the side chains and polypeptide backbone. In addition, they provide a description of the proper way to discuss the concentrations of denaturant and how to apply the model to experimental measures of protein unfolding and the m value; this will be addressed further below.

As stated above, any interpretation of solvent denaturation must satisfy the basic framework of protein unfolding outlined in Fig. 23.1. Specifically, more side chains and backbones are exposed to solvent in the unfolded set of conformations that in the folded ones. In the denaturant binding model, this requirement is fulfilled by Δn, the number of new binding sites. In the transfer models, $\Delta\alpha$ provides the link to the newly exposed side chains and peptide groups. In contrast, it appears that the LEM and the m value have no apparent relationship to the physical unfolding event depicted in Fig. 23.1. This mystery was solved in 1995 when Myers *et al.* (1995) provided evidence that the m value determined from the LEM correlated with the predicted change in solvent accessible surface area (ASA) when a protein unfolds as shown in Fig. 23.3. The original dataset is shown as well as several other proteins that were not available in the original analysis. The entire range spans proteins from 36 to 338 residues that all show reversible two-state equilibrium unfolding. Although there is some scatter, the correlation is remarkably good and provides the connection between the LEM and the physical unfolding event. The best-fit line ($r = 0.94$) provides:

$$m(\text{urea}) = 0.13\Delta\text{ASA} + 243(\text{cal}\,\text{mol}^{-1}\text{M}^{-1})\,\text{with}\,\Delta\text{ASA}\,\text{in}\,\text{Å}^2 \quad (23.9)$$

The relationship shown in Fig. 23.3 between the m value and the amount of newly exposed surface area upon unfolding has several important uses. First, if the structure of a protein is known so the change in ASA can be calculated, the m value can be estimated and compared to the experimental measure. Deviation from a two-state mechanism will reduce the

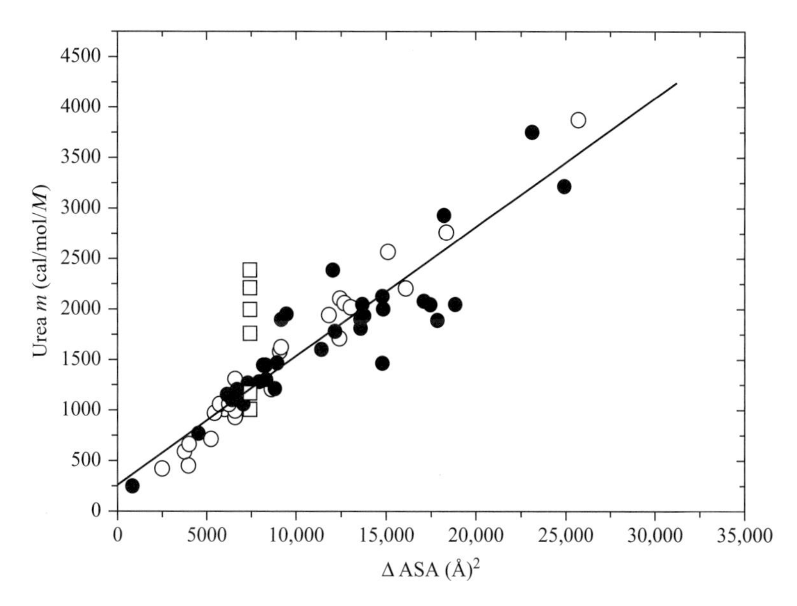

Figure 23.3 The relation between the m values for urea denaturation and the calculated change in solvent accessible surface area for unfolding for a series of proteins. The black circles represent those in the original analysis by Myers *et al.* (1995) and the open circles are proteins that have been added since the original analysis and are discussed in detail in a forthcoming paper. The squares are the RNase Sa variants discussed in the text and illustrate the effects that pH and other alterations in electrostatic properties can have on m values.

experimental m value from that expected and the inequality can be used as an indication that further study is needed to confirm the nature of the unfolding transition. In addition to being able to predict m values from the expected structure through ΔASA, the relationship shown in Fig. 23.3 has been used as a framework to think about structure (or lack thereof) in the denatured state and how mutations in the protein or alterations in the solution conditions could affect the distribution of structures in the denatured ensemble.

This initial interest in m values and their connection to structure was stimulated by a series of papers from the Shortle laboratory (Shortle, 1996; Shortle and Meeker, 1986), including an excellent review (Shortle, 1995). Their studies of variants of staphylococcal nuclease (SNase) showed a wide range of m values. Those with m values 5% or more greater than wild type were designated m^+ mutants (≈25%) and those with m values 5% or more less than wild type were designated m^- mutants (≈50%). Their interpretation was that m^+ mutants unfolded to a greater extent than wild type and to a smaller extent for m^- mutants. These studies were extremely important because they focused attention on the denatured state which had largely been ignored in discussions of protein stability.

We were surprised to find that for RNases A and T1 (Pace *et al.*, 1990) and barnase (Pace *et al.*, 1992) the *m* value increases markedly as the pH is lowered from 7 to 3. This indicates that the denatured states interact more extensively with urea as the pH is lowered, and we suggested that the denatured state ensemble expands at low pH due to electrostatic repulsion among the positive charges. This observation has now been made for some other proteins including SNase (Whitten and Garcia-Moreno, 2000) and we have recently obtained results with RNase Sa that provide further support for this idea.

RNase Sa is an acidic protein with a pI = 3.5 that contains no lysine residues. We have prepared a triple mutant that we call 3K (D1K, D17K, E41K) with a pI = 6.4, and a quintuple mutant that we call 5K (D1K, D17K, D25K, E41K, E74K) with a pI > 9 (Laurents *et al.*, 2003). At pH 3, the estimated net charges are +8 for wild-type RNase Sa, +11 for 3K, and +13 for 5K. The *m* values for the three proteins at pH 7 and pH 3 are shown in Fig. 23.3 (open squares) based on the ΔASA value for the wild-type protein. For all the three proteins, the *m* value is considerably greater at pH 3 than at pH 7. This is consistent with an increase in accessibility to denaturant caused by an expansion of the denatured state due to electrostatic repulsion among the positive charges. As the pH increases from 3 to 7, the carboxyl groups are titrated and both negative and positive charges are present. Now attractive charge–charge interactions are possible, and the decrease in the *m* values suggests that the denatured state ensemble is more compact because of attractive Coulombic interactions. In support of this, the *m* value at pH 7 is lowest for 3K where the number of positive and negative charges is almost equal than for wild-type RNase Sa which has an excess of negative charges or 5K which has an excess of positive charges. This and other evidence led us to conclude that long-range electrostatic interactions are important in determining the denatured state ensemble (Pace *et al.*, 2000).

6. Concluding Remarks

The LEM has proved to be a reliable method for measuring the conformational stability of a protein or the difference in stability between two proteins that differ slightly in structure. In addition, studies of the *m* values of proteins, especially SNase, have focused attention on the denatured state and it is now clear that denatured proteins contain residual structure and that this residual structure can be altered by single mutations.

ACKNOWLEDGMENTS

This work was supported by grants GM-37039 and GM-52483 from the National Institutes of Health (USA), and grants BE-1060 and BE-1281 from the Robert A. Welch Foundation. We thank many colleagues and coworkers over the years who have made important contributions to this work and these protocols.

REFERENCES

Alexander, S. S., and Pace, C. N. (1971). A comparison of the denaturation of bovine-lactoglobulins A and B and goat-lactoglobulin. *Biochemistry* **10,** 2738–2743.

Auton, M., and Bolen, D. W. (2004). Additive transfer free energies of the peptide backbone unit that are independent of the model compound and the choice of concentration scale. *Biochemistry* **43,** 1329–1342.

Auton, M., and Bolen, D. W. (2005). Predicting the energetics of osmolyte-induced protein folding/unfolding. *PNAS* **102,** 15065–15068.

Buck, M. (1998). Trifluoroethanol and colleagues: Cosolvents come of age. Recent studies with peptides and proteins. *Q. Rev. Biophys.* **31,** 2970355.

Courtenay, E. S., Capp, M. W., Saecker, R. M., and Record, M. T. Jr. (2000). Thermodynamic analysis of interactions between denaturants and protein surface exposed on unfolding: Interpretation of urea and guanidinium chloride m-values and their correlation with changes in accessible surface area (ASA) using preferential interaction coefficients and the local-bulk domain model. *Proteins* **Suppl. 4,** 72–85.

Greene, R. F. Jr., and Pace, C. N. (1974). Urea and guanidine hydrochloride denaturation of ribonuclease, lysozyme, alpha-chymotrypsin, and beta-lactoglobulin. *J. Biol. Chem.* **249,** 5388–5393.

Greenstein, J. P. (1938). Sulfhydryl groups in proteins. I. Egg albumin in solutions of urea, guanidine and their derivatives. *J. Biol. Chem.* **125,** 501–513.

Hagel, P., Gerding, J., Fieggen, W., and Bloemendal, H. (1971). Cyanate formation in solutions of urea. I. Calculation of cyanate concentrations at different temperature and pH. *Biochim. Biophys. Acta* **243,** 366–373.

Holthauzen, L. M., and Bolen, D. W. (2007). Mixed osmolytes: The degree to which one osmolyte affects the protein stabilizing ability of another. *Prot. Sci.* **16,** 293–298.

Hong, H. J. N., Bowie, J. U., and Tamm, L. K. (2009). Methods for measuring the thermodynamic stability of membrane proteins. *Methods Enzymol.* **455,** 213–236.

Hu, C. Q., Sturtevant, J. M., Thomson, J. A., Erickson, R. E., and Pace, C. N. (1992). Thermodynamics of ribonuclease T1 denaturation. *Biochemistry* **31,** 4876–4882.

Huyghues-Despointes, B. M., Scholtz, J. M., and Pace, C. N. (1999). Protein conformational stabilities can be determined from hydrogen exchange rates. *Nat. Struct. Biol.* **6,** 910–912.

Ibarra-Molero, B., Loladze, V. V., Makhatadze, G. I., and Sanchez-Ruiz, J. M. (1999). Thermal versus guanidine-induced unfolding of ubiquitin. An analysis in terms of the contributions from charge charge interactions to protein stability. *Biochemistry* **38,** 8138–8149.

Kawahara, K., and Tanford, C. (1966). Viscosity and density of aqueous solutions of urea and guanidine hydrochloride. *J. Biol. Chem.* **241,** 3228–3232.

Lau, F. W., and Bowie, J. U. (1997). A method for assessing the stability of a membrane protein. *Biochemistry* **36,** 5884–5892.

Laurents, D. V., Huyghues-Despointes, B. M., Bruix, M., Thurlkill, R. L., Schell, D., Newsom, S., Grimsley, G. R., Shaw, K. L., Trevino, S., Rico, M., Briggs, J. M.,

Antosiewicz, J. M., *et al.* (2003). Charge–charge interactions are key determinants of the pK values of ionizable groups in ribonuclease Sa (pI = 3.5) and a basic variant (pI = 10.2). *J. Mol. Biol.* **325,** 1077–1092.

Liu, Y., and Bolen, D. W. (1995). The peptide backbone plays a dominant role in protein stabilization by naturally occurring osmolytes. *Biochemistry* **34,** 12884–12891.

Makhatadze, G. I. (1999). Thermodynamics of protein interactions with urea and guanidinium hydrochloride. *J. Phys. Chem. B* **103,** 4781–4785.

Mello, C. C., and Barrick, D. (2003). Measuring the stability of partly folded proteins using TMAO. *Protein Sci.* **12,** 1522–1529.

Monera, O. D., Kay, C. M., and Hodges, R. S. (1994). Protein denaturation with guanidine hydrochloride or urea provides a different estimate of stability depending on the contributions of electrostatic interactions. *Protein Sci.* **3,** 1984–1991.

Myers, J. K., Pace, C. N., and Scholtz, J. M. (1995). Denaturant *m* values and heat capacity changes: Relation to changes in accessible surface areas of protein unfolding. *Protein Sci.* **4,** 2138–2148.

Nicholson, E. M., and Scholtz, J. M. (1996). Conformational stability of the *Escherichia coli* HPr protein: Test of the linear extrapolation method and a thermodynamic characterization of cold denaturation. *Biochemistry* **35,** 11369 11378.

Nozaki, Y. (1972). The preparation of guanidine hydrochloride. *In* "Methods in Enzymology," (C. H. W. Hirs and S. N. Timasheff, eds.), Vol. 26, pp. 43–50. Academic Press, New York.

Pace, C. N., and Laurents, D. V. (1989). A new method for determining the heat capacity change for protein folding. *Biochemistry* **28,** 2520–2525.

Pace, C. N., and Tanford, C. (1968). Thermodynamics of the unfolding of b-lactoglobulin A in aqueous urea solutions between 5 and 55. *Biochemistry* **7,** 198–208.

Pace, C. N., Laurents, D. V., and Thomson, J. A. (1990). pH dependence of the urea and guanidine hydrochloride denaturation of ribonuclease A and ribonuclease T1. *Biochemistry* **29,** 2564–2572.

Pace, C. N., Laurents, D. V., and Erickson, R. E. (1992). Urea denaturation of barnase: pH Dependence and characterization of the unfolded state. *Biochemistry* **31,** 2728–2734.

Pace, C. N., Hebert, E. J., Shaw, K. L., Schell, D., Both, V., Krajcikova, D., Sevcik, J., Wilson, K. S., Dauter, Z., Hartley, R. W., and Grimsley, G. R. (1998). Conformational stability and thermodynamics of folding of ribonucleases Sa, Sa2 and Sa3. *J. Mol. Biol.* **279,** 271–286.

Pace, C. N., Grimsley, G. R., Thomas, S. T., and Makhatadze, G. I. (1999). Heat capacity change for ribonuclease A folding. *Protein Sci.* **8,** 1500–1504.

Pace, C. N., Alston, R. W., and Shaw, K. L. (2000). Charge–charge interactions influence the denatured state ensemble and contribute to protein stability. *Protein Sci.* **9,** 1395–1398.

Pace, C., Grimsley, G. R., and Scholtz, J. M. (2005). Denaturation of proteins by urea and guanidine hydrochloride. *In* "Protein Folding Handbook," (J. A. K. Buchner T., ed.), pp. 45–69. Wiley-VCH Verlag GmbH & Co. KGaA, Hamburg.

Prakash, V., Loucheux, C., Scheufele, S., Gorbunoff, M. J., and Timasheff, S. N. (1981). Interactions of proteins with solvent components in 8 *M* urea. *Arch. Biochem. Biophys.* **210,** 455–464.

Santoro, M. M., and Bolen, D. W. (1992). A test of linear extrapolation of unfolding free energy changes over an extended denaturant concentration range. *Biochemistry* **31,** 4901–4907.

Schellman, J. A. (2003). Protein stability in mixed solvents: A balance of contact interaction and excluded volume. *Biophys. J.* **85,** 108–125.

Schellman, J. A., and Gassner, N. C. (1996). The enthalpy of transfer of unfolded proteins into solutions of urea and guanidinium chloride. *Biophys. Chem.* **59,** 259–275.

Shortle, D. (1995). Staphylococcal nuclease: A showcase of *m*-value effects. *Adv. Protein Chem.* **46,** 217–247.
Shortle, D. (1996). The denatured state (the other half of the folding equation) and its role in protein stability. *FASEB J.* **10,** 27–34.
Shortle, D., and Meeker, A. K. (1986). Mutant forms of staphylococcal nuclease with altered patterns of guanidine hydrochloride and urea denaturation. *Proteins* **1,** 81–89.
Spiro, K. (1900). Uber die beeinflussung der eiweisscoagulation durch stickstoffhaltige substanzen. *Z. Physiol. Chem.* **30,** 182–199.
Stark, G. R. (1965). Reactions of cyanate with functional groups of proteins. 3. Reactions with amino and carboxyl groups. *Biochemistry* **4,** 1030–1036.
Tanford, C. (1964). Isothermal unfolding of globular proteins in aqueous urea solutions. *J. Am. Chem. Soc.* **86,** 2050–2059.
Thomson, J. A., Shirley, B. A., Grimsley, G. R., and Pace, C. N. (1989). Conformational stability and mechanism of folding of ribonuclease T1. *J. Biol. Chem.* **264,** 11614–11620.
Warren, J. R., and Gordon, J. A. (1966). On the refractive indices of aqueous solutions of urea. *J. Phys. Chem.* **70,** 297–300.
Whitten, S. T., and Garcia-Moreno, B. E. (2000). pH dependence of stability of staphylococcal nuclease: Evidence of substantial electrostatic interactions in the denatured state. *Biochemistry* **39,** 14292–14304.
Yu, Y., Makhatadze, G. I., Pace, C. N., and Privalov, P. L. (1994). Energetics of ribonuclease T1 structure. *Biochemistry* **33,** 3312–3319.

CHAPTER TWENTY-FOUR

Measuring Cotranslational Folding of Nascent Polypeptide Chains on Ribosomes

Patricia L. Clark *and* Krastyu G. Ugrinov

Contents

Abstract

Protein folding has been studied extensively *in vitro*, but much less is known about how folding proceeds *in vivo*. A particular distinction of folding *in vivo* is that folding begins while the nascent polypeptide chain is still undergoing synthesis by the ribosome. Studies of cotranslational protein folding are inherently much more complex than classical *in vitro* protein folding studies, and historically there have been few methods available to produce the quantities of

Department of Chemistry and Biochemistry, University of Notre Dame, Notre Dame, Indiana, USA

Methods in Enzymology, Volume 466
ISSN 0076-6879, DOI: 10.1016/S0076-6879(09)66024-9

pure material required for biophysical studies of the nascent chain, or assays to specifically interrogate its conformation. However, the past few years have produced dramatic methodological advances, which now place cotranslational folding studies within reach of more biochemists, enabling a detailed comparison of the earliest stages of protein folding on the ribosome to the wealth of information available for the refolding of full-length polypeptide chains *in vitro*.

1. Introduction

In the cell, proteins are synthesized by the ribosome as linear strings of amino acid residues (Fig. 24.1). For most proteins, this linear chain of amino acid residues must fold into a unique three-dimensional structure; correct folding is a prerequisite for proper protein function. Historically, protein folding has been studied *in vitro*, by diluting full-length polypeptide chains out of a chemical denaturant, and observing the folding process (typically via optical or NMR spectroscopic methods). This homogeneous protein solution, which unfolds and refolds reversibly, permits a detailed biophysical investigation of the thermodynamics, kinetics, conformations, and dynamics of the folding process, as well as the competing off-pathway misfolding and aggregation process. Such studies have contributed greatly to our understanding of proteins.

However, equilibrium thermodynamics can only be applied to proteins that unfold and refold reversibly, and it is often impossible to find conditions under which to study folding/unfolding of a particular protein of interest at equilibrium. A common result is that, upon dilution from chemical denaturant, the polypeptide chain will misfold and aggregate, rather than refold

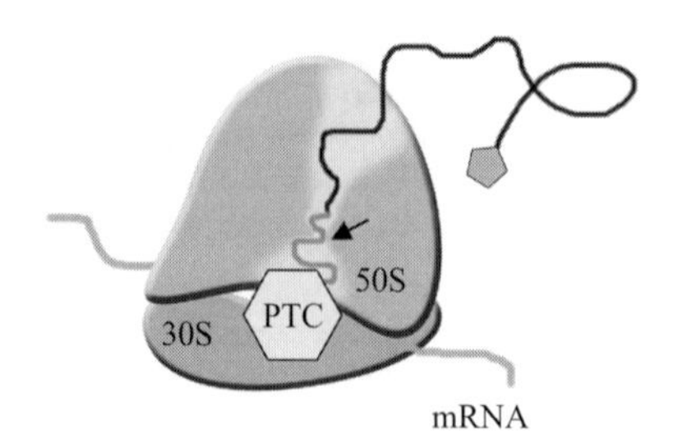

Figure 24.1 The ribosome nascent chain (RNC) complex. Illustration of translating bacterial ribosome with a stalled nascent chain tethered at the peptidyl transferase center (PTC, hexagon). The arrow points to the narrowest part of the ribosomal tunnel, where direct interactions with the nascent stall sequence have been identified (see text). Nascent chain is drawn in black and grey (indicating the polypeptide sequence used to initiate stalling); affinity tag for purification or labeling (optional; pentagon). (See Color Insert.)

to its native structure; the chemical denaturation of GFP is a classic example of this phenomenon (Tsien, 1998). Yet, these same polypeptide chains are synthesized and fold correctly every day, often at high levels, in the crowded environment of the cell.

Early observations of this contrast between the efficiency of protein folding *in vivo* versus inefficient refolding *in vitro* focused attention on what is lost upon removing a protein from its cellular environment. These observations spurred the discovery of molecular chaperones (Hemmingsen *et al.*, 1988), and studies of how these "helper proteins" can suppress aggregation and in some cases actively promote folding to the native structure (Bukau *et al.*, 2006; Hartl and Hayer-Hartl, 2009). Nevertheless, careful proteomics studies have shown that, under normal growth conditions, the majority of proteins do not require an interaction with molecular chaperones in order to fold to the native structure (Hartl and Hayer-Hartl, 2002), casting light onto other fundamental differences between folding *in vivo* and refolding of proteins *in vitro* upon dilution from chemical denaturants.

In vivo, every protein enters the cellular environment vectorially, either entering the cell cytoplasm during synthesis by the ribosome or another cellular compartment upon secretion through a protein pore in a lipid bilayer. There is a profound difference between this vectorial initiation of folding *in vivo* and the refolding of a full-length protein *in vitro*. For example, *in vivo*, N-terminal segments of the polypeptide chain are the first to appear outside the ribosome exit tunnel (Fig. 24.1). These N-terminal segments can start to build interactions with nearby segments of the nascent ribosome-bound polypeptide chain, but are unable to interact with C-terminal segments of the polypeptide chain, which have not yet been synthesized or remain sequestered within the ribosome exit tunnel. One can envision a scenario where parsing out interactions between neighboring parts of the polypeptide chain could reduce the topological frustration experienced by a protein that starts to build all interactions simultaneously upon dilution of a full-length polypeptide chain out of a chemical denaturant (Clark, 2004).

Hence, while numerous studies on diverse proteins have confirmed Anfinsen's theory that proteins fold to their global energy minimum conformation, the route taken to reach that conformation can differ dramatically, depending on whether a polypeptide chain is folding cotranslationally during synthesis by the ribosome *in vivo*, or sifting through myriad possible interactions between all portions of the polypeptide chain simultaneously *in vitro*. The distinctions between cotranslational folding *in vivo* and refolding *in vitro* have been discussed in theory (Clark, 2004; Evans *et al.*, 2005a), explored computationally (Elcock, 2006; Hsiao-Mei and Jie, 2008; Wang and Klimov, 2008), and confirmed experimentally (Evans *et al.*, 2008; Fedorov and Baldwin, 1999; Frydman *et al.*, 1999). Cotranslational folding can indeed influence the folding pathway, often with profound effects on folding efficiency (Evans *et al.*, 2008; Ugrinov and Clark, submitted).

Yet despite the potentially profound effect of cotranslational folding on protein folding mechanisms, cotranslational folding has been studied for only a small handful of proteins, and much remains unknown. This is mostly due to the experimental complexities that must be dealt with in order to study cotranslational folding. The experimental sample is no longer a pure, homogeneous polypeptide chain, but a complex mixture including nascent chains of various lengths at different points during synthesis, ribosomes, and potentially other components (molecular chaperones, the signal recognition particle (SRP), etc.). There are potentially coupled kinetics between nascent chain elongation and chain folding. There are very few experimental methods that can specifically report on the behavior of the nascent chain in such a mixture—although this is changing, and detailed below. Yet despite this complexity, great strides have been made over the past 5 years, making cotranslational measurements of nascent chain conformation increasingly accessible. Here, we discuss current state-of-the-art methods for measuring the cotranslational folding of nascent polypeptide chains. Because of the layered kinetic issue described above, these methods rely heavily on strategies that stall translation, producing a static ribosome: nascent chain complex (see below). Moreover, although this chapter describes results from cotranslational folding studies in both prokaryotic and eukaryotic systems, the detailed methods outlined below focus on cotranslational folding studies on ribosomes derived from *Escherichia coli* cells, and/or translation mixtures.

2. Translation and the Ribosome:Nascent Chain (RNC) Complex

Ribosomes are extremely large ribonucleoproteins, ranging in size from 2 MDa (*E. coli*) to 4 MDa (mammals). *E. coli* ribosomes consist of three separate RNA particles (rRNAs) and >50 proteins, and can be divided into two subunits: a small subunit (30 S) that binds mRNA, and a large subunit (50 S) that forms the tRNA binding sites. These two subunits dock together to produce the functional 70 S ribosome, a roughly spherical complex with the peptidyltransferase center (PTC) active site located near its center. As peptide bonds are formed at the PTC and the nascent chain elongates, it moves through a narrow exit tunnel that spans the width of the large subunit.

For the *E. coli* ribosome, the exit tunnel is approximately 100 Å long and 10–20 Å wide (Schuwirth *et al.*, 2005). Overall, the tunnel is narrowest near the PTC and grows wider as it reaches the outer surface of the large subunit, although a significant constriction occurs approximately 1/3 of the distance between the PTC and the exterior of the ribosome. The narrowness of the exit tunnel places significant restrictions on the extent of nascent chain folding that can occur while the nascent chain resides within the tunnel. While there appears to be enough room for chain compaction into an α-helical conformation (Ban *et al.*, 2000; Woolhead *et al.*, 2004), there does not appear to be enough room for the nascent chain to fold back on itself to form higher-order secondary structure (such as a β-hairpin) or tertiary structure within the tunnel, except perhaps in the broader region immediately adjacent to the ribosome surface (Amit *et al.*, 2005; Lu and Deutsch, 2005a,b). This was confirmed experimentally in a classic study showing that the ribosome exit tunnel protects only the most C-terminal 30–40 aa from proteolytic digestion (Malkin and Rich, 1967).

The surface of the exit tunnel is lined with ribosomal RNA and, to a lesser extent, ribosomal proteins. For decades, the exit tunnel was considered to be a "Teflon tube" (Nissen *et al.*, 2000), devoid of interactions with the nascent polypeptide chain, as it was expected that such interactions would interfere with the smooth progress of protein synthesis and the appearance of the nascent chain. More recently, however, a few amino acid sequences have been identified that interact specifically with both RNA and protein moieties lining the tunnel walls (Nakatogawa and Ito, 2004). These interactions between the nascent chain sequence and the ribosome exit tunnel can control nascent chain conformation (Woolhead *et al.*, 2006), and regulate the conformation of the ribosome and its functions (Rospert, 2004), including protein translation (Nakatogawa and Ito, 2004). Indeed, translational regulation by interactions between the nascent chain and the ribosome exit tunnel have been exploited to produce stalled ribosome:nascent chain (RNC) complexes for detailed biophysical studies of cotranslational folding (see below).

During translation, the C-terminus of the nascent chain is covalently tethered to the PTC center. Hence as the nascent chain grows in length, its N-terminus will exit the ribosome tunnel, but remain held in close proximity to the outer surface of the ribosome. The outer surface of the ribosome immediately adjacent to the exit tunnel is rich in ribosomal proteins (Schuwirth *et al.*, 2005), some of which serve as docking sites for additional proteins such as molecular chaperones and SRP (Buskiewicz *et al.*, 2004; Ullers *et al.*, 2006).

Polypeptide chain synthesis occurs at an average rate of approximately 20 aa/s in eubacteria, and 4–6 aa/s in eukaryotes. Translation therefore occurs much more slowly than early folding steps, such as global collapse and secondary structure formation, which can occur on the order of seconds, milliseconds, or even faster (Kubelka *et al.*, 2004; Roder and Colon, 1997). Hence, for some proteins, the rate of folding might be limited by the rate of chain elongation; these represent two potentially coupled kinetic processes. As a result, a major challenge for cotranslational folding studies is to uncouple the translation and folding processes. Typically, this is accomplished by arresting translation at a specific point, in order to create a homogeneous solution of ribosomes bearing nascent chains of a specific sequence and length, for detailed biophysical studies of nascent chain conformations.

A further complication is that the cellular ribosome concentration is much higher than the mRNA concentration. After translation initiation, as a ribosome moves further down the mRNA, another ribosome can initiate translation right behind the first ribosome. Hence each mRNA typically bears multiple ribosomes, each at a different point in the translation process. These poly-ribosomes (polysomes) further increase sample complexity, as stalling translation at a specific point on an mRNA sequence is likely to produce one ribosome stalled at the stall point, and several more ribosomes (bearing successively shorter nascent chains) stacked up behind it. The resulting sample is therefore not homogeneous, but instead a heterogeneous mixture of nascent chain lengths. Depending on the type of conformational or dynamics analysis performed, this heterogeneity can complicate the interpretation of results. However, there are methods available to separate individual 70 S ribosomes from polysomes and analyze the resulting nascent chain length distribution within a sample (Ugrinov and Clark, submitted).

3. General Approaches for Generating Stalled RNC Complexes

As described above, obtaining a RNC sample that accurately reflects the vectorial synthesis of a nascent chain, while simultaneously providing the homogeneity required for quantitative biophysical assays, is a challenging task. Another challenge is achieving sample concentrations in the range necessary for precise, quantitative biophysical assays (μM-mM). Ribosomes are so large (see above) that the nascent chain, the subject of our investigation, represents only a tiny fraction (typically <0.1%) of the total sample mass (Johnson, 2005). In addition, the strong background signal originating from the nucleic acid and protein component of the ribosome eliminates many common spectroscopic assays from consideration, including tryptophan fluorescence, far-UV circular dichroism, and infra-red spectroscopy. The physical properties of the ribosome require the design of conformational assays with simultaneously high sensitivity and specificity for features of the nascent chain, although sensitivity can be augmented somewhat by large-scale RNC purification strategies (see below).

Currently, two strategies are broadly used to generate RNC complexes: (i) expression of an mRNA sequence lacking a stop codon (truncated mRNA), or (ii) translational stalling induced by a portion of the translating nascent chain itself. Below we briefly describe these two strategies, and discuss some important considerations for the production of RNC complexes created by nascent chain-induced translational stalling for biophysical analyses.

3.1. Truncated mRNA-based production of RNC complexes

Historically, RNC complexes bearing nascent chains of a predetermined length and sequence were produced using an mRNA lacking a stop codon as a template for *in vitro* protein expression (Haeuptle *et al.*, 1986; Hanes and Pluckthun, 1997). Translation of an mRNA lacking a stop codon hinders the recruitment of the translation termination factors responsible for the hydrolysis of the polypeptide chain from the P-site tRNA (Youngman *et al.*, 2008). As a result, a stable complex is formed between the mRNA and the ribosome bearing the synthesized nascent chain.

The elimination of the mRNA stop codon can be achieved in different ways. When an *in vitro* translation system is supplemented with a short DNA fragment complementary to a specific position of the encoding mRNA, a DNA:mRNA hybrid is formed. This double-stranded hybrid sequence is recognized by ribonuclease H (Cerritelli and Crouch, 2009; Tadokoro and Kanaya, 2009), which degrades the DNA:mRNA hybrid stretch, but not the unhybridized portions of the mRNA. This produces a truncated mRNA lacking a stop codon. Alternatively, a DNA plasmid encoding an mRNA with an early transcription terminator sequence (Hanes and Pluckthun, 1997) can be used to produce an mRNA lacking a stop codon. In either case, the mRNA lacking a stop codon is then used as the template in an *in vitro* translation system. Upon reaching the 3′ end of the mRNA, ribosomes will stall, unable to either continue synthesis or terminate translation and release the nascent chain.

While the preparation of mRNA lacking a stop codon is relatively straightforward, the use of a truncated mRNA to produce stable RNC complexes is limited to *in vitro* translation systems, which have limited translational capacity (Underwood *et al.*, 2005). *In vivo* translation of truncated mRNAs is hindered by the challenges of producing or introducing stable truncated mRNAs in live cells. Yet *in vivo*, cellular components are crowded much more closely together than they are *in vitro*, and some *in vitro* translation systems lack a full repertoire of cellular components, including components such as the secretion apparatus, which interacts with some nascent chains.

3.2. Nascent chain sequence-based production of RNC complexes

There are a few specific amino acid sequences that, when present in a newly synthesized polypeptide chain, can control the synthesis of downstream portions of their mRNA sequences, typically via interactions with the ribosome exit tunnel (Lovett and Rogers, 1996). Examples of such translational control, where the nascent chain being synthesized promotes stalling of the translating ribosome, are described in diverse organisms including viruses (Alderete *et al.*, 1999; Janzen *et al.*, 2002), eubacteria (Ambulos *et al.*, 1986; Gong and Yanofsky, 2002; Lovett, 1994), yeast (Delbecq *et al.*, 1994, 2000), fungi (Fang *et al.*, 2000, 2004), plants (Onouchi *et al.*, 2005, 2008), and higher eukaryotes (Mize and Morris, 2001; Parola and Kobilka, 1994). The mechanisms by which these nascent chains stall translation are described in detail elsewhere (Beringer, 2008; Cruz-Vera and Yanofsky, 2008; Ramu *et al.*, 2009; Yap and Bernstein, 2009). The detailed mechanisms of translational stalling, as well as the locations of the effective amino acid sequence within the nascent polypeptide chain sequence, vary considerably; nevertheless, several of these sequences contain significant amino acid similarity at key positions within the ribosome exit tunnel (Fig. 24.2). Significantly, while some nascent chain stall sequences require an additional effector molecule to trigger ribosome stalling (Fang *et al.*, 2004; Gish and Yanofsky, 1995; Werner *et al.*, 1987), other nascent chains alone are sufficient to arrest translation (Nakatogawa and Ito, 2001; Onouchi *et al.*, 2005). Most translation stall sequences are encoded in short leader peptides or upstream open reading frames (uORFs)

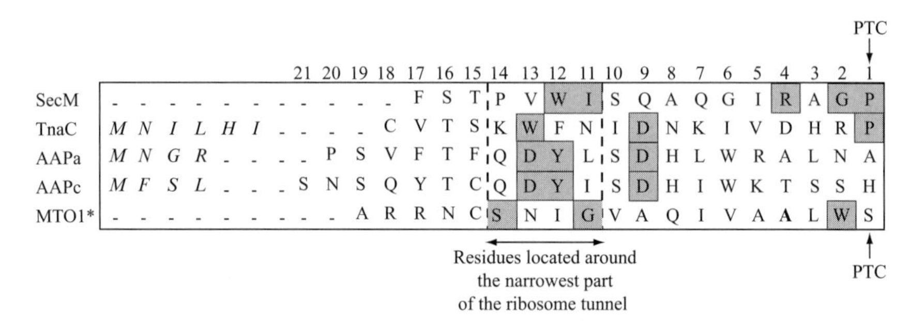

Figure 24.2 Stalling sequences with potential uses for the production of stalled RNCs. Nascent chain amino acid residues are numbered according to their distance from the peptidyl transferase center (PTC, the point of ribosome stalling). The arrows pointing at position 1 mark the residue located at the PTC. The key residues for each stalling sequence are shown in gray boxes. The N-terminal residues of TnaC (the leader peptide from the tryptophanase operon), AAPa (arginine attenuator peptide from Neurospora crassa carbamoyl-phosphate synthetase gene arg2), and AAPc (arginine attenuator peptide from Saccharomyces cerevisiae carbamoyl-phosphate synthetase gene cpa1) peptides, which are not conserved and/or do not affect stalling, are shown in italics (Cruz-Vera and Yanofsky, 2008; Delbecq *et al.*, 2000; Fang *et al.*, 2000; Gong and Yanofsky, 2002; Hood *et al.*, 2007). The MTO1 stalling sequence from cystathionine γ-synthase (marked with an asterisk) includes the sequence RRNCSNIGVAQ, which is highly conserved in multicellular plants; residue Ser94 is located at the PTC (Ominato *et al.*, 2002).

(Lovett and Rogers, 1996; Morris and Geballe, 2000), yet some are found in full-length protein-encoding ORFs (Nakatogawa and Ito, 2001; Ominato *et al.*, 2002).

The bacterial secretion monitor protein SecM (Nakatogawa and Ito, 2001; Sarker *et al.*, 2000) is a 170-aa protein that contains a 17-aa sequence near its C-terminus that is capable of stalling translation when present in the ribosome tunnel, with no requirement for an exogenous effector molecule (Nakatogawa and Ito, 2002) (Fig. 24.2). This SecM stall sequence has been exploited to stall the translation of other, diverse polypeptide chains (Evans *et al.*, 2005b).

The SecM stall sequence stalls translation while lodged in the ribosomal tunnel, which can accommodate the C-terminal 30–40 aa of the nascent chain, depending on the conformation of the polypeptide chain (Ban *et al.*, 2000; Gilbert *et al.*, 2004; Malkin and Rich, 1967). Hence, the 17 aa of the SecM stall sequence spans only one half of the tunnel, closest to the PTC, and presumably has little effect on the conformation and dynamics of the nascent chain on the ribosome surface. The major advantage of the SecM-based translation stalling method over classical truncated mRNA methods is the capacity for *in vivo* production of stable RNC complexes (Evans *et al.*, 2005b). *In vivo* assembly of SecM-stalled RNC complexes, each bearing a nascent chain of defined length and amino acid composition, provides the unique opportunity to mimic the vectorial synthesis of polypeptides by the ribosomal machinery in the context of the native cellular environment. The macromolecular crowding experienced by nascent chains *in vivo* might have a significant impact on the thermodynamic and/or kinetic properties of nascent chains, as crowding has been shown to affect macromolecular diffusion and protein folding/aggregation properties (Ellis, 2001; Ellis and Minton, 2006).

To date, SecM-mediated translation stalling has been used to produce RNC complexes bearing diverse nascent chains (Evans *et al.*, 2005b; Rutkowska *et al.*, 2009; Schaffitzel and Ban, 2007) (Table 24.1). Both polysome and single 70 S RNC complexes are produced, without affecting the overall cellular distribution of ribosomes and polysomes (Brandt *et al.*, 2009; Ugrinov and Clark, submitted). Although prolonged stalling could result in declined ribosome recycling and a consequent decrease of cellular translation efficiency (Nakatogawa and Ito, 2001), no significant reduction in cell viability has been reported so far during *in vivo* production of stalled RNC complexes (Contreras-Martinez and DeLisa, 2007; Evans *et al.*, 2005b; Schaffitzel and Ban, 2007). It appears that the SecM stall sequence can arrest the ongoing translation of virtually any polypeptide chain, underlying the potential of the method for future biophysical studies on ribosome-bound nascent chains.

Table 24.1 Polypeptides nascent chains created as a SecM stall sequence fusion proteins

Nascent chain	Origin	Length(s) (aa)[a]	Translation	References
P22 tailspike	Bacterial phage	248–685	*In vivo/in vitro*	Evans *et al.* (2005b, 2008)
GFP	Eukaryotes	89–303	*In vivo*	Evans *et al.* (2005b), Ugrinov and Clark (submitted)
GFP	Eukaryotes	373	*In vitro*	Uemura *et al.* (2008)
cSNC	Synthetic	39	*In vivo*	Unpublished[c]
cAMP receptor	Eubacteria	197[b]–209	*In vivo*	Sunohara *et al.* (2004)
Spectrin	Eukaryotes	103–367	*In vivo/in vitro*	Hoffmann *et al.* (2006), Merz *et al.* (2008), Rutkowska *et al.* (2008)
FtsQ	Eubacteria	121, 142	*In vivo/in vitro*	Mitra *et al.* (2005), Schaffitzel and Ban (2007), Schaffitzel *et al.* (2006)
ICDH	Eubacteria	35–400	*In vivo/in vitro*	Hoffmann *et al.* (2006), Merz *et al.* (2008), Rutkowska *et al.* (2008)
RpoB	Eubacteria	106–596	*In vivo*	Rutkowska *et al.* (2008, 2009)
Barnase	Eubacteria	177–236	*In vivo*	Rutkowska *et al.* (2008, 2009)
MetK	Eubacteria	112–268	*In vivo*	Rutkowska *et al.* (2008)
Firefly luciferase	Eukaryotes	113–576	*In vivo*	Brandt *et al.* (2009), Rutkowska *et al.* (2008)
Anti β-gal scFv	Eubacteria	308	*In vivo*	Contreras-Martinez and DeLisa (2007)
Anti HEL scFv	Eukaryotes	346	*In vitro*	Ohashi *et al.* (2007)
Dihydrofolate reductase	Eubacteria	284	*In vitro*	Ohashi *et al.* (2007)
Yap65	Eukaryotes	173	*In vitro*	Matsuura *et al.* (2007)
Protein D	Eubacteria	149	*In vitro*	Takahashi *et al.* (2009)

[a] The length of each polypeptide is calculated according to the information provided in the listed reference(s), and references therein. The calculations include the length of the polypeptide, the SecM stall sequence, a ribosome tunnel linker (if used), and an affinity tag (if used). When more than two polypeptide chains of the same protein are used, only the lengths of the shortest and longest are shown. A small variation in the calculated size is not excluded.

[b] Shorter SecM stall sequence was used.

[c] Ugrinov and Clark (unpublished results).

3.3. Caveats and limitations of stalled RNC complexes

A caveat of all ribosome stalling procedures is that the ribosome has stalled, an unnatural condition that does not occur under normal translation conditions. Hence it is quite possible that the nascent chain attached to a stalled ribosome might reach a conformation that would be kinetically inaccessible to a nascent chain on a ribosome that continues to undergo translation of downstream portions of the polypeptide chain. Technically, we are not yet able to simultaneously observe translation and cotranslational folding, although steps are being made in this direction (see Section 7). In the meantime, it is important to remember that a large amount of *in vitro* refolding experimental results collected for equilibrium conformations have assisted with descriptions of *bone fide in vitro* kinetic intermediates. And, care can be taken to design ribosome stall points to mimic endogenous local pauses in translation, resulting from rare codons (Clarke and Clark, 2008) or other effects.

The yield of stalled RNC complexes produced *in vivo* can be reduced by the action of the cellular *trans*-translation (tmRNA, SsrA) mechanism (Withey and Friedman, 2003). The *trans*-translation apparatus clears stalled bacterial ribosomes by disassembling the stalled ribosome from the engaged mRNA. The incomplete nascent chain is tagged with a specific C-terminal sequence, and directed for degradation (Keiler *et al.*, 1996; Withey and Friedman, 1999). To avoid the negative effects of *trans*-translation on SecM-mediated stalling, Δ*ssrA* bacterial strains can be used (Hallier *et al.*, 2004; Komine *et al.*, 1994). Alternatively, when an *in vitro* translation system is used, *trans*-translation can be reduced by supplementing the translation system with an anti-*ssrA* oligonucleotide (Evans *et al.*, 2005b; Schaffitzel and Ban, 2007).

It remains to be determined whether the SecM stalling procedure can be applied to translation systems from organisms other than *E. coli*, particularly eukaryotic translation systems, given the significant difference in architecture between prokaryotic and eukaryotic ribosomes (Morgan *et al.*, 2000). Possible limitations can also arise from the length of the nascent chain that can be stalled on the translating ribosome. Studies performed with P22 tailspike showed that stalling efficiency can decrease with increasing nascent chain length (Evans *et al.*, 2005b). A possible explanation is the viscous drag of a long nascent chain mostly outside the ribosome exit tunnel. In the case of tailspike, the lowest stalling efficiency was determined for a >800 aa nascent chain sequence, >4 times longer than the wild type SecM protein (170 aa).

4. Methods for Preparing RNC Complexes

The detailed protocol described below is based on the experimental work of our laboratory (Evans *et al.*, 2005b; Ugrinov and Clark, submitted), but the procedure is easily adjusted depending on the specific experimental goals and available laboratory equipment. At the end of the procedure, we provide additional comments and suggestions to refine the procedure for specific experimental goals.

4.1. Preliminary considerations

This procedure relies on the ribosome stalling properties of the SecM stall sequence (Evans *et al.*, 2005b; Nakatogawa and Ito, 2001). As a prerequisite, it requires an IPTG-inducible DNA expression plasmid bearing the gene for the nascent chain of interest, followed immediately by the SecM stall sequence. See Evans *et al.* (2005b) for a detailed description of such a plasmid.

4.2. Procedure

1. One liter of sterile Luria–Bertani (LB) broth supplemented with 100 μg/ml ampicillin is inoculated with 20 ml of an overnight stationary phase cell culture of *E. coli* transformed with the plasmid of interest. (The preparation scale can be increased or decreased, as needed. The numerous examples of generated RNC complexes (Table 24.1) demonstrate that the SecM-based methodology is applicable for practically any *E. coli*-based protein expression system.)
2. The cells are grown at 37 °C until reaching an optical density of 0.4–0.5 at 600 nm (approximately 3.5 h).
3. Protein expression is induced for 30 min by addition of IPTG to a final concentration of 0.5 m*M*.
   (There is the potential for cells expressing the SecM stall sequence for prolonged times to experience growth defects (Nakatogawa and Ito, 2004); hence expression time should be kept to a minimum.)
4. Protein expression is halted by chilling the cell culture on ice for ~10 min. All further steps are performed at 4 °C.
5. Cells are harvested by centrifugation and the cells are resuspended gently in 1 ml (for each 500 ml of cell culture) cold R buffer (50 m*M* Tris, pH 7.5; 10 m*M* $MgCl_2$; 150 m*M* KCl).
   (Complete cell resuspension produces more efficient cell lysis and higher RNC yields).
6. The resuspended cells are transferred to 2 ml microcentrifuge tubes and frozen at −80 °C for at least 30 min.
7. The frozen cells are thawed at room temperature and treated with lysozyme for 30 min. The final concentration of lysozyme is 1 mg/ml.
8. The cells are frozen again for ≥30 min at −80 °C.
9. The lysed cells are thawed at room temperature; the sample is supplemented with 50 m*M* $MgSO_4$, and treated with 100 units of DNase I (RNase free) for 30 min.

10. The cell lysate is spun at 14,000 rpm until a solid pellet is formed (40–50 min).
11. The sample supernatant is removed, layered onto a 35% sucrose solution prepared in R buffer, and spun at 229,600×*g* for 40 min (Beckmann 70.1 Ti rotor).
12. The supernatant and the sucrose are removed; the pellet consisting of RNC complexes is gently washed and resuspended in R buffer on a rotary shaker.
    (Mechanical disruption of the pellet by the pipette tip should be avoided).
13. Calculations of ribosome concentration and nascent chain stalling efficiency are described by Evans *et al.* (2005b).

4.3. Additional considerations

A sample purified according the protocol above will consist of RNC complexes and ribosome-associated cellular components. Hence, this sample will permit measurements of nascent chain conformations under conditions that closely resemble the nascent chain conformations and interactions encountered *in vivo*. If further purification of the RNC complexes is required, such as removal of nonspecifically bound cellular components, removal of a labeling reagent, or separation of polysome and monosome (70 S) complexes for detailed quantitative measurements, the RNC complexes can be further refined. Separation of an RNC sample using standard size exclusion chromatography is an efficient way to achieve such separations (Woolhead *et al.*, 2004). Affinity-based chromatography methods are also suitable; however, they require specifically designed nascent chain constructs (Rutkowska *et al.*, 2009; Schaffitzel and Ban, 2007), and care must be taken that neither the affinity tag nor the purification protocol disturbs the nascent chain conformation and/or its interactions.

5. Biophysical Studies with RNC Complexes

The ability to produce large quantities of stalled RNC complexes has opened up a new world of experimental possibilities, expanding studies of protein conformation and dynamics in directions that were previously inaccessible. As mentioned above, the fundamental biophysical principles of protein chemistry have been derived mainly from refolding experiments using simplified experimental systems. Now, these fundamental principles can be readdressed and reevaluated from the perspective of ribosome-bound nascent chains, the true starting point for folding *in vivo*. In addition, although beyond the scope of this chapter, RNC complexes provide an excellent system for examining the complex interplay of interactions between nascent chains and other components of the cell, such as molecular chaperones and the secretory apparatus. Here, we review recent studies of nascent polypeptide chain folding and dynamics. These studies are grouped into broad categories, but it should be noted that some studies span multiple categories, underscoring the complexity of RNC complexes.

5.1. Cotranslational folding studies

The process of cotranslational folding entails acquisition of secondary and/or tertiary structures by the ribosome-bound nascent polypeptide chain prior to its complete synthesis by the ribosome (Evans *et al.,* 2005a,b; Fedorov and Baldwin, 1997; Hardesty *et al.*, 1999; Kolb *et al.*, 1995; Komar, 2009; Lim and Spirin, 1986; Ugrinov and Clark, submitted). There are now many examples of proteins that start to fold cotranslationally, supporting the important role of the vectorial appearance of the polypeptide at the surface of the ribosome, and the role of the synthesizing ribosome itself in correct folding (Clark and King, 2001; Das *et al.*, 1996; Evans *et al.*, 2008; Fedorov and Baldwin, 1997; Frydman *et al.*, 1999; Kudlicki *et al.*, 1997; Singh and Rao, 2002; Svetlov *et al.*, 2006, 2007; Ugrinov and Clark, submitted). As mentioned above, during cotranslational folding, the growing nascent chain experiences significant conformational constraint arising from the walls of the ribosomal tunnel, and numerous macromolecules gating the exit of the tunnel (Kramer *et al.*, 2009; Nissen *et al.*, 2000). In addition, the C-terminus of the nascent chain is immobilized within the bulky ribosome (Svetlov *et al.*, 2007). As a result, the folding landscape for a nascent chain is significantly different from that of a full-length polypeptide chain refolding in a test tube (Clark, 2004; Evans *et al.*, 2005a). Cotranslational folding affects the entropic and enthalpic components of the free energy for folding. Entropically, the accessible conformations of a ribosome-bound nascent chain are restricted versus the conformations of a corresponding free chain. Enthalpically, completion of the network of noncovalent bonds that stabilizes the native structure of the protein is regulated by the vectorial synthesis and appearance of the nascent chain.

The effect of the ribosome tunnel on the conformation of a nascent chain was addressed by Deutsch and coworkers, who experimentally measured the compactness of nascent chains of different lengths inside the ribosome exit tunnel (Kosolapov and Deutsch, 2009; Lu and Deutsch, 2005a,b). The experimental data was used to calculate the free energies of folding and unfolding of the nascent chain in different regions of the ribosome tunnel (Lu and Deutsch, 2005a,b). Based on their analyses, the authors propose that specific zones inside the ribosome exit tunnel play an active role inducing early nascent chain compaction and folding.

In particular, these authors identified the distal portion of the ribosome tunnel, closest to the ribosome surface, as an "entropic window" where N-terminal portions of a nascent chain can explore conformational space (Kosolapov and Deutsch, 2009).

Johnson and coworkers have developed techniques for Förster resonance energy transfer (FRET)-based measurements that have revolutionized the field of cotranslational folding (Johnson, 2005; Woolhead *et al.*, 2004, 2006). FRET provides information on the distance between two fluorescent probes, and so allows evaluation of the conformational ensembles populated by a doubly labeled polypeptide chain in the process of folding/unfolding (Haas, 2005; Stryer, 1978). When the measurements are performed with a series of homogeneous ribosome: nascent chain complexes having specifically labeled nascent chains with a predetermined length, information for the folding of the nascent chain at each step of the biosynthetic process can be gained (Woolhead *et al.*, 2004, 2006). For example, Johnson and coworkers used FRET measurements to determine that transmembrane helical segments of the nascent chain adopt a helical conformation deep within the ribosome exit tunnel, and the ribosome tunnel environment is necessary for formation of this conformation (Woolhead *et al.*, 2004).

Diverse cotranslational folding studies with purified RNC complexes have also revealed the presence of foldable, on-pathway nascent chain conformations on the surface of the ribosome (Clark and King, 2001; Evans *et al.*, 2008; Frydman *et al.*, 1994; Kudlicki *et al.*, 1994, 1995; Ugrinov and Clark, submitted). Studies performed with firefly luciferase and P22 tailspike protein have demonstrated that the conformations of these nascent chains on the surface of the ribosome are distinct from the conformations populated by the protein upon dilution out of a chemical denaturant *in vitro* (Clark and King, 2001; Evans *et al.*, 2008; Frydman *et al.*, 1999). Similarly, for full-length GFP nascent chains, cotranslational conformations have been shown to lead to a much higher folding yield than the conformations populated by full-length free GFP chains unfolded in a chemical denaturant (Ugrinov and Clark, submitted).

5.2. Studies of nascent chain dynamics

Very little is known about the conformational flexibility of the newly synthesized peptide inside the ribosome tunnel, or on the surface of the ribosome. Nascent chain dynamics directly reflect the rigidity of the molecule as well as the conformational space accessible for the growing polypeptide chain. Recently, however, NMR and fluorescence spectroscopy have made possible direct measurements of the dynamics of various nascent chains (Ellis *et al.*, 2008; Hsu *et al.*, 2007, 2009), and highlight the possibility for future advanced thermodynamic and kinetic studies of RNC complexes. Results derived from both NMR relaxation measurements and time resolved fluorescence measurements can be used for evaluation of the conformational entropy of the experimental system and changes in the heat capacity during protein folding and unfolding (Yang and Kay, 1996; Yang *et al.*, 1997).

5.3. Complementary approaches to cotranslational folding

In vitro protein folding studies with truncated free polypeptide chains (no ribosomes present), as well as computational cotranslational folding studies, are briefly discussed here. These studies complement the experimental results on RNC complexes described above, and provide an experimental and theoretical foundation for conformational changes occurring with increasing polypeptide chain length, with or without confinement by a ribosome surface.

A comprehensive *in vitro* structural analysis of C-terminally truncated chains of two small globular proteins, barnase and chymotrypsin inhibitor 2 (CI2), showed that significant stable native-like conformations are present only in polypeptide chains with lengths close to the size of the full-length protein (Neira and Fersht, 1999a,b). Similar results were shown for C-terminally truncated chains of two other small proteins, staphylococcal nuclease and sperm whale apomyoglobin: these polypeptide chains acquired their native state only when their C-terminal residues were present (Chow *et al.*, 2003; Flanagan *et al.*, 1992). Yet these proteins are all quite small, meaning that the C-terminal 20–40 aa of the newly synthesized chain that resides within the ribosome tunnel even at the termination of translation represents a significant portion of the polypeptide chain. Hence it would not be surprising for these four proteins to acquire stable native-like structures only after ribosome release (Chow *et al.*, 2003; Flanagan *et al.*, 1992; Neira and Fersht, 1999a,b). It is important to note, however, that the *in vitro* studies performed with these proteins have shown that even their short truncated chains can form compact structures with some long range (but perhaps non-native) interactions (Chow *et al.*, 2003; Flanagan *et al.*, 1992; Neira and Fersht, 1999a,b). Hence cotranslational formation of similar compact structures cannot be excluded. A direct comparison of the folding of free, truncated polypeptide chains with the folding of nascent chains of various lengths, tethered to the ribosome, would ensure proper understanding of the processes governing vectorial synthesis and folding of these proteins.

Numerous cotranslational folding simulation studies have emerged in recent years (Elcock, 2006; Hsiao-Mei and Jie, 2008; Huard *et al.*, 2006; Lydia *et al.*, 2006; Peiyu and Dmitri, 2008). Comprehensive molecular simulations of the cotranslational folding of CI2, barnase, and Semliki forest virus protein (SFVP) were presented by Elcock (2006). For the purposes of the simulations, the conformations adopted by the polypeptide chains were represented in terms of fraction of native-like contacts between the amino acid residues, and the changes in the energy of folding of the "nascent" polypeptides were monitored as a function of the native-like contacts. Importantly, the values for the folded and unfolded state energies, as well as the acquisition of native-like contacts, were standardized to experimentally calculated values (Neira and Fersht, 1999a,b). These simulations supported the results of the C-terminal truncations described above, suggesting that CI2 and barnase are unlikely to fold to a stable, native-like structure cotranslationally. In contrast, cotranslational folding was observed for the N-terminal domain of the larger multidomain protein SFVP, which is known to fold cotranslationally into an active protease structure (Nicola *et al.*, 1999).

Lattice-based simulations have been used to measure kinetics for cotranslational folding versus refolding of small single domain proteins (Peiyu and Dmitri, 2008). The simulations showed slower, more complex kinetics for "cotranslational" folding, compared to the folding under "*in vitro*" conditions. The different folding mechanisms were attributed to different conformations populated by the free chains and the ribosome-bound chains at the beginning of the reaction and as folding progressed (Peiyu and Dmitri, 2008). Similar observations were reported for HP (hydrophobic/polar) lattice simulation, where the growing nascent chain populated ground states different from the global energy minimum state (Huard *et al.*, 2006).

6. Measuring Nascent Chain Cotranslational Folding and Rigidity by Limited Protease Digestion

Resistance of a protein to a limited protease digestion is commonly used to investigate the compactness of a polypeptide chain (Hubbard, 1998; Park and Marqusee, 2004). The method relies on the higher susceptibility of unstructured and extended conformations to protease cleavage; in contrast, rigid (and hence folded) regions are protected from digestion. Of course, it is also possible that the ribosome itself might provide a steric blockade that shields the nascent chain from the protease. Nevertheless, carefully controlled experiments with diverse RNC complexes have shown that limited protease digestion can provide crucial information on the cotranslational conformations of nascent proteins (Evans *et al.*, 2008; Frydman *et al.*, 1999; Malkin and Rich, 1967; Picking *et al.*, 1992).

6.1. Preliminary considerations

Digestion conditions must be chosen that do not affect the integrity of the ribosome itself. In general, digestion times should be kept as short as possible, to minimize digestion-time dependent changes in the nascent chain conformation. When a nonspecific protease such as proteinase K is used for the limited digest, this method is generally applicable to all RNC complexes. Moreover, the identification of resulting digestion fragments will provide unbiased information on nascent chain compaction, rigidity, and folding (and potentially, ribosome shielding). Alternatively, a specific protease cleavage site (recognized by Factor Xa, enterokinase, TEV protease, etc.) could be introduced at a desired position in the ribosome-bound nascent chain. Cleavage at this position would indicate that the protease cleavage site resides at an unstructured or otherwise accessible location on the nascent chain conformation. The protocol below should be considered as a guideline rather than a detailed procedure, since it is based on the specific aims of our laboratory (Evans *et al.*, 2008).

6.2. Procedure

1. Fresh RNC complexes are purified in R buffer as described above. (Any freezing/thawing prior to the treatment with proteinase K should be avoided. The proteinase K reaction is performed at 4 °C.)
2. Approximately 1 ml of RNC complexes are mixed with an appropriate volume of proteinase K stock solution (prepared in R buffer) to a final volume of 1 ml and final concentration of proteinase K of 1 μg/ml. (The concentration of the RNC complexes, and the [RNC]/[proteinase K] ratio, will vary and is determined by the properties of the specific nascent chain and the goals of the experiment. A final concentration of 1 μg/ml proteinase K and digestion times of <20 min will keep ribosomes intact throughout the experiment.)
3. At various digestion time points, proteinase K is inhibited by addition of 1 m*M* PMSF (phenylmethylsulphonyl fluoride, a serine protease inhibitor) to the reaction mixture.
4. The sample is incubated with PMSF for 15 min and layered onto a 35% sucrose solution prepared in R buffer.
5. The sample is spun at 229,600×*g* for 40 min (Beckmann 70.1 Ti rotor), and the ribosome-free supernatant is collected.
6. The supernatant, which contains the protease resistant fragments, is analyzed by gel electrophoresis, immunoblotting, mass spectrometry, etc. (e.g., see (Evans *et al.*, 2008)).

7. Future Directions

While cotranslational folding studies have made great strides recently, much work remains. For example, an important bottleneck is the current lack of methods suitable to measure cotranslational folding while nascent chain synthesis is underway. An experiment of this sort would remove the caveat that the conformations populated on a stalled ribosome might not resemble the conformations populated if translation were allowed to continue uninterrupted. Recent advances in single molecule techniques to monitor protein translation (Blanchard, 2009; Katranidis *et al.*, 2009; Uemura *et al.*, 2008) hold promise for future technical advances in this direction.

Finally, many of the considerations described here for vectorial, cotranslational folding of nascent polypeptide chains on the ribosome also apply to vectorial, cosecretory folding of polypeptide chains as they are moved across a lipid bilayer from one cellular compartment to another. While landmark studies have provided some information on the direction of chain transport across membranes (Junker *et al.*, 2009), and the effect of premature folding on secretion efficiency (Teschke *et al.*, 1991), the conformations of these polypeptides as they appear in their destination compartment, and the influence of these conformations on final folding yield, are much less well understood.

REFERENCES

Alderete, J. P., *et al.* (1999). Translational effects of mutations and polymorphisms in a repressive upstream open reading frame of the human cytomegalovirus UL4 gene. *J. Virol.* **73,** 8330–8337.

Ambulos, N. P. Jr., *et al.* (1986). Analysis of the regulatory sequences needed for induction of the chloramphenicol acetyltransferase gene *cat-86* by chloramphenicol and amicetin. *J. Bacteriol.* **167,** 842–849.

Amit, M., *et al.* (2005). A crevice adjoining the ribosome tunnel: Hints for cotranslational folding. *FEBS Lett.* **579,** 3207–3213.

Ban, N., *et al.* (2000). The complete atomic structure of the large ribosomal subunit at 2.4 Å resolution. *Science* **289,** 905–920.

Beringer, M. (2008). Modulating the activity of the peptidyl transferase center of the ribosome. *RNA* **14,** 795–801.

Blanchard, S. C. (2009). Single-molecule observations of ribosome function. *Curr. Opin. Struct. Biol.* **19,** 103–109.

Brandt, F., *et al.* (2009). The native 3D organization of bacterial polysomes. *Cell* **136,** 261–271.

Bukau, B., *et al.* (2006). Molecular chaperones and protein quality control. *Cell* **125,** 443–451.

Buskiewicz, I., *et al.* (2004). Trigger factor binds to ribosome-signal-recognition particle (SRP) complexes and is excluded by binding of the SRP receptor. *Proc. Natl. Acad. Sci. USA* **101,** 7902–7906.

Cerritelli, S. M., and Crouch, R. J. (2009). Ribonuclease H: The enzymes in eukaryotes. *FEBS J.* **276,** 1494–1505.

Chow, C. C., *et al.* (2003). Chain length dependence of apomyoglobin folding: Structural evolution from misfolded sheets to native helices. *Biochemistry* **42,** 7090–7099.

Clark, P. L. (2004). Protein folding in the cell: Reshaping the folding funnel. *Trends Biochem. Sci.* **29,** 527–534.

Clark, P. L., and King, J. (2001). A newly synthesized, ribosome-bound polypeptide chain adopts conformations dissimilar from early *in vitro* refolding intermediates. *J. Biol. Chem.* **276,** 25411–25420.

Clarke, T. F. IV., and Clark, P. L. (2008). Rare codons cluster. *PLoS ONE* **3,** e3412.

Contreras-Martinez, L. M., and DeLisa, M. P. (2007). Intracellular ribosome display via SecM translation arrest as a selection for antibodies with enhanced cytosolic stability. *J. Mol. Biol.* **372,** 513–524.

Cruz-Vera, L. R., and Yanofsky, C. (2008). Conserved residues Asp16 and Pro24 of TnaC-tRNAPro participate in tryptophan induction of Tna operon expression. *J. Bacteriol.* **190,** 4791–4797.

Das, B., *et al.* (1996). *In vitro* protein folding by ribosomes from *Escherichia coli*, wheat germ and rat liver: The role of the 50 S particle and its 23 S rRNA. *Eur. J. Biochem.* **235,** 613–621.

Delbecq, P., *et al.* (1994). A segment of mRNA encoding the leader peptide of the CPA1 gene confers repression by arginine on a heterologous yeast gene transcript. *Mol. Cell. Biol.* **14,** 2378–2390.

Delbecq, P., *et al.* (2000). Functional analysis of the leader peptide of the yeast gene CPA1 and heterologous regulation by other fungal peptides. *Curr. Genet.* **38,** 105–112.

Elcock, A. H. (2006). Molecular simulations of cotranslational protein folding: Fragment stabilities, folding cooperativity, and trapping in the ribosome. *PLoS Comput. Biol.* **2,** e98.

Ellis, R. J. (2001). Macromolecular crowding: Obvious but underappreciated. *Trends Biochem. Sci.* **26,** 597–604.

Ellis, R. J., and Minton, A. P. (2006). Protein aggregation in crowded environments. *Biol. Chem.* **387,** 485–497.

Ellis, J. P., *et al.* (2008). Chain dynamics of nascent polypeptides emerging from the ribosome. *ACS Chem. Biol.* **3,** 555–566.

Evans, M. S., *et al.* (2005a). Conformations of co-translational folding intermediates. *Protein Pept. Lett.* **12,** 189–195.

Evans, M. S., *et al.* (2005b). Homogeneous stalled ribosome nascent chain complexes produced *in vivo* or *in vitro*. *Nat. Methods* **2,** 757–762.

Evans, M. S., *et al.* (2008). Cotranslational folding promotes β-helix formation and avoids aggregation *in vivo*. *J. Mol. Biol.* **383,** 683–692.

Fang, P., *et al.* (2000). Evolutionarily conserved features of the arginine attenuator peptide provide the necessary requirements for its function in translational regulation. *J. Biol. Chem.* **275,** 26710–26719.

Fang, P., *et al.* (2004). A nascent polypeptide domain that can regulate translation elongation. *Proc. Natl. Acad. Sci. USA* **101,** 4059–4064.

Fedorov, A. N., and Baldwin, T. O. (1997). Cotranslational protein folding. *J. Biol. Chem.* **272,** 32715–32718.

Fedorov, A. N., and Baldwin, T. O. (1999). Process of biosynthetic protein folding determines the rapid formation of native structure. *J. Mol. Biol.* **294,** 579–586.

Flanagan, J. M., *et al.* (1992). Truncated staphylococcal nuclease is compact but disordered. *Proc. Natl. Acad. Sci. USA* **89,** 748–752.

Frydman, J., *et al.* (1994). Folding of nascent polypeptide chains in a high molecular mass assembly with molecular chaperones. *Nature* **370,** 111–117.

Frydman, J., *et al.* (1999). Co-translational domain folding as the structural basis for the rapid *de novo* folding of firefly luciferase. *Nat. Struct. Biol.* **6,** 697–705.

Gilbert, R. J., *et al.* (2004). Three-dimensional structures of translating ribosomes by cryo-EM. *Mol. Cell.* **14,** 57–66.

Gish, K., and Yanofsky, C. (1995). Evidence suggesting *cis* action by the TnaC leader peptide in regulating transcription attenuation in the tryptophanase operon of *Escherichia coli*. *J. Bacteriol.* **177,** 7245–7254.

Gong, F., and Yanofsky, C. (2002). Instruction of translating ribosome by nascent peptide. *Science* **297,** 1864–1867.
Haas, E. (2005). The study of protein folding and dynamics by determination of intramolecular distance distributions and their fluctuations using ensemble and single-molecule FRET measurements. *Chemphyschem* **6,** 858–870.
Haeuptle, M. T., *et al.* (1986). Translation arrest by oligodeoxynucleotides complementary to mRNA coding sequences yields polypeptides of predetermined length. *Nucleic Acids Res.* **14,** 1427–1448.
Hallier, M., *et al.* (2004). Pre-binding of small protein B to a stalled ribosome triggers trans-translation. *J. Biol. Chem.* **279,** 25978–25985.
Hanes, J., and Pluckthun, A. (1997). *In vitro* selection and evolution of functional proteins by using ribosome display. *Proc. Natl. Acad. Sci. USA* **94,** 4937–4942.
Hardesty, B., *et al.* (1999). Co-translational folding. *Curr. Opin. Struct. Biol.* **9,** 111–114.
Hartl, F. U., and Hayer-Hartl, M. (2002). Molecular chaperones in the cytosol: From nascent chain to folded protein. *Science* **295,** 1852–1858.
Hartl, F. U., and Hayer-Hartl, M. (2009). Converging concepts of protein folding *in vitro* and *in vivo*. *Nat. Struct. Mol. Biol.* **16,** 574–581.
Hemmingsen, S. M., *et al.* (1988). Homologous plant and bacterial proteins chaperone oligomeric protein assembly. *Nature* **333,** 330–334.
Hoffmann, A., *et al.* (2006). Trigger factor forms a protective shield for nascent polypeptides at the ribosome. *J. Biol. Chem.* **281,** 6539–6545.
Hood, H. M., *et al.* (2007). Evolutionary changes in the fungal carbamoyl-phosphate synthetase small subunit gene and its associated upstream open reading frame. *Fungal Genet. Biol.* **44,** 93–104.
Hsiao-Mei, L., and Jie, L. (2008). A model study of protein nascent chain and cotranslational folding using hydrophobic-polar residues. *Proteins: Struct. Funct. Bioinform.* **70,** 442–449.
Hsu, S. T., *et al.* (2007). Structure and dynamics of a ribosome-bound nascent chain by NMR spectroscopy. *Proc. Natl. Acad. Sci. USA* **104,** 16516–16521.
Hsu, S. D., *et al.* (2009). Structure, Dynamics and folding of an immunoglobulin domain of the gelation factor (ABP-120) from *Dictyostelium discoideum*. *J. Mol. Biol.* **388,** 865–879.
Huard, F. P., *et al.* (2006). Modelling sequential protein folding under kinetic control. *Bioinformatics* **22,** e203–e210.
Hubbard, S. J. (1998). The structural aspects of limited proteolysis of native proteins. *Biochim. Biophys. Acta.* **1382,** 191–206.
Janzen, D. M., *et al.* (2002). Inhibition of translation termination mediated by an interaction of eukaryotic release factor 1 with a nascent peptidyl-tRNA. *Mol. Cell. Biol.* **22,** 8562–8570.
Johnson, A. E. (2005). The co-translational folding and interactions of nascent protein chains: A new approach using fluorescence resonance energy transfer. *FEBS Lett.* **579,** 916–920.
Junker, M., *et al.* (2009). Vectorial transport and folding of an autotransporter virulence protein during outer membrane secretion. *Mol. Microbiol.* **71,** 1323–1332.
Katranidis, A., *et al.* (2009). Fast biosynthesis of GFP molecules: A single-molecule fluorescence study. *Angew. Chem. Int. Ed.* **48,** 1758–1761.
Keiler, K. C., *et al.* (1996). Role of a peptide tagging system in degradation of proteins synthesized from damaged messenger RNA. *Science* **271,** 990–993.
Kolb, V. A., *et al.* (1995). Cotranslational folding of proteins. *Biochem. Cell Biol.* **73,** 1217–1220.
Komar, A. A. (2009). A pause for thought along the co-translational folding pathway. *Trends Biochem. Sci.* **34,** 16–24.
Komine, Y., *et al.* (1994). A tRNA-like structure is present in 10Sa RNA, a small stable RNA from *Escherichia coli*. *Proc. Natl. Acad. Sci. USA* **91,** 9223–9227.

Kosolapov, A., and Deutsch, C. (2009). Tertiary interactions within the ribosomal exit tunnel. *Nat. Struct. Mol. Biol.* **16,** 405–411.

Kramer, G., *et al.* (2009). The ribosome as a platform for co-translational processing, folding and targeting of newly synthesized proteins. *Nat. Struct. Mol. Biol.* **16,** 589–597.

Kubelka, J., *et al.* (2004). The protein folding 'speed limit'. *Curr. Opin. Struct. Biol.* **14,** 76–88.

Kudlicki, W., *et al.* (1994). Activation and release of enzymatically inactive, full-length rhodanese that is bound to ribosomes as peptidyl-tRNA. *J. Biol. Chem.* **269,** 16549–16553.

Kudlicki, W., *et al.* (1995). Elongation and folding of nascent ricin chains a peptidyl-tRNA on ribosomes: The effect of amino acid deletion on these processes. *J. Mol. Biol.* **252,** 203–212.

Kudlicki, W., *et al.* (1997). Ribosomes and ribosomal RNA as chaperones for folding of proteins. *Fold. Des.* **2,** 101–108.

Lim, V. I., and Spirin, A. S. (1986). Stereochemical analysis of ribosomal transpeptidation. Conformation of nascent peptide. *J. Mol. Biol.* **188,** 565–574.

Lovett, P. S. (1994). Nascent peptide regulation of translation. *J. Bacteriol.* **176,** 6415–6417.

Lovett, P. S., and Rogers, E. J. (1996). Ribosome regulation by the nascent peptide. *Microbiol. Rev.* **60,** 366–385.

Lu, J., and Deutsch, C. (2005a). Folding zones inside the ribosomal exit tunnel. *Nat. Struct. Mol. Biol.* **12,** 1123–1129.

Lu, J., and Deutsch, C. (2005b). Secondary structure formation of a transmembrane segment in Kv channels. *Biochemistry* **44,** 8230–8243.

Lydia, M. C. M., *et al.* (2006). Protein translocation through a tunnel induces changes in folding kinetics: A lattice model study. *Biotechnol. Bioeng.* **94,** 105–117.

Malkin, L. I., and Rich, A. (1967). Partial resistance of nascent polypeptide chains to proteolytic digestion due to ribosomal shielding. *J. Mol. Biol.* **26,** 329–346.

Matsuura, T., *et al.* (2007). Nascent chain, mRNA, and ribosome complexes generated by a pure translation system. *Biochem. Biophys. Res. Commun.* **352,** 372–377.

Merz, F., *et al.* (2008). Molecular mechanism and structure of trigger factor bound to the translating ribosome. *EMBO J.* **27,** 1622–1632.

Mitra, K., *et al.* (2005). Structure of the *E. coli* protein-conducting channel bound to a translating ribosome. *Nature* **438,** 318–324.

Mize, G. J., and Morris, D. R. (2001). A mammalian sequence-dependent upstream open reading frame mediates polyamine-regulated translation in yeast. *RNA* **7,** 374–381.

Morgan, D. G., *et al.* (2000). A comparison of the yeast and rabbit 80 S ribosome reveals the topology of the nascent chain exit tunnel, inter-subunit bridges and mammalian rRNA expansion segments. *J. Mol. Biol.* **301,** 301–321.

Morris, D. R., and Geballe, A. P. (2000). Upstream open reading frames as regulators of mRNA translation. *Mol. Cell. Biol.* **20,** 8635–8642.

Nakatogawa, H., and Ito, K. (2001). Secretion monitor, SecM, undergoes self-translation arrest in the cytosol. *Mol. Cell.* **7,** 185–192.

Nakatogawa, H., and Ito, K. (2002). The ribosomal exit tunnel functions as a discriminating gate. *Cell.* **108,** 629–636.

Nakatogawa, H., and Ito, K. (2004). Intraribosomal regulation of expression and fate of proteins. *ChemBioChem.* **5,** 48–51.

Neira, J. L., and Fersht, A. R. (1999a). Acquisition of native-like interactions in C-terminal fragments of barnase. *J. Mol. Biol.* **287,** 421–432.

Neira, J. L., and Fersht, A. R. (1999b). Exploring the folding funnel of a polypeptide chain by biophysical studies on protein fragments. *J. Mol. Biol.* **285,** 1309–1333.

Nicola, A. V., *et al.* (1999). Co-translational folding of an alphavirus capsid protein in the cytosol of living cells. *Nat. Cell. Biol.* **1,** 341–345.

Nissen, P., *et al.* (2000). The structural basis of ribosome activity in peptide bond synthesis. *Science* **289,** 920–930.

Ohashi, H., *et al.* (2007). Efficient protein selection based on ribosome display system with purified components. *Biochem. Biophys. Res. Commun.* **352,** 270–276.

Ominato, K., *et al.* (2002). Identification of a short highly conserved amino acid sequence as the functional region required for posttranscriptional autoregulation of the cystathionine gamma-synthase gene in *Arabidopsis. J. Biol. Chem.* **277,** 36380–36386.

Onouchi, H., *et al.* (2005). Nascent peptide-mediated translation elongation arrest coupled with mRNA degradation in the CGS1 gene of *Arabidopsis. Genes Dev.* **19,** 1799–1810.

Onouchi, H., *et al.* (2008). Nascent peptide-mediated translation elongation arrest of *Arabidopsis thaliana* CGS1 mRNA occurs autonomously. *Plant Cell Physiol.* **49,** 549–556.

Park, C., and Marqusee, S. (2004). Probing the high energy states in proteins by proteolysis. *J. Mol. Biol.* **343,** 1467–1476.

Parola, A. L., and Kobilka, B. K. (1994). The peptide product of a 5' leader cistron in the beta 2 adrenergic receptor mRNA inhibits receptor synthesis. *J. Biol. Chem.* **269,** 4497–4505.

Peiyu, W., and Dmitri, K. K. (2008). Lattice simulations of cotranslational folding of single domain proteins. *Proteins: Struct. Funct. Bioinform.* **70,** 925–937.

Picking, W. D., *et al.* (1992). Fluorescence characterization of the environment encountered by nascent polyalanine and polyserine as they exit *Escherichia coli* ribosomes during translation. *Biochemistry* **31,** 2368–2375.

Ramu, H., *et al.* (2009). Programmed drug-dependent ribosome stalling. *Mol. Microbiol.* **71,** 811–824.

Roder, H., and Colon, W. (1997). Kinetic role of early intermediates in protein folding. *Curr. Opin. Struct. Biol.* **7,** 15–28.

Rospert, S. (2004). Ribosome function: Governing the fate of a nascent polypeptide. *Curr. Biol.* **14,** R386–R388.

Rutkowska, A., *et al.* (2008). Dynamics of trigger factor interaction with translating ribosomes. *J. Biol. Chem.* **283,** 4124–4132.

Rutkowska, A., *et al.* (2009). Large-scale purification of ribosome-nascent chain complexes for biochemical and structural studies. *FEBS Lett.* **583,** 2407–2413.

Sarker, S., *et al.* (2000). Revised translation start site for secM defines an atypical signal peptide that regulates *Escherichia coli* secA expression. *J. Bacteriol.* **182,** 5592–5595.

Schaffitzel, C., and Ban, N. (2007). Generation of ribosome nascent chain complexes for structural and functional studies. *J. Struct. Biol.* **158,** 463–471.

Schaffitzel, C., *et al.* (2006). Structure of the *E. coli* signal recognition particle bound to a translating ribosome. *Nature* **444,** 503–506.

Schuwirth, B. S., *et al.* (2005). Structures of the bacterial ribosome at 3.5 Å resolution. *Science* **310,** 827–834.

Singh, R., and Rao, C. M. (2002). Chaperone-like activity and surface hydrophobicity of 70 S ribosome. *FEBS Lett.* **527,** 234–238.

Stryer, L. (1978). Fluorescence energy transfer as a spectroscopic ruler. *Annu. Rev. Biochem.* **47,** 819–846.

Sunohara, T., *et al.* (2004). Ribosome stalling during translation elongation induces cleavage of mRNA being translated in *Escherichia coli. J. Biol. Chem.* **279,** 15368–15375.

Svetlov, M. S., *et al.* (2006). Effective cotranslational folding of firefly luciferase without chaperones of the Hsp70 family. *Protein Sci.* **15,** 242–247.

Svetlov, M. S., *et al.* (2007). Folding of the firefly luciferase polypeptide chain with immobilized C-terminus. *Mol. Biol. (Mosk).* **41,** 96–102.

Tadokoro, T., and Kanaya, S. (2009). Ribonuclease H: Molecular diversities, substrate binding domains, and catalytic mechanism of the prokaryotic enzymes. *FEBS J.* **276,** 1482–1493.

Takahashi, S., *et al.* (2009). Real-time monitoring of cell-free translation on a quartz-crystal microbalance. *J. Am. Chem. Soc.* **131,** 9326–9332.

Teschke, C. M., *et al.* (1991). Mutations that affect the folding of ribose-binding protein selected as suppressors of a defect in export in *Escherichia coli*. *J. Biol. Chem.* **266,** 11789–11796.

Tsien, R. Y. (1998). The green fluorescent protein. *Annu. Rev. Biochem.* **67,** 509–544.

Uemura, S., *et al.* (2008). Single-molecule imaging of full protein synthesis by immobilized ribosomes. *Nucleic Acids Res.* **36,** e70.

Ugrinov, K. G., and Clark, P. L. (submitted). Co-translational folding increases GFP folding yield.

Ullers, R. S., *et al.* (2006). Sequence-specific interactions of nascent *Escherichia coli* polypeptides with trigger factor and signal recognition particle. *J. Biol. Chem.* **281,** 13999–14005.

Underwood, K. A., *et al.* (2005). Quantitative polysome analysis identifies limitations in bacterial cell-free protein synthesis. *Biotechnol. Bioeng.* **91,** 425–435.

Wang, P., and Klimov, D. K. (2008). Lattice simulations of cotranslational folding of single domain proteins. *Proteins* **70,** 925–937.

Werner, M., *et al.* (1987). The leader peptide of yeast gene CPA1 is essential for the translational repression of its expression. *Cell* **49,** 805–813.

Withey, J., and Friedman, D. (1999). Analysis of the role of trans-translation in the requirement of tmRNA for λmmP22 growth in *Escherichia coli*. *J. Bacteriol.* **181,** 2148–2157.

Withey, J. H., and Friedman, D. I. (2003). A salvage pathway for protein structures: tmRNA and trans-translation. *Annu. Rev. Microbiol.* **57,** 101–123.

Woolhead, C. A., *et al.* (2004). Nascent membrane and secretory proteins differ in FRET-detected folding far inside the ribosome and in their exposure to ribosomal proteins. *Cell* **116,** 725–736.

Woolhead, C. A., *et al.* (2006). Translation arrest requires two-way communication between a nascent polypeptide and the ribosome. *Mol. Cell.* **22,** 587–598.

Yang, D., and Kay, L. E. (1996). Contributions to conformational entropy arising from bond vector fluctuations measured from NMR-derived order parameters: Application to protein folding. *J. Mol. Biol.* **263,** 369–382.

Yang, D., *et al.* (1997). Contributions to protein entropy and heat capacity from bond vector motions measured by NMR spin relaxation. *J. Mol. Biol.* **272,** 790–804.

Yap, M. N., and Bernstein, H. D. (2009). The plasticity of a translation arrest motif yields insights into nascent polypeptide recognition inside the ribosome tunnel. *Mol. Cell.* **34,** 201–211.

Youngman, E. M., *et al.* (2008). Peptide release on the ribosome: Mechanism and implications for translational control. *Annu. Rev. Microbiol.* **62,** 353–373.

Author Index

I

J

K

M

N

Subject Index

D

E

K

L

M

U

V

W

X

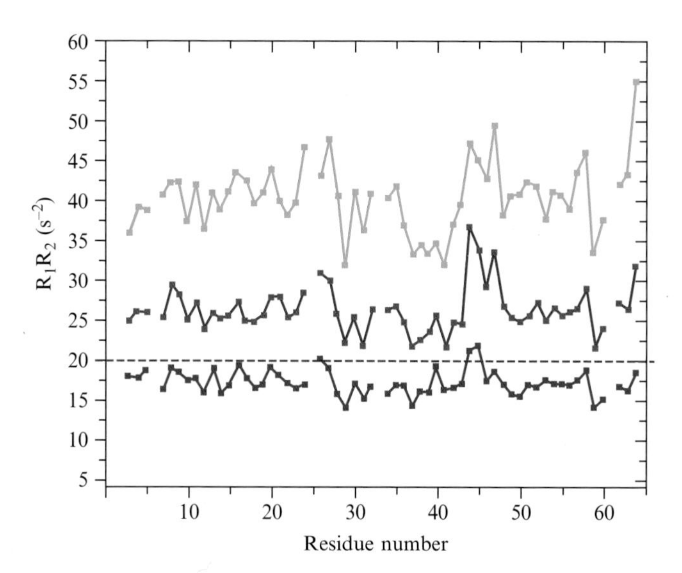

Andrew C. Miklos *et al.*, Figure 1.3 Histograms of R_1R_2 values as a function of residue number for chymotrypsin inhibitor 2 at 25 °C in 42% glycerol (blue) and 200 g/L BSA at pH 5.4 (red) and pH 6.8 (green). The rigid limit is also shown. Reprinted with permission from Li and Pielak (2009). Copyright 2009 American Chemical Society.

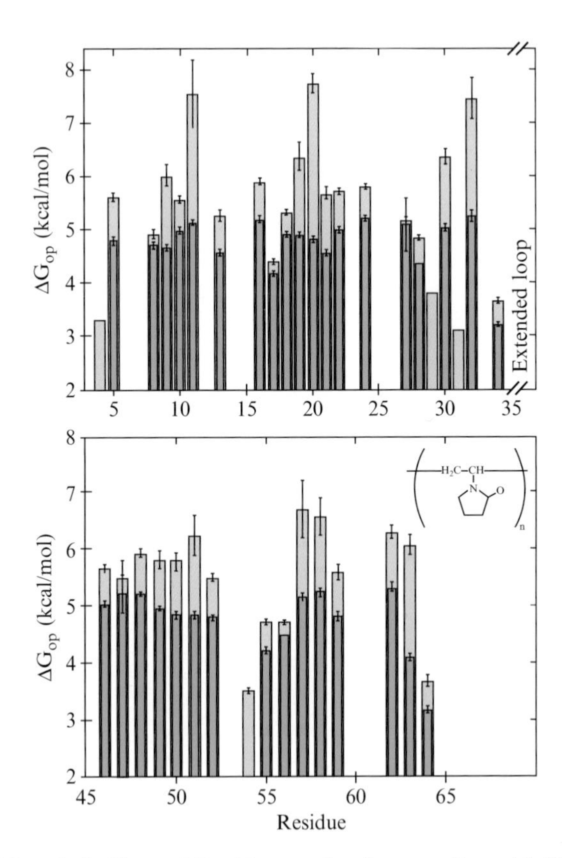

Andrew C. Miklos *et al.*, Figure 1.2 Macromolecular crowding with PVP40 stabilizes the I29A/I37H variant of CI2 relative to dilute solution. Values of $\Delta G^{0'}_{op}$ in 300 g/L PVP40 (cyan) and dilute solution (green) are plotted versus residue number. The height of each bar represents the average from three trials. The error bars represent the standard deviation. *Conditions*: 700–800 μM variant protein, 50 mM acetate buffer in D_2O, pH 5.4 at 37 °C. The inset shows the backbone structure of PVP. Reprinted with permission from Charlton *et al.* (2008). Copyright 2008 American Chemical Society.

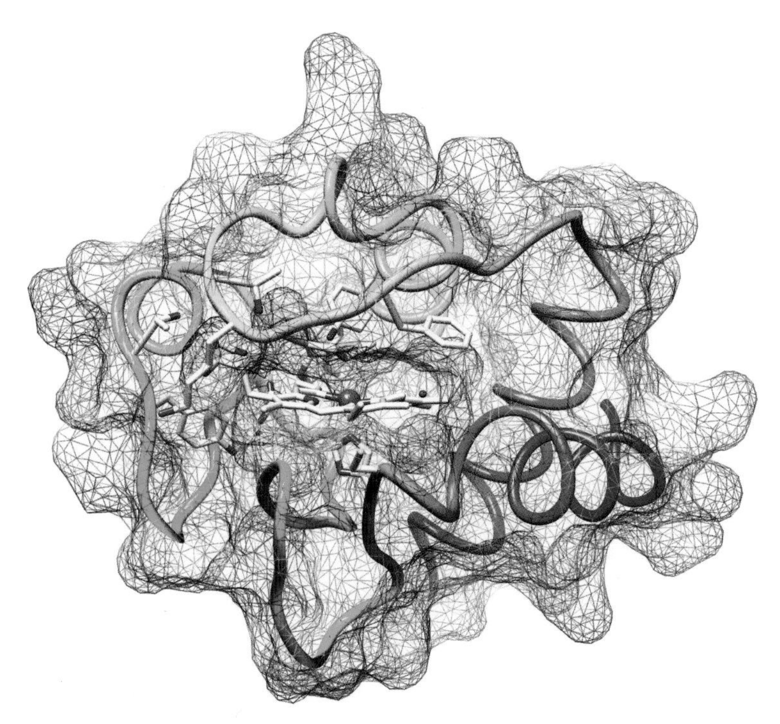

Reinhard Schweitzer-Stenner *et al.*, Figure 6.1 Environment of the functional heme *c* group in horse heart cytochrome *c*. The structure was reported by Bushnell *et al.* (1990), the figure was produced with the VMD software (Humphrey *et al.*, 1996).

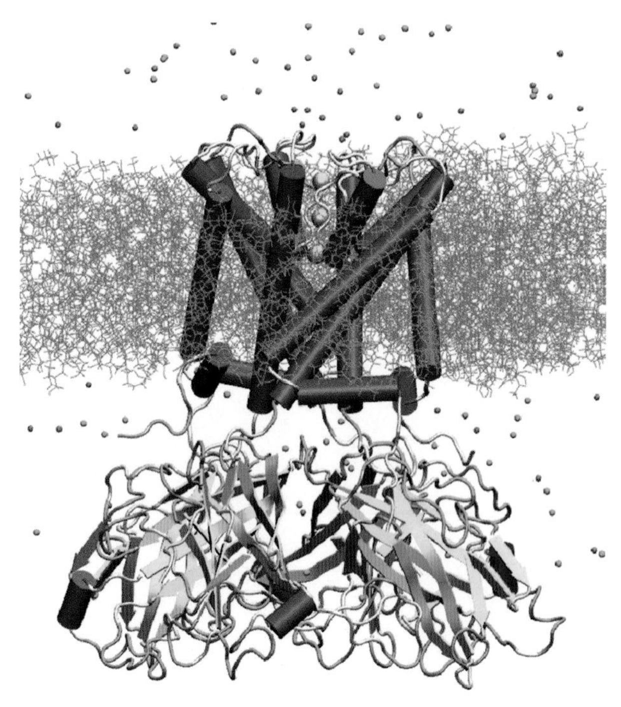

Carmen Domene and Simone Furini, Figure 7.1 The KirBac1.1 K^+ channel structure (PDB 1P7B) embedded in a lipid bilayer with individual lipids rendered as gray chains. The transmembrane α-helices are rendered in purple and the extracellular domain is colored according to its structure (β-sheet: yellow; random coil: white; turns: cyan). K^+ ions inside the selectivity filter of the channel are represented as yellow spheres. Other ions in solution are yellow and orange spheres.

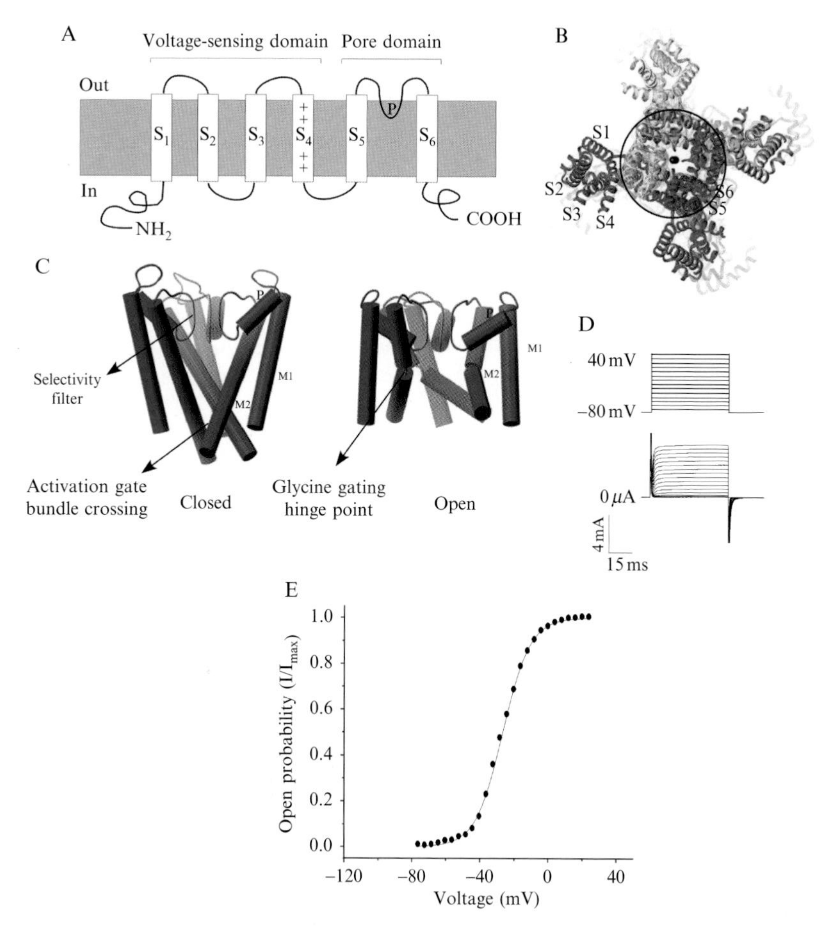

Ofer Yifrach *et al.*, Figure 8.3 The voltage-activated potassium channel model allosteric system. (A) Membrane topology of Kv channels. The alternating basic residues of the S4 transmembrane helix are indicated by a "+" sign. (B) Crystallographic structure of the human Kv1.2 channel viewed from the intracellular side along the channel's fourfold symmetry axis. Each subunit is depicted in a different color. The black circle indicates the ion-conduction pore domain. Adopted with permission from Long *et al.*, 2005. (C) Helix-rod representations of the KcsA (closed) and MthK (open) pore structures, with M1, P, and M2 corresponding to the outer, pore, and inner helices, respectively. Adopted with permission from Yifrach, 2004. (D) Typical voltage protocol for wild-type *Shaker* Kv channel activation and traces of the elicited K^+ currents flowing through the membrane-expressed Kv channels. (E) Voltage-activation curve of the wild-type *Shaker* Kv channel. Smooth curve corresponds to a two-state Boltzmann function.

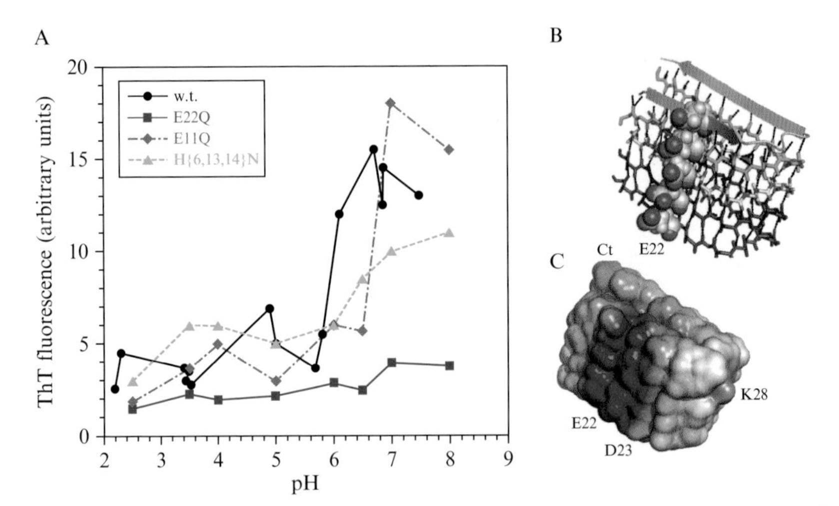

Sarah R. Sheftic *et al.*, Figure 10.4 Roles of charged residues in the pH dependence of Aβ fibrillization. (A) Comparison of the ThT fluorescence plateau of Aβ variants as a function of pH. The lines are not fits but simply included to guide the eye. (B) Structure of the Aβ(1–42) protofilament (Luhrs *et al.*, 2005). The organization of five Aβ monomers in the protofilament is shown, and residue E22 is highlighted with CPK spheres. Note the side-chain carboxylates from E22 (red spheres) form an array of negative charges running along the length of the fibril. (C) Electrostatic surface of the structure in (B). Charged residues in the core of the fibril structure are labeled. The peptides [E11Q]-Aβ(1–40), [E22Q]-Aβ(1–40), and [H(6,13,14)N]-Aβ(1–42) were all from Anaspec (San Jose, CA). All peptides were dissolved according to a literature protocol (Hou *et al.*, 2004) and run through a 100-kDa molecular weight cutoff filter before use, to remove any aggregates at the start of the reactions. All other conditions were as described in the legend to Fig. 10.3.

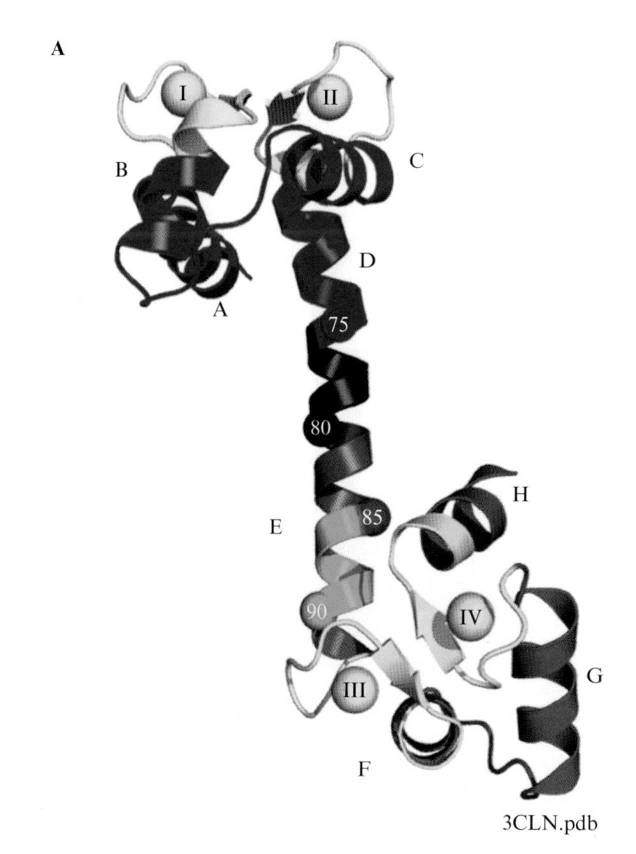

Susan E. O'Donnell *et al.*, Figure 21.1 *Ribbon diagram of CaM and sequence information.* (A) Ribbon diagram of $(Ca^{2+})_4$-CaM (3CLN.pdb). Ca^{2+}-binding sites (yellow) I and II are in the N-domain (blue), and sites III and IV are in the C-domain (red). Residues 76–80 (black), 81–85 (purple), and 86–90 (green) are highlighted. (B) Amino acid sequence for CaM with corresponding color schematic described for (A).

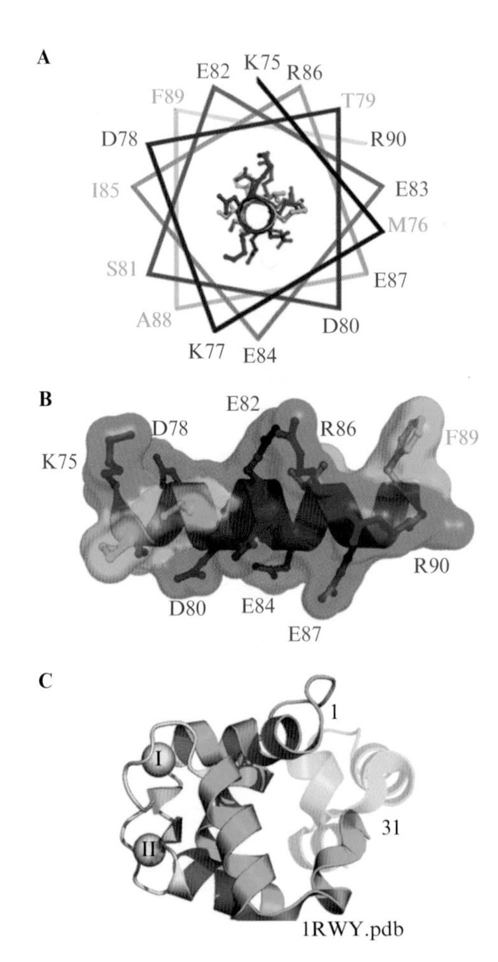

Susan E. O'Donnell *et al.*, Figure 21.2 Helical wheel (A) and surface (B) representation of residues 75–90 of CaM with basic residues shown in blue, acidic residues shown in red, and all other residues shown in green. (C) Ribbon diagram of $(Ca^{2+})_2$-parvalbumin (1RWY.pdb). Ca^{2+}-binding sites are shown in yellow (residues 1–31 are transparent).

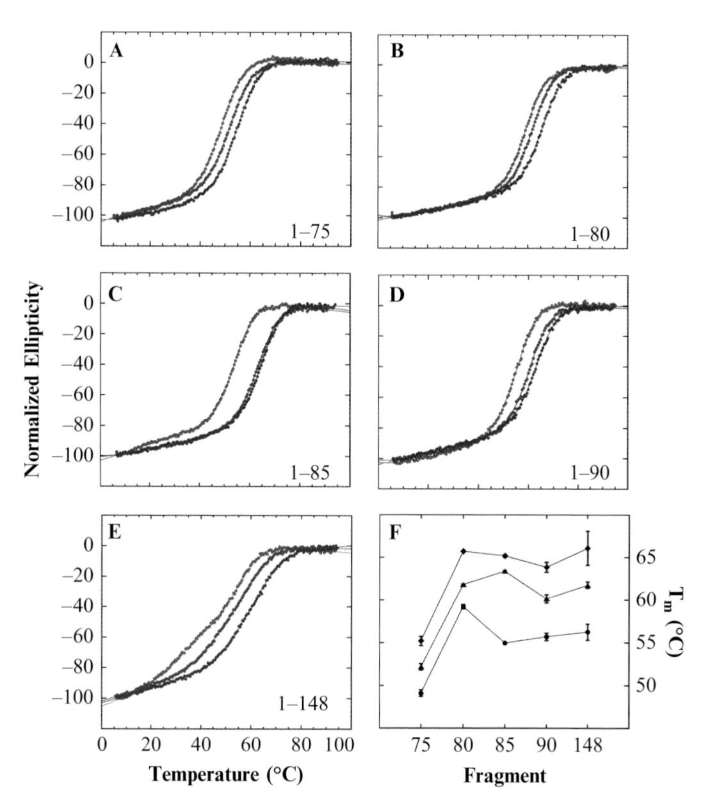

Susan E. O'Donnell *et al.*, Figure 21.4 *Thermal unfolding of CaM fragments.* Thermal unfolding of (A) CaM_{1-75}, (B) CaM_{1-80}, (C) CaM_{1-85}, (D) CaM_{1-90}, and (E) CaM_{1-148} in the presence of 0 m*M* added KCl (red), 100 m*M* KCl (green), or 300 m*M* KCl (purple) as monitored by circular dichroism at 222 nm. Curves were simulated using parameters resolved for unfolding of CaM at each KCl level using a two-state model for unfolding (A–D, Eq. 21.4) or a three-model (E, Eq. 21.5). Comparison of T_m (°C) for each fragment tested (F) at 0 m*M* added KCl (●), 100 m*M* KCl (▲), and 300 m*M* KCl (◆) is shown. Results are summarized in Table 21.2.

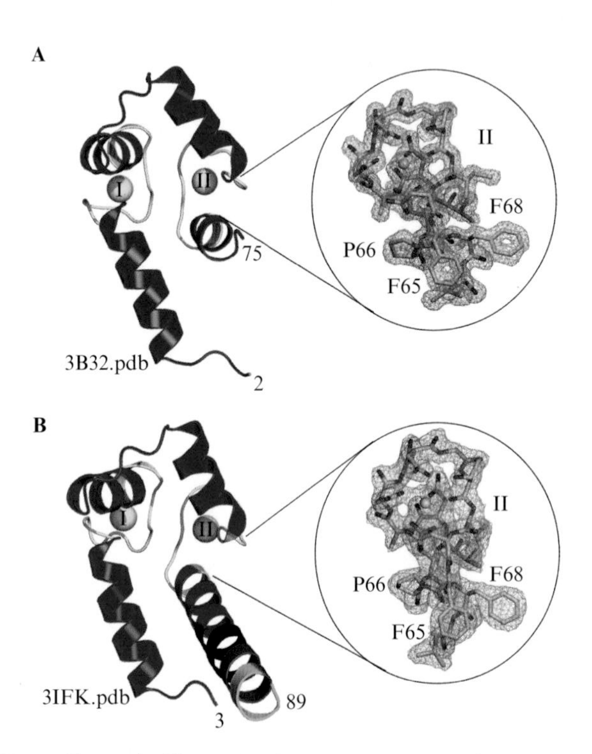

Susan E. O'Donnell *et al.*, Figure 21.6 *Ribbon diagram of* CaM_{1-75} *and* CaM_{1-90}. Ribbon diagram of (A) $(Ca^{2+})_2$-CaM_{1-75} (3B32.pdb) and (B) $(Ca^{2+})_2$-CaM_{1-90} (3IFK.pdb) displayed with electron density for site II. Ca^{2+}-binding sites I and II are shown in yellow. Residues 1–75 (blue, A and B) and residues 76–80 (black), 81–85 (purple), and 86–90 (green) (B) are shown.

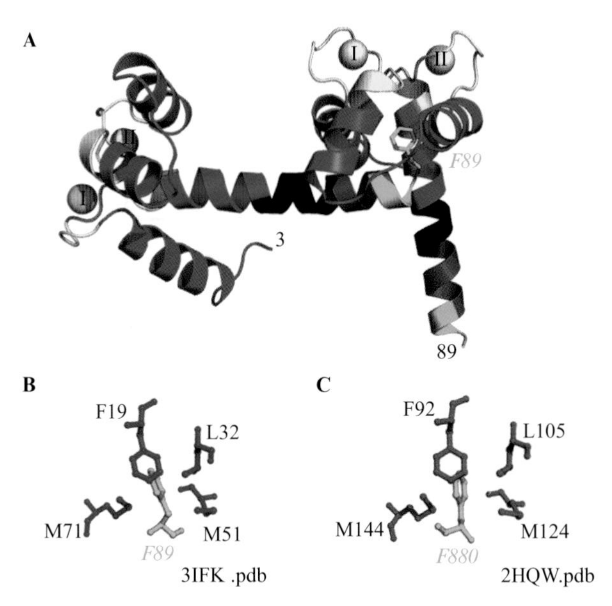

Susan E. O'Donnell *et al.*, Figure 21.7 *Crystal packing-induced dimerization of* $rCaM_{1-90}$ *(3IFK).* (A) Ca^{2+}-binding sites I and II (yellow), residues 1–75 (blue), 76–80 (black), 81–85 (purple), and 86–90 (green) are shown. (B) Occupancy of the N-domain FLMM pocket (blue) of $rCaM_{1-90}$ by F89 (green). (C) Occupancy of the C-domain FLMM pocket (red) of CaM_{1-148} by F880 (green) of the CaM-binding domain of NR1C0 (2HQW.pdb).